D0164611

Geometry

Assume A = area, C = circumference, V = volume, S = surface area, r = radius, h = altitude, l = length, w = width, b (or a) = length of a base, and s = length of a side.

1. **Square** $A = s^2$
2. **Rectangle** $A = lw$
3. **Parallelogram** $A = bh$
4. **Triangle** $A = \frac{1}{2}bh$
5. **Circle** $A = \pi r^2$; $C = 2\pi r$
6. **Trapezoid** $A = \frac{1}{2}(a + b)h$
7. **Cube** $S = 6s^2$; $V = s^3$
8. **Rectangular Box** $S = 2(lw + wh + lh)$; $V = lwh$
9. **Cylinder** $S = 2\pi rh$; $V = \pi r^2 h$
10. **Sphere** $S = 4\pi r^2$; $V = \frac{4}{3}\pi r^3$
11. **Cone** $S = \pi r\sqrt{r^2 + h^2}$; $V = \frac{1}{3}\pi r^2 h$

Tables of Special Values of Trigonometric Functions

Angle θ in degrees	Angle θ in radians	$\sin\theta$	$\cos\theta$	$\tan\theta$	$\cot\theta$	$\sec\theta$	$\csc\theta$
0°	0	0	1	0	—	1	—
30°	$\pi/6$	$1/2$	$\sqrt{3}/2$	$\sqrt{3}/3$	$\sqrt{3}$	$2\sqrt{3}/3$	2
45°	$\pi/4$	$\sqrt{2}/2$	$\sqrt{2}/2$	1	1	$\sqrt{2}$	$\sqrt{2}$
60°	$\pi/3$	$\sqrt{3}/2$	$1/2$	$\sqrt{3}$	$\sqrt{3}/3$	2	$2\sqrt{3}/3$
90°	$\pi/2$	1	0	—	0	—	1
120°	$2\pi/3$	$\sqrt{3}/2$	$-1/2$	$-\sqrt{3}$	$-\sqrt{3}/3$	-2	$2\sqrt{3}/3$
135°	$3\pi/4$	$\sqrt{2}/2$	$-\sqrt{2}/2$	-1	-1	$-\sqrt{2}$	$\sqrt{2}$
150°	$5\pi/6$	$1/2$	$-\sqrt{3}/2$	$-\sqrt{3}/3$	$-\sqrt{3}$	$-2\sqrt{3}/3$	2
180°	π	0	-1	0	—	-1	—
210°	$7\pi/6$	$-1/2$	$-\sqrt{3}/2$	$\sqrt{3}/3$	$\sqrt{3}$	$-2\sqrt{3}/3$	-2
225°	$5\pi/4$	$-\sqrt{2}/2$	$-\sqrt{2}/2$	1	1	$-\sqrt{2}$	$-\sqrt{2}$
240°	$4\pi/3$	$-\sqrt{3}/2$	$-1/2$	$\sqrt{3}$	$\sqrt{3}/3$	-2	$-2\sqrt{3}/3$
270°	$3\pi/2$	-1	0	—	0	—	-1
300°	$5\pi/3$	$-\sqrt{3}/2$	$1/2$	$-\sqrt{3}$	$-3/\sqrt{3}$	2	$-2\sqrt{3}/3$
315°	$7\pi/4$	$-\sqrt{2}/2$	$\sqrt{2}/2$	-1	-1	$\sqrt{2}$	$-\sqrt{2}$
330°	$11\pi/6$	$-1/2$	$\sqrt{3}/2$	$-\sqrt{3}/3$	$-\sqrt{3}$	$2\sqrt{3}/3$	-2
360°	2π	0	1	0	—	1	—

Precalculus

Precalculus

Functions and Graphs

FOURTH EDITION

M. A. MUNEM

J. P. YIZZE

MACOMB COMMUNITY COLLEGE

WORTH PUBLISHERS, INC.

Precalculus: Functions and Graphs Fourth Edition

Printed in the United States of America

Library of Congress Catalog Card No. 84-051926

ISBN: 0-87901-258-7

Third printing, July 1987

Editor: Rosalind Lippel

Production: Jose Fonfrias, Patricia Lawson

Design: Malcolm Grear Designers

Illustrator: ANCO/Boston

Typographer: Syntax International

Printing and Binding: R. R. Donnelley & Sons

Cover: Photograph of Pyrotechnic Burst by Peter Mastrogiannis

Photo Credits: P. 66: Historical Pictures Service, P. 87: Historical Pictures Service, P. 167: The Bettmann Archive, P. 186: Historical Pictures Service, P. 189: The Bettmann Archive, P. 215: UPI/Bettmann Archive, P. 235: The Bettmann Archive, P. 346: © Rae Russel/International Stock Photo, P. 356: NASA, P. 465: t.l., UPI/Bettmann Archive; t.r., Courtesy of the Library Services Department, American Museum of Natural History; b.l., Library of Congress; b.r., Mula & Haramaty/ Phototake, P. 471: top, © Kachaturian/International Stock Photo; b.l., Fundamental Photographs, New York; b.r., The Port Authority of New York & New Jersey, P. 473: top, © Steven Innerfield/International Stock Photo; bot., © Dennis Brack/Black Star

Worth Publishers, Inc.

33 Irving Place

New York, New York 10003

Preface

Each new edition of *Precalculus* provides us with a chance to do all the things we wished we had done in the previous edition. We believe we have taken full advantage of this opportunity here. The principal changes are listed later, and we have made many less obvious improvements, such as adding more graphs, problems, and illustrative examples, and including brief historical notes.

Prerequisites

We assume that students using this book have taken the equivalent of two years of college-preparatory high school mathematics, including algebra and geometry, or an equivalent college course in intermediate algebra. Conscientious students with less preparation can, however, master the contents of this book, particularly if they take advantage of the tutorial help provided by the accompanying *Study Guide*.

Objectives

The aim of the book remains the same—to provide students with the working knowledge of algebra and trigonometry, including functions and graphs, that they will need for their later study. There is added coverage in this edition of topics that will be useful not only in calculus, but also in linear algebra and other courses in business, physics, chemistry, and engineering.

We have written and revised the book with the objective of complementing and enhancing effective classroom teaching. The student is provided here with every opportunity to learn to solve routine problems successfully, to develop computational skills, and to build confidence.

Presentation

Topics are presented in brief sections that progress logically from basic to more difficult mathematical concepts and skills. The discussions of new concepts include clear, succinct explanations and numerous illustrative examples. These examples are worked out in detail, and they provide the student with a strategy or model for working similar problems.

Wherever appropriate, specific problem-solving procedures are also given. To develop students' geometric intuition, there is an emphasis on graphing techniques and illustrations.

Problem Sets

Problems at the end of each section begin with simple drill-type problems and progress gradually to more challenging ones. Most of the problems are in the middle range of difficulty. There is an extensive set of review problems at the end of each chapter, covering all the techniques and concepts in the chapter. Many of the 3,587 problems in this edition are new.

Odd-numbered problems. Most of the odd-numbered problems are similar in scope to the worked-out examples in the text. Answers to all odd-numbered problems, including graphs, are given in the back of the book.

Even-numbered problems. Many of the even-numbered problems are also similar to the worked-out examples. However, some are more challenging and probe for a deeper understanding of concepts.

Applications

The applied problems and examples in this edition cover a wider variety of fields—engineering, geometry, business, economics, medicine, navigation, and the physical and social sciences.

Use of Calculators

Today, most students who study mathematics at the precalculus level own a calculator with function keys. The book contains many examples and problems that make use of a calculator, all of them identified with the symbol [c]. We introduce the calculator in Section 1.1 and offer suggestions for the proper use of this important tool throughout the book. Tables of logarithms, trigonometric functions, and exponential function values are nevertheless included in the Appendix, and their use is carefully explained.

Major Changes

In this Fourth Edition, we have rewritten and reorganized the presentation of some topics and added new material to reflect the needs and interests of students and the goals of the course. The main changes are as follows:

Complex Numbers are now introduced in Chapter 1 where they are used to solve quadratic equations, which have been moved to this chapter.

Polynomial and Rational Inequalities are now covered together in Chapter 1.

Circles and Lines. The function concept is carefully developed after the graphs of circles and straight lines have been considered.

Graphing Techniques, utilizing shrinking, stretching, and shifting, are presented in Chapter 2 and then applied throughout the text. More material on graphing has been added to Chapter 3, including the graphing of power and polynomial functions using synthetic division. The graphs in this book were generated on computers to ensure their accuracy.

Approximating Irrational Zeros of Polynomials. Section 3.5 discusses the bisection method for approximating irrational zeros of polynomials. Several examples illustrate the interplay between Descartes' Rule of Signs and upper and lower bounds for the zeros. In addition, the discussion of complex zeros of polynomials has been moved to Chapter 3.

Mathematical Models and Applications. Many applications and mathematical models in the life sciences, social sciences, economics, and business have been added throughout the book. Sections devoted entirely to applications are 4.4, 5.6, 6.6, 6.10, and 7.6.

Trigonometry. Chapters 5 and 6 on trigonometric functions and analytic trigonometry have been rewritten and reorganized. The trigonometric functions of angles and right triangle trigonometry are introduced early in Chapter 5. The transition from real number domains to angle domains follows immediately in a more integrated fashion. The basic trigonometric identities derived in Chapter 5 are used in Chapter 6 to establish other useful trigonometric identities and formulas for sums, differences, and multiples of angles. The sum and difference formulas are now covered in separate sections, which have been rewritten.

Simple Harmonic Motion. A new subsection on simple harmonic motion has been added to Chapter 5 to illustrate this important application of the trigonometric functions of real numbers.

The Polar Coordinate System is expanded to include graphs of polar equations (Section 6.9) and polar graphs of the conics (Section 8.5). Equations of the conics are also discussed in parametric form (Section 8.5).

Algebra of Matrices. The coverage of matrices in Chapter 7 has been expanded to include the addition, subtraction, and multiplication of matrices. The use of the inverse of a matrix in solving linear systems is also explained.

Partial Fractions. A new section on partial fraction decompositions has been added to Chapter 7.

Rotation of Axes is a new section added to Chapter 8. It allows us to convert a general second degree equation of a rotated conic to standard form.

Arithmetic and Geometric Sequences. A new section on arithmetic and geometric sequences has been added to Chapter 9. Summation notation and combinatorial notation are given more attention.

Student Aids

The *Study Guide* is a supplementary learning resource, offering student tutorial aid on each topic in the textbook. It is a useful source of drill problems for students who need additional reinforcement and practice, who have missed classes and need to catch up, or who want assurance that they are ready for examinations. The *Study Guide* contains study objectives, fill-in statements, and problems (broken down into simple steps), and a self-test for each chapter. Answers to all problems and tests in the *Study Guide* are included.

Instructor Aids

Instructor's Resource Manual. This Manual provides a comprehensive testing program, closely coordinated with the textbook. It includes: (1) a diagnostic examination, (2) a syllabus with teaching suggestions, (3) two examinations for each chapter, one of which is multiple-choice, and (4) a multiple-choice final examination.

Solutions Manual. The step-by-step solution to each problem in the textbook is available in this manual for instructors. A glance at the worked-out solutions will assist instructors in selecting the appropriate problems to assign.

Acknowledgments

In preparing this edition, we have drawn from our own experience in teaching from *Precalculus*, as well as the feedback provided by our students. The suggestions obtained from instructors using the third edition and the appraisals offered by our colleagues at *Macomb Community College* were of great help. We wish to thank all of these people and, in particular, to express our gratitude to the following: James Bright, *Clayton Junior College*, Merle Friel, *Humboldt State University*, Louis Hoelzle, *Bucks County Community College*, Joan Levine, *Kean College of New Jersey*, Thomas Sharp, *West Georgia College*, Alexandra Tauson, *Community College of Allegheny County*, Howard Taylor, *West Georgia College*, Henry Tjoelker, *California State University at Sacramento*, and Ann Wagner, *Towson State University*.

A special tribute is due to Professor Steve Fasbinder of *Oakland University* for his detailed reviewing, proofreading, and checking of all the problems in the book, and to our colleagues Professors S. Hukku and Allen Zerbst for working and checking the problems in the *Solutions Manual*. We also thank the people at Worth Publishers for their assistance.

January, 1985

M.A.M.
J.P.Y.

Contents

Precalculus

CHAPTER 1

Fundamentals of Algebra

The basic concepts of algebra are reviewed in this chapter. The topics covered include sets of real numbers, polynomials, fractions, exponents, radicals, equations, inequalities, and complex numbers.

1.1 Sets of Real Numbers

We often use the term **set** to refer to a collection of numbers. Each number in a given set is called a **member** or **element** of the set. Capital letters such as A, B, C, and D, or braces $\{ \}$ enclosing the elements in a set are used to denote sets. Thus, $A = \{1, 2, 3\}$ denotes a set A consisting of elements 1, 2, and 3.

The set of **real numbers** can be thought of as the collection of all numbers that can be written as decimal numbers. Often the symbol $\mathbb{R}$ is used to represent the set of real numbers. The set $\mathbb{R}$ contains some special subcollections or **subsets** that are defined as follows:

1. The set $\{1, 2, 3, \ldots\}$ consists of all numbers used for counting. This set is referred to as the set of **counting numbers,** or **natural numbers,** or **positive integers;** it is denoted by $\mathbb{N}$.

2. The **set of integers** is represented by I and is defined as

$$\mathrm{I} = \{\ldots, -4, -3, -2, -1, 0, 1, 2, 3, 4, \ldots\}$$

The set I consists of the positive integers, the negative integers, and zero.

3. The set of **rational numbers** consists of all real numbers that *can* be written as the ratio of two integers, that is, numbers that can be written in the form a/b, where a and b are integers and $b \neq 0$.

Examples of rational numbers are

$$-\frac{5}{7}, \qquad 2\frac{1}{2} = \frac{5}{2}, \qquad 3 = \frac{3}{1}, \qquad \text{and} \qquad \frac{53}{100}$$

The set of rational numbers is denoted by the symbol $\mathbb{Q}$. Every rational number can be expressed as a decimal number by using division. For instance,

$$\frac{3}{5} = 0.6000\ldots = 0.6\overline{0}$$

$$\frac{1}{3} = 0.3333\ldots = 0.\overline{3}$$

$$-\frac{1310}{99} = -13.232323\ldots = -13.\overline{23}$$

The decimal forms in the above examples are called **repeating decimals,** and the bar is used to identify the block of digits that repeats itself. When the repeating block consists only of the digit 0, the number is also referred to as a **terminating decimal.** For example,

$$2 = 2.0 \qquad \text{and} \qquad \frac{7}{8} = 0.875$$

are terminating decimals. In summary, a rational number *can* be expressed as a ratio of two integers, as a terminating decimal, or as a nonterminating repeating decimal.

4. The **irrational numbers** are the real numbers that *cannot* be written as the ratio of two integers. A real number is irrational if and only if its decimal representation is **nonterminating** and **nonrepeating.**
   Examples of irrational numbers are $\sqrt{2}$, $\sqrt[3]{5}$, and π.

Whereas the decimal representation of a rational number will either terminate or contain a repeating block of digits, the decimal representation of an irrational number *never* forms a repeating pattern. As a result, the decimal representation of an irrational number is impossible to determine. We can, however, use an arithmetic process or a calculator to obtain approximations of an irrational number. For example, if we use an arithmetic process to approximate $\sqrt{2}$, we obtain successively

$$1.4,\ 1.41,\ 1.414,\ 1.4142,\ \ldots$$

If we use a ten-digit calculator to find an approximate decimal value for $\sqrt{2}$, the number we obtain is 1.414213562.

The Use of Calculators

Most of the calculations done today are performed on scientific calculators with keys for exponential, logarithmic, and trigonometric functions. After acquiring such a calculator, it is important to study the instruction booklet and to practice basic decimal conversions. In particular, practice is required to become proficient at calculations involving more than one step in which storage keys and/or parentheses keys are used. The symbol **C** is used in this text to indicate *when* to use a calculator. Using it to solve problems *not* marked by a **C** may not only waste time but actually hinder your understanding of the underlying concepts.

Normally, we leave real number answers in fractional or radical form, unless approximations are asked for or calculator use is recommended. For instance, an approximate decimal representation of $\sqrt{2}$ is 1.414213562, of $\sqrt[3]{5}$ is 1.709975947, of $1/3$ is 0.333333333, and of π is 3.141592654. In this book, we use the *equal sign* to represent the approximate decimal representation of a real number. Thus, we can write the above approximate decimal representations as

$$\sqrt{2} = 1.414213562, \qquad \sqrt[3]{5} = 1.709975947,$$

$$\frac{1}{3} = 0.333333333, \qquad \text{and} \qquad \pi = 3.141592654$$

Most calculators can display numbers up to 10 digits. We need to decide how many of these digits are *significant*. For example, if the number 120 is *rounded off* to the nearest ten, then 120 has two significant digits; namely, 1 and 2. On the other hand, if 120 is rounded off to the nearest unit, then 0 is significant, and 120 has three significant digits. In this book, we assume the greatest possible number of significant digits unless stated otherwise. For example,

5832 has four significant digits;

702 has three significant digits;

28,000 has five significant digits;

0.00017 has two significant digits;

0.0008 has one significant digit;

1.00007 has six significant digits.

EXAMPLE 1 [c] Change each number to a decimal form, then round off the result to five significant digits.

(a) $\frac{156}{199}$ (b) $\sqrt{237}$
(c) $(5.7)^3$ (d) $\sqrt[3]{25}$

SOLUTION The solutions are given in Table 1. Note that in parts (c) and (d) we utilize a $\boxed{y^x}$ or a comparable key along with the fact that $\sqrt[3]{25} = 25^{\frac{1}{3}}$.

Table 1.

Given Number	Calculator Value	Rounded Off to Five Significant Digits
(a) $\frac{156}{199}$	0.783919598	0.78392
(b) $\sqrt{237}$	15.39480432	15.395
(c) $(5.7)^3$	185.193	185.19
(d) $\sqrt[3]{25}$	2.924017738	2.9240

The Number Line and Absolute Value

The set of real numbers can be represented *geometrically* as the set of all points on the **number line** or **coordinate axis.** By repeating the scale unit used to construct the line and by moving from left to right, starting at the point representing 0, called the **origin,** we can associate the set of positive integers $\mathbb{N} = \{1, 2, 3, 4, \ldots\}$ with equispaced points on the line. Moving from right to left, starting at 0, we can associate the set of negative integers $\{-1, -2, -3, -4, \ldots\}$ with equispaced points on the line (Figure 1). The remaining real numbers can be "located" or "plotted" on the real line by using *decimal representations* as illustrated in Figure 1. The number that is associated with a point on the number line is called the **coordinate** of the point. Note that the number is not the point, nor is the point the number. The point *represents* the number. It is customary, however, to use the words *real number* and *point* interchangeably. Thus we often speak of the *point* $\frac{3}{4}$ rather than the *point corresponding to the real number* $\frac{3}{4}$.

Figure 1

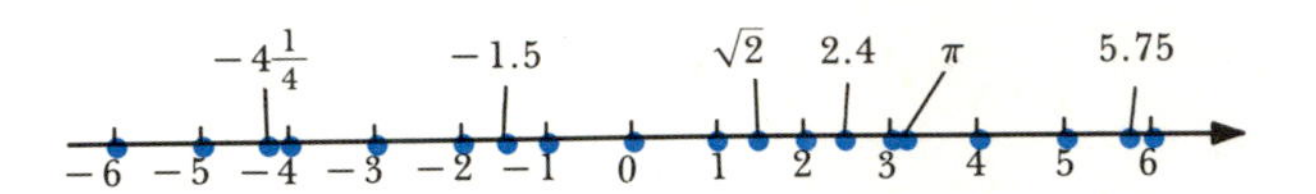

The real numbers to the right of zero are the **positive** numbers (Figure 2a) and the real numbers to the left of zero are the **negative** numbers (Figure 2b). The positive numbers and zero together are called the **nonnegative** numbers.

Figure 2

If x is a real number, then $|x|$ represents the (nonnegative) number of units of distance on the number line between the point with coordinate x and the point with coordinate 0. The nonnegative number $|x|$ is called the **absolute value** of x. For instance, both the point with coordinate -4 and the point with coordinate 4 are 4 units from the origin (Figure 3); that is,

$$\begin{aligned} |-4| &= |4| \\ &= 4 \end{aligned}$$

Figure 3

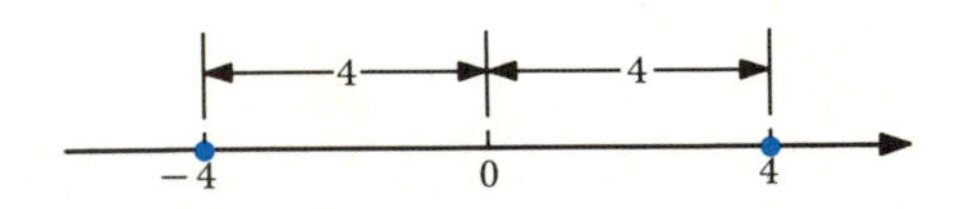

More formally, we have the following definition.

DEFINITION 1 **Absolute Value**

If x is a real number, then the **absolute value** of x, denoted by $|x|$, is defined as follows:

$$|x| = \begin{cases} x \text{ if } x \text{ is positive or zero} \\ -x \text{ if } x \text{ is negative} \end{cases}$$

EXAMPLE 2 Use Definition 1 to find the value of each expression.

(a) $|7|$ (b) $|-7|$ (c) $|0|$ (d) $|3 - \sqrt{3}|$ (e) $|\sqrt{3} - 3|$

SOLUTION

(a) $|7| = 7$ because 7 is positive.

(b) $|-7| = -(-7) = 7$ because -7 is negative.

(c) $|0| = 0$

(d) $|3 - \sqrt{3}| = 3 - \sqrt{3}$ because $3 - \sqrt{3}$ is positive.

(e) $|\sqrt{3} - 3| = -(\sqrt{3} - 3) = -\sqrt{3} + 3 = 3 - \sqrt{3}$ because $\sqrt{3} - 3$ is negative.

Notice in Example 2 that

$$|7| = |-7| \quad \text{and} \quad |3 - \sqrt{3}| = |\sqrt{3} - 3|$$

In general,

$$|x| = |-x| \text{ for every real number } x$$

The concept of absolute value can be used to find the distance between any two points on a number line. For example, consider the points with coordinates 3 and 8 (Figure 4). Note that the distance d between 3 and 8 is 5 units and that

Figure 4

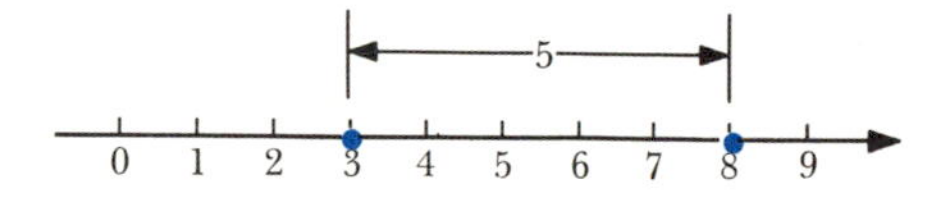

$d = |8 - 3| = 5$ or $d = |3 - 8| = |-5| = 5$. Thus, the distance d between 3 and 8 can be found by determining the absolute value of the difference between 3 and 8 regardless of whether we subtract 3 from 8 or 8 from 3. In general, the distance d between two points on the number line whose coordinates are x and y is given by

$$d = |x - y|$$

The statement holds regardless of which point is to the left of the other.

EXAMPLE 3 Find the distance d between the two points on the number line whose coordinates are the given numbers.

(a) -3 and -5

(b) -5 and 4

(c) 4 and a

SOLUTION Using the formula for d, we have

(a) $d = |-3 - (-5)| = |-3 + 5| = |2| = 2$

(b) $d = |-5 - 4| = |-9| = 9$

(c) $d = |4 - a| = |a - 4|$

Basic Algebraic Properties of Real Numbers

We assume that you are familiar with the operations of addition and multiplication on the set of real numbers. The basic *algebraic* properties of real numbers are:

Let a, b, and c be real numbers.

1. The Closure Properties

(i) $a + b$ is a real number (ii) $a \cdot b$ is a real number

2. The Commutative Properties

(i) $a + b = b + a$ (ii) $a \cdot b = b \cdot a$

3. The Associative Properties

(i) $(a + b) + c = a + (b + c)$ (ii) $(a \cdot b) \cdot c = a \cdot (b \cdot c)$

4. The Distributive Properties

(i) $a \cdot (b + c) = a \cdot b + a \cdot c$ (ii) $(b + c) \cdot a = b \cdot a + c \cdot a$

5. The Identity Properties

(i) $a + 0 = 0 + a = a$ (ii) $a \cdot 1 = 1 \cdot a = a$

6. The Inverse Properties

(i) For each real number a, there is a real number called the **additive inverse,** denoted by $-a$, such that

$$a + (-a) = (-a) + a = 0$$

(ii) For each real number $a \neq 0$, there is a real number called the **multiplicative inverse** or **reciprocal,** denoted by $1/a$ or a^{-1}, such that

$$a \cdot \frac{1}{a} = \frac{1}{a} \cdot a = 1$$

A set that satisfies the above properties is referred to as a **field.** For this reason, Properties 1–6 are sometimes called the **field properties** of real numbers. The following important properties can be derived from the field properties. Assume that a, b, x, and y are real numbers.

7. The Cancellation Properties

(i) If $a + x = a + y$, then $x = y$ (ii) If $ax = ay$ and $a \neq 0$, then $x = y$

8. The Zero-Factor Properties

(i) $a \cdot 0 = 0 \cdot a = 0$
(ii) $a \cdot b = 0$ if and only if $a = 0$ or $b = 0$ or both

9. The Properties of Negation

(i) $-(-a) = a$ (ii) $(-a)(b) = a(-b) = -(ab)$
(iii) $(-a)(-b) = ab$

EXAMPLE 4 State the property that justifies each of the following equalities.

(a) $\frac{5}{3} \cdot (-\frac{7}{2}) = (-\frac{7}{2}) \cdot \frac{5}{3}$
(b) $(2 \cdot 5) \cdot a = 2 \cdot (5a)$
(c) $14 \cdot (3 + y) = 14 \cdot 3 + 14y$
(d) $11 \cdot \frac{1}{11} = 1$
(e) $7 + (-7) = 0$
(f) $9 \cdot 0 = 0$
(g) $-(-4) = 4$
(h) $(-3)(-b) = 3b$
(i) If $3 + y = 3 + z$, then $y = z$.

SOLUTION

(a) $\frac{5}{3} \cdot (-\frac{7}{2}) = (-\frac{7}{2}) \cdot \frac{5}{3}$ [Commutative property 2(ii)]
(b) $(2 \cdot 5) \cdot a = 2 \cdot (5a)$ [Associative property 3(ii)]
(c) $14 \cdot (3 + y) = 14 \cdot 3 + 14y$ [Distributive property 4(i)]
(d) $11 \cdot \frac{1}{11} = 1$ [Multiplicative inverse property 6(ii)]
(e) $7 + (-7) = 0$ [Additive inverse property 6(i)]
(f) $9 \cdot 0 = 0$ [Zero-factor property 8(i)]
(g) $-(-4) = 4$ [Negation property 9(i)]
(h) $(-3)(-b) = 3b$ [Negation property 9(iii)]
(i) If $3 + y = 3 + z$, then $y = z$. [Cancellation property 7(i)]

The operations of subtraction and division are defined in terms of addition and multiplication, respectively.

DEFINITION 2 **Subtraction and Division**

Assume a and b are real numbers.

(i) The **difference** between a and b is denoted by $a - b$ and is defined by

$$a - b = a + (-b)$$

(ii) The **quotient** of a and b, denoted by $a \div b$ or a/b, where $b \neq 0$, is defined by

$$\frac{a}{b} = a \cdot \frac{1}{b}$$

For example,

$$4 - (-5) = 4 + 5 = 9$$

and

$$4 \div 5 = 4 \cdot \frac{1}{5} = \frac{4}{5}$$

Division by zero is *not* defined in the real number system.

PROBLEM SET 1.1

1. List all of the elements in the given set.
(a) $A = \{-1, 8, 0, 9\}$
(b) B is the set of all odd integers between -4 and 6.
(c) C is the set of all integers between $-\frac{13}{2}$ and $\frac{7}{3}$.
(d) D is the set of all even integers between π and 2π.

2. Locate the numbers in each given set on a number line.
(a) A is the set of positive integers less than 6.
(b) B is the set of integers greater than -3 and less than 4.

3. Indicate whether each statement is true or false.

(a) 0 is an element of $\mathbb{N}$.
(b) $\frac{1}{\pi}$ is an element of $\mathbb{R}$.
(c) -5 is an element of $\mathbb{Q}$.
(d) $\frac{3}{\frac{1}{8}}$ is an element of $\mathbb{Q}$.
(e) $\sqrt{9}$ is an element of I.
(f) $\frac{5}{0}$ is an element of $\mathbb{R}$.
(g) $\frac{3}{4} + \sqrt{3}$ is an element of $\mathbb{Q}$.
(h) $\frac{0}{7}$ is an element of I.
(i) 0.0314 is an element of $\mathbb{Q}$.
(j) Every integer is a rational number.
(k) $\sqrt{13} - \sqrt{4}$ is an element of I.
(l) Every radical is an irrational number.

4. Explain why $\frac{22}{7}$ could not be the exact value of π.

In problems 5–13, express each rational number in decimal form.

5. $\frac{3}{4}$ **6.** $-\frac{7}{4}$ **7.** $-\frac{5}{6}$
8. $\frac{5}{9}$ **9.** $\frac{7}{3}$ **10.** $-\frac{6}{7}$
11. $17\frac{1}{4}$ **12.** $-6\frac{5}{33}$ **13.** $\frac{7}{330}$

14. Interest rates, profits, and losses are frequently expressed as a **percent.** The percent symbol % means "per hundred." For example,

$$7\% = \frac{7}{100} = 0.07$$

Write each percent in a decimal form.
(a) 37% (b) 400% (c) 66% (d) 33%

15. Write each terminating decimal as a ratio of two integers. Reduce each answer to lowest terms.
(a) 0.27 (b) 1.71 (c) -0.125 (d) -0.008

16. Convert each percent to a ratio of two integers. Reduce each answer to lowest terms.
(a) $33\frac{1}{3}\%$ (b) $66\frac{2}{3}\%$ (c) $5\frac{1}{2}\%$

17. A repeating decimal can be written as the ratio of two integers. For example, to convert $0.\overline{35}$ to the rational number form $\frac{a}{b}$ we can set $n = 0.\overline{35}$ and then proceed as follows:

$$\begin{array}{rl} 100n = & 35.\overline{35} \\ -\quad n = & -0.\overline{35} \\ \hline 99n = & 35 \end{array}$$

Thus,

$$n = \frac{35}{99}$$

Use this technique to write each decimal in the form $\frac{a}{b}$, where a and b are integers. Reduce each answer to lowest terms.
(a) $-3.\overline{5}$ (b) $1.\overline{71}$ (c) $23.\overline{189}$ (d) $0.\overline{9}$

18. A suit whose original price was $240 was discounted by 40%. What is the price of the suit after the discount?

In problems 19–29, rewrite each expression without the absolute value symbol.

19. $|-5|$ **20.** $|-22 + 1|$ **21.** $|-22| + |1|$ **22.** $|-25 - 4|$
23. $|-25| - |4|$ **24.** $|-7| - |-3|$ **25.** $5 - |-5|$
26. $|\pi - 5|$ **27.** $|\sqrt{2} - \pi|$ **28.** $|\sqrt{7} - 2|$ **29.** $\dfrac{-5}{|-5|}$

30. Use $x = -7$ and $y = 4$ to verify that:

(a) $|xy| = |x||y|$ (b) $\left|\dfrac{x}{y}\right| = \dfrac{|x|}{|y|}$ (c) $|x - y| = |y - x|$

In problems 31–36, find the distance between each pair of points.

31. -3 and -7 **32.** 17 and $5a$ **33.** -4 and b
34. 18 and $-4b$ **35.** -9 and $-2a$ **36.** $-7.3a$ and $5.7b$

C In problems 37–48, change each number to a decimal. Round off the result to four significant digits.

37. $\frac{126}{99}$ **38.** $-\frac{9}{111}$ **39.** $-\frac{497}{345}$ **40.** $\frac{66}{201}$
41. $-\dfrac{\sqrt{3}}{2}$ **42.** $-\dfrac{\pi}{4}$ **43.** $\dfrac{180}{\pi}$ **44.** $\left(\dfrac{\pi}{180}\right)^2$
45. $(2.718)^6$ **46.** $\left(\dfrac{\sqrt{2}}{2}\right)^3$ **47.** $\sqrt[5]{226}$ or $(226)^{\frac{1}{5}}$ **48.** $(\sqrt[3]{1111})^4$ or $(1111)^{\frac{4}{3}}$

In problems 49–64, state the basic algebraic property of real numbers that justifies each statement.

49. $5 + \sqrt{7}$ is a real number.
50. $5\sqrt{7}$ is a real number.
51. $\frac{7}{8} \cdot (-\frac{2}{3}) = (-\frac{2}{3}) \cdot \frac{7}{8}$
52. $\frac{9}{16} + (-\frac{8}{11}) = (-\frac{8}{11}) + \frac{9}{16}$
53. $a + (7 + 16) = (a + 7) + 16$
54. $(\frac{2}{7} \cdot c) \cdot d = \frac{2}{7} \cdot (c \cdot d)$
55. $8 \cdot (y + b) = 8 \cdot y + 8 \cdot b$
56. $1 \cdot c = c$
57. $(5 + 3) \cdot a = 5 \cdot a + 3 \cdot a$
58. $\sqrt{3} + (-\sqrt{3}) = 0$
59. $(-2) \cdot (-3) = 6$
60. If $5a = 0$, then $a = 0$.
61. If $3x = 3y$, then $x = y$.
62. If $3x + y = 3x + 5z$, then $y = 5z$.
63. $5 \cdot 0 = 0$
64. $-(-3) = 3$

1.2 Polynomials

In this section we discuss polynomials and algebraic expressions. We begin by reviewing some concepts of *algebra*. When a letter is used to represent any one of the members of a set of real numbers, it is called a **variable.** Letters used to designate fixed but unspecified numbers are called **constants;** specific numbers are also considered to be constants.

Positive integer exponents are used as a shorthand notation for representing repeated multiplication. The **exponential notation** x^n denotes a product in which x is used as a factor n times, that is,

$$x^n = \overbrace{x \cdot x \cdot x \cdots x}^{n \text{ factors}}$$

x^n is read as the ***n*th power** of x; x is called the **base** and n is called the **exponent** of the expression x^n.

For example, $5^3 = 5 \cdot 5 \cdot 5$; 5 is the base, and 3 is the exponent. Similarly, $(x + y)^2 = (x + y)(x + y)$; $x + y$ is the base, and 2 is the exponent. Also

$$x^1 = x$$

Expressions formed by any combination of addition, subtraction, multiplication, exponentiation, or extraction of roots of numbers and variables are called **algebraic expressions.** For example,

$$-2, \qquad 5x + 7, \qquad y^2 + 5y - 3, \qquad \frac{6x^3y^2 + 4}{y^3 - x}, \qquad \text{and} \qquad \sqrt{x^2 + y^2}$$

are algebraic expressions, where x and y are variables. If specific numbers are substituted for the variables in an algebraic expression, the resulting real number is called the *value* of the expression for these numbers. For example, if we replace x by -1 and y by 2 in the expression $\frac{6x^3y^2 + 4}{y^3 - x}$, we obtain

$$\frac{6(-1)^3(2^2) + 4}{2^3 - (-1)} = \frac{-24 + 4}{8 + 1} = \frac{-20}{9}$$

An algebraic expression that results from applying the operations of addition, subtraction, and multiplication *only* on a set of real numbers and variables is called a **polynomial.** Real numbers are also considered to be polynomials. The expressions $5x + 2$, $3x^2 - 2x + 7$, and $4xy^4 - 3xy + 5x + 6y^2$ are examples of polynomials. However,

$$\frac{6x^3 + 4}{2x^2 + 5} \qquad \text{and} \qquad 5y^2 + \sqrt{y} - 7$$

are not polynomials, since the first expression involves division by a variable expression and the second expression contains a root of a variable.

The expression $5x + 2$ is a polynomial *in one variable*. In this case the variable is x and the polynomial is the sum of the two **terms** $5x$ and 2. The number 5 in the term $5x$ is called its **coefficient,** and the term 2, which contains no variable, is called the **constant term.** The polynomial $3x^2 - 2x + 7$ in one variable is the sum of three terms $3x^2$, $-2x$, and 7; the coefficients are 3, -2, and 7, respectively. The polynomial $4xy^4 - 3xy + 5x + 6y^2$ *in two variables* is the sum of four terms $4xy^4$, $-3xy$, $5x$, and $6y^2$ whose *numerical* coefficients are 4, -3, 5, and 6, respectively. The **degree of a polynomial** in one variable is the highest exponent of that variable in any term of the polynomial. For instance, the polynomial $3x^2 + x - 3$ is of degree 2 in x; the polynomial $\frac{1}{2}x^5 + 1$ is of degree 5 in x; and the polynomial $z^5 - 3z^4 + 5z^2 + z + 3$ is of degree 5 in z.

In the polynomial of two variables

$$3xy^2 - xy + \frac{1}{2}x - \sqrt{2}y$$

the numerical coefficients are 3, -1, $\frac{1}{2}$, and $-\sqrt{2}$; and the degree is 3, the highest sum of the degrees of the variables for any one of the four terms in the expression.

A polynomial containing one term is called a **monomial;** a polynomial of two terms is called a **binomial;** a polynomial of three terms is called a **trinomial.** For example,

$3x^2 + x - 3$ is a trinomial;

$\frac{1}{2}x^5 + 1$ is a binomial;

$x^3y^4 - 3xy + 5x$ is a trinomial in two variables.

Terms of a polynomial, such as $5x^2y$ and $-7x^2y$, which differ only in their numerical coefficients, are called **like terms.**

Polynomials can be *added* and *subtracted* by using the algebraic properties in Section 1.1 to group and combine like terms.

EXAMPLE 1 Perform each operation.

(a) $(4x^3 + 7x - 13) + (-2x^3 + 5x + 17)$

(b) $(2x^3 + 3x^2 - 5x + 11) - (4x^3 - 5x^2 + 9)$

SOLUTION (a) We apply the associative and commutative properties of addition to group the like terms. Then we combine the like terms by using the distributive property to get

$$\begin{aligned}
&(4x^3 + 7x - 13) + (-2x^3 + 5x + 17)\\
&\quad = (4x^3 - 2x^3) + (7x + 5x) + (-13 + 17) && \text{(Rearrange and regroup)}\\
&\quad = (4 - 2)x^3 + (7 + 5)x + (-13 + 17) && \text{(Distributive property)}\\
&\quad = 2x^3 + 12x + 4
\end{aligned}$$

(b) First, we use the definition of subtraction to rewrite the subtraction problem as an addition problem, and then we proceed as explained below.

$$\begin{aligned}
&(2x^3 + 3x^2 - 5x + 11) - (4x^3 - 5x^2 + 9)\\
&\quad = (2x^3 + 3x^2 - 5x + 11) + (-4x^3 + 5x^2 - 9) && \text{(Definition of subtraction)}\\
&\quad = (2x^3 - 4x^3) + (3x^2 + 5x^2) - 5x + (11 - 9) && \text{(Rearrange and regroup)}\\
&\quad = -2x^3 + 8x^2 - 5x + 2 && \text{(Combine like terms)}
\end{aligned}$$

To *multiply* two or more monomials, we use the commutative and associative properties of multiplication and the following properties of exponents.

Product Properties of Exponents

Suppose x and y are real numbers. If m and n are positive integers, then

(i) $x^m \cdot x^n = x^{m+n}$ (ii) $(x^m)^n = x^{mn}$ (iii) $(xy)^n = x^n y^n$

We verify property (i) as follows:

$$x^m \cdot x^n = \overbrace{(x \cdot x \cdots x)}^{m \text{ factors}} \cdot \overbrace{(x \cdot x \cdots x)}^{n \text{ factors}}$$

$$= \overbrace{x \cdot x \cdot x \cdot x \cdots x \cdot x}^{m + n \text{ factors}} = x^{m+n}$$

The verifications of properties (ii) and (iii) are left as exercises (Problem 28).

EXAMPLE 2 Determine each of the given products of polynomials.

(a) $(-2x)(7x^3)$ (b) $(-2x^2y)^2(3x^3y^2)^3$

SOLUTION We apply the associative and commutative properties of multiplication along with the properties of positive integer exponents.

(a) $(-2x)(7x^3) = (-2 \cdot 7)(x \cdot x^3) = -14x^4$

(b) $(-2x^2y)^2(3x^3y^2)^3 = [(-2)^2(x^2)^2(y)^2] \cdot [(3)^3(x^3)^3(y^2)^3]$ (Property iii)

$= (4x^4y^2)(27x^9y^6)$ (Property ii)

$= (4 \cdot 27)(x^4 \cdot x^9)(y^2 \cdot y^6) = 108x^{13}y^8$ (Property i)

To multiply polynomials with more than one term, we use the distributive property, together with the properties of exponents, and then we combine like terms.

EXAMPLE 3 Perform each multiplication.

(a) $(2x^2y^3)(4xy^2 - 3x^2y^5 - 8)$ (b) $(x^2 - 2x + 1)(x^2 + x + 2)$

SOLUTION (a) $(2x^2y^3)(4xy^2 - 3x^2y^5 - 8)$

$= (2x^2y^3)(4xy^2) + (2x^2y^3)(-3x^2y^5) + (2x^2y^3)(-8)$ (Distributive property)

$= 8x^3y^5 - 6x^4y^8 - 16x^2y^3$ (Properties of exponents)

(b) It is sometimes easier to perform the multiplication of polynomials with more than *two* terms by using the following type of *vertical arrangement*:

$$\begin{array}{rrrrr}
 & & x^2 & -2x & +1 \\
 & & x^2 & +x & +2 \\
\hline
x^4 & -2x^3 & +x^2 & & \\
 & x^3 & -2x^2 & +x & \\
 & & 2x^2 & -4x & +2 \\
\hline
x^4 & -x^3 & +x^2 & -3x & +2
\end{array}$$

Certain types of products of polynomials occur often enough in algebra to be worthy of special consideration. We list these types of products below.

Special Products

Assume that a and b represent real numbers.

1. $(a + b)^2 = a^2 + 2ab + b^2$
2. $(a - b)^2 = a^2 - 2ab + b^2$
3. $(a - b)(a + b) = a^2 - b^2$
4. $(a + b)(a^2 - ab + b^2) = a^3 + b^3$
5. $(a - b)(a^2 + ab + b^2) = a^3 - b^3$
6. $(a + b)^3 = a^3 + 3a^2b + 3ab^2 + b^3$
7. $(a - b)^3 = a^3 - 3a^2b + 3ab^2 - b^3$

EXAMPLE 4 Multiply each expression by using the special products.

(a) $(x + 2y)^2$

(b) $(2x - 5y)^3$

(c) $(2x^2 + 3y)(2x^2 - 3y)$

SOLUTION (a) We use Special Product 1 to obtain

$$(x + 2y)^2 = x^2 + 2(x)(2y) + (2y)^2 = x^2 + 4xy + 4y^2$$

(b) Using Special Product 7, we have

$$\begin{aligned}(2x - 5y)^3 &= (2x)^3 - 3(2x)^2(5y) + 3(2x)(5y)^2 - (5y)^3 \\ &= 8x^3 - 60x^2y + 150xy^2 - 125y^3\end{aligned}$$

(c) Using Special Product 3, we have

$$(2x^2 + 3y)(2x^2 - 3y) = (2x^2)^2 - (3y)^2 = 4x^4 - 9y^2$$

Factoring Polynomials

If a polynomial is expressed as the product of two or more polynomials, then each polynomial in the product is called a **factor** of the original polynomial. The process is known as **factoring** a polynomial. A polynomial which has no factors other than itself and 1 or its negative and -1 is called a **prime** polynomial. To factor a polynomial, we express it as the product of prime polynomials, or as the product of a monomial and prime polynomials. Once such a factorization is accomplished, we refer to the final result as a **complete factorization.** Unless otherwise specified, we accept as factors of any polynomial with integral coefficients only those prime factors that also contain integral coefficients.

The most familiar kind of factoring is removing the **greatest common factor.**

This process is accomplished by using the distributive property as illustrated below.

EXAMPLE 5 Factor each polynomial by removing the greatest common factor.

(a) $2x^2 + 5x$

(b) $3x^2y^5 - 5xy^3$

SOLUTION

(a) $2x^2 + 5x = x(2x + 5)$

(b) $3x^2y^5 - 5xy^3 = xy^3(3xy^2) + xy^3(-5) = xy^3(3xy^2 - 5)$

Some factoring depends on recognizing polynomials that fit the forms of the special products listed on page 14. The next example illustrates the use of these formulas.

EXAMPLE 6 Factor each of the given polynomials by using the special products.

(a) $16x^2 - 49y^2$ (b) $8x^3 + y^3$ (c) $x^3 - 27y^3$

SOLUTION

(a) We apply Special Product 3 to get

$$16x^2 - 49y^2 = (4x)^2 - (7y)^2 = (4x - 7y)(4x + 7y)$$

(b) By using Special Product 4, we obtain

$$8x^3 + y^3 = (2x)^3 + y^3 = (2x + y)(4x^2 - 2xy + y^2)$$

(c) By using Special Product 5, we have

$$x^3 - 27y^3 = x^3 - (3y)^3 = (x - 3y)(x^2 + 3xy + 9y^2)$$

It is often possible to factor *trinomials* of the form

$$ax^2 + bx + c$$

where a, b, and c are integers and $a \neq 0$, into the product of two binomials. This is accomplished by trying various combinations until one is found that satisfies the equation

$$ax^2 + bx + c = (sx + p)(rx + q)$$

such that

$$ax^2 + bx + c = (sr)\,x^2 + \underbrace{(sq + pr)}\,x + (pq)$$

We try different numerical factors s and r (including 1 and -1) to obtain the product a, in combination with the various numerical factors p and q to obtain c, until we find the combination that yields the "middle term" b. This is illustrated in the following example. We should note that sometimes such a combination may not exist in the real number system.

EXAMPLE 7 Factor each trinomial.

(a) $3x^2 + x - 2$ (b) $12x^2 + x - 20$

SOLUTION (a) The only integer factors of 3 are ± 1 and ± 3, and the only integer factors of -2 are ± 1 and ± 2. After trying different combinations, we discover that

$$3x^2 + x - 2 = (3x - 2)(x + 1)$$

(b) Here 12 can be factored as $1 \cdot 12$, $2 \cdot 6$, or $3 \cdot 4$, and 20 can be factored as $1 \cdot 20$, $2 \cdot 10$, or $4 \cdot 5$. After trying various combinations with different algebraic signs, we find that

$$12x^2 + x - 20 = (4x - 5)(3x + 4)$$

Sometimes the terms of a given polynomial have to be *rearranged and grouped* before they can be factored. This is illustrated in the next example.

EXAMPLE 8 Factor the polynomial

$$x^2 + x - y^2 - y$$

SOLUTION We rearrange, group, and then factor as shown below:

$$\begin{aligned} x^2 + x - y^2 - y &= (x^2 - y^2) + (x - y) \\ &= (x + y)(x - y) + (x - y) \\ &= [(x + y) + 1]\,(x - y) \\ &= (x + y + 1)\,(x - y) \end{aligned}$$

PROBLEM SET 1.2

In problems 1–4, write each expression in equivalent exponential form.

1. $6 \cdot 6 \cdot 6 \cdot 6 \cdot 6 \cdot 6$

2. $x \cdot x \cdot x \cdot x \cdot x$

3. $2x \cdot x \cdot x \cdot y \cdot y - 3x \cdot x \cdot x$

4. $5x \cdot x \cdot x \cdot y \cdot y + 7y \cdot z \cdot z$

In problems 5–14, identify which of the algebraic expressions are polynomials. For each polynomial determine the degree and the nonzero numerical coefficients, and specify whether the polynomial is a monomial, binomial, or trinomial.

5. $4x^2 - x - 7$

6. 6

7. $3x^5 - x^3 + \dfrac{1}{x}$

8. $\dfrac{x^2 - 3x + 1}{2}$

9. $-11x^4$

10. $-8x + x^2$

11. $\frac{1}{2}x^5 + \pi x$

12. $\frac{4}{5}x^2 - x^4 + \sqrt{2}x$

13. $(-2x^5)^4$

14. $\sqrt{x + 2}$

In problems 15–18, find the value of the algebraic expression for the given values of the variables.

15. $2x^3y + 5xy^5$; $x = -2$ and $y = 2$

16. $\dfrac{7xy^2 + 8x^3y}{3x^2 - 1}$; $x = 2$ and $y = -1$

[C] **17.** $\sqrt{\dfrac{3x^3 - y^2}{5xy + 7}}$, to four decimal places; $x = 3.21$ and $y = 7.4$

[C] **18.** $\left(1 + \dfrac{r}{t}\right)^{tn}$, to four decimal places; $r = 0.08$, $t = 4$, and $n = 6$

In problems 19–27, perform the specified operations.

19. $5xy^2 - 3x^2y^2 - 2xy^2 + x^2y^2$

20. $(2x^2 - 3x - 1) + (5x^2 - 2x + 4)$

21. $(-7x^3 + 4x + 3) + (3x^3 + 5x^2 - 7x - 8)$

22. $(4x^2 - 7x - 8) - (-2x^2 + 4x - 2)$

23. $(3x^2 + 5x + 7) - (x^2 + x + 21)$

24. $(3x^4 - 4x^3 + 6x^2 + x - 1) - (4 - x + 2x^2 - 3x^3 - x^4)$

25. $(2x^3 - 5x^2 + 7x - 1) + (x^3 + 6x^2 - 7x + 2) - (3x^3 + 4x - 1)$

26. $(3x^4 - 8x^3 + x - 7) - (-2x^4 + x^3 - 5x - 1) - (5x^4 - 2x^3 + 8x - 3)$

27. $2(6x^3 - 8x^2 + 5x + 1) - 3(-3x^3 + 6x^2 - 3x + 2)$

28. (a) Verify that $(x^m)^n = x^{mn}$, where m and n are positive integers.
(b) Verify that $(xy)^n = x^ny^n$, where n is a positive integer.

In problems 29–38, use the properties of positive integer exponents to multiply the monomials.

29. $(11x^4)(-2x^7)$

30. $(2x^4)(-3x^4)(7x)$

31. $(-2x^2y^3)(3xy^5)(-5x^2y^2)$

32. $(4x^{2m})(2x^{5m})$

33. $(3x^2yz^3)^3$

34. $(-x^{11})^2(x^2)^{11}$

35. $(-5x^4y^2)^2(7xy^2)^3$

36. $(2x^n)^3(-3x^n)^2$

37. $(6xy)^2(-2xy)^3(-x^3y^5)^2$

38. $[(-5x)^2]^2[(-x)^3]^3$

In problems 39–48, determine each product.

39. $(3x - 1)(2x + 3)$

40. $(9x + 7y)(5x - 4y)$

41. $(6x - 5y)(4x + 3y)^2$

42. $3(10x - 7)(5x + 8)$

43. $2x(5x + 7)(-3x + 1)$

44. $-8x(2x + y)(y - 2x)$

45. $(2x - 1)(x^2 + 3x + 7)$

46. $(x^2 - 5)(x^2 + 8x - 11)$

47. $(x^2 - 5x + 6)(x^2 + 4x + 9)$

48. $(x^2 + 2y)^2(x^2 - xy + 4y^2)$

In problems 49–58, use the special products to perform each operation.

49. $(2x + y)^2$

50. $(3x - 5)^2$

51. $(8y - 8z)^2$

52. $(7t + 3)^2$

53. $(2p - q^2)^3$

54. $(2x^2 + 3y)^3$

55. $(5r - 7s)(5r + 7s)$

56. $(1 - 10x)(1 + 10x)$

57. $(3 + y)(9 - 3y + y^2)$

58. $(3x - 4y)(9x^2 + 12xy + 16y^2)$

In problems 59–90, factor each expression completely.

59. $9x^2 + 3x$

60. $17x^3y^2 - 34x^2y$

61. $81x^3y - 12x^2y^5$

62. $4xy^2z + x^2y^2z^2 - x^3y^3$

63. $m(x + y) + (x + y)^2$

64. $x(y - z) + (z - y)$

65. $9y^2 - 1$

66. $25 - 4a^2$

67. $16x^2 - 25y^2$

68. $x^2z^2 - 144$

69. $x^4 - 81y^4$

70. $625a^4 - 81b^4$

71. $x^3 - 8$

72. $(3x + 2y)^2 - 25b^2$

73. $125y^3 + 64$

74. $27 - y^3$

75. $(w + 1)^3 + 27$

76. $(1 - x)^3 - 64$

77. $x^2 - 16x + 63$

78. $x^2 - 9xy - 10y^2$

79. $3x^2 + 5x - 2$

80. $16 - x^2 - 6x$

81. $12x^2 + 17x - 5$

82. $10x^2 - 19x + 6$

83. $-2x^3 - 5x^2 + 12x$

84. $42x^2 + x - 30$

85. $24x^2 - 8x^3 - 2x^4$

86. $12xy + 10y^2 - 16x^2$

87. $2y - 3xy - 4 + 6x$

88. $x^2 + 2x - y^2 + 2y$

89. $25x^2 - 2y - 4y^2 + 5x$

90. $7x^3 - 28x^2 - x + 4$

1.3 Fractions

In Section 1.1, we indicated that a rational number can be represented in the form a/b, where a and b are integers and $b \neq 0$. The form a/b is called a **fraction** with **numerator** a and **denominator** b. If the numerator and denominator of a fraction are polynomials, the fraction is called a **rational expression.** Examples of rational expressions are

$$\frac{2}{3}, \quad \frac{1}{x}, \quad \frac{x+2}{x-5}, \quad \frac{7}{t^2-4}, \quad \text{and} \quad \frac{4y^3+1}{8y^4+13y^2-8}$$

Since division by zero is *not* defined, it is always understood that *the denominator of a rational expression cannot represent zero.* Thus, for

$$\frac{x+2}{x-5}, \qquad x \neq 5$$

and for

$$\frac{3x^5}{x(x-1)}, \qquad x \neq 0 \quad \text{and} \quad x \neq 1$$

For $\frac{3x^5}{x(x-1)}$, we say that $x \neq 0$ and $x \neq 1$ are the restrictions on x under which this rational expression is defined. Whenever we use a rational expression, we assume that the variables involved are restricted to numerical values that will give a nonzero denominator.

Rational expressions, as with rational numbers, can be written in **lowest terms** in which the numerator and the denominator have no common factor other than 1 and -1. This procedure is called **simplifying** or **reducing** the rational expression. The simplification technique is based on the *fundamental principle of fractions*, which holds for all rational expressions.

Fundamental Principle of Fractions

$$\frac{PK}{QK} = \frac{P}{Q} \qquad \text{where } Q \neq 0 \quad \text{and} \quad K \neq 0$$

EXAMPLE 1 Reduce each rational expression.

(a) $\frac{12x^2y}{9xy^2}$ (b) $\frac{2x^2+5x-3}{10x^2+9x-7}$

SOLUTION

(a) $\frac{12x^2y}{9xy^2} = \frac{4x(3xy)}{3y(3xy)} = \frac{4x}{3y}$

(b) $\frac{2x^2+5x-3}{10x^2+9x-7} = \frac{(2x-1)(x+3)}{(2x-1)(5x+7)} = \frac{x+3}{5x+7}$

In calculus, we often encounter rational expressions that have to be simplified, as the next example shows.

EXAMPLE 2 Simplify the expression

$$\frac{(3x-1)(2)-(2x+1)(3)}{5(3x-1)^2}$$

SOLUTION

$$\frac{(3x-1)(2)-(2x+1)(3)}{5(3x-1)^2} = \frac{(6x-2)-(6x+3)}{5(3x-1)^2}$$

$$= \frac{-\cancel{5}}{\cancel{5}(3x-1)^2}$$

$$= \frac{-1}{(3x-1)^2}$$

The negative of a rational expression can be written in the following ways:

$$\frac{-P}{Q} = \frac{P}{-Q} = -\frac{P}{Q}$$

For example,

$$\frac{-4}{3-x} = \frac{4}{-(3-x)} = \frac{4}{x-3}$$

$$= -\frac{4}{3-x}$$

Also note that

$$\frac{a-b}{b-a} = \frac{-(-a+b)}{b-a} = \frac{-\cancel{(b-a)}}{\cancel{b-a}}$$

$$= -1$$

Multiplication and Division of Fractions

The following rules are used to *multiply* and *divide* fractions.

Let P/Q and R/S be rational expressions.

Rule for Multiplication of Fractions

$$\frac{P}{Q}\cdot\frac{R}{S} = \frac{PR}{QS}$$

Rule for Division of Fractions

$$\frac{P}{Q} \div \frac{R}{S} = \frac{P}{Q} \cdot \frac{S}{R} = \frac{PS}{QR}, \text{ provided } \frac{R}{S} \neq 0$$

The key to writing the results of these operations in simplified form is factoring, as illustrated in the following example.

EXAMPLE 3 Perform each operation and simplify the result.

(a) $\dfrac{x^2 + 4x + 4}{2x^2 + 2x - 4} \cdot \dfrac{2x - 2}{x^2 + 2x}$ (b) $\dfrac{x^2 - 10x + 25}{x^2 - 100} \div \dfrac{x^2 - 7x + 10}{x^2 + 12x + 20}$

SOLUTION

(a) $$\frac{x^2 + 4x + 4}{2x^2 + 2x - 4} \cdot \frac{2x - 2}{x^2 + 2x} = \frac{(x^2 + 4x + 4)(2x - 2)}{(2x^2 + 2x - 4)(x^2 + 2x)}$$
$$= \frac{(x + 2)^2(2)(x - 1)}{(2)(x + 2)(x - 1)(x)(x + 2)} = \frac{1}{x}$$

(b) $$\frac{x^2 - 10x + 25}{x^2 - 100} \div \frac{x^2 - 7x + 10}{x^2 + 12x + 20} = \frac{x^2 - 10x + 25}{x^2 - 100} \cdot \frac{x^2 + 12x + 20}{x^2 - 7x + 10}$$
$$= \frac{(x^2 - 10x + 25)(x^2 + 12x + 20)}{(x^2 - 100)(x^2 - 7x + 10)}$$
$$= \frac{(x - 5)^2(x + 2)(x + 10)}{(x - 10)(x + 10)(x - 2)(x - 5)}$$
$$= \frac{(x - 5)(x + 2)}{(x - 10)(x - 2)}$$

Addition and Subtraction of Fractions

The following rules are used to *add* and *subtract* fractions.

Let P/Q and R/Q be rational expressions.

Rule for Addition of Fractions

$$\frac{P}{Q} + \frac{R}{Q} = \frac{P + R}{Q}$$

Rule for Subtraction of Fractions

$$\frac{P}{Q} - \frac{R}{Q} = \frac{P - R}{Q}$$

EXAMPLE 4 Perform each operation and simplify the result.

(a) $\dfrac{3}{2x+4} + \dfrac{5}{2x+4}$ (b) $\dfrac{x}{4-x^2} - \dfrac{2}{4-x^2}$

SOLUTION

(a) $$\frac{3}{2x+4} + \frac{5}{2x+4} = \frac{8}{2x+4} = \frac{\overset{4}{\cancel{8}}}{\cancel{2}(x+2)} = \frac{4}{x+2}$$

(b) $$\frac{x}{4-x^2} - \frac{2}{4-x^2} = \frac{x-2}{4-x^2}$$

$$= \frac{\overset{-1}{\cancel{x-2}}}{(2+x)\cancel{(2-x)}} = \frac{-1}{2+x}$$

To add or subtract fractions with different denominators, it is necessary *first* to convert the fractions to equivalent forms that have the same denominators. Only then can the addition and subtraction rules be applied. To convert a fraction to a "built-up" equivalent form, we reverse the reducing process by applying the fundamental principle of fractions in the form

$$\frac{P}{Q} = \frac{PK}{QK}$$

For instance,

$$\frac{3}{x+1} = \frac{3(x+2)}{(x+1)(x+2)} = \frac{3x+6}{(x+1)(x+2)}$$

Normally, it is desirable to find the **least common denominator** (LCD) of fractions in order to make the denominators the same. The LCD is found in the following way:

Step 1. Factor each denominator completely.

Step 2. List each factor with the largest exponent it has in any factored denominator.

The product of the factors so listed is the LCD.

This process is shown in the next example.

EXAMPLE 5 Perform each operation and simplify the result.

(a) $\dfrac{6}{x^2-2x-8} + \dfrac{x}{x+2}$ (b) $\dfrac{x}{x^2+6x+5} - \dfrac{2}{x^2+2x+1}$

SOLUTION (a) First we determine the LCD. The following table is helpful.

Denominators	Prime Factors
$x^2 - 2x - 8$	$(x-4)(x+2)$
$x + 2$	$(x+2)$
LCD	$(x-4)(x+2)$

Next we convert both fractions to a form that has the LCD as the denominator, and then we apply the rule for addition.

$$\begin{aligned}\frac{6}{x^2 - 2x - 8} + \frac{x}{x + 2} &= \frac{6}{(x - 4)(x + 2)} + \frac{x}{x + 2}\\ &= \frac{6}{(x - 4)(x + 2)} + \frac{x(x - 4)}{(x - 4)(x + 2)}\\ &= \frac{6 + x(x - 4)}{(x - 4)(x + 2)}\\ &= \frac{x^2 - 4x + 6}{(x - 4)(x + 2)}\end{aligned}$$

(b) Again we use a table to determine the LCD.

Denominators	Prime Factors
$x^2 + 6x + 5$	$(x + 1)(x + 5)$
$x^2 + 2x + 1$	$(x + 1)^2$
LCD	$(x + 1)^2(x + 5)$

Now we convert each fraction to a form with the LCD in the denominator, and then we apply the subtraction rule.

$$\begin{aligned}\frac{x}{x^2 + 6x + 5} - \frac{2}{x^2 + 2x + 1} &= \frac{x}{(x + 1)(x + 5)} - \frac{2}{(x + 1)^2}\\ &= \frac{x(x + 1)}{(x + 1)^2(x + 5)} - \frac{2(x + 5)}{(x + 1)^2(x + 5)}\\ &= \frac{(x^2 + x) - (2x + 10)}{(x + 1)^2(x + 5)}\\ &= \frac{x^2 - x - 10}{(x + 1)^2(x + 5)}\end{aligned}$$

Complex Fractions

Sometimes the numerator or denominator (or both) of a fraction contains one or more fractions. In such a case, the fraction is called a **complex fraction.** Examples of complex fractions are

$$\frac{\frac{1}{2}}{3}, \qquad \frac{\frac{1}{x}}{\frac{3}{x + 1} + 7}, \qquad \text{and} \qquad \frac{x - \frac{1}{x}}{\frac{2}{1 + x} - \frac{x}{1 - x}}$$

To *simplify* a complex fraction means to write the fraction in a reduced form without any fraction in the numerator or denominator. The next example illustrates one procedure for simplifying complex fractions by using division. Recall that

$$\frac{P}{Q} = P \div Q = P \cdot \frac{1}{Q}$$

EXAMPLE 6 Simplify the complex fraction

$$\frac{\frac{1}{x-y}}{\frac{1}{x+y} + \frac{1}{x-y}}$$

SOLUTION

$$\frac{\frac{1}{x-y}}{\frac{1}{x+y} + \frac{1}{x-y}} = \frac{\frac{1}{x-y}}{\frac{x-y+x+y}{x^2-y^2}} = \frac{\frac{1}{x-y}}{\frac{2x}{x^2-y^2}} = \frac{1}{x-y} \div \frac{2x}{x^2-y^2}$$

$$= \frac{1}{x-y} \cdot \frac{x^2-y^2}{2x} = \frac{1}{x-y} \cdot \frac{(x-y)(x+y)}{2x} = \frac{x+y}{2x}$$

A second method for simplifying a complex fraction is to multiply the numerator and the denominator by the LCD of all fractions that appear in both the numerator and the denominator.

EXAMPLE 7 Simplify the complex fraction

$$\frac{\frac{1}{x+h} - \frac{1}{x}}{h}$$

SOLUTION After multiplying the numerator and the denominator by the LCD, $x(x + h)$, we obtain

$$\frac{\frac{1}{x+h} - \frac{1}{x}}{h} = \frac{\left(\frac{1}{x+h} - \frac{1}{x}\right) \cdot x(x+h)}{h \cdot x(x+h)}$$

$$= \frac{\frac{1}{x+h} \cdot x(x+h) - \frac{1}{x} \cdot x(x+h)}{hx(x+h)}$$

$$= \frac{x-(x+h)}{hx(x+h)} = \frac{-h}{hx(x+h)} = \frac{-1}{x(x+h)}$$

PROBLEM SET 1.3

In problems 1–10, specify the restriction on x under which the given rational expression is defined.

1. $\dfrac{5x}{x-7}$ **2.** $\dfrac{x-6}{x-2}$ **3.** $2-\dfrac{1}{x+8}$ **4.** $\dfrac{3}{x^2+4}$

5. $\dfrac{-1}{(x+2)(x+3)}$ **6.** $\dfrac{1}{x}+\dfrac{1}{x-1}$ **7.** $\dfrac{4}{x(x-1)^2(x+2)^3}$ **8.** $\dfrac{x^2-4}{x+2}$

9. $\dfrac{5xy}{x-y}$ **10.** $\dfrac{x-3}{3-x}$

In problems 11–20, reduce each rational expression.

11. $\dfrac{3x^3y^7a}{15x^4y^2a^3}$ **12.** $\dfrac{15x^2y^5c-15x^2y^5d}{45x^3y^2c-45x^3y^2d}$ **13.** $\dfrac{4x^2-9}{9x-6x^2}$ **14.** $\dfrac{x^3+4x^2}{x^3+6x^2+8x}$

15. $\dfrac{7x^2-5xy}{49x^3-25xy^2}$ **16.** $\dfrac{-x^2+8x+9}{x^2-14x+45}$ **17.** $\dfrac{x^2-4x-32}{x^2-10x+16}$ **18.** $\dfrac{1-x^2}{x^3-1}$

19. $\dfrac{xz+xy-z^2-yz}{xy+xw-yz-zw}$ **20.** $\dfrac{(t+z)^2-4t^2z^2}{t^2+zt-2t^2z}$

In problems 21–26, simplify each rational expression.

21. $\dfrac{(3x)(2)-(2x+3)(3)}{(3x)^2}$ **22.** $\dfrac{(x+h)^3-x^3}{h}$

23. $\dfrac{(x-1)^2(2)-(2x+1)(2)(x-1)}{(x-1)^4}$ **24.** $\dfrac{(x+1)^3-(x)(3)(x+1)^2}{[(x+1)^3]^2}$

25. $\dfrac{[(x+h)^2+5(x+h)+2]-(x^2+5x+2)}{(2x+h+5)^2}$

26. $\dfrac{(3x-4)^3(2)(5x+1)(5)-(5x+1)^2(3)(3x-4)^2(3)}{[(3x-4)^3]^2}$

In problems 27–36, perform the indicated operations and simplify the result.

27. $\dfrac{3x+6}{5x+5}\cdot\dfrac{x+1}{x^2+5x+6}$ **28.** $\dfrac{x+2}{x^2+8x-9}\cdot\dfrac{2x+18}{x^2-4}$

29. $\dfrac{a^2-1}{a+1}\cdot\dfrac{7a^2-5a-2}{a^2-2a+1}$ **30.** $\dfrac{a^2-9b^2}{a^2-b^2}\cdot\dfrac{5a-5b}{a^2+6ab+9b^2}$

31. $\dfrac{x^2-11x+10}{9x^2-25}\div\dfrac{x^2-8x-20}{12x^2+20x}$ **32.** $\dfrac{a^2+8a+16}{a^2-8a+16}\div\dfrac{a^3+4a^2}{a^2-16}$

33. $\dfrac{x-1}{x^2-1}\cdot\dfrac{2x+2}{x^2-4}\div\dfrac{3x+3}{x^2+4x+4}$ **34.** $\dfrac{x-3}{x^2+2x-3}\cdot\dfrac{x^2-5x+6}{x^2-2x-3}\div\dfrac{x^2-9}{x^2-1}$

35. $\dfrac{4-x^2}{x^2-5x+6}\div\dfrac{x+2}{x+6}\cdot\dfrac{x^2+4}{2x^2+12x}$ **36.** $\dfrac{(x+4)^2-9y^2}{9-25x^2}\div\dfrac{(x-3y+4)^2}{25x^2+30x+9}$

In problems 37–46, perform the indicated operations and simplify the result.

37. $\dfrac{x^2}{x+3} - \dfrac{9}{x+3}$

38. $\dfrac{3x^2}{x-2} - \dfrac{13x}{x-2} + \dfrac{14}{x-2}$

39. $\dfrac{x}{x^2-25} + \dfrac{1}{x+5}$

40. $\dfrac{x}{x-1} - \dfrac{1}{x^2-x}$

41. $\dfrac{x}{x^2-9} - \dfrac{x-1}{x^2-5x+6}$

42. $\dfrac{x}{x^2+5x-6} + \dfrac{3}{x+6}$

43. $\dfrac{4}{2y^2+y-1} - \dfrac{3}{2-3y-2y^2}$

44. $\dfrac{5b-1}{3b^2-2b-8} - \dfrac{3b+2}{2b^2-3b-2}$

45. $\dfrac{x}{x+2} - \dfrac{x}{x-2} - \dfrac{x^2}{x^2-4}$

46. $\dfrac{x-3}{x+3} - \dfrac{x+3}{3-x} + \dfrac{x^2}{9-x^2}$

In problems 47–50, perform all operations and simplify the result.

47. $\dfrac{3}{x+1} - \dfrac{2}{x-1} \cdot \dfrac{x^2-1}{8x^3}$

48. $\left(\dfrac{5}{x^2+2x} + \dfrac{6}{x^2-2x-8}\right) \div \dfrac{33x-60}{x^2+4x+4}$

49. $\dfrac{1}{x-1} \cdot \dfrac{5}{x} - \dfrac{2}{x-1} \cdot \dfrac{2}{x+1}$

50. $\left(\dfrac{2}{x-3} + \dfrac{8}{x+3} \cdot \dfrac{x}{x-3}\right) \cdot \dfrac{9-x^2}{28}$

In problems 51–60, simplify the complex fraction.

51. $\dfrac{x + \dfrac{3}{x}}{1 + \dfrac{3}{x^2}}$

52. $\dfrac{\dfrac{x-1}{x+1} - 3}{\dfrac{x-1}{x+1} + 4}$

53. $\dfrac{\dfrac{x}{y} - \dfrac{y}{x}}{\dfrac{x}{y} + \dfrac{y}{x}}$

54. $\dfrac{x + \dfrac{y}{y-x}}{y + \dfrac{x}{x-y}}$

55. $\dfrac{\dfrac{1}{x+5} - \dfrac{1}{x-5}}{\dfrac{3}{x^2-25}}$

56. $\dfrac{\dfrac{x-1}{x+1} - \dfrac{x+1}{x-1}}{\dfrac{x-1}{x+1} + \dfrac{x+1}{x-1}}$

57. $\dfrac{\dfrac{1}{(x+h)^2} - \dfrac{1}{x^2}}{h}$

58. $\dfrac{\dfrac{1}{(x+h+1)^2} - \dfrac{1}{(x+1)^2}}{h}$

59. $\dfrac{\dfrac{1}{6(x+h)^3} - \dfrac{1}{6x^3}}{h}$

60. $\dfrac{\dfrac{2}{x-2} + 3}{\dfrac{-1}{x^3-8}}$

1.4 Exponents, Radicals, and Complex Numbers

In Section 1.2, page 13, we reviewed the product properties of positive integer exponents. We now extend the definition of exponents to include zero and negative integers so that those properties continue to hold.

DEFINITION 1 **Zero and Negative Integer Exponents**

If a is any *nonzero* real number and n is a positive integer,

(i) $a^0 = 1$ (ii) $a^{-n} = \frac{1}{a^n}$

If $a = 0$ in Definition 1(i), we obtain 0^0, *which is undefined.*

Since

$$\left(\frac{a}{b}\right)^{-1} = \frac{1}{a/b} = 1 \cdot \frac{b}{a} = \frac{b}{a}$$

it follows that

$$\left(\frac{a}{b}\right)^{-1} = \frac{b}{a}$$

In addition to the product properties, the following properties of exponents also hold for *all* integer exponents m and n and all nonzero real numbers a and b.

Quotient Properties of Exponents

(i) $\left(\frac{a}{b}\right)^n = \frac{a^n}{b^n}$ (ii) $\frac{a^m}{a^n} = a^{m-n}$ (iii) $\frac{a^{-n}}{b^{-n}} = \frac{b^n}{a^n}$

EXAMPLE 1 Rewrite each expression so that it contains nonnegative exponents and simplify each result.

(a) $(2x^0 + 7)^{-2}$ (b) $(\frac{1}{2})^{-3}$ (c) $(2x^{-3}y^2)^{-4}$

(d) $\left(\frac{x^{-4}}{y^3}\right)^{-5}$ (e) $\frac{7x^{-2}(x+y)^3}{21x^5(x+y)^{-4}}$ (f) $\frac{x^{-2}+y^{-2}}{(xy)^{-1}}$

SOLUTION By applying the properties and definitions of exponents, we obtain

(a) $(2x^0 + 7)^{-2} = (2 \cdot 1 + 7)^{-2} = 9^{-2} = \frac{1}{9^2} = \frac{1}{81}$

(b) $(\frac{1}{2})^{-3} = (2^{-1})^{-3} = 2^3 = 8$

(c) $(2x^{-3}y^2)^{-4} = 2^{-4}(x^{-3})^{-4}(y^2)^{-4} = \frac{1}{2^4} \cdot x^{12}y^{-8} = \frac{x^{12}}{16y^8}$

(d) $\left(\frac{x^{-4}}{y^3}\right)^{-5} = \frac{(x^{-4})^{-5}}{(y^3)^{-5}} = \frac{x^{20}}{y^{-15}} = x^{20}y^{15}$

(e) $\dfrac{7x^{-2}(x+y)^3}{21x^5(x+y)^{-4}} = \dfrac{x^{-2-5}(x+y)^{3+4}}{3} = \dfrac{x^{-7}(x+y)^7}{3} = \dfrac{(x+y)^7}{3x^7}$

(f) $\dfrac{x^{-2}+y^{-2}}{(xy)^{-1}} = \dfrac{\dfrac{1}{x^2}+\dfrac{1}{y^2}}{\dfrac{1}{xy}} \cdot \dfrac{x^2y^2}{x^2y^2} = \dfrac{y^2+x^2}{xy}$

Problems similar to the following example occur often in calculus.

EXAMPLE 2 Simplify the expression, that is, write the expression without negative exponents.

$$-4y^4(2y-1)^{-5}(2) + 4y^3(2y-1)^{-4}$$

SOLUTION

$$\begin{aligned} &-4y^4(2y-1)^{-5}(2) + 4y^3(2y-1)^{-4} \\ &\quad = 4y^3(2y-1)^{-5}[-2y+(2y-1)] \qquad [\text{Factor out } 4y^3(2y-1)^{-5}] \\ &\quad = 4y^3(2y-1)^{-5}(-1) = -4y^3(2y-1)^{-5} = \frac{-4y^3}{(2y-1)^5} \end{aligned}$$

Exponents can also be extended to include rational numbers so that all the properties of exponents we have studied so far continue to hold. If n is a positive integer, we define an *nth root* of a as follows:

> b is an ***n*th root** of a if $b^n = a$.

When $n = 2$, we say b is a square root of a; when $n = 3$, we say b is a cube root of a. For instance, both 2 and -2 are square roots of 4, since

$$2^2 = (-2)^2 = 4$$

Also, -3 is a cube root of -27, since $(-3)^3 = -27$.

Let's consider what meaning can be given to $a^{1/n}$ where n is a positive integer. If the property $(b^m)^n = b^{mn}$ is to hold, then we want

$$(a^{1/n})^n = a^{n/n} = a^1 = a$$

Thus we would like $a^{1/n}$ to be an nth root of a, and in addition, we want $a^{1/n}$ to be unique so that we have the following definition.

DEFINITION 2 **Principal *n*th Root**

> Let a be a real number and let n be a positive integer. Then $a^{1/n}$, **the principal *n*th root** of a, is defined to be the nth root of a which has the same sign as a. Thus $b = a^{1/n}$ if $b^n = a$ and
>
> (i) b is nonnegative when a is nonnegative.
> (ii) b is negative when a is negative and n is odd.

In summary, $a^{1/n} = b$ is equivalent to $b^n = a$, where b has the same sign as a. It should be emphasized that if a is negative and n is even, then $a^{1/n}$ is not defined in the set of real numbers $\mathbb{R}$. For example,

$$4^{1/2} = 2 \text{ (not } -2) \qquad \text{since} \quad 2^2 = 4, \text{ and both 4 and 2 are positive;}$$

$$(-8x^3)^{1/3} = -2x \qquad \text{since} \qquad (-2x)^3 = -8x^3;$$

$(-9)^{1/2}$ is not defined in the real number system, since there is *no real number* x that can satisfy $x^2 = -9$.

Radical notation is commonly used to indicate the principal root. We write

$$\sqrt[n]{a} \text{ to represent } a^{1/n}$$

In the expression $\sqrt[n]{a}$, the symbol $\sqrt[n]{\ }$ is called the **radical.** The positive integer n is called the **index,** and the real number a under the radical is called the **radicand.** If the index is not written, it is understood to be 2. Thus, $\sqrt{25} = 25^{1/2} = 5$.

If m/n is a rational number, we define $a^{m/n}$ as follows.

DEFINITION 3 **Rational Exponents**

Let a be a nonzero real number. Suppose that m and n are integers, that n is positive, and that the fraction m/n is *reduced to lowest terms.* Then, if each of the roots exists,

$$a^{m/n} = (a^{1/n})^m = (\sqrt[n]{a})^m = \sqrt[n]{a^m}$$

For instance,

$$32^{2/5} = (\sqrt[5]{32})^2 = 2^2 = 4 \qquad \text{and} \qquad 81^{-3/4} = (\sqrt[4]{81})^{-3} = 3^{-3} = \frac{1}{3^3} = \frac{1}{27}$$

The properties of exponents considered so far are also true for rational exponents. For convenience these properties are listed in the front cover of the book.

EXAMPLE 3 Simplify each expression and write the answer so that it contains only positive exponents. Assume that the variables are restricted to values for which all expressions are defined.

(a) $x^{-1/2} \cdot x^{3/2}$

(b) $(x^{-5/7})^{-14}$

(c) $\left(\dfrac{x^{-9}}{y^{-6}}\right)^{-2/3}$

(d) $(x^{3/2} - x^{-3/2})^2$

(e) $\dfrac{4}{3}x^{1/3} + \dfrac{1}{3}x^{-2/3}$

(f) $3y^2(3y^2 - 1)^{-1/3} - \frac{1}{3}(3y^2 - 1)^{-4/3}(6y)(y^3)$

SOLUTION

(a) $x^{-1/2} \cdot x^{3/2} = x^{-1/2+3/2} = x^{2/2} = x$

(b) $(x^{-5/7})^{-14} = x^{(-5/7)(-14)} = x^{10}$

(c) $\left(\dfrac{x^{-9}}{y^{-6}}\right)^{-2/3} = \dfrac{(x^{-9})^{-2/3}}{(y^{-6})^{-2/3}} = \dfrac{x^{(-9)(-2/3)}}{y^{(-6)(-2/3)}} = \dfrac{x^6}{y^4}$

(d)
$$\begin{aligned}(x^{3/2} - x^{-3/2})^2 &= (x^{3/2})^2 - 2(x^{3/2})(x^{-3/2}) + (x^{-3/2})^2 \\ &= x^3 - 2x^{(3/2)-(3/2)} + x^{-3} \\ &= x^3 - 2 + \frac{1}{x^3} = \frac{x^6 - 2x^3 + 1}{x^3}\end{aligned}$$

(e) $\dfrac{4}{3}x^{1/3} + \dfrac{1}{3}x^{-2/3} = \dfrac{1}{3}x^{-2/3}(4x + 1) = \dfrac{4x + 1}{3x^{2/3}}$

(f)
$$\begin{aligned}&3y^2(3y^2 - 1)^{-1/3} - \tfrac{1}{3}(3y^2 - 1)^{-4/3}(6y)(y^3) \\ &= y^2(3y^2 - 1)^{-4/3}[3(3y^2 - 1) - 2y^2] \\ &= y^2(3y^2 - 1)^{-4/3}(9y^2 - 3 - 2y^2) \\ &= y^2(3y^2 - 1)^{-4/3}(7y^2 - 3) = \frac{y^2(7y^2 - 3)}{(3y^2 - 1)^{4/3}}\end{aligned}$$

Radical Expressions

We indicated earlier that b is the *principal* nth root of a if $b^n = a$ and b has the same sign as a. Using the notation for radicals, we have

$$\sqrt[n]{a} = b \qquad \text{is equivalent to} \qquad b^n = a$$

where b has the same sign as a. For example, $\sqrt[3]{8} = 2$; $\sqrt{9} = 3$, *not* -3; $\sqrt{-4}$ is not defined in the real number system. Note that $\sqrt[4]{(-2)^4} = \sqrt[4]{16} = 2$, and *not* -2. This shows us that $\sqrt[n]{a^n}$ is *not* always the same as a (Problem 30b). However, if n is any positive integer and $\sqrt[n]{a}$ is defined, then

$$\sqrt[n]{a^n} = (\sqrt[n]{a})^n = a$$

Since radicals are just another way of writing exponents, the properties of radicals can be derived from the properties of exponents (Problem 44).

Properties of Radicals

Let a and b be real numbers and suppose that m and n are positive integers. Then, provided that all expressions are defined,

(i) $\sqrt[n]{ab} = \sqrt[n]{a}\,\sqrt[n]{b}$ (ii) $\sqrt[n]{\dfrac{a}{b}} = \dfrac{\sqrt[n]{a}}{\sqrt[n]{b}}$ (iii) $\sqrt[n]{\sqrt[m]{a}} = \sqrt[nm]{a}$

To *simplify* a radical expression, we use the properties of radicals and write the expression in a form that satisfies the following conditions:

1. The power of any term under the radical is less than the index of the radical, that is, in $\sqrt[n]{a^m}$, $m < n$.

2. The exponent of any term under the radical and the index of the radical have no common factor other than 1 or -1, that is, in $\sqrt[n]{a^m}$, m and n have no common factors.

3. The radicand contains no fractions.

EXAMPLE 4 Use the properties of radicals to simplify each expression. Assume that all variables are positive.

(a) $\sqrt{63}$ (b) $\sqrt{125x^2}$

(c) $\sqrt[4]{\dfrac{32x^9}{81y^4}}$ (d) $\sqrt[5]{\sqrt[3]{y^{30}}}$

SOLUTION

(a) $\sqrt{63} = \sqrt{9 \cdot 7} = \sqrt{9}\sqrt{7} = 3\sqrt{7}$

(b) $\sqrt{125x^2} = \sqrt{25x^2 \cdot 5} = \sqrt{25x^2}\sqrt{5} = 5x\sqrt{5}$

(c) $\sqrt[4]{\dfrac{32x^9}{81y^4}} = \dfrac{\sqrt[4]{32x^9}}{\sqrt[4]{81y^4}} = \dfrac{\sqrt[4]{16x^8 \cdot 2x}}{3y} = \dfrac{2x^2\sqrt[4]{2x}}{3y}$

(d) $\sqrt[5]{\sqrt[3]{y^{30}}} = \sqrt[15]{y^{30}} = y^{30/15} = y^2$

To *add* or *subtract* radicals, we simplify each of the radicals and then combine like terms, as we did with polynomials.

EXAMPLE 5 Perform the following operations.

$$5x\sqrt[3]{x} - \sqrt[3]{64x^4} + 7\sqrt[3]{x^4}$$

SOLUTION

$$\begin{aligned}5x\sqrt[3]{x} - \sqrt[3]{64x^4} + 7\sqrt[3]{x^4} &= 5x\sqrt[3]{x} - \sqrt[3]{64 \cdot x^3 \cdot x} + 7\sqrt[3]{x^3 \cdot x}\\ &= 5x\sqrt[3]{x} - \sqrt[3]{64}\sqrt[3]{x^3}\sqrt[3]{x} + 7\sqrt[3]{x^3} \cdot \sqrt[3]{x}\\ &= 5x\sqrt[3]{x} - 4x\sqrt[3]{x} + 7x\sqrt[3]{x} = 8x\sqrt[3]{x}\end{aligned}$$

The properties of real numbers and the special products on page 14 hold when we multiply expressions containing radicals. Note that if x is nonnegative, then

$$\sqrt{x}\sqrt{x} = \sqrt{x^2} = x$$

EXAMPLE 6 Perform each multiplication and simplify the result.

(a) $2\sqrt{3}(\sqrt{6} - \sqrt{5})$ (b) $(2\sqrt{3} + \sqrt{5})(\sqrt{3} - 4\sqrt{5})$

(c) $(3\sqrt{x} + 5\sqrt{y})(3\sqrt{x} - 5\sqrt{y})$

SOLUTION

(a) $2\sqrt{3}(\sqrt{6} - \sqrt{5}) = 2\sqrt{3} \cdot \sqrt{6} - 2\sqrt{3} \cdot \sqrt{5} = 2\sqrt{3 \cdot 6} - 2\sqrt{3 \cdot 5}$

$$= 2\sqrt{18} - 2\sqrt{15} = 2\sqrt{9 \cdot 2} - 2\sqrt{15} = 2 \cdot 3\sqrt{2} - 2\sqrt{15}$$

$$= 6\sqrt{2} - 2\sqrt{15}$$

(b) $(2\sqrt{3} + \sqrt{5})(\sqrt{3} - 4\sqrt{5}) = 2\sqrt{3}(\sqrt{3} - 4\sqrt{5}) + \sqrt{5}(\sqrt{3} - 4\sqrt{5})$

$$= 2\sqrt{3} \cdot \sqrt{3} - 8\sqrt{3} \cdot \sqrt{5} + \sqrt{5} \cdot \sqrt{3} - 4\sqrt{5} \cdot \sqrt{5}$$

$$= 2 \cdot 3 - 8\sqrt{15} + \sqrt{15} - 4 \cdot 5$$

$$= 6 - 7\sqrt{15} - 20 = -14 - 7\sqrt{15}$$

(c) $(3\sqrt{x} + 5\sqrt{y})(3\sqrt{x} - 5\sqrt{y}) = (3\sqrt{x})^2 - (5\sqrt{y})^2 = 9x - 25y$

In Example 6(c), the product of two radical expressions contains no radical. In such a case, we say the two expressions are **rationalizing factors** for each other. We often write a fraction so that there are no radicals in the denominator by multiplying the numerator and denominator by a rationalizing factor for the denominator. This process is called **rationalizing the denominator,** as illustrated in the next example.

EXAMPLE 7 Rationalize the denominator of each fraction.

(a) $\dfrac{1}{\sqrt{3}}$ (b) $\dfrac{4}{2\sqrt{x} - 3\sqrt{y}}$

SOLUTION

(a) $\dfrac{1}{\sqrt{3}} = \dfrac{1}{\sqrt{3}} \cdot \dfrac{\sqrt{3}}{\sqrt{3}} = \dfrac{\sqrt{3}}{3}$

(b) $\dfrac{4}{2\sqrt{x} - 3\sqrt{y}} = \dfrac{4}{2\sqrt{x} - 3\sqrt{y}} \cdot \dfrac{2\sqrt{x} + 3\sqrt{y}}{2\sqrt{x} + 3\sqrt{y}}$

$$= \frac{4(2\sqrt{x} + 3\sqrt{y})}{(2\sqrt{x})^2 - (3\sqrt{y})^2} = \frac{8\sqrt{x} + 12\sqrt{y}}{4x - 9y}$$

EXAMPLE 8 Rewrite

$$\frac{\sqrt{x + 3} - \sqrt{x}}{3}$$

by rationalizing the *numerator.*

SOLUTION

Since a rationalizing factor for $\sqrt{x + 3} - \sqrt{x}$ is $\sqrt{x + 3} + \sqrt{x}$, it follows that

$$\frac{\sqrt{x + 3} - \sqrt{x}}{3} = \frac{\sqrt{x + 3} - \sqrt{x}}{3} \cdot \frac{\sqrt{x + 3} + \sqrt{x}}{\sqrt{x + 3} + \sqrt{x}}$$

$$= \frac{x + 3 - x}{3(\sqrt{x + 3} + \sqrt{x})} = \frac{3}{3(\sqrt{x + 3} + \sqrt{x})} = \frac{1}{\sqrt{x + 3} + \sqrt{x}}$$

Complex Numbers

Earlier we saw that symbols such as $(-9)^{1/2}$, $\sqrt{-4}$, and $\sqrt{-x}$, where x is positive, do not represent real numbers because there is no *real number* whose square is negative. In order to consider square roots of negative numbers, we must use a set of numbers that contains the set of real numbers $\mathbb{R}$ as a subset, and also contains square roots of negative numbers. We call this set of numbers the set of *complex numbers* and denote it by $\mathbb{C}$.

A number of the form

$$a + bi, \text{ where } a \text{ and } b \text{ are real numbers}$$

is called a **complex number.** The symbol i denotes a number whose square is -1; that is, we define i by the equation

$$i^2 = -1$$

and we write $\sqrt{-1} = i$.

If the properties of radicals and exponents in the real number system are extended to include the complex numbers, then

$$\sqrt{-4} = \sqrt{4(-1)} = \sqrt{4} \cdot \sqrt{-1} = 2i$$

and, if x is nonnegative,

$$-\sqrt{-25x^2} = -\sqrt{25x^2(-1)} = -\sqrt{25x^2} \cdot \sqrt{-1} = -5xi$$

For the complex number $z = a + bi$, a is called the **real part** of z and b is called the **imaginary part** of z. For example, the complex numbers

$$3 + 5i, \qquad \sqrt{5} - 2i, \qquad \text{and} \qquad -\sqrt{3} - 7i$$

have, respectively, real parts 3, $\sqrt{5}$, and $-\sqrt{3}$, and imaginary parts 5, -2, and -7.

If i is raised to successive positive integer powers, we obtain results such as

$$i^3 = i^2 \cdot i = -i, \qquad i^4 = i^2 \cdot i^2 = (-1)(-1) = 1, \qquad \text{and} \qquad i^5 = i^4 \cdot i = i$$

Since any real number a can be expressed in the complex number form $a + 0 \cdot i$, we consider the real number set to be a subset of the complex number set.

The algebraic properties listed in Section 1.1 for the real number system also hold for complex numbers. In fact, the rules for the *addition, subtraction*, and *multiplication* for complex numbers are the same as those for polynomials, where i is treated like a variable with the added property that $i^2 = -1$. The next example illustrates the similarities.

EXAMPLE 9 Perform each operation and express the result in the form $a + bi$.

(a) $(3 + 2i) + (5 - 8i)$

(b) $(4 - 7i) - (1 + 2i)$

(c) $(3 - 5i)(2 + 3i)$

SOLUTION

(a) $(3 + 2i) + (5 - 8i) = (3 + 5) + (2 - 8)i = 8 - 6i$

(b) $(4 - 7i) - (1 + 2i) = (4 - 1) + (-7 - 2)i = 3 - 9i$

(c) $$\begin{aligned}(3 - 5i)(2 + 3i) &= 6 - 10i + 9i - 15i^2 \\ &= 6 - i - 15(-1) \\ &= 6 + 15 - i = 21 - i\end{aligned}$$

We divide complex numbers by applying a process similar to the one we used to rationalize the denominators of radical expressions. In this case, the factor that plays the role of the rationalizing factor is called the *complex conjugate*.

If $z = a + bi$, then the **complex conjugate** is denoted by $\bar{z} = \overline{a + bi}$, and it is defined as

$$\bar{z} = \overline{a + bi} = a - bi$$

For example,

if $z = 3 + 2i$, then its complex conjugate $\bar{z} = \overline{3 + 2i} = 3 - 2i$;

if $z = -5 - 7i$, then $\bar{z} = -5 + 7i$.

Complex conjugates have the following properties which can be easily verified (Problem 90).

Properties of Complex Conjugates

1. $\overline{z_1 + z_2} = \bar{z}_1 + \bar{z}_2$
2. $\overline{z_1 z_2} = \overline{z_1} \cdot \overline{z_2}$
3. $\overline{\left(\dfrac{z_1}{z_2}\right)} = \dfrac{\overline{z_1}}{\overline{z_2}}$
4. $\overline{z^n} = \bar{z}^n$
5. $z \cdot \bar{z}$ is a real number

In fact, if $z = a + bi$, then

$$z\bar{z} = (a + bi)(a - bi) = a^2 - b^2 i^2 = a^2 - b^2(-1) = a^2 + b^2$$

The *division* of two complex numbers is accomplished by multiplying the numerator and the denominator by the conjugate of the denominator. This results in a fraction that contains a real number in the denominator.

EXAMPLE 10 Perform each division and express the answer in the form $a + bi$.

(a) $\dfrac{1}{5 + 3i}$ (b) $\dfrac{5 - 10i}{3 - 4i}$

SOLUTION

(a) $$\frac{1}{5 + 3i} = \frac{1}{5 + 3i} \cdot \frac{5 - 3i}{5 - 3i} = \frac{5 - 3i}{25 + 9} = \frac{5 - 3i}{34} = \frac{5}{34} - \frac{3}{34}i$$

(b) $$\begin{aligned}\frac{5 - 10i}{3 - 4i} &= \frac{5 - 10i}{3 - 4i} \cdot \frac{3 + 4i}{3 + 4i} = \frac{55 - 10i}{9 + 16} = \frac{55 - 10i}{25} \\ &= \frac{5(11 - 2i)}{25} = \frac{11 - 2i}{5} = \frac{11}{5} - \frac{2}{5}i\end{aligned}$$

PROBLEM SET 1.4

In problems 1–24, simplify the expression; that is, write each expression without negative exponents. Assume that the variables are restricted to values for which the expressions are defined.

1. $\dfrac{3 \cdot 2^{-1} \cdot 4^{-1}}{2^2}$

2. $\dfrac{2^3 \cdot 2^4 \cdot 6^{-2}}{6^2}$

3. $\dfrac{2^{-3} \cdot 5^{-2}}{10^{-1} \cdot 16}$

4. $\dfrac{3^{-5} \cdot 4^{-2} \cdot 2^{-1}}{2^3 \cdot 3^{-2}}$

5. $16^{-2}[(2^{-1})(2)(2^5)]^4$

6. $[(5^{-1})(5^2 \cdot 5^{-3})]^{-1}$

7. $\left(\dfrac{3^2 \cdot 3^4 \cdot 9^{-1}}{4^3 \cdot 3^{-2} \cdot 5^0}\right)^{-1}$

8. $\left(\dfrac{3^0 \cdot 2^{-6}}{2^{-2} \cdot 4 \cdot 7^0}\right)^{-2}$

9. $\dfrac{x^{-4}y^2z^{-4}}{(xy)^{-2}(yz)^{-4}}$

10. $\left[\dfrac{(ab)^{-2}(bc)^{-3}}{(ac)^3(cd)^{-2}}\right]^{-2}$

11. $\dfrac{x^2 \cdot x^3 \cdot x^{-1} \cdot (x^{-2})^3}{x^{-3}}$

12. $\left[\dfrac{x^{-4}y^2z^{-3}}{x^3(yz)^{-2}}\right]^{-4}$

13. $\dfrac{a^{-1} - b^{-1}}{a - b}$

14. $a^{-1}b + ab^{-1}$

15. $\dfrac{a^{-1} + b^{-1}}{a^{-1} - b^{-1}}$

16. $\dfrac{x^{-2} - y^{-2}}{x^{-1} - y^{-1}}$

17. $x^{2/3} \cdot x^{-1/2} \cdot x^{1/6}$

18. $(a^{7/9} \cdot a^{-1/3})^{-18}$

19. $\dfrac{a^{1/3} \cdot a^{3/8}}{a^{-7/2}}$

20. $\left(\dfrac{x^{-1} \cdot y^{-2/3}}{z^{-2}}\right)^{-3}$

21. $(x^{-3/8})^{-8/3} \cdot (y^{-1/3})^{3/2}$

22. $(x^{5/7} \cdot y^{3/14})^{-14} \cdot (x^{2/3})^6$

23. $\left(\dfrac{81x^{-12}}{y^{16}}\right)^{-1/4}$

24. $\left(\dfrac{x^{-1/4}y^{-5/2}}{x^3y^{-3}}\right)^4$

In problems 25–28, simplify the expression.

25. $\dfrac{(2x + 3)^2(3)(4x + 1)^2(4) - (4x + 1)^3(2)(2x + 3)(2)}{[(2x + 3)^2]^2}$

26. $\dfrac{(x + 7)^3(-5)(3x + 1)^{-6}(3) - (3x + 1)^{-5}(3)(x + 7)^2}{[(x + 7)^3]^2}$

27. $\dfrac{(4x + 5)^{3/2}(-\frac{1}{2})(2x - 1)^{-3/2}(2) - (2x - 1)^{-1/2}(\frac{3}{2})(4x + 5)^{1/2}(4)}{[(4x + 5)^{3/2}]^2}$

28. $\dfrac{(x + 1)(\frac{1}{2})(2x - 3x^2)^{-1/2}(2 - 6x) - (2x - 3x^2)^{1/2}}{(x + 1)^2}$

29. Rewrite each expression in simplified radical form. Assume that all variables represent positive real numbers.

(a) $(4x)^{1/2}$ (b) $4x^{1/2}$ (c) $(x^{1/2} + y^{1/2})^2$
(d) $(x^2 + y^2)^{1/2}$ (e) $(x - y)^{2/3}$ (f) $(3x^2 - 5x + 7)^{-1/2}$
(g) $(x^3 + 2y^3)^{-4/3}$ (h) $\left(\dfrac{1}{x^2 + 9}\right)^{-3/2}$ (i) $\left(\dfrac{5 - x}{x + 4}\right)^{-5/3}$

30. (a) If x is a positive real number, verify that for any positive integer n

$$\sqrt[n]{x^n} = x$$

(b) If x is a negative number, verify that $\sqrt{x^2} = -x$.
(c) Verify that $\sqrt{x^2} = |x|$, where x is any real number.

In problems 31–41, simplify each expression. Assume all variables are restricted to values for which the radical expressions are defined.

31. $\sqrt[5]{-64}$
32. $5\sqrt[3]{-16}$
33. $\sqrt{25x^3}$
34. $\sqrt[3]{(x^3 + y^3)^6}$
35. $2xy^2\sqrt[4]{x^7y^5}$
36. $\sqrt{\sqrt{32x^5}}$
37. $\sqrt[3]{250x^4y^7}$
38. $\sqrt[3]{-x^3y^4t^5z^{27}}$
39. $\dfrac{\sqrt[3]{200uv^7}}{\sqrt[3]{25u^4v}}$
40. $\dfrac{\sqrt{9r} \cdot \sqrt{3r^4}}{\sqrt{3r^3}}$
41. $\dfrac{\sqrt[3]{9t^5} \cdot \sqrt[3]{27t^2}}{\sqrt[3]{216t^4}}$

In problems 42 and 43, simplify each algebraic expression.

42. $\dfrac{\dfrac{x^2}{\sqrt{x^2 - 4}} - \sqrt{x^2 - 4}}{x^2 - 4}$
43. $\dfrac{3y^2\sqrt{1 - y^2} + \dfrac{y^4}{\sqrt{1 - y^2}}}{1 - y^2}$

44. Verify the properties of radicals on page 29 with the aid of Definition 3 and the properties of exponents.

In problems 45–52, simplify each sum. Assume each variable is a positive real number.

45. $-3\sqrt{32} + \sqrt{8} - \sqrt{2}$
46. $5\sqrt[3]{54} - 2\sqrt[3]{16}$
47. $9\sqrt{27x^2} - 5x - 3\sqrt{12x^2}$
48. $\sqrt{\frac{1}{2}} - \sqrt{\frac{1}{8}}$
49. $\sqrt[3]{8x^4} + 2\sqrt[3]{-125x^4} - x\sqrt[3]{x}$
50. $\sqrt{(x + 3)^2} + 2x - 6$
51. $\sqrt[3]{(2x + 5)^6} - \sqrt[3]{(5x - 4)^3}$
52. $\sqrt[3]{-x^6} + \sqrt{x^4} - x^2$

In problems 53–62, perform each multiplication and simplify the result.

53. $\sqrt{2}(3\sqrt{6} - 2\sqrt{14} + \sqrt{2})$
54. $\sqrt[3]{7x^2}(\sqrt[3]{49x} - \sqrt[3]{x^2})$
55. $(5\sqrt{2} - \sqrt{6})(-2\sqrt{2} + \sqrt{3})$
56. $(4\sqrt{x} - 9\sqrt{y})^2$
57. $(\sqrt{3} - 5\sqrt{2})^2$
58. $(\sqrt{x} + \sqrt{y})^3$
59. $(2\sqrt{5} - \sqrt{3})(2\sqrt{5} + \sqrt{3})$
60. $(\sqrt[3]{3} + \sqrt[3]{2})^3$
61. $(3\sqrt{x} - \sqrt{y})(3\sqrt{x} + \sqrt{y})$
62. $(\sqrt{2} + \sqrt{3} + \sqrt{5})(\sqrt{2} - \sqrt{3} - \sqrt{5})$

In problems 63–71, rationalize the denominator and simplify the result.

63. $\dfrac{2}{\sqrt{2}}$
64. $\dfrac{-6}{\sqrt{14}}$
65. $\dfrac{10x}{3\sqrt{5x}}$
66. $\dfrac{x^2 - 4}{\sqrt{x + 2}}$
67. $\dfrac{5 + \sqrt{2}}{\sqrt{3} - \sqrt{7}}$
68. $\dfrac{3}{\sqrt{x} + 1}$
69. $\dfrac{6\sqrt{2} - 12\sqrt{3}}{3 + \sqrt{3}}$
70. $\dfrac{\sqrt{y + 3} + \sqrt{y}}{\sqrt{y + 3} - \sqrt{y}}$
71. $\dfrac{1}{\sqrt{t} + \sqrt{t - 1}}$

72. Rewrite $\dfrac{\sqrt{x + h} - \sqrt{x}}{h}$ by rationalizing the *numerator*.

In problems 73–89, perform the indicated operation and write the answer in the form $a + bi$.

73. $(2 + 3i) + (4 + 5i)$
74. $(-1 + 2i) + (3 + 5i)$
75. $(2 + i) - (4 + 3i)$
76. $(-4 + 2i) - (3 + 2i)$
77. $(6 + i)(5 + 3i)$
78. $(3 - 4i)(7 - 3i)$
79. $(-7 + 2i)(-7 - 2i)$
80. $(-2 + 3i)(-2 - 3i)$
81. $i^{14} - 3i^5$
82. $i^{102} + 5i^{51}$
83. $(3 - i)^2$
84. $(1 + 3i)^2$
85. $\sqrt{-49x} + \sqrt{-121x}$
86. $\sqrt{-16xy^5} - y\sqrt{-25xy^3}$, y is positive
87. $\dfrac{7 - 3i}{5i}$
88. $\dfrac{5 + 8i}{3i}$
89. $\dfrac{3 + 5i}{4 - 3i}$

90. (a) Verify Properties 1–3 of complex conjugates on page 33.
(b) Verify Property 4 for $n = 2$.
(c) Use $z_1 = 4 + 3i$, $z_2 = 2 - 5i$, and $n = 2$ to illustrate the five properties.

1.5 Equations

An **equation** is a statement that two mathematical expressions are equal. Some examples of equations in *one variable* are

$$3x + 8 = 0, \qquad x^2 + 5x - 6 = 0, \qquad \text{and} \qquad \sqrt{5y - 7} = 11$$

To **solve** an equation in one variable we determine all values which, when substituted for the variable, make the statement of equality a true statement. These values are called **solutions** or **roots** of the equation. For example, if we substitute -1 for the variable x in the equation

$$3x + 8 = 5$$

we have

$$3(-1) + 8 = 5, \text{ which is a true statement.}$$

Therefore, -1 is a solution to the equation.

Two equations are **equivalent** if they both have the same solutions. We can convert an equation to an equivalent equation by applying the following properties.

Assume P, Q, and R represent numbers.

(i) **Addition Property**
If $P = Q$, then $P + R = Q + R$ and $P - R = Q - R$.

(ii) **Multiplication Property**
If $P = Q$ and $R \neq 0$, then $PR = QR$ and $\dfrac{P}{R} = \dfrac{Q}{R}$.

Linear Equations

A **linear** or **first-degree** equation is a polynomial equation in which no variable appears with an exponent other than 1 and in which there are no products of variables. For instance,

$$4x - 7 = 1, \qquad 2 - 7y = y + 1, \qquad \text{and} \qquad 2x - 5y = 4 - 3x$$

are linear equations. Although the precise steps taken in order to solve an equation depend upon the nature of the equation, the two properties on page 36 are needed to convert equations to equivalent forms when solving equations. The next two examples illustrate this process.

EXAMPLE 1 Solve the equation

$$7x - 5 = 3(x - 2) + 5(1 - 3x)$$

SOLUTION First, we simplify the right-hand side of the equation and then apply the addition and multiplication properties:

$$7x - 5 = 3(x - 2) + 5(1 - 3x)$$

$$7x - 5 = 3x - 6 + 5 - 15x \qquad \text{(Distributive property)}$$

$$7x - 5 = -12x - 1 \qquad \text{(Combine like terms)}$$

$$7x + 12x = -1 + 5 \qquad \text{(Add } 12x + 5 \text{ to each side)}$$

$$19x = 4 \qquad \text{(Combine like terms)}$$

$$x = \frac{4}{19} \qquad \text{(Divide each side by 19)}$$

The solution of the equation is $\frac{4}{19}$.

EXAMPLE 2 The formula

$$A = P + Prt$$

gives A, the total amount of money due, if P dollars is invested for t years at a **simple interest** rate r. Solve the equation for P, that is, express P in terms of A, r, and t.

SOLUTION The equation can be solved as follows:

$$P + Prt = A$$

$$P(1 + rt) = A \qquad \text{(Factor out } P\text{)}$$

$$P = \frac{A}{1 + rt} \qquad \text{(Divide both sides by } 1 + rt\text{)}$$

If an equation involves *fractions*, we begin by multiplying both sides of the equation by the LCD of the fractions and then we solve the resulting equation. The proposed solution should always be checked. This is illustrated in the next example.

EXAMPLE 3 Solve each equation.

(a) $\frac{5}{x-5} + 6 = \frac{x}{x-5}$ (b) $\frac{3}{x} - \frac{1}{2x} = \frac{2}{x-1}$

SOLUTION (a) We multiply both sides of the equation by the LCD $x - 5$ to obtain

$$(x-5)\left(\frac{5}{x-5} + 6\right) = (x-5)\left(\frac{x}{x-5}\right)$$

$$(x-5)\left(\frac{5}{x-5}\right) + (x-5)6 = (x-5)\left(\frac{x}{x-5}\right) \quad \text{(Distributive property)}$$

$$5 + (x-5)6 = x \quad \text{(Simplify each side)}$$

$$5 + 6x - 30 = x \quad \text{(Distributive property)}$$

$$6x - 25 = x \quad \text{(Combine like terms)}$$

$$5x = 25 \quad \text{(Add } 25 - x \text{ to each side)}$$

$$x = 5 \quad \text{(Divide each side by 5)}$$

We check the *proposed solution* by substituting $x = 5$ in the original equation to obtain

$$\frac{5}{5-5} + 6 = \frac{5}{5-5}$$

Because division by zero is not defined, neither side of the equation is defined. Thus, the *proposed solution* does not satisfy the original equation. In fact there is no solution. This "misleading root" 5 is called an **extraneous root.**

(b) After multiplying both sides of the equation by the LCD $2x(x - 1)$, we get

$$2x(x-1)\left(\frac{3}{x} - \frac{1}{2x}\right) = 2x(x-1)\left(\frac{2}{x-1}\right)$$

$$2x(x-1)\left(\frac{3}{x}\right) - 2x(x-1)\left(\frac{1}{2x}\right) = 2x(x-1)\left(\frac{2}{x-1}\right)$$

$$6(x-1) - (x-1) = 4x$$

$$6x - 6 - x + 1 = 4x$$

$$5x - 5 = 4x$$

$$x = 5$$

Next we check the proposed solution $x = 5$ to obtain

$$\frac{3}{5} - \frac{1}{10} \stackrel{?}{=} \frac{2}{4} \quad \text{or} \quad \frac{6-1}{10} \stackrel{?}{=} \frac{2}{4} \quad \text{or} \quad \frac{5}{10} = \frac{2}{4}$$

which is true. Therefore, the solution is $x = 5$.

The next example is an application that utilizes a linear equation.

EXAMPLE 4 A chemist has one solution containing a 60% concentration of acid and a second solution containing a 75% concentration of acid. How many liters of each should be mixed to obtain 10 liters of a solution containing a 65% concentration of acid?

SOLUTION Let x equal the number of liters poured from the first solution, so $10 - x$ equals the number of liters from the second solution in the mixture. The following table summarizes the given information:

	Number of Liters of Solution	Percentage of Acid Concentration	Number of Liters of Acid in Solution
First solution	x	0.60	$0.60x$
Second solution	$10 - x$	0.75	$0.75(10 - x)$
Mixture	10	0.65	$0.65(10) = 6.5$

Since the amount of acid in the mixture is the sum of the amounts of acid in the two other solutions, we have

$$0.60x + 0.75(10 - x) = 6.5$$

$$60x + 75(10 - x) = 650$$

$$60x + 750 - 75x = 650$$

$$-15x = -100$$

$$x = \frac{-100}{-15} = \frac{20}{3} = 6\tfrac{2}{3}$$

Hence, $6\frac{2}{3}$ liters of the first solution and $10 - 6\frac{2}{3} = 3\frac{1}{3}$ liters of the second solution should be mixed to obtain the desired concentration of acid in the mixture.

Equations Involving Absolute Values

We note from Section 1.1 that

$$|u| = |-u|$$

and we use this result to solve **equations involving absolute values.** For instance, if

$$|x| = 3, \text{ then } x = 3 \text{ or } x = -3$$

so that the solutions are 3 and -3.

EXAMPLE 5 Solve each absolute value equation.

(a) $|3x - 4| = 5$

(b) $|y + 2| = |y - 7|$

SOLUTION (a) $|3x - 4| = 5$ implies that

$$3x - 4 = 5 \quad \text{or} \quad 3x - 4 = -5$$
$$3x = 9 \quad \Big| \quad 3x = -1$$
$$x = 3 \quad \Big| \quad x = -\frac{1}{3}$$

Hence, the solutions are 3 and $-\frac{1}{3}$.

(b) $|y + 2| = |y - 7|$ means

$$y + 2 = y - 7 \quad \text{or} \quad y + 2 = -(y - 7)$$
$$2 = -7 \quad \Big| \quad y + 2 = -y + 7$$
$$2y = 5$$
$$y = \frac{5}{2}$$

The first equation $y + 2 = y - 7$ leads to a false statement because $2 \neq -7$. Thus the only solution of the original equation is $\frac{5}{2}$.

Nonlinear Equations

Now we examine **nonlinear equations,** including quadratic equations, radical equations, and equations involving rational exponents.

Any equation that can be written in the form

$$ax^2 + bx + c = 0, \qquad a \neq 0$$

is called a **second-degree** or **quadratic equation** in one variable. This form is referred to as the **standard form** of a quadratic equation. For instance,

$$2x^2 - 3x + 1 = 0 \qquad \text{and} \qquad 4x = x^2 + 7$$

are quadratic equations because the first can be expressed as $2x^2 + (-3)x + 1 = 0$ and the second as $x^2 + (-4)x + 7 = 0$.

We examine three methods for solving quadratic equations with real number coefficients.

Solving Quadratic Equations by Factoring

To solve a quadratic equation by factoring, we use the methods of factoring developed in Section 1.2 and the zero-factor property:

$$a \cdot b = 0 \text{ if and only if } a = 0 \text{ or } b = 0, \text{ or both } a \text{ and } b = 0.$$

EXAMPLE 6 Use factoring to solve $2x^2 - 5x = -2$.

SOLUTION First we convert the equation to standard form. Then we proceed as follows:

$$2x^2 - 5x + 2 = 0$$

$$(2x - 1)(x - 2) = 0$$

$$2x - 1 = 0 \quad \Big| \quad x - 2 = 0$$

$$x = \frac{1}{2} \quad \Big| \quad x = 2$$

Thus, the roots are $\frac{1}{2}$ and 2.

EXAMPLE 7 Solve the equation

$$\frac{12}{y} - 7 = \frac{12}{1 - y}$$

SOLUTION We multiply both sides of the equation by $y(1 - y)$, the LCD of the fractions, and then proceed as follows:

$$y(1 - y)\left[\frac{12}{y} - 7\right] = y(1 - y)\left[\frac{12}{1 - y}\right]$$

$$y(1 - y)\left(\frac{12}{y}\right) - 7y(1 - y) = y(1 - y)\left(\frac{12}{1 - y}\right)$$

$$12 - 12y - 7y + 7y^2 = 12y$$

$$7y^2 - 31y + 12 = 0$$

After factoring the left side of the equation, we have $(7y - 3)(y - 4) = 0$, so that

$$7y - 3 = 0 \quad \Big| \quad y - 4 = 0$$

$$7y = 3 \quad \Big| \quad y = 4$$

$$y = \frac{3}{7}$$

To check the solution, we substitute $\frac{3}{7}$ and 4 in the original equation. We find that both numbers do satisfy the equation. Therefore, the solutions are $\frac{3}{7}$ and 4.

Solving Quadratic Equations by Completing the Square

Suppose we are asked to solve a quadratic equation that is not readily factorable, $3x^2 - 2x - 2 = 0$, for example. This quadratic equation can be solved by a process known as **completing the square,** which proceeds as follows:

First, we isolate the x terms of $3x^2 - 2x - 2 = 0$ by adding 2 to both sides of the equation to get

$$3x^2 - 2x = 2$$

Next, we change the resulting equation to an equivalent equation that has 1 as the coefficient of the x^2 term by dividing both sides by 3, to obtain

$$x^2 - \frac{2}{3}x = \frac{2}{3}$$

Finally, we make the left-hand side a "perfect square" by adding the appropriate number. In order to form a perfect square on the left side, we take one-half the coefficient of x, square it, and then we add the result to *both* sides of the equation. In this situation, we add $[\frac{1}{2}(-\frac{2}{3})]^2 = (-\frac{1}{3})^2$ to get

$$x^2 - \frac{2}{3}x + \left(-\frac{1}{3}\right)^2 = \frac{2}{3} + \left(-\frac{1}{3}\right)^2$$

$$x^2 - \frac{2}{3}x + \frac{1}{9} = \frac{7}{9}$$

$$\left(x - \frac{1}{3}\right)^2 = \frac{7}{9}$$

This last equation implies, then, that

$$x - \frac{1}{3} = \sqrt{\frac{7}{9}} = \frac{\sqrt{7}}{3}$$

or

$$x - \frac{1}{3} = -\sqrt{\frac{7}{9}} = -\frac{\sqrt{7}}{3}$$

so that the solutions are

$$x = \frac{1}{3} + \frac{\sqrt{7}}{3} = \frac{1 + \sqrt{7}}{3}$$

and

$$x = \frac{1}{3} - \frac{\sqrt{7}}{3} = \frac{1 - \sqrt{7}}{3}$$

EXAMPLE 8 Use the method of completing the square to solve $2x^2 - 2x - 1 = 0$.

SOLUTION

$2x^2 - 2x = 1$ (Isolate x terms by adding 1 to both sides of the equation)

$x^2 - x = \frac{1}{2}$ (Make the coefficient of x^2 equal to 1 by dividing each side of the equation by 2)

$x^2 - x + \frac{1}{4} = \frac{1}{2} + \frac{1}{4}$ (Complete the square by adding $[\frac{1}{2}(-1)]^2 = \frac{1}{4}$ to each side of the equation)

$\left(x - \frac{1}{2}\right)^2 = \frac{3}{4}$ (Factor and simplify)

Thus

$$x - \frac{1}{2} = -\sqrt{\frac{3}{4}} \qquad \bigg| \qquad x - \frac{1}{2} = \sqrt{\frac{3}{4}}$$

$$x - \frac{1}{2} = -\frac{\sqrt{3}}{2} \qquad \bigg| \qquad x - \frac{1}{2} = \frac{\sqrt{3}}{2}$$

$$x = \frac{1}{2} - \frac{\sqrt{3}}{2} \qquad \bigg| \qquad x = \frac{1}{2} + \frac{\sqrt{3}}{2}$$

$$x = \frac{1 - \sqrt{3}}{2} \qquad \bigg| \qquad x = \frac{1 + \sqrt{3}}{2}$$

so that the roots are

$$\frac{1 - \sqrt{3}}{2} \quad \text{and} \quad \frac{1 + \sqrt{3}}{2}$$

We can generalize the method of completing the square to arrive at a formula that enables us to solve any quadratic equation with relative ease. Proof of this formula is left as an exercise (Problem 46).

THEOREM 1 **Quadratic Formula**

If $ax^2 + bx + c = 0$, with a, b, and c real numbers and $a \neq 0$, then the roots of the equation can be determined by the formula

$$x = \frac{-b \pm \sqrt{b^2 - 4ac}}{2a}$$

EXAMPLE 9 Solve each quadratic equation by the quadratic formula.

(a) $2x^2 - 8x + 3 = 0$

(b) $5x^2 + 2x + 1 = 0$

SOLUTION (a) Here $a = 2$, $b = -8$, and $c = 3$, so that by using the quadratic formula we have

$$\begin{aligned} x &= \frac{-b \pm \sqrt{b^2 - 4ac}}{2a} \\ &= \frac{-(-8) \pm \sqrt{64 - 4(2)(3)}}{2(2)} = \frac{8 \pm \sqrt{64 - 24}}{4} \\ &= \frac{8 \pm \sqrt{40}}{4} = \frac{8 \pm 2\sqrt{10}}{4} \\ &= \frac{2(4 \pm \sqrt{10})}{4} = \frac{4 \pm \sqrt{10}}{2} \end{aligned}$$

Therefore, the solutions are

$$2 + \frac{\sqrt{10}}{2} \qquad \text{and} \qquad 2 - \frac{\sqrt{10}}{2}$$

(b) Using the quadratic formula with $a = 5$, $b = 2$, and $c = 1$, we have

$$x = \frac{-2 \pm \sqrt{4 - 4(5)(1)}}{10}$$

$$= \frac{-2 \pm \sqrt{-16}}{10} = \frac{-2 \pm 4i}{10}$$

$$= \frac{2(-1 \pm 2i)}{10} = \frac{-1 \pm 2i}{5}$$

Hence, the solutions are $\frac{-1}{5} + \frac{2}{5}i$ and $\frac{-1}{5} - \frac{2}{5}i$

The following example shows how quadratic equations may be used in solving applied problems.

EXAMPLE 10 Suppose that a live stock rancher wants to enclose a rectangular feed lot on his property. One side of the lot lies along a river and does not need to be fenced. The other three sides require fencing. He needs an area of 87,500 square feet to accommodate all of his cattle, and he has 950 feet of fencing. What dimensions should the lot be in order to use as much of the river front as possible?

SOLUTION Let x equal the width of the rectangular lot in feet. It follows that $(950 - 2x)$ feet of fencing remains for the length of the lot (Figure 1). Since the area to be enclosed is 87,500 square feet, we have

$$x(950 - 2x) = 87{,}500$$

$$2x^2 - 950x + 87{,}500 = 0$$

$$x^2 - 475x + 43{,}750 = 0$$

$$(x - 125)(x - 350) = 0$$

so that $x = 125$ or $x = 350$. If we use $x = 350$, however, the length of the lot turns out to be only $950 - 2(350) = 250$. If $x = 125$ is used, the length is $950 - 2(125) = 700$. Since the rancher wants to use as much river front as possible, we must select $x = 125$ so that the lot should be 125 feet wide and 700 feet long.

Figure 1

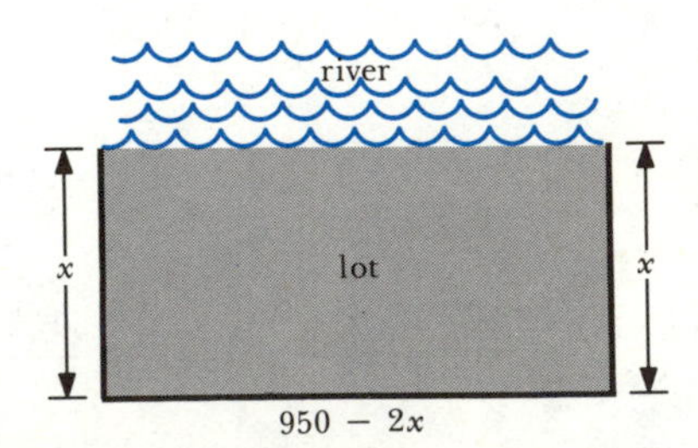

Equations Involving Radicals

Next we consider real number solutions of algebraic equations that contain *radicals* or *rational exponents*. Examples of these equations are

$$\sqrt{x} = 5, \qquad \sqrt{5y + 1} = 4, \qquad \sqrt[3]{4t + 1} = 5$$

and

$$(3u + 1)^{1/3} = u + 1$$

A method of solving equations of this type is to isolate one of the radicals and then raise each side of the equation to the same positive integer power that eliminates the isolated radical. It may be necessary to repeat this step until all the radicals are eliminated. Then we simplify the resulting equation and solve it. For example, to solve the equation

$$\sqrt{x} = 5$$

we square each side of the equation to obtain

$$(\sqrt{x})^2 = 5^2$$

$$x = 25$$

By substituting 25 into the original equation, we see that $\sqrt{25} = 5$. Hence 25 is the solution of the equation.

It is important to realize that this process can introduce an *extraneous root*. Therefore, it is necessary to check all the proposed solutions in the original equation whenever a radical with an *even* index is involved to determine whether the solutions should be *accepted* or *rejected*. For example, to solve the equation

$$\sqrt{x} = -3$$

we square both sides of the equation to get

$$(\sqrt{x})^2 = (-3)^2$$

$$x = 9$$

If we substitute $x = 9$ in the original equation, we have $\sqrt{9} = -3$, which is false. Therefore 9 is *not* a solution to the original equation. The number 9 is an *extraneous solution*, and the equation has no solution.

EXAMPLE 11 Solve each equation.

(a) $\sqrt[3]{t^2 - 1} - 2 = 0$ (b) $\sqrt{x + 4} - 2 = \sqrt{x}$

SOLUTION (a) We add 2 to both sides of the equation to isolate the radical,

$$\sqrt[3]{t^2 - 1} = 2$$

Next we raise both sides of the equation to the third power and obtain

$$(\sqrt[3]{t^2 - 1})^3 = 2^3 \qquad \text{or} \qquad t^2 - 1 = 8$$

so that $t^2 = 9$ and $t = -3$ or $t = 3$. The solutions are -3 and 3.

(b) We begin by squaring both sides of the equation to obtain

$$(\sqrt{x+4}-2)^2 = (\sqrt{x})^2$$
$$x + 4 - 4\sqrt{x+4} + 4 = x$$
$$-4\sqrt{x+4} + 8 = 0$$
$$-4\sqrt{x+4} = -8$$
$$\sqrt{x+4} = 2$$

We repeat the process by squaring each side of the latter equation to obtain

$$(\sqrt{x+4})^2 = 2^2$$
$$x + 4 = 4$$
$$x = 0$$

Check For $x = 0$,

$$\sqrt{0+4} - 2 \stackrel{?}{=} \sqrt{0}$$
$$2 - 2 = 0$$

Therefore, the solution is 0.

EXAMPLE 12 Solve the equation $(3x + 1)^{-1/5} = 2$.

SOLUTION First we rewrite the equation as

$$\frac{1}{\sqrt[5]{3x+1}} = 2$$

Next we raise each side at the equation to the fifth power,

$$\left(\frac{1}{\sqrt[5]{3x+1}}\right)^5 = 2^5$$
$$\frac{1}{3x+1} = 32$$
$$1 = 32(3x + 1)$$
$$1 = 96x + 32$$
$$-31 = 96x$$
$$x = \frac{-31}{96}$$

Thus, the root is $\frac{-31}{96}$

Some algebraic equations that are not quadratic can be transformed into a *quadratic form* by making an appropriate substitution. The solution of the "new" quadratic equation can be used to obtain the solution of the original equation.

EXAMPLE 13 Solve the equation $(x + 2)^{2/3} + (x + 2)^{1/3} - 2 = 0$.

SOLUTION The equation has an underlying quadratic form. If we let $u = (x + 2)^{1/3}$, the equation becomes

$$u^2 + u - 2 = 0 \text{ or } (u + 2)(u - 1) = 0$$

$u + 2 = 0$	$u - 1 = 0$
$u = -2$	$u = 1$
$(x + 2)^{1/3} = -2$	$(x + 2)^{1/3} = 1$
$\sqrt[3]{x + 2} = -2$	$\sqrt[3]{x + 2} = 1$
$x + 2 = (-2)^3$	$x + 2 = 1^3$
$x + 2 = -8$	$x + 2 = 1$
$x = -10$	$x = -1$

Therefore, the solutions are -1 and -10.

PROBLEM SET 1.5

In problems 1–28, solve each equation.

1. $34 - 3x = 4(3 - 2x) + 23$

2. $8(5x - 1) + 36 = -3(x + 5)$

3. $3y - 2(y + 1) = 2(y - 1)$

4. $7(t - 3) = 4(t + 5) - 47$

5. $5 + 8(u + 2) = 23 - 2(2u - 5)$

6. $11 - 7(1 - 2p) = 9(p + 1)$

7. $8 = 3v - 8(7 - v) + 23$

8. $6(y - 10) + 3(2y - 7) = -45$

9. $P = 2l + 2w$, for l

10. $5F = 9C + 100$, for C

11. $\dfrac{3x - 2}{3} + \dfrac{x - 3}{2} = \dfrac{5}{6}$

12. $\dfrac{x - 14}{5} + 4 = \dfrac{x + 16}{10}$

13. $\dfrac{1}{y} + \dfrac{2}{y} = 3 - \dfrac{3}{y}$

14. $\dfrac{2}{3u} + \dfrac{1}{6u} = \dfrac{1}{4}$

15. $\dfrac{2t}{t - 2} = \dfrac{4}{t - 2} + 1$

16. $\dfrac{9}{5y - 3} = \dfrac{5}{3y + 7}$

17. $\dfrac{a + b}{x} + \dfrac{a - b}{x} = 2a$, for x

18. $\dfrac{1}{R_1} + \dfrac{1}{R_2} = \dfrac{1}{R}$, for R

19. $|3x + 2| = 5$

20. $|2t + 6| = 18$

21. $|1 - \frac{2}{3}x| = 3$

22. $|1 - 7y| = 22$

23. $|7x + 4| = -8$

24. $|4 - 3t| = -12$

25. $|u - 2(u - 1)| = 3$

26. $|v - 2(v + 5)| = 1$

27. $|y - 7| = |3y + 1|$

28. $|2t - 9| = |5t - 3|$

In problems 29–36, solve each equation by factoring.

29. $x^2 + 5x = 0$

30. $x^2 = x + 2$

31. $y^2 - 6y + 8 = 0$

32. $-2t^2 + 19t + 33 = 0$

33. $-6u^2 + 5u - 1 = 0$

34. $y^4 - 16 = 0$

35. $18t^3 + 15t^2 - 12t = 0$

36. $6u^2 - u^4 = u^3$

In problems 37–45, solve each quadratic equation by completing the square.

37. $x^2 - 2x - 2 = 0$ **38.** $x^2 + 10x + 3 = 0$ **39.** $7y^2 - 4y - 1 = 0$
40. $2p^2 + 6p - 7 = 0$ **41.** $9t^2 - 30t = -21$ **42.** $5m^2 - 8m = -17$
43. $2r^2 + 3r = -2$ **44.** $4y^2 + 7y = -5$ **45.** $-gt^2 + vt = s$, for t

46. Prove Theorem 1 on page 43. (*Hint*: Use the method of completing the square to solve $ax^2 + bx + c = 0$, where $a \neq 0$.)

In problems 47–53, solve each equation by the quadratic formula.

47. $2x^2 + x - 1 = 0$ **48.** $3x^2 - 5x + 1 = 0$
49. $2y^2 - 6y + 3 = 0$ **50.** $3u^2 + 4u - 4 = 0$
c **51.** $1.5x^2 - 7.5x - 4.3 = 0$, to two decimal places c **52.** $-6.7m^2 - 4.2m - 5.3 = 0$, to two decimal places
53. $LI^2 + RI + \frac{1}{C} = 0$, for I, where C is positive

54. The expression $D = b^2 - 4ac$, which appears under the radical sign in the quadratic formula

$$x = \frac{-b \pm \sqrt{b^2 - 4ac}}{2a}$$

is called the **discriminant.** We can use the algebraic sign of the discriminant to determine the number and kind of roots of a quadratic equation.

(i) If D is positive, the equation has two distinct real roots. Why?

(ii) If $D = 0$, the equation has one real root. Why?

(iii) If D is negative, the equation has two complex roots. Why?

Use the discriminant D to determine the number and kind of roots of each equation.

(a) $2x^2 - 4x + 1 = 0$ (b) $x^2 - 6x + 9 = 0$ (c) $2x^2 + 3x + 2 = 0$

In problems 55–78, solve each equation.

55. $\sqrt[3]{6x - 3} = 5$ **56.** $\sqrt[3]{3x - 1} = 5$ **57.** $5 + \sqrt{y - 5} = 4$
58. $8 + \sqrt{t - 1} = 6$ **59.** $\sqrt{u^2 + 3u} = u + 1$ **60.** $\sqrt{r^2 + 5r + 2} = r$
61. $\sqrt{x + 12} = 2 + \sqrt{x}$ **62.** $\sqrt{y} = \sqrt{y + 16} - 2$ **63.** $\sqrt{m + 3} = \sqrt{m - 4} + 1$
64. $\sqrt{\sqrt{c + 4} + c} = 4$ **65.** $(4x + 5)^{-1/4} = 2$ **66.** $4 - (2t + 1)^{3/2} = 5$
c **67.** $8(1 - 3u)^{-2/3} = 4$, to two decimal places **68.** $(y + 2)^{1/3} + 5 = 4$
69. $(t^2 + 6t)^{1/3} = 3$ **70.** $(5x^2 + 7x - 3)^{3/2} = -1$
71. $y^{-2} + y^{-1} = 2$, use the substitution $u = y^{-1}$
72. $x + 2 + \sqrt{x + 2} - 6 = 0$, use the substitution $u = \sqrt{x + 2}$
73. $x^4 - 13x^2 + 36 = 0$ **74.** $x^3 - 9x^{3/2} + 8 = 0$
75. $(x^2 + 1)^2 - 3(x^2 + 1) + 2 = 0$ **76.** $(2x^2 + 7x)^2 - 3(2x^2 + 7x) = 10$
77. $\left(3x - \frac{2}{x}\right)^2 + 6\left(3x - \frac{2}{x}\right) + 5 = 0$ **78.** $\left(t - \frac{5}{t}\right)^2 - 2t + \frac{10}{t} = 8$

79. A medicine contains 25% alcohol. How much water should be added to 120 cubic centimeters of the medicine if the final mixture is to contain only 20% alcohol?

80. An investor wishes to borrow \$39,000. She is able to obtain an annual interest rate of 13.5% on part of the loan, but must pay 15% on the other part. If the annual interest on both parts totals \$5670, how much money was borrowed at each rate?

81. A rectangle has a perimeter of 64 centimeters. Find the dimensions if the length is 2 more than 3 times the width.

82. The perimeter of an isosceles triangle (a triangle with two sides of equal length) is 86 centimeters. If the third side of the triangle is 5 more than the length of either of the other two sides, find the length of each side of the triangle.

83. An electronic computer can be used to generate a large stock portfolio analysis in 6 minutes. With the help of a new computer, the analysis can be run in 2 minutes. How long would it take the new computer to run the analysis alone?

84. A drain pipe can empty a full tank of liquid waste in 15 minutes. The chemical factory that produces the waste has two feeder pipes—one can fill the tank by itself in 18 minutes and the other by itself in 24 minutes. With all three pipes open, how long before the tank will be filled?

85. A physicist has determined that the altitude h in feet of a rocket t seconds after firing is given by $h = -16t^2 + 96t$. Find how long the rocket will take to (a) reach a height of 144 feet. (b) hit the ground.

86. If a jogger runs at a speed of x miles per hour and burns $6x^2 + 20x + 25$ calories per hour, how fast should the jogger run to burn 505 calories in one hour?

1.6 Inequalities

The number line enables us to establish **order relationships** between real numbers. If the point with coordinate a lies to the left of the point with coordinate b (Figure 1), we say that a is **less than** b (or equivalently, that b is **greater than** a) and we write $a < b$ (or $b > a$). More formally, we have the following definition.

DEFINITION 1 **Order**

Assume that a and b are real numbers. Then $a < b$ (or $b > a$) means that $b - a$ is a positive number.

Figure 1

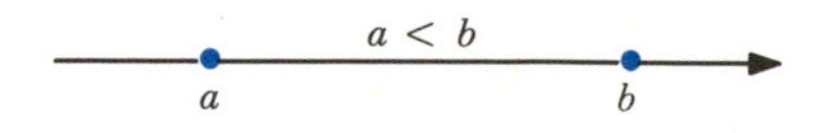

For instance, $-5 < 2$ since $2 - (-5) = 7$, which is positive. A statement of the form $a < b$ (or $b > a$) is called an **inequality.** The inequality symbol $\leq$ means **less than or equal to;** the symbol $\geq$ means **greater than or equal to.** Table 1 describes both the algebraic and geometric interpretations of inequality symbols.

Table 1

Inequality Notation	Algebraic Statement	Geometric Statement
$a > b$	$a - b$ is positive	a lies to the right of b
$a < b$	$a - b$ is negative	a lies to the left of b
$a \geq b$	$a - b$ is nonnegative	a coincides with or lies to the right of b
$a \leq b$	$a - b$ is nonpositive	a coincides with or lies to the left of b
$a > 0$	a is positive	a lies to the right of the origin
$a < 0$	a is negative	a lies to the left of the origin

The set of positive numbers is **closed under addition;** that is, the sum of any two positive numbers is always a positive number. The set of positive numbers is also **closed under multiplication;** that is, the product of two positive numbers is always a positive number.

The notation $a < x < b$, called a **compound inequality,** means

$$a < x \quad \text{and} \quad \text{simultaneously} \quad x < b \quad \text{(Figure 2)}$$

Figure 2

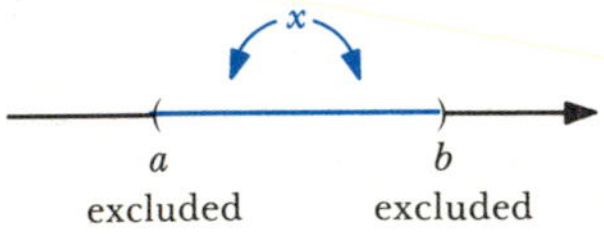

x

a excluded $\quad$ b excluded

The set of all real numbers x that satisfy $a < x < b$ is called an **open interval** with **endpoints** a and b. Note that points a and b do *not* belong to this set. The classifications of other types of intervals, as well as the notation sometimes used to denote intervals, are shown in Table 2.

Table 2

Terminology	Interval Notation	Inequality Notation	Number Line Representation
Bounded intervals:			
Open interval	(a, b)	$a < x < b$	 a b
Closed interval	$[a, b]$	$a \leq x \leq b$	 a b
Half-open interval	$[a, b)$	$a \leq x < b$	 a b
Half-open interval	$(a, b]$	$a < x \leq b$	 a b
Unbounded intervals:			
	(a, ∞)	$a < x$	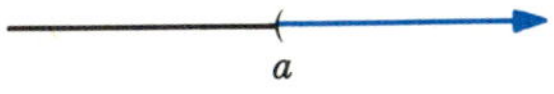 a
	$[a, \infty)$	$a \leq x$	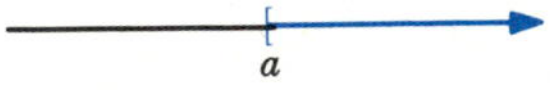 a
	$(-\infty, a)$	$x < a$	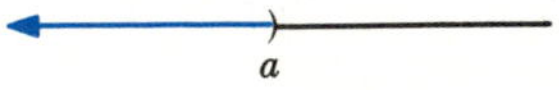 a
	$(-\infty, a]$	$x \leq a$	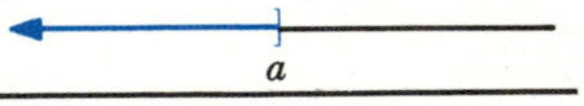 a

We use the notation $(-\infty, \infty)$ to denote the *entire real number line*. Note that ∞ ("infinity") and $-\infty$ are convenient symbols. They are *not* real numbers.

EXAMPLE 1 Express each set in interval notation and represent the set on the number line.

(a) $2 < x \leq 4$
(b) $x < 3$
(c) $x \geq -1$
(d) $x < -2$ or $x \geq 1$

SOLUTION

(a) $(2, 4]$ (Figure 3a)
(b) $(-\infty, 3)$ (Figure 3b)
(c) $[-1, \infty)$ (Figure 3c)
(d) x is in interval $(-\infty, -2)$ or in interval $[1, \infty)$ (Figure 3d).

Figure 3

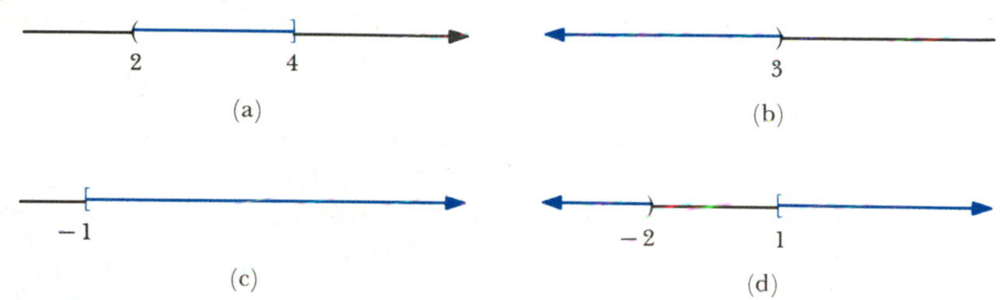

EXAMPLE 2 Express each interval in inequality notation and represent the set on the number line.

(a) $(2, 5)$ (b) $[-5, \infty)$ (c) $(-\infty, \frac{1}{2}]$

SOLUTION

(a) $(2, 5)$ includes all x such that $2 < x < 5$ (Figure 4a).
(b) $[-5, \infty)$ includes all x such that $-5 \leq x$ (Figure 4b).
(c) $(-\infty, \frac{1}{2}]$ includes all x such that $x \leq \frac{1}{2}$ (Figure 4c).

Figure 4

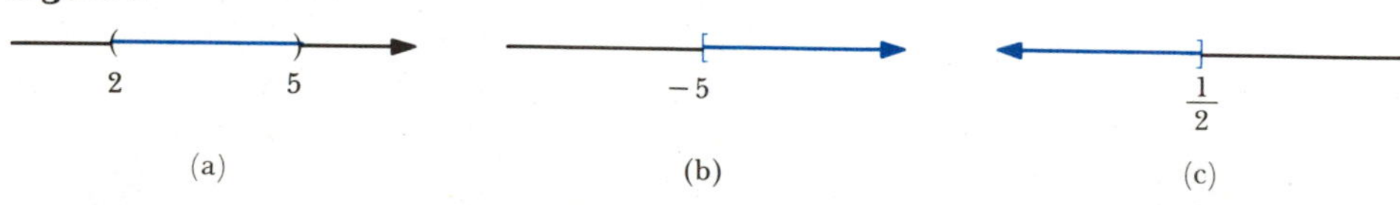

Much of the terminology of equations carries over to inequalities. A **solution** of an inequality is a value of the variable or unknown that satisfies the inequality. The set of *all* solutions of an inequality is called the **solution** or the **solution set** of the inequality. Two inequalities are **equivalent** if they have the same solution set. To solve an equality is to find its solution set. As with equations, we perform operations on inequalities to produce equivalent inequalities and continue the process until we obtain an inequality whose solution set is obvious. This process is based on the following properties of inequalities.

Properties of Inequalities

Assume a, b, and c are real numbers.

1. **The Trichotomy Property**
   One and only one of the following is true:

$$a < b \quad \text{or} \quad b < a \quad \text{or} \quad a = b$$

2. **The Transitive Property**
   If $a < b$ and $b < c$, then $a < c$
3. **The Addition Property**
   If $a < b$, then $a + c < b + c$
4. **The Multiplication Properties**
   (i) If $a < b$ and $c > 0$, then $ac < bc$
   (ii) If $a < b$ and $c < 0$, then $ac > bc$

We include here only the proofs of the multiplication properties. (See Problem 12.)

PROOFS

4(i) By Definition 1, $a < b$ means that $b - a = p$ is a positive number. But $c > 0$ and $p > 0$ implies that $pc > 0$, since the positive numbers are closed under multiplication. Hence $(b - a)c = pc$ is positive, so that $bc - ac$ is positive. Thus, by Definition 1,

$$ac < bc$$

4(ii) Since $a < b$, $b - a = p$ is a positive number. But c is negative, so that by the trichotomy principle, $-c$ is positive. Hence, $p(-c)$ is a positive number because the position numbers are closed under multiplication. Consequently,

$$p(-c) = (b - a)(-c) = ac - bc \quad \text{is positive}$$

and

$$bc < ac$$

Note that Property 3, together with the fact that $a - c = a + (-c)$, implies that if the same number is *subtracted* from both sides of an equality, the order of the inequality remains the same, that is,

$$\text{if } a < b, \text{ then } a - c < b - c \text{ for any real number } c$$

In addition, using the fact that $a \div c = a(1/c)$ for $c \neq 0$, Property 4 also holds for division.

EXAMPLE 3 Which of the properties of inequalities justifies each statement?

(a) If $x < 2$, then $x + 3 < 2 + 3$.
(b) If $x < z$ and $z < 3$, then $x < 3$.
(c) If $x < -4$, then $-3x > 12$.
(d) If $y < -5$, then $7y < -35$.
(e) If $-5t < 35$, then $t > -7$.

SOLUTION (a) Addition Property (b) Transitive Property
(c) Multiplication Property (ii) (d) Multiplication Property (i)
(e) Multiplication Property (ii)

Linear Inequalities

Inequalities such as

$$3x + 2 \le 5 \qquad \text{and} \qquad 5x - 2 \ge 7x + 11$$

are called **linear inequalities.** To solve a linear inequality, we try to isolate the variable on one side of the inequality as we did with linear equations. The crucial difference is that when we multiply or divide all members by a negative number, we have to reverse the direction of the inequality.

EXAMPLE 4 Solve each inequality, express the answer in interval notation, and display the solution on a number line.

(a) $2x - 8 > 7$ (b) $\frac{3}{4} - \frac{2}{5}y \le y - \frac{9}{20}$ (c) $-4 \le 3t + 1 < 5$

SOLUTION (a) $2x - 8 > 7$

$2x > 15$ (Add 8 to both sides)

$x > \frac{15}{2}$ (Divide both sides by 2)

Thus the solution in interval form is $(\frac{15}{2}, \infty)$ (Figure 5a).

(b) $\frac{3}{4} - \frac{2}{5}y \le y - \frac{9}{20}$

$20(\frac{3}{4} - \frac{2}{5}y) \le 20(y - \frac{9}{20})$ (Multiply both sides by the LCD 20 to clear fractions)

$15 - 8y \le 20y - 9$

$-28y \le -24$ (Subtract 15 from both sides and subtract $20y$ from both sides)

$y \ge \frac{6}{7}$ (Divide both sides by -28 and reverse the direction of the inequality)

Hence, the solution set in interval form is $[\frac{6}{7}, \infty)$ (Figure 5b).

(c) Here, we actually have two inequalities:

$$-4 \le 3t + 1 \qquad \text{and} \qquad 3t + 1 < 5$$

However, we can leave them in the original form and proceed as follows:

$-4 \le 3t + 1 < 5$

$-5 \le 3t < 4$ (Subtract 1 from both sides of both inequalities)

$-\frac{5}{3} \le t < \frac{4}{3}$ (Divide both sides of both inequalities by 3)

Thus the interval form of the solution set is $[-\frac{5}{3}, \frac{4}{3})$ (Figure 5c).

Figure 5

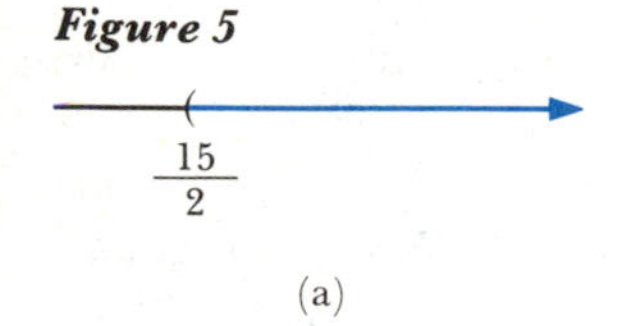

(a)

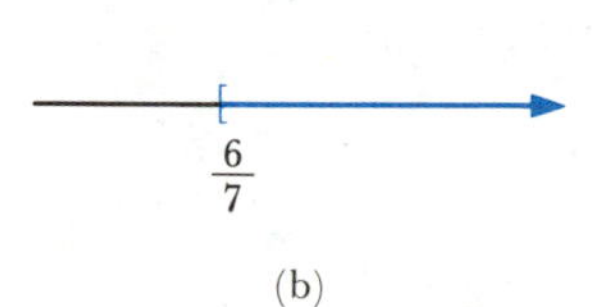

(b)

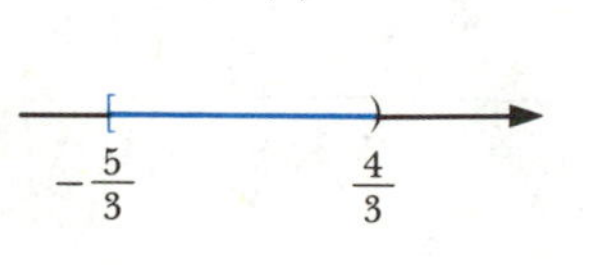

(c)

The next example shows how inequalities can be used in practical situations.

EXAMPLE 5 In filing her income tax, a taxpayer has two choices:

(1) Pay a tax equal to 30% of her gross income.
(2) Pay a tax equal to 35% of the difference between her gross income and \$5000.

Above what level of income should she elect to pay 30% of her gross income?

SOLUTION Let x equal the taxpayer's gross income, then her choices are:

(1) Pay 30% of her gross income, which is $0.30x$.
(2) Pay 35% of her gross income above \$5000, which is $0.35(x - 5000)$.

To determine when the first choice results in less tax than the second choice, we solve the inequality

$$0.30x < 0.35(x - 5000)$$

$$0.30x < 0.35x - 1750$$

$$-0.05x < -1750$$

$$x > \frac{-1750}{-0.05} = 35{,}000$$

Therefore, she should elect to pay the 30% rate when her gross income is greater than \$35,000.

Inequalities Involving Absolute Values

To solve inequalities involving absolute values, such as

$$|x| < 2 \qquad \text{and} \qquad |x| > 2$$

we can use the number line. The inequality $|x| < 2$ is satisfied by a number x if the distance from the origin to the point that represents x is less than 2. Figure 6a indicates that x must lie between -2 and 2, that is, $-2 < x < 2$. On the other hand, the inequality $|x| > 2$ is satisfied by a number x if the distance from the origin to the point x is greater than 2. Figure 6b illustrates that x must lie to the right of 2, or it must lie to the left of -2; that is, $x > 2$ or $x < -2$. This example is generalized in the following properties.

Figure 6

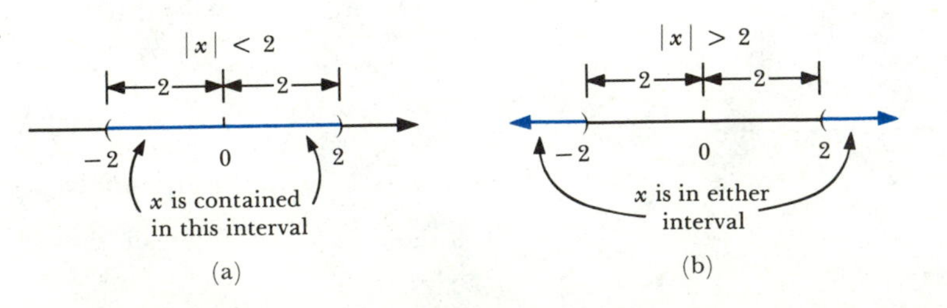

Properties of Inequalities Involving Absolute Values

> If $a > 0$, then
>
> (i) $|u| < a$ if and only if $-a < u < a$
>
> (ii) $|u| > a$ if and only if $u < -a$ or $u > a$

EXAMPLE 6 Solve each absolute value inequality, express the solution in interval form, and display the solution on a number line.

(a) $|3x - 2| \leq 8$ (b) $|2x - 3| > 5$

SOLUTION (a) By Property (i) above, we can write the inequality $|3x - 2| \leq 8$ as

$$-8 \leq 3x - 2 \leq 8$$

$$-6 \leq 3x \leq 10$$

$$-2 \leq x \leq \tfrac{10}{3}$$

Thus, the solution set in interval form is $[-2, \frac{10}{3}]$ (Figure 7a).

(b) From Property (ii), we can write $|2x - 3| > 5$ as

$$2x - 3 < -5 \quad \text{or} \quad 2x - 3 > 5$$

so that

$$2x < -2 \quad \Big| \quad 2x > 8$$

$$x < -1 \quad \Big| \quad x > 4$$

Therefore, the solution set consists of all numbers in intervals $(-\infty, -1)$ or $(4, \infty)$ (Figure 7b).

Figure 7

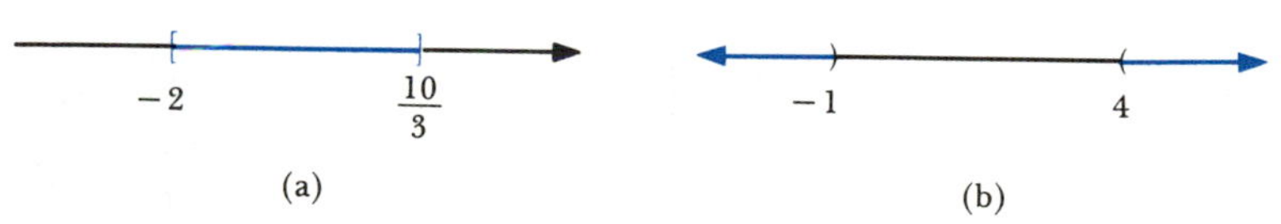

Polynomial Inequalities

A **polynomial inequality** is an inequality that contains only polynomials. For instance, $x^2 + 2x < 15$, $x^2 \geq 2x + 3$, and $x(x + 4)(x - 1) \leq 0$ are polynomial inequalities. A polynomial inequality is in standard form when it is written so that 0 is on one side of the inequality. Thus, standard forms for the above inequalities are $x^2 + 2x - 15 < 0$, $x^2 - 2x - 3 \geq 0$, and $x(x + 4)(x - 1) \leq 0$, respectively. Note that an inequality can always be converted to standard form by adding (or subtracting) an appropriate expression on both sides of the inequality. When solving polynomial inequalities, we use standard forms.

Let us consider the *cut point method* for solving the polynomial inequality

$$x^2 - 2x > 3$$

First, we convert the inequality $x^2 - 2x > 3$ to standard form by subtracting 3 from both sides to obtain

$$x^2 - 2x - 3 > 0$$

Next, we solve the *associated equation* $x^2 - 2x - 3 = 0$:

$$x^2 - 2x - 3 = 0$$

$$(x - 3)(x + 1) = 0$$

$$x = 3 \quad | \quad x = -1$$

so that the roots are -1 and 3.

It can be shown that the algebraic sign of the polynomial $x^2 - 2x - 3$ remains the same for all values of x in each of the three intervals defined by using the roots of $x^2 - 2x - 3 = 0$ as *division* points of the number line (Figure 8). Because we are using the roots as division points of the number line, we refer to the values of the roots as the *cut points* of the inequality.

Figure 8

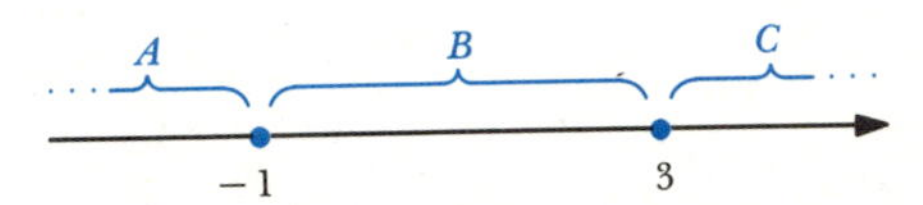

$x^2 - 2x - 3$ retains the same algebraic sign within each interval: $A = (-\infty, -1)$, $B = (-1, 3)$, and $C = (3, \infty)$

In other words, $x^2 - 2x - 3$ is always positive or always negative on each of the three intervals defined by the cut points -1 and 3. Consequently, it is sufficient to test *any* number in each interval to decide whether *all values* in the interval yield positive or negative values for $x^2 - 2x - 3$. The table below summarizes the results.

Interval	Test Number x	Test Value of $x^2 - 2x - 3$	Sign of $x^2 - 2x - 3$
A: $x < -1$ or $(-\infty, -1)$	-2	$(-2)^2 - 2(-2) - 3 = 5$	+
B: $-1 < x < 3$ or $(-1, 3)$	0	$0^2 - 2 \cdot 0 - 3 = -3$	−
C: $3 < x$ or $(3, \infty)$	5	$5^2 - 2(5) - 3 = 12$	+

Thus $x^2 - 2x - 3 = (x - 3)(x + 1) > 0$ if $x < -1$ or $x > 3$ and the solution set of $x^2 - 2x > 3$ includes all numbers in intervals $(-\infty, -1)$ or $(3, \infty)$ (Figure 9).

Figure 9

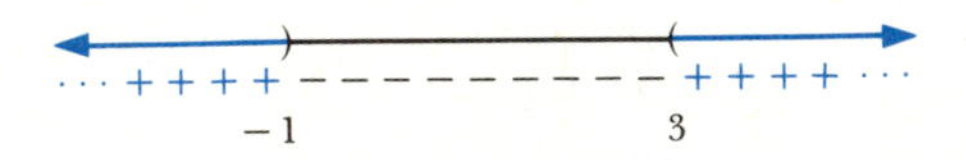

It is always true that *any polynomial expression* retains its algebraic sign for *all* numbers within an interval defined by the roots or cut points. Thus, the procedure used above can be applied to *any* polynomial inequality. This procedure is summarized as follows:

Procedure for Solving a Polynomial Inequality: Cut Point Method

Step 1. Convert the inequality to *standard form.*
Step 2. Solve the *associated equation* that results from the standard form.
Step 3. Use the roots from step 2 as **cut points.** Arrange these roots in increasing order on a number line. The roots will divide the number line into open intervals; the algebraic sign of the polynomial cannot change over any of these intervals.
Step 4. Test each interval obtained in step 3 by selecting a number within each interval and substituting it for the variable in the inequality. The algebraic sign of the resulting value is the sign of the polynomial over the entire interval.

We can see the solution set of the original inequality by displaying the appropriate algebraic signs (pluses or minuses) over each interval, as shown in the following example.

EXAMPLE 7 Solve each of the given polynomial inequalities by the cut point method. Express the solution set in interval notation and display it on the number line.

(a) $6x^2 + 7x \leq 3$ (b) $x(x + 4)(x - 1) < 0$

SOLUTION We apply the procedure outlined above.

(a) Step 1. The standard form is obtained by subtracting 3 from both sides of the inequality to get

$$6x^2 + 7x - 3 \leq 0$$

Step 2. The associated equation $6x^2 + 7x - 3 = 0$ is solved as follows:

$$6x^2 + 7x - 3 = 0$$

$$(3x - 1)(2x + 3) = 0$$

so that the roots are $\frac{1}{3}$ and $-\frac{3}{2}$.
Step 3. The cut points are, *in order*, $-\frac{3}{2}$ and $\frac{1}{3}$. They define three intervals A, B, and C (Figure 10).
Step 4. We test each interval as follows:

Interval	Test Number	Test Value of $6x^2 + 7x - 3$	Sign of $6x^2 + 7x - 3$
A: $x < -\frac{3}{2}$ or $(-\infty, -\frac{3}{2})$	-2	$6(-2)^2 + 7(-2) - 3 = 7$	$+$
B: $-\frac{3}{2} < x < \frac{1}{3}$ or $(-\frac{3}{2}, \frac{1}{3})$	0	$6 \cdot 0^2 + 7 \cdot 0 - 3 = -3$	$-$
C: $\frac{1}{3} < x$ or $(\frac{1}{3}, \infty)$	1	$6 \cdot 1^2 + 7 \cdot 1 - 3 = 10$	$+$

Figure 10

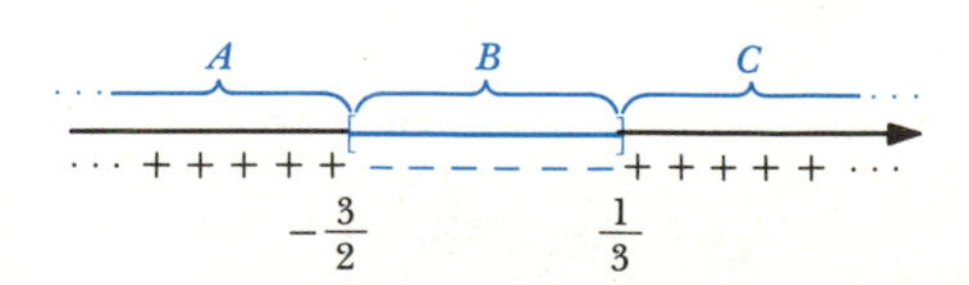

Thus the solution includes all x such that $-\frac{3}{2} \leq x \leq \frac{1}{3}$. The interval form of the solution is $[-\frac{3}{2}, \frac{1}{3}]$ (Figure 10). Note that the cut points are *included* because the original inequality includes equality.

(b) Step 1. We leave the inequality in factored form in anticipation of step 2.
Step 2. The associated equation $x(x + 4)(x - 1) = 0$ has roots 0, -4, and 1.
Step 3. The cut points, in order, are -4, 0, and 1. They define four intervals A, B, C, and D (Figure 11).
Step 4. We test each interval as follows:

Interval	Test Number	Test Value of $x(x + 4)(x - 1)$	Sign of $x(x + 4)(x - 1)$
A: $x < -4$ or $(-\infty, -4)$	-5	$-5(-1)(-6) = -30$	$-$
B: $-4 < x < 0$ or $(-4, 0)$	-1	$-1(3)(-2) = 6$	$+$
C: $0 < x < 1$ or $(0, 1)$	$\frac{1}{2}$	$\frac{1}{2}(\frac{9}{2})(-\frac{1}{2}) = -\frac{9}{8}$	$-$
D: $1 < x$ or $(1, \infty)$	2	$2(6)(1) = 12$	$+$

Figure 11

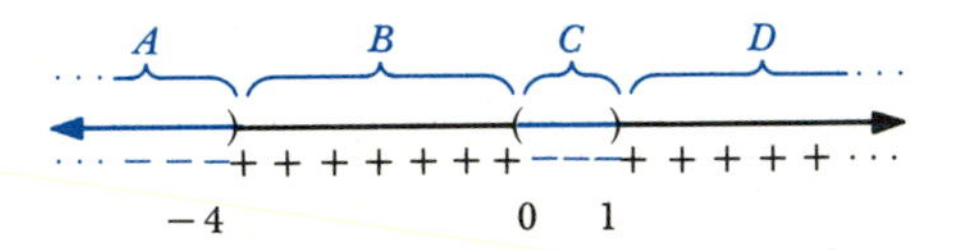

Thus the solution includes all x such that $x < -4$ or $0 < x < 1$. In interval form, the solution set includes all numbers in the intervals $(-\infty, -4)$ or $(0, 1)$ (Figure 11).

Rational Inequalities

A **rational inequality** is an inequality that contains rational expressions. The following are examples of rational inequalities:

$$\frac{x - 1}{x} \geq 0 \quad \text{and} \quad \frac{x + 1}{3x - 4} < \frac{2x - 1}{x^2 - 1}$$

We convert the rational inequality

$$\frac{x}{x + 1} \leq \frac{x - 1}{x - 2}$$

to *standard form* by subtracting $\frac{x - 1}{x - 2}$ from both sides of the inequality to obtain

$$\frac{x}{x + 1} - \frac{x - 1}{x - 2} \leq 0 \quad \text{or} \quad \frac{-2x + 1}{(x + 1)(x - 2)} \leq 0$$

so that the right side of the inequality is zero.

Notice that we do *not* multiply both sides by the LCD $x(x-1)$ because we do not know whether we are multiplying by a positive or a negative number.

To solve a rational inequality in standard form, we use the values for which the numerator or the denominator is zero as the *cut points*; and then we proceed as we did for polynomial inequalities.

EXAMPLE 8 Solve

$$\frac{x}{x+1} \leq \frac{x-1}{x-2}$$

Express the solution set in interval notation and display it on the number line.

SOLUTION Step 1. We convert to standard form

$$\frac{-2x+1}{(x+1)(x-2)} \leq 0$$

Step 2. Next we solve the associated polynomial equations, one for the numerator and one for the denominator, in order to determine the cut points:

$$-2x+1=0 \qquad\Big|\qquad (x+1)(x-2)=0$$
$$x=\tfrac{1}{2} \qquad\Big|\qquad x=-1 \quad \text{or} \quad x=2$$

Step 3. The cut points, in order, are -1, $\frac{1}{2}$, 2. They define four intervals A, B, C, and D (Figure 12).

Step 4. Each interval is tested as follows:

Interval	Test Number	Test Value of $\frac{-2x+1}{(x+1)(x-2)}$	Sign of $\frac{-2x+1}{(x+1)(x-2)}$
A: $x < -1$ or $(-\infty, -1)$	-2	$\frac{5}{(-1)(-4)} = \frac{5}{4}$	+
B: $-1 < x < \frac{1}{2}$ or $(-1, \frac{1}{2})$	0	$\frac{1}{(1)(-2)} = -\frac{1}{2}$	−
C: $\frac{1}{2} < x < 2$ or $(\frac{1}{2}, 2)$	1	$\frac{-1}{(2)(-1)} = \frac{1}{2}$	+
D: $2 < x$ or $(2, \infty)$	3	$\frac{-5}{(4)(1)} = -\frac{5}{4}$	−

Figure 12

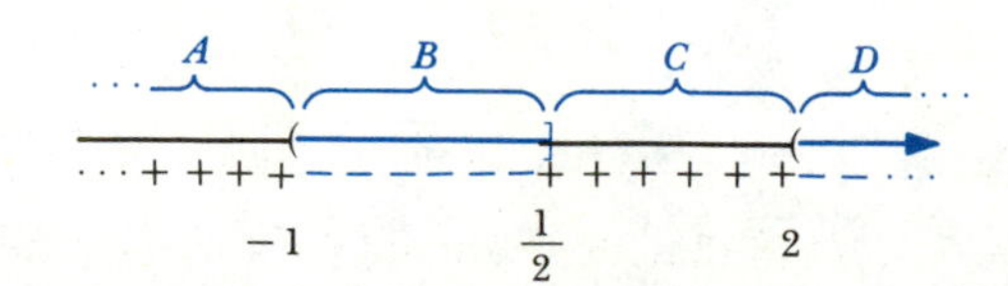

Since

$$\frac{-2x+1}{(x+1)(x-2)} = 0$$

if $x = \frac{1}{2}$, we include this number in interval B to get the solution set, which includes all x in intervals $(-1, \frac{1}{2}]$ or $(2, \infty)$ (Figure 12).

PROBLEM SET 1.6

In problems 1–5, express each set in interval notation and represent the set on the number line.

1. $-1 \leq x < 3$

2. $-\frac{5}{2} \leq t \leq -\frac{1}{2}$

3. $x > \frac{3}{4}$ or $x < 0$

4. $x > 0$ or $x < 0$

5. $\frac{1}{\pi} < x < \pi$ or $x \geq 4$

6. Use the transitive property to explain why it is incorrect to write $2 < x < 1$. In general, if $a < x < b$, then what must be the order relation between a and b?

In problems 7–11, rewrite in inequality notation each set that is given in interval notation, and display each set on the number line.

7. $[0, \frac{1}{2}]$

8. $[0, \infty)$

9. $(-\infty, 1)$

10. $(-\infty, 0)$ or $(0, \infty)$

11. x is in interval $(-2, 4)$ or in interval $[5, 8)$

12. (a) Prove the transitive property on page 52.
(b) Prove the addition property on page 52.

In problems 13–20, indicate the property that justifies each statement.

13. If $x < 5$, then $x + 7 < 5 + 7$.

14. If $x > -3$, then $x + 4 > -3 + 4$.

15. If $y \geq -4$, then $y - 5 \geq -4 - 5$.

16. If $t \leq 6$, then $t + 10 \leq 6 + 10$.

17. If $a < 4$ and $b > 4$, then $a < b$.

18. If $c \geq 5$ and $d \leq 5$, then $c \geq d$.

19. If $x \geq 7$, then $4x \geq 28$.

20. If $x < y$, then $x/(-10) > y/(-10)$.

In problems 21–37, express the solution set of each inequality in both inequality notation and interval notation, and then represent the solution set on the number line.

21. $3x < 9$

22. $-21w \leq -63$

23. $4x + 3 \geq 12$

24. $3x - 2 > 7$

25. $2t - 5 \geq 3$

26. $-8t - 4 \leq -16 - 2t$

27. $-9x - 2 < 16 + 4x$

28. $-(8 + x) - 5 + 4x \geq 1$

29. $3(x + 2) - 5x \leq 4x$

30. $5 - x < -x + 3$

31. $x + 6 \leq 4 - 3x$

32. $-4 < \frac{4x - 2}{3} \leq 6$

33. $-6 < \frac{1}{2}t + 1 < 8$

34. $0.5 \leq 0.75(t - 1) \leq 1.5$

35. $x(x - 1) \leq x(x + 1)$

36. $\frac{3x - 7}{6} - 13 \geq 1 - \frac{x}{2}$

37. $\frac{x}{3} + 2 \leq \frac{x}{4} - 2x$

38. The **triangle inequality** asserts that $|a + b| \leq |a| + |b|$ for any real numbers a and b. Use the triangle inequality to verify each of the following statements.
(a) If $|x| < 3$ and $|y| < 1$, then $|x + y| < 4$.
(b) If $|x - y| < \frac{1}{10}$ and $|y - t| < \frac{1}{10}$, then $|x - t| < \frac{1}{5}$.

In problems 39–80, solve each inequality. Express the solution set in interval notation and represent the solution on a number line. For problems 55–80, use the cut point method.

39. $|x| + 3 \leq 7$
40. $|2x| - 1 \leq 5$
41. $|x - 1| \leq 3$
42. $|3 - 2x| \leq 5$
43. $|2x + 3| < 1$
44. $|3x - 6| \leq 0$
45. $|5p| - 1 \geq 14$
46. $|6x| - 3 > 9$
47. $|x + 2| \geq 5$
48. $|t + 1| > 7$
49. $|4x - 3| > 9$
50. $|5x - 3| - 2 \geq 10$
51. $|2(t - 1) - 2| \geq 3$
52. $\sqrt{x^2 - \frac{1}{3}x + \frac{1}{36}} > \frac{2}{3}$
53. $|1 - 2(y + 3)| \leq 2$
54. $|x - 4| < |2x - 2|$
55. $(x + 3)(x - 1) < 0$
56. $x^2 - 7x \leq -6$
57. $t^2 + 5t \geq 6$
58. $-y^2 < -6y + 8$
59. $2x^2 + 3x - 9 \leq 0$
60. $t^2 - 4t + 4 \geq 0$
61. $-6t^2 + 5t + 4 > 0$
62. $8x^2 + 5x + 4 < 2x^2 - 5x - 8$
c **63.** $-3.1x^2 + 5.2x + 7.2 \leq 0$; two decimal places
c **64.** $1.4x^2 \geq -3.4x - 5.7$; two decimal places
65. $(y - 1)(y - 2)(y + 3) < 0$
66. $(t + 1)^2(t - 3) \geq 0$
67. $(2t + 4)(3t - 1)(t - 7) \geq 0$
68. $(x + 3)(x^2 + x - 2) \leq 0$
69. $(r^2 - r - 6)(2r^2 - 5r - 3) < 0$
70. $(x - 4)^3(x^2 + 7x) > 0$
71. $\dfrac{x - 1}{x - 4} < 0$
72. $\dfrac{2x - 1}{x + 2} < 0$
73. $\dfrac{1 - y}{y} \leq 1$
74. $\dfrac{1}{r + 2} > 3$
75. $\dfrac{4}{t^2 - t - 12} > 0$
76. $\dfrac{x^2 - 4}{x^2 - 9} > 0$
77. $\dfrac{x + 1}{x - 3} \geq \dfrac{x - 4}{x + 1}$
78. $\dfrac{(y - 3)(1 - y)}{y(y + 2)} \leq 0$
79. $-4 \leq \dfrac{t + 2}{t - 3} \leq 4$
80. $0 < \dfrac{1}{x - 2} < \dfrac{4}{x + 3}$

81. A small business had a choice between two schedules of billing for its telephone services: 1) A fixed \$250 monthly charge for unlimited calls or 2) a base rate of \$70 per month plus 7¢ per message unit. After how many message units does the second option become more expensive than the first one?

82. A stock worth \$60 per share pays an 11% dividend each year. In how many years will each share of stock have paid more than \$58 in dividends?

83. If one of the walls of a rectangular room is 13 feet long and the room's floor area is less than 432 square feet, what can be concluded about the length of the other wall?

84. A tire manufacturing company shows a *profit*, that is, the revenue obtained from the sales of its tires is greater than the cost of producing them. If each tire costs \$30 in materials, and if it costs \$7500 a week in labor and overhead to produce and sell the tires, how many tires must be sold each week at \$85 each in order for the company to show a profit?

85. Wildlife biologists predict that the population N of a certain endangered species after t years will be given by the equation $N = 50t^2 + 200t + 250$. In how many years will the population N be at least 5050?

86. In economics the supply equation relates the price per item p and the quantity supplied q by the equation $q = ap + b$, where a and b are constants depending on the particular situation. The total profit from selling q items equals the profit per item times q. Suppose that it costs a wheat farmer \$4 per bushel to grow wheat and that the supply equation is $q = 8p + 90$. What price should the farmer charge per bushel to realize a profit of at least \$315?

REVIEW PROBLEM SET, CHAPTER 1

[C] **1.** Express each of the given real numbers in decimal form, rounded off to three decimal places.

(a) $\frac{7}{41}$ (b) $-\frac{2}{11}$ (c) $3\frac{1}{8}$ (d) $\frac{7}{6}$
(e) $-\frac{131}{22}$ (f) $\sqrt{237}$ (g) $\sqrt[4]{19}$ (h) 17%
(i) $\frac{1}{\pi}$ (j) 0.05%

2. Indicate whether each statement is true or false.

(a) $\pi = 3.14$ (b) 0 is a positive number.
(c) 7% is an element of $\mathbb{Q}$. (d) $\sqrt{6} + \sqrt{10} = 4$
(e) $3 - 2|-5| = 5$ (f) $\frac{1}{(\frac{1}{3})}$ is an element of I.
(g) $\sqrt{x^2} = x$ (h) Every real number is a complex number.
(i) $\sqrt{4} + \sqrt{9}$ is an element of $\mathbb{Q}$. (j) $-|x| \le \sqrt{x^2} \le |x|$

3. Write each of the given rational numbers as the ratio of two integers. Reduce each answer to lowest terms.

(a) 0.14 (b) −0.0035 (c) $0.\overline{31}$
(d) $-2.5\overline{37}$ (e) $5\frac{1}{3}\%$ (f) $\frac{7.84}{23.95}$

4. Solve each equation.

(a) $|x| = x$ (b) $\sqrt{x^2} = -x$

In problems 5–14, justify each statement by giving the appropriate property. Assume that all variables represent real numbers.

5. $x + 5$ is a real number.
6. $a + (b + 3) = (a + b) + 3$
7. $3x = x \cdot 3$
8. $2(xy) = (2x)y$
9. $1 \cdot 4 = 4$
10. $5 + 0 = 5$
11. If $2x = 2y$, then $x = y$.
12. $x \cdot (1/x) = 1$, for $x \ne 0$
13. If $7a = 0$, then $a = 0$.
14. If $x + 9 = y + 9$, then $x = y$.

In problems 15–26, perform the indicated operation.

15. $(2x^2 + 3x - 4) + (x^2 - 5x + 7)$
16. $(3x^2 + 7x + 8) - (2x^2 + 3x + 2)$
17. $(5x^2 - 3x + 2) + (2x^2 + 5x - 7) - (3x^2 - 4x - 1)$
18. $(x - y)(x^2 - 2xy + y^2)$
19. $(x^2 - x + 1)(2x^2 - 3x + 2)$
20. $(2x - 3)(x + 5)$
21. $(7 - 5xy)(4 + 3xy)$
22. $(3x - 2)^2$
23. $(x^2 + 4)(x^2 - 4)$
24. $(1 + 7y)^2$
25. $(x^2 - y)(x^4 + x^2y + y^2)$
26. $(2x + 7)(4x^2 - 14x + 49)$

In problems 27–36, factor each polynomial completely.

27. $26x^3y^2 + 39x^5y^4 - 52x^2y^3$
28. $7xy - 7yz + 14y^2z - 14xy^2$
29. $y^2 - 121$
30. $y^3 - 216$
31. $25y^2 - 81z^2$
32. $x^8 - 256$
33. $x^3 + 64$
34. $x^4 - 10x^2 + 9$
35. $x^2 - x - 56$
36. $6x^2 - 29x + 35$

In problems 37–46, perform the indicated operations and simplify the result.

37. $\dfrac{x^2-16}{x^2-4x}\cdot\dfrac{x-4}{x+4}$

38. $\dfrac{x^2+5x-6}{x^2+x-2}\cdot\dfrac{x^2+3x-4}{x^2+7x+12}$

39. $\dfrac{x^2-x-2}{x^2-x-6}\div\dfrac{x^2-2x}{2x+x^2}$

40. $\dfrac{y^3+1}{x^2-4y^2}\div\dfrac{y^2-y+1}{x-2y}$

41. $\dfrac{3}{x^2-7x+12}-\dfrac{2}{x^2-5x+4}$

42. $\dfrac{5}{x^2+8x+15}+\dfrac{4}{x^2+2x-3}$

43. $\dfrac{1-\dfrac{1}{x}}{x-2+\dfrac{1}{x}}$

44. $\dfrac{\dfrac{y}{y^2-1}-\dfrac{1}{y+1}}{\dfrac{y}{y-1}+\dfrac{1}{y+1}}$

45. $\dfrac{x^{-2}-y^{-2}}{(x+y)^2}$

46. $\dfrac{x^2-25}{x+3}\div\dfrac{5-x}{x^2-9}\cdot\dfrac{1}{2x+10}$

In problems 47–58, simplify each expression. Assume that all variables represent positive real numbers.

47. $(x^{-2}y^{-3}z^0)^{-3}$

48. $\dfrac{(xy^{-1}z)^{-2}}{(xy^{-1}z)^{-6}}$

49. $\dfrac{(x^{-1}y^{-2/3})^{-3}}{(x^{-1}y^{-2/3})^{-5}}$

50. $\dfrac{x^{-2}-y^{-2}}{x^{-1}+y^{-1}}$

51. $3\sqrt{8x^2}-x\sqrt{2}-7\sqrt{18x^2}$

52. $(5\sqrt{3}-2)(2\sqrt{3}+3)$

53. $\sqrt{32x^5y^{10}z^{15}}$

54. $\sqrt[3]{81x^4z^5w^6}$

55. $\sqrt[6]{128x^{13}y^{25}}$

56. $\sqrt{98(x+2y)^2}$

57. $\dfrac{\sqrt{7}-\sqrt{6}}{\sqrt{7}+\sqrt{6}}$

58. $\dfrac{\sqrt{x}}{3-\sqrt{y}}$

In problems 59–70, perform each operation and write the answer in the form $a + bi$.

59. $(3-2i)+(7-3i)$

60. $(3-\sqrt{7}i)(3+\sqrt{7}i)$

61. $(5+12i)+(-5-3i)-(1-3i)$

62. $(4-i)-(7-3i)+(3-i)$

63. $(5+3i)(3-5i)$

64. $\dfrac{3+7i}{2-3i}$

65. $\dfrac{4-\sqrt{3}i}{2+\sqrt{3}i}$

66. $\dfrac{3-5i}{4i}$

67. $\sqrt{-25xy^3}-7yi\sqrt{xy}$, $x>0$ and $y>0$

68. $3i^5+\dfrac{2}{i^3}$

69. $(\overline{5-2i})^2(5-2i)$

70. $\sqrt{-63x}+17i\sqrt{7x}$, $x>0$

In problems 71–88, solve each equation.

71. $12(x-2)+8=5(x-1)+2x$

72. $x-7(4+x)=5x-6(3-4x)$

73. $\dfrac{2}{x}+\dfrac{x-1}{3x}=\dfrac{2}{5}$

74. $\dfrac{10-y}{y}+\dfrac{3y+3}{3y}=3$

75. $\dfrac{2}{y-1}-\dfrac{3}{y+2}=\dfrac{4}{y^2+y-2}$

76. $x+5=\dfrac{-6}{x}$

77. $(t+1)^2=7$

78. $x^2+2x-3=0$

79. $-2y^2+y+1=0$

80. $3x^2 - x - 1 = 0$

81. $4x^2 + 6x - 1 = 0$

82. $6y^2 - 7y - 5 = 0$

83. $2x^2 + 2x + 1 = 0$

84. $v^2 = 1 - v$

85. $\sqrt[3]{2t + 5} = 3$

86. $2\sqrt{3x - 1} - 1 = 3$

87. $\sqrt{1 - w} = 2 - \sqrt{1 - 5w}$

88. $\sqrt{4y^2 - 3} = 2y + 1$

89. A monthly phone bill includes a charge of \$9.80 for local calls plus an additional charge of \$0.60 for each out-of-zone call placed within a certain area. Suppose that a tax of 9% of all charges is added to the total bill and assume that all the out-of-zone calls were within the \$0.60 area. How many out-of-zone calls were made if the total bill, including taxes, is \$34.88?

90. The manager of a parking garage agreed to pay a troop of Boy Scouts a fixed amount for washing some cars, enough to give each boy in the troop \$10. When 25 boys failed to show up, those already present agreed to do the work, which meant that each boy who worked got \$15. How many boys are in the troop?

91. A projectile is fired from the ground in such a way that it is h feet above the ground t seconds after the firing, where $h = 88t - 16t^2$. Find

(a) t when $h = 0$.

(b) how long it takes for the projectile to reach a height of 121 feet.

92. An apple orchard now has 30 trees per acre, and the average yield is 400 apples per tree. For each additional tree planted per acre, the average yield per tree is reduced by approximately 10 apples. How many additional trees per acre will yield a crop of 12,250 apples?

In problems 93 and 94, solve each equation.

93. $8x^{2/3} + 7x^{1/3} - 1 = 0$

94. $(5x - 1)^{3/2} = 8$

95. Indicate which of the following statements are true for *all* real numbers a and b with $a < b$.

(a) $2a > -(-2)b$

(b) $5/a < 5/b$

(c) $a - c < b - c$

(d) $a + 3 < b + 4$

96. Indicate the appropriate property that justifies each statement. Assume that all variables represent real numbers.

(a) If $x > y$ and $z > 0$, then $x/z > y/z$.

(b) If $x < y$, then $x/(-3) > y/(-3)$.

(c) If $-7y < 35$, then $y > -5$.

(d) If $a < 5$ and $5 < d$, then $a < d$.

In problems 97–102, find the solution set of each inequality and represent the solution set on the number line. Write the solution using interval notation.

97. $5x - 9 > 2x + 3$

98. $\frac{2x}{3} + \frac{1}{5} > \frac{7}{15} + \frac{4x}{5}$

99. $5t - 1 \leq 8t - 5$

100. $5y - 2 \geq 6y + 5$

101. $2 < 3x - 8 \leq 5$

102. $-3 \leq 5 - 2x < 2$

103. In a fund-raising gathering, a woman sold five times as many tickets as a man, but the woman could not have sold more than 35 tickets. How many tickets might the man have sold?

104. Compute $\left|\frac{x}{|x|}\right|$ if x is a nonzero real number.

In problems 105–114, solve each equation or inequality and represent the solution set of each inequality on the number line.

105. $|3x + 4| = 12$

106. $|5 - 4x| = 11$

107. $|2x + 5| \leq 9$

108. $|x - 2| = |2 - x|$

109. $|3t + 1| > 8$

110. $|x - 5| \geq \frac{3}{2}$

111. $|y - 1| = 2y$

112. $|3t - 7| + 4 < 5$

113. $|w - 2| > -5$

114. $|y + 5| < |y - 1|$

In problems 115–123, solve the inequality. Show the solution set on the number line and write the solution using interval notation.

115. $x^2 - 4x - 5 \leq 0$

116. $-t^2 + 3t + 4 < 0$

117. $3y^2 - 5 > 2y$

118. $(x - 2)^2(3x - 5) \leq 0$

119. $x(x + 1)(x - 3) \geq 0$

120. $\frac{5}{y} < \frac{10}{3y}$

121. $\frac{5w - 1}{2w + 2} \geq 0$

122. $\frac{2x - 1}{(x + 7)^2} < 0$

123. $\frac{3}{y^2 - 25} < \frac{2}{y^2 + 10y + 25}$

CHAPTER 2

Functions and Graphs

Many applications in mathematics, science, economics, medicine, and technology deal with relationships between variable quantities. For example, in geometry, the area A of a square can be completely determined by the length of one of its sides x, and this relationship defines the *function*

$$A = x^2$$

In this chapter, we develop the concept of a function as a means of studying such relationships. The chapter focuses on the graphs of functions, the algebra of functions, and inverse functions. Since graphs are a powerful means of exhibiting relationships, we begin by introducing the Cartesian coordinate system, and then we graph equations of circles and lines.

2.1 The Cartesian Coordinate System and Circles

René Descartes

We saw in Section 1.1 that a point on a number line can be represented by a single *real number* called its *coordinate.* This *one-dimensional* coordinate system is extended to *two dimensions* by using the **Cartesian coordinate system,** named in honor of the French mathematician René Descartes (1596–1650). In the Cartesian coordinate system, a point in the plane is represented by a *pair of real numbers.*

The Cartesian coordinate system consists of two perpendicular number lines, one horizontal and one vertical, called the **coordinate axes.** These axes intersect at a point called the **origin.** Traditionally, the horizontal number line is called the **x axis** and the vertical number line is called the **y axis** (Figure 1).

The numerical coordinates increase to the right along the x axis and upward along the y axis (Figure 1). The positive portion of the x axis is to the right of the origin, and the negative portion is to the left of the origin; the positive portion of the y axis is above the origin, and the negative portion is below the origin.

If P is a point in the Cartesian coordinate system, the coordinates of P are the x and y values of the points where the perpendiculars to the respective axes from P meet the two axes (Figure 2). The x coordinate is called the **abscissa** of P, and the y coordinate is called the **ordinate** of P. The **coordinates** of P are traditionally

Figure 1

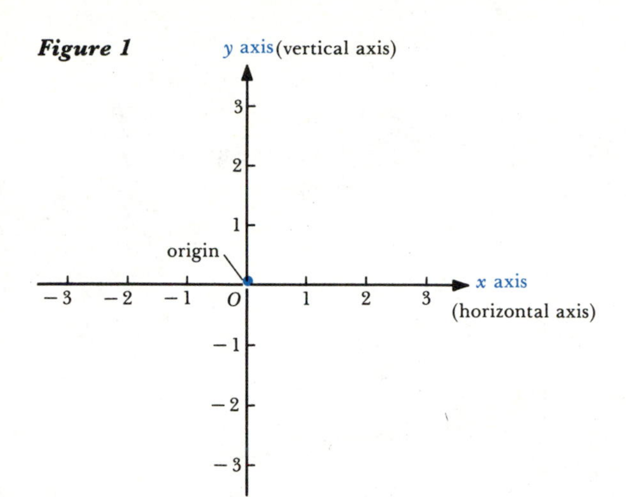

Figure 2

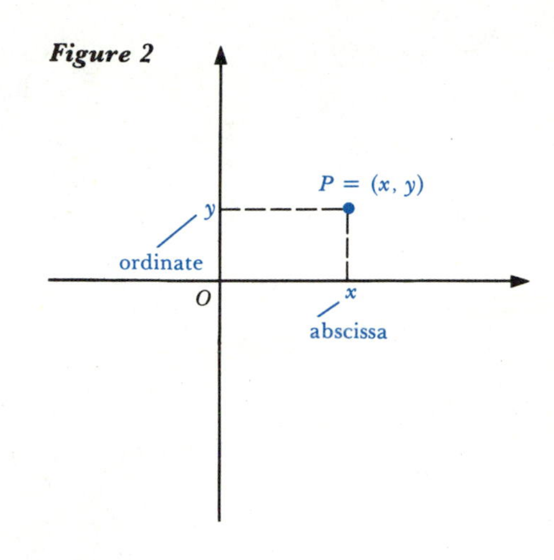

written as the **ordered pair** (x, y). To *plot* or *locate* the point P with coordinates (x, y) means to draw Cartesian coordinate axes and to place a dot representing P at the point with abscissa x and ordinate y. We can think of the ordered pair (x, y) as the numerical "address" of P. The correspondence between P and (x, y) seems so natural that in practice we identify the point P with its "address" (x, y) by writing $P = (x, y)$. With this identification in mind, we refer to an ordered pair of real numbers (x, y) as a *point,* and we refer to the set of all such ordered pairs as the **Cartesian plane** or the ***xy* plane.**

The x and y axes divide the plane into four regions called **quadrants I, II, III,** and **IV** (Figure 3). Notice that the quadrants are numbered in the *counterclockwise* direction from one to four and that a point on any one of the coordinate axes does not belong to any quadrant.

Figure 3

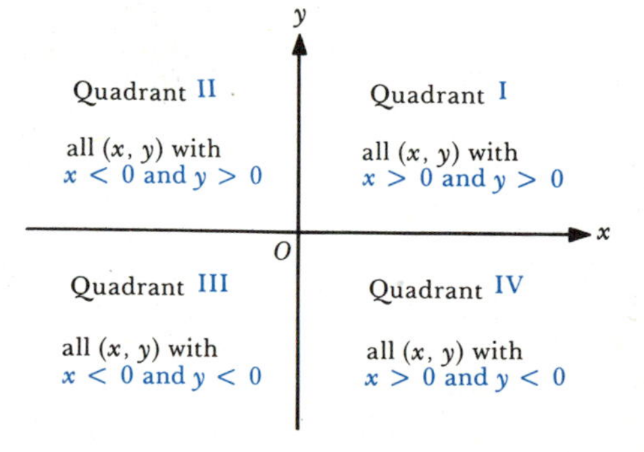

EXAMPLE 1 Plot each point and indicate which quadrant or coordinate axis contains the point.

(a) $(2, 3)$ (b) $(-3, 4)$ (c) $(-2, -6)$ (d) $(3, -5)$

(e) $(4, 0)$ (f) $(0, -5)$ (g) $(0, 0)$ (h) $(0, \frac{1}{2})$

SOLUTION

The points are plotted in Figure 4.

(a) $(2, 3)$ lies in quadrant I.
(b) $(-3, 4)$ lies in quadrant II.
(c) $(-2, -6)$ lies in quadrant III.
(d) $(3, -5)$ lies in quadrant IV.
(e) $(4, 0)$ lies on the positive x axis.
(f) $(0, -5)$ lies on the negative y axis.
(g) $(0, 0)$, the origin, lies on both axes.
(h) $(0, \frac{1}{2})$ lies on the positive y axis.

Figure 4

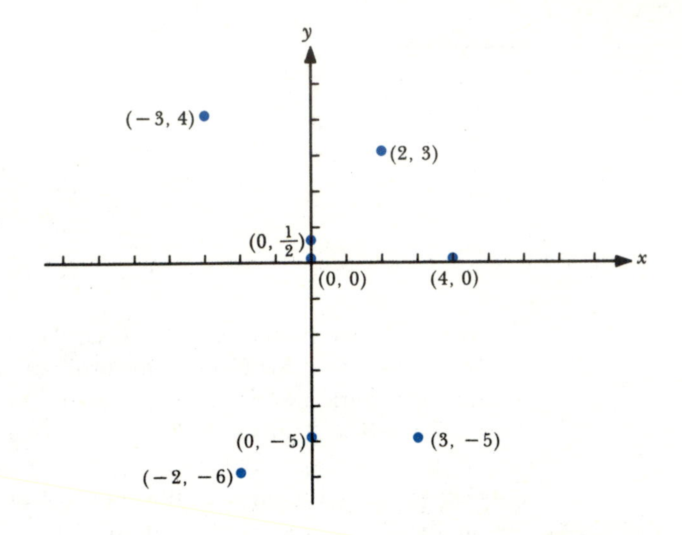

EXAMPLE 2

Plot the point $P = (-2, 3)$ and give the coordinates of point Q if the line segment $\overline{PQ}$ is perpendicular to the y axis and bisected by it.

SOLUTION

By examining Figure 5, we see that Q has the same y coordinate as P. Since it must be located two units to the right of the y axis, its x coordinate must be 2. Therefore, $Q = (2, 3)$.

Figure 5

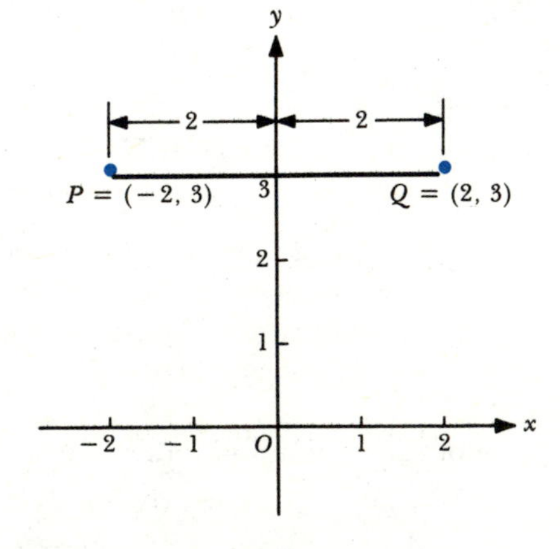

EXAMPLE 3 Shade the region in the xy plane that contains all points (x, y) for which $x > 1$ and $y \leq -2$.

SOLUTION All the points that satisfy $x > 1$ and $y \leq -2$ are contained in the shaded region in Figure 6. Note that the solid line is part of the region, whereas the dashed line is not part of the region.

Figure 6

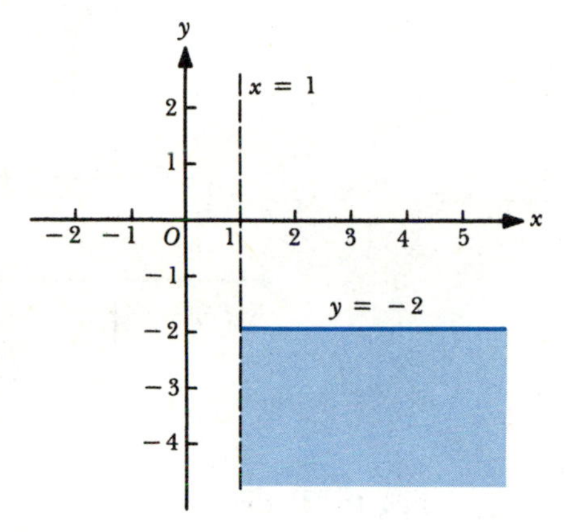

The Distance Formula

We often need to calculate the distance between two points in the Cartesian plane by using their coordinates. If two points are given in the Cartesian plane, then the distance between the points can be evaluated by using the following formula.

THEOREM 1 **The Distance Formula**

Assume P_1 and P_2 are points such that $P_1 = (x_1, y_1)$ and $P_2 = (x_2, y_2)$, then the distance d between P_1 and P_2, denoted by $d = |\overline{P_1P_2}|$, is given by

$$d = |\overline{P_1P_2}| = \sqrt{(x_1 - x_2)^2 + (y_1 - y_2)^2}$$

PROOF The distance d between P_1 and P_2 in terms of coordinates x_1, y_1, x_2, and y_2 can be derived by considering three cases:

Case 1. If the two points P_1 and P_2 lie on the same vertical line, that is, if $x_1 = x_2$, then the vertical distance d between P_1 and P_2 is the same as the distance between the points with coordinates y_1 and y_2 on the y axis, so that $d = |y_1 - y_2|$ (Figure 7a).

Case 2. If the two points P_1 and P_2 lie on the same horizontal line, that is, if $y_1 = y_2$, then the horizontal distance d between P_1 and P_2 is given by $d = |x_1 - x_2|$ (Figure 7b).

Case 3. If the two points lie on a line that is neither horizontal nor vertical, then a right triangle, that is, a triangle P_1PP_2 with a 90° angle at P, can be constructed as shown in Figure 7c.

Figure 7

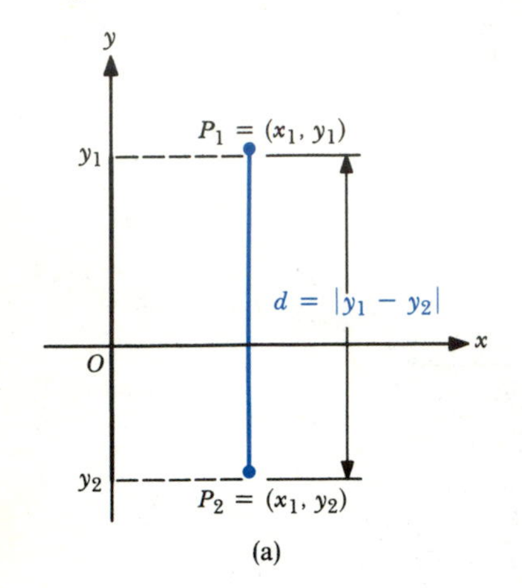

(a)

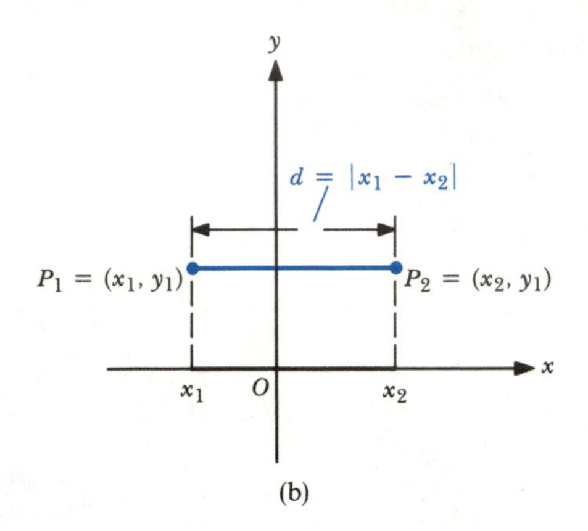

(b)

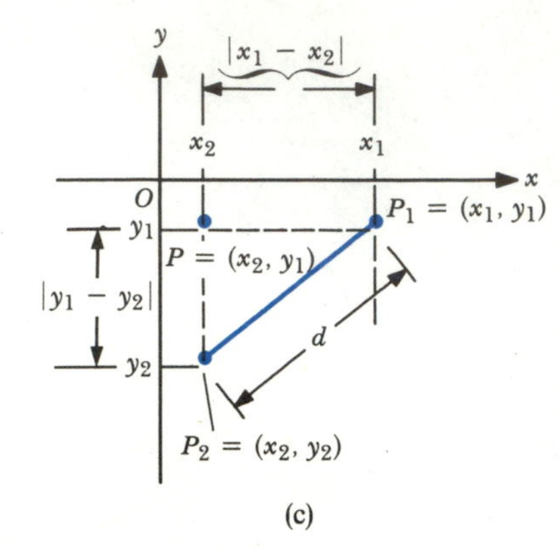

(c)

Using the Pythagorean theorem, we get

$$d^2 = |\overline{PP_1}|^2 + |\overline{PP_2}|^2$$

so that

$$d^2 = |x_1 - x_2|^2 + |y_1 - y_2|^2$$

or

$$d^2 = (x_1 - x_2)^2 + (y_1 - y_2)^2$$

Hence,

$$d = \sqrt{(x_1 - x_2)^2 + (y_1 - y_2)^2}$$

Because $(x_1 - x_2)^2 = (x_2 - x_1)^2$ and $(y_1 - y_2)^2 = (y_2 - y_1)^2$, the distance formula can also be written as

$$d = |\overline{P_1P_2}| = \sqrt{(x_2 - x_1)^2 + (y_2 - y_1)^2}$$

EXAMPLE 4 Let $P_1 = (-1, -2)$ and $P_2 = (3, -4)$. Find the distance $|\overline{P_1P_2}|$.

SOLUTION The distance $d = |\overline{P_1P_2}|$ is given by

$$d = \sqrt{[3 - (-1)]^2 + [-4 - (-2)]^2} = \sqrt{4^2 + (-2)^2} = \sqrt{20} = 2\sqrt{5}$$

EXAMPLE 5 [C] Let $P = (71.37, -27.04)$ and $Q = (14.86, 11.73)$. Find $|\overline{PQ}|$ and round off the answer to four significant digits.

SOLUTION The distance $d = |\overline{PQ}|$ is given by

$$d = \sqrt{(71.37 - 14.86)^2 + (-27.04 - 11.73)^2} = 68.53$$

EXAMPLE 6 Use the distance formula to show that the triangle whose vertices are $(-2,6)$, $(-1,-1)$, and $(1,0)$ is a right triangle.

SOLUTION Let $A = (-2,6)$, $B = (-1,-1)$, and $C = (1,0)$ be the vertices of the triangle. Then

Figure 8

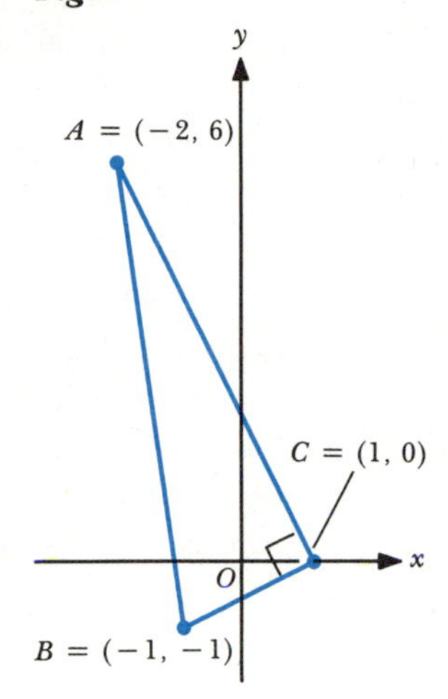

$$|\overline{AB}| = \sqrt{(-1+2)^2 + (-1-6)^2} = \sqrt{1+49} = \sqrt{50} = 5\sqrt{2}$$
$$|\overline{BC}| = \sqrt{(1+1)^2 + (0+1)^2} = \sqrt{4+1} = \sqrt{5}$$
$$|\overline{AC}| = \sqrt{(1+2)^2 + (0-6)^2} = \sqrt{9+36} = \sqrt{45} = 3\sqrt{5}$$

Figure 8 leads us to suspect that the angle at vertex C is a right angle. To confirm this, we use the converse of the Pythagorean theorem; that is, we check to see if $|\overline{AC}|^2 + |\overline{BC}|^2 = |\overline{AB}|^2$. Since

$$|\overline{AC}|^2 + |\overline{BC}|^2 = 45 + 5 = 50 = |\overline{AB}|^2$$

triangle ACB is, indeed, a right triangle.

Graphs of Circles in the Cartesian Plane

The **graph** of an equation in two variables x and y is defined to be the set of all points $P = (x, y)$ in the Cartesian plane whose coordinates x and y satisfy the equation. Many equations in x and y have graphs that are smooth curves. For instance, consider the equation

Figure 9

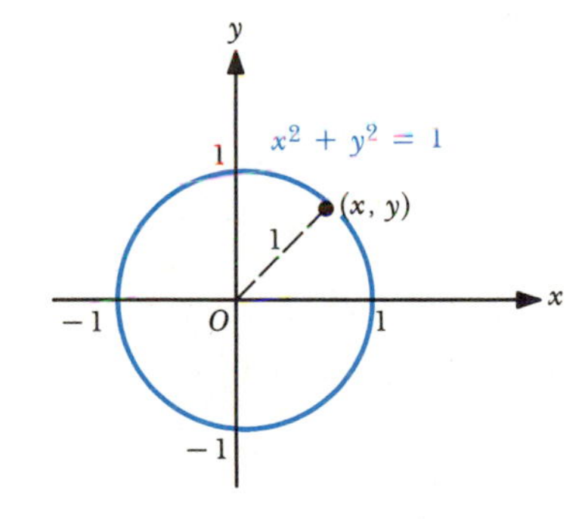

$$x^2 + y^2 = 1$$

We can write this equation as

$$\sqrt{x^2 + y^2} = \sqrt{1}$$

or as

$$\sqrt{(x-0)^2 + (y-0)^2} = 1$$

By the distance formula, this last equation holds if and only if the point $P = (x, y)$ is one unit from the origin $(0,0)$. Therefore, the graph of $x^2 + y^2 = 1$ consists of all the points that lie on a circle of radius 1 with its center at the origin O (Figure 9).

We often use an equation for a circle to designate the circle; for instance, if we speak of "the circle $x^2 + y^2 = 1$," we mean "the circle is the graph of $x^2 + y^2 = 1$." This circle, with its center at the origin and radius equal to one, is called the **unit circle.**

Figure 10

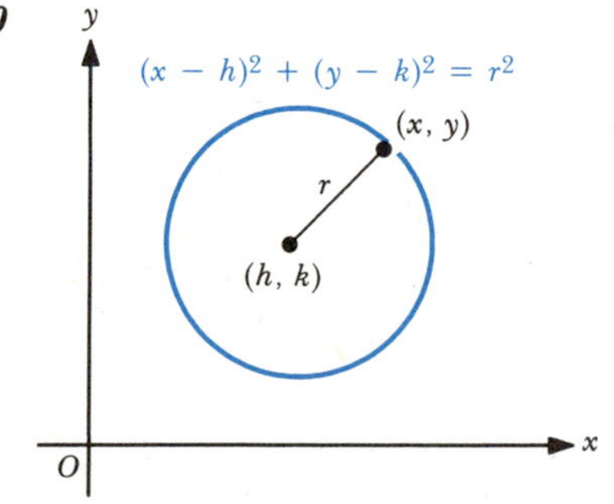

In general, if $r > 0$, then the circle of radius r with center (h, k) consists of all points (x, y) such that the distance between (x, y) and (h, k) is r units (Figure 10).

Using the distance formula, we can write an equation for this circle as

$$\sqrt{(x-h)^2 + (y-k)^2} = r$$

or equivalently, as

$$(x-h)^2 + (y-k)^2 = r^2$$

This latter equation is called the **standard form** for an equation of the circle with radius r and center (h, k).

EXAMPLE 7 Find an equation for the circle of radius 4 with center at the point $(4, -1)$ and sketch the graph.

SOLUTION Here $r = 4$ and $(h, k) = (4, -1)$ so, in standard form, an equation of the circle is

$$(x-4)^2 + [y-(-1)]^2 = 4^2$$

or

$$(x-4)^2 + (y+1)^2 = 16$$

The circle is graphed in Figure 11.

Note that if we expanded the squares and combined like terms on the left side, the equation would take the form

$$x^2 + y^2 - 8x + 2y + 1 = 0$$

Figure 11

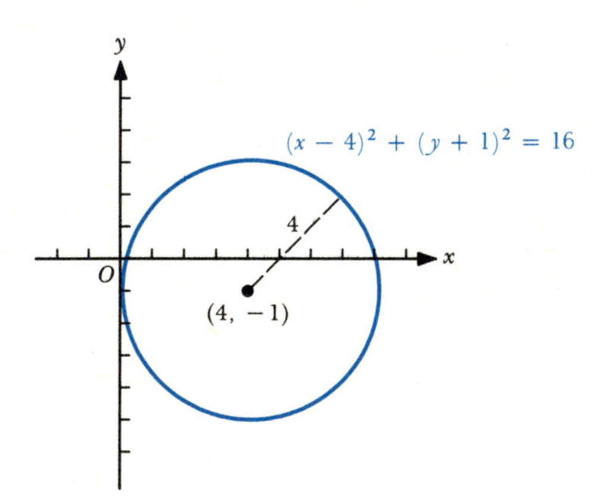

EXAMPLE 8 Show that the graph of the equation

$$x^2 + y^2 + 6x - 2y - 15 = 0$$

is a circle and determine the radius and the center of the circle. Sketch the graph.

SOLUTION First, we rewrite the equation in the form

$$(x^2 + 6x) + (y^2 - 2y) = 15$$

Figure 12

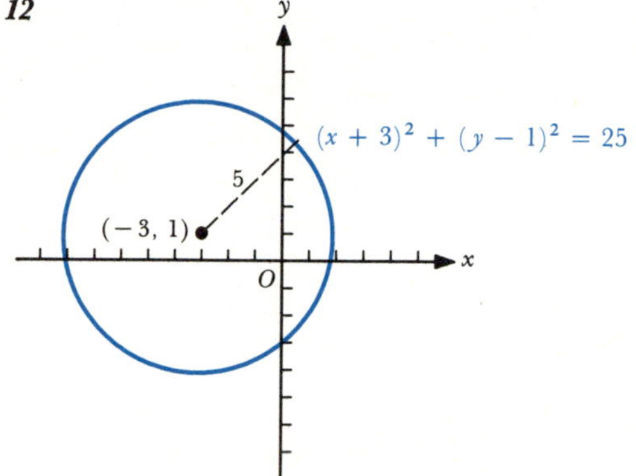

Next, we complete the square in each of the parentheses by adding 9 to the terms in x and 1 to the terms in y. Since we are adding 9 and 1 to the left side of the equation, we must also add 9 and 1 to the right side. The equation now has the form

$$(x^2 + 6x + 9) + (y^2 - 2y + 1) = 15 + 9 + 1$$

or

$$(x + 3)^2 + (y - 1)^2 = 25$$

We recognize this as an equation of the form

$$(x - h)^2 + (y - k)^2 = r^2$$

with $h = -3$, $k = 1$, and $r = 5$. Therefore, the graph is a circle of radius $r = 5$ with center $(h, k) = (-3, 1)$. The circle is graphed in Figure 12.

PROBLEM SET 2.1

1. Plot each of the given points and indicate which quadrant or coordinate axis contains the point.
(a) $(3, 3)$ (b) $(-2, 4)$ (c) $(5, -1)$ (d) $(3, -2)$
(e) $(0, 7)$ (f) $(0, -3)$ (g) $(-3, 0)$ (h) $(-1, -5)$

2. (a) Give the coordinates of any five points on the x axis.
(b) What is common to the coordinates of all points on the x axis?

In problems 3–8, locate the point P on a Cartesian coordinate system and give the coordinates of the points Q, R, and S that satisfy the given conditions.
(a) The line segment $\overline{PQ}$ is perpendicular to the x axis and is bisected by it.
(b) The line segment $\overline{PR}$ is perpendicular to the y axis and is bisected by it.
(c) The line segment $\overline{PS}$ is bisected by the origin.

3. $P = (1, 4)$
4. $P = (-4, -2)$
5. $P = (-3, 2)$
6. $P = (4, -1)$
7. $P = (2, -3)$
8. $P = (2, 0)$

In problems 9–21, use the distance formula to find the distance between the two given points.

9. $(1, 2)$ and $(7, 10)$
10. $(-3, -4)$ and $(-5, -7)$
11. $(1, 1)$ and $(-3, 2)$
12. $(-2, 5)$ and $(3, -1)$
13. $(-\frac{2}{5}, -\frac{1}{5})$ and $(\frac{1}{5}, \frac{3}{5})$
14. $(t, 8)$ and $(t, 7)$
15. $(2, 3)$ and $(-\frac{1}{2}, 1)$
16. $(t, u + 1)$ and $(t + 1, u)$
17. $(5, -t)$ and $(7, t)$
18. $(2t - 1, (t - 1)^2)$ and $(t^2, 0)$
19. $(-\sqrt{2}, -\sqrt{6})$ and $(\sqrt{3}, -\sqrt{2})$

c **20.** $(-3.651, 8.215)$ and $(4.734, 5.324)$, to three decimal places

c **21.** $(\pi, \frac{47}{4})$ and $(-\sqrt{19}, \frac{189}{3})$, to two decimal places

22. If $A = (-2, -1)$, $B = (2, 2)$, and $C = (5, -2)$, determine whether or not triangle ABC is isosceles.

In problems 23–26, use the distance formula and the converse of the Pythagorean theorem to show that the triangle with the given vertices is a right triangle.

23. $(0, 0)$, $(-3, 0)$, and $(-3, 4)$
24. $(-3, 1)$, $(3, 1)$, and $(3, 10)$
25. $(1, 1)$, $(5, 1)$, and $(5, 7)$
26. $(-2, -2)$, $(0, 0)$, and $(3, -3)$

27. (a) Find all real numbers x for which the distance between $(4, 2)$ and $(4, x)$ is five units.
(b) Shade the region in the xy plane that contains all points (x, y) for which $|x| \leq 3$ and $y > 1$.

28. Find a point $(x, 1)$ in the first quadrant for which the triangle whose vertices are $A = (1, 1)$, $B = (4, 7)$, and $C = (x, 1)$ is an isosceles triangle, such that $|\overline{AB}| = |\overline{AC}|$.

29. Let $A = (-2, -1)$, $B = (-4, 5)$, $C = (3, 5)$, and $D = (3, -1)$. Find the perimeter of the quadrilateral $ABCD$.

30. If P_1, P_2, and P_3 are points in the plane, then P_2 lies on the line segment $\overline{P_1P_3}$ if and only if $|\overline{P_1P_3}| = |\overline{P_1P_2}| + |\overline{P_2P_3}|$. Illustrate this fact with a sketch. When this occurs, the points P_1, P_2, and P_3 are said to be **collinear.** Use the distance formula to prove that $P_1 = (x, y)$, $P_2 = (x + 1, y + 2)$, and $P_3 = (x + 2, y + 4)$ are collinear.

In problems 31–33, use problem 30 and the distance formula to determine whether or not the points P_1, P_2, and P_3 are collinear.

31. $P_1 = (-3, -2)$, $P_2 = (1, 2)$, $P_3 = (3, 4)$
32. $P_1 = (-1, 2)$, $P_2 = (1, 1)$, $P_3 = (5, -1)$
33. $P_1 = (1, 5)$, $P_2 = (-1, 2)$, $P_3 = (-2, -2)$

34. Show that the point

$$P_2 = \left(\frac{a + c}{2}, \frac{b + d}{2}\right)$$

is the **midpoint of the line segment** joining the points $P_1 = (a, b)$ and $P_3 = (c, d)$. [*Hint:* Use the condition in problem 30 to show that P_2 actually belongs to the line segment $\overline{P_1P_3}$, then show that $|\overline{P_1P_2}| = |\overline{P_2P_3}|$.]

35. Use the result of problem 34 to find the coordinates of the midpoint P_2 of the line segment joining P_1 and P_3 if:
(a) $P_1 = (-3, 5)$ and $P_3 = (-8, -4)$
(b) $P_1 = (1, 7)$ and $P_3 = (4, -1)$
(c) $P_1 = (1, -3)$ and $P_3 = (5, 8)$
(d) $P_1 = (-2, -4)$ and $P_3 = (-1, -2)$

36. Verify that each of the given points lies on the graph of the unit circle, which has equation $x^2 + y^2 = 1$, by showing that the coordinates of the point satisfy the equation.

(a) $\left(\frac{\sqrt{2}}{2}, -\frac{\sqrt{2}}{2}\right)$ (b) $\left(\frac{-\sqrt{3}}{2}, \frac{-1}{2}\right)$ (c) $\left(-\frac{3}{5}, \frac{4}{5}\right)$ (d) $\left(\frac{15}{17}, \frac{8}{17}\right)$

In problems 37–40, find the standard form of the equation of the circle with center C and radius r; also sketch the graph.

37. $r = 5; C = (3, -2)$

38. $r = 2; C = (-1, 3)$

39. $r = 4; C = (-3, 1)$

40. $r = \frac{3}{2}; C = (-2, 0)$

In problems 41–48, show that the graph of each equation is a circle; find the center C and the radius r of the circle, and sketch the circle.

41. $(x - 1)^2 + (y + 2)^2 = 25$

42. $(x + 2)^2 + (y - 1)^2 = 16$

43. $x^2 + y^2 - 3x + 4y + 4 = 0$

44. $x^2 + y^2 - 4x - 6y = 3$

45. $x^2 + y^2 + 6x - 2y + 6 = 0$

46. $4x^2 + 4y^2 + 4x - 4y + 1 = 0$

47. $x^2 + y^2 - 2x + \frac{4}{3}y - 12 = 0$

48. $4x^2 + 4y^2 + 12x + 20y + 25 = 0$

2.2 Lines and Their Slopes

In this section we derive some forms for equations whose graphs are straight lines. If a line is neither vertical nor horizontal, then it is said to *slant* either upward or downward as we move from the left to the right (Figure 1a and b). To measure the amount of slant a line possesses we introduce the concept of *slope*.

We know from plane geometry that there is one and only one line L containing two distinct points P_1 and P_2. If the coordinates of P_1 and P_2 are given, we define the slope of L as follows.

DEFINITION 1 **Slope of a Line**

If $P_1 = (x_1, y_1)$ and $P_2 = (x_2, y_2)$ are two *different* points on a nonvertical line, the **slope** m of the line is defined by

$$m = \frac{y_2 - y_1}{x_2 - x_1}$$

If the line is vertical, then the slope is not defined.

In finding the slope of a line through the points P_1 and P_2, it is not important which point is called $P_1 = (x_1, y_1)$ and which point is called $P_2 = (x_2, y_2)$, since

$$\frac{y_2 - y_1}{x_2 - x_1} = \frac{-(-y_2 + y_1)}{-(-x_2 + x_1)} = \frac{y_1 - y_2}{x_1 - x_2}$$

EXAMPLE 1 In each case, sketch the line containing the points A and B and find its slope m.

(a) $A = (-2, -1), B = (2, 5)$ (b) $A = (-1, 3), B = (4, 1)$

(c) $A = (-2, 4), B = (3, 4)$ (d) $A = (4, -2), B = (4, 3)$

SOLUTION The lines are sketched in Figure 1. Using Definition 1, we have

(a) $m = \dfrac{y_2 - y_1}{x_2 - x_1} = \dfrac{5 - (-1)}{2 - (-2)} = \dfrac{5 + 1}{2 + 2} = \dfrac{6}{4} = \dfrac{3}{2}$

(b) $m = \dfrac{y_2 - y_1}{x_2 - x_1} = \dfrac{1 - 3}{4 - (-1)} = \dfrac{1 - 3}{4 + 1} = -\dfrac{2}{5}$

(c) $m = \dfrac{y_2 - y_1}{x_2 - x_1} = \dfrac{4 - 4}{3 + 2} = \dfrac{0}{5} = 0$

(d) Since $x_2 - x_1 = 0$, m is undefined.

Figure 1

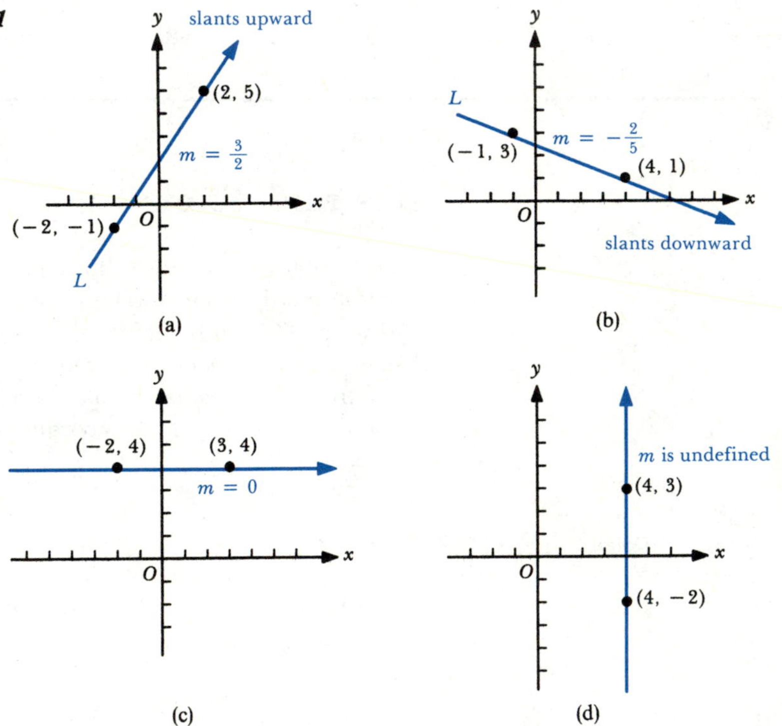

Consider the line L containing the points P_1 and P_2 in Figure 2. The "directed" vertical distance $y_2 - y_1$ between P_1 and P_2 is called the **rise.** The directed horizontal distance $x_2 - x_1$ between P_1 and P_2 is called the **run.** Thus, the slope m of the line L is given by

$$m = \frac{\text{rise}}{\text{run}}$$

Figure 2

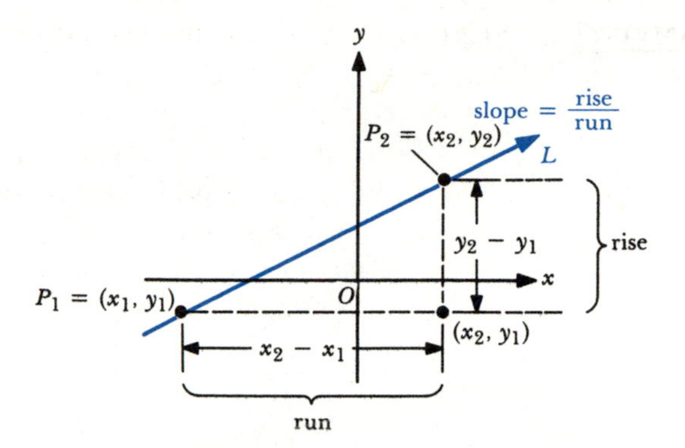

Suppose that the line L contains the points P_1 and P_2. If the line L slants upward to the right, its slope is positive (Figure 3a). If the line L slants downward to the right, its slope is negative (Figure 3b). If the line L is horizontal, its rise is zero and so its slope is zero (Figure 3c).

Figure 3

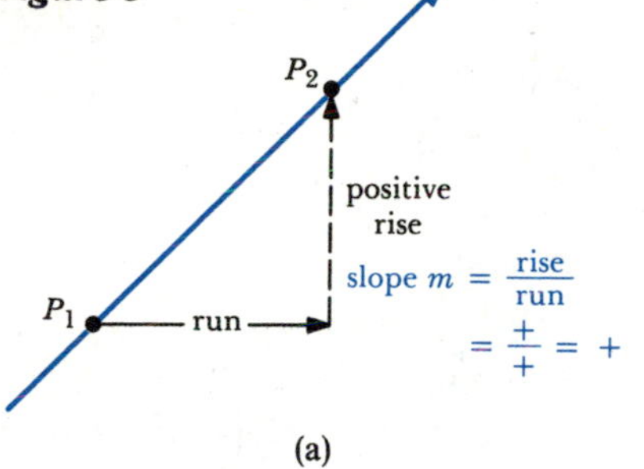

(a)

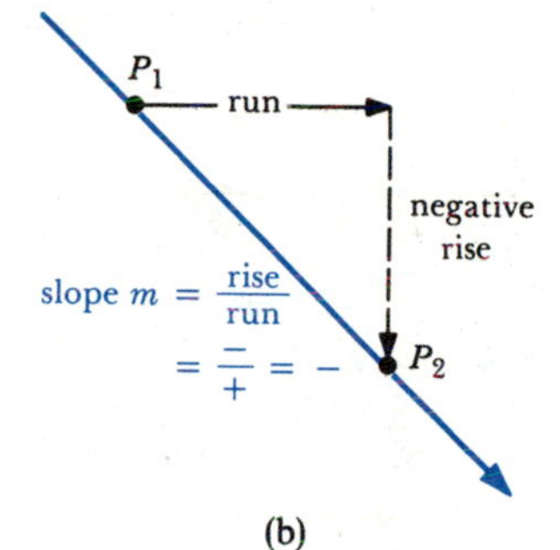

(b)

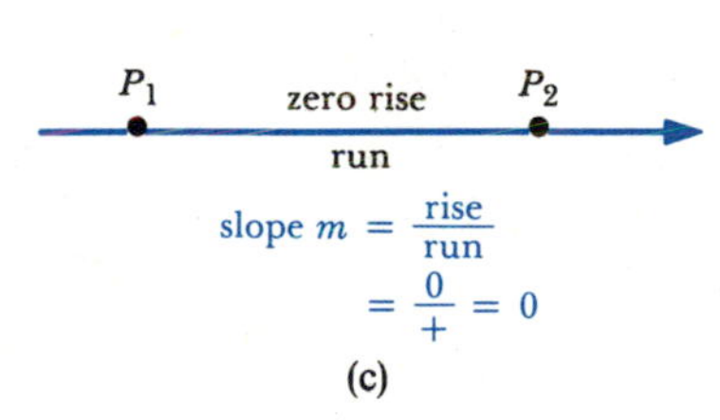

(c)

The value of the slope of a line is the same regardless of which points on the line are chosen to compute the slope. For example, if we choose the two points $P_1 = (x_1, y_1)$ and $P_2 = (x_2, y_2)$ on a line L, and then choose two other points $P'_1 = (x'_1, y'_1)$ and $P'_2 = (x'_2, y'_2)$ on L (Figure 4), the right triangles $P_1P_2P_3$ and $P'_1P'_2P'_3$ are similar (why?), so that the ratios of the lengths of the corresponding sides are equal.

Figure 4

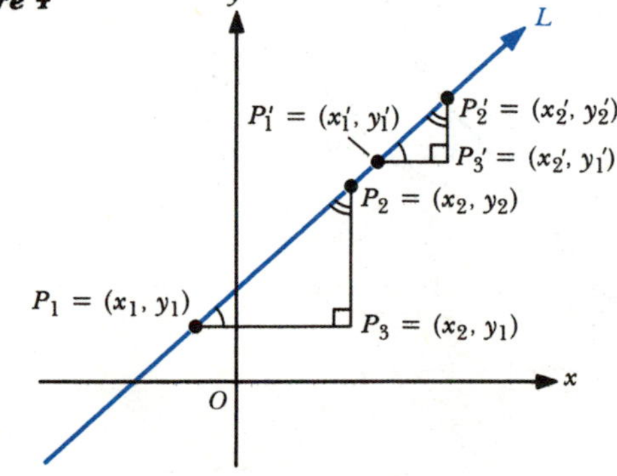

Thus,

$$\frac{|\overline{P_2P_3}|}{|\overline{P_1P_3}|} = \frac{|\overline{P'_2P'_3}|}{|\overline{P'_1P'_3}|}$$

or

$$\frac{y_2 - y_1}{x_2 - x_1} = \frac{y'_2 - y'_1}{x'_2 - x'_1}$$

EXAMPLE 2 Sketch the line L that contains the point $P = (2, 3)$ and has the given slope.

(a) $m = \frac{3}{4}$ (b) $m = -\frac{3}{4}$

SOLUTION (a) The condition $m = \frac{3}{4}$ means that for every four units we move to the right (the run) from a point on L, we must move up three units (the rise) to obtain another point on L (Figure 5a).

(b) Here the slope is $m = -\frac{3}{4} = \frac{-3}{4}$, so that for every run of four units we move to the right from a point on L, there is a corresponding "rise" of -3 units to obtain another point on L (Figure 5b).

Figure 5

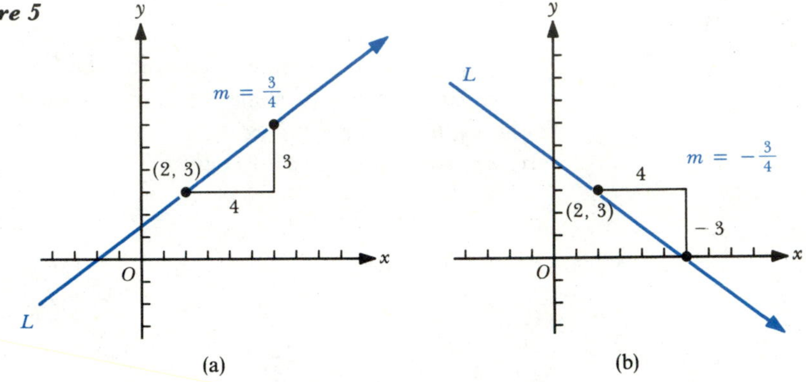

Equations of Lines

Consider a nonvertical line L having slope m and containing the point $P_1 = (x_1, y_1)$ (Figure 6). If $P = (x, y)$ represents any other point on L, then the slope of the line containing P_1 and P is

$$\frac{y - y_1}{x - x_1} = m$$

This equation may be written as

$$y - y_1 = m(x - x_1)$$

Figure 6

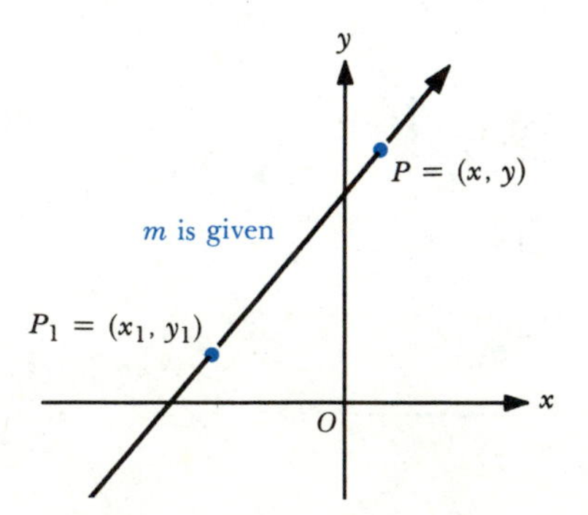

Notice that this latter equation holds even if $P = P_1$, when it simply reduces to $0 = 0$. Hence *any* point $P = (x, y)$ whose coordinates satisfy the above equation lies on L. Conversely, any point that lies on L has coordinates that satisfy the equation. This equation for line L is referred to as the *point-slope form* of an equation of a line.

Point-Slope Form

> An equation for the line L that contains the point $P_1 = (x_1, y_1)$ and has slope m is given by
> $$y - y_1 = m(x - x_1)$$

In Examples 3 and 4, find an equation in point-slope form for the given line L.

EXAMPLE 3 L contains the point $(2, 3)$ and has slope 5.

SOLUTION Substituting $x_1 = 2$, $y_1 = 3$, and $m = 5$ in the equation $y - y_1 = m(x - x_1)$, we have

$$y - 3 = 5(x - 2)$$

EXAMPLE 4 L contains the points $(-2, 5)$ and $(3, -4)$.

SOLUTION By Definition 1, the slope of the line L is given by

$$m = \frac{5 - (-4)}{-2 - 3} = -\frac{9}{5}$$

Since the point $(-2, 5)$ belongs to L, we can use

$$(x_1, y_1) = (-2, 5)$$

in the point-slope form equation to get an equation of the line

$$y - 5 = -\frac{9}{5}[x - (-2)] \quad \text{or} \quad y - 5 = -\frac{9}{5}(x + 2)$$

Figure 7

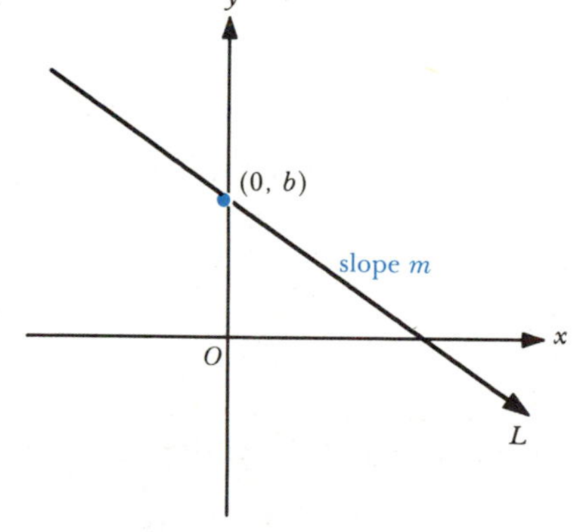

Suppose that L is a nonvertical line with slope m. Since L is not parallel to the y axis, it must cross this axis at some point, say $(0, b)$ (Figure 7). The ordinate b of the intersection point is called the ***y* intercept** of L. Since $(0, b)$ belongs to L, a point-slope form equation of L is

$$y - b = m(x - 0)$$

which simplifies to $y = mx + b$. The latter equation is called the *slope-intercept form* of an equation for line L.

Slope-Intercept Form

The graph of the equation

$$y = mx + b$$

is a line with slope m and y intercept b.

For instance, $y = 3x + 5$ is an equation of a line which has slope 3 and y intercept 5.

If we replace m by 0 in the equation $y = mx + b$, we have $y = 0x + b$ or

$$y = b$$

which is an equation of a **horizontal line** (Figure 8).

The slope of a **vertical line** is undefined. Therefore, we *cannot* write an equation for a vertical line in *the slope-intercept form*. However, because all points on a vertical line have the same abscissa or x coordinate, an equation of such a line (Figure 9) takes the form

$$x = a$$

where a is a constant.

Figure 8

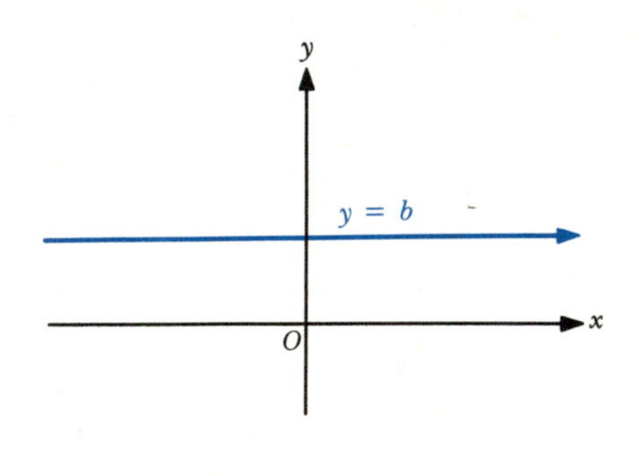

Figure 9

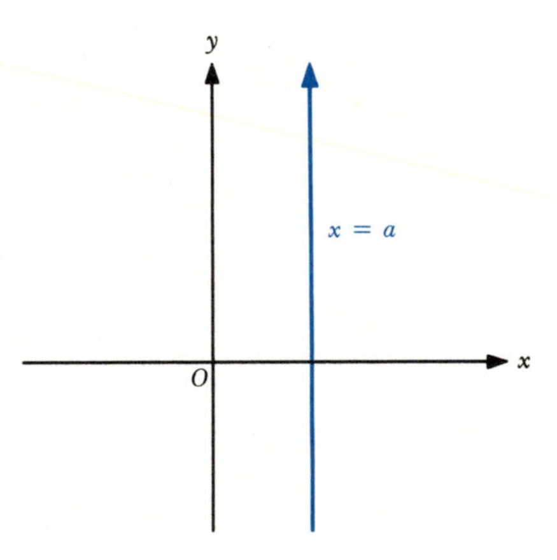

We have shown that every line is the graph of an equation in one of the above special forms. Any of these equations can also be represented in the form

$$Ax + By + C = 0$$

where A, B, and C are constants; and A and B are not both zero. We call such an equation the **general form of a linear equation** in the variables x and y.

EXAMPLE 5 Express each of the given linear equations in the general form $Ax + By + C = 0$.

(a) $x = 7$

(b) $y = -3$

(c) $y = 2x + 1$

SOLUTION (a) $x = 7$ can be written in the general form

$$1 \cdot x + 0 \cdot y + (-7) = 0$$

or

$$x - 7 = 0$$

(b) $y = -3$ can be written in the general form

$$0 \cdot x + 1 \cdot y + 3 = 0$$

or

$$y + 3 = 0$$

(c) $y = 2x - 1$ can be written in the general form

$$2x + (-1) \cdot y + (-1) = 0$$

or

$$2x - y - 1 = 0$$

Table 1 summarizes various situations that occur according to whether A or B is nonzero in the general form $Ax + By + C = 0$.

Table 1

$Ax + By + C = 0$		
Condition on A and B	Converted Form of the Equation	Graph of the Equation
$A \neq 0, B = 0$	$x = -C/A$	Vertical line
$A = 0, B \neq 0$	$y = -C/B$	Horizontal line
$A \neq 0, B \neq 0$	$y = (-A/B)x + (-C/B)$	A line with slope $-A/B$ and y intercept $-C/B$

The ***x* intercept** of a line is the value of the x coordinate of the point where the line intersects the x axis. For example, the x intercept of the graph of the line $3x + 6y = 12$ can be found by setting $y = 0$ to get $x = 4$. Thus, the graph of $3x + 6y = 12$ intersects the x axis at the point $(4, 0)$.

EXAMPLE 6 Find the x intercept, y intercept, and the slope of the line with general equation $x + 2y - 2 = 0$. Also sketch the graph.

SOLUTION We begin by solving the equation for y in terms of x.

$$x + 2y - 2 = 0$$
$$2y = -x + 2$$
$$y = -\tfrac{1}{2}x + 1$$

Figure 10

x	y
0	1
2	0

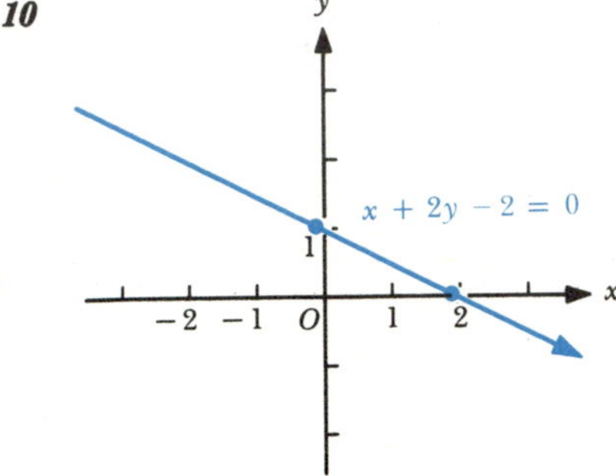

Thus, we have an equation of the line in slope-intercept form with slope $m = -\frac{1}{2}$ and y intercept $b = 1$. To find the x intercept, we set $y = 0$, to obtain

$$0 = x - 2 \quad \text{or} \quad x = 2$$

Therefore, the x intercept is 2. The graph is displayed in Figure 10.

EXAMPLE 7 The *normal* weight w in kilograms of individuals 153 centimeters tall or taller is related to their height h in centimeters by a linear equation. Assume that the normal weight for a person 153 centimeters tall is 50 kilograms and that the normal weight for a person 178 centimeters tall is 75 kilograms. Write an equation that expresses weight w in terms of height h. Use the equation to find the normal weight of a person whose height is 162 centimeters.

SOLUTION Here $w = 50$ when $h = 153$, and $w = 75$ when $h = 178$. The slope m of the line containing the points $(h_1, w_1) = (153, 50)$ and $(h_2, w_2) = (178, 75)$ is given by

$$m = \frac{w_2 - w_1}{h_2 - h_1} = \frac{75 - 50}{178 - 153} = \frac{25}{25} = 1$$

Substituting $w_1 = 50$, $h_1 = 153$, and $m = 1$ into the point-slope equation

$$w - w_1 = m(h - h_1)$$

we obtain

$$w - 50 = 1(h - 153)$$

or

$$w = h - 103$$

If $h = 162$, then

$$w = 162 - 103 = 59$$

Therefore, the normal weight of a person 162 centimeters tall is 59 kilograms.

Parallel and Perpendicular Lines

If two *different* lines are graphed on the same coordinate system, either they are parallel or they intersect at one point. We can use the slopes of two nonvertical lines to determine whether they are parallel or intersect. For instance, the graphs of the linear equations $y = 3x - 1$ and $y = 3x + 2$ suggest that the two lines are *parallel* (Figure 11a), because both lines have the same slope, 3. By contrast, the graphs of the linear equations $y = 3x - 1$ and $y = -\frac{1}{3}x + 2$ show that the two lines intersect (Figure 11b). [The point of intersection is $(\frac{9}{10}, \frac{17}{10})$. The point at which two nonparallel lines intersect can be found by solving the equations of the two lines simultaneously. Methods for solving systems of linear equations will be studied in Chapter 7.] In this latter situation, the two lines have different slopes, 3 and $-\frac{1}{3}$.

Figure 11

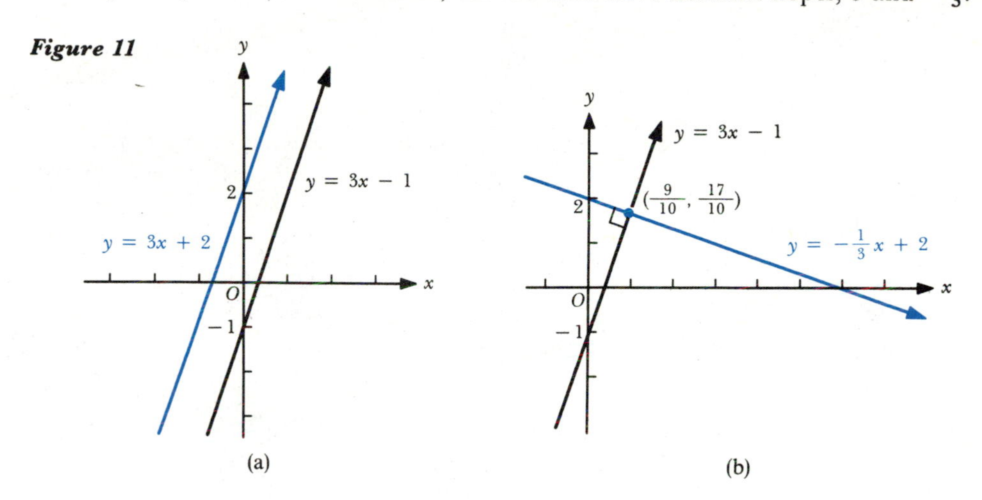

More specifically, the two lines are *perpendicular*. Note that the product of the slopes 3 and $-\frac{1}{3}$ is $(3)(-\frac{1}{3}) = -1$. These observations lead to the following property.

Property of Slopes

(i) Two different lines with slopes m_1 and m_2 are **parallel** if and only if $m_1 = m_2$.
(ii) Two different lines with slopes m_1 and m_2 intersect if and only if $m_1 \neq m_2$. They are **perpendicular** if and only if $m_1 m_2 = -1$.

The proof of Property (ii) is left as an exercise (Problem 44). It should be noted that Property (ii) does not apply if either line is horizontal or vertical. (Recall that if a line is vertical, its slope is undefined; if it is horizontal, its slope is zero.)

EXAMPLE 8 Determine whether each pair of lines is parallel or perpendicular.

(a) $3x - 2y - 3 = 0$ and $3x - 2y + 5 = 0$

(b) $-3x + 2y + 3 = 0$ and $2x + 3y - 10 = 0$

SOLUTION (a) We write both equations in the slope-intercept form:

$$3x - 2y - 3 = 0 \qquad\qquad 3x - 2y + 5 = 0$$
$$-2y = -3x + 3 \qquad\qquad -2y = -3x - 5$$
$$y = \frac{3}{2}x - \frac{3}{2} \qquad\qquad y = \frac{3}{2}x + \frac{5}{2}$$

The two lines are parallel since both of their slopes equal $\frac{3}{2}$ (Figure 12a).

(b) We write both equations in the slope-intercept form to obtain

$$-3x + 2y + 3 = 0 \qquad\qquad 2x + 3y - 10 = 0$$
$$2y = 3x - 3 \qquad\qquad 3y = -2x + 10$$
$$y = \frac{3}{2}x - \frac{3}{2} \qquad\qquad y = -\frac{2}{3}x + \frac{10}{3}$$

Thus the slope of the first line is $m_1 = \frac{3}{2}$, and the slope of the second line is $m_2 = -\frac{2}{3}$. Since $m_1m_2 = (\frac{3}{2})(-\frac{2}{3}) = -1$, the two lines are perpendicular (Figure 12b).

Figure 12

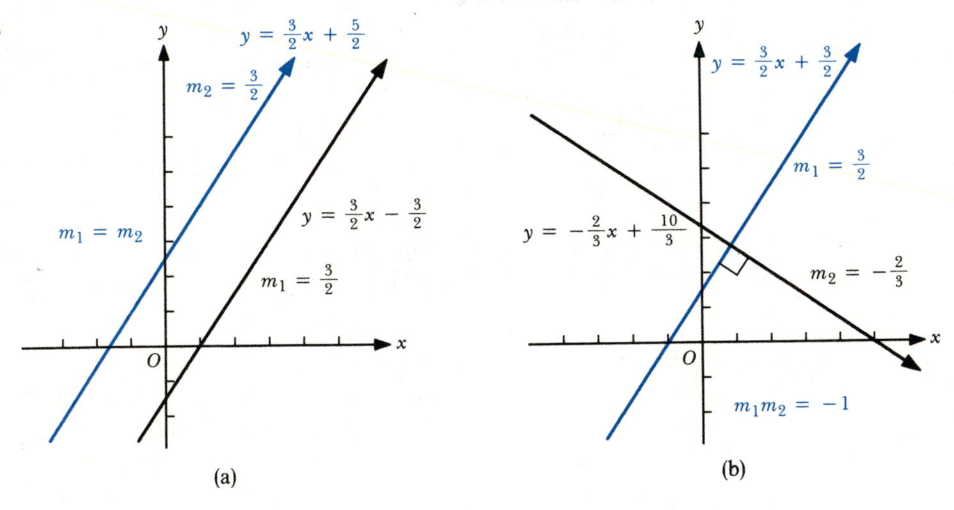

EXAMPLE 9 Let L be the line whose equation is given by $2x - y + 6 = 0$. Find an equation of the line, in slope-intercept form, that contains the point $(3, 2)$ and is (a) perpendicular to L, (b) parallel to L.

SOLUTION Writing the equation of L in slope-intercept form, we have

$$y = 2x + 6$$

so that the slope of L is 2.

(a) The slope of a line perpendicular to L is $-\frac{1}{2}$ since $(2)(-\frac{1}{2}) = -1$. Hence an equation of the line perpendicular to L and containing the point $(3, 2)$ is given by

$$y - 2 = -\frac{1}{2}(x - 3)$$

Solving the equation for y in terms of x, we obtain the slope-intercept form,

$$y = -\frac{1}{2}x + \frac{7}{2}$$

(b) The slope of a line parallel to L is 2. Thus an equation of the line parallel to L and containing the point $(3, 2)$ is given by

$$y - 2 = 2(x - 3)$$

After we solve the equation for y in terms of x, we obtain the slope-intercept form,

$$y = 2x - 4$$

PROBLEM SET 2.2

In problems 1–10, find the slope of the line containing the given pair of points and sketch the line.

1. $A = (2, 3)$, $B = (-1, -2)$
2. $A = (4, 1)$, $B = (3, -1)$
3. $A = (-\frac{2}{3}, 1)$, $B = (-1, \frac{5}{3})$
4. $A = (0, 1)$, $B = (-1, 3)$
5. $A = (0, 0)$, $B = (3, 7)$
6. $A = (1, 1)$, $B = (2, 0)$
7. $A = (-1, 3)$, $B = (2, 3)$
8. $A = (-3, 1)$, $B = (2, 1)$
9. $A = (2, -2)$, $B = (2, 1)$
10. $A = (\frac{5}{7}, -\frac{2}{7})$, $B = (\frac{5}{7}, \frac{2}{7})$

In problems 11–16, sketch the line L that contains the points P and has the slope m.

11. $P = (4, -3)$, $m = 3$
12. $P = (-2, 3)$, $m = 0$
13. $P = (-3, 1)$, $m = -\frac{2}{3}$
14. $P = (-2, 0)$, $m = -\frac{1}{2}$
15. $P = (-1, -2)$, $m = -1$
16. $P = (5, -1)$, $m = \frac{3}{5}$

17. A particle, initially located at the point $P = (4, 5)$, moves along a line of slope -2 to a new position (x, y).
(a) Find y if $x = 6$ (b) Find x if $y = 8$.

18. Find the slopes of the sides of the triangle with vertices $A = (-4, -2)$, $B = (2, -8)$, and $C = (4, 6)$. Use the slopes to determine whether triangle ABC is a right triangle.

In problems 19–22, find an equation in point-slope form of the given line L.

19. L contains the point $(-4, 1)$ and has slope $m = -3$.
20. L contains the point $(2, 8)$ and has slope $m = \frac{2}{3}$.
21. L contains the points $(3, 4)$ and $(2, -5)$.
22. L contains the point $(7, -1)$ and has slope $m = 0$.

In problems 23–32, rewrite each equation in slope-intercept form. Find the slope, the y intercept, the x intercept, and sketch the graph.

23. $3x - 5y = 0$
24. $y - 8 = 0$
25. $y + 1 = -2(x + 1)$
26. $x = 3y + 2$
27. $4x + 5y + 7 = 0$
28. $6x - 9y - 7 = 0$
29. $5x + 12y - 36 = 0$
30. $2x + 5y - 15 = 0$
31. $9x - y - 23 = 0$
32. $\frac{y}{2} - \frac{3x}{5} = 1$

In problems 33–41, find an equation of the line L in (a) point-slope form, (b) slope-intercept form, and (c) general form.

33. L contains the point $(1, 1)$ and has slope $m = 4$.
34. L contains the point $(-3, -2)$ and has slope $m = 0$.
35. L has slope $m = -3$ and y intercept 2.
36. L contains the points $(-2, 5)$ and $(2, -3)$.
37. L contains the points $(3, 3)$ and $(1, -2)$.
38. L contains the point $(2, 1)$ and is parallel to the line $2x - y = 6$.
39. L contains the point $(-1, 4)$ and is parallel to the line $3x + 4y - 5 = 0$.
40. L contains the point $(-3, 2)$ and is perpendicular to the line $4x - 2y - 7 = 0$.
41. L contains the point $(-3, -5)$ and is perpendicular to the line $3x + 2y = -2$.
42. (a) Show that an equation of a line with y intercept $b \neq 0$ and with x intercept $a \neq 0$ can be written in the form

$$\frac{x}{a} + \frac{y}{b} = 1$$

This equation is called the **intercept form** of an equation of a line.
(b) Use the result of part (a) to find an equation of the line whose x intercept is 4 and y intercept is -6.
43. Find an equation of the line that is parallel to the y axis and contains the point $(-3, 2)$.
44. Prove that if the product of the slopes of two lines is -1, then the two lines are perpendicular. (*Hint:* Use the distance formula.)
45. The equation that relates the Celsius C and Fahrenheit F temperatures is linear. Find an equation for F in terms of C if 0°C corresponds to 32°F and 100°C corresponds to 212°F. Graph the equation and use the equation to find the Fahrenheit measure of 45°C.
46. An apartment building was built in 1969 at a cost of $350,000. What will be its value (for tax purposes) in 1990 if it is being depreciated linearly over 40 years according to the formula $v = c - (c/N)n$, where c (in dollars) is the original cost of the property, N is the fixed number of years over which the property is being depreciated, and v is the value of the depreciated balance at the end of n years?
47. A state has an income tax of 7% on all income over $3000. Write a linear equation that determines a person's tax T in terms of income I for someone earning more than $3000. At what income level will a person owe $840 in state tax?

48. A projectile fired straight up attains a velocity of v feet per second after t seconds of flight, and the equation that relates the numbers v and t is linear. If the projectile is fired at a velocity of 200 feet per second and reaches a velocity of 90 feet per second after 2 seconds of flight, express v in terms of t. How soon after the projectile is fired does its velocity equal 0?

49. Let the temperature k meters above the surface of the earth be t degrees Celsius, and assume that the equation that relates t and k is linear. If the temperature on the surface of the earth is 30°C and the temperature at 2500 meters is 16°C, what is the temperature at 5000 meters?

50. A veterinarian wants to prepare a special diet for animals. Two food mixes A and B are available. If mix A contains 18% protein and mix B contains 9% protein, write a linear equation relating the amount x of A and the amount y of B to yield a combination of exactly 25 grams of protein. What combinations of each mix will provide 25 grams of protein?

2.3 Functions

Gottfried Wilhelm von Leibniz

The term "function" was first used by Leibniz in 1673 to denote the dependence of one quantity on another. The notion of functional dependence is encountered quite often in everyday living. For example, the number of textbooks to be ordered for a course depends on the number of students enrolled, and the gravitational attraction between two material bodies depends on the distance between them.

In mathematics, the general idea of a function is simple. Suppose that one variable quantity, say, y, depends in a definite way on another variable quantity, say, x. Then for each particular value of x, there is *one* corresponding value of y. Such a correspondence defines a *function*, and we say that (the variable) y is a function of (the variable) x. For example, if x is used to denote the radius of a circle and y is used to denote the area of this circle, then y depends on x in a definite way, namely, $y = \pi x^2$. Thus, we say that the area y of a circle is a function of its radius x. In a sense, the value of y depends on the value assigned to x. For this reason, we sometimes refer to x as the *independent variable* and y as the *dependent variable*. If $x = 5$, $y = 25\pi$; if $x = 7$, $y = 49\pi$; if $x = 10$, $y = 100\pi$.

More formally, we have the following definition.

DEFINITION 1 **Function**

A **function** is a rule or correspondence that assigns to each member in a certain set, called the **domain** of the function, one and only one member in a second set, called the **range** of the function. The **independent variable** of the function can take on any value in the domain of the function. The set of all possible corresponding values that the **dependent variable** assumes is the range of the function.

In order to discuss functions that relate the values of two variable quantities, the Swiss mathematician Leonhard Euler (1707–1783) developed the notation of using letters of the alphabet, such as f, g, h, F, G, H, etc., to designate functions. By writing

$$y = f(x)$$

(read "y equals f of x"), we convey the idea that y is a function of x in the sense that the value of y (the dependent variable) *depends* on the value of x (the independent variable). Thus, instead of writing

"y is a function of x and $y = 3x + 2$"

we can use the notation

$$y = f(x) = 3x + 2$$

If x is an element in the domain of a function f, then the element in the range that f associates with x is denoted by the symbol $f(x)$, and is called the **image** of x under the function f. The image is also referred to as the **value** of f at x. For example, if $f(x) = 3x + 2$,

the image of $x = 2$ is $3(2) + 2 = 8$

the image of $x = 0$ is $3(0) + 2 = 2$

the image of $x = -1$ is $3(-1) + 2 = -1$

We write

$$f(2) = 8, \qquad f(0) = 2, \qquad \text{and} \qquad f(-1) = -1$$

In other words, when a function f is defined by an equation, we determine the value $f(x)$ when $x = a$ by simply substituting a for x in the equation to obtain $y = f(a)$. For instance, if f is defined by

$$f(x) = 2x^2 + 1$$

then

$$f(1) = 2(1)^2 + 1 = 3$$

$$f(2) = 2(2)^2 + 1 = 9$$

$$f(-3) = 2(-3)^2 + 1 = 19$$

$$f(b) = 2b^2 + 1$$

$$f(k + 3) = 2(k + 3)^2 + 1 = 2k^2 + 12k + 19$$

The particular letters used to denote the dependent and independent variables are of no importance in themselves—the important thing is the rule by which a definite value of the dependent variable is assigned to each value of the independent variable. In many applications, variables other than x and y are used because physical and geometrical quantities are designated by conventional symbols. For instance, if A is the function that gives the area of a circle in terms of its radius r, then we write

$$A(r) = \pi r^2$$

EXAMPLE 1 Let f be the function defined by $f(x) = 7x + 2$. Find the indicated values.

(a) $f(1)$ (b) $f(-2)$ (c) $f(t^2)$
(d) $[f(4)]^2$ (e) $\sqrt{f(2)}$ (f) $f(t + h)$

SOLUTION Since $f(x) = 7x + 2$, we have

(a) $f(1) = 7(1) + 2 = 7 + 2 = 9$
(b) $f(-2) = 7(-2) + 2 = -14 + 2 = -12$
(c) $f(t^2) = 7(t^2) + 2 = 7t^2 + 2$
(d) $[f(4)]^2 = [7(4) + 2]^2 = (28 + 2)^2 = 30^2 = 900$
(e) $\sqrt{f(2)} = \sqrt{7(2) + 2} = \sqrt{14 + 2} = \sqrt{16} = 4$
(f) $f(t + h) = 7(t + h) + 2 = 7t + 7h + 2$

If f is a function and x represents a member of the domain, then $f(x)$ represents the corresponding member of the range. Note that $f(x)$ is *not* the function f. In the interest of brevity, however, we often use the phrase "the function $y = f(x)$." There is no great harm in this practice as long as it is understood that the phrase "the function $y = f(x)$" actually means "f is a function defined by the equation $y = f(x)$." Although we shall avoid this practice when absolute precision is desired, we shall indulge in it whenever it seems convenient.

In calculus, it is necessary to deal with expressions of the form

$$\frac{f(x + h) - f(x)}{h}, \qquad h \neq 0$$

which is called the **difference quotient** of a function f.

EXAMPLE 2 Find the difference quotient for each function.

(a) $f(x) = 2x + 1$ (b) $f(x) = \sqrt{x}$

SOLUTION (a) Since $f(x + h) = 2(x + h) + 1$, it follows that

$$\frac{f(x + h) - f(x)}{h} = \frac{[2(x + h) + 1] - (2x + 1)}{h}$$

$$= \frac{2x + 2h + 1 - 2x - 1}{h} = \frac{2h}{h} = 2$$

(b) Here, $$\frac{f(x + h) - f(x)}{h} = \frac{\sqrt{x + h} - \sqrt{x}}{h}$$

However, this expression can be rewritten by rationalizing the numerator, so that

$$\frac{f(x + h) - f(x)}{h} = \frac{(\sqrt{x + h} - \sqrt{x})(\sqrt{x + h} + \sqrt{x})}{h(\sqrt{x + h} + \sqrt{x})}$$

$$= \frac{(x + h) - x}{h(\sqrt{x + h} + \sqrt{x})} = \frac{\not{h}}{\not{h}(\sqrt{x + h} + \sqrt{x})} = \frac{1}{\sqrt{x + h} + \sqrt{x}}$$

Certain functions are defined by using different equations for different parts of the domains. These **piecewise-defined** functions are illustrated in the next example.

EXAMPLE 3 Let

$$g(x) = \begin{cases} 3 + x^2 & \text{if } x < -2 \\ 2x & \text{if } -2 \le x < 1 \\ 11 - x^2 & \text{if } x \ge 1 \end{cases}$$

Find

(a) $g(-3)$ (b) $g(-1)$

(c) $g(-2)$ (d) $g(4)$

SOLUTION (a) When $x < -2$, $g(x) = 3 + x^2$, so that

$$g(-3) = 3 + (-3)^2 = 3 + 9 = 12$$

(b) When $-2 \le x < 1$, $g(x) = 2x$, so that

$$g(-1) = 2(-1) = -2$$

(c) $g(-2) = 2(-2) = -4$ since $g(x) = 2x$ if $-2 \le x < 1$.

(d) When $x \ge 1$, $g(x) = 11 - x^2$, so that

$$g(4) = 11 - (4)^2 = 11 - 16 = -5$$

EXAMPLE 4 An ecologist investigating the effect of air pollution on plant life finds that the percentage, $p(s)$ percent, of damaged trees and shrubs at a distance s yards from an industrial plant is given by the function

$$p(s) = 32 - \frac{3s}{50}$$

Determine the percentage of affected trees and shrubs if

(a) $s = 50$ (b) $s = 100$ (c) $s = 200$

SOLUTION (a) $p(50) = 32 - \dfrac{3(50)}{50} = 32 - 3 = 29\%$

(b) $p(100) = 32 - \dfrac{3(100)}{50} = 32 - 6 = 26\%$

(c) $p(200) = 32 - \dfrac{3(200)}{50} = 32 - 12 = 20\%$

We have indicated in Definition 1 that the domain of a function is the set of values the independent variable can take on. We shall limit our study to functions whose domains are sets of real numbers. When a function is described by an equation and the domain is not explicitly specified, it is to be understood that the domain is the set of all real numbers for which the equation produces a single real number. We must always be alert to potential problems, such as division by zero and square roots (or even roots) of negative numbers. For example, the domain of $f(x) = 1/x$ consists of all real numbers *except zero,* since division by zero is not defined; and, the domain of $f(x) = \sqrt{x}$ includes all *nonnegative* real numbers, because the square root is defined only for nonnegative numbers in the real number system.

In Examples 5 and 6, determine the domain of the given function in interval notation.

EXAMPLE 5 $f(x) = \dfrac{5}{x+2}$

SOLUTION The expression $\dfrac{5}{x+2}$ represents a real number for all real values of x except values for which the denominator $x + 2$ is zero. Hence, the domain of f is the set of all real numbers *except* -2. That is, it consists of the interval $(-\infty, -2)$ together with the interval $(-2, \infty)$.

EXAMPLE 6 $G(x) = \sqrt{3 - x}$

SOLUTION The expression $\sqrt{3 - x}$ represents a real number if and only if $3 - x \geq 0$; that is, if and only if $x \leq 3$. Therefore, the domain of G is the interval $(-\infty, 3]$.

Occasionally one of the notations

$$x \overset{f}{\mapsto} y, \qquad f{:}\,x \mapsto y, \qquad \text{or} \qquad x \mapsto f(x)$$

is used to indicate that there is a function f such that for each value of x there corresponds a uniquely determined value of y. In this case, we say that each value of x is **mapped** onto a corresponding value of y by f, or f **maps** x onto y. For example, the function $f(x) = 7x + 2$ maps x onto $7x + 2$ and we write

$$f{:}\,x \mapsto 7x + 2$$

Another way to denote the function association is to use *ordered pair* notation. If x is an element in the domain of a function f, then we can form the **ordered pair** of numbers

$$(x, f(x))$$

by matching x with the value of f at x. For example, if $f(x) = 2x^2 + 1$, a few such ordered pairs would be

$$(0, 1), \qquad (1, 3), \qquad (-2, 9), \qquad \text{and} \qquad (\tfrac{1}{2}, \tfrac{3}{2})$$

EXAMPLE 7 Let $g(x) = x^3 - x$.

(a) Find $g(-1)$, $g(1)$, $g(2)$, and $g(3)$.

(b) Represent these function values using function notation, mapping notation, and ordered pair notation.

SOLUTION (a) Since $g(x) = x^3 - x$, we have

$$g(-1) = (-1)^3 - (-1) = -1 + 1 = 0$$

$$g(1) = 1^3 - 1 = 0$$

$$g(2) = 2^3 - 2 = 8 - 2 = 6$$

$$g(3) = 3^3 - 3 = 27 - 3 = 24$$

(b) Thus we have the representations:

Function Notation $g(x)$	Mapping $g: x \mapsto y$	Ordered Pairs (x, y) for g
$g(-1) = 0$	$g: -1 \mapsto 0$	$(-1, 0)$
$g(1) = 0$	$g: 1 \mapsto 0$	$(1, 0)$
$g(2) = 6$	$g: 2 \mapsto 6$	$(2, 6)$
$g(3) = 24$	$g: 3 \mapsto 24$	$(3, 24)$

The **graph** of a function f is defined to be the graph of the corresponding equation $y = f(x)$. In other words, the graph of f is the set of all points $(x, f(x))$ in the Cartesian plane such that x is in the domain of f.

Suppose that we are given a curve in the plane. How can we tell whether that curve represents the graph of a function? According to Definition 1 of a function, each domain number has *one* associated range number. If we use ordered pair notation, this means that no two ordered pairs formed from a function can have the same domain value. Thus, *on the graph of a function, we cannot have* two points (x, y_1) and (x, y_2) with the same abscissa x and different ordinates y_1 and y_2. Hence, we have the next result.

Vertical Line Test

A set of points in the Cartesian plane is the graph of a function if and only if no vertical line intersects the set more than once.

EXAMPLE 8 Which of the curves in Figure 1 is the graph of a function?

SOLUTION By the vertical line test, the curve in Figure 1a is the graph of a function, since no more than one point (x, y) lies on any vertical line intersecting the curve. The curve in Figure 1b is not the graph of a function, since there is a vertical line containing two points (x, y_1) and (x, y_2) on the curve.

Figure 1

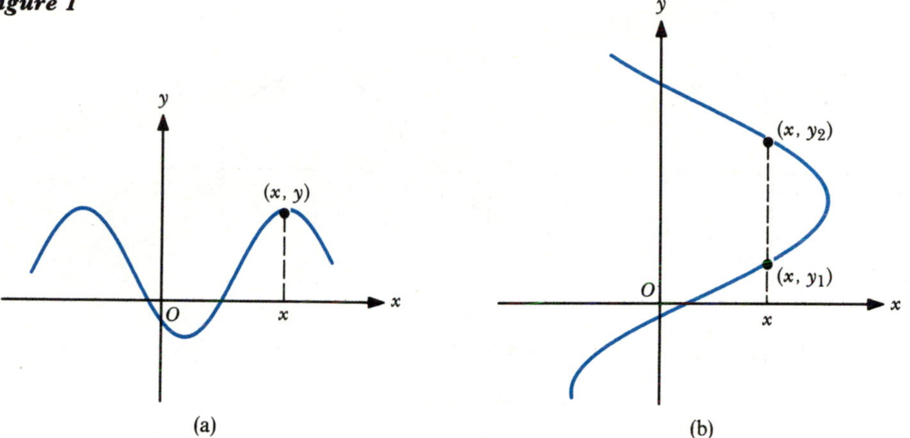

EXAMPLE 9 Use the vertical line test to determine whether or not each of the given equations is a function.

(a) $x^2 + y^2 = 9$

(b) $y = \sqrt{9 - x^2}$

(c) $y = -\sqrt{9 - x^2}$

SOLUTION (a) The graph of the equation $x^2 + y^2 = 9$ is a circle with its center at $(0, 0)$ and radius 3 (see page 72). A vertical line can intersect the graph at two points (Figure 2a). Hence the graph of the equation does not represent a function.

(b) and (c) Solving the equation $x^2 + y^2 = 9$ for y yields

$$y^2 = 9 - x^2 \qquad \text{or} \qquad y = \pm\sqrt{9 - x^2}$$

The nonnegative ordinates of points above and on the x axis are given by the equation

$$y = \sqrt{9 - x^2}$$

and the negative and zero ordinates of points below and on the x axis are given by the equation

$$y = -\sqrt{9 - x^2}$$

Therefore, we can think of the upper half of the circle as the graph of the equation $y = \sqrt{9 - x^2}$, and the lower half as the graph of $y = -\sqrt{9 - x^2}$. Clearly, no vertical line can intersect either graph at more than one point, so each curve in Figure 2b and Figure 2c is the graph of a function.

Figure 2

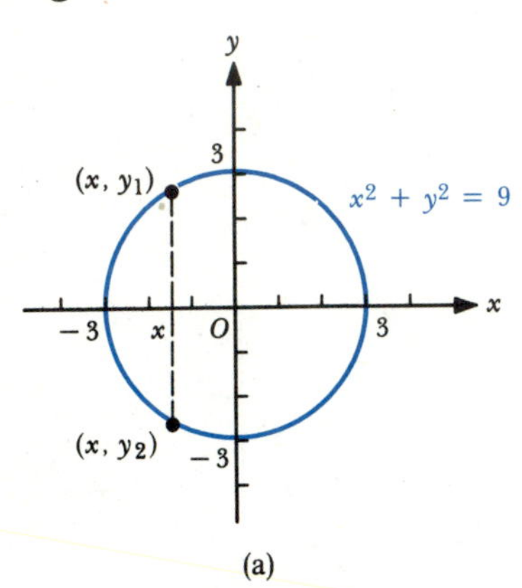

(a)

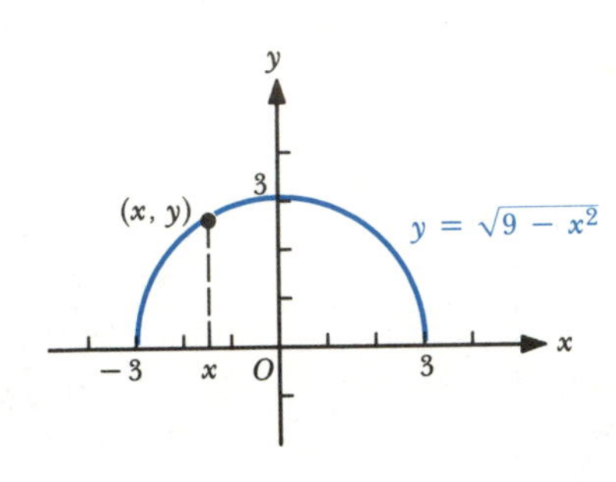

(b)

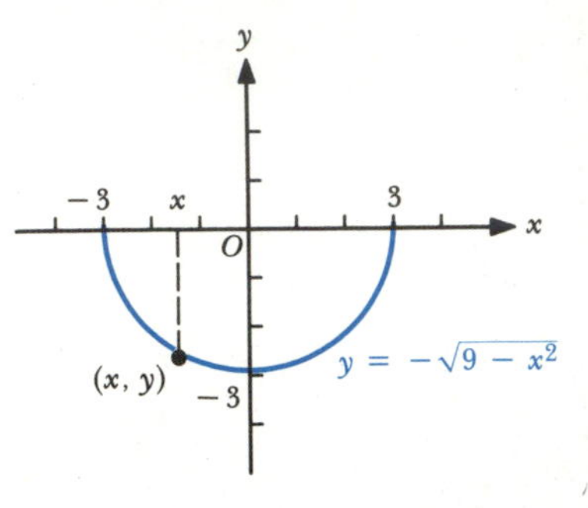

(c)

PROBLEM SET 2.3

In problems 1–52, let $f(x) = 2x + 5$, $g(x) = x^2 - 2x + 8$, $h(x) = (x + 3)/(x - 2)$, $k(x) = 1 + x$, $H(x) = |3 - 4x|$, and $F(x) = \sqrt[3]{x^3 + 8}$. Find each of the indicated values. In problems 48–51, round off each answer to three decimal places.

1. $f(0)$ **2.** $k(0)$ **3.** $h(0)$ **4.** $H(-1)$
5. $F(0)$ **6.** $k(81)$ **7.** $f(3)$ **8.** $g(3)$
9. $h(3)$ **10.** $k(1)$ **11.** $k(4)$ **12.** $H(-4)$
13. $\sqrt{F(-2)}$ **14.** $\sqrt{f(-2)}$ **15.** $\sqrt{g(-2)}$ **16.** $\sqrt{h(6)}$
17. $F(\sqrt[3]{-35})$ **18.** $f(a) + 2$ **19.** $h(a) + 2$ **20.** $k(a) + 2$
21. $f(a + 2)$ **22.** $h(a + 2)$ **23.** $k(a + 2)$ **24.** $f(a) + f(2)$
25. $H(x + \frac{3}{4})$ **26.** $k(a) + k(2)$ **27.** $f(a + b) - f(a)$ **28.** $h(a + b) - h(a)$
29. $g(a + b) - g(a)$ **30.** $f(7x)$ **31.** $h(7x)$ **32.** $g(7x)$
33. $7f(x)$ **34.** $7h(x)$ **35.** $7g(x)$ **36.** $f\left(\frac{1}{x + a}\right)$
37. $h\left(\frac{1}{x + a}\right)$ **38.** $g\left(\frac{1}{x + a}\right)$ **39.** $f(-x)$ **40.** $h(-x)$
41. $g(-x)$ **42.** $f(x^2)$ **43.** $H(x^2)$ **44.** $g(x^2)$
45. $[f(x)]^2$ **46.** $[h(x)]^2$ **47.** $[g(x)]^2$ c **48.** $F(1.460)$
c **49.** $g(-3.712)$ c **50.** $F(2.341)$ c **51.** $h(-3.471)$ **52.** $g(\sqrt{7})$

In problems 53–60, find the difference quotient for each function.

53. $f(x) = 3x + 5$

54. $f(x) = 2$

55. $f(x) = -5x + 1$

56. $f(x) = x^2 - 3$

57. $f(x) = \frac{1}{x}$

58. $f(x) = \sqrt{2x - 1}$

59. $f(x) = \frac{1}{x^2}$

60. $f(x) = x(x - 2)$

In problems 61–63, find the specified values for each function.

61. $f(x) = \begin{cases} -2 & \text{if } x < 1 \\ 3x & \text{if } 1 \le x < 3 \\ 2 - x^2 & \text{if } x \ge 3 \end{cases}$ (a) $f(0)$ (b) $f(2)$ (c) $f(3)$

62. $g(x) = \begin{cases} x + 2 & \text{if } x \le 3 \\ x + 4 & \text{if } x > 3 \end{cases}$ (a) $g(-1)$ (b) $g(3)$ (c) $g(4)$

63. $h(x) = \begin{cases} 1 - x & \text{if } 0 < x \le 1 \\ 3 + x & \text{if } 1 < x \le 2 \\ -1 & \text{if } 2 < x \le 3 \end{cases}$ (a) $h(\frac{1}{2})$ (b) $h(2)$ (c) $h(\frac{5}{2})$

64. Express y explicitly as a function of x if

$$x = \frac{2 - y}{1 + y}$$

65. Let $f(x) = 3x^2 - x$.
(a) Determine each of the values $f(-2)$, $f(-1)$, $f(0)$, $f(1)$, and $f(2)$.
(b) Represent the values in part (a) using function notation, mapping notation, and ordered pair notation.

66. Let $f(x) = x + 2$, $g(x) = 2x - 1$, and $h(x) = x - 1$. Determine which of the following equations hold for all values of x.

(a) $f\left(\frac{1}{x}\right) = \frac{1}{f(x)}$
(b) $g(x) \cdot \left[\frac{1}{g(x)}\right] = 1$
(c) $h(x - 1) = h(x) - h(1)$
(d) $f(x + 2) = f(x) + f(2)$
(e) $g(2x) = 2g(x)$

In problems 67–74, find the domain of each function.

67. $f(x) = 5x^2 + 1$

68. $g(x) = 5x^3 - 7$

69. $h(x) = \frac{3}{x + 1}$

70. $F(x) = \frac{5}{x - 2}$

71. $G(x) = \sqrt{3x - 2}$

72. $f(x) = \frac{1}{\sqrt{3x + 2}}$

73. $g(x) = \frac{x^2 - 4}{x^2 - 1}$

74. $h(x) = \frac{2}{x^2 + 1}$

In problems 75–80, graph the equation and determine whether the graph represents a function.

75. $y = 3x + 1$

76. $3x^2 + 3y^2 = 75$

77. $y = \sqrt{25 - x^2}$

78. $y = -\sqrt{25 - x^2}$

79. $(x - 1)^2 + (y + 2)^2 = 4$

80. $x = 1$

81. Which of the graphs in Figure 3 represent functions?

Figure 3

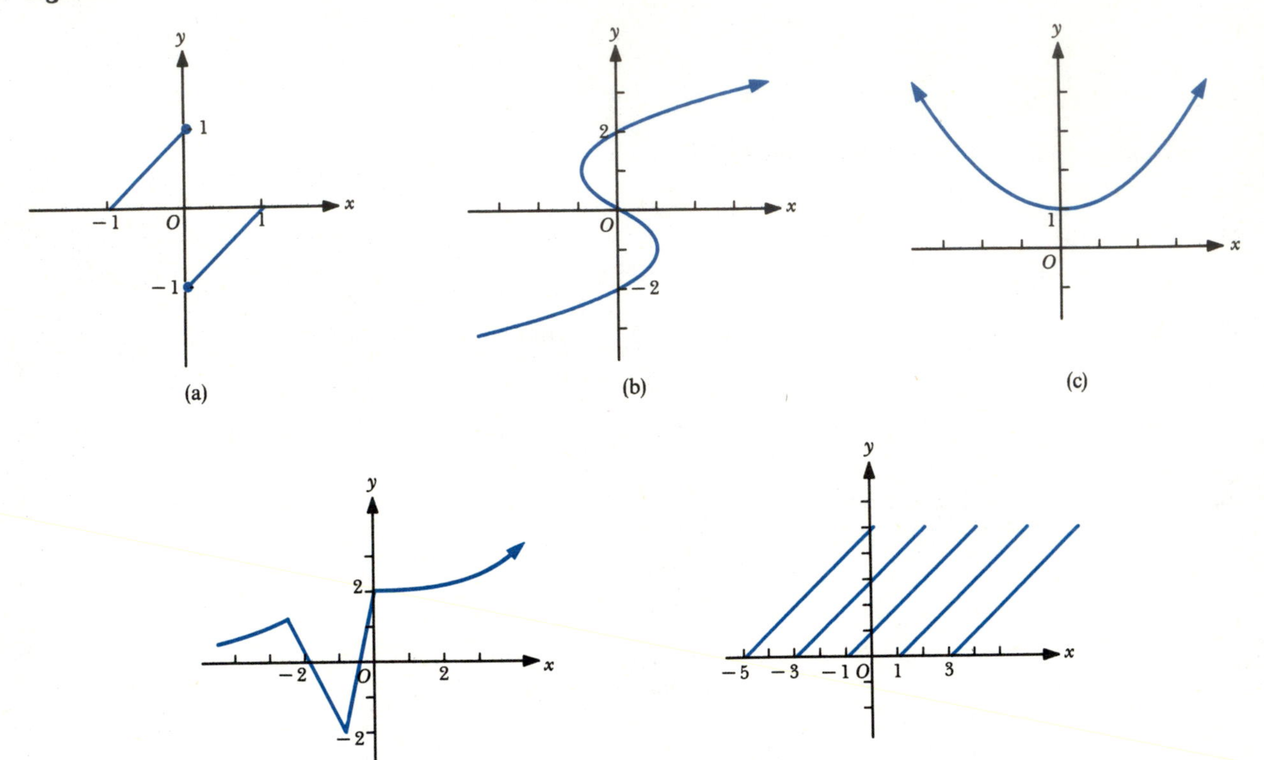

82. Suppose that a rectangular-shaped box has a height of 3 centimeters and a square base with a length of x centimeters. Express the volume V of the box as a function of x.

c **83.** A rock falls from a height of 40 meters, and its height h after t seconds is given by the function

$$h(t) = 40 - 13t^2$$

Determine the height of the rock above the ground level (rounded to two decimal places) when
(a) $t = 1$ (b) $t = 1.2$ (c) $t = 1.75$

c **84.** The velocity V of blood, in centimeters per second, at x centimeters from the center of a given artery is given by the function

$$V(x) = 1.28 - 20{,}000x^2$$

Determine the velocity of the blood (rounded to four decimal places) when
(a) $x = 0.0020$ (b) $x = 0.0040$ (c) $x = 0.0053$

85. The function C for converting temperature from Fahrenheit degrees F to Celsius (centigrade) degrees is given by

$$C(\mathrm{F}) = \tfrac{5}{9}(\mathrm{F} - 32)$$

Convert each of the following from Fahrenheit to Celsius.
(a) 32°F (b) 86°F (c) 0°F (d) −13°F

86. A closed box with a square base of x centimeters has a volume of 400 cubic centimeters. Express the total surface area A of the exterior of the box as a function of x.

87. An outdoor track is to be constructed in the shape shown in Figure 4 and is to have a perimeter of P meters. Express the total area A enclosed by the track as a function of P.

Figure 4

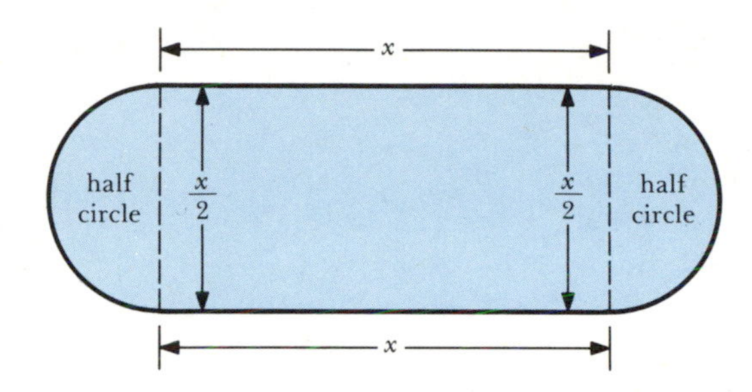

88. A manufacturer's sales representative receives a weekly salary of $325 plus a commission of 10 percent of the sales made that week. Express the representative's weekly salary S as a function of the amount of the weekly sales x, where x is in dollars.

89. Equal squares are cut from the four corners of a rectangular piece of cardboard that is 10 inches by 14 inches. An open box is then formed by folding up the flaps (Figure 5). Express the volume V of the box as function of x, where x is the length of the side of the squares removed.

90. The demand for a certain kind of candy bar is such that the product of the demand d (in thousands of cartons) and the price p (in dollars per carton) is always equal to 250,000, so long as the price is not less than $2.00 nor more than

Figure 5

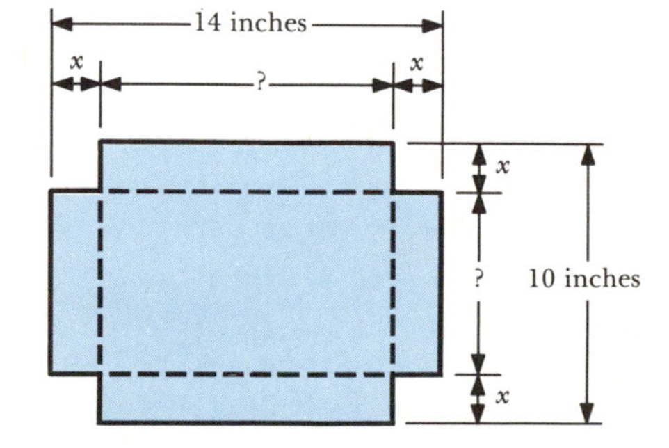

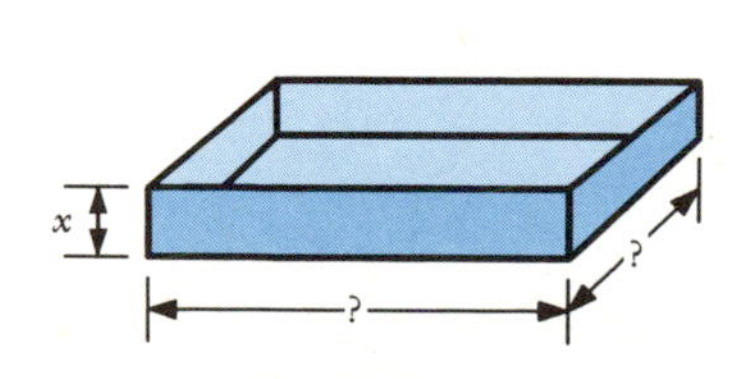

\$3.50. Express the demand d as a function of the price p. Compute the demand when the price is
(a) \$2.00 (b) \$2.50 (c) \$3.00 (d) \$3.50

91. Water is being pumped into a cylindrical tank of radius $r = 8$ feet (Figure 6). Express the volume V of the water in the tank as a function of the depth h of the water. Find the volume when $h = 10$ feet.

Figure 6

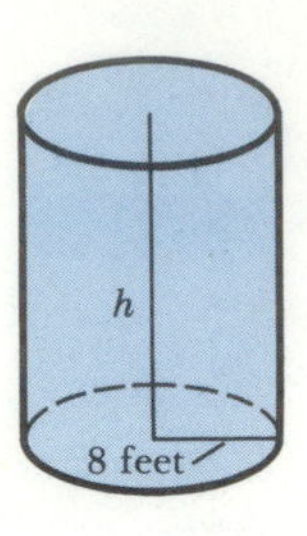

2.4 Graphs of Functions

In Section 2.3, we used the vertical line test to determine whether or not a curve is the graph of a function. In this section, we explore methods for sketching graphs of functions and discuss some of the properties of these graphs. To sketch a graph of a function $y = f(x)$, we select several values of x in the domain of f, calculate the corresponding values of $f(x)$, plot the resulting points $(x, f(x))$, and then connect the points with a smooth curve.

In some cases, the process of graphing a function is relatively easy. For instance, the graph of any function of the form $f(x) = mx + b$, where m and b are constants, is the same as the graph of the linear equation $y = mx + b$. It follows from the slope-intercept form of the equation of a line that the graph of $f(x) = mx + b$ is a straight line with slope m and y intercept b. Thus, any function f of the form

> $f(x) = mx + b$ is called a **linear function.**

EXAMPLE 1 Sketch the graph of $f(x) = 2x - 1$.

SOLUTION Since the domain of the linear function f is the set of all real numbers $\mathbb{R}$, we begin by selecting a few real number values for x and then calculating corresponding values of $f(x) = 2x - 1$ to get points $(x, f(x))$ on the graph.

The table at the top of page 99 exhibits the coordinates of some points on the graph. After plotting the points given in the table and drawing a line containing them, we obtain the graph (Figure 1).

To sketch the graph of a function requires us either to know or to guess the shape of the curve that connects the points plotted. If the function is complicated, it may require advanced methods that are studied in calculus to obtain an accurate sketch. However, we can use the geometric properties of graphs to develop techniques that enable us to sketch a variety of functions without plotting many points.

Geometric Properties of Graphs

Consider the graphs in Figure 2. If the graph of f in Figure 2a is "folded" along the y axis, then the portion of the graph to the right of the y axis coincides with that portion to the left of it. We say that the graph of f is *symmetric with respect to the y axis.*

Figure 1

x	$f(x) = 2x - 1$
-2	-5
0	-1
3	5

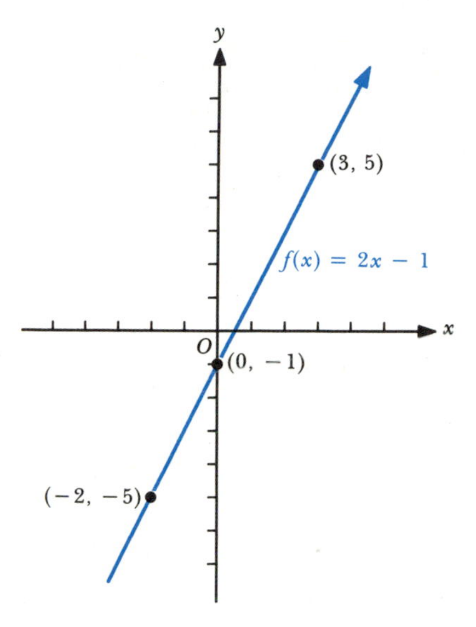

More specifically, the graph of f is **symmetric with respect to the *y* axis** if, whenever the point (x, y) belongs to the graph of f, the point $(-x, y)$ also belongs to the graph of f; that is,

$$f(-x) = f(x)$$

for all x in the domain of f.

The graph of g in Figure 2b is **symmetric with respect to the origin** because, if the point (x, y) belongs to the graph of g, then so does the point $(-x, -y)$; that is,

$$g(-x) = -g(x)$$

for all x in the domain of g.

Figure 2

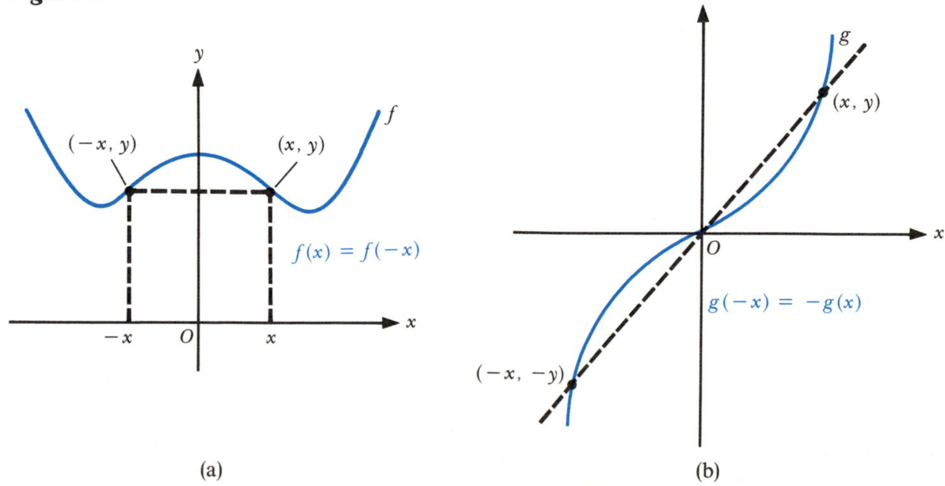

A function whose graph is symmetric about the y axis is called an *even function;* a function whose graph is symmetric about the origin is called an *odd function.* More formally, we have the following definition.

DEFINITION 1 **Even and Odd Functions**

(i) A function f is said to be **even** if, for every number x in the domain of f, $-x$ is also in the domain of f and

$$f(-x) = f(x)$$

(ii) A function f is said to be **odd** if, for every number x in the domain of f, $-x$ is also in the domain of f and

$$f(-x) = -f(x)$$

EXAMPLE 2 Determine which of the functions graphed in Figure 3 are even, odd, or neither.

SOLUTION Figure 3a indicates that the graph of f is symmetric with respect to the y axis; thus $f(-x) = f(x)$ and f is even.

Figure 3b indicates that the graph of g is symmetric with respect to the origin; thus $g(-x) = -g(x)$ and g is odd.

Figure 3c indicates that the graph of h is not symmetric with respect to the y axis nor with respect to the origin; so h is neither even nor odd.

Figure 3

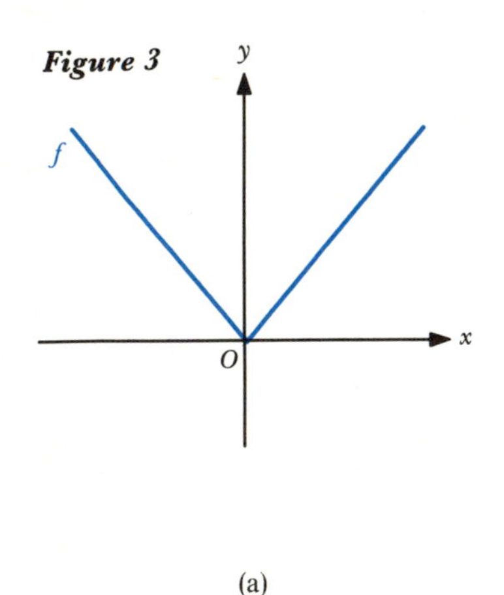

(a)

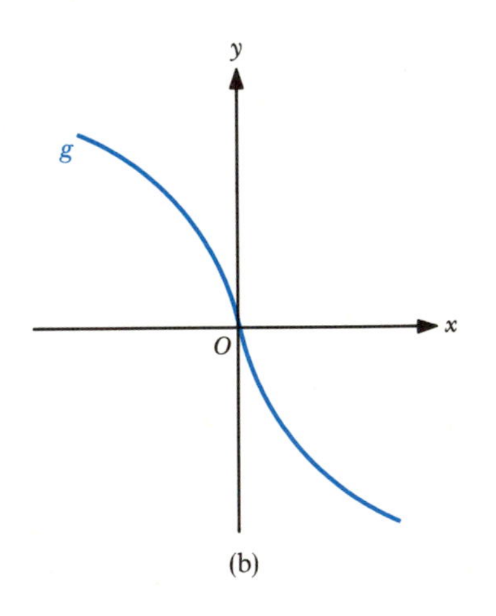

(b)

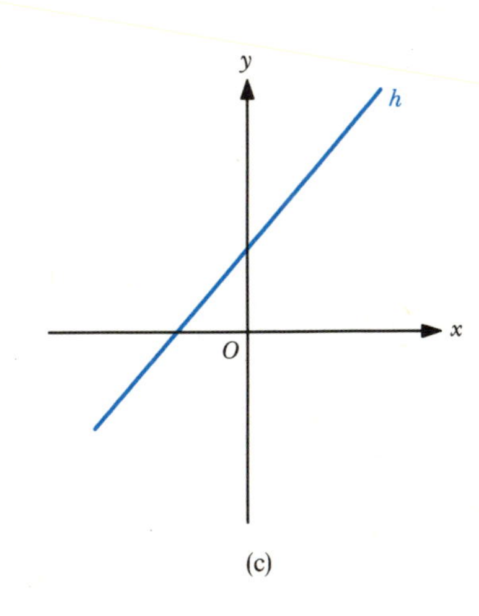

(c)

EXAMPLE 3 Without graphing, classify each function as even or odd or neither.

(a) $f(x) = 3x^4$

(b) $g(x) = -2x^5$

(c) $h(x) = 3x + 1$

SOLUTION By applying Definition 1 we have

(a) $f(-x) = 3(-x)^4 = 3x^4 = f(x)$, so f is an even function.

(b) $g(-x) = -2(-x)^5 = -2(-x^5) = 2x^5 = -g(x)$, so g is an odd function.

(c) $h(-x) = 3(-x) + 1 = -3x + 1$, while $-h(x) = -(3x + 1) = -3x - 1$. Since $-3x + 1 \neq 3x + 1$ and $-3x + 1 \neq -3x - 1$, then $h(-x) \neq h(x)$ and $h(-x) \neq -h(x)$. Thus, h is neither even nor odd.

The domain and range of a function may be determined by examining the graph of the function. Figure 4 illustrates that the *domain* of a function is the set of all abscissas of points on its graph (Figure 4a) and the *range* of a function is the set of all ordinates of points on its graph (Figure 4b).

Figure 4

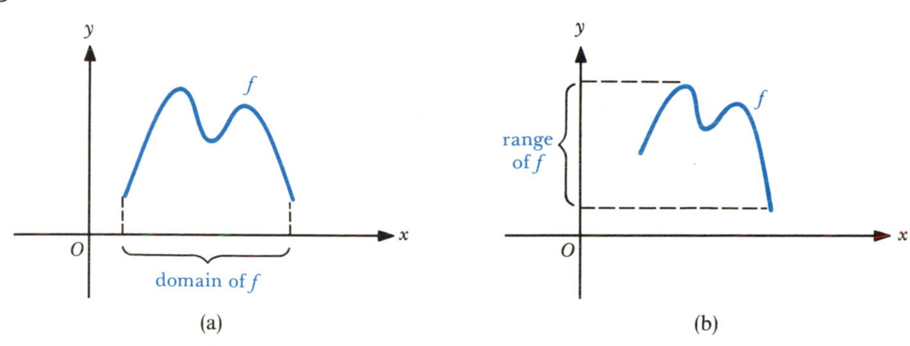

Consider the graph of the linear functions in Figure 5. As we view each graph *from left to right,* we see that the graph in Figure 5a is always rising as we move to the right, a geometric indication that f is *increasing;* that is, as x increases, so do the values $f(x)$. On the other hand, the graph in Figure 5b is always falling as we move to the right, indicating that the function g is *decreasing;* that is, as x increases, the values of $g(x)$ decrease. In Figure 5c, the graph doesn't rise or fall, indicating that h is a *constant* function whose values $h(x)$ do not change as x increases. These illustrations lead us to the next definition.

Figure 5

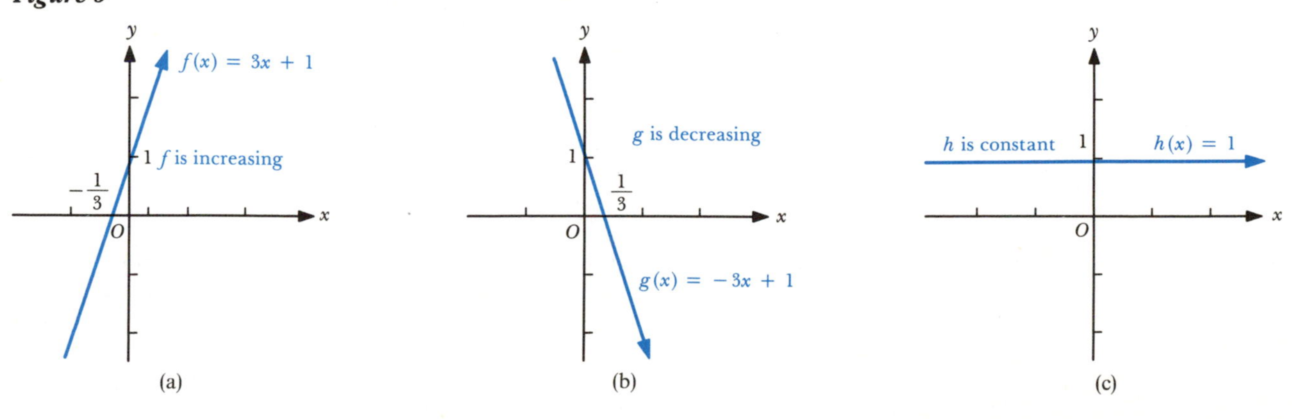

DEFINITION 2 **Increasing and Decreasing Functions**

(i) A function f is said to be an **increasing** function on an interval I if $f(a) < f(b)$ whenever $a < b$, and a and b are in the interval I.

(ii) A function f is said to be a **decreasing** function on an interval I if $f(a) > f(b)$ whenever $a < b$, and a and b are in the interval I.

EXAMPLE 4 Indicate the intervals over which f is increasing, the intervals over which f is decreasing, and the intervals over which f is constant for the function f whose graph is shown in Figure 6. Also, find the domain and range of f.

Figure 6

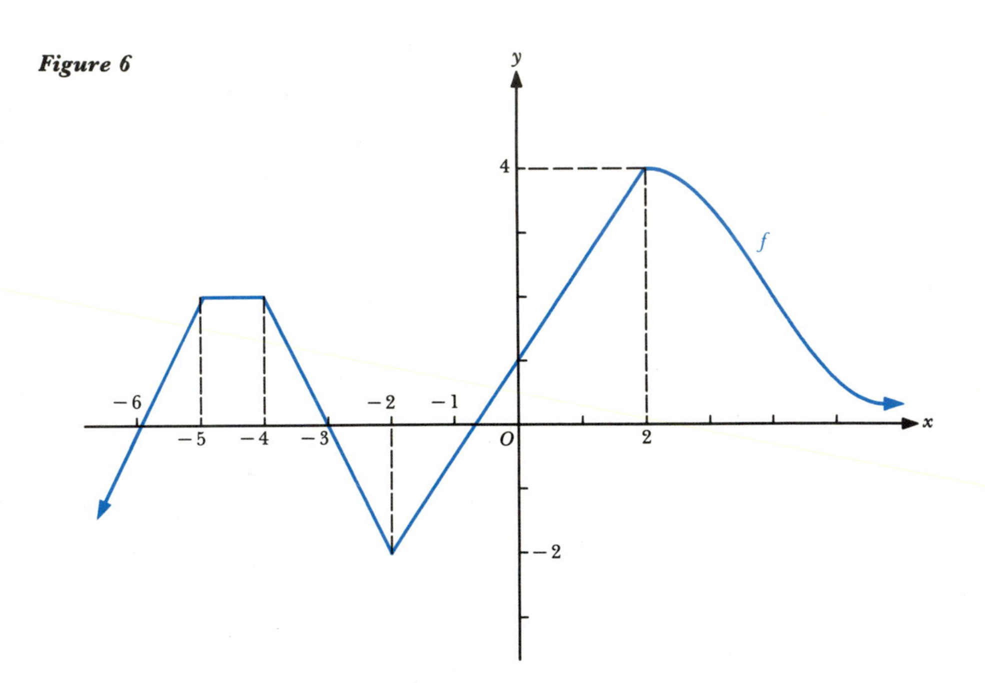

SOLUTION As we move from left to right, the function f is increasing on the intervals where the graph is rising and decreasing on the intervals where the graph is falling. Thus, assuming the function continues indefinitely to the right and to the left in the directions indicated by the arrowheads, we conclude that f is increasing on the intervals $(-\infty, -5]$ and $[-2, 2]$, and that f is decreasing on the intervals $[-4, -2]$ and $[2, \infty)$. On the interval $[-5, -4]$, f is constant. Note that for any selected value of x, there is a point on the graph of f with abscissa equal to x, so the graph suggests that the domain of f is the set $\mathbb{R}$. Assuming that the graph keeps dropping as we move to the left of -5 on the x axis, the graph suggests that the range is the interval $(-\infty, 4]$.

Special Functions

Now we examine "special" functions, which play important roles in the study of mathematics. These examples help to demonstrate the properties and graphs of functions.

In Examples 5–10, in whatever order seems most convenient, find the domain of f and sketch its graph. Discuss the symmetry of the graph and indicate whether f is even or odd. Determine the range of f, and the intervals where f is increasing or decreasing.

EXAMPLE 5 $f(x) = x$ (the **identity function**)

SOLUTION This is a special case of a linear function $f(x) = mx + b$, with slope $m = 1$ and y intercept $b = 0$ (Figure 7). Since $f(-x) = -x = -f(x)$, it follows that f is an odd function and that its graph is symmetric with respect to the origin. The domain and range of f is the set of real numbers $\mathbb{R}$, and f is increasing in $\mathbb{R}$.

Figure 7

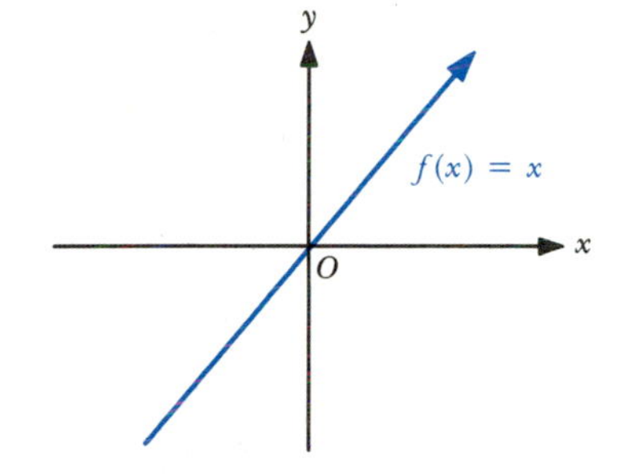

Figure 8

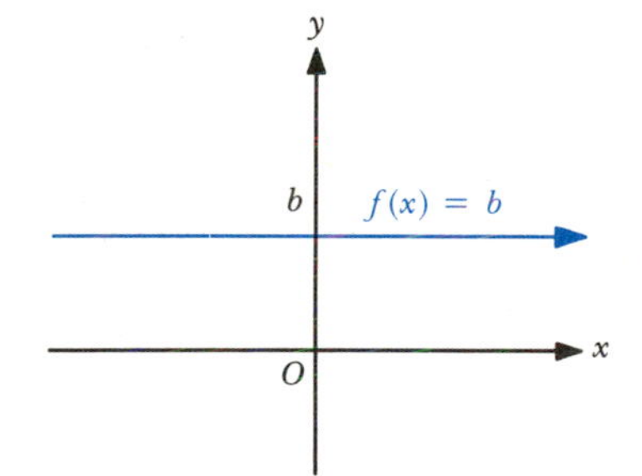

EXAMPLE 6 $f(x) = b$, where b is a constant (a **constant function**)

SOLUTION This is another special case of a linear function $f(x) = mx + b$, with slope $m = 0$ and y intercept b. The graph of $f(x) = b$ is the x axis if $b = 0$ or a line parallel to the x axis if $b \neq 0$; the graph contains the point $(0, b)$ on the y axis. Figure 8 displays a typical graph, where b is positive. Since $f(-x) = b = f(x)$, the function f is even; that is, its graph is symmetric with respect to the y axis. The domain of f is the set of real numbers $\mathbb{R}$ and the range of f is the set $\{b\}$. The function f is neither increasing nor decreasing.

EXAMPLE 7 $f(x) = |x|$ (the **absolute value function**)

SOLUTION For $x \geq 0$, $f(x) = x$. Thus the portion of the graph to the right of the y axis is the same as the graph of the identity function for x in the interval $[0, \infty)$ (Figure 7).

If $x < 0$, $f(x) = -x$, so the portion of the graph to the left of the y axis is a line with slope -1 containing the point $(-1, 1)$. Since $f(-x) = |-x| = |x| = f(x)$, the function f is even, so its graph is symmetric with respect to the y axis. A table of values helps us to sketch the graph (Figure 9). The domain of f is the set of real numbers $\mathbb{R}$. The graph indicates that the range is the interval $[0, \infty)$ and that the function is increasing in the inteval $[0, \infty)$ and decreasing in the interval $(-\infty, 0]$.

Figure 9

x	$y = f(x) = \|x\|$
-2	2
-1	1
0	0
1	1
2	2

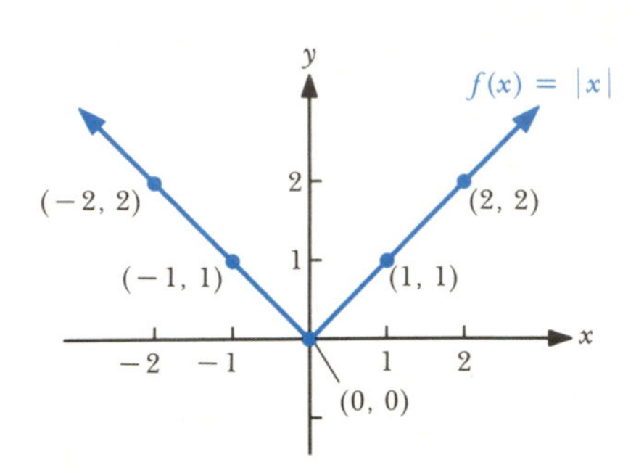

EXAMPLE 8 $f(x) = x^2$ (the **squaring function**)

SOLUTION The domain of f is the set of all real numbers $\mathbb{R}$. The table lists some points $(x, f(x))$ on the graph. By plotting these points and connecting them with a smooth curve, we obtain the graph (Figure 10). Since $f(-x) = (-x)^2 = x^2 = f(x)$, f is an even function; that is, the graph is symmetric with respect to the y axis. The graph indicates that the range of f is the interval $[0, \infty)$, and that f is increasing in the interval $[0, \infty)$ and decreasing in the interval $(-\infty, 0]$.

Figure 10

x	$y = f(x) = x^2$
-2	4
-1	1
0	0
1	1
2	4

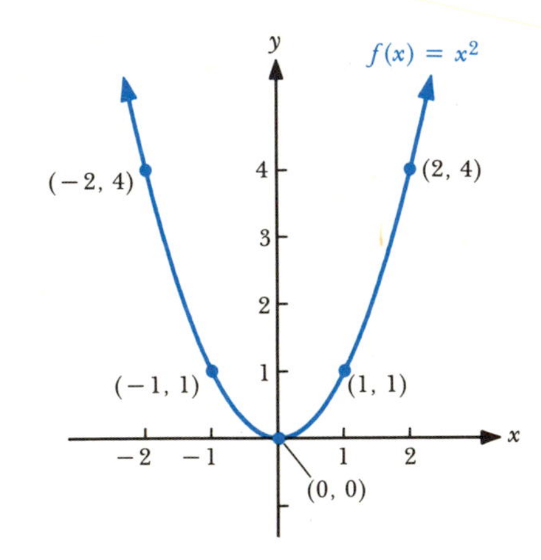

EXAMPLE 9 $f(x) = \sqrt{x}$ (the **square root function**)

SOLUTION Since $\sqrt{x}$ is not defined for $x < 0$, the domain of the function f is the interval $[0, \infty)$. The table lists some points $(x, f(x))$ on the graph. By plotting these points and connecting them with a smooth curve, we obtain the graph (Figure 11). The function is neither even nor odd, so the graph is not symmetric with respect to either the y axis or the origin. The range of f is the interval $[0, \infty)$ and f is an increasing function in the interval $[0, \infty)$.

Figure 11

x	$y = f(x) = \sqrt{x}$
0	0
1	1
4	2
9	3

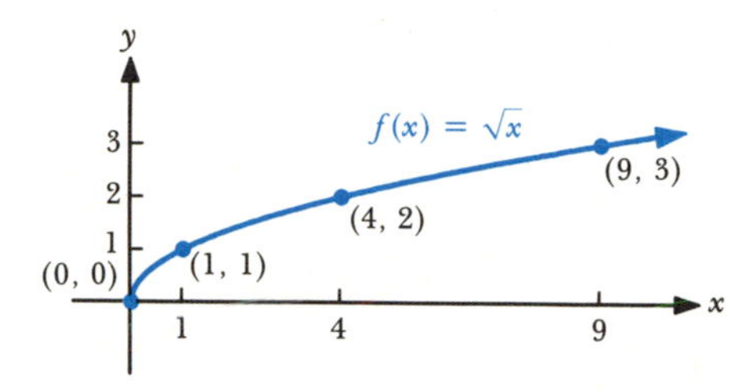

EXAMPLE 10 $f(x) = x^3$ (the **cubing function**)

SOLUTION The domain of f is the set of real numbers $\mathbb{R}$. Since $f(-x) = (-x)^3 = -x^3 = -f(x)$, the graph is symmetric with respect to the origin. Therefore, f is an odd function. The table lists some points $(x, f(x))$ on the graph. By plotting these points and connecting them with a smooth curve, we obtain the graph (Figure 12). The range of f is the set of real numbers $\mathbb{R}$ and f is increasing in $\mathbb{R}$.

Figure 12

x	$y = f(x) = x^3$
−2	−8
−1	−1
0	0
1	1
2	8

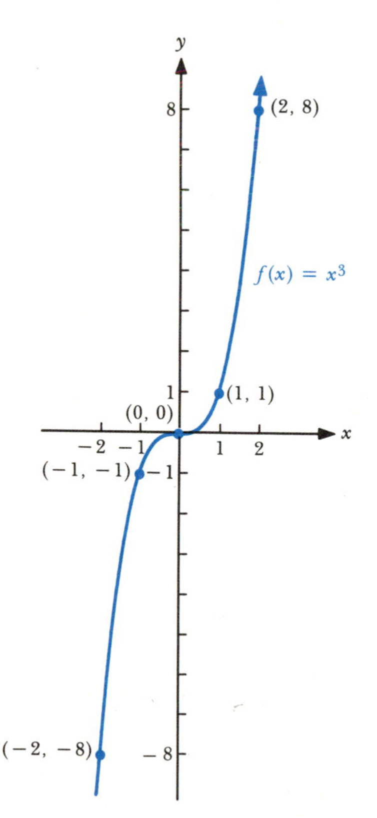

EXAMPLE 11 Sketch the graph of the function f defined by

$$f(x) = \begin{cases} x^2 & \text{if } x < 1 \\ x & \text{if } x \geq 1 \end{cases}$$

SOLUTION If $x < 1$, then $f(x) = x^2$. This portion of the graph coincides with the graph of the squaring function for x in the interval $(-\infty, 1)$ (Figure 10). For $x \geq 1$, the graph of f will coincide with the graph of the identity function restricted to the interval $[1, \infty)$ (Figure 7). Figure 13 displays the graph of the given piecewise-defined function f.

Figure 13

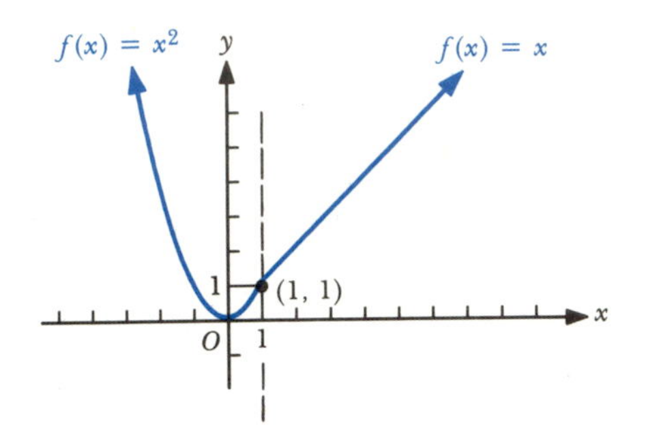

Aids in Graphing

We can use the graphs of the special functions, together with the techniques of geometric shifting, stretching, shrinking, and reflecting, to obtain graphs of other functions. For example, in order to graph the functions

$$g(x) = x^2 - 1 \quad \text{and} \quad h(x) = x^2 + 2$$

first we graph $f(x) = x^2$. Then we subtract one unit from every y value of $f(x) = x^2$ to obtain the graph of $g(x) = x^2 - 1$; that is, we *shift* the graph of $f(x) = x^2$ one unit vertically downward (Figure 14). Analogously, we obtain the graph of $h(x) = x^2 + 2$ by adding two units to every y value of $f(x) = x^2$; that is, we *shift* the graph of $f(x) = x^2$ vertically upward two units (Figure 14).

In general, we have the following result.

Figure 14

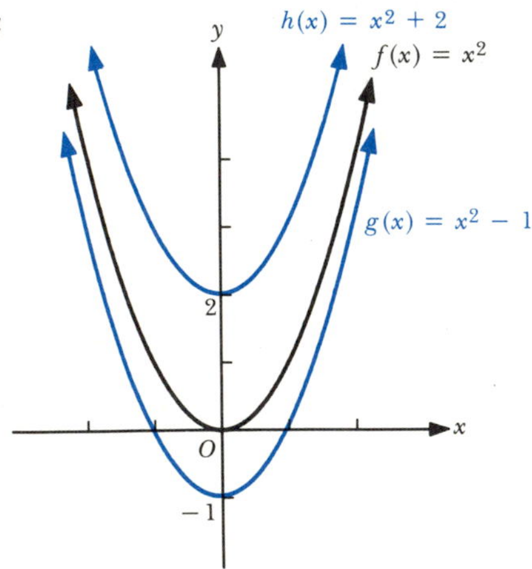

If f is a function and c is a constant, then the graph of the function F defined by

$$F(x) = f(x) + c$$

can be obtained by **shifting** the graph of f **vertically** by $|c|$ units, **upward** if $c > 0$, and downward if $c < 0$.

EXAMPLE 12 Use the graph of $f(x) = |x|$ to sketch the graph $F(x) = |x| - 3$.

SOLUTION Figure 9 shows the graph of $f(x) = |x|$. Because of the "-3", we shift the graph of f vertically downward by three units to obtain the graph of $F(x) = |x| - 3$ (Figure 15).

Figure 15

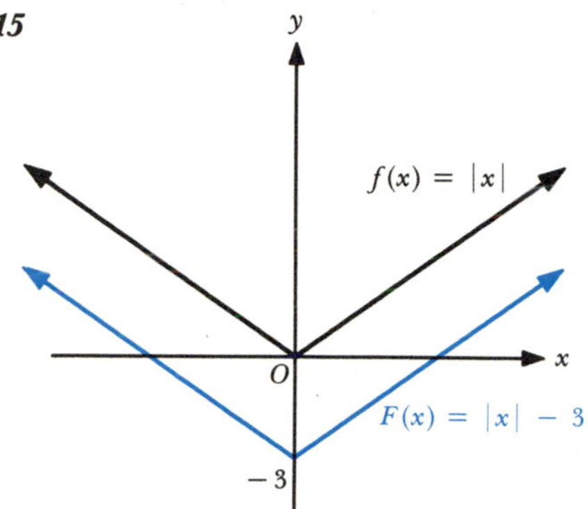

Now we graph the functions

$$g(x) = (x + 1)^2 \qquad \text{and} \qquad h(x) = (x - 2)^2$$

by using the graph of $f(x) = x^2$. Notice that $y = 0$ when $x = 0$ in the function $f(x) = x^2$, and $y = 0$ when $x = -1$ in $g(x) = (x + 1)^2$. Thus, the point $(0, 0)$ belongs to the graph of $f(x) = x^2$ if and only if the point $(-1, 0)$ belongs to the graph of $g(x) = (x + 1)^2$. Geometrically, if (a, b) is a point on the graph of f, then $(a - 1, b)$ will be a point on the graph of g. Thus, the graph of $g(x) = (x + 1)^2$ can be obtained by *shifting* the graph of $f(x) = x^2$ one unit to the *left* (Figure 16). Analogously, since the point $(a + 2, b)$ belongs to the graph of $h(x) = (x - 2)^2$ whenever the point (a, b) belongs to the graph of $f(x) = x^2$, we observe that the graph of $h(x) = (x - 2)^2$ is obtained geometrically by *shifting* the graph of $f(x) = x^2$ two units to the *right* (Figure 16).

Figure 16

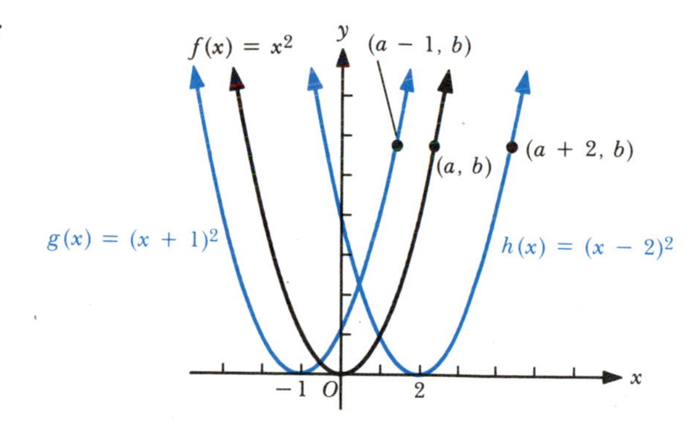

In general, we have the following result.

If f is a function and c is a *positive constant*, then the graph of the function F defined by

$$F(x) = f(x - c)$$

can be obtained by **shifting** the graph of f **horizontally** by c units to the **right.** The graph of the function G defined by

$$G(x) = f(x + c)$$

can be obtained by **shifting** the graph of f **horizontally** by c units to the **left.**

EXAMPLE 13 Use the graph of $g(x) = \sqrt{x}$ to sketch the graph of $G(x) = \sqrt{x + 1}$.

SOLUTION The graph of $g(x) = \sqrt{x}$ is shown in Figure 17. The graph of G is obtained geometrically by shifting the graph of $g(x) = \sqrt{x}$ horizontally one unit to the left (Figure 17).

Finally, if we sketch the graphs of the functions

$$f(x) = x^2, \qquad g(x) = 2x^2, \qquad \text{and} \qquad h(x) = \tfrac{1}{2}x^2$$

Figure 17

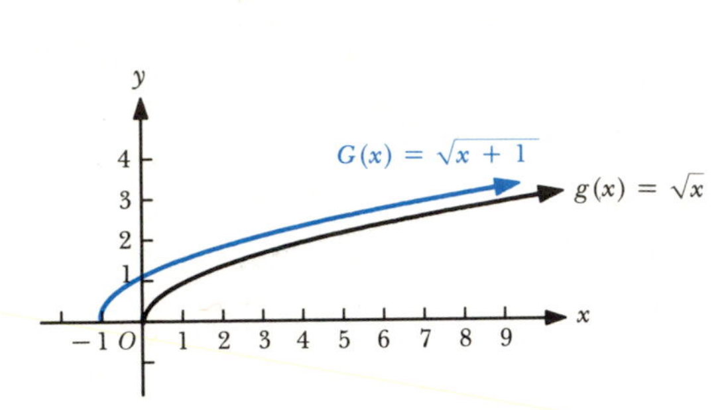

Figure 18

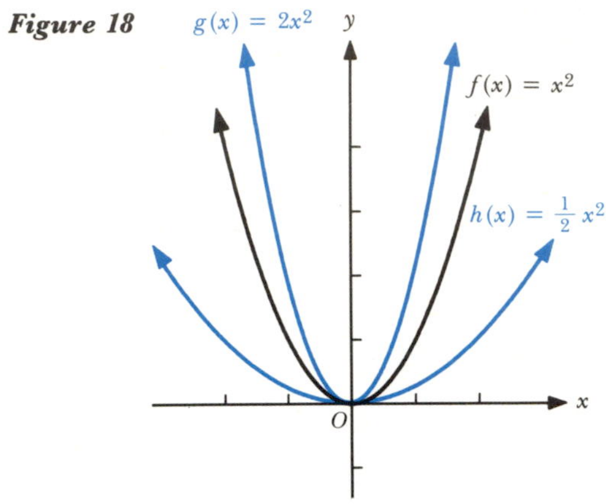

we note that the graph of $g(x) = 2x^2$ rises more rapidly than that of $f(x) = x^2$. In fact, the graph of g can be obtained by doubling the ordinates of points on the graph of f. Geometrically, this means that the graph of $g(x) = 2x^2$ can be formed by "vertically stretching" the graph of $f(x) = x^2$ by a multiple of 2 (Figure 18). Similarly, the graph of $h(x) = \frac{1}{2}x^2$ can be obtained by multiplying all the ordinates of the points on the graph of $f(x) = x^2$ by $\frac{1}{2}$. This is geometrically equivalent to vertically "shrinking" the graph of f by a multiple of $\frac{1}{2}$ to obtain the graph of h (Figure 18).

In general, we have the following result.

> If f is a function and c is a *positive constant*, then the graph of the function F defined by
>
> $$F(x) = c\,f(x)$$
>
> can be obtained from the graph of f by **"stretching"** the graph of f vertically away from the x axis if $c > 1$, or by **"shrinking"** or **"flattening"** the graph of f vertically toward the x axis if $0 < c < 1$. If $c = -1$, the graph of F is obtained by **reflecting** the graph of f across the x axis.

In Examples 14–16, use the graph of f to sketch the graph of each given function.

EXAMPLE 14 $F(x) = c|x|$ where $f(x) = |x|$ if (a) $c = 3$ (b) $c = \frac{1}{3}$

SOLUTION

(a) To obtain the graph of $F(x) = 3|x|$, we refer to the graph of $f(x) = |x|$ (Figure 9) and multiply the ordinate of each point by 3. This gives us the graph of F that is stretched vertically away from the x axis (Figure 19).

(b) Similarly, the graph of $F(x) = \frac{1}{3}|x|$ can be obtained by multiplying ordinates of points on the graph of $f(x) = |x|$ by $\frac{1}{3}$. This gives us a graph that is flattened toward the x axis (Figure 19).

Figure 19

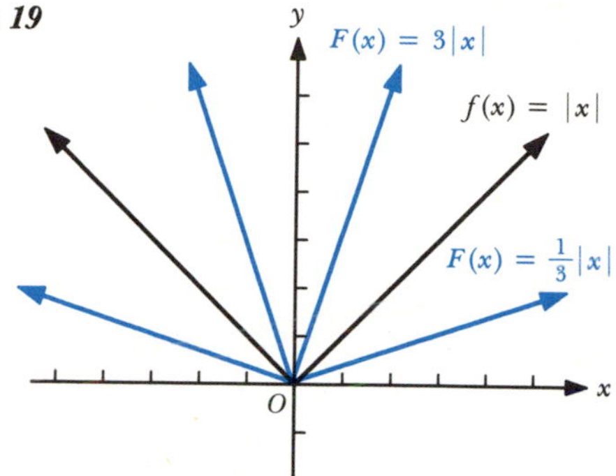

Figure 20

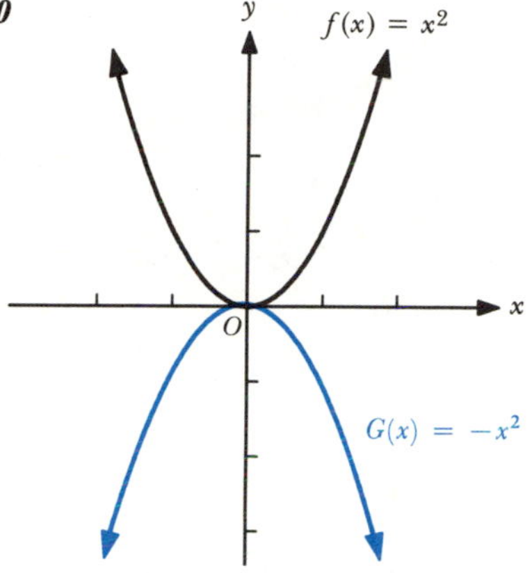

EXAMPLE 15 $G(x) = -x^2$ where $f(x) = x^2$.

SOLUTION

The graph of $G(x) = -x^2$ can be obtained by reflecting the graph of $f(x) = x^2$ (Figure 10) across the x axis so that each point on the graph of G has an ordinate opposite in algebraic sign to the ordinate of the point on the graph of f with the same abscissa value (Figure 20).

EXAMPLE 16 $F(x) = 3(x + 4)^2 + 2$ where $f(x) = x^2$.

SOLUTION

To obtain the graph of $F(x) = 3(x + 4)^2 + 2$ we begin by horizontally shifting the graph of $f(x) = x^2$ (Figure 10) four units to the left. Then we multiply the ordinates of points of the resulting graph by three. (This is equivalent to stretching the resulting graph vertically away from the x axis.) Finally we shift the latter graph two units upward (Figure 21).

Figure 21

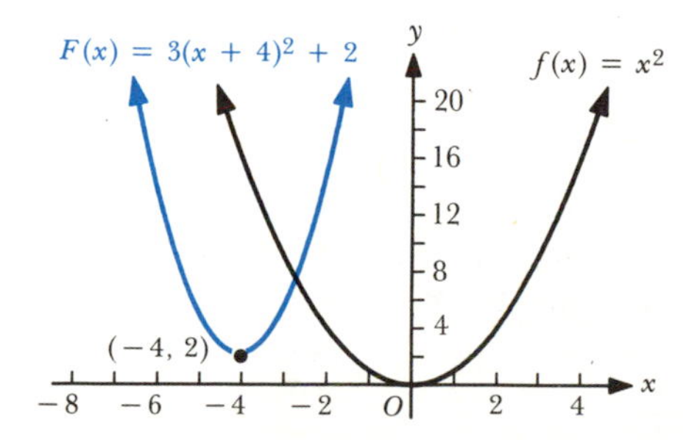

PROBLEM SET 2.4

In problems 1–12, sketch the graph of each function.

1. $f(x) = 3x$

2. $g(x) = -2x$

3. $h(x) = -4x$

4. $f(x) = 3x + 2$

5. $F(x) = -5x + 10$

6. $G(x) = -7x + 1$

7. $f(x) = |x| - x$

8. $H(x) = |x| + x$

9. $f(x) = \dfrac{x}{|x|}$

10. $G(x) = -4$

11. $F(x) = \sqrt{4 - x^2}$

12. $H(x) = -\sqrt{4 - x^2}$

13. Classify each of the functions whose graphs are shown in Figure 22 as even, odd, or neither. Find the domain and range of each function and indicate the intervals over which the function is increasing, decreasing, and constant.

Figure 22

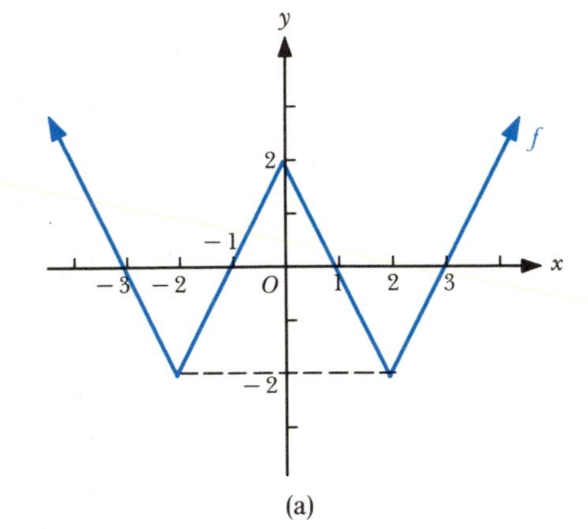

(a)

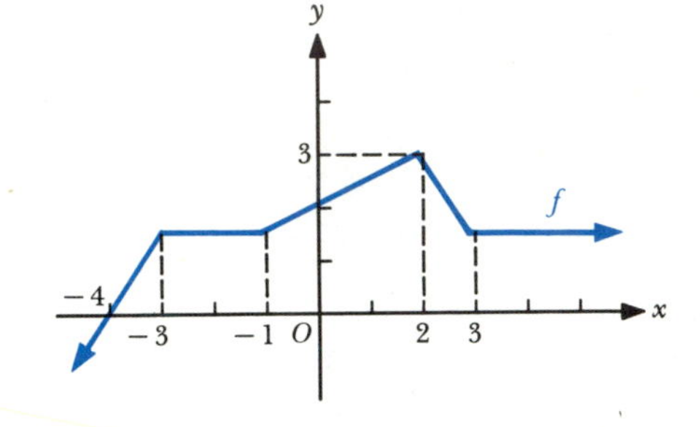

(b)

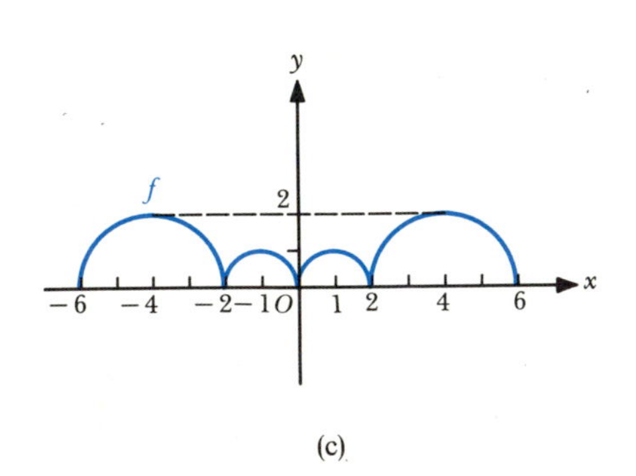

(c)

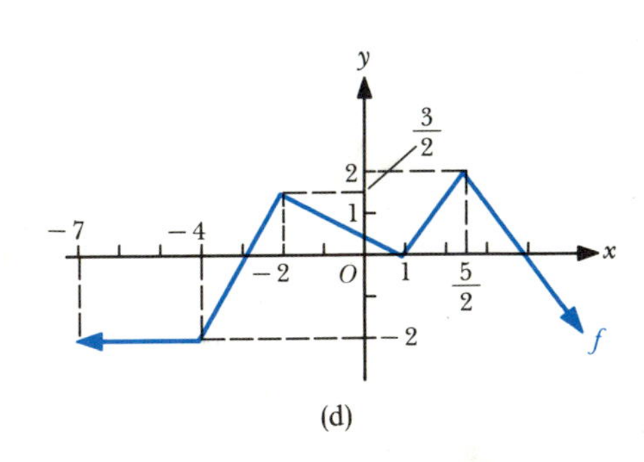

(d)

14. The function f defined by $f(x) = [\![x]\!]$, where $[\![x]\!]$ represents the integer n that satisfies $n \le x < n + 1$ for any given real number x, is called the **greatest integer function.** Use the sketch of the graph of f in Figure 23 to find the ordinates $f(x)$ of points on the graph corresponding to the values of x indicated in the table on page 111.

Figure 23

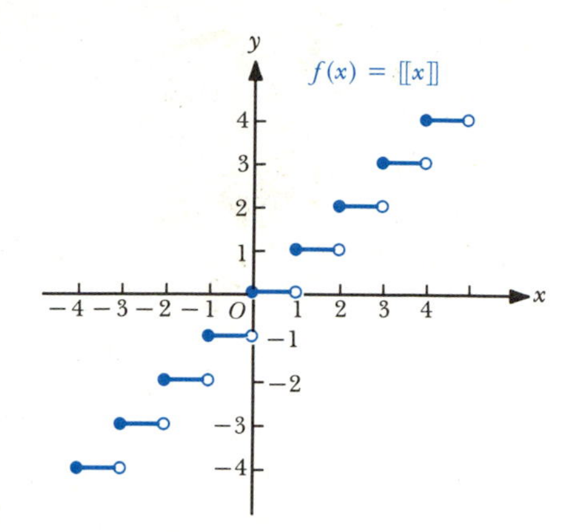

x	$-2 \le x < -1$	$-1 \le x < 0$	$0 \le x < 1$	$1 \le x < 2$	$2 \le x < 3$
Value of $f(x)$					

15. Figure 24 displays the graphs of $f(x) = x^4$ and $g(x) = \sqrt[3]{x}$.

(a) Determine if f and g have symmetry with respect to the y axis or the origin.

(b) Find the reflections of the given points on the graph of f across the y axis in Figure 24a and on the graph of g across the origin in Figure 24b.

(c) Indicate the intervals over which f and g are increasing and decreasing.

Figure 24

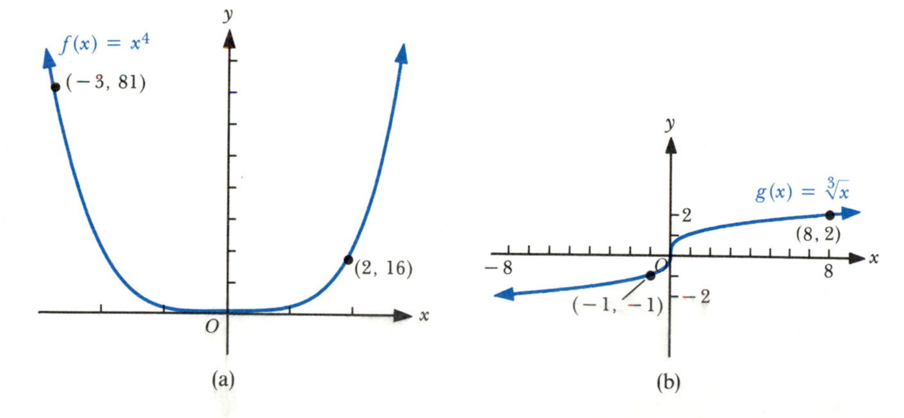

Figure 25

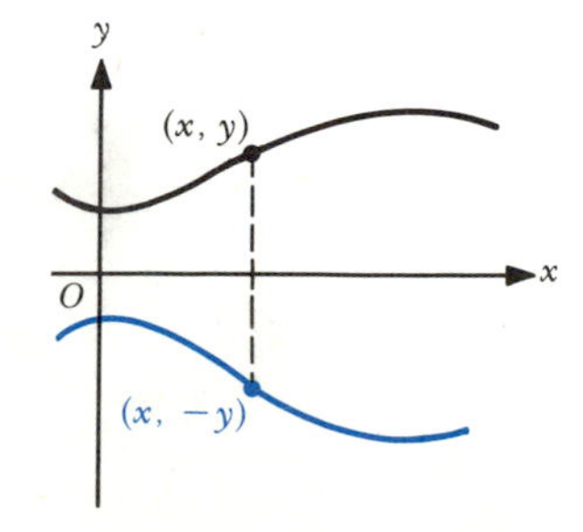

16. Figure 25 shows a graph of an equation that is *not* a function. The graph of an equation is **symmetric with respect to the *x* axis** if, whenever a point (x, y) is on the graph of the equation, then $(x, -y)$ is also on the graph. Show that the graph of $y^2 = x$ is symmetric about the x axis and sketch the graph of the equation.

In problems 17–24, determine without graphing whether each function is even, odd, or neither, and whether the graph is symmetric with respect to the y axis or origin.

17. $f(x) = 8x^2 + 3$
18. $g(x) = -4x^3 + x$
19. $h(x) = -x^4 + 3$
20. $F(x) = \sqrt{7x^4 + 3}$
21. $f(x) = 5x^3 - 2x$
22. $f(x) = 5x^2 + x^3$
23. $f(x) = -x^2 - |x|$
24. $G(x) = \dfrac{\sqrt{x^2 + 4}}{|x|}$

In problems 25–28, sketch the graph of each piecewise-defined function.

25. $f(x) = \begin{cases} x + 3 & \text{if } x \leq 1 \\ 4x^2 & \text{if } x > 1 \end{cases}$

26. $g(x) = \begin{cases} 2x + 1 & \text{if } x < 3 \\ 10 - x & \text{if } x \geq 3 \end{cases}$

27. $h(x) = \begin{cases} 3x + 1 & \text{if } x < -2 \\ 0 & \text{if } -2 \leq x \leq 2 \\ -2x + 3 & \text{if } x > 2 \end{cases}$

28. $f(x) = \begin{cases} -x & \text{if } x < 0 \\ 4 - 2x & \text{if } 0 \leq x < 3 \\ x^2 & \text{if } x \geq 3 \end{cases}$

In problems 29–34, sketch the graphs of the two given functions on the same coordinate axes.

29. $f(x) = x$ and $F(x) = x + 3$
30. $g(x) = |x|$ and $G(x) = |x| + 2$
31. $h(x) = x^2$ and $H(x) = (x + 2)^2$
32. $p(x) = x^3$ and $P(x) = (x + 1)^3$
33. $f(x) = \sqrt{x}$ and $F(x) = -\sqrt{x}$
34. $h(x) = x^2$ and $H(x) = -2x^2$

In problems 35–50, sketch the graph of each function by shifting, stretching, or reflecting a known special graph.

35. $f(x) = x + 3$
36. $g(x) = \sqrt{x} - 2$
37. $G(x) = 1 - x^3$
38. $h(x) = (x + 3)^2$
39. $f(x) = \sqrt{x + 4}$
40. $g(x) = |x + 3|$
41. $F(x) = 2|x - 1|$
42. $h(x) = 2|x + 1| - 1$
43. $H(x) = -5\sqrt{x}$
44. $f(x) = 7 + |x - 2|$
45. $f(x) = -\frac{1}{3}(x + 1)^3$
46. $g(x) = |x^3 + 2|$
47. $g(x) = 4x^2 - 1$
48. $f(x) = 2(x - 1)^2 + 1$
49. $h(x) = 2(x - 3)^2 + 4$
50. $h(x) = (x + 1)^3 - 2$

2.5 Algebra of Functions and Composition of Functions

In this section we examine how functions can be added, subtracted, multiplied, and divided to form new functions. We also introduce the concept of the *composition* of functions.

Algebra of Functions

If $f(x) = x^2 - 1$ (Figure 1a) and $g(x) = 2x + 1$ (Figure 1b), then we can form a new function h simply by adding $f(x)$ and $g(x)$ to get

$$h(x) = f(x) + g(x) = (x^2 - 1) + (2x + 1) = x^2 + 2x$$

Figure 1

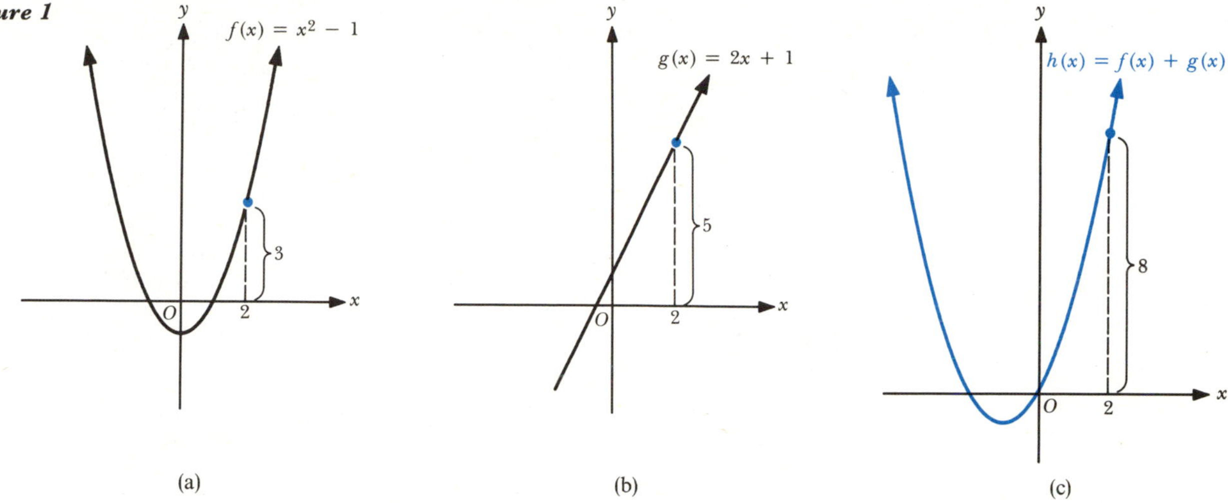

We refer to the function h as the sum of the functions f and g and we write $h = f + g$. Note that the graph of h can be obtained from the graphs of f and g by adding ordinates; for instance, in Figure 1c

$$h(2) = f(2) + g(2) = 3 + 5 = 8$$

It is also possible to subtract, multiply, and divide functions as specified in the next definition.

DEFINITION 1 **Sum, Difference, Product, and Quotient Functions**

Let f and g be any two functions. We define the functions $f + g$, $f - g$, $f \cdot g$, and f/g as follows:

(i) **Sum function:** $(f + g)(x) = f(x) + g(x)$
(ii) **Difference function:** $(f - g)(x) = f(x) - g(x)$
(iii) **Product function:** $(f \cdot g)(x) = f(x) \cdot g(x)$
(iv) **Quotient function:** $\left(\frac{f}{g}\right)(x) = \frac{f(x)}{g(x)}$, for $g(x) \neq 0$

In each case, the domain of the defined function consists of all values of x *common* to the domains of f and g, except for the quotient function, in which case the values of x for which $g(x) = 0$ are also excluded.

EXAMPLE 1 Let $f(x) = 3x^3 + 7$ and $g(x) = x^2 - 1$. Find

(a) $(f + g)(x)$ (b) $(f - g)(x)$ (c) $(f \cdot g)(x)$

(d) $\left(\frac{f}{g}\right)(x)$ and the domain of f/g

SOLUTION

(a) $(f + g)(x) = f(x) + g(x) = (3x^3 + 7) + (x^2 - 1) = 3x^3 + x^2 + 6$

(b) $(f - g)(x) = f(x) - g(x) = (3x^3 + 7) - (x^2 - 1) = 3x^3 - x^2 + 8$

(c) $(f \cdot g)(x) = f(x) \cdot g(x) = (3x^3 + 7)(x^2 - 1) = 3x^5 - 3x^3 + 7x^2 - 7$

(d) $\left(\frac{f}{g}\right)(x) = \frac{f(x)}{g(x)} = \frac{3x^3 + 7}{x^2 - 1}$

The expression $\frac{3x^3 + 7}{x^2 - 1}$ represents a real number for all real values of x except values for which the denominator $x^2 - 1 = 0$. Thus the domain of f/g includes all real numbers *except* 1 and -1.

EXAMPLE 2

In business, if $P(x)$ represents the total dollar *profit* obtained by producing and selling x units of a commodity, then

$$P(x) = R(x) - C(x)$$

where $R(x)$ is the total dollar *revenue* from the sale of x units and $C(x)$ is the total dollar *cost* of producing x units. Suppose that a manufacturer of calculators finds the total cost of manufacturing x calculators per week to be given by the function

$$C(x) = 5000 + 2x$$

and the total revenue for selling x calculators per week to be given by the function

$$R(x) = 10x - \frac{x^2}{1000} \quad \text{for} \quad 0 \leq x \leq 8000$$

(a) Write an expression for the profit function P if x calculators are manufactured and sold in a week.

(b) What is the profit if 4000 calculators are manufactured and sold in a week?

SOLUTION

(a) The profit function P is given by

$$\begin{aligned} P(x) &= R(x) - C(x) \\ &= \left(10x - \frac{x^2}{1000}\right) - (5000 + 2x) \\ &= 8x - \frac{x^2}{1000} - 5000 \end{aligned}$$

(b) If $x = 4000$, then

$$P(4000) = 8(4000) - \frac{16{,}000{,}000}{1000} - 5000 = 11{,}000$$

so that the profit is \$11,000.

Composition of Functions

The **composition** of two functions can be thought of as a "chain reaction" in which the functions occur one after the other. Let us consider a specific example.

We know from solid geometry that the volume V of a sphere of radius r is given by the formula

$$V = \frac{4}{3}\pi r^3$$

Now, suppose that air is being pumped into a spherical balloon so that at the end of t seconds, the radius r satisfies the equation

$$r = t^2 + 1$$

The equation

$$V = \frac{4}{3}\pi r^3$$

defines a function

$$f(r) = \frac{4}{3}\pi r^3$$

whereas the equation

$$r = t^2 + 1$$

defines a function

$$g(t) = t^2 + 1$$

Thus, the first two equations can be represented as the *functions* $V = f(r)$ and $r = g(t)$, respectively. We can express V in terms of t by using substitution to obtain

$$V = \frac{4}{3}\pi r^3 = \frac{4}{3}\pi(t^2 + 1)^3$$

that is,

$$V = f(r) = f(t^2 + 1) = f(g(t))$$

The equation $V = f(g(t))$ defines a new function

$$h(t) = f(g(t))$$

In order to prevent a "pile up" of parentheses, we often replace the outer parentheses in the latter equation by square brackets and write

$$h(t) = f[g(t)]$$

The function h, which was obtained by "chaining" g *followed* by f, is called the *composition of* g by f and is sometimes written as $h = f \circ g$.

DEFINITION 2 **Composition of Functions**

Let f and g be two functions satisfying the condition that at least one number in the range of g belongs to the domain of f. Then the **composition** of g by f is the function $f \circ g$ defined by the equation

$$(f \circ g)(x) = f[g(x)]$$

$f \circ g$ is often called a **composite function.**

The domain of the composite function $f \circ g$ is the set of all values x in the domain of g such that $g(x)$ belongs to the domain of f. The range of $f \circ g$ is the set of all possible values of $f[g(x)]$.

Two important observations regarding the composition of functions are:

1. The composition $f \circ g$ is *different* from the product function $f \cdot g$.

2. In general, the composition function $f \circ g$ is *not* the same as the composition function $g \circ f$ (watch the order!).

For example, if $f(x) = 1/x$ and $g(x) = x - 5$, then

$$(f \circ g)(x) = f[g(x)] = f[x - 5] = \frac{1}{x - 5}$$

whereas

$$(f \cdot g)(x) = f(x) \cdot g(x) = \frac{1}{x} \cdot (x - 5) = \frac{x - 5}{x}$$

and

$$(g \circ f)(x) = g[f(x)] = g\left[\frac{1}{x}\right] = \frac{1}{x} - 5 = \frac{1 - 5x}{x}$$

EXAMPLE 3 Let $f(x) = x^2$ and $g(x) = 2x - 3$. Find

(a) $(f \circ g)(2)$ (b) $(g \circ f)(2)$ (c) $(f \circ g)(x)$
(d) $(g \circ f)(x)$ (e) $(f \circ f)(x)$ C (f) $(f \circ g)(2.008)$

SOLUTION

(a) $(f \circ g)(2) = f[g(2)] = f(1) = 1$
(b) $(g \circ f)(2) = g[f(2)] = g(4) = 5$
(c) $(f \circ g)(x) = f[g(x)] = f(2x - 3) = (2x - 3)^2 = 4x^2 - 12x + 9$
(d) $(g \circ f)(x) = g[f(x)] = g(x^2) = 2x^2 - 3$
(e) $(f \circ f)(x) = f[f(x)] = f(x^2) = (x^2)^2 = x^4$
(f) $(f \circ g)(2.008) = f[g(2.008)] = [2(2.008) - 3]^2 = 1.032256$

EXAMPLE 4 Let $f(x) = 5x - 1$ and $g(x) = \frac{1}{5}(x + 1)$. Find

(a) $f[g(x)]$ (b) $g[f(x)]$

SOLUTION

(a) $f[g(x)] = f[\frac{1}{5}(x + 1)] = 5 \cdot [\frac{1}{5}(x + 1)] - 1 = x + 1 - 1 = x$

(b) $g[f(x)] = g(5x - 1) = \frac{1}{5}[(5x - 1) + 1] = \frac{1}{5} \cdot 5x = x$

EXAMPLE 5 C A damaged tanker leaks oil, and the oil spill is spreading on the ocean surface in the form of a circle of radius r (in kilometers). Suppose that r is expressed as a function of time t (in hours) by the equation $r(t) = 0.75 + 2.25t$, where $t = 0$ corresponds to the time at which the radius of the spill is 0.75 kilometer.

(a) Express the area A of the oil spill as a function of t; that is, find $(A \circ r)(t)$ explicitly.

(b) What is the area of the spill after 3 hours?

SOLUTION

(a) The area A of the oil spill is given by the formula $A = \pi r^2$ and r is given by the function $r(t) = 0.75 + 2.25t$. Thus, the expression for the composite function $A \circ r$ is given by

$$\begin{aligned}(A \circ r)(t) &= A[r(t)] = A(0.75 + 2.25t) \\ &= \pi(0.75 + 2.25t)^2 \\ &= \pi(0.5625 + 3.375t + 5.0625t^2)\end{aligned}$$

(b) If 3.14 is used as an approximate value for π, then the approximate area of the oil spill when $t = 3$ is given by

$$\begin{aligned}(A \circ r)(3) &= A[r(3)] = \pi[0.5625 + 3(3.375) + 9(5.0625)] \\ &= 3.14(56.25) = 176.625 \text{ square kilometers}\end{aligned}$$

If we use the mapping notation of functions, then the composition $f \circ g$ can be illustrated schematically (Figure 2).

Figure 2

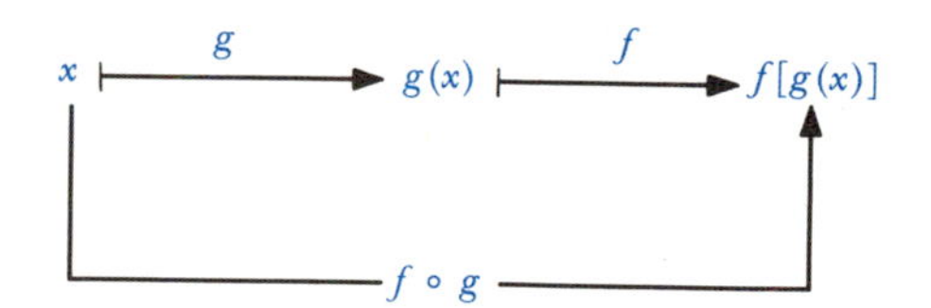

In this scheme,

$$x \overset{g}{\mapsto} g(x) \overset{f}{\mapsto} f[g(x)]$$

indicates that each value of x is *first* mapped by the g function to the image $g(x)$. Next, $g(x)$ is mapped by the f function to the image $f[g(x)]$. The composite function maps each value of x directly to the image $f[g(x)]$ and this is shown by

$$x \xmapsto{f \circ g} f[g(x)]$$

In calculus, it is often necessary to express a given function as a composition of two other functions. This process is illustrated in the next example.

EXAMPLE 6 Express each function h as the composition of two functions f and g so that $h = f \circ g$.

(a) $h(x) = (2x + 1)^3$ (b) $h(x) = \sqrt{x^2 + 5}$

SOLUTION The mapping scheme can be used to determine functions f and g.

(a) Since

$$x \xmapsto{\text{1st}} (2x + 1) \xmapsto{\text{2nd}} (2x + 1)^3$$

we define $g(x) = 2x + 1$ (first function) and $f(u) = u^3$ (second function) so that

$$(f \circ g)(x) = f[g(x)] = f(2x + 1) = (2x + 1)^3$$

and $h = f \circ g$.

(b) The order of the mappings is

$$x \xmapsto{\text{1st}} (x^2 + 5) \xmapsto{\text{2nd}} \sqrt{x^2 + 5}$$

If we define $g(x) = x^2 + 5$ and $f(u) = \sqrt{u}$ then

$$(f \circ g)(x) = f[g(x)] = f[x^2 + 5] = \sqrt{x^2 + 5}$$

and $h = f \circ g$.

PROBLEM SET 2.5

In problems 1–9, using the given pair of functions, find an expression in terms of x for each of the following four functions.

(a) $(f + g)(x)$ (b) $(f - g)(x)$ (c) $(f \cdot g)(x)$ (d) $(f/g)(x)$

What is the domain of f/g?

1. $f(x) = 3x + 1$ and $g(x) = 3x - 7$

2. $f(x) = x$ and $g(x) = -5$

3. $f(x) = 4x - 5$ and $g(x) = x + 6$

4. $f(x) = \dfrac{2}{x - 5}$ and $g(x) = \dfrac{x}{3 - 4x}$

5. $f(x) = \dfrac{2x + 3}{x - 5}$ and $g(x) = \dfrac{2 - 7x}{3x + 1}$

6. $f(x) = x^2 - 3x$ and $g(x) = 4x + 1$

7. $f(x) = 7x + 1$ and $g(x) = -3x + 8$

8. $f(x) = x^3$ and $g(x) = 2x^2$

9. $f(x) = x^2 + 5$ and $g(x) = -x^2 + x$

10. Let $f(x) = \sqrt{x}$ and $g(x) = x^2$, find $(f \circ f)(4)$ and $(g \circ g)(-2)$.

In problems 11–20, use $f(x) = 2x^2 + 6$ and $g(x) = 7x + 2$ to find each of the values.

11. $(f \circ g)(2)$ **12.** $(g \circ f)(2)$ **13.** $(f \circ f)(2)$ **14.** $(g \circ g)(2)$ **15.** $g[f(4)]$

16. $f[g(3)]$ **17.** $f[g(5)]$ **18.** $g[f(5)]$ **19.** $f[f(-1)]$ **20.** $g[g(-1)]$

In problems 21–28, find

(a) $f[g(x)]$ and the domain of $f \circ g$ (b) $g[f(x)]$ and the domain of $g \circ f$

21. $f(x) = 2x$ and $g(x) = 5x - 3$

22. $f(x) = 2x^2 + 5$ and $g(x) = 7x$

23. $f(x) = 3x$ and $g(x) = -3x$

24. $f(x) = x^3 + 1$ and $g(x) = \sqrt[3]{x - 1}$

25. $f(x) = 11x + 2$ and $g(x) = \dfrac{x}{11} - \dfrac{2}{11}$

26. $f(x) = \dfrac{1}{3x + 2}$ and $g(x) = \dfrac{3}{2x - 5}$

27. $f(x) = \dfrac{x}{3x - 5}$ and $g(x) = \dfrac{1 - x}{3 + 2x}$

28. $f(x) = \sqrt{2x + 1}$ and $g(x) = x^2 + 9$

In problems 29–34, express the given function h as the composition $h = f \circ g$ of two functions f and g.

29. $h(x) = (5x - 3)^3$ **30.** $h(x) = (x^2 + 5x)^4$ **31.** $h(t) = (t^2 - 2)^{-2}$

32. $h(s) = \left(\frac{s+1}{s-1}\right)^3$ **33.** $h(x) = \sqrt[3]{x + x^{-1}}$ **34.** $h(x) = (x^2 + 30)^{-6}$

35. Let $F(x) = 3x - 7$ and $G(x) = 2x + k$. Determine the value of k so that $F[G(x)] = G[F(x)]$.

36. Use $f(x) = 5x - 7$ and $g(x) = x^2$ to compare the composition $(f \circ g)(x)$ and the product $(f \cdot g)(x)$.

37. Let $f(x) = x^2 + 1$ and $g(x) = (x + 1)^2$. Form $f \circ g$ and $g \circ f$.

38. Let $F(x) = x$ and $G(x) = 1/x$. Find the domains of $F \circ G$ and $G \circ F$.

39. Let $f(x) = 3x + 1$ and $g(x) = -5x + 2$.
(a) Find $f[g(x)]$. (b) For what values of x does $f[g(x)] = 2$?

40. Verify that $f[f[f(x)]] = x$ if $f(x) = 1 - (1/x)$.

41. Let $f(x) = x^3 + 2$ and $g(x) = \sqrt[3]{x + 7}$.
(a) Find $f[g(x)]$. (b) For what values of x does $f[g(x)] = 13$?

42. Suppose that $f(x) = x^{1/3}$, $g(x) = (x^9 + x^6)^{1/2}$, and that $h(x) = x(x + 1)^{1/2}$. Show that $g[f(x)] = h(x)$.

43. Let $f(x) = ax + 1$. Find the value of a such that $f[f(x)] = x$.

44. Let $f(x) = x$, $g(x) = |x|$, and $h(x) = \sqrt{x}$. Answer true if the given equation is true for *all* real values; answer false if the equation is false for *any* real value. If your answer is false, give an example to support your claim.
(a) $g(x^3) = [g(x)]^3$ (b) $g(x + y) = g(x) + g(y)$
(c) $g(xy) = g(x) \cdot g(y)$ (d) $f(x^3) = [f(x)]^3$
(e) $f(x + y) = f(x) + f(y)$ (f) $f(xy) = f(x) \cdot f(y)$
(g) $h(x^3) = [h(x)]^3$ (h) $h(x + y) = h(x) + h(y)$
(i) $h(xy) = h(x) \cdot h(y)$

45. A company manufactures and sells x microwave ovens per month. Assume that the cost and the revenue functions are given by $C(x) = 72{,}000 + 60x$ and $R(x) = 200x - (x^2/30)$, respectively, for $0 \leq x \leq 6000$. Write an equation for the profit function P in terms of x. Find the profit if 1500 microwave ovens are sold in a month.

46. A baseball diamond is a square, 90 feet long on each side. A ball is hit down the third-base line at the rate of 50 feet per second. Let y denote the distance in feet between the ball and first base, let x denote its distance in feet from home plate, and let t denote the elapsed time in seconds after the ball was hit. Express y as a function f of x and x as a function g of t. Find $(f \circ g)(t)$ explicitly.

47. Suppose that C is the circumference of a circle of radius r. Assume that the circle is shrinking in size and that the radius is given by

$$r = f(t) = \frac{1}{2 + t^2}$$

at a time t, where t is measured in minutes and r is measured in centimeters.
(a) Express C as a function of t; that is, find $(C \circ f)(t)$.
(b) Find the circumference when $t = 4$ minutes.

48. A manufacturer of electric appliances finds that the production cost C (in dollars per unit) for its deluxe model can opener is a function of the number x of can openers produced, defined by

$$C(x) = \frac{x^2 + 120x + 8000}{20x}$$

The selling price S (in dollars) of each can opener, which is a function of the production cost C per unit, is given by $S(C) = 1.02C$. Express the selling price as a function of the number of can openers produced; that is, find $(S \circ C)(x)$ explicitly. What is the selling price per unit if 1000 can openers are produced?

2.6 Inverse Functions

Consider the functions

$$f(x) = x - 1 \qquad \text{and} \qquad g(x) = x + 1$$

By forming the composite functions $f \circ g$ and $g \circ f$ for these particular functions we find that

$$(f \circ g)(x) = f[g(x)] = f(x + 1) = (x + 1) - 1 = x$$

and

$$(g \circ f)(x) = g[f(x)] = g(x - 1) = (x - 1) + 1 = x$$

We can say that each function "undoes" what the other function "does" or that the second function "returns" the image values of the first function back to the domain number x. Schematically, we can display this behavior as follows:

$$x \overset{g}{\longmapsto} (x + 1) \overset{f}{\longmapsto} x \qquad \text{and} \qquad x \overset{f}{\longmapsto} (x - 1) \overset{g}{\longmapsto} x$$

Two functions, such as f and g above, that are related in such a way that each "undoes" what the other "does" are said to be *inverse* functions of one another.

DEFINITION 1 **Inverse Functions**

Two functions f and g are **inverses** of each other if and only if

$$f[g(x)] = x$$

for every value of x in the domain of g and

$$g[f(x)] = x$$

for every value of x in the domain of f.
A function f, for which such a function g exists, is said to be **invertible.**

In Examples 1 and 2, show that the functions f and g are inverses of each other.

EXAMPLE 1 $f(x) = 5x$ and $g(x) = \dfrac{x}{5}$

SOLUTION Since

$$f[g(x)] = f\left(\frac{x}{5}\right) = 5\left(\frac{x}{5}\right) = x$$

and

$$g[f(x)] = g(5x) = \frac{1}{5}(5x) = x$$

for all x, we conclude, by Definition 1, that f and g are inverses of one another.

EXAMPLE 2 $f(x) = 3x + 2$ and $g(x) = \dfrac{1}{3}x - \dfrac{2}{3}$

SOLUTION

$$\begin{aligned} f[g(x)] &= f\left(\frac{1}{3}x - \frac{2}{3}\right) \\ &= 3\left(\frac{1}{3}x - \frac{2}{3}\right) + 2 = x \end{aligned}$$

and

$$\begin{aligned} g[f(x)] &= g(3x + 2) \\ &= \frac{1}{3}(3x + 2) - \frac{2}{3} = x \end{aligned}$$

so that f and g are inverses of each other.

When f and g are inverses of each other, we refer to g as the *inverse* function of f and vice versa, and we write

$$g = f^{-1} \quad \text{and} \quad g^{-1} = f$$

The notation f^{-1} is read "f inverse." [Care must be taken not to confuse $y = f^{-1}(x)$, which is the *inverse function* of function f, with $[f(x)]^{-1} = 1/f(x)$, which is the reciprocal of $f(x)$.]

Thus in Example 2 above

$$g(x) = \frac{1}{3}x - \frac{2}{3}$$

is the inverse of $f(x) = 3x + 2$ and vice versa, and we write

$$f^{-1}(x) = \frac{1}{3}x - \frac{2}{3} \qquad \text{and} \qquad g^{-1}(x) = 3x + 2$$

Notice that the conditions stated in Definition 1 can be restated as:

$$f^{-1}[f(x)] = x \qquad \text{for every } x \text{ in the domain of } f$$

and

$$f[f^{-1}(x)] = x \qquad \text{for every } x \text{ in the domain of } f^{-1}$$

Consider the functions

$$f(x) = 3x + 2 \qquad \text{and} \qquad f^{-1}(x) = \frac{1}{3}x - \frac{2}{3}$$

in Example 2. If we graph f and f^{-1} on the same coordinate axes, we see that the graph of f^{-1} is the reflection of the graph of f across the line $y = x$ (Figure 1). For instance, the reflections across the line $y = x$ of the points $(2, 8)$, $(1, 5)$, $(0, 2)$, and $(-2, -4)$ on the graph of f are, respectively, the points $(8, 2)$, $(5, 1)$, $(2, 0)$, and $(-4, -2)$ on the graph of f^{-1}.

Figure 1

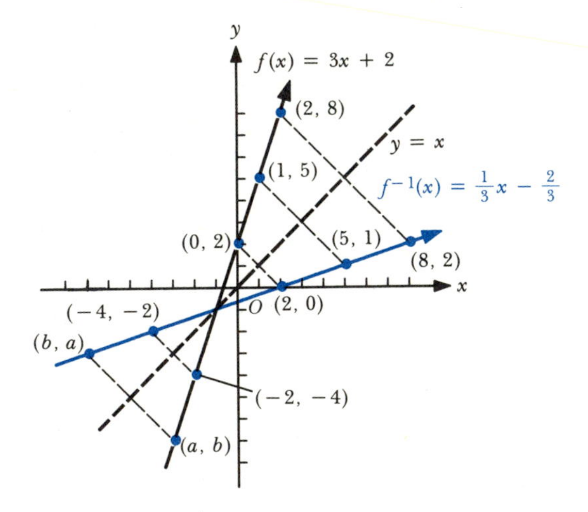

In general, if f is an invertible function and the point (a, b) is on the graph of f, then

$$b = f(a) \qquad \text{and} \qquad f^{-1}(b) = f^{-1}[f(a)] = a$$

so that (b, a) belongs to the graph of f^{-1}. In other words, the graph of f^{-1} is the reflection of the graph of f across the line $y = x$ and vice versa.

Existence of the Inverse Function

Not every function has an inverse. To determine the condition that ensures that a function has an inverse, consider the graph of the function f in Figure 2. The mirror image of the graph of f across the line $y = x$ is *not* the graph of a function, because there is a vertical line l that intersects the graph more than once. Thus f is *not* invertible. Notice that the *horizontal line* L obtained by reflecting l across the line $y = x$ intersects the graph of f more than once. This observation provides the basis for using the graph of a function to determine whether it has an inverse.

Figure 2

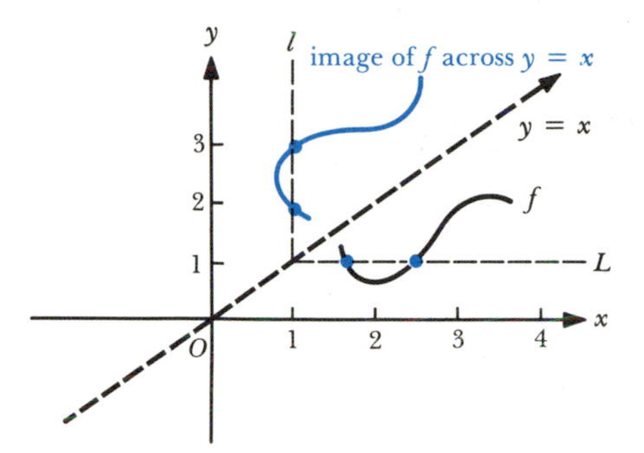

Horizontal Line Test

> A function f is invertible if and only if no horizontal straight line intersects its graph more than once.

EXAMPLE 3 Use the horizontal line test to determine whether or not the functions graphed in Figure 3 are invertible.

Figure 3

(a)

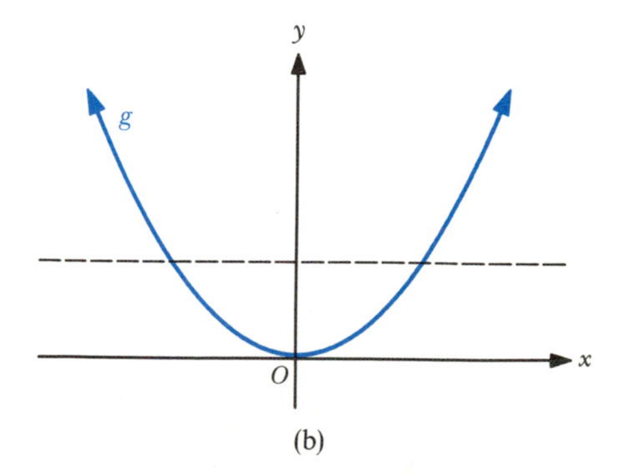

(b)

SOLUTION

(a) No horizontal line intersects the graph of f more than once; hence, f is invertible.

(b) Any horizontal line drawn above the x axis will intersect the graph of g twice. Therefore, g is not invertible.

The horizontal line test enables us to use the graph of a given function to answer the question: Does the function have an inverse?

We now turn our attention to answering the question: If a given function has an inverse, how do we determine an equation that defines the inverse function?

Earlier we saw that if (a, b) is on the graph of f, then (b, a) is on the graph of f^{-1}. This means that f^{-1} is the function formed by switching the roles of the domain and range variables in the equation used to define f.

Procedure for Finding f^{-1}

Step 1. Write the equation $y = f(x)$ that defines f.
Step 2. Interchange variables x and y in the equation obtained in step 1, so that the equation becomes $x = f(y)$.
Step 3. Solve the equation in step 2 for y in terms of x to get $y = f^{-1}(x)$. This latter equation defines f^{-1}.

In Examples 4 and 5, use the above procedure to determine f^{-1}. Sketch the graphs of f and f^{-1} on the same coordinate system.

EXAMPLE 4 $f(x) = 2x - 3$

SOLUTION Step 1. $y = 2x - 3$

Step 2. Interchange x and y in the equation in step 1, so that

$$x = 2y - 3$$

Step 3. Solving the equation $x = 2y - 3$ for y, we obtain

$$y = \frac{x + 3}{2}$$

so that $f^{-1}(x) = \dfrac{x + 3}{2}$. By graphing

$$f(x) = 2x - 3$$

and

$$f^{-1}(x) = \frac{x + 3}{2}$$

on the same coordinate system (Figure 4), we observe that the graph of the inverse function f^{-1} is a reflection of the graph of f across the line $y = x$.

Figure 4

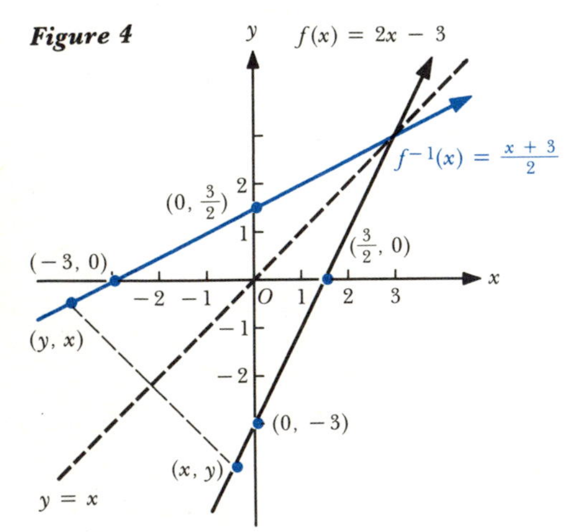

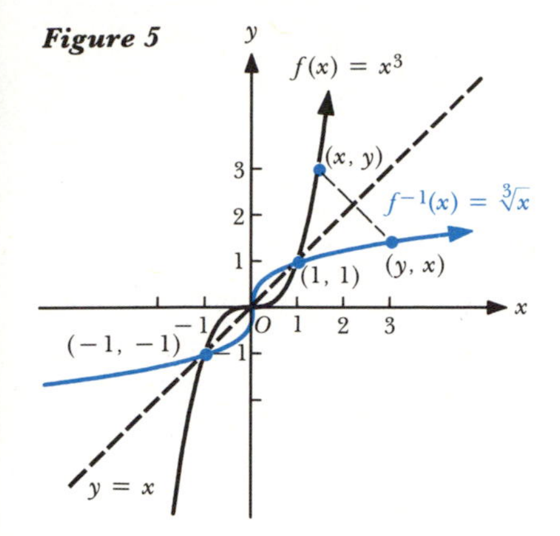

Figure 5

EXAMPLE 5 $f(x) = x^3$

SOLUTION Step 1. $y = x^3$

Step 2. $x = y^3$

Step 3. Solving the equation $x = y^3$ for y, we obtain $y = \sqrt[3]{x}$. Therefore, $f^{-1}(x) = \sqrt[3]{x}$. When both $f(x) = x^3$ and $f^{-1}(x) = \sqrt[3]{x}$ are graphed on the same coordinate system (Figure 5), we see that the graph of f^{-1} is a reflection of the graph of f across the line $y = x$.

EXAMPLE 6 Let $f(x) = 4x - 5$.

(a) Find $f^{-1}(x)$.

(b) Verify that (i) $f^{-1}[f(x)] = x$ and (ii) $f[f^{-1}(x)] = x$.

SOLUTION (a) Step 1. $y = 4x - 5$

Step 2. $x = 4y - 5$

Step 3.
$$\begin{aligned} 4y - 5 &= x \\ 4y &= x + 5 \\ y &= \frac{1}{4}(x + 5) \end{aligned}$$

therefore,

$$f^{-1}(x) = \frac{1}{4}(x + 5)$$

(b) (i) $f^{-1}[f(x)] = f^{-1}[4x - 5] = \frac{1}{4}[(4x - 5) + 5] = \frac{4x}{4} = x$

and

(ii) $f[f^{-1}(x)] = f\left[\frac{1}{4}(x + 5)\right] = 4\left[\frac{1}{4}(x + 5)\right] - 5 = (x + 5) - 5 = x$

PROBLEM SET 2.6

In problems 1–6, show that the functions f and g are inverses of each other by verifying that $f[g(x)] = x$ and $g[f(x)] = x$.

1. $f(x) = 7x - 2$ and $g(x) = \frac{x}{7} + \frac{2}{7}$

2. $f(x) = 1 - 5x$ and $g(x) = \frac{1}{5} - \frac{x}{5}$

3. $f(x) = x^4$, where $x \geq 0$, and $g(x) = \sqrt[4]{x}$

4. $f(x) = \dfrac{1}{x-2}$ and $g(x) = \dfrac{1}{x} + 2$

5. $f(x) = \dfrac{1}{x}$ and $g(x) = \dfrac{1}{x}$

6. $f(x) = x^3$ and $g(x) = \sqrt[3]{x}$

7. Use the horizontal line test to determine whether each of the functions whose graphs are given in Figure 6 has an inverse.

Figure 6

(a)

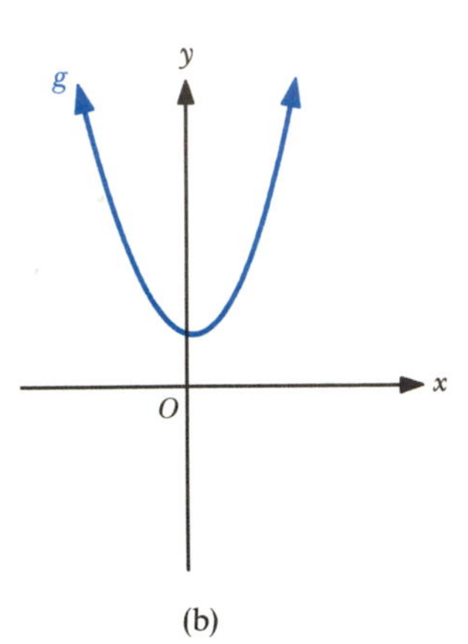

(b)

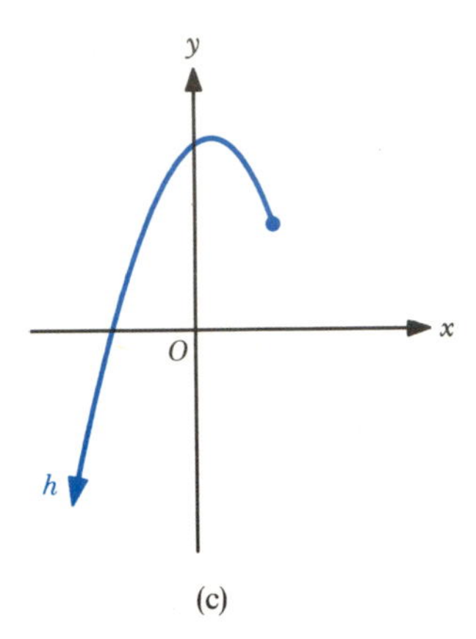

(c)

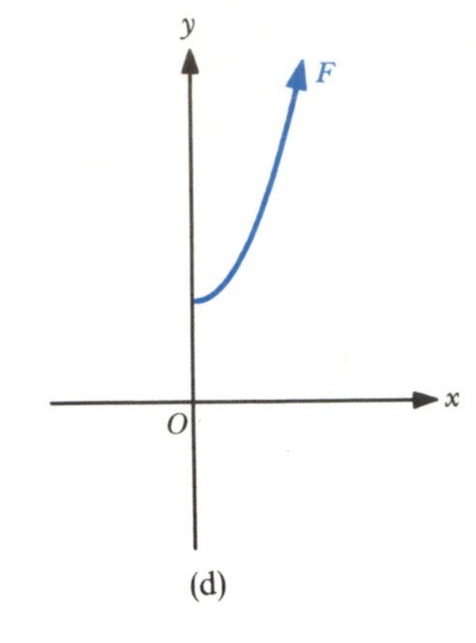

(d)

8. Suppose that $f(x) = 2x + 4$ and $g(x) = 3x - 1$.
(a) Determine f^{-1}, g^{-1}, and $(f \circ g)^{-1}$. (b) Show that $(f \circ g)^{-1}(x) = (g^{-1} \circ f^{-1})(x)$.

9. The functions graphed in Figure 7 are invertible. Sketch the graph of each inverse function by reflecting the given graph across the line $y = x$.

Figure 7

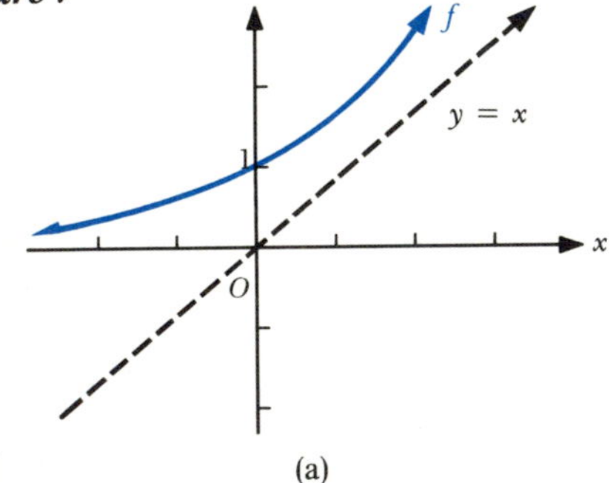

(a)

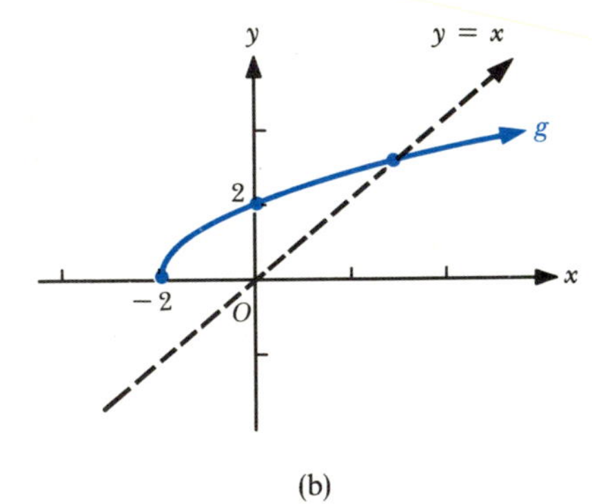

(b)

10. Use the results from problem 9 to answer the following:
(a) Determine the domains and ranges of f and f^{-1} (Figure 7a).
(b) Determine the domains and ranges of g and g^{-1} (Figure 7b).
(c) What can be said in general about the domain and range of a function in comparison to the domain and range of the inverse of the function?

In problems 11–18, use the step-by-step procedure on page 124 to find $f^{-1}(x)$ and sketch the graphs of f and f^{-1} on the same coordinate system.

11. $f(x) = 7x + 5$

12. $f(x) = x^2 - 4$, $x \geq 0$

13. $f(x) = 1 - 3x$

14. $f(x) = x^3 + 5$

15. $f(x) = x^3 - 8$

16. $f(x) = \sqrt{1 - x^2},\ 0 \le x \le 1$

17. $f(x) = x^2,\ x \ge 0$

18. $f(x) = \sqrt{x + 2}$

In problems 19–28, (a) use the step-by-step procedure on page 124 to find $f^{-1}(x)$; (b) verify that $f[f^{-1}(x)] = x$; (c) verify that $f^{-1}[f(x)] = x$.

19. $f(x) = 3x - 7$

20. $f(x) = 5 - 11x$

21. $f(x) = \frac{3}{4}x + 5$

22. $f(x) = x^5$

23. $f(x) = \dfrac{1}{x} - 1$

24. $f(x) = \dfrac{1}{x + 1}$

25. $f(x) = -\dfrac{5}{x}$

26. $f(x) = x^2,\ x < 0$

27. $f(x) = \sqrt{1 + x}$

28. $f(x) = \sqrt{3 - 2x}$

29. Show that the given functions are inverses of each other.

$$f(x) = \frac{3x - 7}{x + 1} \text{ for } x \neq -1 \qquad \text{and} \qquad g(x) = \frac{7 + x}{3 - x} \text{ for } x \neq 3$$

30. (a) Use the graph of $f(x) = -3x^2 + 1$ to show that f^{-1} does not exist.
(b) Show that $f(x) = -3x^2 + 1$ for $x \le 0$ has an inverse function and find the inverse function f^{-1}.
(c) Explain why the function in part (b) is invertible whereas the function in part (a) is not.

31. Does a constant function have an inverse? Examine the graph of g to determine whether g^{-1} exists if $g(x) = 3$.

32. Examine the graph of $f(x) = |x + 1|$ to determine whether f^{-1} exists.

33. A function f is said to be **one-to-one** if, whenever a and b are in the domain of f and $f(a) = f(b)$, then $a = b$. Use the horizontal line test to illustrate graphically that f has an inverse if it is one-to-one.

34. Use the horizontal line test to determine whether the function f defined by

$$f(x) = \begin{cases} 2x - 4 & \text{for } x < 0 \\ -(x + 2)^2 & \text{for } x \ge 0 \end{cases}$$

has an inverse.

35. Let f be the function representing the conversion from Fahrenheit to Celsius, and let g be the function representing the conversion from Celsius to Fahrenheit. Then

$$f(x) = \frac{5}{9}(x - 32) \qquad \text{and} \qquad g(x) = \frac{9}{5}x + \frac{160}{5}$$

Show that f and g are inverses of each other.

c **36.** Suppose that

$$f(x) = -2x^5 + \frac{9}{4}$$

Find (a) $f^{-1}(3.01)$, (b) $f^{-1}(7.45)$, and (c) $f^{-1}(-3.06)$. Round off each answer to two decimal places.

REVIEW PROBLEM SET, CHAPTER 2

1. Locate each of the given points and indicate in which quadrant, if any, the point is found.
(a) $(1, 2)$ (b) $(-1, 1)$ (c) $(2, -1)$ (d) $(-1, -1)$ (e) $(1, -\frac{1}{2})$ (f) $(4, 0)$ (g) $(-2, 0)$ (h) $(0, -4)$
2. Locate the point $P = (-1, 4)$ on a Cartesian coordinate system and then give the coordinates of Q and S if
(a) Line segment $\overline{PQ}$ is perpendicular to and bisected by the x axis.
(b) Line segment $\overline{PS}$ is bisected by the origin.

In problems 3–6, use the distance formula to find (a) the distance between each given pair of points and (b) the coordinates of the midpoint between each given pair of points (see problem 34 on page 74).

3. $(2, 1)$ and $(4, -5)$
4. $(-3, 2)$ and $(6, -1)$
5. $(-6, -3)$ and $(2, 1)$
6. [c] $(5.82, -3.71)$ and $(0, 2\pi)$, to two decimal places
7. Show that the triangle whose vertices are $P_1 = (-2, 4)$, $P_2 = (-5, 1)$, and $P_3 = (-6, 5)$ is an isosceles triangle.
8. The abscissa of a point P is 3 and its distance from the point $(3, 4)$ is $2\sqrt{13}$. Find the possible ordinates of P.

In problems 9 and 10, show that the graph of the given equation is a circle. Determine the radius and the center of the circle. Sketch the graph.

9. $x^2 + y^2 - 6x + 4y - 12 = 0$
10. $x^2 + y^2 - 6x + 2y - 26 = 0$

In problems 11–14, find the slope of the line containing the given pair of points P_1 and P_2, and find an equation in point-slope form of the line.

11. $P_1 = (2, 3)$ and $P_2 = (-1, 4)$
12. $P_1 = (1, -5)$ and $P_2 = (-4, 0)$
13. $P_1 = (4, 6)$ and $P_2 = (-1, -5)$
14. $P_1 = (3, -2)$ and $P_2 = (-3, -2)$
15. Sketch the line L that contains the point $(4, 1)$ and has slope $m = -3/7$.
16. Suppose that $a > 0$ and $a + k > 0$. Show that the line L containing the points $(a, \sqrt{a})$ and $(a + k, \sqrt{a + k})$ has slope

$$\frac{1}{\sqrt{a + k} + \sqrt{a}}$$

In problems 17–24, find an equation of each line in the slope-intercept form and sketch its graph.

17. Slope $m = 3$; contains the point $(4, 5)$
18. Slope $m = 0$; contains the point $(-2, 3)$
19. Contains the points $(3, -2)$ and $(5, 6)$
20. Contains the points $(0, 2)$ and $(3, 0)$
21. Parallel to the line $3x + y = 5$ and containing the point $(-2, 3)$
22. Containing the point $(-3, -5)$ and parallel to the line containing points $(5, 8)$ and $(-1, 3)$
23. Perpendicular to the line $8x + 7y = -3$ and containing the point $(2, -4)$
24. Perpendicular to the line containing the points $(-3, 7)$ and $(-1, -1)$, and with x intercept -2
25. The three vertices of a right triangle are located at points $(9, 3)$, $(5, 9)$, and $(-7, 1)$. Which one is the vertex of the right angle?

26. Two opposite vertices of a rectangle are located at points (6, 2) and (−5, 4). Two sides of the rectangle are parallel to the line $8x - 6y + 5 = 0$. Find equations of the lines containing these two sides.

In problems 27–38, assume that $f(x) = x^3$, $g(x) = 2x + 2$, and $h(x) = x^2 - 5x$. Find each value.

27. $f(-1)$ **28.** $g(-1)$ **29.** $h(-5)$ **30.** $f(\sqrt{2})$

31. $f(\sqrt[3]{2})$ **32.** $h(\sqrt{2})$ **33.** $f(\frac{1}{3})$ **34.** $g(a + b) - g(a)$

35. $\dfrac{g(a+b) - g(a)}{b}$ **36.** $f(b^2)$ **37.** $[f(\sqrt{x})]^2$ **38.** $g\left(\dfrac{1}{x+a}\right)$

In problems 39 and 40, find the domain of each function.

39. $f(x) = \dfrac{1}{(x-2)^2}$ **40.** $g(x) = \sqrt{x^2 - 9}$

41. Which of the graphs in Figure 1 represent functions?

Figure 1

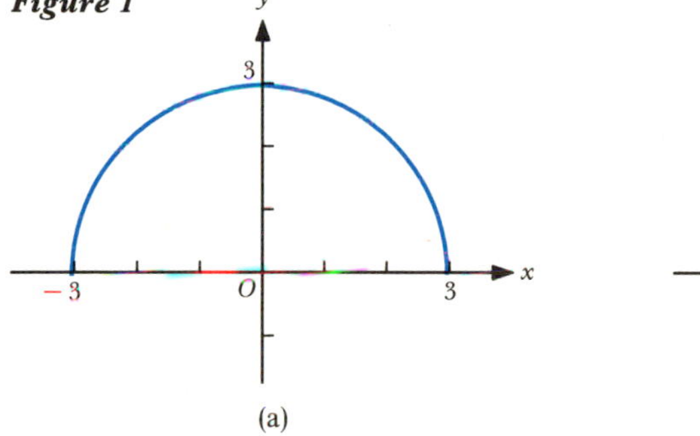

(a)

y
O
x
(b)

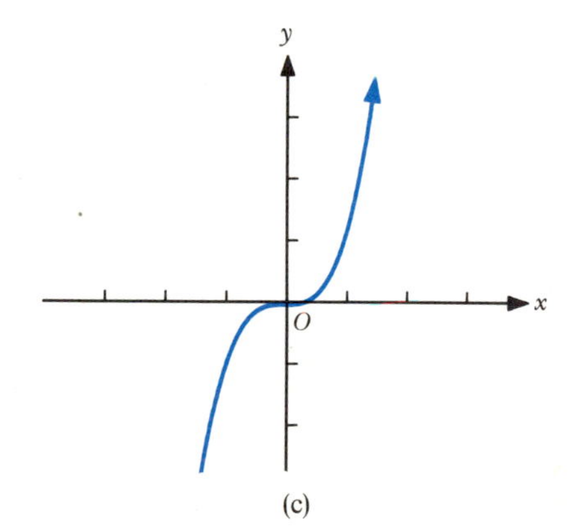

(c)

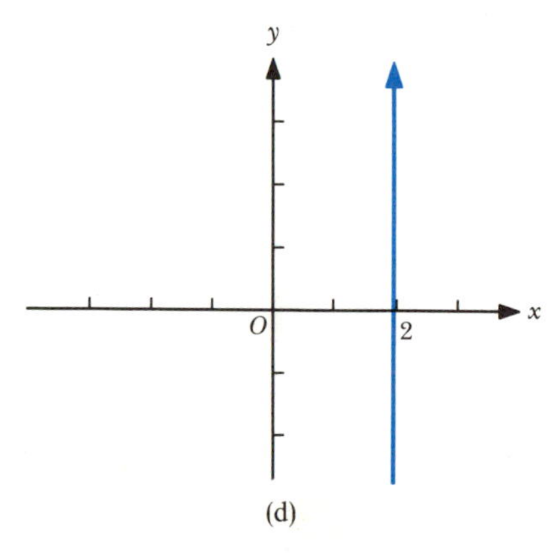

(d)

42. Determine which of these points lie on the graph of $f(x) = |5x - 1|$: $(-1, 6)$, $(0, 2)$, $(3, 14)$, $(-\frac{1}{2}, -\frac{7}{2})$, and $(\frac{1}{3}, \frac{2}{3})$. What is the domain of f?

43. Let $f(x) = \sqrt{x^2 + 5}$.

(a) Determine each of the values $f(-2)$, $f(2)$, and $f(0)$.

(b) Describe the function values using function notation, mapping notation, and ordered pair notation.

44. Suppose that $f(x) = \dfrac{3+x}{3-x}$; find and simplify $\dfrac{f(a) - f(-a)}{1 + f(a)f(-a)}$.

45. A duplicating machine produces y copies in x hours. The relationship between y and x is linear. If the machine has produced 120 copies after running for $\frac{1}{2}$ hour and 470 copies after running for $1\frac{1}{2}$ hours, find a function f such that $y = f(x)$.

46. A projectile fired upward attains a speed of v feet per second after t seconds of flight, and the relationship between the two numbers v and t is linear. If the projectile is fired at a speed of 2000 feet per second and reaches a speed of 90 feet per second after 2 seconds of flight, express v as a function of t.

47. Write a formula that expresses the area A of an equilateral triangle as a function of the length x of a side.

48. Let $f(x) = |x - 6| - |x + 6|$. Sketch the graph of f. Determine the intervals in which f is increasing and the intervals in which f is decreasing.

49. A rectangle with sides $2x$ and $2y$ units long is inscribed in a circle of radius 3 units, whose center is at the origin. Express y as a function of x. Express the area A of the rectangle as a function of x.

50. Let $g(x) = x^4 + 5x^2$. Which of the following holds for all real numbers?
(a) $g(-x) = g(x)$ (b) $g(-x) = -g(x)$

In problems 51–54, find the domain of each function. Determine whether the function is even or odd or neither.

51. $f(x) = \sqrt{x^3 - 1}$

52. $g(x) = \dfrac{x^2}{x^2 + 1}$

53. $h(x) = \dfrac{1}{1 + x^2 + x^4 + x^6}$

54. $G(x) = \sqrt[3]{x^2 - 3x + 2}$

In problems 55–66, find the domain of the function and sketch its graph. Is the function even or odd? Does the graph of the function have symmetry with respect to the y axis or the origin? From the graph of the function, find its range and indicate the intervals where the function is increasing and the intervals where it is decreasing.

55. $f(x) = -2x + 3$

56. $f(x) = 2x^2 + 5$

57. $g(x) = |2x| - 2x$

58. $f(x) = |3x| + 3x$

59. $h(x) = -1$

60. $h(x) = 2\sqrt[3]{x}$

61. $H(x) = \begin{cases} -x^2 + 2 & \text{if } x \geq 1 \\ 2x - 1 & \text{if } x < 1 \end{cases}$

62. $F(x) = \begin{cases} x^4 + 1 & \text{if } x \geq -1 \\ 5 + 3x & \text{if } x < -1 \end{cases}$

63. $F(x) = -2x^3$

64. $f(x) = 5$

65. $f(x) = \begin{cases} x^3 & \text{if } x > 0 \\ x^2 & \text{if } x \leq 0 \end{cases}$

66. $g(x) = \begin{cases} x & \text{if } x < 0 \\ 2x & \text{if } 0 \leq x \leq 1 \\ 3x^2 - 1 & \text{if } x > 1 \end{cases}$

67. Use the graph of $y = x^2$ and the techniques in Section 2.4 to sketch the graph of each of the functions.
(a) $h(x) = x^2 + 4$ (b) $w(x) = (x - \frac{1}{2})^2$ (c) $f(x) = -2(x - \frac{1}{3})^2$

68. Figure 2 shows the graph of a function f. Use this graph to sketch the graph of g.
(a) $g(x) = f(x) - 3$ (b) $g(x) = f(x - 3)$ (c) $g(x) = -3f(x)$

Figure 2

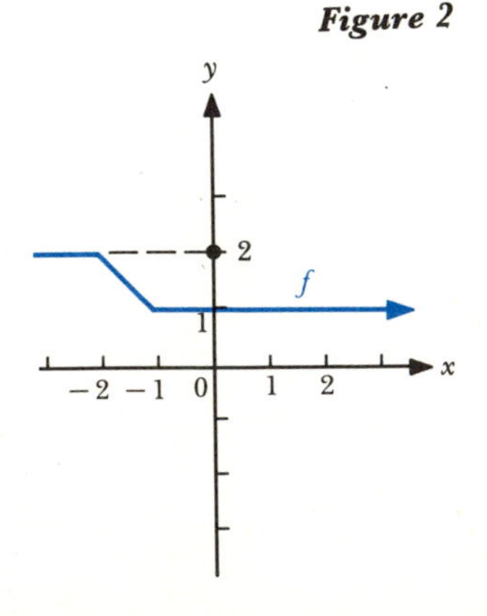

In problems 69–78, find an expression for (a) $(f + g)(x)$, (b) $(f - g)(x)$, (c) $(f \cdot g)(x)$, (d) $(f \circ g)(x)$, (e) $(g \circ f)(x)$, and (f) $(f/g)(x)$. Indicate the domain of f/g.

69. $f(x) = x + 2$ and $g(x) = x - 1$

70. $f(x) = x^3$ and $g(x) = -x$

71. $f(x) = x^2$ and $g(x) = 2x + 1$

72. $f(x) = 2x^2 - 1$ and $g(x) = 2x^2 + 1$

73. $f(x) = \dfrac{2}{1 - x}$ and $g(x) = \dfrac{1}{1 + 5x}$

74. $f(x) = 4$ and $g(x) = -7$

75. $f(x) = 7 - x^2$ and $g(x) = 1 + 5x$

76. $f(x) = \sqrt{x + 1}$ and $g(x) = x^4$

77. $f(x) = |x|$ and $g(x) = |x - 3|$

78. $f(x) = \sqrt[3]{x + 1}$ and $g(x) = \sqrt[4]{x - 1}$

[c] In problems 79 and 80, let $f(x) = 3 + x^5$ and $g(x) = 3 - \sqrt{x}$. Find each value and round off to four decimal places.

79. $(f \circ g)(3.7814)$

80. $[(g \circ f) \circ f](3.8195)$

81. Express the given function h as the composition of two other functions f and g so that $h = f \circ g$.

(a) $h(x) = (7x + 2)^5$ (b) $h(t) = \sqrt{t^2 + 17}$

82. Suppose that

$$f(x) = \begin{cases} x^2 & \text{if } x > 0 \\ x^3 & \text{if } x \le 0 \end{cases} \quad \text{and} \quad g(x) = \begin{cases} \sqrt{x} & \text{if } x > 0 \\ \sqrt{-x} & \text{if } x \le 0 \end{cases}$$

Find (a) $(f \circ g)(x)$ (b) $(g \circ f)(x)$.

83. Let $f(x) = 2x + 4$ and $g(x) = \frac{1}{2}x - 2$. Show that

$$(f \circ g)(x) = (g \circ f)(x) = x$$

84. Let $f(x) = ax + b$ and $g(x) = cx + d$. Find conditions on a, b, c, and d in order that $(f \circ g)(x) = (g \circ f)(x) = x$.

85. Let $f(x) = -4x + 5$. Find a function g such that $(f \circ g)(x) = (g \circ f)(x) = x$.

86. A city estimates that its population during the next 6 years will be approximated by the function $p(t) = 100t^2 + 20{,}000$, where t is the time in years and $t = 0$ corresponds to the present year. The city administration also estimates that its average daily pollution index I is a function of the population p of the city and is given by the function $I(p) = 20 + p/1000$. Express the pollution index I as a function of t explicitly; that is, find $I[p(t)]$.

In problems 87–92, determine which function has an inverse. If the function f has an inverse, find f^{-1}, verify that $f[f^{-1}(x)] = f^{-1}[f(x)] = x$, and sketch the graph of f and f^{-1} on the same coordinate system.

87. $f(x) = 7 - 13x$

88. $f(x) = \frac{1}{3}x + \frac{7}{5}$

89. $f(x) = 3\sqrt{x} - 1$

90. $f(x) = \dfrac{x^3}{2}$

91. $f(x) = -3|x|$

92. $f(x) = 2x^4$

CHAPTER 3

Polynomial and Rational Functions

In Chapter 2, we examined linear functions and graphed the squaring and cubing functions. These functions belong to a wider class of functions known as *polynomial functions,* which we study in this chapter. They arise directly in many applications and are used in numerical approximations. Our analysis will include methods for determining the real and complex "zeros" of polynomial functions. We begin by investigating the graphs of the quadratic and power functions, which are the basic building blocks for polynomial functions. Then, we use synthetic division as a shorthand technique for dividing polynomials and for evaluating polynomial functions. Finally, we turn our attention to the graphs of rational functions.

3.1 Quadratic Functions

A function of the form

$$f(x) = ax^2 + bx + c$$

where a, b, and c are constants and $a \neq 0$, is called a **quadratic function.** These functions arise in many applications. We know from physics, for instance, that if a ball is thrown straight up from a height of 65 feet with an initial speed of 64 feet per second, after t seconds the height h of the ball in feet is given by the quadratic function

$$h(t) = -16t^2 + 64t + 65$$

Other examples of quadratic functions are

$$h(x) = x^2, \qquad f(x) = 3x^2 - 1, \qquad \text{and} \qquad g(x) = \tfrac{1}{2}x^2 - x$$

The simplest quadratic function is the squaring function $f(x) = x^2$, whose graph was shown on page 104. The graph of functions of the form

$$f(x) = ax^2$$

can be obtained from the graph of $y = x^2$ by vertically stretching or flattening the graph of f by the factor a, if a is positive. Furthermore, the graph of $f(x) = ax^2$ for negative values of a is obtained by reflecting the graph of $y = |a|x^2$ across the x axis. Figure 1 shows the graph of $f(x) = ax^2$ for selected values of a.

Figure 1

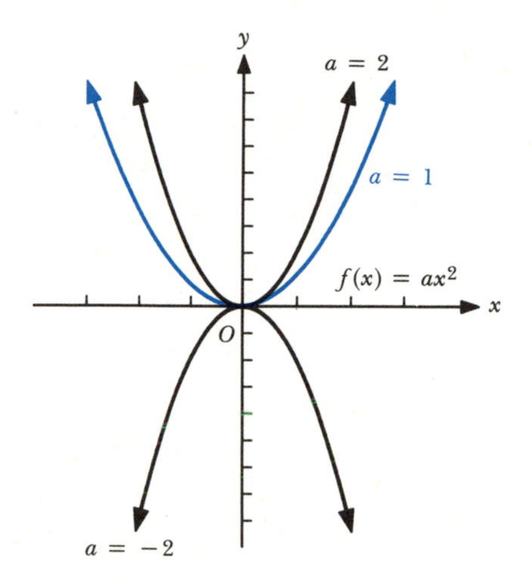

The graphs of equations of the form $y = ax^2$ (Figure 1) are examples of curves called **parabolas.** These parabolas *open upward* and have a low point, called the *minimum point,* at (0, 0) if $a > 0$, and they *open downward* and have a high point, called the *maximum point,* at (0, 0) if $a < 0$.

Figure 2 shows graphs of two quadratic functions of the form

$$y = ax^2 + bx + c$$

Figure 2a shows a parabola that opens upward and has a minimum point at $(\frac{5}{2}, -\frac{9}{4})$, whereas Figure 2b shows a parabola that opens downward and has a maximum point at (1, 1).

Figure 2

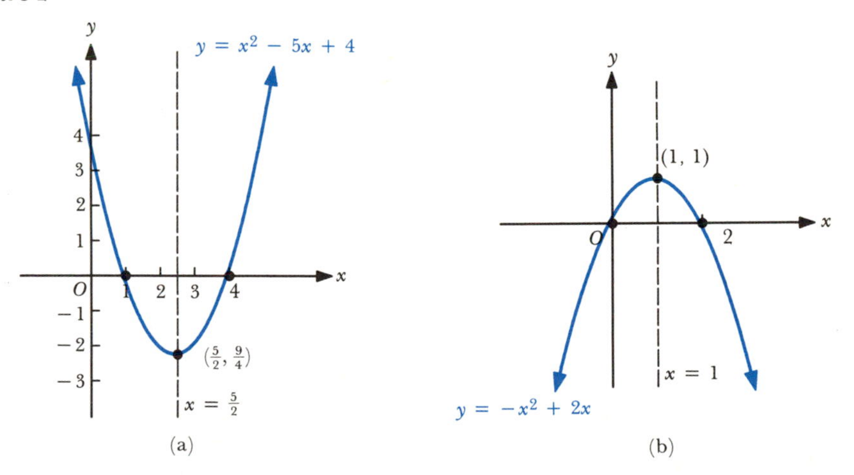

The maximum point or minimum point of a parabola is also referred to as the **extreme point** or the **vertex** of the parabola. The graph of a quadratic function has symmetry with respect to a vertical line through the vertex of the parabola.

Thus, the graph of the quadratic function in Figure 2a has a line of symmetry $x = \frac{5}{2}$, whereas the graph in Figure 2b has a line of symmetry $x = 1$.

In general, the graph of a quadratic function is a parabola that opens either upward or downward and that has a vertical line of symmetry passing through the extreme point. (A detailed discussion of parabolas is given in Chapter 8.)

EXAMPLE 1 Sketch the graph of each quadratic function with the given intercepts and extreme point.

(a) y intercept: 3; x intercepts: 1 and 3; minimum point or vertex at $(2, -1)$

(b) y intercept: $\frac{3}{2}$; x intercepts: $-\frac{1}{2}$ and 3; maximum point or vertex at $(\frac{5}{4}, \frac{49}{16})$

SOLUTION (a) The graph of the function is shown in Figure 3a. Notice that the graph opens upward and has a line of symmetry $x = 2$.

(b) Figure 3b shows the graph of the function. Here the parabola opens downward with a line of symmetry $x = \frac{5}{4}$.

Figure 3

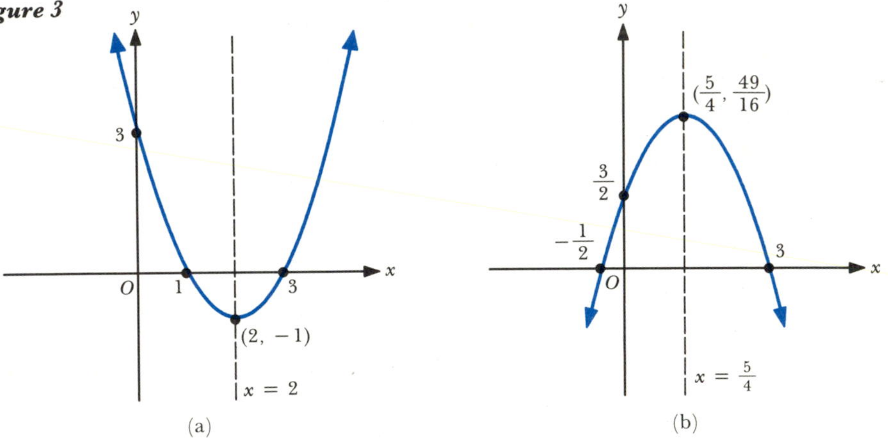

A technique for graphing a quadratic function of the form

$$y = ax^2 + bx + c$$

is outlined in the following step-by-step procedure.

Step 1. *Locate the y intercept* by substituting $x = 0$ into the equation to get $y = c$.

Step 2. *Locate the x intercepts* by setting $y = 0$ and then solving the resulting quadratic equation $ax^2 + bx + c = 0$.

Step 3. *Locate the extreme point* or *vertex*. We shall develop an efficient way to do this later.

Step 4. Connect the points located in the previous three steps with a smooth *parabolic-shaped curve* similar to the graph of $y = x^2$ or $y = -x^2$.

EXAMPLE 2 Sketch the graph of the function $y = x^2 - 3x + 2$ by using this four-step process. Use the graph to determine the range of the function.

SOLUTION

Step 1. We locate the y intercept by setting $x = 0$ to get $y = 2$. Thus, the point $(0, 2)$ is on the graph of the function.

Step 2. The x intercepts are found by substituting $y = 0$ into the equation to get $x^2 - 3x + 2 = (x - 2)(x - 1) = 0$, so that 2 and 1 are the x intercepts and the points $(2, 0)$ and $(1, 0)$ are on the graph.

Step 3. To find the extreme point (the process used here to determine the extreme point is generalized in Theorem 1), we first rewrite the equation as

$$y = (x^2 - 3x \qquad) + 2$$

Next we complete the square of the expression $x^2 - 3x$ to obtain

$$y = \left(x^2 - 3x + \frac{9}{4}\right) + 2 - \frac{9}{4}$$

$$y = \left(x - \frac{3}{2}\right)^2 - \frac{1}{4}$$

Since $(x - \frac{3}{2})^2 \geq 0$, it follows that $y \geq -\frac{1}{4}$. This means that $-\frac{1}{4}$ is a minimum value for y, which indicates that the parabola opens upward. In fact, this minimum y value occurs when $x - \frac{3}{2} = 0$, that is, when $x = \frac{3}{2}$, so that $(\frac{3}{2}, -\frac{1}{4})$ is the minimum point of the parabola.

Step 4. The graph is obtained by connecting the points given in the table with a smooth parabolic curve (Figure 4). From the graph, it is clear that the range of the function is the interval $[-\frac{1}{4}, \infty)$. Notice that $x = \frac{3}{2}$ is the line of symmetry.

Figure 4

x	y
0	2
2	0
1	0
$\frac{3}{2}$	$-\frac{1}{4}$

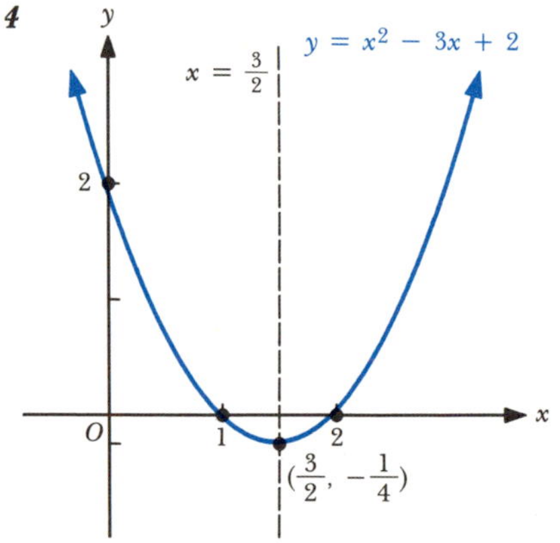

The following theorem generalizes the process of completing the square, used in the preceding example, to find the extreme point and provides us with a simple way of locating the extreme points of graphs of quadratic functions.

THEOREM 1 **Extreme Point of a Parabola**

The graph of the quadratic function defined by $y = f(x) = ax^2 + bx + c$ has an extreme point

$$\left(\frac{-b}{2a}, f\left(\frac{-b}{2a}\right)\right)$$

If $a > 0$, the extreme point is a **minimum** and the parabola opens upward. If $a < 0$, the extreme point is a **maximum** and the parabola opens downward.

PROOF First, we factor a from the two terms involving x in the equation $y = ax^2 + bx + c$ to get

$$y = a\left(x^2 + \frac{b}{a}x\right) + c$$

Next we complete the square of the expression in parentheses, $x^2 + (b/a)x$, by adding to it

$$\left[\frac{1}{2}\left(\frac{b}{a}\right)\right]^2 = \frac{b^2}{4a^2}$$

However, because there is an a outside the parentheses, this is the same as adding $b^2/(4a)$ to the right side of the equation. By also adding the expression $b^2/(4a)$ to the left side of the equation we obtain the equivalent equation

$$y + \frac{b^2}{4a} = a\left(x^2 + \frac{b}{a}x\right) + \frac{b^2}{4a} + c$$

$$= a\left(x^2 + \frac{b}{a}x + \frac{b^2}{4a^2}\right) + c$$

Thus

$$y = a\left(x^2 + \frac{b}{a}x + \frac{b^2}{4a^2}\right) + c - \frac{b^2}{4a}$$

$$= a\left(x + \frac{b}{2a}\right)^2 + c - \frac{b^2}{4a}$$

Figure 5

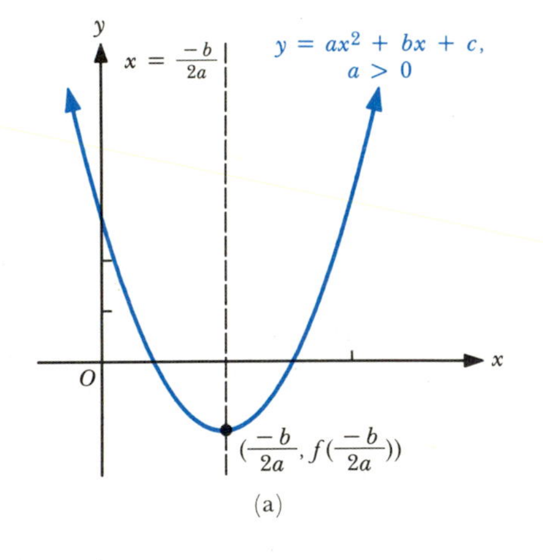

(a)

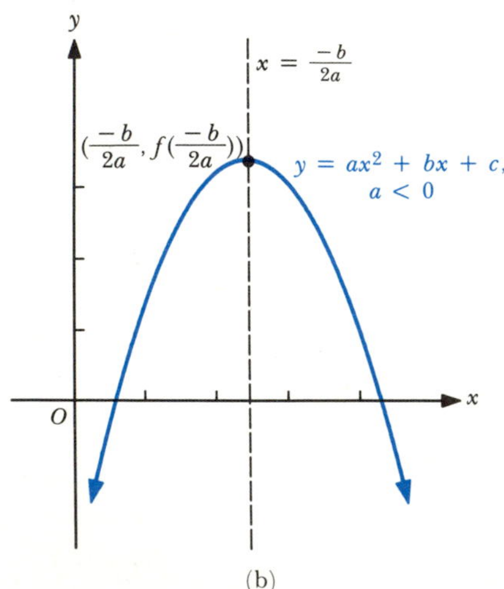

(b)

Notice that $(x + (b/(2a)))^2 \geq 0$ for all values of x.

If $a > 0$, then $a(x + (b/(2a)))^2 \geq 0$, so that

(i) $y = a\left(x + \dfrac{b}{2a}\right)^2 + c - \dfrac{b^2}{4a}$

$y \geq c - \dfrac{b^2}{4a}$ (Figure 5a)

If $a < 0$, then $a(x + (b/(2a)))^2 \leq 0$, so that

(ii) $y \leq c - \dfrac{b^2}{4a}$ (Figure 5b)

Now, if we substitute $x = -b/(2a)$ into the function $y = f(x) = ax^2 + bx + c$, we get

$$y = a\left(\frac{b^2}{4a^2}\right) + b\left(\frac{-b}{2a}\right) + c = \frac{b^2}{4a} - \frac{b^2}{2a} + c$$

$$= c - \frac{b^2}{4a}$$

which is the same as the expression on the right side of both inequalities (i) and (ii).

Consequently, if $a > 0$, $f(-b/(2a))$ is the minimum value of y, and the graph of f, a parabola, opens upward as in Figure 5a. If $a < 0$, $f(-b/(2a))$ is the maximum value of y, and the graph of f, a parabola, opens downward as in Figure 5b. In either case, the value of the function $f(x) = ax^2 + bx + c$ at $x = -b/(2a)$ gives the extreme value of $y = f(x)$. Thus the extreme point of the graph is given by

$$\left(\frac{-b}{2a}, f\left(\frac{-b}{2a}\right)\right)$$

Since the line of symmetry for the graph of a quadratic function is a vertical line passing through the vertex of the parabola, it follows that an equation of the line of symmetry is

$$x = \frac{-b}{2a}$$

In Examples 3–5, sketch the graph of the quadratic function by locating the x and y intercepts and the extreme point (vertex) of the parabola. Specify an equation of the line of symmetry and use the graph to determine the range of the function.

EXAMPLE 3 $f(x) = -x^2 + 2x$

SOLUTION Using the step-by-step procedure, we have

Step 1. Since $f(0) = 0$, the y intercept is 0.

Step 2. The x intercepts are found by letting $f(x) = 0$, so that

$$-x^2 + 2x = 0$$
$$x(-x + 2) = 0$$

Therefore, 0 and 2 are the x intercepts.

Step 3. We have $a = -1$ and $b = 2$, so that by Theorem 1 the extreme point occurs when

$$x = \frac{-b}{2a} = \frac{-2}{-2} = 1$$

Since $f(1) = -1 + 2 = 1$, then the vertex is $(1, 1)$. Because the coefficient of x^2 is negative, the extreme point is a maximum point. Also, $x = 1$ is the line of symmetry.

Step 4. Finally, the graph is obtained by sketching a parabola through the points determined and by recognizing the fact that $x = 1$ is the line of symmetry (Figure 6). From the graph we see that the range of f is the interval $(-\infty, 1]$.

x	y
0	0
2	0
1	1

Figure 6

EXAMPLE 4 $y = 3x^2 - 5x + 2$

SOLUTION Step 1. If $x = 0$, then $y = 2$, so that the y intercept is 2.

Step 2. In order to find the x intercepts, we let $y = 0$ to obtain the quadratic equation

$$3x^2 - 5x + 2 = (3x - 2)(x - 1) = 0$$

so that $x = \frac{2}{3}$ or $x = 1$. Thus, the x intercepts are $\frac{2}{3}$ and 1.

Step 3. By using Theorem 1, we find that the extreme point occurs when

$$x = \frac{-b}{2a} = \frac{-(-5)}{2(3)} = \frac{5}{6}$$

Substituting $x = \frac{5}{6}$ in the equation yields $y = -\frac{1}{12}$. Since $a > 0$, the extreme point is a minimum point. Also, $x = \frac{5}{6}$ is the line of symmetry.

Step 4. The graph of f is obtained by locating the intercepts and minimum point and then sketching a parabola that contains the points and is symmetric with respect to $x = \frac{5}{6}$ (Figure 7). The graph shows that the range of f is the interval $[-\frac{1}{12}, \infty)$.

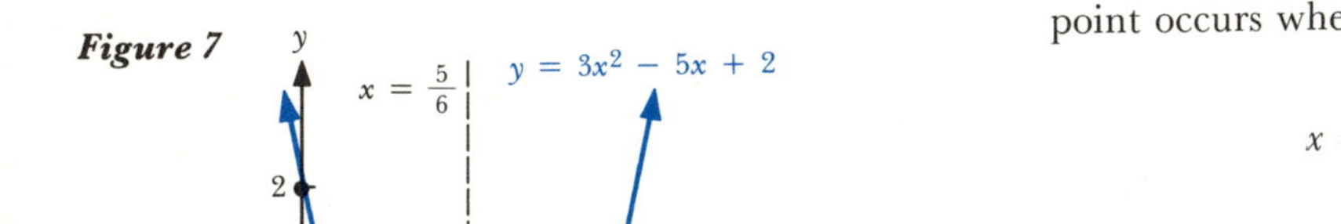

Figure 7

x	y
0	2
$\frac{2}{3}$	0
1	0
$\frac{5}{6}$	$-\frac{1}{12}$

EXAMPLE 5 $y = x^2 + 2x + 3$

SOLUTION Step 1. Setting $x = 0$ yields the y intercept 3.

Step 2. To find the x intercepts, we let $y = 0$ to get $x^2 + 2x + 3 = 0$. Recall (from problem 54 on page 48) that the discriminant of a quadratic equation is given by $b^2 - 4ac$, which in this case is $4 - 12 = -8$. Since the discriminant is negative, the equation has no real number roots and the graph has no x intercepts.

Step 3. The extreme point, which is a minimum because the coefficient of x^2 is positive, occurs when

$$x = \frac{-b}{2a} = \frac{-2}{2} = -1$$

so that $x = -1$ is the line of symmetry. Substituting $x = -1$ into the function, we get $y = 2$, so that the vertex is $(-1, 2)$.

Step 4. The graph and the line of symmetry are displayed in Figure 8. The range is $[2, \infty)$.

Figure 8

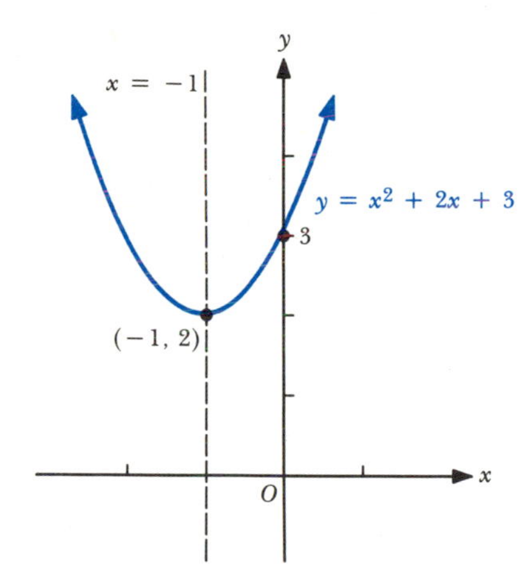

EXAMPLE 6 A rocket is fired vertically upward and is h feet above the ground t seconds after being fired, where $h = 96t - 16t^2$.

(a) Find the height of the rocket after 1 second.

(b) Find the maximum height to which the rocket ascends.

(c) Sketch the graph of the equation.

SOLUTION (a) At $t = 1$, $h = 96(1) - 16(1)^2 = 80$ feet.

(b) $h = -16t^2 + 96t$ is a quadratic function with $a = -16$ and $b = 96$, so the maximum value of h occurs when

$$t = \frac{-b}{2a} = \frac{-96}{-32} = 3$$

Figure 9

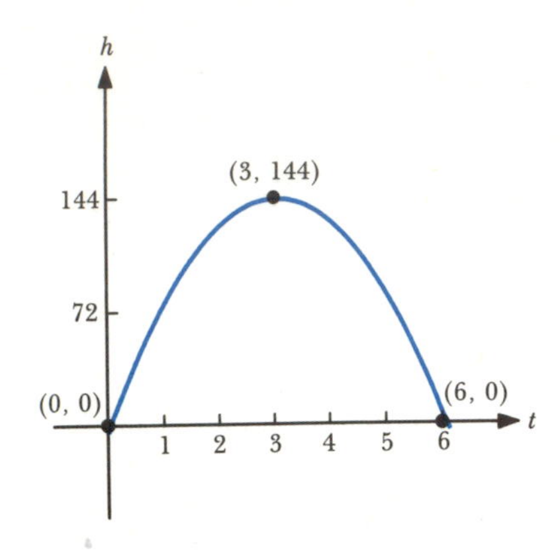

Substituting $t = 3$ into $h = 96t - 16t^2$, we obtain

$$h = 96(3) - 16(9) = 144$$

Hence, h has a maximum value of 144 feet when $t = 3$ seconds.

(c) The t intercepts are found by setting $h = 0$, to obtain

$$96t - 16t^2 = -16t(t - 6) = 0$$

so $t = 0$ or $t = 6$. Notice that the h intercept is 0. The graph is determined by constructing the parabola that contains the points $(0, 0)$ and $(6, 0)$, and the maximum point $(3, 144)$ (Figure 9). The graph does not extend beyond the interval $[0, 6]$ on the t axis because h is defined as "feet above the ground."

EXAMPLE 7 A woman with a budget of \$120 wants to plant a rectangular vegetable garden and put an ornamental fence around it. The fencing for three sides of the garden costs \$2 per foot, and for the fourth side costs \$3 per foot. What dimensions should she give the garden to maximize its area?

SOLUTION Let x feet be the length of the side of the rectangle for which the fencing costs \$3 per foot. Let y feet be the length of the sides perpendicular to this side. The total cost of the fencing in dollars will be $3x + 2x + 2y + 2y$ (Figure 10) or

$$5x + 4y = 120$$

Figure 10

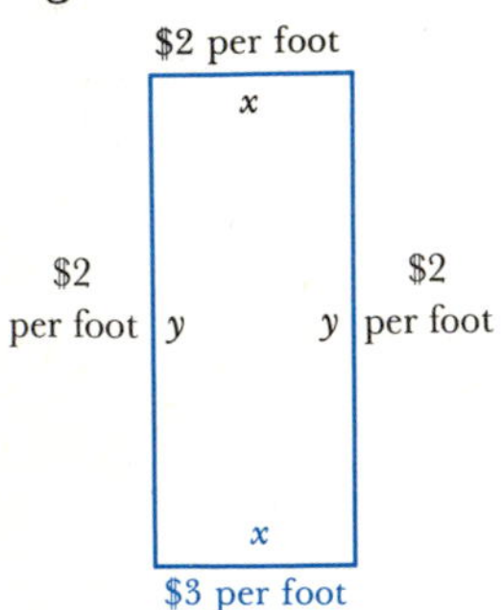

so that

$$y = \frac{120 - 5x}{4}$$

The area A of the garden is given by

$$A = xy = x\left(\frac{120 - 5x}{4}\right)$$

$$= -\frac{5}{4}x^2 + 30x$$

Thus, A is a quadratic function with $a = -\frac{5}{4}$ and $b = 30$. Hence, the maximum value of the area A is obtained when

$$x = -\frac{b}{2a} = \frac{-30}{-\frac{5}{2}} = 12 \qquad \text{and} \qquad y = \frac{120 - 5(12)}{4} = 15$$

Therefore, the dimensions of the garden that give a maximum area are 12 feet by 15 feet. The \$3 per foot fencing should be used for one of the 12-foot sides.

The graphs of quadratic functions can be used to solve quadratic inequalities, as illustrated in the following examples.

EXAMPLE 8 Use the graph of $y = x^2 - 2x - 3$ in Figure 11 to solve $x^2 - 2x - 3 \geq 0$.

SOLUTION If $y = x^2 - 2x - 3$, the inequality is satisfied by all values of x such that $y \geq 0$. The graph (Figure 11) indicates that $y \geq 0$ whenever $x \leq -1$ or $x \geq 3$. Thus the solution set of the inequality is all values of x in the intervals $(-\infty, -1]$ or $[3, \infty)$.

Figure 11

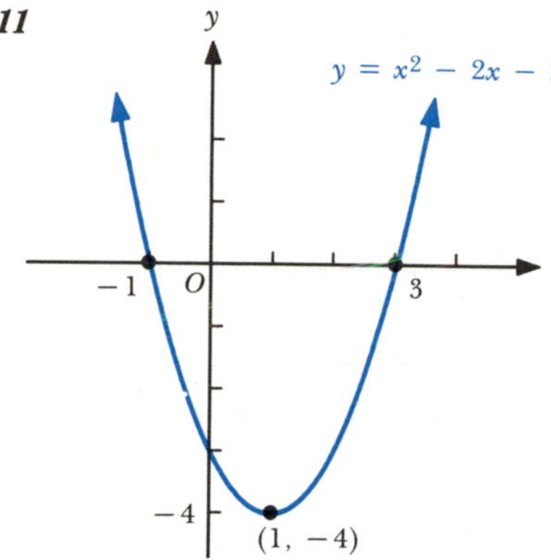

Figure 12

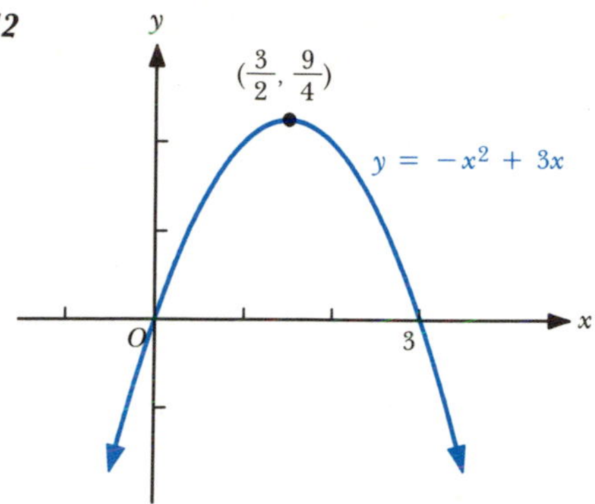

EXAMPLE 9 Use the graph of $y = -x^2 + 3x$ shown in Figure 12 to solve $-x^2 + 3x > 0$.

SOLUTION The graph of the function $y = -x^2 + 3x$ shows that $y = -x^2 + 3x > 0$ is satisfied if $0 < x < 3$. Thus the solution set is the interval $(0, 3)$.

EXAMPLE 10 Use the graph of $y = x^2 + 1$ in Figure 13 to solve $x^2 < -1$.

SOLUTION The inequality $x^2 < -1$ is equivalent to $x^2 + 1 < 0$. The graph of $y = x^2 + 1$ (Figure 13) shows that $x^2 + 1 > 0$ for all real numbers. Thus, $x^2 + 1 < 0$ has no real number solutions, and so $x^2 < -1$ has no real number solutions.

Figure 13

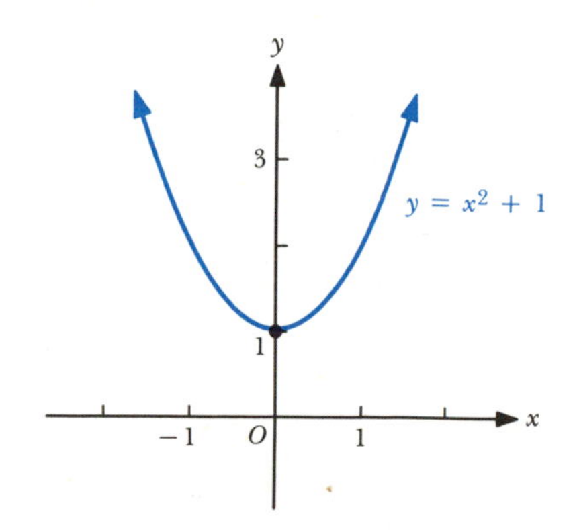

PROBLEM SET 3.1

1. Graph on the same coordinate system the quadratic function $y = ax^2$, for $a = -7, -4, -1, -\frac{1}{3}, \frac{1}{3}, 1, 4,$ and 7.
(a) For what values of a does the graph open upward?
(b) For what values of a does the graph open downward?
(c) For what positive values of a does the graph lie below the graph of $y = x^2$?
(d) For what positive values of a does the graph lie above the graph of $y = x^2$?

2. Use the graph of $y = x^2$ to obtain the graph of each function by using an appropriate horizontal shift.
(a) $f(x) = (x - 1)^2$
(b) $g(x) = (x + 2)^2$

3. Sketch the graph of the following four functions on the same coordinate system by using the graph of $y = x^2$ and suitable vertical shifts.

$$f(x) = x^2 - 2 \qquad h(x) = x^2 - 1 \qquad g(x) = x^2 + 1 \qquad F(x) = x^2 + 2$$

4. Use the graph of $y = x^2$ to obtain the graph of each function.
(a) $f(x) = \frac{1}{2}(x - 3)^2 + 2$
(b) $G(x) = -2(x + 4)^2 - 3$

In problems 5–9, sketch the graph of the quadratic function with the given intercepts and extreme point.

5. y intercept: -8; x intercepts: $-4, 2$; minimum point: $(-1, -9)$
6. y intercept: 6; x intercepts: $-3, 3$; maximum point: $(0, 6)$
7. y intercept: -4; x intercepts: $1, 2$; maximum point: $(\frac{3}{2}, \frac{1}{2})$
8. y intercept: 5; x intercept: 1; minimum point: $(1, 0)$
9. y intercept: 5; x intercepts: none; minimum point: $(1, 3)$
10. Explain why quadratic functions do *not* have inverse functions.

In problems 11–16, use the method of completing the square illustrated in Example 2 on page 135 to find the extreme point. Indicate whether the extreme point is a maximum or minimum point.

11. $y = x^2 - 4x - 5$
12. $y = x^2 + x + 1$
13. $y = -x^2 + 6x$
C **14.** $y = -0.53x^2 + 1.41x$
15. $y = 5x^2 - x - 3$
16. $y = \frac{1}{3}x^2 + \frac{1}{5}x - \frac{3}{7}$

In problems 17–29, sketch the graph of the quadratic function by locating the x and y intercepts and the extreme point of the parabola. Use the graph to determine the range. Give an equation of the line of symmetry.

17. $f(x) = x^2 - 9x$
18. $g(x) = x^2 + 2x$
19. $g(x) = x^2 + 5x - 14$
20. $f(x) = x^2 + 3x - 9$
21. $f(x) = -x^2 - 4x - 4$
22. $h(x) = -7x^2 - 28x + 3$
23. $h(x) = -4x^2 + 4x + 3$
24. $g(x) = -\frac{1}{3}x^2 + 3$
25. $g(x) = 2x^2 + 4x - 3$
26. $f(x) = 3x^2 - x - 3$
27. $f(x) = -2x^2 + 3x - 1$
28. $H(x) = 2x^2 - 5x - 3$
C **29.** $G(x) = 0.4x^2 - 2.35x - 7.2$

30. Given the quadratic function $y = ax^2 + bx + c$, assume that D is the value of the *discriminant* $b^2 - 4ac$ (see page 46, problem 48).

(a) If $D < 0$, how many real number roots does the quadratic equation have? How many x intercepts does the function have? Sketch a parabola that displays this situation.
(b) Answer the questions in part (a) if $D = 0$.
(c) Answer the questions in part (a) if $D > 0$.

In problems 31–40, graph the quadratic function determined by the given equation, and then use the graph to solve the corresponding inequality. Display the solution set on the number line.

31. $y = x^2 - 9x + 14$; $x^2 - 9x + 14 > 0$

32. $y = -x^2 + x + 20$; $x^2 - x - 20 \leq 0$

33. $y = -2x^2 + 5x - 3$; $-2x^2 + 5x - 3 \leq 0$

34. $y = 2x^2 - x - 1$; $2x^2 - x - 1 > 0$

35. $y = (x - 1)^2$; $x^2 - 2x < -1$

36. $y = x^2 + 6x + 8$; $-x^2 - 6x - 8 \leq 0$

37. $y = x^2 + 2$; $x^2 + 2 > 0$

38. $y = 2x^2 - 7x + 6$; $2x^2 - 7x + 6 > 0$

39. $y = 3x^2 - 5x + 1$; $3x^2 - 5x + 1 < 0$

40. $y = \frac{1}{3}x^2 - \frac{1}{2}x - \frac{3}{2}$; $\frac{1}{3}x^2 - \frac{1}{2}x - \frac{3}{2} > 0$

41. A rocket is fired vertically upward from a balloon in such a way that it is h feet above the ground t seconds after the firing, where $h = -16t^2 + 96t + 256$.
(a) Find the height h of the rocket when $t = 0$.
(b) Find the maximum height reached by the rocket.
(c) Determine the value of t when the rocket strikes the ground.
(d) Sketch the graph of the equation.

c **42.** A rocket is fired vertically upward and it is h meters above the ground t seconds after being fired, where $h = 190t - 4.9t^2$.
(a) Find how long it takes for the rocket to reach its maximum height.
(b) Determine the value of t at which the rocket strikes the ground.
(c) Sketch the graph of the equation.

43. A rectangular field adjacent to a river must be fenced on three sides but not on the river bank. What is the largest area that can be enclosed if 50 yards of fencing are used?

44. A travel agency advertises all-expenses-paid trips to the World Series for special groups. Transportation is by charter bus, which seats 48 passengers, and the charge per person is \$80 plus \$2 for each empty seat. (If there are four empty seats, each person has to pay \$88; if there are six empty seats, each person has to pay \$92 and so on.) If there are x empty seats, how many passengers are there on the bus? How much does each passenger have to pay? Determine and then graph the function that relates the travel agency's total receipts to the number of empty seats.

3.2 Graphs of Polynomial Functions

We have discussed the constant, linear, and quadratic functions given, respectively, by equations of the forms

$$f(x) = b, \qquad f(x) = mx + b, \qquad \text{and} \qquad f(x) = ax^2 + bx + c$$

These functions are special cases of *polynomial functions*. In general, a function f of the form

$$f(x) = a_n x^n + a_{n-1} x^{n-1} + a_{n-2} x^{n-2} + \cdots + a_1 x + a_0$$

where n is a nonnegative integer and $a_n, a_{n-1}, a_{n-2}, \ldots, a_1$, and a_0 are real numbers with $a_n \neq 0$, is called a **polynomial function of degree n in x.** The numbers $a_n, a_{n-1}, a_{n-2}, \ldots, a_1$, and a_0 are called the **coefficients** of the polynomial function. The function $f(x) = 0$ is called the **zero polynomial function,** and no degree is assigned to it.

For example, $f(x) = 2x^3 - 5x^2 + 3$ is a polynomial function of degree 3, since it can be written in the form

$$f(x) = 2x^3 + (-5)x^2 + 0x + 3$$

The coefficients of this function are

$$2, \quad -5, \quad 0, \quad \text{and} \quad 3$$

where $a_3 = 2$, $a_2 = -5$, $a_1 = 0$, and $a_0 = 3$. On the other hand,

$$h(x) = \frac{2x^3 + 1}{x^2} = 2x + x^{-2}$$

is *not* a polynomial function because of the exponent -2.

Using this definition of a polynomial function, it follows that a constant function other than $f(x) = 0$ has degree 0, a linear function has degree 0 or 1, and a quadratic function has degree 2.

In this section, we discuss polynomial functions of the form $f(x) = ax^n$, where $a \neq 0$ and $n \geq 3$, and polynomial functions written in factored form with first-degree factors.

A polynomial function f of the form

$$f(x) = ax^n$$

where $a \neq 0$, is called a **power function.**

For example,

$$f(x) = x^5, \qquad g(x) = 3x^4, \qquad \text{and} \qquad h(x) = -\tfrac{1}{2}x^6$$

are power functions of degrees 5, 4, and 6, respectively.

Figure 1a displays the graphs of

$$y = x^2 \qquad \text{and} \qquad f(x) = x^4$$

and Figure 1b displays the graphs of

$$g(x) = x^6 \qquad \text{and} \qquad h(x) = x^8$$

These functions are of the form $f(x) = x^n$, where n is an *even* integer.

Figure 1

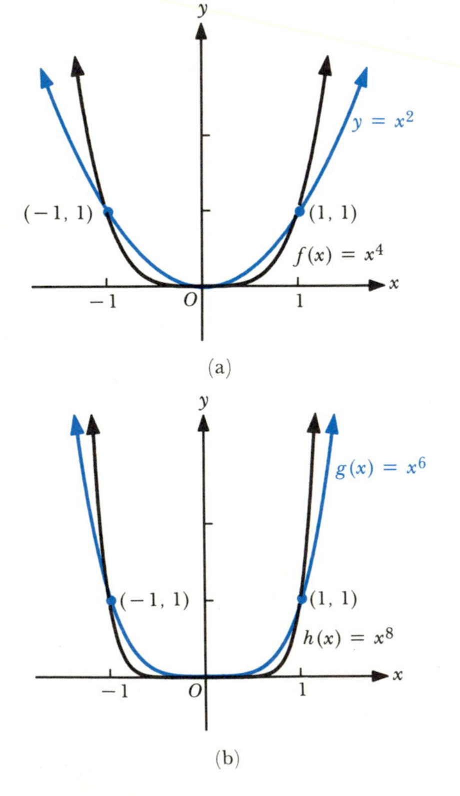

The characteristics of the graphs of these functions can be summarized as follows:

1. All the graphs contain the points $(-1, 1)$, $(0, 0)$, and $(1, 1)$.

2. Since n is even,

$$f(-x) = (-x)^n$$
$$= x^n = f(x)$$

so that f is even and the graphs are symmetric with respect to the y axis.

3. Since $x^n \geq 0$, the graphs lie in quadrants I and II.

4. For $n > 2$, the portions of the graphs on *intervals* $(-1, 0)$ and $(0, 1)$ fall below the graph of $y = x^2$. For increasing values of n, the graphs on these intervals get flatter and flatter because increasing even powers of numbers less than 1 in absolute value yield smaller and smaller results. (For example, $(\frac{1}{2})^2 = \frac{1}{4} > (\frac{1}{2})^4 = \frac{1}{16} > (\frac{1}{2})^6 = \frac{1}{64}$, and so forth.)

5. For increasing values of n, the portions of the graphs on intervals $(-\infty, -1)$ and $(1, \infty)$ tend to rise more rapidly because increasing even powers of numbers greater than 1 in absolute value yield larger and larger results. (For example, $2^2 = 4 < 2^4 = 16 < 2^6 = 64 < 2^8 = 256$, and so forth.)

We can use the graphs of power functions, together with the techniques of shifting, stretching, flattening, and reflecting, to obtain graphs of other functions.

In Examples 1 and 2, use the graph of f to sketch the graph of g.

EXAMPLE 1 $g(x) = 3x^4$ and $f(x) = x^4$

SOLUTION We begin by graphing $f(x) = x^4$. We then graph $g(x) = 3x^4$ by multiplying the ordinates of f by 3. Geometrically, we obtain the graph of g by vertically stretching the graph of f by a multiple of 3 (Figure 2).

Figure 2

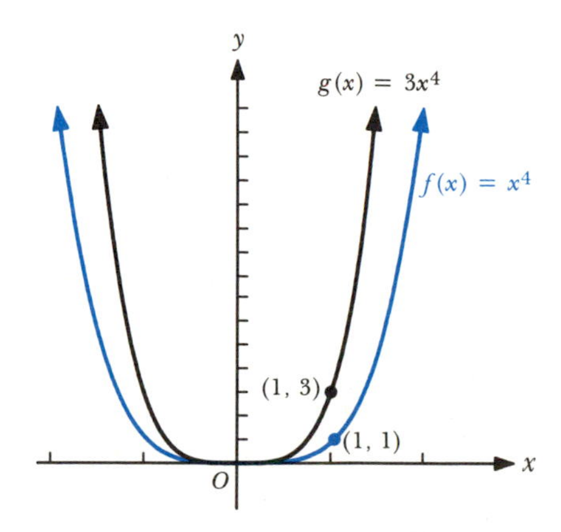

EXAMPLE 2 $g(x) = -\frac{1}{2}x^4 + 1$ and $f(x) = x^4$

SOLUTION After graphing $f(x) = x^4$ (Figure 3), we can obtain the graph of g from f.

First, we flatten the graph of f by the multiple $\frac{1}{2}$ to get the graph of $y = \frac{1}{2}x^4$.

Second, we reflect the resulting graph across the x axis (because of the negative multiplier) to obtain the graph of $F(x) = -\frac{1}{2}x^4$.

Third, we shift the graph resulting from the second step one unit vertically upward to get the graph of g (Figure 3).

Figure 3

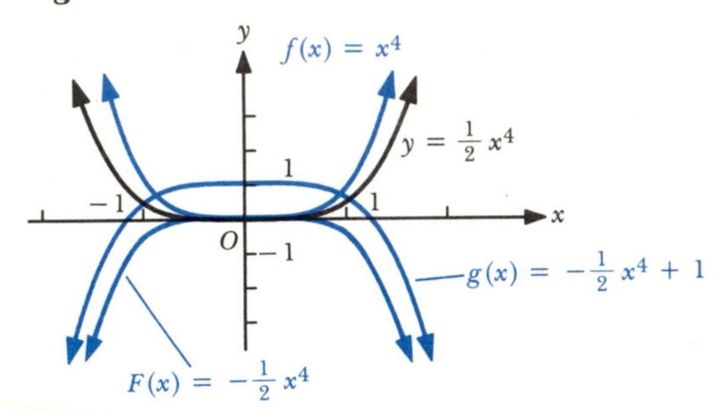

Figure 4

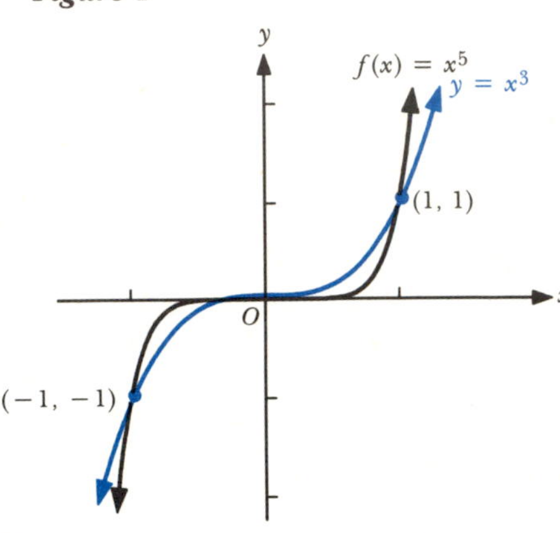

Figure 4 displays the graphs of

$$y = x^3 \quad \text{and} \quad f(x) = x^5$$

which are of the form $f(x) = x^n$, where n is an *odd* integer.

The characteristics of these graphs are summarized as follows:

1. The graphs contain the points $(-1, -1)$, $(0, 0)$, and $(1, 1)$.

2. Since n is odd, $f(-x) = (-x)^n = -x^n = -f(x)$, so that the graphs are symmetric with respect to the origin.

3. Since $x^n > 0$ if $x > 0$, and $x^n < 0$ if $x < 0$, it follows that the graphs lie in quadrants I and III.

4. For $n > 3$, the portions of the graphs on *intervals* $(-1, 0)$ and $(0, 1)$ fall *between* the graph of $y = x^3$ and the x axis. For increasing values of n the graphs get flatter and flatter on these intervals because increasing odd powers of numbers less than 1 in absolute value yield smaller and smaller absolute value results.

5. For increasing values of n, the portions of the graphs on the interval $(-\infty, -1)$ tend to fall more rapidly, and the portions on the interval $(1, \infty)$ tend to rise more rapidly, because increasing odd powers of numbers greater than 1 in absolute value yield larger and larger absolute value results.

EXAMPLE 3 Use the graph of $f(x) = x^5$ to sketch the graph of $g(x) = 2(x + 3)^5 - 1$.

SOLUTION We can obtain the graph of g from the graph of f.

First, we shift the graph of f (Figure 5) horizontally three units to the left because of the factor $(x + 3)$.

Second, we vertically stretch the resulting graph by the multiple 2.

Third, we shift the resulting graph one unit vertically downward to obtain the graph of g (Figure 5).

Figure 5

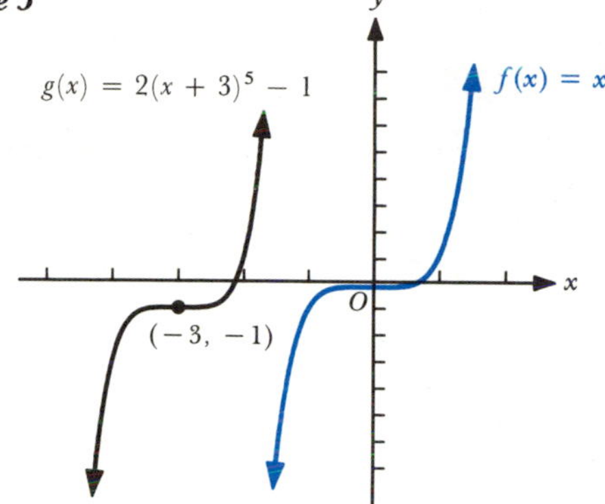

When graphing functions, it is helpful to locate the points where the graph intersects the x axis. These points, of course, are the x intercepts of the graph. We already know that the x intercepts of a given function $y = f(x)$ are found by first setting $y = 0$ and then solving the resulting equation $f(x) = 0$. The values of x that satisfy the latter equation are also called the *zeros* of the function f. Thus, the **zeros of the function** f are the *solutions* or *roots* of the equation $f(x) = 0$.

The zeros of linear and quadratic functions are relatively easy to determine because the techniques for solving linear and quadratic equations are well established. For example, the only zero of $f(x) = x - 2$ is 2. The zeros of $g(x) = x^2 - 4x - 5$ are found by solving $x^2 - 4x - 5 = 0$ or $(x - 5)(x + 1) = 0$ so that the roots of this equation are 5 and -1. Thus, 5 and -1 are the zeros of g.

The polynomial function $g(x) = x^2 - 4x - 5$ can be written in *linear* (first-degree) factored form as

$$g(x) = (x - 5)(x + 1)$$

Sometimes, a polynomial can be written in a linear factored form in which a linear factor may appear more than once. For example, we can verify by direct multiplication that

$$f(x) = 3x^3 + 12x^2 + 15x + 6 = 3(x + 1)(x + 1)(x + 2)$$

or more concisely,

$$f(x) = 3(x + 1)^2(x + 2)$$

In this case, we say that -1 is a *double zero* or a *zero of multiplicity 2* of the function f, and -2 is a *zero of multiplicity 1* of f.

In general, if

$$f(x) = (x - c)^s Q(x) \qquad \text{and} \qquad Q(c) \neq 0, \qquad \text{where } s \text{ is a positive integer,}$$

we say that c is a **zero of multiplicity *s*** of the function f.

For example, $f(x) = (x - 1)^2 (x - 2)$ has 1 as a zero of multiplicity 2, and 2 as a zero of multiplicity 1. The zeros of $g(x) = x(2x - 1)^2 (x + 1)^3 (x - 4)$ are 0, $\frac{1}{2}$, -1, and 4 of multiplicities 1, 2, 3, and 1, respectively. As you can see, the zeros of polynomial functions that have been written in linear factored form are easy to determine. This is useful in locating the x intercepts of the graphs of such functions.

Before explaining a technique for graphing polynomial functions that have been written in linear factored form, we must state two general characteristics of these graphs. First, the graph of a polynomial function is **continuous,** that is, the graph is a smooth curve without breaks or jumps. (It can be sketched without lifting the pencil.) Second, the graph of a polynomial function of degree n, where $n > 1$, will have no more than $n - 1$ "peaks" (high points) and "valleys" (low points).

If $y = f(x)$ is a polynomial function written in factored form, we can first locate the x intercepts of the graph of f. Then we can determine whether the graph lies above or below the x axis by selecting values of x between consecutive x intercepts and plotting the points $(x, f(x))$. Because of the two characteristics of polynomial functions, we can come up with a rough sketch of the graph by connecting the points with a smooth continuous curve that intersects the x axis at each intercept.

In Examples 4 and 5, sketch the graph of each function by locating a few points. Identify the x and y intercepts, and then use the graph to find the solution set of the corresponding inequality.

EXAMPLE 4 $f(x) = (x + 1)(x - 2)(x - 3); \qquad (x + 1)(x - 2)(x - 3) \leq 0$

SOLUTION The graph of f intersects the x axis only at -1, 2, and 3. Some additional points on the graph of f are given in the table accompanying Figure 6.

The portion of the graph below or on the x axis indicates the values of x that satisfy the inequality. Thus the solution set of the inequality consists of all values of x such that $x \leq -1$ or $2 \leq x \leq 3$, that is, all values of x in intervals $(-\infty, -1]$ and $[2, 3]$.

Figure 6

x	$f(x)$
-1	0
$-\frac{1}{2}$	$\frac{35}{8}$
0	6
1	4
2	0
$\frac{5}{2}$	$-\frac{7}{8}$
3	0
$\frac{7}{2}$	$\frac{27}{8}$

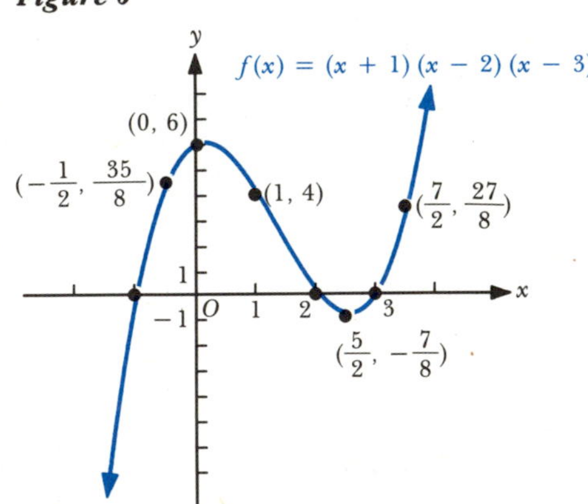

EXAMPLE 5 $g(x) = (2x - 5)(x - 1)^2; \qquad (2x - 5)(x - 1)^2 > 0$

SOLUTION Clearly, 1 and $\frac{5}{2}$ are the zeros of g so that 1 and $\frac{5}{2}$ are the x intercepts of the graph. The table in Figure 7 gives some additional points for values of x between the

Figure 7

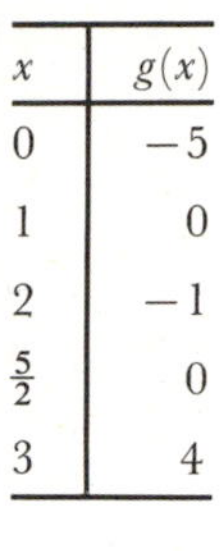

x	$g(x)$
0	-5
1	0
2	-1
$\frac{5}{2}$	0
3	4

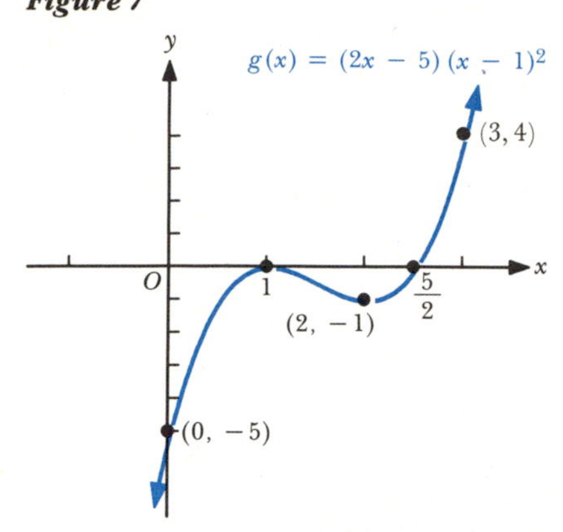

x intercepts. After connecting these points with a smooth curve, we obtain a sufficiently accurate sketch of the graph to solve the inequality. Note that since $(x-1)^2 \geq 0$ for all values of x, it follows that $g(x) \leq 0$ only when $2x - 5 \leq 0$, that is, when $x \leq \frac{5}{2}$. Thus, the graph is below or on the x axis if $x \leq \frac{5}{2}$, and the graph is above the x axis if $x > \frac{5}{2}$. Consequently, the values of x that satisfy $g(x) > 0$ are all numbers in the interval $(\frac{5}{2}, \infty)$.

PROBLEM SET 3.2

In problems 1–10, identify whether or not the function is a polynomial function. Specify the degree and identify the nonzero coefficients of each polynomial function.

1. $f(x) = 4x^3 - 2x^2 + x - 5$

2. $g(x) = 1 - x^3$

3. $h(x) = \frac{1}{3}$

4. $f(x) = (2x^3)^{-1} + x^2 - 5x + 7$

5. $g(x) = 5x^4 + \dfrac{x^3}{2} - 2x^2$

6. $h(x) = \dfrac{x}{5}$

7. $f(x) = (3x^2 + 2x - 1)^{-1}$

8. $g(x) = 4^{-3}x^3 + \frac{1}{2}x^2 - \sqrt{2}x + 1$

9. $g(x) = x^3 - 3x^5 + 8 - x$

10. $f(x) = \dfrac{x^2 - 4}{x - 2}$

In problems 11–24, use the techniques of shifting, stretching, flattening, and reflecting to graph the function using an appropriate power function graph. Locate the x and y intercepts.

11. $g(x) = \frac{1}{3}x^4$

12. $h(x) = -5x^4$

13. $f(x) = -2x^5$

14. $g(x) = \frac{1}{2}x^5$

15. $h(x) = x^5 + 3$

16. $f(x) = x^6 - 7$

17. $f(x) = 2(x-1)^4 + 3$

18. $h(x) = \frac{1}{3}(x-2)^5 + 2$

19. $g(x) = -\frac{1}{5}(x+1)^4$

20. $f(x) = 4(x-4)^5$

21. $h(x) = -4(x-2)^3 + 1$

22. $g(x) = -\frac{1}{2}(x-2)^4 - 2$

23. $f(x) = \frac{1}{4}(x - \frac{1}{3})^5 - \frac{1}{5}$

24. $h(x) = \pi(x + \sqrt{2})^6 + \sqrt{3}$

In problems 25–34, find all real number zeros of the polynomial function and indicate the multiplicity of each zero.

25. $f(x) = -3x + 2$

26. $g(x) = 3x^2 + x - 2$

27. $h(x) = (x-1)(x^2 - 3x + 2)^2$

28. $f(x) = (x-1)^3(x-2)(x-5)^2$

29. $h(x) = x^4 - 16$

30. $g(x) = (x-1)^2 - 9$

31. $f(x) = x^3 - 9x$

32. $f(x) = (x^4 - x^3)(x - 1)$

33. $g(x) = (x+1)(x-5)^2(x^2 - 25)$

34. $h(x) = (2x-5)(3x+1)^5(x^2 - 5x)$

In problems 35–43, sketch the graph of the function by locating the x and y intercepts, and plotting a few points. Use the graph to solve the accompanying inequality.

35. $f(x) = x(x-1)(x+2)$; $x(x-1)(x+2) > 0$
36. $g(x) = (2x+1)^4$; $(2x+1)^4 > 0$
37. $h(x) = (x+1)(x+2)(x+3)$; $(x+1)(x+2)(x+3) \leq 0$
38. $g(x) = (x-1)^3(x+2)^3$; $(x-1)^3(x+2)^3 \geq 0$
39. $h(x) = (x-2)^2(x-3)$; $(x-2)^2(x-3) < 0$
40. $f(x) = x^3(x+1)^2$; $x^3(x+1)^2 \geq 0$
41. $h(x) = (x-4)^3(x-5)$; $(x-4)^3(x-5) \geq 0$
42. $g(x) = x(x-1)^2(x-3)$; $x(x-1)^2(x-3) < 0$
43. $f(x) = (x+1)^2(x-2)^2$; $(x+1)^2(x-2)^2 \geq 0$

3.3 Division of Polynomials

In Section 3.2, we graphed some polynomial functions that had been written in a linear factored form. To aid us in writing a polynomial in a linear factored form, we review the division process for polynomials and describe the procedure known as *synthetic division.*

In long division, we normally organize the work needed to divide 6741 by 23 as follows:

$$\begin{array}{rr|rl} & & 293 & \longleftarrow \text{quotient} \\ \text{divisor} \longrightarrow & 23 & 6741 & \longleftarrow \text{dividend} \\ & & 46 & \\ & & 214 & \\ & & 207 & \\ & & 71 & \\ & & 69 & \\ & & 2 & \longleftarrow \text{remainder} \end{array}$$

The result of this calculation can be expressed as

$$\frac{6741}{23} = 293 + \frac{2}{23}$$

or

$$6741 = (23)(293) + 2$$

In general,

$$\boxed{\text{dividend} = (\text{divisor})(\text{quotient}) + \text{remainder}}$$

To divide one polynomial by another polynomial, we use a method similar to the long division method in arithmetic. For example, to divide the polynomial expression $2x^4 - 3x^3 + 5x^2 + 2x + 7$ by $x^2 - x + 1$, we note that both the divisor $x^2 - x + 1$ and the dividend $2x^4 - 3x^3 + 5x^2 + 2x + 7$ are arranged in descending powers of x. (If they weren't we would begin by rewriting them so that they were.) We proceed as follows:

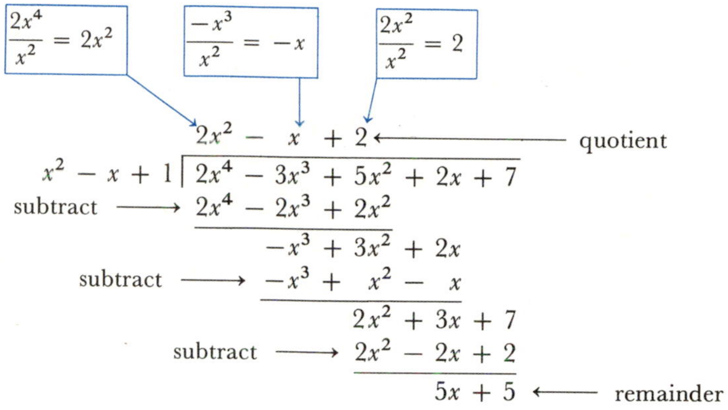

$\frac{2x^4}{x^2} = 2x^2$ $\frac{-x^3}{x^2} = -x$ $\frac{2x^2}{x^2} = 2$

$$
\begin{array}{rrrrrrl}
 & 2x^2 & -\ x & +\ 2 & & & \leftarrow \text{quotient} \\
x^2 - x + 1 \,\big) & 2x^4 - 3x^3 & +\ 5x^2 & +\ 2x & +\ 7 & & \\
\text{subtract} \longrightarrow & 2x^4 - 2x^3 & +\ 2x^2 & & & & \\
\hline
 & -x^3 & +\ 3x^2 & +\ 2x & & & \\
\text{subtract} \longrightarrow & -x^3 & +\ x^2 & -\ x & & & \\
\hline
 & & 2x^2 & +\ 3x & +\ 7 & & \\
\text{subtract} \longrightarrow & & 2x^2 & -\ 2x & +\ 2 & & \\
\hline
 & & & 5x & +\ 5 & & \leftarrow \text{remainder}
\end{array}
$$

The division results in the quotient $2x^2 - x + 2$ and the remainder $5x + 5$. Note that the division process ends either when the remainder is a constant or when the degree of the remainder is smaller than the degree of the divisor. The result can be written as

$$\frac{2x^4 - 3x^3 + 5x^2 + 2x + 7}{x^2 - x + 1} = (2x^2 - x + 2) + \frac{5x + 5}{x^2 - x + 1}$$

or as

$$2x^4 - 3x^3 + 5x^2 + 2x + 7 = (x^2 - x + 1)(2x^2 - x + 2) + (5x + 5)$$

The preceding example illustrates the next property.

Division Algorithm

If $f(x)$ and $D(x)$ are nonconstant polynomials such that the degree of $f(x)$ is greater than or equal to the degree of $D(x)$, then there exist unique polynomials $Q(x)$ and $R(x)$ such that

$$f(x) = D(x) \cdot Q(x) + R(x)$$

where the degree of $R(x)$ is less than the degree of $D(x)$ (note that $R(x)$ may be 0). The expression $D(x)$ is called the **divisor,** $f(x)$ is the **dividend,** $Q(x)$ is the **quotient,** and $R(x)$ is the **remainder.**

EXAMPLE 1 Suppose that $f(x) = 3x^3 - x^2 + 2x - 1$ and $D(x) = x - 3$. Use long division to find $Q(x)$ and $R(x)$ such that $3x^3 - x^2 + 2x - 1 = (x - 3)Q(x) + R(x)$.

SOLUTION We arrange this division in the following manner:

$$
\begin{array}{r|rrrr}
 & 3x^2 & +\,8x & +\,26 & \\
\cline{2-5}
x - 3 & 3x^3 & -\;\;x^2 & +\;\;2x & -\;\;1 \\
 & 3x^3 & -\;9x^2 & & \\
\cline{2-3}
 & & 8x^2 & +\;\;2x & \\
 & & 8x^2 & -\;24x & \\
\cline{3-4}
 & & & 26x & -\;\;1 \\
 & & & 26x & -\;78 \\
\cline{4-5}
 & & & & 77
\end{array}
$$

Hence

$$Q(x) = 3x^2 + 8x + 26 \qquad \text{and} \qquad R(x) = 77$$

and

$$3x^3 - x^2 + 2x - 1 = (x - 3)(3x^2 + 8x + 26) + 77$$

In this example, the divisor is a first-degree polynomial of the form of $x - c$, where c is a constant. Whenever this occurs, the remainder will be a constant. A shorthand method known as *synthetic division* can be used to perform the division of a polynomial by a first-degree binomial.

Synthetic Division

Synthetic division, which is a streamlined version of the long division process, can be used *only* when the divisor has the special form $x - c$. In order to explain this process, let us use the division in Example 1.

The basic idea behind synthetic division is to leave out all the variables involved and just copy down the coefficients in their positions. The coefficients from the dividend are written in the order corresponding to decreasing powers of x. Zero coefficients must be supplied to account for any missing terms. The division process for Example 1 can be rewritten as follows:

$$
\begin{array}{r|rrrr}
 & \boxed{3} & \boxed{8} & \boxed{26} & \\
\hline
-3 & \boxed{3} & -1 & 2 & -1 \\
 & & -9 & & \\
\cline{3-3}
 & & 8 & 2 & \\
 & & & -24 & \\
\cline{4-4}
 & & & 26 & -1 \\
 & & & & -78 \\
\cline{5-5}
 & & & & 77
\end{array}
$$

In this format, the quotient and remainder are obtained through the following steps:

Step 1. The first coefficient of the dividend, 3, is the first coefficient of the quotient, and we write it directly above the dividend.

Step 2. Multiply the 3 obtained in step 1 by -3 (in the divisor) to obtain

$$3 \cdot (-3) = -9$$

Then subtract -9 from -1, the second coefficient of the dividend, to obtain

$$-1 - (-9) = 8$$

The number 8 is the second coefficient of the quotient.

Step 3. Multiply the 8 obtained in step 2 by -3 (in the divisor) to obtain

$$8 \cdot (-3) = -24$$

Then subtract -24 from 2, the third coefficient of the dividend, to obtain

$$2 - (-24) = 26$$

The number 26 is the third coefficient of the quotient.

Step 4. Multiply the 26 obtained in step 3 by -3 (in the divisor) to obtain

$$26 \cdot (-3) = -78$$

Then subtract -78 from -1, the fourth coefficient of the dividend, to obtain

$$-1 - (-78) = 77$$

The number 77 is the remainder in this division.

A convenient format for setting up this process is shown below. In this representation, the coefficients of the quotient appear in the bottom row, together with the remainder in the last position.

$$\begin{array}{r|rrrr} -3 & 3 & -1 & 2 & -1 \\ & & -9 & -24 & -78 \\ \hline & 3 & 8 & 26 & 77 \end{array}$$

Step	Procedure	Column
1	Bring down the 3	1
2	$3 \cdot (-3) = -9; -1 - (-9) = 8$	2
3	$8 \cdot (-3) = -24; 2 - (-24) = 26$	3
4	$26 \cdot (-3) = -78; -1 - (-78) = 77$	4

Now, we can change the sign of -3 to $+3$ so that it is no longer necessary to use subtraction to obtain the bottom row entries. All we need do is to *add* algebraically to get the coefficients of the quotient and the remainder, as displayed in the following diagram.

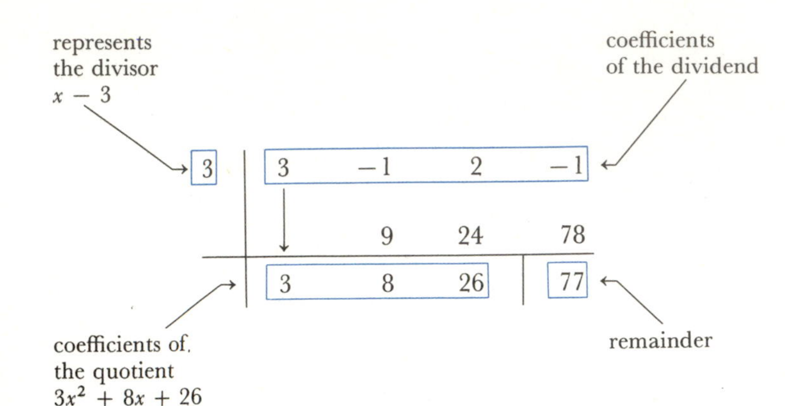

Step	Procedure	Column
1	Bring down the 3	1
2	$3 \cdot 3 = 9$; $(-1) + 9 = 8$	2
3	$3 \cdot 8 = 24$; $2 + 24 = 26$	3
4	$3 \cdot 26 = 78$; $(-1) + 78 = 77$	4

This final form of the long division is called **synthetic division.** Again we obtain

$$3x^3 - x^2 + 2x - 1 = (x - 3)(3x^2 + 8x + 26) + 77$$

EXAMPLE 2 Use synthetic division to find the quotient $Q(x)$ and the remainder R if the dividend is given by $f(x) = 3x^3 - 8x + 1$ and the divisor is $D(x) = x + 2$.

SOLUTION First, we write the divisor $x + 2$ in the form $x - c$ as $x - (-2)$. Then we use -2 as the divisor in the synthetic division.

$$\begin{array}{r|rrrr} -2 & 3 & 0 & -8 & 1 \\ & \downarrow & -6 & 12 & -8 \\ \hline & 3 & -6 & 4 & \multicolumn{1}{|r}{-7} \end{array}$$

Notice 0 is used as the coefficient of the "missing" x^2 term

Notice again the pattern in the synthetic division:

First, we bring down the leading coefficient, 3.

Then we multiply 3 by -2 to get -6, and then *add* -6 and 0 to get -6.

Next, we multiply -6 by -2 to get 12, and then *add* 12 and -8 to get 4.

Finally, we multiply 4 by -2 to get -8, and then *add* -8 and 1 to get the remainder -7.

The entries in the last row, namely, 3, -6, 4, and -7, indicate the quotient

$$Q(x) = 3x^2 - 6x + 4$$

and the remainder $R = -7$.

Thus, $3x^3 - 8x + 1 = (x + 2)(3x^2 - 6x + 4) + (-7)$, or

$$\frac{3x^3 - 8x + 1}{x + 2} = (3x^2 - 6x + 4) - \frac{7}{x + 2}$$

EXAMPLE 3 Use synthetic division to divide $4x^5 - 15x^4 - 20x^3 - 350x$ by $x - 6$.

SOLUTION By synthetic division, we have

6	4	−15	−20	0	−350	0
		24	54	204	1224	5244
	4	9	34	204	874	5244

Hence

$$4x^5 - 15x^4 - 20x^3 - 350x = (x - 6)(4x^4 + 9x^3 + 34x^2 + 204x + 874) + 5244$$

or

$$\frac{4x^5 - 15x^4 - 20x^3 - 350x}{x - 6} = 4x^4 + 9x^3 + 34x^2 + 204x + 874 + \frac{5244}{x - 6}$$

Synthetic division can be used to find values of polynomial functions. For example, if $f(x) = x^5 - 4x^3 - 6x^2 - 9$, we can find $f(-3)$ as follows.

We begin by dividing $f(x)$ by $x + 3$, using synthetic division, to get

−3	1	0	−4	−6	0	−9
		−3	9	−15	63	−189
	1	−3	5	−21	63	−198

Next, $f(x)$ can be written as

$$\begin{aligned} f(x) &= x^5 - 4x^3 - 6x^2 - 9 \\ &= (x^4 - 3x^3 + 5x^2 - 21x + 63)(x + 3) + (-198) \end{aligned}$$

so that, by direct substitution, we have

Note that this is 0

$$\begin{aligned} f(-3) &= [(-3)^4 - 3(-3)^3 + 5(-3)^2 - 21(-3) + 63] \cdot \boxed{(-3 + 3)} + (-198) \\ &= 0 + (-198) \end{aligned}$$

Thus, by comparing the function value $f(-3)$ to the remainder R, when dividing $f(x)$ by $x - (-3)$, we see that

$$f(-3) = -198 = R$$

This result is generalized in the following theorem.

THEOREM 1 **Remainder Theorem**

If a polynomial $f(x)$ of degree $n > 0$ is divided by $x - c$, the remainder R is a constant and is equal to the value of the polynomial when the constant c is substituted for x; that is, $f(c) = R$.

PROOF Let $Q(x)$ be the quotient and $R(x)$ be the remainder obtained from the division algorithm when $f(x)$ is divided by $x - c$. Thus

$$f(x) = (x - c)Q(x) + R(x)$$

Since the remainder $R(x)$ is of degree less than the divisor $x - c$, $R(x)$ must be constant and we denote it as R. The equation

$$f(x) = (x - c)Q(x) + R$$

holds for all x, and if we set $x = c$, we find that

$$f(c) = (c - c)Q(c) + R = 0 \cdot Q(c) + R = R$$

so that $f(c) = R$.

In Examples 4 and 5, use the remainder theorem and synthetic division to find the value of each polynomial.

EXAMPLE 4 $f(2)$ if $f(x) = 3x^5 - 5x^3 + 1$

SOLUTION By the remainder theorem $f(2)$ is the remainder when $f(x)$ is divided by $x - 2$. Using synthetic division, we have

$$\begin{array}{r|rrrrrr} 2 & 3 & 0 & -5 & 0 & 0 & 1 \\ & & 6 & 12 & 14 & 28 & 56 \\ \hline & 3 & 6 & 7 & 14 & 28 & 57 = R \end{array}$$

Therefore, $f(2) = 57$.

EXAMPLE 5 $f(-2)$ if $f(x) = x^3 - 2x^2 + 5x + 26$

SOLUTION By the remainder theorem $f(-2)$ is the remainder R when $f(x)$ is divided by $x - (-2)$ or $x + 2$. Using synthetic division, we have

$$\begin{array}{r|rrrr} -2 & 1 & -2 & 5 & 26 \\ & & -2 & 8 & -26 \\ \hline & 1 & -4 & 13 & 0 = R \end{array}$$

Therefore, $f(-2) = 0$.

In Example 5, since the remainder R is 0 when $f(x)$ is divided by $x + 2$, we can write

$$f(x) = x^3 - 2x^2 + 5x + 26 = (x^2 - 4x + 13)(x + 2)$$

so that $x + 2$ is a *factor* of $f(x)$. This example illustrates an important special case of the remainder theorem called the *factor theorem.*

THEOREM 2 **Factor Theorem**

If $f(x)$ is a polynomial of degree $n > 0$ and $f(c) = 0$, then $x - c$ is a factor of the polynomial $f(x)$; conversely, if $x - c$ is a factor of $f(x)$, then c is a zero of f.

PROOF Assume that $f(x) = (x - c)Q(x) + R$. If $f(c) = 0$, then the remainder theorem indicates that $R = 0$. Thus, $f(x) = Q(x)(x - c)$ and $(x - c)$ is a factor of $f(x)$. Conversely, if $x - c$ is a factor of $f(x)$, then $f(x) = (x - c)Q(x)$, so that $f(c) = 0 \cdot Q(c) = 0$ and c is a zero of f.

EXAMPLE 6 Let $f(x) = 3x^5 - 38x^3 + 5x^2 - 720$. Use the factor theorem to determine

(a) whether $x - 4$ is a factor of $f(x)$ (b) whether $x + 1$ is a factor of $f(x)$

SOLUTION (a) The expression $x - 4$ is of the form $x - c$ with $c = 4$. Using synthetic division, we obtain

$$\begin{array}{r|rrrrr|r} 4 & 3 & 0 & -38 & 5 & 0 & -720 \\ & & 12 & 48 & 40 & 180 & 720 \\ \hline & 3 & 12 & 10 & 45 & 180 & 0 = R \end{array}$$

It follows from the factor theorem that $x - 4$ is a factor of $f(x)$. In fact,

$$3x^5 - 38x^3 + 5x^2 - 720 = (x - 4)(3x^4 + 12x^3 + 10x^2 + 45x + 180)$$

(b) We have $x + 1 = x - (-1)$, which is of the form $x - c$ with $c = -1$. Using synthetic division, we get

$$\begin{array}{r|rrrrr|r} -1 & 3 & 0 & -38 & 5 & 0 & -720 \\ & & -3 & 3 & 35 & -40 & 40 \\ \hline & 3 & -3 & -35 & 40 & -40 & -680 = R \end{array}$$

Since $R = -680 \neq 0$, it follows from the factor theorem that $x + 1$ is *not* a factor of $f(x)$.

EXAMPLE 7 Find the value of k that makes $x - 3$ a factor of $f(x) = 3x^3 - 4x^2 - kx - 33$.

SOLUTION Using synthetic division, we have

$$\begin{array}{r|rrr|r} 3 & 3 & -4 & -k & -33 \\ & & 9 & 15 & -3k + 45 \\ \hline & 3 & 5 & -k + 15 & -3k + 12 \end{array}$$

If $x - 3$ is to be a factor of $f(x)$, it follows from the factor theorem that $f(3) = 0$, and from the remainder theorem that the remainder $-3k + 12$ must be zero; that is, $-3k + 12 = 0$ and $k = 4$.

PROBLEM SET 3.3

In problems 1–6, use the long division process to find the quotient $Q(x)$ and the remainder $R(x)$ if $f(x)$ is divided by $D(x)$.

1. $f(x) = 4x^3 - 2x^2 + 7x - 1;\ D(x) = x^2 - 2x + 3$
2. $f(x) = 8x^3 - x^2 - x + 5;\ D(x) = x^2 + x$
3. $f(x) = 5x^3 + 7x^2 - 2x + 9;\ D(x) = x^2 + 3x - 4$
4. $f(x) = 3x^4 - 2x^3 + 4x^2 + 3x + 7;\ D(x) = 2x^2 - x + 3$
5. $f(x) = 2x^5 + 3x^4 - x^2 + x - 4;\ D(x) = 2x^3 - x^2 + x + 2$
6. $f(x) = 6x^4 + 10x^2 + 7;\ D(x) = 3x^2 - 1$

In problems 7–11, complete the synthetic division, and then express each result in the form $f(x) = D(x) \cdot Q(x) + R(x)$.

7.

2	1	−4	5	7
		2	−4	
	1	−2	1	

8.

−4	1	0	−2	0
		−4		
	1	−4		

9.

−1	2	−1	0	5	−3
		−2	3		
	2	−3	3		

10.

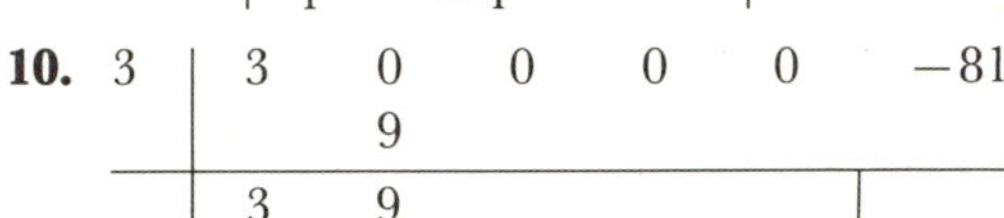

3	3	0	0	0	0	−81
		9				
	3	9				

C 11.

2.3	3.8	−7.3	−2.1

12. In the division algorithm, explain why the degree of the remainder $R(x)$ is less than the degree of the divisor $D(x)$.

In problems 13–22, use synthetic division to find the quotient $Q(x)$ and the remainder R if the first polynomial is divided by the second.

13. $x^2 - 4x - 5$ by $x + 1$
14. $x^3 - 2x^2 - 1$ by $x + 2$
15. $4x^4 - 3x^3 + 2x^2 + 5$ by $x - 1$
16. $x^8 + 1$ by $x - 1$
17. $5x^3 - 2x^2 + 3x - 4$ by $x - 3$
18. $2x^4 - 5x^2 - 1$ by $x + 1$
19. $5x^5 + x^2 + 3$ by $x + 2$
20. $2x^4 - 3x^3 + 5x^2 + 6x - 3$ by $x + 2$
21. $-4x^6 - 5x^3 + 3x^2 + x + 7$ by $x - 2$
22. $-2x^4 + 3x^3 - 3x^2 + x - 1$ by $x + 4$

In problems 23–27, use synthetic division to write each expression in the form

$$\frac{f(x)}{D(x)} = Q(x) + \frac{R}{D(x)}$$

23. $\dfrac{x^2 + 5x + 1}{x - 2}$
24. $\dfrac{x^3 - x^2 - 3}{x + 1}$
25. $\dfrac{4x^3 - 2x^2 + x - 5}{x + 2}$
26. $\dfrac{x^4 - 1}{x - 1}$
27. $\dfrac{5x^5 - 5}{x - 1}$

28. Use synthetic division to find the remainder if $f(x) = x^3 - 4x^2 + 5x + 3$ is divided by
    (a) x (b) $x - 1$ (c) $x + 1$ (d) $x - 2$ (e) $x + 2$

In problems 29–37, use synthetic division and the remainder theorem to find $f(c)$.

29. $f(x) = x^2 - 7x + 11$ and $c = -2$

30. $f(x) = 3x^5 - 5x^3 + 7$ and $c = 1$

31. $f(x) = -x^3 - 3x^2 + 2$ and $c = -3$

32. $f(x) = -2x^4 - 3$ and $c = 3$

33. $f(x) = 3x^3 + 6x^2 - 10x + 7$ and $c = 2$

34. $f(x) = 3x^3 + 4x^2 - 7x + 16$ and $c = -1$

35. $f(x) = 2x^3 - 5x^2 + 5x + 11$ and $c = \frac{1}{2}$

36. $f(x) = -2x^4 + 3x^3 + 5x - 13$ and $c = 3$

37. $f(x) = -3x^4 - 3x^3 + 3x^2 + 2x - 4$ and $c = -2$

38. Use synthetic division to show that $x - 3$ is a factor of $f(x) = x^3 - x^2 - 3x + 18$.

In problems 39–43, use synthetic division to show that $D(x)$ is a factor of $f(x)$.

39. $f(x) = x^2 + 5x + 6$ and $D(x) = x + 3$

40. $f(x) = x^3 - 5x^2 + 8x - 6$ and $D(x) = x - 3$

41. $f(x) = 2x^4 + 3x - 26$ and $D(x) = x + 2$

42. $f(x) = x^3 - 8$ and $D(x) = x - 2$

43. $f(x) = x^3 - 4x^2 - 3x + 18$ and $D(x) = x - 3$

44. (a) If $f(x) = 2x^3 - 6x^2 + x + k$, find k so that $f(3) = -2$.
(b) Find k so that $x - 2$ is a factor of $f(x) = 3x^3 + 4x^2 + kx - 20$.
(c) Show that $x - 1$ is a factor of $f(x) = 14x^{99} - 65x^{56} + 51$.

45. If $f(x) = x^3 + 2x^2 - 13x + 10$, use synthetic division to determine $f(-5)$, $f(-4)$, $f(-3)$, $f(-2)$, $f(-1)$, $f(1)$, $f(2)$, $f(3)$, $f(4)$, and $f(5)$. What are the factors of $f(x)$?

C **46.** If $g(x) = -4.3x^3 + 1.7x - 5.3$, use synthetic division to determine $g(-2.1)$, $g(-1.5)$, $g(3.3)$, and $g(4.9)$ to two decimal places.

3.4 Rational Zeros of Polynomials

So far, we have been able to find the zeros of linear, quadratic, and linear factored polynomial functions. We now consider general techniques for determining all the rational numbers that are zeros of polynomial functions. These numbers are called **rational zeros** of polynomial functions.

By the factor theorem, each zero of a polynomial function f corresponds to a first-degree factor of $f(x)$. Because the polynomial $f(x)$ can't have more first-degree factors than its degree, we have the following property.

PROPERTY 1 **Maximum Number of Zeros of a Polynomial Function**

If f is a polynomial function of degree n, then f cannot have more than n zeros.

We saw in Section 3.2 that the zeros of a function have the same values as the x intercepts of its graph. Therefore, the graph of a polynomial function f cannot have more x intercepts than the degree of the polynomial $f(x)$. It should be noted that some of the zeros of a polynomial function f may be multiple zeros, or some may be complex numbers. (This latter situation is discussed in Section 3.5.) For instance, if

$$f(x) = x(x - 3)^2(x^2 + 4)$$

then $f(x)$ has degree 5 and f possesses four zeros: 0, 3, $2i$, and $-2i$, where 3 has multiplicity 2.

If complex zeros are counted, and if zeros of multiplicity m are counted as m zeros, we can state the following property.

PROPERTY 2 **Number of Zeros of a Polynomial Function**

> If f is a polynomial function of degree $n > 0$, and if a zero of multiplicity m is counted m times, then f has *precisely* n zeros.

We shall see that the possible number of positive real zeros, negative real zeros, and complex zeros of a polynomial function are limited. Consider the polynomial

$$f(x) = 3x^5 - 2x^4 + 4x^2 + 6x - 7$$

This polynomial is arranged in descending powers of x. Notice that there are *three variations in the signs* of the coefficients, reading left to right:

$$+3x^5 - 2x^4 + 4x^2 + 6x - 7$$

① ② ③

If we replace x by $-x$ in $f(x)$, we obtain

$$\begin{aligned} f(-x) &= 3(-x)^5 - 2(-x)^4 + 4(-x)^2 + 6(-x) - 7 \\ &= -3x^5 - 2x^4 + 4x^2 - 6x - 7 \end{aligned}$$

Here, we have *two variations of signs*:

$$-3x^5 - 2x^4 + 4x^2 - 6x - 7$$

① ②

Descartes discovered a connection between the number of variations in the signs and the potential number of positive and negative real number roots of a polynomial. Recall that the zeros of a polynomial function f are the same values as the roots of the polynomial equation $f(x) = 0$.

Descartes' Rule of Signs

> Let $f(x)$ be a polynomial arranged in descending powers of x, with real coefficients and a nonzero constant term.
>
> (i) The number of *positive* real roots of the equation $f(x) = 0$ either is equal to the number of variations in the signs of $f(x)$ or is less than the number of variations by an even number.
>
> (ii) The number of *negative* real roots of the equation $f(x) = 0$ either is equal to the number of variations in the signs of $f(-x)$ or is less than the number of variations by an even number.

If it is determined that a polynomial $f(x)$ of degree n has c real number zeros, then by Property 2, the remaining $n - c$ zeros must be complex numbers.

EXAMPLE 1 Use Descartes' rule of signs to determine the possible number of positive real zeros, negative real zeros, and complex zeros of

$$f(x) = 3x^5 - 2x^4 + 3x^3 + 2x^2 + 7x - 3$$

SOLUTION There are three variations of signs in $f(x)$. Therefore, $f(x)$ has either three positive zeros or one positive zero. Since

$$\begin{aligned} f(-x) &= 3(-x)^5 - 2(-x)^4 + 3(-x)^3 + 2(-x)^2 + 7(-x) - 3 \\ &= -3x^5 - 2x^4 - 3x^3 + 2x^2 - 7x - 3 \end{aligned}$$

$f(-x)$ has two variations in signs. This means that $f(x)$ has either two negative zeros or no negative zero.

Since f has precisely 5 zeros (with a zero of multiplicity k counted as k zeros), the possible number of zeros of f as determined by Descartes' rule of signs are summarized in the following table.

Number of Positive Zeros	Number of Negative Zeros	Number of Complex Zeros
3	2	0
3	0	2
1	2	2
1	0	4

Rational Zeros

It may be difficult to find the zeros of a polynomial function of degree higher than 2. However, if all the coefficients are integers or rational numbers, we can use a method for finding the *rational* number zeros, if they exist, by applying the next theorem.

THEOREM 1 **Rational Zero Theorem**

Assume that $f(x) = a_nx^n + a_{n-1}x^{n-1} + \cdots + a_1x + a_0$, where $a_n \neq 0$ and n is a positive integer. If the coefficients are integers and if p/q is a rational number zero in lowest terms, then p is a factor of a_0 and q is a factor of a_n.

It should be noted that -1 and 1 are acceptable factors for a_0 and a_n.

The rational zero theorem can be used to find the rational zeros of a polynomial function whose coefficients are integers by carrying out the following procedure:

Step 1. List all factors of the constant term a_0 in the polynomial. This list provides us with all *possible* values of p.

Step 2. List all factors of the coefficients of the term of the highest degree, a_n, in the polynomial. This list provides us with all *possible* values of q.

Step 3. Use the lists from step 1 and step 2 to form the list of all possible rational number zeros p/q.

Step 4. Check each of the possible rational zeros produced in step 3. If none of them is a root of $f(x) = 0$, we conclude that f has no rational zeros.

In Examples 2 and 3, find the rational zeros of each function.

EXAMPLE 2 $f(x) = x^3 - 2x^2 - x + 2$

SOLUTION We know from Property 2 that there are three zeros. We obtain the possible number of positive and negative real zeros from Descartes' rule of signs. These are listed in the following table.

Number of Positive Zeros	Number of Negative Zeros
2	1
0	1

Next we apply the above procedure.

Step 1. The factors of $a_0 = 2$ provide the possibilities for p which are

$$p: \quad -1, 1, -2, 2$$

Step 2: The possible values for q are the factors of $a_3 = 1$ and they are

$$q: \quad -1, 1$$

Step 3: The possible values for p/q are

$$\frac{p}{q}: \quad -1, 1, -2, 2$$

Step 4. We test the possible values for p/q by using synthetic division to get

$$\begin{array}{r|rrr|r} -1 & 1 & -2 & -1 & 2 \\ & & -1 & 3 & -2 \\ \hline & 1 & -3 & 2 & 0 \end{array}$$

so that $f(-1) = 0$ and -1 is a zero of f.

The result of the synthetic division also indicates that

$$x^3 - 2x^2 - x + 2 = (x + 1)(x^2 - 3x + 2)$$

Thus, the remaining zeros are found by solving

$$x^2 - 3x + 2 = 0 \qquad \text{or}$$

$$(x - 2)(x - 1) = 0 \qquad \text{so that} \qquad x = 1 \qquad \text{and} \qquad x = 2$$

Hence the zeros of f are -1, 1, and 2.

EXAMPLE 3 $f(x) = 4x^3 + 16x^2 + 9x - 9$

SOLUTION There are three zeros of f, and by Descartes' rule of signs we can find the possible number of positive and negative real zeros.

Number of Positive Zeros	Number of Negative Zeros
1	2
1	0

We follow the step-by-step procedure.

Step 1. The possible values of p are

$$p: \quad -1, 1, -3, 3, -9, 9$$

Step 2. The possible values of q are

$$q: -1, 1, -2, 2, -4, 4$$

Step 3: The possible values of p/q are

$$\frac{p}{q}: \quad -1, 1, -3, 3, -9, 9, -\tfrac{1}{2}, \tfrac{1}{2}, -\tfrac{1}{4}, \tfrac{1}{4}, -\tfrac{3}{2}, \tfrac{3}{2}, -\tfrac{3}{4}, \tfrac{3}{4}, -\tfrac{9}{2}, \tfrac{9}{2}, -\tfrac{9}{4}, \tfrac{9}{4}$$

Step 4. After using synthetic division to test various values, we find

$$\begin{array}{c|cccc} \tfrac{1}{2} & 4 & 16 & 9 & -9 \\ & & 2 & 9 & 9 \\ \hline & 4 & 18 & 18 & 0 \end{array}$$

Thus,

$$\begin{aligned} 4x^3 + 16x^2 + 9x - 9 &= (x - \tfrac{1}{2})(4x^2 + 18x + 18) \\ &= (x - \tfrac{1}{2})2(2x^2 + 9x + 18) \\ &= 2(x - \tfrac{1}{2})(x + 3)(2x + 3) \end{aligned}$$

so that the zeros of f are $\frac{1}{2}$, -3, and $-\frac{3}{2}$.

If the polynomial equation $f(x) = 0$ has rational numbers that are not integers as coefficients, then we first multiply both sides of the equation by the least common denominator (LCD) of the coefficients to convert the equation $f(x) = 0$ to an equivalent form containing only integer coefficients. Then the rational zero theorem can be applied. This technique is illustrated in the next example.

EXAMPLE 4 Find the rational zeros and the other zeros, if possible, for $f(x) = \frac{1}{2}x^3 - 3x^2 + 5x - \frac{3}{2}$.

SOLUTION The zeros of f are the same as the roots of

$$\frac{1}{2}x^3 - 3x^2 + 5x - \frac{3}{2} = 0$$

After multiplying both sides of the equation by the LCD, 2, we obtain the equivalent equation

$$x^3 - 6x^2 + 10x - 3 = 0$$

The possible number of positive and negative real roots are listed in the table below.

Number of Positive Roots	Number of Negative Roots
3	0
1	0

We apply the rational zero theorem to the latter equation.

Step 1. Possibilities for p include

$$p: \quad -1, 1, -3, 3$$

Step 2. Possibilities for q include

$$q: \quad -1, 1$$

Step 3. Possibilities for p/q are

$$\frac{p}{q}: \quad -1, 1, -3, 3$$

Step 4. After using synthetic division to try the various possibilities, we find

$$\begin{array}{r|rrrr} 3 & 1 & -6 & 10 & -3 \\ & & 3 & -9 & 3 \\ \hline & 1 & -3 & 1 & 0 \end{array}$$

so that 3 is a root of the equation, and

$$x^3 - 6x^2 + 10x - 3 = (x - 3)(x^2 - 3x + 1)$$

Therefore, 3 is a zero of f. The remaining zeros of f are found by solving

$$x^2 - 3x + 1 = 0$$

by the quadratic formula, to obtain

$$x = \frac{3 + \sqrt{5}}{2} \qquad \text{or} \qquad x = \frac{3 - \sqrt{5}}{2}$$

Hence, the zeros of f are

$$3, \quad \frac{3 + \sqrt{5}}{2}, \quad \text{and} \quad \frac{3 - \sqrt{5}}{2}$$

It is important to realize that the rational zero theorem has only limited usefulness in determining roots of polynomial equations. For example, the roots of the equation $x^4 + 5x^2 + 6 = 0$ cannot be determined by this method, because none of the roots of this equation are rational numbers.

By using the rational zero theorem and the factor theorem, we can often factor a polynomial function completely as a product of linear factors in the set of real numbers. When this is possible, we can use the techniques discussed in Section 3.2 to sketch a rough graph of a polynomial function.

EXAMPLE 5 Write the polynomial function

$$f(x) = 2x^4 - 3x^3 - 12x^2 + 7x + 6$$

in a linear factored form in the set of real numbers. Locate the x and y intercepts and sketch the graph.

SOLUTION After applying the rational zero theorem to the function f, with $a_0 = 6$ and $a_4 = 2$, we get the possibilities:

$$p: \quad -1, 1, -2, 2, -3, 3, -6, 6$$

$$q: \quad -1, 1, -2, 2$$

$$\frac{p}{q}: \quad -1, 1, -2, 2, -3, 3, -6, 6, -\tfrac{1}{2}, \tfrac{1}{2}, -\tfrac{3}{2}, \tfrac{3}{2}$$

We test the potential rational zeros of f by repeatedly using synthetic division to find that

1	2	−3	−12	7	6
		2	−1	−13	−6
−2	2	−1	−13	−6	0
		−4	10	6	
	2	−5	−3	0	

This synthetic division implies that $f(x) = (x - 1)(2x^3 - x^2 - 13x - 6)$

This synthetic division implies that $2x^3 - x^2 - 13x - 6 = (x + 2)(2x^2 - 5x - 3)$

so that

$$f(x) = (x - 1)(x + 2)(2x^2 - 5x - 3)$$

or

$$f(x) = (x - 1)(x + 2)(2x + 1)(x - 3)$$

x	y
$-\frac{5}{2}$	$\frac{77}{2} = 38.5$
-2	0
$-\frac{3}{2}$	$-\frac{45}{4} = -11.25$
-1	-8
$-\frac{1}{2}$	0
0	6
1	0
$\frac{3}{2}$	$-\frac{21}{2} = -10.5$
2	-20
$\frac{5}{2}$	$-\frac{81}{4} = -20.25$
3	0

From the factored form of f, we can see that the x intercepts of the graph of f occur when $x = 1$, $x = -2$, $x = -\frac{1}{2}$, and $x = 3$. The y intercept is found by setting $x = 0$ to get $y = 6$. A rough graph of f is determined by plotting a few points and then connecting them with a smooth unbroken curve (Figure 1).

Figure 1

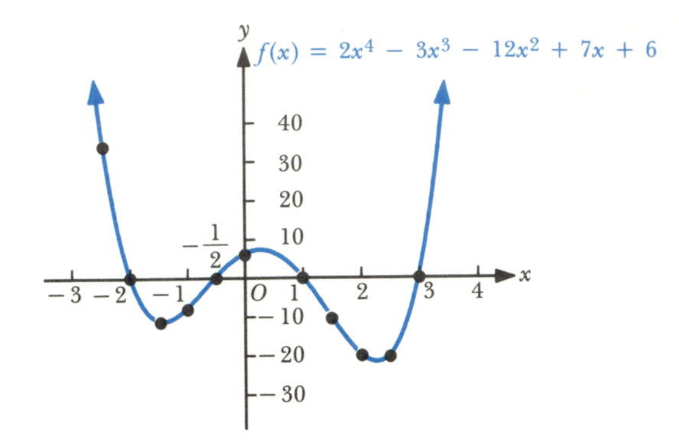

PROBLEM SET 3.4

In problems 1–10, use Descartes' rule of signs to determine the possible number of positive real zeros, negative real zeros, and complex zeros for the given function.

1. $f(x) = x^2 - 8x - 2$
2. $g(x) = x^3 + x^2 + x + 1$
3. $g(x) = 3x^3 - 4x + 1$
4. $h(x) = -x^3 + x^2 - 2x + 7$
5. $f(x) = x^5 - 6x - 5$
6. $g(x) = 7x^4 - x^3 + 11x - 1$
7. $h(x) = 2x^4 + x^3 + 5x - 1$
8. $f(x) = 2x^4 - 5x^3 - 6x^2 - 8$
9. $f(x) = x^4 - 3x^3 + 7x^2 - x + 3$
10. $h(x) = x^5 + 4x^4 - 3x^2 + x$

In problems 11–26, use the rational zero theorem to list all possible rational zeros of the given function. Use synthetic division to test the possibilities in order to find the rational zeros.

11. $f(x) = x^3 - x^2 - 4x + 4$
12. $g(x) = x^3 + 2x - 12$
13. $h(x) = 5x^3 - 12x^2 + 17x - 10$
14. $f(x) = x^3 - x^2 - 14x + 24$
15. $f(x) = 3x^3 - 7x^2 + 8x - 2$
16. $h(x) = 4x^3 - 13x + 6$
17. $g(x) = 10x^3 + x^2 - 7x + 2$
18. $f(x) = 4x^4 - 15x^2 + 5x + 6$
19. $h(x) = 4x^4 - 4x^3 - 7x^2 + 4x + 3$
20. $g(x) = x^4 + 9x^2 + 20$
21. $f(x) = \frac{x^4}{2} + x^3 + x^2 - 2x - 4$
22. $h(x) = 2x^3 + x^2 + x$
23. $f(x) = x^3 - \frac{7}{2}x^2 + 3x + \frac{5}{2}$
24. $g(x) = x^3 - \frac{x^2}{2} - 4x + 2$
25. $f(x) = \frac{2}{3}x^4 + \frac{x^3}{2} - \frac{5}{4}x^2 - x - \frac{1}{6}$
26. $h(x) = \frac{2}{3}x^3 - x^2 - \frac{8}{3}x - 1$

In problems 27–35, write each polynomial function in a linear factored form in the set of real numbers. Locate the x and y intercepts and sketch the graph of the function.

27. $f(x) = x^3 - x^2 - 10x - 8$
28. $g(x) = 2x^3 - 3x^2 - 17x + 30$
29. $h(x) = x^3 - 4x^2 + x + 6$
30. $f(x) = 6x^3 + 19x^2 + x - 6$
31. $g(x) = 9x^4 + 15x^3 - 20x^2 - 20x + 16$
32. $h(x) = x^4 - 5x^2 + 4$
33. $h(x) = 2x^3 - 7x^2 - 10x + 24$
34. $g(x) = 2x^3 - 9x^2 + 12x - 5$
35. $f(x) = 3x^4 - 19x^3 + 20x^2$

3.5 Complex and Irrational Zeros of Polynomial Functions

Until now we have focused our attention on determining the *rational* zeros of polynomial functions with rational number coefficients. In this section, we extend these results to search for the *complex* zeros of polynomial functions with *complex* coefficients. In addition, we study a numerical technique for approximating the values of *irrational* zeros of polynomial functions with real coefficients.

Complex Polynomials

Karl Friedrich Gauss

If x represents a *complex number*, then a polynomial function of the form

$$f(x) = a_nx^n + a_{n-1}x^{n-1} + \cdots + a_1x + a_0$$

where n is a nonnegative integer and the coefficients a_n, a_{n-1}, . . . , a_1, and a_0 are complex numbers, is called a **complex polynomial** in x. If $a_n \neq 0$, we say that the polynomial has *degree n* and we call a_n the leading coefficient. Note that real numbers are included in the set of complex numbers.

The factor theorem also holds for complex polynomials, and the relationship between the zeros and the factors of a complex polynomial can tell us how many zeros to expect. This fact depends on an important theorem, often referred to as Gauss' theorem or as the *fundamental theorem of algebra*, which was proved by the German mathematician Karl Friedrich Gauss (1777–1855).

THEOREM 1 **The Fundamental Theorem of Algebra**

Every complex polynomial of degree $n \geq 1$ has at least one complex zero.

By combining Theorem 1 with the factor theorem, we obtain the following result.

THEOREM 2 **The Complete Linear Factorization Theorem**

> If $f(x)$ is a complex polynomial of degree $n \geq 1$, then there is a nonzero complex number a_n and there are complex numbers $c_1, c_2, \ldots, c_n$ such that
>
> $$f(x) = a_n(x - c_1)(x - c_2) \cdots (x - c_n)$$

PROOF According to the fundamental theorem of algebra, f has a zero c_1, so that, by the factor theorem (see Theorem 2 on page 157), we have

$$f(x) = (x - c_1)Q_1(x)$$

where $Q_1(x)$ is a polynomial of degree $n - 1$. $Q_1(x)$ has a zero c_2 if $n - 1 \geq 1$, and, as above,

$$Q_1(x) = (x - c_2)Q_2(x)$$

so that

$$f(x) = (x - c_1)(x - c_2)Q_2(x)$$

where $Q_2(x)$ has degree $n - 2$. Continuing the process, we get

$$f(x) = (x - c_1)(x - c_2) \cdots (x - c_n)Q_n(x)$$

where $Q_n(x)$ is a constant, which we write simply as Q_n. Multiplying this expression out for $f(x)$, it is seen that the coefficient of x^n is Q_n; hence, $Q_n = a_n$ and the theorem is proved.

It must be emphasized that the factors in the above theorem may not all be distinct, which is equivalent to saying that the c_k are not necessarily distinct. Each factor $(x - c_k)$ corresponds to a root of the equation $f(x) = 0$, according to the factor theorem. We therefore have the following theorem, which is a generalization of the result we stated for polynomials with real coefficients.

THEOREM 3

> If f is a complex polynomial function of degree $n \geq 1$, then f has at most n zeros.

PROOF By the factorization theorem,

$$f(x) = a_n(x - c_1)(x - c_2) \cdots (x - c_n)$$

Clearly, the numbers $c_1, c_2, \ldots c_n$ are zeros of f. Also, if $f(c) = 0$, for some number $c \neq c_i$ and $i = 1, \ldots, n$, then

$$f(x) = a_n(x - c_1)(x - c_2) \cdots (x - c_n)(x - c)$$

so that the degree of $f(x)$ is $n + 1$. However, this contradicts our assumption that $f(x)$ is a polynomial of degree n. So there can be no more than n zeros.

Consider the polynomial function $f(z) = z^2 - 6z + 25$. We find the zeros of f, by using the quadratic formula, to be

$$z = 3 + 4i \qquad \text{and} \qquad z = 3 - 4i$$

Note that the two complex zeros of f are complex conjugates of each other. Similarly, the complex zeros $-i$ and i of $g(z) = z^2 + 1$ are complex conjugates of each other. The following theorem generalizes these observations.

THEOREM 4 **Conjugate Root Theorem**

> If $f(z)$ is a complex polynomial of degree $n \geq 1$ with *real* coefficients and if $f(z_0) = 0$, where $z_0 = a + bi$, then $f(\bar{z}_0) = 0$, where $\bar{z}_0 = a - bi$.

PROOF Let $f(z) = a_n z^n + a_{n-1}z^{n-1} + \cdots + a_1 z + a_0$ be a polynomial with real coefficients. Since $z_0 = a + bi$ is a root of $f(z) = 0$, then

$$f(z_0) = a_n z_0{}^n + a_{n-1} z_0{}^{n-1} + \cdots + a_1 z_0 + a_0 = 0$$

Recall from Section 1.4, page 33, that the conjugate of the sum of complex numbers is the same as the sum of the conjugates of the complex numbers. Thus,

$$\begin{aligned}\overline{f(z_0)} &= \overline{a_n z_0{}^n + a_{n-1} z_0{}^{n-1} + \cdots + a_1 z_0 + a_0} \\ &= \overline{a_n z_0{}^n} + \overline{a_{n-1} z_0{}^{n-1}} + \cdots + \overline{a_1 z_0} + \overline{a_0} = \overline{0} = 0\end{aligned}$$

Also, recall that the conjugate of a product of complex numbers is the same as the product of the conjugates of the complex numbers so that

$$\overline{a_n z_0{}^n} = \bar{a}_n \bar{z}_0{}^n,\ \overline{a_{n-1} z_0{}^{n-1}} = \overline{a_{n-1}}\, \bar{z}_0{}^{n-1}, \ldots, \overline{a_1 z_0} = \bar{a}_1 \bar{z}_0$$

Since the conjugate of a real number is the real number itself, we have $\bar{a}_0 = a_0$, $\bar{a}_1 = a_1, \ldots,$ and $\bar{a}_n = a_n$. Hence,

$$\overline{f(z_0)} = a_n \bar{z}_0{}^n + a_{n-1} \bar{z}_0{}^{n-1} + \cdots + a_1 \bar{z}_0 + a_0 = f(\bar{z}_0) \qquad \text{and} \qquad \overline{f(z_0)} = 0$$

so that $f(\bar{z}_0) = 0$ and $\bar{z}_0$ is also a root of $f(z) = 0$.

EXAMPLE 1 Form a polynomial function f of degree 5 with real coefficients that has the following numbers as zeros: $-\frac{1}{2}$, $1 + i$, and 1 as a double zero.

SOLUTION Since $1 + i$ is a root of the equation $f(x) = 0$, it follows from the conjugate root theorem that $1 - i$ must also be a root; therefore, by the factorization theorem

$$f(x) = (x + \tfrac{1}{2})(x - 1)^2[x - (1 + i)][x - (1 - i)]$$

After multiplying the above factors, we obtain

$$f(x) = x^5 - \tfrac{7}{2}x^4 + 5x^3 - \tfrac{5}{2}x^2 - x + 1$$

Note that there are many other polynomials with the given zeros. However, if we seek a polynomial of degree 5, then the polynomial is either this $f(x)$ or $c \cdot f(x)$, where c is a nonzero constant.

EXAMPLE 2 Given that $f(x) = x^4 - 4x^3 + 5x^2 - 4x + 4$ has as one of its zeros i with a multiplicity 1, find the other zeros of f and determine their multiplicities.

SOLUTION Since i is a zero of f, it follows that $x = i$ is a root of the equation $f(x) = 0$, so that from the conjugate root theorem, $-i$ is also a root of $f(x) = 0$. By the factorization theorem, we get

$$\begin{aligned} f(x) &= x^4 - 4x^3 + 5x^2 - 4x + 4 \\ &= (x - i)(x + i)Q(x) \\ &= (x^2 + 1)Q(x) \end{aligned}$$

After dividing $f(x)$ by $x^2 + 1$, using long division, we get

$$\begin{aligned} f(x) &= (x^2 + 1)(x^2 - 4x + 4) \\ &= (x - i)(x + i)(x - 2)^2 \end{aligned}$$

so that f has zeros i and $-i$, each of multiplicity 1, and 2 is a double zero.

Approximating Irrational Zeros

We can approximate the value of an irrational zero of a polynomial function with *real* coefficients by successively enlarging the graph of the function on smaller and smaller intervals that are known to contain the zero until the desired approximation is obtained. Although there is more than one way of doing this, we examine a technique called the *bisection method,* based on the next property.

Intermediate-Value Property

> Suppose that f is a polynomial function with real coefficients. If the values of $f(a)$ and $f(b)$ have opposite algebraic signs, where $a < b$, then there is a number c in the interval (a, b) such that $f(c) = 0$.

Figure 1 illustrates the intermediate-value property for two cases. The **bisection method,** which can be used to approximate the value of an irrational zero, utilizes the following recursive process (refer to Figure 2).

Assume that f is a polynomial function with real coefficients.

Step 1. Find values for a and b where $a < b$ and the values $f(a)$ and $f(b)$ have opposite signs. (A graph of the function is helpful here.)

Figure 1

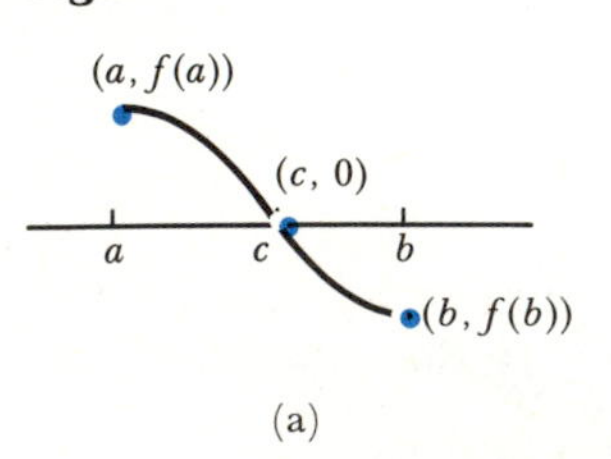

(a)

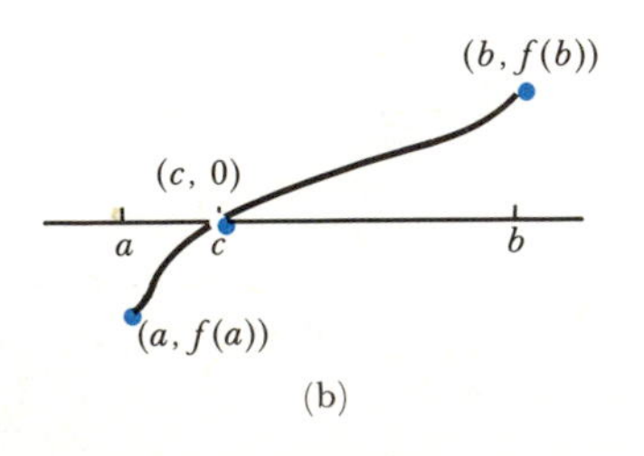

(b)

Figure 2

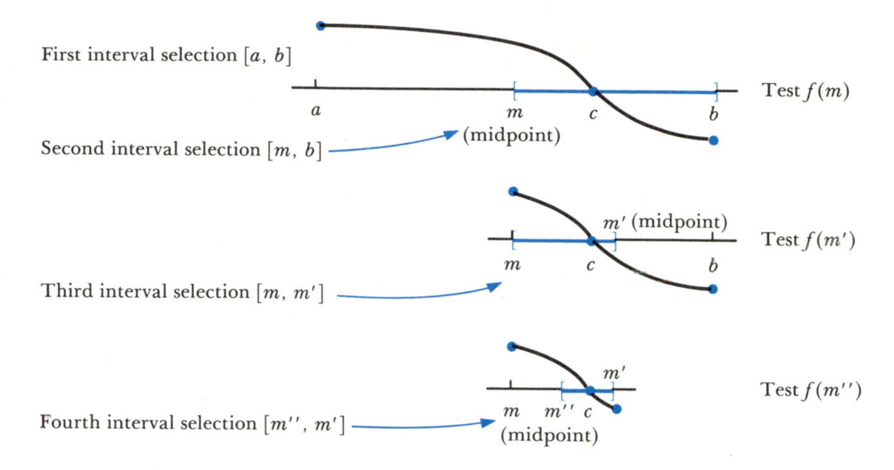

Step 2. Determine the midpoint m of interval $[a, b]$, namely, $m = (a + b)/2$.

Step 3. From the intervals $[a, m]$ and $[m, b]$, select the interval where the values of the given function f at the endpoints have opposite algebraic signs (one is positive and the other is negative).

Step 4. Repeat steps 2 and 3 on the interval selected above. Continue this process and evaluate $f(m)$ in each repetition until the value of $f(m)$ differs from 0 by less than an amount decided upon at the start.

Notice that each bisection point (midpoint of the interval) is getting closer to the value c where $f(c) = 0$.

The following property helps to find the intervals where $f(x)$ changes sign.

Upper and Lower Bound Property

Suppose that $f(x)$ is a polynomial with real coefficients and a positive leading coefficient and that $f(x)$ is divided by $x - c$ using synthetic division.

(i) If $c > 0$ and the numbers in the last row of the synthetic division process are nonnegative (positive or zero), then c is an **upper bound** for the real roots of the equation $f(x) = 0$, that is, any root is less than or equal to c.

(ii) If $c < 0$ and the numbers in the last row of the synthetic division process are alternately nonnegative (positive or zero) and nonpositive (negative or zero), then c is a **lower bound** for the real roots of the equation $f(x) = 0$, that is, any root is greater than or equal to c.

This property is helpful in locating real zeros because if we find upper and lower bounds for the real zeros of a polynomial function, we can then focus our search on numbers that fall between the bounds.

EXAMPLE 3 Find upper and lower integer bounds for the real zeros of $f(x) = x^3 + 3x^2 - 2x - 5$.

SOLUTION Our strategy will be to check for an upper bound by beginning with 1 and working upward to 2, then to 3, and so forth. The results of the synthetic divisions by $x - 1$ and $x - 2$ are

$$\begin{array}{r|rrrr} 1 & 1 & 3 & -2 & -5 \\ & & 1 & 4 & 2 \\ \hline & 1 & 4 & 2 & -3 \end{array} \quad \text{and} \quad \begin{array}{r|rrrr} 2 & 1 & 3 & -2 & -5 \\ & & 2 & 10 & 16 \\ \hline & 1 & 5 & 8 & 11 \end{array}$$

Since all numbers in the third row of the synthetic division by $x - 2$ are positive, it follows that 2 is an upper bound for the real roots of $f(x) = 0$. This means that we needn't bother to check whether any number greater than 2 is a root of the equation $f(x) = 0$. By trial and error, we use -3 and -4 to check for a lower bound.

$$\begin{array}{r|rrrr} -3 & 1 & 3 & -2 & -5 \\ & & -3 & 0 & 6 \\ \hline & 1 & 0 & -2 & 1 \end{array} \quad \text{and} \quad \begin{array}{r|rrrr} -4 & 1 & 3 & -2 & -5 \\ & & -4 & 4 & -8 \\ \hline & 1 & -1 & 2 & -13 \end{array}$$

We see that the numbers in the third row of the synthetic division by $x - (-4)$ are alternately positive and negative; it follows that -4 is a lower bound for the real zeros of f. Thus any real zero of f would have to be in the interval $(-4, 2)$.

Once the upper and lower bounds are determined, we can apply the rational zero theorem to find the possible rational number zeros that fall between the bounds. Finally, if there are irrational zeros remaining, we can use a graph of the function (sketched on the interval defined by the upper and lower bounds), along with the bisection method, to approximate these values.

EXAMPLE 4 [c] Use the graph of $f(x) = x^3 + 3x^2 - 2x - 5$, along with the bisection method to approximate the value of the largest zero so that $|f(x)| < 0.005$.

SOLUTION Using Descartes' rule of signs, we list the possible number of positive and negative real zeros. This is the same function that was used in Example 3. Thus we know that all real zeros for f must be in the interval $(-4, 2)$. After applying the rational zero theorem and testing the possible zeros, -1 and 1 in the interval $(-4, 2)$, we find that there are no rational zeros.

Number of positive zeros	Number of negative zeros
1	0
1	2

Next we apply the bisection method:

Step 1. The computer-generated graph of f on the interval $(-4, 2)$ (Figure 3) shows that the largest zero lies in the interval $[1, 2]$, since $f(x)$ changes algebraic signs at the endpoints. In fact $f(1) = -3$ and $f(2) = 11$.

Step 2. The midpoint of interval $[1, 2]$ is 1.5 and, by using a calculator, we find that

$$f(1.5) = 2.125$$

Step 3. Since $f(1) = -3$ and $f(1.5) = 2.125$, we select interval $[1, 1.5]$ to do the next bisection.

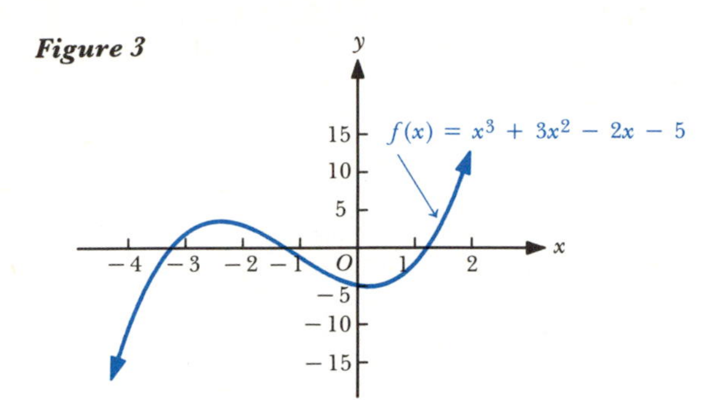

Figure 3

Step 4. The midpoint of interval [1, 1.5] is 1.25 and $f(1.25) = -0.8594$ (approximately).

Next, we select interval [1.25, 1.5] because $f(1.25)$ is negative and $f(1.5)$ is positive. The midpoint is 1.375 and $f(1.375) = 0.5215$ (approximately).

Next, we select interval [1.25, 1.375] because $f(1.25)$ is negative and $f(1.375)$ is positive. The midpoint is 1.3125 and $f(1.3125) = -0.1960$ (approximately).

Then we select interval [1.3125, 1.375] because $f(1.3125)$ is negative and $f(1.375)$ is positive. The midpoint is approximately 1.3438 and $f(1.3438) = 0.1564$ (approximately).

Continuing this process will yield more accurate approximations of the zero of f between 1 and 2. The progression of improved approximations is shown in the following table.

Interval	Midpoint m (4 decimal places)	$f(m)$ (4 decimal places)
[1, 2]	1.5000	2.1250
[1, 1.5]	1.2500	−0.8594
[1.25, 1.5]	1.3750	0.5215
[1.25, 1.375]	1.3125	−0.1960
[1.3125, 1.375]	1.3438	0.1564
[1.3125, 1.3438]	1.3282	−0.0210
[1.3282, 1.3438]	1.3360	0.0673
[1.3282, 1.3360]	1.3321	0.0231
[1.3282, 1.3321]	1.3302	0.0016

Admittedly the bisection method entails tedious computations; however, with the increasing availability of programmable calculators and appropriate microcomputer software, much of the tedium can be eliminated. Other methods of approximating irrational zeros are available, but they make use of calculus.

PROBLEM SET 3.5

In problems 1–10, find a polynomial function of the specified degree with real coefficients that has the given numbers as its zeros.

1. $1 + i$, $1 - i$; degree 2

2. $3, -3, 3i, -3i$; degree 4

3. -1, $2 + i$; degree 3

4. $1 + 5i$, $1 - 5i$, 2 (double zero); degree 4

5. $-2, 3, i$ (double zero); degree 6

6. $0, 1, 2i, i$ (double zero); degree 8

7. $1 - 3i$ (double zero); degree 4

8. $\frac{1}{2}i$, $1 - 2i$; degree 4

9. $-\frac{1}{2} + \frac{\sqrt{3}}{2}i$, 2 (double zero); degree 4

10. $\sqrt{3}$, $-\sqrt{2}$, $1 + \sqrt{5}i$; degree 4

In problems 11–14, use the given information to find all other zeros of f and express $f(x)$ in completely factored form in the complex number system.

11. $f(x) = x^4 + 13x^2 + 36$, where $x = 2i$ is a zero of f
12. $f(x) = 2x^6 + x^5 + 2x^3 - 6x^2 + x - 4$, where $x = i$ is a double zero of f
13. $f(x) = x^4 + 2x^3 - 4x - 4$, where $x = -1 + i$ is a zero of f
14. $f(x) = x^5 - 2x^4 + 2x^3 + 8x^2 - 16x + 16$, where $x = 1 - i$ is a zero of f

In problems 15–19, find upper and lower integer bounds for the real zeros of f.

15. $f(x) = x^3 - 4x^2 + x + 6$
16. $f(x) = 2x^3 - 7x^2 - 7x + 5$
17. $f(x) = 2x^4 - x^3 - 11x^2 + 4x + 12$
18. $f(x) = x^4 + 5x^3 - 64x - 320$
19. $f(x) = x^5 - 2x^4 - 9x^3 + 22x^2 + 4x - 24$
20. Explain why the bisection method cannot be applied to determine approximations for the zeros of $f(x) = x^4 + 8x^2 + 16$.

Figure 4

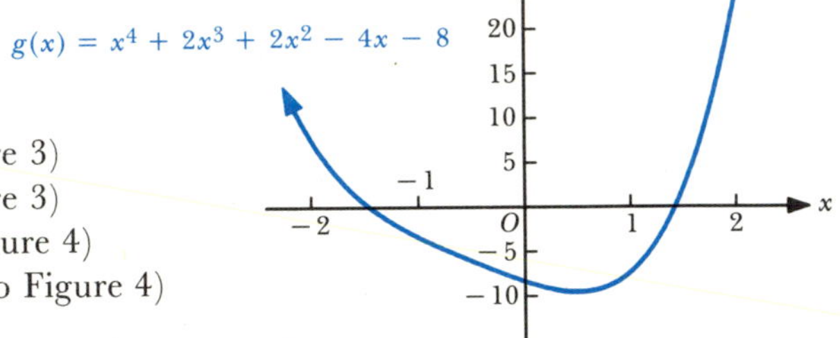

c In problems 21–24, use the bisection method to approximate the value of the zero on the specified interval so that $|f(x)| < 0.005$.

21. $f(x) = x^3 + 3x^2 - 2x - 5$, $[-2, -1]$ (Refer to Figure 3)
22. $f(x) = x^3 + 3x^2 - 2x - 5$, $[-4, -3]$ (Refer to Figure 3)
23. $f(x) = x^4 + 2x^3 + 2x^2 - 4x - 8$, $[1, 2]$ (Refer to Figure 4)
24. $f(x) = x^4 + 2x^3 + 2x^2 - 4x - 8$, $[-2, -1]$ (Refer to Figure 4)

c In problems 25–28, graph the function and use the bisection method to approximate the values of the x intercepts that are irrational numbers so that $|f(x)| < 0.005$.

25. $f(x) = x^3 + x - 3$
26. $f(x) = x^4 - 7x^2 + 10$
27. $f(x) = x^4 + 2x^3 - 4x - 4$
28. $f(x) = x^3 - 9x + 3$

3.6 Rational Functions

A function f of the form

$$f(x) = \frac{g(x)}{h(x)}$$

where $g(x)$ and $h(x)$ are polynomials, is called a **rational function.** The domain of f consists of all real numbers x except those for which $h(x) = 0$, since division by zero is undefined.

Examples of rational functions are

$$f(x) = \frac{1}{x}, \qquad g(x) = \frac{x+1}{x-1}, \qquad \text{and} \qquad H(x) = \frac{3x^3 + 5x}{x^2 - 4}$$

EXAMPLE 1 Find the domain of each function.

(a) $f(x) = \dfrac{1}{x}$ (b) $g(x) = \dfrac{x+1}{x-1}$ (c) $H(x) = \dfrac{3x^3 + 5x}{x^2 - 4}$

SOLUTION (a) The domain of $f(x) = 1/x$ is the set of all real numbers except 0.

(b) The domain of

$$g(x) = \frac{x+1}{x-1}$$

is the set of all real numbers except 1.

(c) The domain of

$$H(x) = \frac{3x^3 + 5x}{x^2 - 4}$$

is the set of all real numbers except -2 and 2.

If $f(x) = g(x)/h(x)$ is a rational function, where $h(a) = 0$ and $g(a) \neq 0$, then the graph of f displays some interesting behavior for values of x close to a, as we shall see in the next example.

EXAMPLE 2 Graph $f(x) = \dfrac{1}{x}$.

SOLUTION The domain of f includes all real numbers *except* $x = 0$. Upon examining the values in Table 1, we notice that as the values of x get closer to 0 "from the right" on the x axis, the corresponding values of $f(x)$ become very large.

Table 1

x	3	2	1	$\frac{1}{2}$	$\frac{1}{4}$	$\frac{1}{10}$	$\frac{1}{100}$	$\frac{1}{1000}$
$f(x)$	$\frac{1}{3}$	$\frac{1}{2}$	1	2	4	10	100	1000

This situation occurs because the denominator x is getting very close to zero, so that the corresponding values of $1/x$ increase without bound.

Table 2 shows that as x gets closer to 0 "from the left" on the x axis, the corresponding values of $f(x)$, which are negative numbers, become very large in absolute value.

Table 2

x	-2	-1	$-\frac{1}{2}$	$-\frac{1}{4}$	$-\frac{1}{10}$	$-\frac{1}{100}$	$-\frac{1}{1000}$
$f(x)$	$-\frac{1}{2}$	-1	-2	-4	-10	-100	-1000

x	$f(x)$
-2	$-\frac{1}{2}$
-1	-1
$-\frac{1}{4}$	-4
$\frac{1}{2}$	2
1	1
2	$\frac{1}{2}$
3	$\frac{1}{3}$

Figure 1

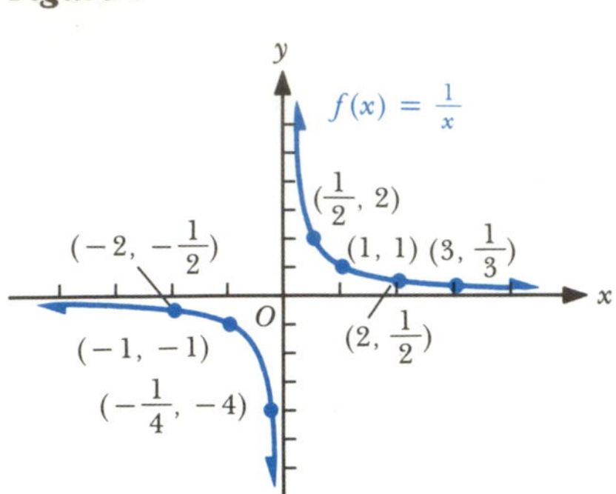

By plotting a few points from both Tables 1 and 2 and connecting the points, noting that there can be no point on the graph corresponding to $x = 0$, we come up with two separate curves that together constitute the graph of f (Figure 1). Notice that since $f(-x) = 1/(-x) = -1/x = -f(x)$, the graph of f is symmetric with respect to the origin.

The behavior of the graph of $f(x) = 1/x$, as x gets closer to 0, is described by saying that the graph of f is getting closer to the line $x = 0$ "asymptotically," and the line $x = 0$ is called a *vertical asymptote.* More formally, we have the following criteria.

Vertical Asymptote Criteria

A rational function $f(x) = g(x)/h(x)$ has a **vertical asymptote** at $x = a$ if $h(a) = 0$ and $g(a) \neq 0$. The value $|g(x)/h(x)|$ increases without bound as x gets closer to a.

EXAMPLE 3 Graph $g(x) = \dfrac{1}{x^2}$ and identify any vertical asymptotes.

SOLUTION Clearly, the domain of g consists of all real numbers except $x = 0$. In fact, the line $x = 0$ is a vertical asymptote since, when $x = 0$, the numerator 1 is not zero and the denominator x^2 is zero. Also, since

$$g(-x) = \frac{1}{(-x)^2} = \frac{1}{x^2} = g(x)$$

g is even and thus the graph is symmetric with respect to the y axis. The table of values next to Figure 2 is used to sketch the part of the graph of g that lies in quadrant I. The other branch of the graph in quadrant II is obtained by using symmetry with respect to the y axis.

We use the graphs of $f(x) = 1/x$ and $g(x) = 1/x^2$, together with the techniques of shifting, stretching, flattening, and reflecting, to obtain graphs of other functions.

Figure 2

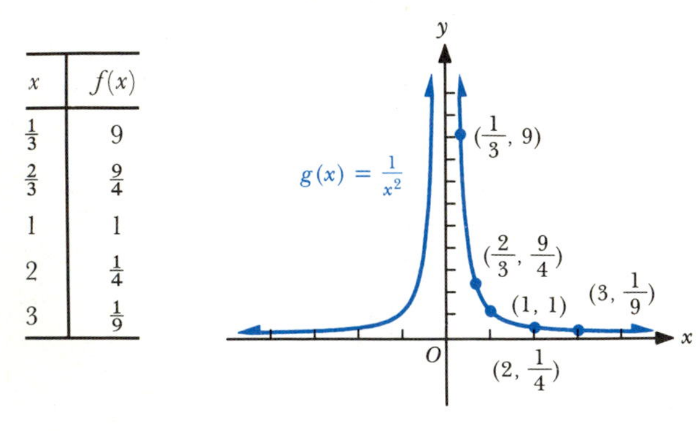

x	$f(x)$
$\frac{1}{3}$	9
$\frac{2}{3}$	$\frac{9}{4}$
1	1
2	$\frac{1}{4}$
3	$\frac{1}{9}$

In Examples 4 and 5, graph the function as specified. Locate the x and y intercepts and any vertical asymptotes.

EXAMPLE 4 Use the graph of $y = \dfrac{1}{x}$ to graph $g(x) = \dfrac{1}{x-3}$.

SOLUTION We can obtain the graph of $g(x) = 1/(x - 3)$ geometrically from the graph of $y = 1/x$ by horizontally shifting the graph three units to the right (Figure 3). Thus g has a vertical asymptote at $x = 3$. The graph of g has a y intercept at $y = -\frac{1}{3}$ that is found by substituting $x = 0$ into the function. There is no x intercept.

Figure 3

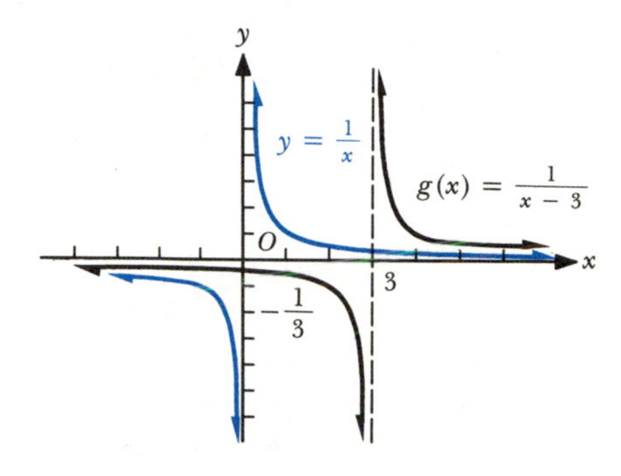

EXAMPLE 5 Use the graph of $y = \dfrac{1}{x^2}$ to graph $f(x) = -\dfrac{2}{x^2}$.

SOLUTION The graph of $f(x) = -2/x^2 = -(2)(1/x^2)$ can be obtained from the graph of $y = 1/x^2$ as follows:

Figure 4

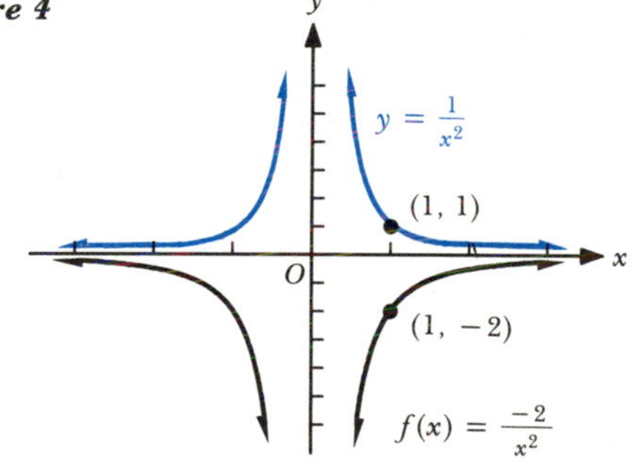

First, we vertically stretch the graph of $y = 1/x^2$ by the multiple 2.

Second, because of the negative sign, we reflect this graph to get the graph of f (Figure 4). There are no x or y intercepts and the y axis is a vertical asymptote.

Consider the graph of the rational function

$$f(x) = \frac{6x}{x - 2}$$

The domain of f includes all real numbers except 2. In fact, $x = 2$ is a vertical asymptote. Both the x and y intercepts equal 0. Using the division algorithm, we can write $f(x)$ as

$$f(x) = \frac{6x}{x - 2} = 6 + \frac{12}{x - 2}$$

Notice here that the degree of the numerator is at least as large as the degree of the denominator. This comparison of the degrees of the numerator and the denominator is what prompts the use of the division algorithm. Although it is possible to use $6 + 12/(x - 2)$ to obtain the graph of f by shifting and stretching the graph of $y = 1/x$, let us consider another point of view.

x	$f(x)$
-2	3
$-\frac{3}{2}$	$\frac{18}{7}$
-1	2
0	0
$\frac{1}{2}$	-2
1	-6
3	18

Figure 5

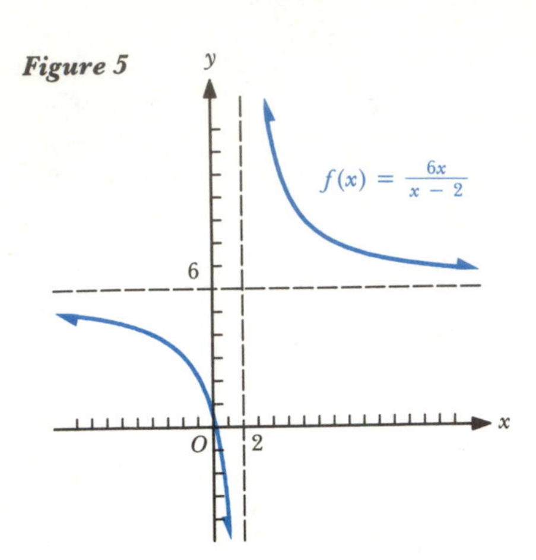

As x becomes larger in absolute value, the rational expression $12/(x-2)$ approaches zero. As that happens, $f(x) = 6 + 12/(x-2)$ gets very close to 6. In other words, as x becomes very large in absolute value, $f(x)$ approaches 6 "asymptotically." The line $y = 6$ is referred to as a *horizontal asymptote*. Notice also that if $x > 2$, then $f(x) > 6$; and if $x < 2$, then $f(x) < 6$. The graph of f illustrates this behavior (Figure 5).

In general, we have the following criteria.

Horizontal Asymptote Criteria

> A rational function $f(x) = g(x)/h(x)$ has a **horizontal asymptote** at $y = b$ if the values of $f(x)$ approach the value b as $|x|$ becomes increasingly large.

Note that the x axis is a horizontal asymptote for $f(x) = 1/x$ (Figure 1), $g(x) = 1/x^2$ (Figure 2), $g(x) = 1/(x-3)$ (Figure 3), and $f(x) = -2/x^2$ (Figure 4).

In Examples 6–8, find the domain, the x and y intercepts, and the vertical and the horizontal asymptotes. Sketch the graph.

EXAMPLE 6 $\quad f(x) = \dfrac{3x + 17}{x + 5}$

SOLUTION The domain of f includes all real numbers *except* -5. Since, for $x = -5$, $x + 5 = 0$ and $3x + 17 \neq 0$, it follows that $x = -5$ is a vertical asymptote. When $x = 0$, $f(x) = \frac{17}{5}$, so that the y intercept is $\frac{17}{5}$. Setting $f(x) = 0$ yields

$$\frac{3x + 17}{x + 5} = 0$$

so that $-\frac{17}{3}$ is the x intercept. Since the degree of the denominator equals the degree of the numerator, we use the division algorithm to write $f(x)$ as

$$f(x) = \frac{3x + 17}{x + 5} = 3 + \frac{2}{x + 5}$$

As $|x|$ becomes increasingly large, $2/(x+5)$ approaches the value 0, and $f(x)$, in turn, approaches the value 3, so that $y = 3$ is a horizontal asymptote. The graph of f is given in Figure 6.

Figure 6

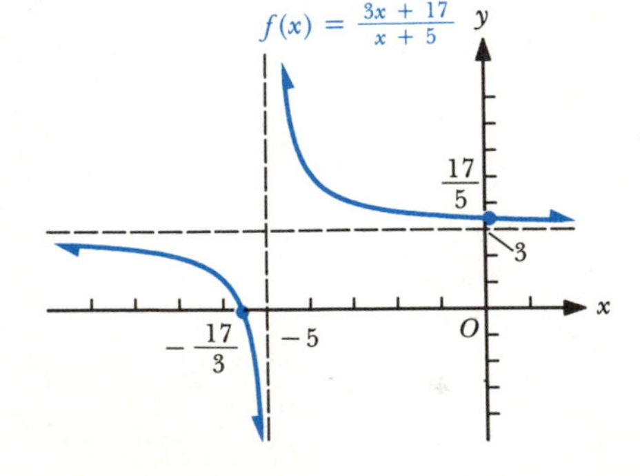

EXAMPLE 7 $f(x) = \dfrac{5}{x^2 + 4}$

SOLUTION The domain of f consists of all real numbers. When $x = 0$, $f(x) = \frac{5}{4}$ so that the y intercept is $\frac{5}{4}$. Since $5/(x^2 + 4) = 0$ has no solution, there is no x intercept. Also, since $x^2 + 4 = 0$ has no real number solution, f has no vertical asymptote. Notice that

$$f(-x) = \frac{5}{(-x)^2 + 4} = \frac{5}{x^2 + 4} = f(x)$$

so that the graph is symmetric with respect to the y axis. If we allow $|x|$ to become increasingly large, the fraction $5/(x^2 + 4)$ approaches the value 0, and so that $y = 0$ is a horizontal asymptote. The graph of f is displayed in Figure 7.

Figure 7

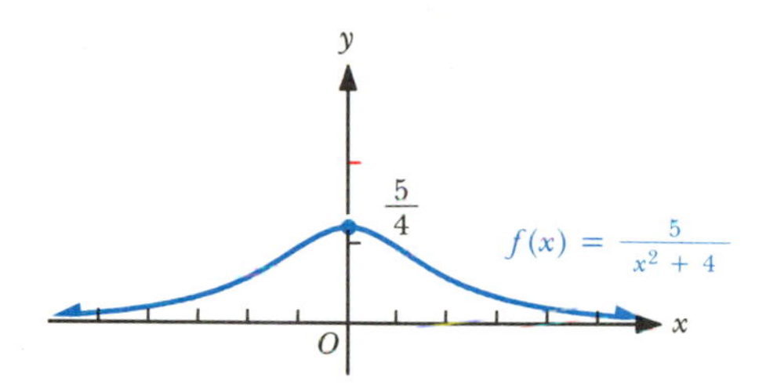

EXAMPLE 8 $f(x) = \dfrac{2x^2 + 1}{2x^2 - 3x}$

SOLUTION Since

$$\frac{2x^2 + 1}{2x^2 - 3x} = \frac{2x^2 + 1}{x(2x - 3)}$$

it follows that the domain of f includes all real numbers *except* $x = 0$ and $x = \frac{3}{2}$. Since neither of these two values of x cause the expression $2x^2 + 1$ to equal 0, it follows that $x = 0$ and $x = \frac{3}{2}$ are vertical asymptotes. Also f has no y intercept because $f(0)$ is not defined; and f has no x intercept because $2x^2 + 1 = 0$ does not have a solution in the real number system. Since the degree of the numerator is the same as the degree of the denominator, we use the division algorithm to get

$$f(x) = \frac{2x^2 + 1}{2x^2 - 3x} = 1 + \frac{3x + 1}{2x^2 - 3x}$$

To determine the horizontal asymptote, we divide both the numerator and the denominator by x^2 to obtain

$$\frac{3x + 1}{2x^2 - 3x} = \frac{\dfrac{3}{x} + \dfrac{1}{x^2}}{2 - \dfrac{3}{x}}$$

x	$f(x)$
-4	$\frac{3}{4}$
-1	$\frac{3}{5}$
$-\frac{1}{2}$	$\frac{3}{4}$
$-\frac{1}{3}$	1
$\frac{1}{2}$	$-\frac{3}{2}$
1	-3
2	$\frac{9}{2}$
4	$\frac{33}{20}$

Figure 8

We notice that as $|x|$ becomes very large,

$$\frac{\frac{3}{x} + \frac{1}{x^2}}{2 - \frac{3}{x}}$$

approaches $(0 + 0)/(2 - 0) = 0$, so that, in turn, $f(x) = 1 + (3x + 1)/(2x^2 - 3x)$ approaches 1 asymptotically; that is, $y = 1$ is a horizontal asymptote. Using this information, along with plotting a few points, we obtain the graph (Figure 8).

Notice that the graph of f *intersects the horizontal asymptote.* In fact, if we set

$$\frac{2x^2 + 1}{2x^2 - 3x} = 1$$

we obtain

$$2x^2 + 1 = 2x^2 - 3x$$

$$1 = -3x$$

$$x = -\frac{1}{3}$$

so that $f(-\frac{1}{3}) = 1$, as shown in the graph.

Some rational functions in which the numerator and denominator have a common factor are often used in calculus. Such functions have "holes" in their graphs and therefore are not *continuous.*

EXAMPLE 9 Sketch the graph of the function $f(x) = \dfrac{x^2 - 4}{x - 2}$.

SOLUTION We observe that

$$f(x) = \frac{x^2 - 4}{x - 2} = \frac{\cancel{(x - 2)}(x + 2)}{\cancel{x - 2}}$$

$$= x + 2, \qquad \text{for} \qquad x \neq 2$$

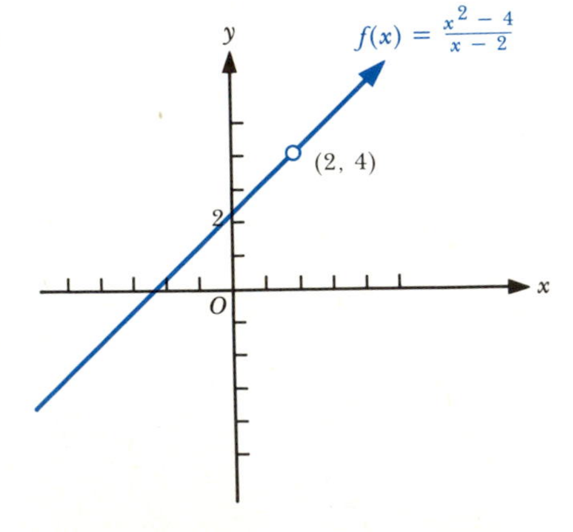

Figure 9

Thus, the graph of the function f coincides with the line $y = x + 2$, with the exception that f is undefined when $x = 2$ (Figure 9). Notice here that $x = 2$ is *not* a vertical asymptote; *both* the numerator and denominator of $f(x)$ equal 0 when $x = 2$.

PROBLEM SET 3.6

In problems 1–10, determine the domain and all vertical asymptotes of each rational function.

1. $f(x) = \dfrac{-2}{x + 7}$

2. $h(x) = 5x^{-1}$

3. $g(x) = \dfrac{1}{(x - 4)(x + 7)^2}$

4. $f(x) = \dfrac{x^3 - 8}{x - 2}$

5. $h(x) = \dfrac{1}{x^2 + 4x - 5}$

6. $g(x) = \dfrac{2}{x} - \dfrac{1}{x - 1}$

7. $g(x) = \dfrac{x^2 - 9}{x - 3}$

8. $h(x) = \dfrac{x^3 - 2x}{x^3 + 5x^2 + 6x}$

9. $f(x) = \dfrac{3}{x^2 + 25}$

10. $g(x) = 5 - \dfrac{1}{x - 5}$

In problems 11–26, use the graph of either $y = 1/x$ or $y = 1/x^2$ along with the techniques of shifting, stretching, flattening, and reflecting to obtain a graph of the function. Determine the vertical and horizontal asymptotes.

11. $f(x) = \dfrac{2}{x + 4}$

12. $g(x) = \dfrac{1}{x} + 3$

13. $g(x) = \dfrac{1}{(x - 3)^2}$

14. $h(x) = \dfrac{4}{x^2} - 1$

15. $f(x) = \dfrac{-3}{(x + 1)^2}$

16. $g(x) = \dfrac{4}{2x + 3}$

17. $h(x) = \dfrac{3x + 2}{x}$

18. $f(x) = -\dfrac{5}{x^2}$

19. $g(x) = \dfrac{x^2 - 2}{x^2}$

20. $g(x) = \dfrac{3}{x - 2} + 2$

21. $f(x) = \dfrac{2}{(x + 3)^2} - 3$

22. $h(x) = \dfrac{-1}{3(x + 2)^2} - 4$

23. $g(x) = \dfrac{1}{5(x - 2)} + 2$

24. $f(x) = \dfrac{2x^2 + 4x + 3}{x^2 + 2x + 1}$

25. $h(x) = \dfrac{5x - 1}{x + 1}$

26. $f(x) = \dfrac{2x^2 + 3}{x^2} - 1$

In problems 27–42, find the domain, the x and y intercepts, the vertical and horizontal asymptotes, and sketch the graph of each function.

27. $f(x) = \dfrac{x^2 - 9}{x - 3}$

28. $g(x) = \dfrac{x^2 - 1}{x - 1}$

29. $g(x) = \dfrac{2x}{x + 1}$

30. $h(x) = \dfrac{-3x}{x + 4}$

31. $h(x) = \dfrac{x^2}{x^2 + 1}$

32. $f(x) = \dfrac{x^2 - 4}{x^2 + 9}$

33. $g(x) = \dfrac{x^2 + 2}{x^2 + x}$

34. $f(x) = \dfrac{3}{x^2 - 2x - 3}$

35. $g(x) = \dfrac{x^2 - 4}{x^2 - 1}$

36. $h(x) = \dfrac{x^2 + 1}{x^2 - 9}$

37. $g(x) = \dfrac{2x + 8}{(2x - 3)(x + 5)}$

38. $F(x) = \dfrac{x + 4}{(x + 5)(2x - 3)}$

39. $H(x) = \dfrac{4}{(x - 1)(x + 2)^2}$

40. $h(x) = \dfrac{(2x - 3)^2}{(x + 4)(x + 5)}$

41. $f(x) = \dfrac{x^2 - 3x - 4}{x^2 \;[\text{illegible}]\; x - 2}$

42. $g(x) = \dfrac{x^2 - 3x - 4}{x^2 - x + 2}$

43. Figure 10a displays the graphs of $y = 1/x$ and $f(x) = 1/x^3$ on the same coordinate system, and Figure 10b displays the graphs of $y = 1/x^2$ and $g(x) = 1/x^4$ on the same coordinate system.
(a) Find the domains of f and g.
(b) Discuss the symmetry of the graphs of f and g.
(c) Find the horizontal and vertical asymptotes of f and g.

Figure 10

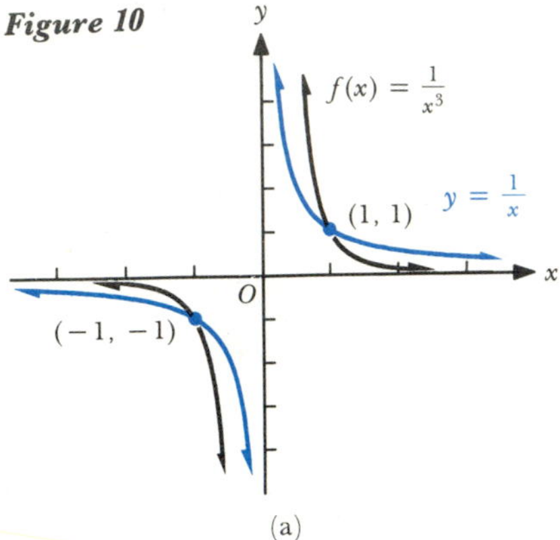

(a)

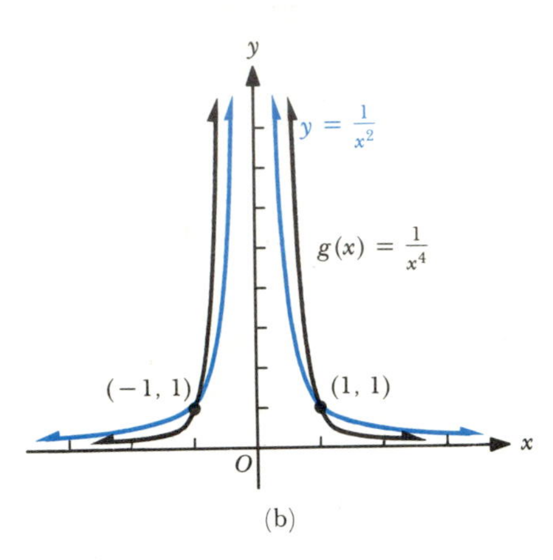

(b)

44. Can the graph of a rational function intersect a vertical asymptote? Can it intersect a horizontal asymptote? Explain.

In problems 45–50, use the graphs in Figure 10 of problem 43 to sketch the graph of each rational function. Find the horizontal and vertical asymptotes.

45. $g(x) = \dfrac{1}{x^4} + 3$

46. $f(x) = \dfrac{4}{x^3} - 1$

47. $f(x) = -\dfrac{2}{x^3}$

48. $g(x) = \dfrac{-3}{x^4}$

49. $h(x) = \dfrac{2}{(x-2)^3} + 1$

50. $g(x) = \dfrac{2}{x^3 + 3x^2 + 3x + 1}$

51. The cost $c(x)$ (in dollars) of producing x computer interface cables for a color graphics printer is found to be:

$$c(x) = \begin{cases} 90 & \text{if} \quad 1 \le x \le 100 \\ \dfrac{9000}{x} & \text{if} \quad 100 < x \le 200 \\ 45 & \text{if} \quad x > 200 \end{cases}$$

Graph the function and interpret the graph.

52. Living neural tissue can be excited by an electric current only if the current across the tissue reaches or exceeds a certain threshold, which we denote by I. The threshold I depends on the duration t of current flow. **Weiss' law** expresses I as a function of t by the equation

$$I = \frac{a}{t} + b$$

where a and b are positive constants.

(a) Graph the rational function $g(t) = 4/t$ for $t > 0$.
(b) Use the graph of g to graph $I = (4/t) + 3$ for $t > 0$.
(c) Use the graph of part (b) to describe the behavior of the threshold I as t approaches 0 (from the right of 0), and as t becomes very large.

53. A psychologist estimates that the percent P of the information remembered by a subject in an experiment t months after the subject learned the material is given by the function $P = 100/(1 + t)$. Sketch the graph of P. Use the graph to interpret the values of P as t gets larger.

REVIEW PROBLEM SET, CHAPTER 3

1. Use the graph of $y = x^2$ to graph each function.
(a) $f(x) = (x + 5)^2$
(b) $g(x) = (x - \frac{1}{2})^2$
(c) $h(x) = 2(x - 1)^2 + 4$
(d) $f(x) = -\frac{1}{3}(x + 2)^2 - 1$

2. Use the method of completing the square illustrated on page 135 to find the extreme point of each of the following quadratic functions.
(a) $g(x) = x^2 + 4x - 7$
(b) $f(x) = 2x^2 + 5x - 1$
(c) $h(x) = -3x^2 + 6x + 9$
(d) $g(x) = -x^2 + \frac{1}{2}x + 1$

In problems 3–10, sketch the graph by locating the x and y intercepts, the extreme point, and the axis of symmetry of the parabola. Use the graph to determine the range.

3. $f(x) = x^2 + 14x + 49$
4. $g(x) = -2x^2 + 1$
5. $h(x) = -2x^2 + x + 3$
6. $f(x) = 3x^2 + 7x + 2$
7. $g(x) = -x^2 - 3x - 1$
8. $h(x) = x^2 - x + 1$
9. $h(x) = 5x^2 + 7x + 2$
10. $f(x) = -\frac{1}{2}x^2 + x - 3$

In problems 11–15, graph each of the quadratic functions by locating the x and y intercepts and the extreme point. Then use the graph to solve the given associated inequality.

11. $f(x) = -x^2 + 2$; $-x^2 + 2 \geq 0$
12. $g(x) = 2x^2 - x - 1$; $2x^2 - x - 1 < 0$
13. $f(x) = 6x^2 - 5x - 4$; $6x^2 - 5x - 4 \leq 0$
14. $h(x) = -3 + 10x - 8x^2$; $-3 + 10x - 8x^2 < 0$
15. $g(x) = -x^2 + 2x - 1$; $-x^2 + 2x - 1 < 0$

16. A projectile is fired vertically upward from the ground in such a way that it is h feet above the ground t seconds after the firing, where $h = 88t - 16t^2$.
(a) Find t when $h = 0$.
(b) Find the maximum height reached by the projectile.
(c) Sketch the graph of the equation.

17. A manufacturer can sell x units of a product at a price of p cents per unit, where $p = 300 - 0.25x$. What number of units should be sold to achieve maximum income?

18. An apple orchard now has 30 trees per acre, and the average yield is 400 apples per tree. For each additional tree planted per acre, the average yield is reduced by approximately 10 apples. How many trees per acre will give the largest crop of apples?

In problems 19–24, use the graph of f to graph g.

19. $g(x) = 2x^3$ and $f(x) = x^3$

20. $g(x) = -\frac{1}{2}x^4$ and $f(x) = x^4$

21. $g(x) = \frac{1}{3}(x - 2)^4$ and $f(x) = x^4$

22. $g(x) = 4x^3 - 2$ and $f(x) = x^3$

23. $g(x) = -3(x + 2)^3 + 1$ and $f(x) = x^3$

24. $g(x) = -2x^5 + 3$ and $f(x) = x^5$

In problems 25–30, find all real number zeros of the function and indicate the multiplicity of each zero.

25. $h(x) = \frac{1}{3}x + \frac{1}{2}$

26. $g(t) = t^4 - t$

27. $f(w) = (w - 1)(w^2 - 5w + 4)$

28. $h(y) = y^2 - 5y$

29. $F(x) = (x^2 - 4)(x^2 - 3x + 2)$

30. $g(x) = (2x^2 + 5x)(6x^2 - 11x - 10)$

In problems 31–34, sketch the graph of the function by locating the x and y intercepts and plotting a few points. Use the graph to solve the accompanying inequality.

31. $f(x) = x(x + 3)(x - 5)$; $x(x + 3)(x - 5) < 0$

32. $g(x) = x(3x + 5)^2$; $x(3x + 5)^2 \geq 0$

33. $h(x) = x(x - 1)^2(x + 2)$; $x(x - 1)^2(x + 2) \leq 0$

34. $f(x) = x^3(x + 4)^2$; $x^3(x + 4)^2 > 0$

In problems 35–38, use synthetic division to determine the quotient $Q(x)$ and the remainder R so that $f(x) = (x - c)Q(x) + R$.

35. $f(x) = 3x^3 + 5x^2 + 7x - 3$; $c = 2$

36. $f(x) = 5x^4 - 2x^3 + 11x^2 + 5x + 36$; $c = 1$

37. $f(x) = 2x^3 + 4x^2 + 3x - 18$; $c = -1$

38. $f(x) = x^7 - 5$; $c = 2$

39. Let $f(x) = x^5 + x^4 - 3x^3 - 5x^2 + x + 7$. Use synthetic division to find each of the following values.
(a) $f(2)$ (b) $f(3)$ (c) $f(-1)$ (d) $f(-2)$ (e) $f(\frac{1}{2})$

40. Use synthetic division to find k so that $x + 2$ is a factor of $5x^3 - 2x^2 + 4x + k$.

In problems 41–44, use Descartes' rule of signs to determine the possible number of positive real zeros, negative real zeros, and complex zeros.

41. $f(x) = x^3 + 2x^2 - 3x - 11$

42. $h(x) = 3x^4 + 3x^3 - 13x - 6$

43. $g(x) = 2x^4 - 5x^2 + 8x - 3$

44. $f(x) = x^3 + 3x^2 - 4x - 12$

In problems 45–48, find all rational zeros of each function.

45. $f(x) = x^3 - 4x^2 + x + 6$

46. $g(x) = x^4 - 7x^3 + 18x^2 - 20x + 8$

47. $h(x) = 2x^4 + x^3 - 19x^2 - 9x + 9$

48. $f(x) = 4x^4 + 4x^3 - 3x^2 - 2x + 1$

In problems 49–52, write the polynomial function in linear factored form and then sketch the graph of the function by locating the x and y intercepts and plotting a few points.

49. $f(x) = x^4 - 10x^2 + 9$

50. $g(x) = x^3 + 2x^2 - 5x - 6$

51. $h(x) = x^3 - x^2 - 5x - 3$

52. $f(x) = x^4 + 5x^2 - 36$

In problems 53–56, find a polynomial function with real coefficients and of the given degree that has the given numbers as its zeros.

53. $-2, -2, -3, 3$; degree is 4

54. $1, 3, -2, 5$ (multiplicity of 2); degree is 5

55. $2, 2 - i$; degree is 3

56. $i, 1 + i$; degree is 4

In problems 57–60, find upper and lower integer bounds for the real zeros of f.

57. $f(x) = 2x^3 - 7x^2 - 7x + 5$

58. $f(x) = x^3 - 6x^2 + 12x - 8$

59. $f(x) = x^3 - 3x^2 - 4x + 12$

60. $f(x) = x^4 - 9x^2 + 7x + 4$

In problems 61 and 62, use the bisection method to approximate the value of the zero on the specified interval so that the absolute value of each function is less than 0.005.

61. $h(x) = x^3 - 5x^2 + 8x - 3$; $[0, 1]$

62. $g(x) = 2x^3 - 7x^2 - 7x + 5$; $[4, 5]$

In problems 63–68, determine the domain and all vertical asymptotes of each rational function.

63. $f(x) = \dfrac{3}{x + 1}$

64. $g(x) = \dfrac{x^2 - 25}{x + 5}$

65. $h(x) = \dfrac{-2x}{x^2 - 4}$

66. $f(x) = \dfrac{-8x}{x^3 - 9x}$

67. $g(x) = \dfrac{x + 7}{x^2 - 5x - 6}$

68. $h(x) = \dfrac{5x}{(x^3 - 8)(x^2 - 49)}$

69. Use the graph of f to graph g in each situation.

(a) $g(x) = \dfrac{3}{(x - 1)^2}$ and $f(x) = \dfrac{1}{x^2}$

(b) $g(x) = \dfrac{-2}{(x + 1)^3} + 1$ and $f(x) = \dfrac{1}{x^3}$

(c) $g(x) = \dfrac{-3}{(x - 4)^4} - 2$ and $f(x) = \dfrac{1}{x^4}$

70. Explain why the graphs of $f(x) = x - 4$ and $g(x) = (x^2 - 16)/(x + 4)$ are different.

In problems 71–75, determine the domain, the x and y intercepts, and the horizontal and vertical asymptotes, then sketch the graph of each rational function.

71. $f(x) = \dfrac{2}{x - 5}$

72. $g(x) = \dfrac{x}{x + 3}$

73. $h(x) = \dfrac{x - 2}{x^2 - 9}$

74. $f(x) = \dfrac{5x^2}{x^2 + 4}$

75. $G(x) = \dfrac{x^2 + 2}{x^2 + 3}$

76. According to Boyle's law, at a fixed temperature the pressure p and the volume v of a confined gas satisfy the equation $p = 3000/v$. The pressure is measured in pounds per square inch and the volume in cubic inches. Sketch the graph of the equation $p = 3000/v$ for $v > 0$.

CHAPTER 4

Exponential and Logarithmic Functions

John Napier

In this chapter, we introduce two types of functions that are closely related—*exponential* and *logarithmic* functions. Both are widely used to model situations in biology, chemistry, business and economics, as well as in engineering. Logarithms were first developed in the late sixteenth century by the Scottish mathematician, John Napier (1550–1617), who also seems to have been the first person to use the decimal point in arithmetic as we do today. By shortening the process of computation, logarithms served for many years to speed up the work of astronomers and engineers. Now, however, the widespread use of calculators and microcomputers has virtually eliminated the need to refer to tables of logarithms, or even to use logarithmic computations. Nevertheless, logarithms are still important in many varied applications, as we will see in this chapter.

4.1 Exponential Functions

In Chapter 1, we introduced the basic properties of exponents. Now let us see how to use exponential expressions to define functions that are appropriate for modeling population growth. Suppose, for instance, that a biologist is studying a colony of bacteria. She wishes to determine how the number of bacteria in the culture changes with time. She discovers that it takes one day for the number of bacteria to triple. If there are initially x_0 bacteria present, then there are $3x_0$ present after 1 day, $3(3x_0) = 3^2x_0$ after 2 days, $3(3^2x_0) = 3^3x_0$ after 3 days, and so on; so that after n days there are 3^nx_0 bacteria present. This phenomenon of bacteria growth can be represented by the function

$$f(t) = 3^t x_0$$

where t represents the number of days after the experiment begins and x_0 represents the number of bacteria present initially. Although it is true that the actual experiment supplies data only when t is a *positive integer*, we can assume that the function $f(t) = 3^tx_0$ is *continuous* so that $f(t)$ indicates the number of bacteria present after t days, where t is *any* positive real number. For example, after 2 days, 4 hours ($2\frac{1}{6}$ or $\frac{13}{6}$ days), there would be $3^{13/6}x_0$ bacteria present. It is our purpose here to examine the properties of functions of the type $f(t) = 3^tx_0$.

First we need to extend the definition of exponents, given in Section 1.4 for any *rational number*, so that we will have a meaning for the symbol b^x, where b is a positive number and x is any *real number*. Let us consider the function $f(x) = 2^x$. For each rational value for x we can find a value for 2^x, as shown in Table 1.

Table 1

x	-4	-3	-2	-1	0	$\frac{1}{2}$	1	2	3	4
2^x	$\frac{1}{16}$	$\frac{1}{8}$	$\frac{1}{4}$	$\frac{1}{2}$	1	1.414	2	4	8	16

The points determined by the pairs $(x, 2^x)$ in Table 1 are plotted in Figure 1a. We could, of course, plot as many points as we please by taking more and more rational values for x. However, there would not be any points corresponding to such values of x as $\sqrt{2}$, $-\sqrt{3}$, or π since $2^{\sqrt{2}}$, $2^{-\sqrt{3}}$, and 2^{π} have not been defined. In fact, there would be no points corresponding to any *irrational* values of x. We can fill in the "gaps" of the curve in Figure 1a by making the following agreement: $f(x) = 2^x$ is the function whose domain is the set of *all real* numbers and whose graph is the smooth curve drawn through the points $(x, 2^x)$, where x is rational (Figure 1b).

Figure 1

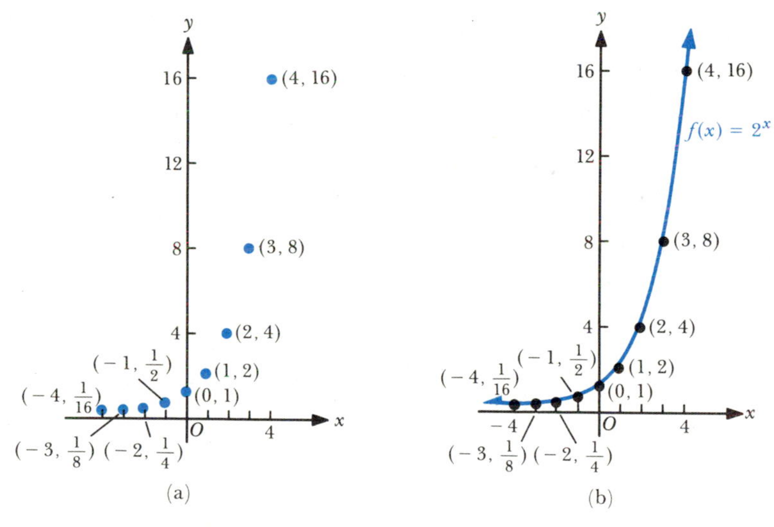

We generalize this idea in the following definition.

DEFINITION 1

Exponential Function

A function f of the form

$$f(x) = b^x$$

where b is a positive real number, is called an **exponential function** with base b.

The graph of $f(x) = b^x$ is the smooth continuous curve drawn through the points (x, b^x) obtained by taking rational values of x. Thus, the domain of *every* exponential function is the set $\mathbb{R}$ of real numbers.

For example, $f(x) = 3^x$, $h(t) = (\frac{1}{3})^t$, $g(x) = 2^{-x} = (2^{-1})^x = (\frac{1}{2})^x$ are all exponential functions with bases 3, $\frac{1}{3}$, and $\frac{1}{2}$, respectively. If we examine the graph of $f(x) = 2^x$ in Figure 1b, we see that the range of this function is the set of all *positive* real numbers and that the x axis is a horizontal asymptote for the graph. In fact the range of *every* exponential function with $b \neq 1$ is the set of positive real numbers and the x axis is a horizontal asymptote for the graph.

EXAMPLE 1 Sketch the graphs of $f(x) = 3^x$ and $g(x) = (\frac{1}{3})^x$.

SOLUTION We begin by calculating values of $f(x) = 3^x$ and $g(x) = (\frac{1}{3})^x$ for integer values of x. Then we plot the points corresponding to the pairs we have found and connect the points with smooth curves to obtain the graphs of $f(x) = 3^x$ (Figure 2a) and $g(x) = (\frac{1}{3})^x$ (Figure 2b).

x	$f(x) = 3^x$	$g(x) = (\frac{1}{3})^x$
-3	$\frac{1}{27}$	27
-2	$\frac{1}{9}$	9
-1	$\frac{1}{3}$	3
0	1	1
1	3	$\frac{1}{3}$
2	9	$\frac{1}{9}$

Figure 2

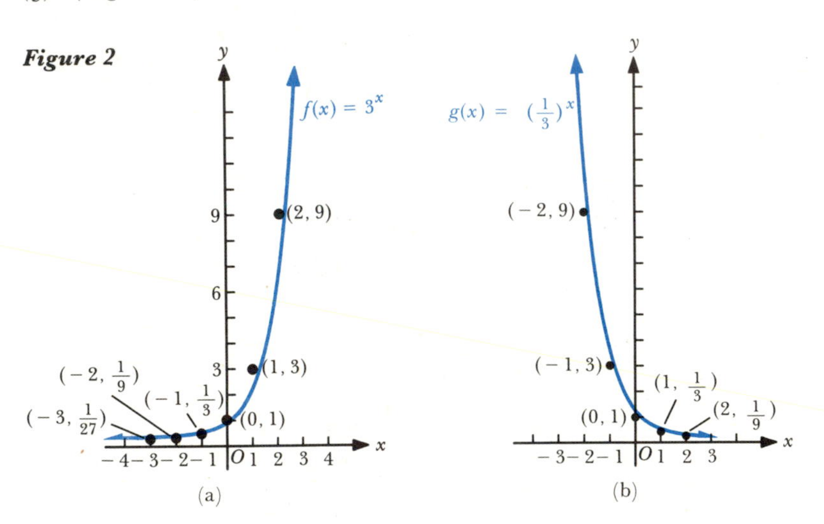

The graphs illustrate the fact that the domain of each of the functions f and g is $\mathbb{R}$, and the range of each function is $(0, \infty)$. Also we note that f is an increasing function, reflecting the fact that the base of f is *greater than* 1; g is a decreasing function, reflecting the fact that the base of g is *less than* 1. Because

$$g(x) = \left(\frac{1}{3}\right)^x = \frac{1}{3^x} = 3^{-x} = f(-x)$$

these curves are reflections of each other across the y axis.

In general, $f(x) = b^x$ *will be an increasing function if* $b > 1$, *and it will be a decreasing function if* $b < 1$. *If* $b = 1$, *the graph of* $f(x) = b^x = 1^x = 1$ *is the line whose equation is* $y = 1$.

The techniques of sketching graphs that were presented in Section 2.4 can be applied in graphing exponential functions, as illustrated in the next example.

EXAMPLE 2 Use the graphs of f and g in Example 1 to sketch the graph of the given function.

(a) $F(x) = 3(3^x)$ (b) $G(x) = -(\frac{1}{3})^x + 1$

SOLUTION

(a) The graph of F is obtained from the graph of $f(x) = 3^x$ by multiplying each ordinate of the graph of f by 3 (Figure 3a). Notice that, since

$$F(x) = 3(3^x) = 3^{x+1} = f(x+1)$$

the graph of F could also be obtained by shifting the graph of f one unit to the left. The domain of F is $\mathbb{R}$ and the range is $(0, \infty)$.

(b) First we obtain the graph of $y = -(\frac{1}{3})^x$ by reflecting the graph of $g(x) = (\frac{1}{3})^x$ across the x axis (Figure 3b). Then the graph of $G(x) = -(\frac{1}{3})^x + 1$ is obtained by shifting the graph of $y = -(\frac{1}{3})^x$ one unit upward (Figure 3b). The domain of G is $\mathbb{R}$ and the range is $(-\infty, 1)$. Notice that $y = 1$ is a horizontal asymptote.

Figure 3

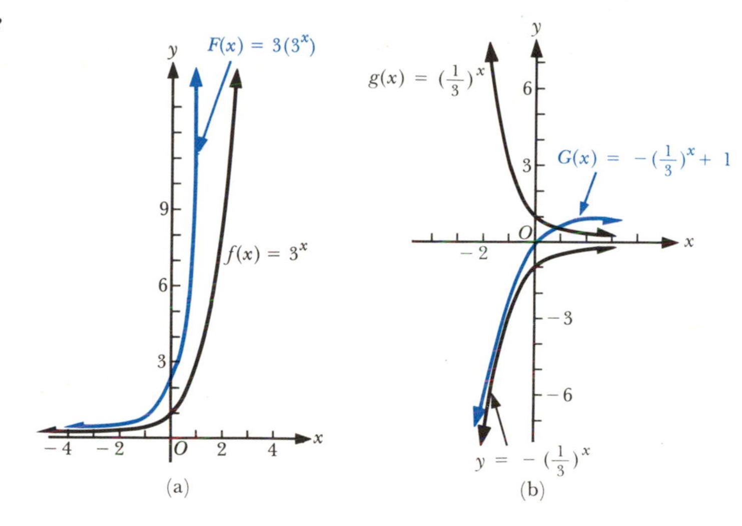

The graphs of exponential functions suggest that b^x is defined for *any* real number x, if $b > 0$. We can approximate the value of b^x for any *irrational* number x by rounding off the decimal representation of x to a finite number of places. Such an approximation for b^x can be obtained by using a calculator with a $\boxed{y^x}$ key. For example, on a nine-digit calculator with a $\boxed{y^x}$ key, we have

$$3^{\sqrt{2}} = 4.72880439$$

$$4^{\pi} = 77.8802336$$

and

$$\pi^{-\sqrt{3}} = 0.137692777$$

The Exponential Function with Base *e*

Leonard Euler

Although any positive number can be used as a base for an exponential function, some bases occur more often than others in applications. The base most commonly used is the number denoted by e, which was first used by Leonhard Euler(1707–1783). The number e is an *irrational* number whose nine decimal place approximation is given by

$$e = 2.718281828$$

Since $e > 1$, the exponential function with base e,

$$f(x) = e^x$$

is an increasing function and its graph (Figure 4) resembles that of $y = 3^x$. The domain is $\mathbb{R}$ and the range is $(0, \infty)$. Notice that $y = 0$ is a horizontal asymptote.

Figure 4

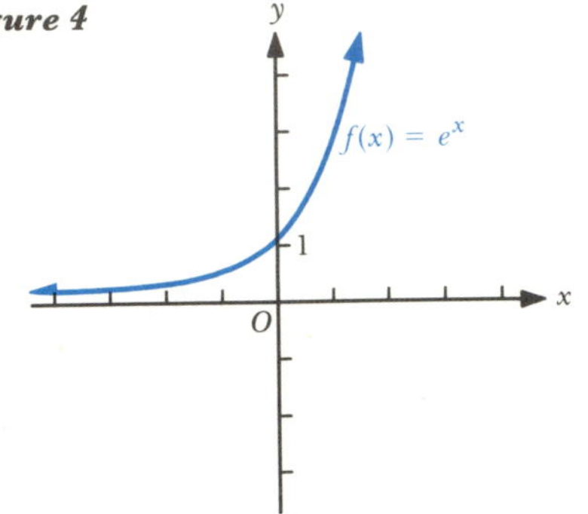

Figure 5

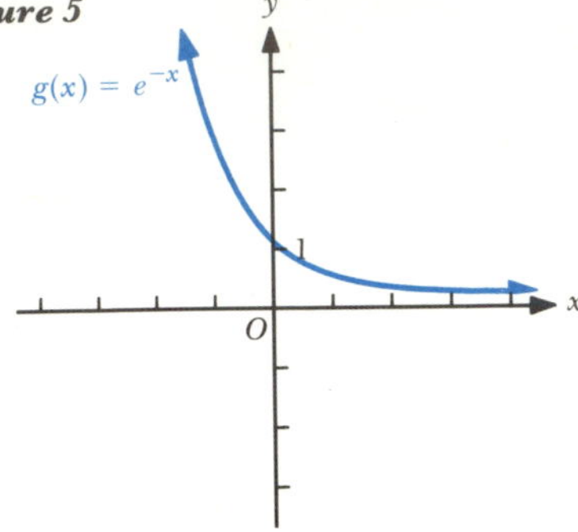

EXAMPLE 3 Use the graph of $f(x) = e^x$ to sketch the graph of $g(x) = e^{-x}$.

SOLUTION The effect of replacing x by $-x$ in the function $f(x) = e^x$ is to reflect the graph of f about the y axis (Figure 5). Since $e^{-1} < 1$, the function $g(x) = e^{-x}$ is a decreasing function, and its graph resembles that of $y = (\frac{1}{3})^x = 3^{-x}$. The domain of g is $\mathbb{R}$ and the range is $(0, \infty)$. Notice that $y = 0$ is a horizontal asymptote.

Appendix Table V contains values of e^x for selected values of x. However, values of e^x can be found easily by using a calculator that has the key $\boxed{e^x}$. If a calculator does not have the $\boxed{e^x}$ key, we use the $\boxed{\text{INV}}$ key, followed by the $\boxed{\ln}$ key.

EXAMPLE 4 [C] Use a calculator to approximate each given number.

(a) e^1 (b) $e^{2.3}$ (c) $e^{\sqrt{3}}$ (d) $e^{-\sqrt{5}}$

SOLUTION On a nine-digit calculator, we have

(a) $e^1 = 2.71828183$ (b) $e^{2.3} = 9.97418246$

(c) $e^{\sqrt{3}} = 5.65223367$ (d) $e^{-\sqrt{5}} = 0.106877925$

Business Applications

Exponential functions occur in many financial transactions involving compound interest, annuities, installment payments, and the like. Consider what happens when a sum of money is deposited in a savings account or invested in a certificate of deposit, so that it earns interest that is "compounded" at stated intervals of time. This means that the amount in the account is increased by a certain percentage at the end of each interval of time called the **compounding period.**

Compound interest is interest paid on the interest previously earned as well as on the initial deposit. The **compound interest formula** will enable us to compute the amount of money in a savings account after any length of time with any initial deposit and any rate of interest and compounding period.

Suppose P dollars, called the **principal,** are deposited in an account with a **nominal annual interest rate** r (expressed in decimal form) compounded *n times* a year. (Notice that the *interest rate per compounding period* is r/n.) The final accumulated amount of S dollars in the account at the end of *k compounding periods* is given by

$$S = P\left(1 + \frac{r}{n}\right)^k$$

At the end of *t years* the number of compounding periods will be $n \cdot t$, and the compound interest formula can be written as

$$S = P\left(1 + \frac{r}{n}\right)^{nt}$$

EXAMPLE 5 [c] If \$5250 is invested in an account paying nominal annual interest at 8%, find the amount S in the account at the end of one year if the interest is compounded (a) Quarterly (b) Monthly

SOLUTION (a) In this case, $P = 5250$, $r = 8\% = 0.08$, $n = 4$, and $t = 1$. The interest rate per period is 0.08/4, and the number of periods is 4, so that

$$S = 5250\left(1 + \frac{0.08}{4}\right)^4 = 5682.7688$$

Thus the amount in the account at the end of the year is \$5682.77.

(b) Here $P = 5250$, $r = 8\% = 0.08$, $n = 12$, and $t = 1$. The interest rate per period is 0.08/12, and the number of periods is 12, so that

$$S = 5250\left(1 + \frac{0.08}{12}\right)^{12} = 5685.7474$$

The amount in the account at the end of one year is \$5685.75.

In Example 5b the amount of interest earned during the year is \$435.75. If the interest were not compounded, a higher rate of interest would be required in order for \$5250 to earn \$435.75 in interest in a year. If we let R represent that higher rate, the value of R can be found from the equation

$$(5250)R = 435.75$$

so that

$$R = \frac{435.75}{5250} = 0.083 \qquad \text{that is} \qquad 8.3\%$$

The rate R is called the **effective annual interest rate** corresponding to the nominal interest rate r. The relation between the nominal interest rate r, the number n of compounding periods per year, and the effective interest rate R is given by the formula

$$R = \left(1 + \frac{r}{n}\right)^n - 1$$

EXAMPLE 6 C Find the effective annual interest rate R corresponding to a nominal interest rate of 10% compounded monthly.

SOLUTION Here $r = 10\% = 0.10$ and $n = 12$, so that

$$R = \left(1 + \frac{r}{n}\right)^n - 1$$

$$= \left(1 + \frac{0.10}{12}\right)^{12} - 1 = 0.10471307$$

Therefore the effective annual interest rate is approximately 10.47%.

The **present value** of a sum of money due to be paid to an investor in the future is the amount which, if it were on hand today, would grow with interest to equal the future sum. Thus, P dollars in hand now is worth S dollars to be received t years in the future. Solving the compound interest formula for P, we obtain the *present value formula*, which gives the present value, P dollars, of the amount, S dollars, due in t years, if investments during this time earn a nominal interest rate r compounded n times a year,

$$P = \frac{S}{\left(1 + \frac{r}{n}\right)^{nt}} = S\left(1 + \frac{r}{n}\right)^{-nt}$$

EXAMPLE 7 C Find the present value of \$8000 to be paid 4 years in the future, if investments during this period earn a nominal interest rate of 9% compounded monthly.

SOLUTION Here $S = 8000$, $r = 9\% = 0.09$, $n = 12$, and $t = 4$ so that

$$P = 8000\left(1 + \frac{0.09}{12}\right)^{-(12)(4)} = 5588.91309$$

Thus the present value is \$5588.91.

In these examples we have used compounding periods of three months (quarterly) and one month. Compounding periods can be of any length; the most common now in use by savings institutions is one day; that is, the amount in the account is compounded 365 times a year. We could go further and compound interest every minute, every second, every half-second, and so on. If the number of compounding periods increases without bound, the interest is said to be compounded *continuously*. When interest is compounded continuously for t years, with nominal rate r, the amount S in the account at the end of the t years can be expressed in terms of an exponential function with base e.

If the principal P is invested in an account with interest rate r **compounded continuously,** the amount S in the account at the end of t years is given by

$$S = Pe^{rt}$$

EXAMPLE 8 [C] If $20,000 is invested in an account that pays 8.5% compounded continuously, what will be the amount S in the account after 10 years?

SOLUTION Here, $P = 20{,}000$, $r = 0.085$, and $t = 10$, so that

$$\begin{aligned} S &= Pe^{rt} \\ &= (20{,}000)e^{(0.085)(10)} \\ &= 46{,}792.9370 \end{aligned}$$

Therefore, the amount is $46,792.94.

Special Exponential Equations

Suppose that the number of bacteria in a certain culture after t hours is given by the function $Q(t) = 2000(3^t)$. One might ask how long it will take to grow 486,000 bacteria in the culture; that is, when is $Q(t) = 486{,}000$? Such a question is answered by solving an equation involving exponents called an **exponential equation.** For instance,

$$2000(3^t) = 486{,}000 \qquad \text{or equivalently,} \qquad 3^t = 243 = 3^5$$

is an example of an exponential equation. Other examples are

$$5^x = 125, \qquad 3^{2x-1} = 9, \qquad 7^{x^2+x} = 49, \qquad \text{and} \qquad 10(1.07)^x = 33$$

Later, in Section 4.3, we will solve such equations by using logarithms. However, we can deal with some *special* exponential equations now. A technique for solving these special equations entails first writing each side of the equation in terms of the same base and then using the following property, where $a > 0$ and $a \neq 1$.

$$a^u = a^v \qquad \text{if and only if} \qquad u = v$$

EXAMPLE 9 Solve each exponential equation.

(a) $5^x = 125$ (b) $3^{2x-1} = \frac{1}{9}$ (c) $7^{x^2+x} = 49$

SOLUTION (a) For $5^x = 125$, since $125 = 5^3$, we write $5^x = 5^3$. Therefore, $x = 3$.

(b) Since $\frac{1}{9} = 3^{-2}$, we write $3^{2x-1} = \frac{1}{9}$ as

$$3^{2x-1} = 3^{-2}$$

so that

$$2x - 1 = -2$$
$$2x = -1$$
$$x = -\tfrac{1}{2}$$

(c) Since $7^{x^2+x} = 49$ and $49 = 7^2$, we write $7^{x^2+x} = 7^2$ or $x^2 + x = 2$, so that $x^2 + x - 2 = 0$ or $(x + 2)(x - 1) = 0$ and

$x + 2 = 0$	$x - 1 = 0$
$x = -2$	$x = 1$

PROBLEM SET 4.1

1. Suppose that $f(x) = 4^{-x/2}$. Find each value.
(a) $f(-1)$ (b) $f(-2)$ (c) $f(2)$
c (d) $f(0.3)$ c (e) $f(-1.2)$ c (f) $f(\sqrt{3})$

2. Suppose that $g(x) = (\frac{2}{3})^x$. Find each value.
(a) $g(-1)$ (b) $g(-2)$ (c) $g(3)$
c (d) $g(1.7)$ c (e) $g(-3.4)$ c (f) $g(-\sqrt{7})$

In problems 3–20, sketch the graph of the given function, determine its domain, its range, and indicate whether the function is increasing or decreasing. Also identify any horizontal asymptote.

3. $f(x) = 5^x$ **4.** $h(x) = (\sqrt{3})^x$ **5.** $h(x) = (0.7)^x$
6. $f(x) = (0.1)^x$ **7.** $g(x) = (\frac{1}{7})^x$ **8.** $f(x) = (\frac{3}{4})^x$
9. $f(x) = (\frac{1}{3})^{-x}$ **10.** $g(x) = (\frac{1}{6})^{-x}$ **11.** $g(x) = -4^x$
12. $h(x) = -2^x$ **13.** $f(x) = 5^{x+1}$ **14.** $h(x) = 4^{x-2}$
15. $f(x) = 2^x + 3$ **16.** $h(x) = 4^x - 1$ **17.** $f(x) = 4 \cdot 5^x$
18. $h(x) = 7(\frac{1}{2})^x + 1$ **19.** $f(x) = 2^{-x} - 3$ **20.** $F(x) = 3 - 4^x$

In problems 21–24, each of the curves in Figures 6–9 is a graph of an exponential function of the form $y = c \cdot b^x$ that contains the given point. Find the value of the base b and the value of c if $c = 1$ or $c = -1$.

21.

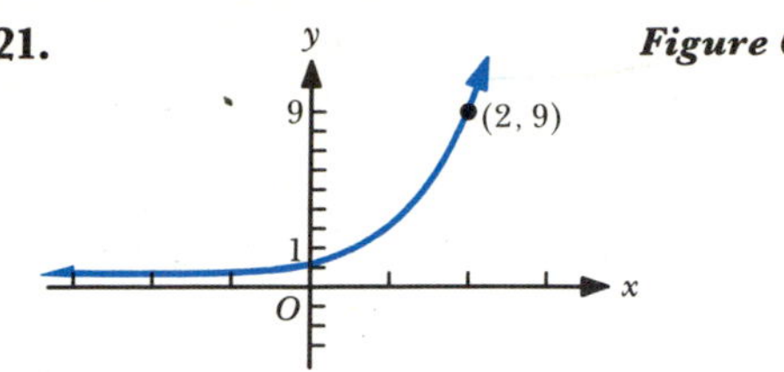

Figure 6

22.

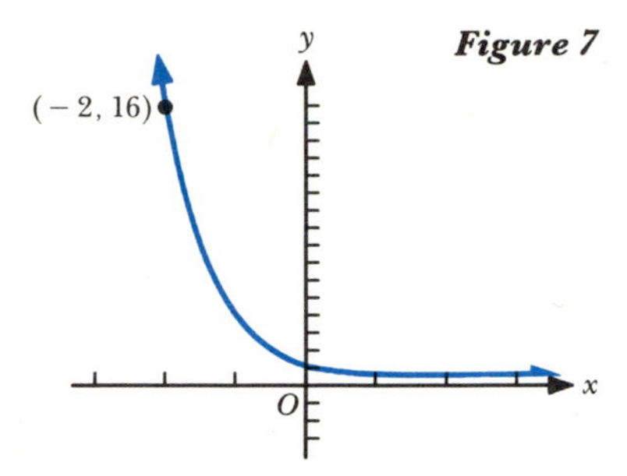

Figure 7

23.

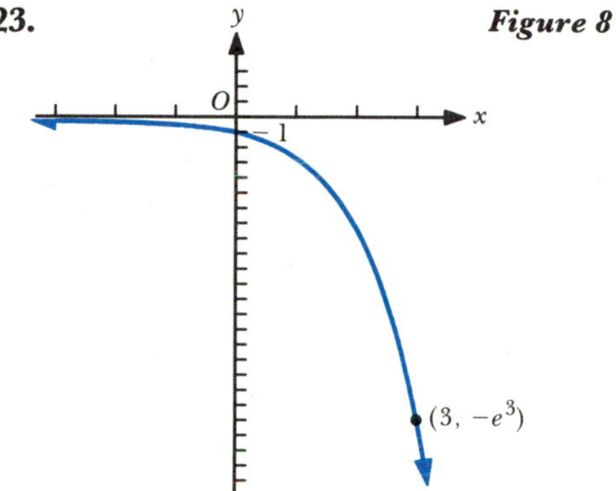

Figure 8

24.

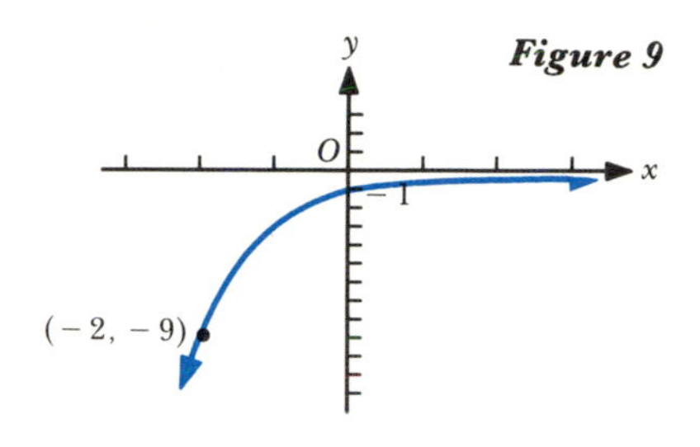

Figure 9

C In problems 25–28, use a calculator with a $\boxed{y^x}$ key to verify each equation for the indicated values of the variables.

25. $a^x \cdot a^y = a^{x+y}$, for $a = 2.062$, $x = 3.182$, $y = 1.081$

26. $(a^x)^y = a^{xy}$, for $a = 3.081$, $x = 4.315$, $y = 2.132$

27. $a^{x-y} = a^x/a^y$, for $a = 1.718$, $x = -2.041$, $y = 4.325$

28. $(ab)^x = a^x \cdot b^x$, for $a = 3.141$, $b = 2.718$, $x = 3.172$

29. Find the zeros of

$$f(x) = x^2(3e^{3x}) + 2xe^{3x} + 4e^{3x} + 5xe^{3x}$$

C **30.** Use a calculator with a $\boxed{y^x}$ key to approximate the value of each expression to seven significant digits.

(a) $\sqrt{3}^{\sqrt{3}}$ (b) $3^{-\sqrt{3}}$ (c) 3^{π} (d) $\pi^{-\pi}$

In problems 31–34, use the graph of $y = e^x$ to sketch the graph of each function. Find the domain and the range of each function and indicate whether the function is increasing or decreasing.

31. $f(x) = 1 + e^x$ **32.** $g(x) = e^{1-x}$ **33.** $h(x) = 2 - 3e^x$ **34.** $g(x) = 2e^{-x}$

35. Simplify the expression

$$\frac{(e^x + e^{-x})(e^x + e^{-x}) - (e^x - e^{-x})(e^x - e^{-x})}{(e^x + e^{-x})^2}$$

36. Simplify the expression

$$\frac{(3^x - 3^{-x})^2 - (3^x + 3^{-x})^2}{(3^x + 3^{-x})^2}$$

c **37.** Approximate each number to seven significant digits.
(a) $e^{3.4}$ (b) e^{π} (c) $e^{\sqrt{2}}$ (d) e^e
(e) $e^{-2.5}$ (f) $e^{-3/7}$ (g) $e^{-\sqrt{3}}$ (h) $e^{-\sqrt{0.5}}$

c **38.** To obtain an approximation for the number e, we can evaluate the following expression

$$\left(1 + \frac{1}{x}\right)^x$$

for large positive values of x. Find the value of the expression for
(a) $x = 100$ (b) $x = 1000$ (c) $x = 10{,}000$

In problems 39–48, solve each exponential equation.

39. $3^{3x} = 27$
40. $6^{2x-1} = 216$
41. $2^{x+1} = \frac{1}{8}$
42. $4^{3x-1} = 16$
43. $16^{3x} = 8^{2x-1}$
44. $7^{5x+4} = 1$
45. $5^{x^2+3x} = 625$
46. $2^{2y+2} + 2^{y+2} = 3$ (Let $x = 2^y$)
47. $3^{x^2+x} = 1$
48. $2^{15t^2+14t} = 256$

c In problems 49–54, find the final value of each investment.

49. \$3000 is invested for 4 years at 8% compounded semiannually.
50. \$8000 is invested for 5 years at 9% compounded quarterly.
51. \$20,000 is invested for 3 years at 8.5% compounded continuously.
52. \$4000 is invested for 30 months at 7.5% compounded weekly.
53. \$10,000 is invested for 6 years at 11.2% compounded quarterly.
54. \$85,000 is invested for 6 months at 15.4% compounded continuously.

c In problems 55–58, find the present value of each amount due.

55. \$10,000 to be paid after 4 years; investments during this period earn a nominal interest rate of 11% compounded quarterly.
56. \$12,000 to be paid after 5 years; investments during this period earn a nominal interest rate of 10% compounded monthly.
57. \$14,000 to be paid after 7 years; investments during this period earn a nominal interest rate of 10.5% compounded monthly.
58. \$40,000 to be paid after 3 years; investments during this period earn a nominal interest rate of 13% compounded semiannually.

c In problems 59–62, find the *effective* annual interest rate corresponding to the given *nominal* annual interest rate.

59. 9% compounded semiannually
60. 13% compounded monthly
61. 10% compounded quarterly
62. 15% compounded semiannually

c **63.** Suppose that the number of employees N of a certain company is related to its annual sales x (in millions of dollars) by the function $N = 50e^{0.06x}$, $x \geq 0$. Find the number of employees when annual sales reach 10 million dollars.

c **64.** A vertical beam of light with initial intensity I_0 at the surface of a lake has intensity I at a depth of d meters, given by the function

$$I = I_0 e^{-1.4d}$$

Compare the intensity at the surface to the intensity at depths of
(a) 0.75 meter (b) 1.85 meters

c **65.** Medical researchers estimate that the number of cancer cells present in a certain tissue sample after t days is given by the function

$$N(t) = 200(2^{0.12t})$$

Find the number of cancer cells present in the tissue sample after 8 days.

c **66.** A department store's annual profit from the sale of a certain toy is given by the function

$$y = 8000 + 30{,}000(2^{-0.4x})$$

where y is the annual profit in dollars and x denotes the number of years the toy has been on the market.
(a) Calculate the store's annual profit for $x = 1, 2, 3, 5, 10$, and 15.
(b) Use the results of part (a) to sketch the graph of this equation.

c **67.** A power supply for a lunar sensor uses a radioisotope whose power P is related to the number t of days of operation by the function

$$P = 40e^{-0.006t}$$

Find the power output after 150 days of operation to two decimal places.

4.2 Logarithmic Functions

An exponential function $g(x) = b^x$ is an increasing function when $b > 1$ and is a decreasing function when $0 < b < 1$. Therefore, for $b \neq 1$, no horizontal line can intersect the graph of g more than once, and we conclude that the exponential function $g(x) = b^x$ is *invertible* if $b \neq 1$. The inverse of the function $g(x) = b^x$ for $b \neq 1$ is called the *logarithmic function with base b*. To find the inverse of an exponential function, we use the following three-step procedure.

Step 1. First we write $y = g(x) = b^x$.

Step 2. Next we interchange x and y in the equation to obtain $x = b^y$.

Step 3. We attempt to solve this equation in step 2 for y in terms of x. However, we do not have the tools to solve the equation $x = b^y$ for y, in order to produce a formula for the dependent variable y in terms of the independent variable x.

To maintain the convention of expressing the dependent variable in terms of the independent variable, we introduce the terminology in the following definition.

DEFINITION 1 **Logarithm**

If b is a positive real number *not equal to* 1 and x is any *positive* real number, then the exponent y such that $b^y = x$ is called the **logarithm of x with base b** and is denoted by $\log_b x$. That is,

$$y = \log_b x \quad \text{is equivalent to} \quad b^y = x$$

For example,

$$3 = \log_2 8 \quad \text{because} \quad 2^3 = 8$$
$$2 = \log_5 25 \quad \text{because} \quad 5^2 = 25$$
$$-2 = \log_7(\tfrac{1}{49}) \quad \text{because} \quad 7^{-2} = \tfrac{1}{49}$$

EXAMPLE 1 Rewrite each exponential equation as an equivalent logarithmic equation.

(a) $3^2 = 9$ (b) $e^{-a} = b$ (c) $10^0 = 1$

SOLUTION Using Definition 1, we have

(a) $3^2 = 9$ is equivalent to $2 = \log_3 9$

(b) $e^{-a} = b$ is equivalent to $-a = \log_e b$

(c) $10^0 = 1$ is equivalent to $0 = \log_{10} 1$

EXAMPLE 2 Rewrite each logarithmic equation as an equivalent exponential equation.

(a) $\log_{10} 10 = 1$ (b) $\log_e\left(\dfrac{1}{c}\right) = -t$ (c) $\log_e e = 1$

SOLUTION Using Definition 1, we have

(a) $\log_{10} 10 = 1$ is equivalent to $10^1 = 10$

(b) $\log_e\left(\dfrac{1}{c}\right) = -t$ is equivalent to $e^{-t} = \dfrac{1}{c}$

(c) $\log_e e = 1$ is equivalent to $e^1 = e$

The next example establishes a pair of *identities* for logarithms.

EXAMPLE 3 Show that

(a) $\log_b b = 1$ (b) $\log_b 1 = 0$

SOLUTION

(a) If $t = \log_b b$, then $b^t = b$ and so $t = 1$; thus, $\log_b b = 1$.

(b) If $t = \log_b 1$, then $b^t = 1$ and so $t = 0$; thus, $\log_b 1 = 0$.

In view of Definition 1, the **logarithmic function f with base b,** which is the inverse of the exponential function $g(x) = b^x$, can be expressed as

$$f(x) = \log_b x \qquad \text{where} \quad b > 0 \quad \text{and} \quad b \neq 1$$

To summarize, we have the following result.

> The inverse of the exponential function $g(x) = b^x$ with $b \neq 1$ is the logarithmic function $f(x) = \log_b x$.

As we saw in Section 4.1, the domain of $g(x) = b^x$ is $\mathbb{R}$ and the range is $(0, \infty)$; therefore, the inverse function $f(x) = \log_b x$ has domain $(0, \infty)$ and range $\mathbb{R}$. Thus, we can only evaluate the logarithm of a positive number in $\mathbb{R}$.

The relation between logarithms and exponentials given in Definition 1, along with the solutions of special exponential equations, can often be used to find values of logarithmic functions.

EXAMPLE 4 Let $f(x) = \log_5 x$. Find

(a) $f(25)$ (b) $f(5)$ (c) $f(\frac{1}{5})$ (d) $f(\frac{1}{25})$ (e) $f(1)$

SOLUTION

(a) Let $u = f(25) = \log_5 25$. Since $u = \log_5 25$ is equivalent to $5^u = 25$, then $u = 2$ and $\log_5 25 = 2$.

(b) Let $u = f(5) = \log_5 5$. Then $5^u = 5$ so that $u = 1$ and $\log_5 5 = 1$.

(c) Let $u = f(\frac{1}{5}) = \log_5(\frac{1}{5})$. Then $5^u = \frac{1}{5}$. Thus, $u = -1$ and $\log_5 (\frac{1}{5}) = -1$.

(d) Let $u = f(\frac{1}{25}) = \log_5(\frac{1}{25})$. Then $5^u = \frac{1}{25}$ so that $u = -2$ and $\log_5(\frac{1}{25}) = -2$.

(e) Let $u = f(1) = \log_5 1$. Then $5^u = 1$. Thus $u = 0$ and $\log_5 1 = 0$.

EXAMPLE 5 Find the domain of each function.

(a) $g(x) = \log_5(-x)$ (b) $f(x) = \log_8(2x - 1)$ (c) $h(x) = \log_7(x^2 - 4)$

SOLUTION

(a) $\log_5(-x)$ is defined only if $-x > 0$. But $-x > 0$ only if $x < 0$, so the domain of g is the interval $(-\infty, 0)$.

(b) $\log_8(2x - 1)$ is defined only if $2x - 1 > 0$. But $2x - 1 > 0$ only if $x > \frac{1}{2}$, so the domain of f is the interval $(\frac{1}{2}, \infty)$.

(c) $\log_7(x^2 - 4)$ is defined only if $x^2 - 4 > 0$. But $x^2 - 4 > 0$ only if $x < -2$ or if $x > 2$. Therefore, the domain of h includes all of the numbers in the intervals $(-\infty, -2)$ and $(2, \infty)$.

Graphs of Logarithmic Functions

By using the idea that the graph of the inverse function f^{-1} is a reflection of the graph of f across the line $y = x$, we can sketch the graph of a logarithmic function by reflecting the graph of an exponential function across the line $y = x$.

For example, to sketch the graph of

$$F(x) = \log_2 x$$

we first sketch the graph of $f(x) = 2^x$; then we reflect the graph of f across the line $y = x$ to obtain the graph of $F(x) = \log_2 x$ (Figure 1). We can also sketch the graph of $y = F(x) = \log_2 x$ *directly* by first converting the equation $y = \log_2 x$ to the equivalent form $x = 2^y$ and then forming a table of values. By plotting the points determined by the entries in the table and connecting them with a smooth curve we obtain the graph of F (Figure 1).

$x = 2^y$	y
$\frac{1}{4}$	-2
$\frac{1}{2}$	-1
1	0
2	1
4	2
8	3

Figure 1

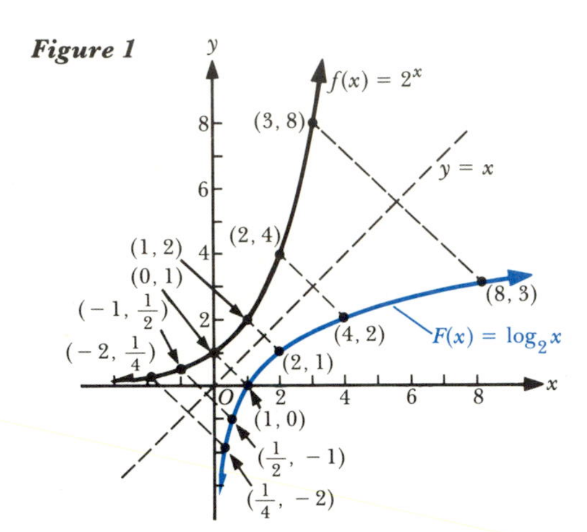

EXAMPLE 6 Graph each function directly, without using symmetry.

(a) $f(x) = \log_4 x$ (b) $g(x) = \log_{1/4} x$

SOLUTION

(a) To determine the graph of f we first consider $y = f(x) = \log_4 x$ in its equivalent form $4^y = x$ and then form a table of values for this exponential equation. Next we plot the points whose coordinates are given in the table and then we join the points with a smooth curve, as shown in Figure 2a.

(b) The graph of g can be determined by considering $y = g(x) = \log_{1/4} x$ in its equivalent form $(\frac{1}{4})^y = x$ and forming a table of values for this latter equation. We plot the points whose coordinates are given in the table and then connect them with a smooth curve to obtain the graph of g (Figure 2b).

Notice that the graph of f in Figure 2a is increasing, whereas the graph of g in Figure 2b is decreasing. In each case, the y axis is a vertical asymptote.

Figure 2

$x = 4^y$	y
$\frac{1}{4}$	-1
1	0
4	1
16	2
64	3

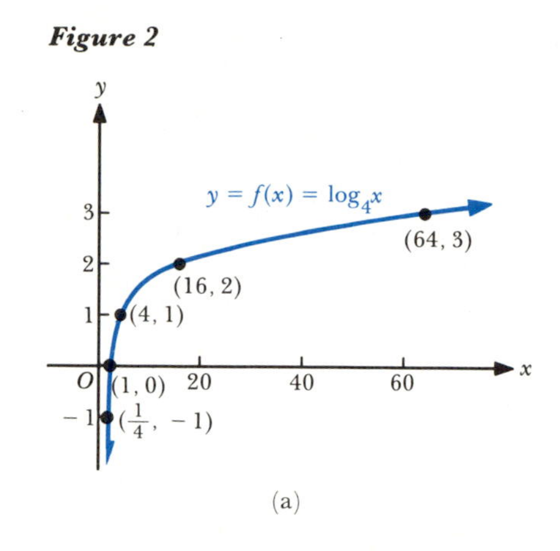

(a)

$x = (\frac{1}{4})^y$	y
$\frac{1}{2}$	$\frac{1}{2}$
1	0
2	$-\frac{1}{2}$

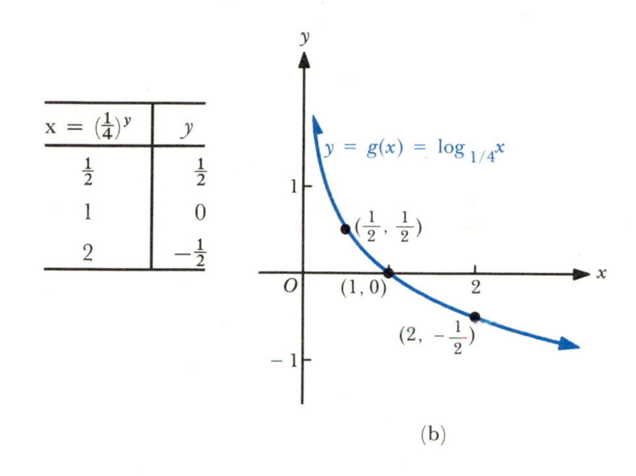

(b)

The techniques of graph sketching presented in Section 2.4 can be applied to functions involving logarithms as illustrated in the next two examples.

In Examples 7 and 8, use the specified graph to sketch the graph of each function, determine the domain and range of the function, and indicate whether the function is increasing or decreasing. Find the equation of any vertical asymptote.

EXAMPLE 7 $f(x) = \log_2(-x)$ by using $F(x) = \log_2 x$.

SOLUTION The domain of f consists of values of x for which $-x > 0$ or $x < 0$, that is, the interval $(-\infty, 0)$. By reflecting the graph of $F(x) = \log_2 x$ (Figure 1) across the y axis, we obtain the graph of f (Figure 3). From the graph, we see that the range of f is the set of real numbers $\mathbb{R}$, and that the function f is decreasing over its entire domain. Notice that the y axis, that is, $x = 0$, is a vertical asymptote of the graph of f.

Figure 3

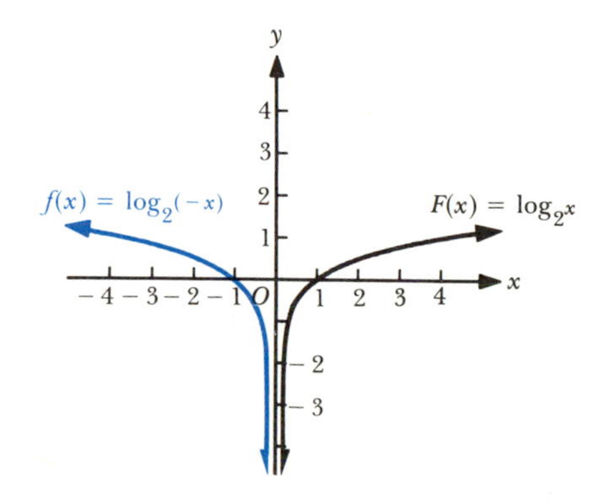

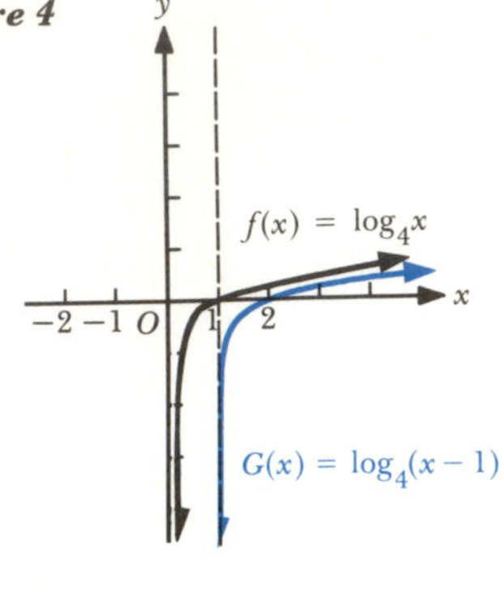

Figure 4

EXAMPLE 8 $G(x) = \log_4(x - 1)$ by using $f(x) = \log_4 x$.

SOLUTION The domain of G consists of all values of x for which $x - 1 > 0$; that is, the interval $(1, \infty)$. After shifting the graph of $f(x) = \log_4 x$ (Figure 2a) one unit to the right, we obtain the graph of G (Figure 4). From the graph, we see that the range of G is the set of real numbers $\mathbb{R}$, and that the function G is increasing over its entire domain. Notice that the line $x = 1$ is a vertical asymptote for G.

The Natural and Common Logarithms

Although the base of a logarithm may be any positive number except 1, in practice, the bases that are normally used are 10 and e. Logarithms with base 10 are called **common logarithms** and are denoted by

$$\log x = \log_{10} x$$

Logarithms with base e are called **natural logarithms** and are denoted by

$$\ln x = \log_e x$$

Natural logarithms are used throughout the study of calculus and have many applications in science, engineering, business, and technology.

EXAMPLE 9 Sketch the graph of $f(x) = \ln x$.

SOLUTION Because the natural logarithm function is the inverse of the exponential function with base e, the graph of

$$f(x) = \ln x$$

can be obtained by reflecting the graph of $y = e^x$ across the line $y = x$ (Figure 5). The domain of f is the interval $(0, \infty)$, the range is the set of real numbers $\mathbb{R}$, and f is increasing over its entire domain. The line $x = 0$ is a vertical asymptote for f.

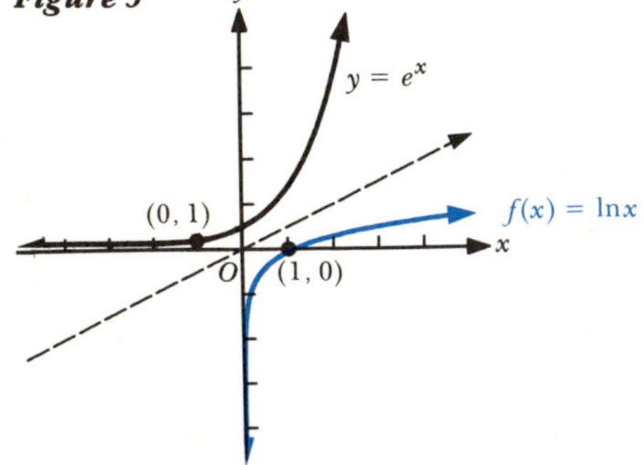

Figure 5

For limited situations, it is possible to use Definition 1 to evaluate logarithms. For instance,

$$\log 100 = 2 \qquad \text{since} \qquad 10^2 = 100$$

and

$$\log 0.001 = -3 \qquad \text{since} \qquad 10^{-3} = 0.001$$

In other situations, however, such as the evaluation of log 4819 or ln 0.03714, we need another method. In the past, logarithm tables were used to facilitate numerical calculations involving common or natural logarithms. Now, modern calculators and computers provide an easier way of determining the approximate values of logarithms.

For example, using a nine-digit calculator with a $\boxed{\log}$ key and an $\boxed{\ln}$ key, we have

$$\log 4819 = 3.68295693 \qquad \text{and} \qquad \ln 0.03714 = -3.29306072$$

On the other hand, by using a calculator with a $\boxed{y^x}$ and an $\boxed{e^x}$ key we find that

$$10^{3.68295693} = 4819 \qquad \text{and} \qquad e^{-3.29306072} = 0.03714$$

confirming the fact that

$$\log 4819 = 3.68295693 \qquad \text{means that} \qquad 10^{3.68295693} = 4819 \text{ and}$$

$$\ln 0.03714 = -3.29306072 \qquad \text{means that} \qquad e^{-3.29306072} = 0.03714$$

EXAMPLE 10 $\boxed{c}$ Use a calculator with a $\boxed{\log}$ key and an $\boxed{\ln}$ key to evaluate each expression.

(a) log 5310 (b) log 0.003864 (c) ln 5834 (d) ln 0.04183

SOLUTION On a nine-digit calculator, we obtain

(a) $\log 5310 = 3.72509452$ (b) $\log 0.003864 = -2.41296288$

(c) $\ln 5834 = 8.67145815$ (d) $\ln 0.04183 = -3.17414149$

Now consider the problem of finding the value x when the value of $\log x$ or $\ln x$ is given. A calculator can also be used to solve this type of problem. For example, we can use a calculator with a $\boxed{y^x}$ key to find the value x such that

$$\log x = 2.7435$$

Writing the above equation in exponential form, we have

$$x = 10^{2.7435}$$

Rounding off to four significant digits, we obtain

$$x = 10^{2.7435} = 554.0$$

The value $x = 554.0$ is called the **antilogarithm** of 2.7435.

EXAMPLE 11 [c] Use a calculator with a $\boxed{y^x}$ key and an $\boxed{e^x}$ key or $\boxed{\text{INV}}$ – $\boxed{\text{ln}}$ keys to find the value of x to four significant digits.

(a) $\log x = 0.7235$ (b) $\ln x = 1.4674$

SOLUTION

(a) $\log x = 0.7235$ is equivalent to $x = 10^{0.7235}$, which equals 5.291, rounded off to four significant digits.

(b) $\ln x = 1.4674$ is equivalent to $x = e^{1.4674}$, which equals 4.338, rounded off to four significant digits.

EXAMPLE 12 [c] A company's sales S (in dollars) are related to its advertising expenditures x (in dollars) by the function

$$S(x) = 100{,}000 + 8100 \log(x + 1)$$

Calculate the sales associated with the given advertising expenditure.

(a) \$250 (b) \$471 (c) \$1148

SOLUTION

Using a calculator, we obtain

(a) $S(250) = 100{,}000 + 8100 \log(250 + 1) = \$119{,}437.36$

(b) $S(471) = 100{,}000 + 8100 \log(471 + 1) = \$121{,}658.93$

(c) $S(1148) = 100{,}000 + 8100 \log(1148 + 1) = \$124{,}788.59$

PROBLEM SET 4.2

In problems 1–6, write each exponential equation as an equivalent logarithmic equation.

1. $5^3 = 125$ **2.** $4^{-2} = \frac{1}{16}$ **3.** $\sqrt{9} = 3$

4. $(\frac{1}{3})^{-2} = 9$ **5.** $\sqrt[5]{32} = 2$ **6.** $\pi^t = w$

In problems 7–12, write each logarithmic equation as an equivalent exponential equation.

7. $\log_9 81 = 2$ **8.** $\log_{10} 0.0001 = -4$ **9.** $\log_{1/3} 9 = -2$

10. $\log_{10} \frac{1}{10} = -1$ **11.** $\log_{36} 216 = \frac{3}{2}$ **12.** $\log_\pi s = t$

In problems 13–21, evaluate each expression.

13. $\log_2 \frac{1}{8}$ **14.** $\log_5 \frac{1}{25}$ **15.** $\log_{10} \frac{1}{10}$

16. $\log_{1/2} 8$ **17.** $\log_5 5$ **18.** $\log_{35} 35$

19. $\log_7 1$ **20.** $\log_8 1$ **21.** $\log_{10} 0.0001$

22. Given $f(x) = b^x$ and $f^{-1}(x) = \log_b x$, use the fact that $f[f^{-1}(x)] = x$ to verify

$$b^{\log_b x} = x$$

In problems 23–28, use $f(x) = \log_7 x$, $g(x) = \log_8 x$, and $h(x) = \log_{1/6} x$ to find the function value.

23. $f(49)$ **24.** $g(64)$ **25.** $h(216)$ **26.** $f(1)$ **27.** $g(\frac{1}{64})$ **28.** $h(\frac{1}{36})$

In problems 29–32, each of the curves in Figures 6–9 is the graph of a logarithmic function $y = \log_b x$. Find the base b if its graph contains the given point.

29.

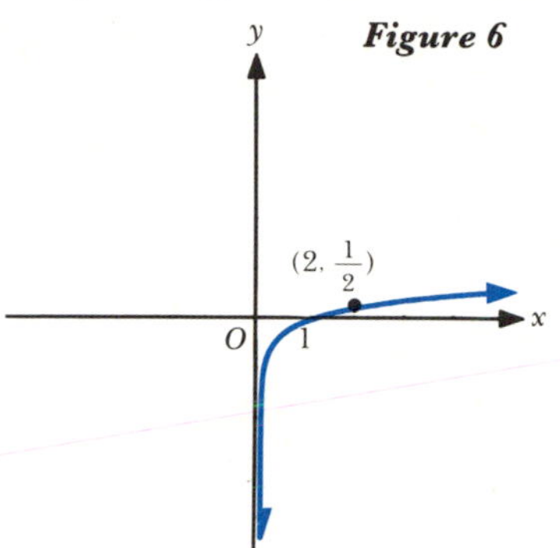

Figure 6

30.

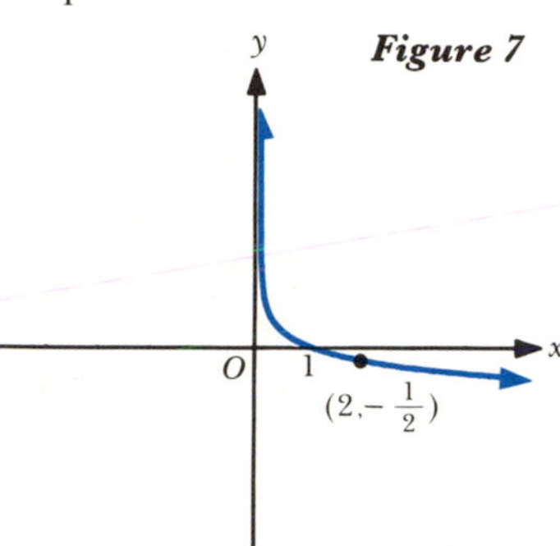

Figure 7

31.

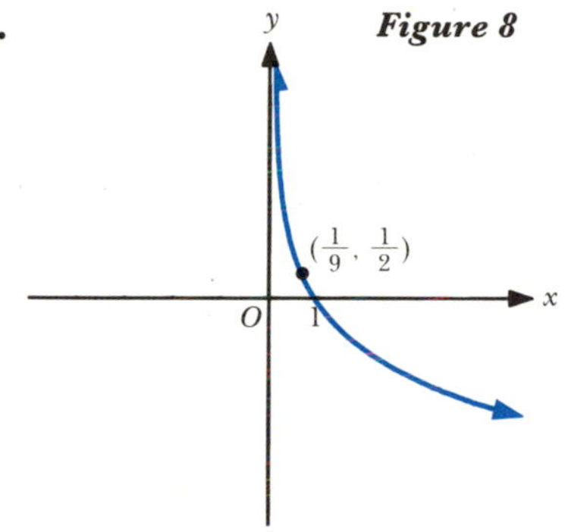

Figure 8

32.

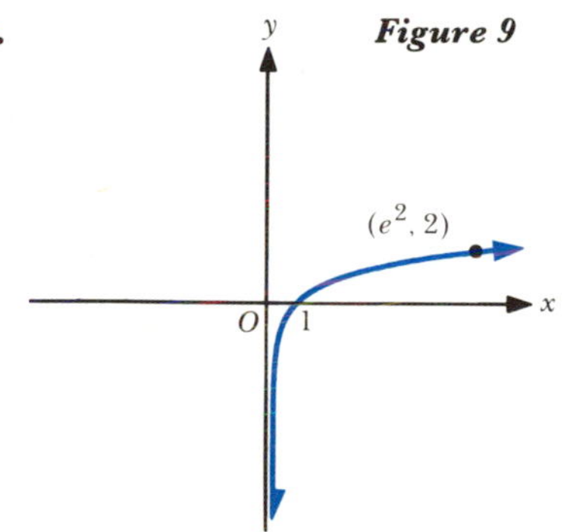

Figure 9

In problems 33–40, find the domain of each function.

33. $f(x) = \log(x + 1)$ **34.** $g(x) = \log_2(2 - x)$ **35.** $h(x) = \log_6(1 - x)$

36. $F(x) = \log_5\left(\frac{1}{x}\right)$ **37.** $K(x) = \ln(x^2 + 1)$ **38.** $G(x) = \log_7|x|$

39. $f(x) = \ln(x^2 - 3x + 2)$ **40.** $G(x) = \log_3(x^2 - 5x + 4)$

In problems 41–50, sketch the graph of each function. Determine the domain and range, and indicate whether the function is increasing or decreasing. Find an equation of any vertical asymptote.

41. $f(x) = \log_6(x + 1)$ **42.** $g(x) = \log_5|x|$ **43.** $h(x) = 1 + \ln x$

44. $f(x) = \ln(1 - x)$ **45.** $f(x) = \frac{1}{3}\log\left(\frac{1}{x}\right)$ **46.** $g(x) = |\log_3 x|$

47. $g(x) = \log_5 x^2$ **48.** $f(x) = \log_{1/2}(2 - x)$ **49.** $f(x) = 2 - \log_3 x$

50. $h(x) = \log_{1/3}|3 - x|$

c In problems 51–62, use a calculator with a log key and an ln key to evaluate each expression to seven significant digits

51. $\log 39.2$ **52.** $\log 9614$ **53.** $\log 1732.8$ **54.** $\log 0.1438$
55. $\log 0.005812$ **56.** $\log(4.835 \cdot 10^{-7})$ **57.** $\ln 961$ **58.** $\ln \sqrt[3]{583}$
59. $\ln 0.752$ **60.** $\ln 4765$ **61.** $\ln 0.00419$ **62.** $\ln(1.87 \cdot 10^{-5})$

c In problems 63–68, use a calculator with a y^x key and an e^x key or INV – ln keys to find the value of x. Round off the answer to four significant digits.

63. $\log x = 0.4187$ **64.** $\log x = 1.4371$ **65.** $\ln x = 2.6143$
66. $\ln x = -1.3712$ **67.** $\log x = -2.4507$ **68.** $\ln x = 0.01481$

c **69.** Use a calculator to verify each equation.
(a) $\ln e^{\pi} = \pi$
(b) $e^{\ln \pi} = \pi$

70. Solve the inequality $2 \leq \log_2 x \leq 3$ by using the graph of $y = \log_2 x$.

71. Graph $f(x) = \log_b x$, where $b = 2$, 3, 4, and 5 on the same coordinate system. What is the relationship among these graphs?

72. Graph $f(x) = \log_5 x$ and $g(x) = 5^x$ on the same coordinate system. Explain the symmetry with respect to the line $y = x$.

73. Sketch the graphs of $f(x) = \log_3 x$ and $g(x) = \log_{1/3} x$. Compare the graphs of f and g. In general, how do the graphs of logarithmic functions with base less than 1 compare to the graphs of logarithmic functions with base greater than 1?

74. Let $f(x) = 10^x$ and $g(x) = \log_{10} x$. Find $f[g(x)]$ and $g[f(x)]$. Simplify your answer by using the fact that $g(x) = \log_{10} x$ is equivalent to $10^{g(x)} = x$.

75. Graph $y = \log_{10} x$.
(a) For what values of x is $y < 0$?
(b) For what values of x is $y > 0$?
(c) For what value of x is $y = 0$?
(d) If $x_1 < x_2$, how does $\log_{10} x_1$ compare to $\log_{10} x_2$?

76. A company's weekly profit P in dollars is estimated by $P(x) = 914 \log_9 x$, where x is the number of parts produced each minute. Find the company's weekly profit if it produced 729 parts per minute.

c **77.** A company's sales S in dollars are related to advertising expenditures x in dollars by the function $S(x) = 200{,}000 + 11{,}000 \ln x$. Calculate the sales associated with each of the following advertising expenditures.
(a) \$241
(b) \$542
(c) \$1325

c **78.** A lake polluted by coliform bacteria is treated with bactericidal agents. Environmentalists estimate that t days after the treatment, the number N of viable bacteria per milliliter will be given by

$$N(t) = 10t - 100 \ln \frac{t}{10} - 30, \qquad \text{for} \quad 1 \leq t \leq 12$$

Find the number N of viable bacteria per milliliter t days after the treatment if
(a) $t = 3$ (b) $t = 4.5$ (c) $t = 7.5$

4.3 Properties of Logarithms

Since logarithms are exponents, their properties are derived from the properties of exponents. These properties are listed in the front cover.

THEOREM 1 **Logarithm Properties**

Suppose that M, N, and b are positive real numbers, where $b \neq 1$, and r is any real number. Then

(i) $\log_b MN = \log_b M + \log_b N$ (ii) $\log_b \dfrac{M}{N} = \log_b M - \log_b N$

(iii) $\log_b N^r = r \log_b N$

PROOF (i) Since $M = b^{\log_b M}$ and $N = b^{\log_b N}$ (Problem Set 4.2, Problem 22),

$$MN = b^{\log_b M} \cdot b^{\log_b N} = b^{\log_b M + \log_b N}$$

Thus, by Definition 1 of a logarithm given in Section 4.2

$$\log_b MN = \log_b M + \log_b N$$

(ii) Since $M = (M/N)N$, we can use Property (i) above to get

$$\log_b M = \log_b \frac{M}{N} + \log_b N$$

so that

$$\log_b \frac{M}{N} = \log_b M - \log_b N$$

(iii) Since $N = b^{\log_b N}$, then

$$N^r = (b^{\log_b N})^r = b^{r \log_b N}$$

so that

$$\log_b N^r = r \log_b N$$

These properties enable us to perform certain operations involving logarithms.

EXAMPLE 1 Use Theorem 1 to write each expression as a sum or difference of multiples of logarithms.

(a) $\log_8 7x$ (b) $\log_8 \frac{17}{5}$ (c) $\log_8(x^7y^{11})$ (d) $\ln \dfrac{(y+7)^3}{\sqrt{y}}$

SOLUTION We assume that all numbers whose logarithms are taken are positive, so that

(a) $\log_8 7x = \log_8 7 + \log_8 x$ [Theorem 1 (i)]

(b) $\log_8 \frac{17}{5} = \log_8 17 - \log_8 5$ [Theorem 1 (ii)]

(c) $\log_8(x^7y^{11}) = \log_8 x^7 + \log_8 y^{11}$ [Theorem 1 (i)]
$= 7\log_8 x + 11\log_8 y$ [Theorem 1 (iii)]

(d) $\ln \dfrac{(y+7)^3}{\sqrt{y}} = \ln(y+7)^3 - \ln\sqrt{y}$ [Theorem 1 (ii)]
$= 3\ln(y+7) - \frac{1}{2}\ln y$ [$\sqrt{y} = y^{1/2}$ and Theorem 1 (iii)]

EXAMPLE 2 Use Theorem 1 to write each expression as a single logarithm.

(a) $\log_3 7 + \log_3 4$ (b) $2\ln x - 3\ln y$
(c) $\log_a(c^2 - cd) - \log_a(2c - 2d)$ (d) $\log_7 \sqrt[5]{7}$

SOLUTION We assume that all numbers whose logarithms are taken are positive. Then

(a) $\log_3 7 + \log_3 4 = \log_3[(7)(4)]$ [Theorem 1 (i)]
$= \log_3 28$

(b) $2\ln x - 3\ln y = \ln x^2 - \ln y^3$ [Theorem 1 (iii)]
$= \ln \dfrac{x^2}{y^3}$ [Theorem 1 (ii)]

(c) $\log_a(c^2 - cd) - \log_a(2c - 2d) = \log_a\left(\dfrac{c^2 - cd}{2c - 2d}\right) = \log_a \dfrac{c(c-d)}{2(c-d)} = \log_a \dfrac{c}{2}$

(d) $\log_7 \sqrt[5]{7} = \log_7 7^{1/5} = \frac{1}{5}\log_7 7 = \frac{1}{5}$, since $\log_7 7 = 1$

EXAMPLE 3 Assume that $\log_b 2 = 0.35$, $\log_b 3 = 0.55$, and $\log_b 5 = 0.82$. Use Theorem 1 to find each of the following values.

(a) $\log_b \frac{2}{3}$ (b) $\log_b 2^3$ (c) $\dfrac{\log_b 2}{\log_b 3}$ C (d) $(\log_b 2)^3$

(e) $\log_b 24$ (f) $\log_b \sqrt{\frac{2}{3}}$ (g) $\log_b\left(\frac{60}{b}\right)$ (h) $\log_b 0.6$

SOLUTION

(a) $\log_b \frac{2}{3} = \log_b 2 - \log_b 3 = 0.35 - 0.55 = -0.20$

(b) $\log_b 2^3 = 3\log_b 2 = 3(0.35) = 1.05$

(c) $\dfrac{\log_b 2}{\log_b 3} = \dfrac{0.35}{0.55} = \dfrac{7}{11}$

(d) $(\log_b 2)^3 = (0.35)^3 = 0.042875$

(e) $\log_b 24 = \log_b(2^3 \cdot 3) = \log_b 2^3 + \log_b 3$
$= 3\log_b 2 + \log_b 3 = 3(0.35) + 0.55 = 1.60$

(f) $\log_b\sqrt{\frac{2}{3}} = \log_b\left(\frac{2}{3}\right)^{1/2} = \frac{1}{2}\log_b\left(\frac{2}{3}\right) = \frac{1}{2}(\log_b 2 - \log_b 3)$
$= \frac{1}{2}(0.35 - 0.55) = -0.10$

(g) $\log_b\left(\frac{60}{b}\right) = \log_b 60 - \log_b b = \log_b(2^2 \cdot 3 \cdot 5) - \log_b b$
$= 2\log_b 2 + \log_b 3 + \log_b 5 - \log_b b$
$= 2(0.35) + 0.55 + 0.82 - 1 = 0.70 + 0.55 + 0.82 - 1 = 1.07$

(h) $\log_b 0.6 = \log_b \frac{3}{5} = \log_b 3 - \log_b 5 = 0.55 - 0.82 = -0.27$

Logarithmic Equations

Logarithmic equations can be solved by changing them to equivalent exponential forms. However, at times it is necessary first to use the basic properties of logarithms to simplify one side of an equation.

In Examples 4–6, solve each equation.

EXAMPLE 4 $\log_4(3x - 2) = 2$

SOLUTION The equation $\log_4(3x - 2) = 2$ is equivalent to $3x - 2 = 4^2$ so that

$$3x - 2 = 16$$

$$3x = 18$$

$$x = 6$$

Therefore, 6 is the solution.

EXAMPLE 5 $\log_3(x + 1) + \log_3(x + 3) = 1$

SOLUTION Applying Theorem 1(i), we have

$$\log_3(x + 1) + \log_3(x + 3) = \log_3[(x + 1)(x + 3)]$$

so that

$$\log_3[(x + 1)(x + 3)] = 1$$

or

$$(x + 1)(x + 3) = 3^1$$

That is,

$$x^2 + 4x + 3 = 3$$

$$x^2 + 4x = 0$$

$$x(x + 4) = 0$$

so $x = 0$ or $x = -4$. Since $x + 1$ and $x + 3$ must be positive numbers for $\log_3(x + 1)$ and $\log_3(x + 3)$ to be defined, we can eliminate $x = -4$ as an *extraneous* root. Therefore, the solution is 0.

EXAMPLE 6 $\log_4(t^2 - t - 2) - \log_4(t + 1) = 2$

SOLUTION Applying Theorem 1(ii), we can rewrite the given equation

$$\log_4(t^2 - t - 2) - \log_4(t + 1) = 2$$

as

$$\log_4 \frac{t^2 - t - 2}{t + 1} = 2$$

so that

$$\frac{t^2 - t - 2}{t + 1} = 4^2$$

$$\frac{(t - 2)(t + 1)}{t + 1} = 16$$

$$t - 2 = 16$$

$$t = 18$$

Therefore, the solution is 18.

Change of Logarithm Base

Most scientific calculators are programmed to provide common and natural logarithms only. Microcomputers, on the other hand, are frequently programmed to provide *only* natural logarithms. Therefore, in order to be able to work with bases different than 10 or e, it is necessary to be able to write a logarithm that is given in one base, say b, as a logarithm in terms of another base, say a. This is accomplished by the following formula.

THEOREM 2 **Change of Base Formula**

$$\log_b x = \frac{\log_a x}{\log_a b}$$

where a and b are positive real numbers different from 1, and x is positive.

PROOF Let $y = \log_b x$ then $b^y = x$ and it follows that, after taking the logarithm of each side of the latter equation to base a, we obtain

$$\log_a b^y = \log_a x$$

Using Theorem 1(iii), we have

$$y \log_a b = \log_a x \qquad \text{or} \qquad y = \frac{\log_a x}{\log_a b}$$

Therefore,

$$\log_b x = \frac{\log_a x}{\log_a b}$$

By replacing x by a in the above formula and using the fact that $\log_a a = 1$, we obtain the formula

$$\log_b a = \frac{1}{\log_a b}$$

EXAMPLE 7 [c] Use Theorem 2 to find the value of each expression by converting to (a) natural logarithms and then to (b) common logarithms. Round off each answer to four significant digits.

(i) $\log_3 7$
(ii) $\log_2 10$

SOLUTION Using a nine-digit calculator and rounding off to four significant digits, we get

$$\text{(a)}\quad \text{(i)}\ \log_3 7 = \frac{\ln 7}{\ln 3} = \frac{1.94591015}{1.09861229} = 1.771$$

$$\text{(ii)}\ \log_2 10 = \frac{\ln 10}{\ln 2} = \frac{2.30258509}{0.69314718} = 3.322$$

$$\text{(b)}\quad \text{(i)}\ \log_3 7 = \frac{\log 7}{\log 3} = \frac{0.84509804}{0.47712125} = 1.771$$

$$\text{(ii)}\ \log_2 10 = \frac{\log 10}{\log 2} = \frac{1}{0.301029995} = 3.322$$

Solving Exponential Equations

Recall from Section 4.1 on page 193 that *special* types of exponential equations were solved by applying the properties of exponents, together with the techniques we have already studied for solving equations. Now we can solve *any* exponential equation, such as $5^x = 7$ or $e^{-3t} = 0.5$, by taking the logarithms of both sides of the equation and using the properties of logarithms to simplify the resulting equation. Depending on the situation, either common or natural logarithms are used.

[c] *In Examples 8 and 9, solve each exponential equation. Round off the answer to two decimal places.*

EXAMPLE 8 $41^{2x-1} = 3^x$

SOLUTION By taking the common logarithms of both sides, we have

$$\log 41^{2x-1} = \log 3^x$$

$$(2x - 1)\log 41 = x \log 3$$

$$2x \log 41 - \log 41 = x \log 3$$

$$2x \log 41 - x \log 3 = \log 41$$

$$x(2 \log 41 - \log 3) = \log 41$$

$$x = \frac{\log 41}{2 \log 41 - \log 3} = 0.59$$

EXAMPLE 9 $e^{-3t} = 0.5$

SOLUTION By taking the natural logarithms of both sides, we have

$$\ln e^{-3t} = \ln 0.5$$

$$-3t \ln e = \ln 0.5$$

$$t = -\frac{\ln 0.5}{3} \qquad (\ln e = 1)$$

$$t = 0.23$$

EXAMPLE 10 [C] Suppose that \$1000 is invested at a 6% nominal interest rate compounded every 4 months. How long will it take for this investment to double?

SOLUTION The final value of S dollars for this investment after t years is given by

$$S = 1000\left(1 + \frac{0.06}{3}\right)^{3t}$$

If t is the time required to double the money, then

$$2000 = 1000(1.02)^{3t} \qquad \text{or} \qquad (1.02)^{3t} = 2$$

Taking the common logarithms of both sides of the latter equation, we have

$$3t \log 1.02 = \log 2$$

so that

$$t = \frac{\log 2}{3 \log 1.02} = 11.6675963$$

Hence, it takes about 11.7 years or approximately 11 years and 8 months to double the investment.

PROBLEM SET 4.3

In problems 1–20, write each expression as a sum or difference of multiples of logarithms. All necessary assumptions about the values of the variables should be made so that all logarithms are defined.

1. $\log_3 x(x + 1)$

2. $\log_3 \dfrac{18}{x + 2}$

3. $\log_b x^2 y^3$

4. $\log_b \dfrac{x^9}{y^7}$

5. $\log_b (x + 3)^4$

6. $\log_b x\sqrt{2x + 1}$

7. $\log_b \sqrt[6]{\dfrac{x^5}{y^2}}$

8. $\log_b \sqrt{x\sqrt{xy}}$

9. $\log \sqrt[3]{x\sqrt[3]{y}}$

10. $\ln \dfrac{x^5 y^2}{\sqrt[5]{z}}$

11. $\ln[y(3x + 1)^{2/3}]$

12. $\ln(x^2 + 7x)$

13. $\ln \sqrt[3]{x^2 - y^2}$

14. $\ln \dfrac{1}{\sqrt{7x + 1}}$

15. $\ln \sqrt[3]{\dfrac{4x + 3}{x^4}}$

16. $\log_b \dfrac{\sqrt[5]{1 + 3x}}{(2x + 3)^3}$

17. $\log_b \dfrac{5x(x^2 + 1)^2}{(x + 1)\sqrt{7x + 3}}$

18. $\ln \dfrac{4x^3}{(x + 1)^{2/3}(x + 2)^5}$

19. $\ln \dfrac{1}{\sqrt[7]{6x^2 - 7x - 3}}$

20. $\ln \left[\dfrac{\log_b x}{\ln x}\right]$

In problems 21–34, use Theorem 1 to write each expression as a single logarithm. All necessary assumptions about the values of the variables should be made so that all logarithms are defined.

21. $\log_5 \frac{5}{7} + \log_5 \frac{40}{25}$

22. $\log_2 \frac{32}{11} + \log_2 \frac{121}{16} - \log_2 \frac{4}{5}$

23. $\log_3 \frac{3}{4} - \log_3 \frac{3}{8}$

24. $\log_2\left(a + \dfrac{a}{b}\right) - \log_2\left(c + \dfrac{c}{b}\right)$

25. $\log_c(a^2 - ab) - \log_c(7a - 7b)$

26. $\frac{1}{2}\log_a(x - 1) - 3\log_a(x + 2)$

27. $\ln(m^2 + m) - \ln(m^2 - m)$

28. $\ln(x^2 - 9) - \ln(x^2 - 6x - 9)$

29. $\ln(4u^2 - 9) - \ln(8u^3 - 27)$

30. $\ln(3x^2 + 7x + 4) - \ln(3x^2 - 5x - 12)$

31. $\log_7\left(\dfrac{1}{4} - \dfrac{1}{x^2}\right) - \log_7\left(\dfrac{1}{2} - \dfrac{1}{x}\right)$

32. $\log_a\left(\dfrac{a}{\sqrt[3]{x}}\right) - \log_a\left(\dfrac{\sqrt[3]{x}}{a}\right)$

33. $3\log_e\left(\dfrac{a^2 b}{c^2}\right) + 2\log_e\left(\dfrac{bc^2}{a^4}\right) + 2\log_e\left(\dfrac{abc}{2}\right)$

34. $3\ln(x + 2) - \frac{1}{3}\ln x - \frac{1}{3}\ln(1 - 2x)$

In problems 35–41, evaluate each expression without the use of a calculator.

35. $\log_2 \sqrt{256}$

36. $\log_{10} \sqrt[7]{0.001}$

37. $\ln \sqrt[4]{e^3}$

38. $\log_3 \frac{1}{243}$

39. $\log_2 16^{2/5}$

40. $\log_7\left(\dfrac{49^{2/3}}{7^{3/2}}\right)$

41. $\log_2(16^{1/5} \cdot 64^{1/4})$

42. Use $x_1 = 10{,}000$, $x_2 = 10$, $b = 10$, and $p = 3$ to show that each statement is *false*.

(a) $\log_b\left(\dfrac{x_1}{x_2}\right) = \dfrac{\log_b x_1}{\log_b x_2}$

(b) $\dfrac{\log_b x_1}{\log_b x_2} = \log_b x_1 - \log_b x_2$

(c) $\log_b x_1 \cdot \log_b x_2 = \log_b x_1 + \log_b x_2$

(d) $\log_b(x_1 x_2) = \log_b x_1 \cdot \log_b x_2$

(e) $\log_b(x_1^p) = (\log_b x_1)^p$

(f) $(\log_b x_1)^p = p \log_b x_1$

43. Suppose that $\log_b 2 = 0.39$, $\log_b 3 = 0.61$, $\log_b 5 = 0.90$, and $\log_b 7 = 1.09$. Use Theorem 1 to find each of the given values. Round off each answer to two decimal places.

(a) $\log_b 6$ (b) $\log_b 21$ (c) $\log_b 35$ (d) $\log_b \frac{14}{5}$

(e) $\log_b \sqrt[4]{5}$ (f) $\log_b \sqrt[3]{42}$ (g) $\log_b \sqrt[5]{27}$

44. Let $\log_b 2 = A$, $\log_b 3 = B$, and $\log_b 5 = C$. Express $\log_b 0.012$ in terms of A, B, and C.

In problems 45–65, solve each equation without the use of a calculator.

45. $\log_3 x = 2$

46. $\log_{27} x = \frac{1}{3}$

47. $\log_{10} x^2 = -\frac{1}{2}$

48. $\log_x 4 = \frac{2}{5}$

49. $\log_3(x + 1) = 2$

50. $\log_{125} x = \frac{2}{3}$

51. $\log_7(2x - 7) = 2$

52. $\log_2(3x - 1) = 3$

53. $\log_5(5x - 1) = -2$

54. $\log_8 \sqrt{\dfrac{3x + 4}{x}} = 0$

55. $\log_2(x^2 + 3x + 4) = 1$

56. $\log_2|4x - 3| = 1$

57. $\log 2x = \log 3 + \log(x - 1)$

58. $\log_5(y^2 - 4y) = 1$

59. $\ln x + \ln(x - 2) = \ln(x + 4)$

60. $2\ln(x + 1) - \ln(x + 4) = \ln(x - 1)$

61. $\log_2(x^2 - 9) - \log_2(x + 3) = 2$

62. $\log(x + 1) - \log x = 1$

63. $\log_4 x + \log_4(6x + 10) = 1$

64. $\log_3 x + \log_3(x - 6) = \log_3 7$

65. $\log(x^2 - 144) - \log(x + 12) = 1$

66. Show that

$$\ln(x + \sqrt{x^2 - 1}) = -\ln(x - \sqrt{x^2 - 1})$$

c In problems 67–70, use Theorem 2 to find the value of each expression by converting to (a) natural logarithms and then to (b) common logarithms. Round off each answer to four significant digits.

67. $\log_2 5$

68. $\log_7 9$

69. $\log_4 11$

70. $\log_8 13$

c In problems 71–80, solve each equation. Round off the answer to two decimal places.

71. $6^x = 12$

72. $3^{5x} = 2$

73. $7^{3x-1} = 5$

74. $3^{x+1} = 17^{2x}$

75. $3^{2x-1} = 5^x$

76. $10^{x+1} = 4$

77. $e^{-5x} = 7$

78. $e^{x+1} = 10$

79. $e^{2x-1} = 10^x$

80. $e^x = 10^{x+1}$

c **81.** Suppose that \$1500 is invested at 8% nominal interest rate. Find how many years it will take for the money to triple if the interest is compounded
(a) Semiannually (b) Quarterly (c) Monthly (d) Continuously
(Use the formulas on pages 191 and 193.)

c **82.** Suppose that \$10,000 is invested at 10% nominal interest rate. Find how many years it will take for the money to double if the interest is compounded
(a) Semiannually (b) Quarterly (c) Monthly (d) Continuously
(Use the formulas on pages 191 and 193.)

4.4 Applications and Models Involving Exponential and Logarithmic Functions

This section discusses several applications of exponential and logarithmic functions, as well as some mathematical models based upon these functions.

Applications of Logarithms in the Physical Sciences

The use of logarithmic functions occurs frequently in chemistry, earth science, and physics.

EXAMPLE 1 [c] In *chemistry*, the **pH** of a substance is defined by

$$\text{pH} = -\log[\text{H}^+]$$

where $[\text{H}^+]$ is the concentration of hydrogen ions in the substance, measured in moles per liter. Find the pH of tuna fish if its $[\text{H}^+]$ concentration is given by $9.55 \cdot 10^{-7}$ moles per liter. Round off the answer to two decimal places.

SOLUTION

$$\begin{aligned}\text{pH} = -\log[\text{H}^+] &= -\log[9.55 \cdot 10^{-7}] = -(-6.02)\\ &= 6.02\end{aligned}$$

EXAMPLE 2 In *physics*, the basic unit of sound measurement is called a *bel* (in honor of Alexander Graham Bell). One-tenth of a bel is called a **decibel,** which is the smallest difference noticeable by the average person in the loudness of two sounds. Experiments in psychology suggest that the relationship of the "loudness" L of a sound and the "intensities" I and I_0 is given by the **Weber-Fechner law,** in honor of the German physicist Gustav Fechner (1801–1887) and the German physiologist Ernst Weber (1795–1878). The law is given by the formula

$$L = 10 \log \frac{I}{I_0}$$

where L is measured in decibels, I_0 is the intensity of the threshold of human hearing, which is given by 10^{-12} watt per square meter, and I is the intensity of the sound in question measured in watts per square meter. Find the loudness L (in decibels) of the noise of a car passing in the street if the sound wave intensity I of the car is 10^{-5} watt per square meter.

SOLUTION Here, $I = 10^{-5}$ and $I_0 = 10^{-12}$. Thus we have

$$\begin{aligned}L = 10 \log \frac{I}{I_0} &= 10 \log \frac{10^{-5}}{10^{-12}}\\ &= 10 \log 10^7\\ &= 10(7) = 70 \text{ decibels}\end{aligned}$$

Damage from an earthquake in Southern California, 1971

In *earth science*, the *magnitude* of an earthquake can be measured in terms of its intensity on the **Richter scale,** devised by the American seismologist Charles R. Richter (1900–1984). Various Richter scale readings are indicated in the following table.

Magnitude	Resulting Physical Damage
2.0	Not noticed
4.5	Felt by most people, destruction in a limited area
6.0	Hazardous, serious destruction in a limited area
7.5	Significant damage over a wide area
8.0	Serious damage to buildings
8.9	Maximum recorded damage

EXAMPLE 3 The **magnitude** M of an earthquake is given by

$$M = \log \frac{I}{I_0}$$

where I_0 represents a standard minimum intensity and I is the intensity of the earthquake being measured. Find M if an earthquake has intensity $10^{6.4}$ times that of I_0.

SOLUTION Here $I = 10^{6.4} I_0$, so that

$$M = \log \frac{I}{I_0} = \log \frac{10^{6.4} I_0}{I_0}$$

$$= \log 10^{6.4} = 6.4 \log 10 = 6.4$$

Exponential Functions as Mathematical Models

A **mathematical model** is an equation or a set of equations in which variables represent real-world quantities. In Section 4.1, we sketched the graphs of exponential functions. These graphs show that exponential functions are either increasing or decreasing if the base is different from 1.

Biological Growth Model

It has been observed experimentally that the number of bacteria in certain cultures grows exponentially. Suppose that on a given day there are x_0 bacteria present; from then on, the number N of bacteria increases so that after t days the number of bacteria present is given by the model

$$N = x_0 \cdot a^t \qquad \text{where} \quad a > 1$$

EXAMPLE 4 [c] Suppose that a specific bacteria is known to triple its population every day. If 10,000 bacteria are present initially, find the number of bacteria present at the end of $2\frac{1}{2}$ days.

SOLUTION There will be 3(10,000) bacteria one day after the experiment begins, 90,000, or $3^2(10{,}000)$ after two days, and so on. The number N of bacteria present after t days is given by

$$N = 10{,}000 \cdot 3^t$$

Substituting $t = 2\frac{1}{2}$ in the equation, we have

$$N = 10{,}000 \cdot 3^{2.5} = 155{,}884.573$$

Therefore, the number of bacteria present at the end of $2\frac{1}{2}$ days is approximately 155,885.

Radioactive Decay

Radioactive substances disintegrate by spontaneously emitting particles. Specifically, the mass of the radioactive substance y is related to the time elapsed t by the model

$$y = y_0 e^{-kt}$$

where y_0 is the original mass at $t = 0$, and the positive constant k is associated with the specific radioactive substance being considered.

EXAMPLE 5 [c] Suppose a radioactive substance decays according to the equation

$$y = 2000e^{-0.75t}$$

where y represents the mass in grams after t hours. Calculate the mass of the substance after $3\frac{1}{2}$ hours have elapsed. Round off to two decimal places.

SOLUTION Substituting $t = 3.5$ in the equation, we have

$$\begin{aligned} y &= 2000e^{-0.75(3.5)} \\ &= 144.879514 \end{aligned}$$

Therefore, the mass of the substance after $3\frac{1}{2}$ hours is approximately 144.88 grams.

To find the **half-life** T of a given radioactive substance, we use the equation

$$y = y_0 e^{-kT}$$

and replace y by $\frac{1}{2}y_0$, so that

$$\tfrac{1}{2}y_0 = y_0 e^{-kT}$$

or

$$\tfrac{1}{2} = e^{-kT}$$

that is,

$$e^{kT} = 2$$

After taking the natural logarithm of each side of this equation, we obtain

$$\begin{aligned} kT \ln e &= \ln 2 \\ kT &= \ln 2 \end{aligned}$$

Therefore,

$$T = \frac{\ln 2}{k}$$

EXAMPLE 6 [c] Assume that the half-life of radium is 1600 years.

(a) If y_0 milligrams of radium are initially present, write a formula for the number of milligrams y present after t years.

(b) How much of 2 milligrams will remain after 8100 years? Round off to three decimal places.

SOLUTION (a) Since $T = 1600$, it follows from the equation $T = (\ln 2)/k$ that

$$k = \frac{\ln 2}{T} = \frac{\ln 2}{1600}$$
$$= 0.000433217$$

Therefore,

$$y = y_0 e^{-0.000433217t}$$

(b) Substituting $y_0 = 2$ and $t = 8100$ into the latter equation, we obtain

$$y = 2e^{-(0.000433217)(8100)}$$
$$= 0.059850199$$

Therefore, about 0.060 milligram remains after 8100 years.

Learning Curve

In 1930, the American psychologist L. L. Thurstone invented a mathematical model for the learning process called the **learning curve.** The model reflects the fact that when a person is learning to perform a new task, the rate of learning is rapid at first. As time passes, this rate tends to taper off as the level of learning approaches an upper limit. More precisely, if y denotes the number of tasks mastered during the tth unit of time, then the learning curve is given by

$$y = c - ce^{-kt}$$

where k and c are positive constants.

EXAMPLE 7 [c] Suppose that learning vocabulary words in a foreign language follows the learning curve expressed by

$$y = 40 - 40e^{-0.2t}$$

where y is the number of words a student can learn during the tth day of study. Find how many words a student would be expected to learn on the fifth day of studying a foreign language. Sketch the graph of the function.

SOLUTION If $t = 5$, then

$$y = 40 - 40e^{-0.2(5)} = 25.2848224$$

Therefore, on the fifth day of study, the typical student learns approximately 25 words (Figure 1).

Figure 1

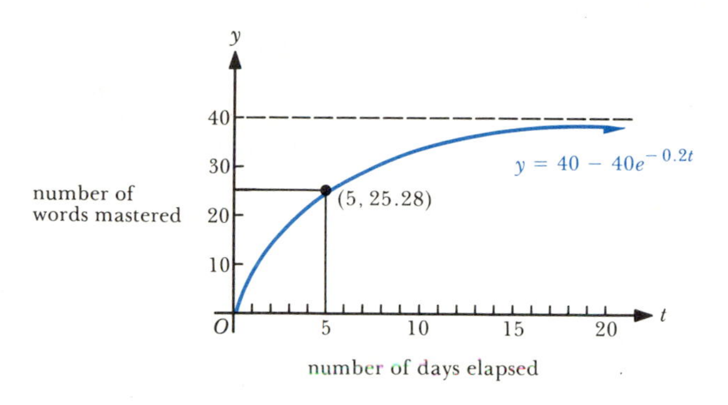

Newton's Law of Cooling

Newton's law of cooling states that the temperature of a body warmer than the surrounding air decreases at a rate that is proportional to the difference between the two temperatures. If T is the temperature of a cooling object after t units of time, and T_0 is the temperature of the medium surrounding the cooling object, then the **law of cooling** is given by

$$T = T_0 + Ce^{-kt}$$

where k and C are positive constants associated with the cooling object.

EXAMPLE 8 [c] A cake baked at a temperature of 180°F is placed on a dinner table where the room temperature is 70°F. After 5 minutes, the temperature of the cake is 100°F. Find its temperature after 8 minutes to one decimal place.

SOLUTION Using the law of cooling

$$T = T_0 + Ce^{-kt}$$

with $T = 180°\text{F}$ and $T_0 = 70°\text{F}$ when $t = 0$, we obtain

$$180 = 70 + C \qquad \text{or} \qquad C = 110$$

so that

$$T = 70 + 110e^{-kt}$$

Also, since $T = 100°\text{F}$ when $t = 5$, we get

$$100 = 70 + 110e^{-5k}$$

$$e^{-5k} = \frac{3}{11}$$

$$-5k = \ln\left(\frac{3}{11}\right)$$

$$k = \frac{\ln 3 - \ln 11}{-5} = 0.26$$

so that

$$T = 70 + 110e^{-0.26t}$$

Therefore, when $t = 8$, we have

$$T = 70 + 110e^{-0.26(8)} = 83.7423233$$

Hence, the temperature of the cake after 8 minutes is about 83.7°F.

PROBLEM SET 4.4

In this problem set, round off all approximated numerical answers to two decimal places.

C **1.** Find the pH of each substance.
(a) Beer: $[H^+] = 3.16 \cdot 10^{-3}$ moles per liter
(b) Eggs: $[H^+] = 1.6 \cdot 10^{-8}$ moles per liter
(c) Milk: $[H^+] = 4 \cdot 10^{-7}$ moles per liter

C **2.** What is the hydrogen ion concentration $[H^+]$ in moles per liter of each substance?
(a) Calcium hydroxide: pH = 13.2
(b) Vinegar: pH = 3.1
(c) Tomatoes: pH = 4.2

3. At take-off, a supersonic jet transport produces a sound wave of intensity 0.3 watt per square meter. Taking I_0 to be 10^{-12} watt per square meter, find the loudness in decibels of the take-off.

4. Suppose that a supersonic plane on take-off produces 120 decibels of sound. How many decibels of sound would be produced by two such planes taking off side by side?

5. Find the magnitude of an earthquake on the Richter scale if its intensity I is 1000 times that of I_0.

6. The Richter scale is also given by the equation

$$R = \log \frac{A}{t}$$

where A is the amplitude of the tremor in microns and t represents how long the tremor lasts, in seconds. Find the magnitude of an earthquake with an amplitude of 10,000 microns and a period of 0.1 second.

C **7.** The **barometric equation**

$$h = (30T + 8000) \ln \frac{P_0}{P}$$

relates the height h in meters above sea level, the air temperature T in degrees Celsius, the atmospheric pressure P_0 in centimeters of mercury at sea level, and the atmospheric pressure P in centimeters of mercury at height h. On a certain day, assume that the atmospheric pressure at a height of h meters measures 38 centimeters of mercury and the atmospheric pressure at sea level is 76 centimeters of mercury. Find the height h if the average air temperature is 4°C.

c **8.** The equation

$$\ln 10W \doteq 4.4974 + 3.135 \ln h$$

is used to express the relationship between the weight W of a person in kilograms and the person's height h in meters. Find the weight W of a person whose height h is

(a) 2 meters (b) 1.8 meters

c **9.** The decibel voltage gain G of an amplifier is given by the formula

$$G = 10 \log \frac{P_o}{P_i}$$

where P_o is the power output of the amplifier and P_i is the power input. If an amplifier produces an output of 35 watts when driven by an input signal of 0.12 watt, what is the amplifier's voltage gain?

10. The **Beer-Lambert law** relates the absorption of light traveling through a material to the concentration and thickness of the material. The formula

$$D = -\frac{1}{k} \log \frac{I}{I_0}$$

relates the length D of the path followed by a beam of light through the material, to the intensities I_0 and I of light of a particular wavelength before and after passing through the material, where k is a positive constant called the **absorption coefficient.** If the absorption coefficient k of seawater is 0.2, and if the eye can perceive light of intensity I smaller than that of sunlight by a factor of 10^{-18}, what is the greatest depth below the ocean surface at which light can be seen?

c **11.** A biologist finds that the number N of bacteria in a certain culture after t hours is given by

$$N = 600 \cdot 3^{t/2}$$

(a) How many bacteria were present when $t = 0$?
(b) How many bacteria are present after 12 hours?

c **12.** A biologist finds that the number N of bacteria present in a culture after t hours is given by $N = Pe^{0.2t}$, where P represents the number of bacteria present initially. How long will it take for the number of bacteria to triple?

c **13.** Suppose that the population N of the United States grows in t years according to the **Malthusian model,** named after the English economist Thomas Malthus (1766–1834). The model is given by the formula

$$N = N_0 e^{kt}$$

where N_0 is the population at $t = 0$, and $k = \ln(1 + k_1)$, where k_1 is the yearly percentage increase and t is the number of elapsed years. According to the U.S. Census Bureau, the population of the United States in 1982 was $N_0 = 231$ million. Suppose that the population grows at 1.1% per year from 1982. Estimate the population of the United States in the year

(a) 1990 (b) 2000

c **14.** The world population is growing according to the Malthusian model (in Problem 13). In 1962 it was 3,150,000,000 and in 1978 it was 4,238,000,000.
(a) Determine the yearly percentage increase of the population.
(b) Estimate the world population in 1992.

c **15.** If the population of the United States grows according to the Malthusian model (in problem 13) at 1.1% per year, in approximately how many years will it double?

c **16.** The area A of a wound that remains unhealed after t days is given by the formula

$$A = A_0 e^{-rt}$$

where A_0 is the original area of the wound and r is the rate of healing. If $r = 0.12$, find the time required for a wound to be half-healed.

c **17.** A radioactive material decays according to the equation $y = 3000e^{-0.6t}$, where y represents the mass, in grams, after t years. What amount will be present in 5 years?

c **18.** Suppose that the amount of radium R present after t years of radioactive decay is given by $R = Pe^{-kt}$, where P represents the number of grams present initially and k is a constant. Find k if 30% of the radium disappears in 100 years.

c **19.** The half-life of radioactive carbon 14, denoted by ^{14}C, is 5570 years. Assume that ^{14}C decays according to the equation

$$y = y_0 e^{-kt}$$

If 100 milligrams of ^{14}C are present at $t = 0$, determine the amount of ^{14}C present after 2000 years.

c **20.** Iodine 131, which is used in the treatment of cancer of the thyroid gland, has a half-life of 8.14 days, where the model is given by the equation

$$y = y_0 e^{-kt}$$

If 100 milligrams of iodine 131 are present at $t = 0$, determine the amount of iodine present after two days.

c **21.** The number y of new tasks mastered by an assembly-line worker during the tth day after a training period begins is given by the learning-curve model

$$y = 50 - 50e^{-0.3t}$$

(a) How many new tasks are learned during the sixth day after the start of the training period?
(b) Predict on what day after the training period 22 tasks will be learned.

c **22.** The growth of physical quantities is often described by the **Gompertz growth model**

$$N = Ce^{-(Ae^{-Bt})}$$

where $A = \ln(C/N_0)$ and B is a positive constant. Find the value of N for the special case $N_0 = 300$, $C = 700$, $B = 0.07$, and
(a) $t = 1$ (b) $t = 2$ (c) $t = 3$ (d) $t = 4$
Do the values of N increase or decrease for increasing values of t?

c **23.** A kettle of water has an initial temperature of 100°C. The room temperature is 20°C. After 10 minutes, the temperature of the kettle is 80°C.
(a) What is the temperature of the kettle after 20 minutes?
(b) When will the kettle's temperature be 40°C?

c **24.** A roast that has cooled to a room temperature of 70°F is placed in a refrigerator with a constant temperature of 35°F. If the temperature of the roast after 2 hours is 45°F, what is its temperature after 4 hours?

c **25.** The electric current I in amperes flowing in a series circuit having an inductance L henrys, a resistance R ohms, and an electromotive force E volts, is given by the equation

$$I = \frac{E}{R}(1 - e^{-Rt/L})$$

where t is the time in seconds after the current begins to flow. If $E = 12$ volts, $R = 5$ ohms, and $L = 0.03$ henry, find the amount of current that flows in the circuit after 3 seconds.

c **26.** A simple model for inhibited population growth (that is, when constraints have a greater and greater effect as time passes) is called the **logistic growth model.** The population N in t years is given by

$$N = \frac{C}{1 + C_0 e^{-kt}}$$

where C_0, C, and k are constants. Also $C_0 = (C - N_0)/N_0$; N_0 is the initial population, and $C - N_0$ is positive. Suppose that the population of the United States has been growing according to the logistic model, where t is the time in years elapsed since 1780 and $k = 0.03$. If the population in 1780 was 3 million and in 1880 was 50 million, find N in the year 1988. Round off the population to the nearest 10 million.

4.5 Tables of Logarithms

We have been using a calculator to determine the values of common logarithms and natural logarithms. Before the widespread availability of computers and scientific calculators, these types of calculations were accomplished by the use of logarithmic tables.

Common Logarithm Table

We can use logarithms to base 10 in computational work. First we observe that any positive real number x can be written in the form

$$x = s \cdot 10^n$$

where $1 \leq s < 10$ and n is an integer. This form is referred to as **scientific notation.** For example, 3721 and 300,000,000 can be expressed in scientific notation as follows:

$$3721 = 3.721 \cdot 10^3$$

$$300{,}000{,}000 = 3 \cdot 10^8$$

If we apply Property (i) of Theorem 1 in Section 4.3 to $x = s \cdot 10^n$ we have

$$\begin{aligned} \log x &= \log[s \cdot 10^n] \\ &= \log s + \log 10^n \end{aligned}$$

Since $\log 10^n = n$, we obtain

$$\log x = \log s + n$$

This last equation is called the **standard form of log *x*.** The number $\log s$, where $1 \leq s < 10$, is called the **mantissa of log *x*,** and the integer n is called the **characteristic of log *x*.**

Notice that for $1 \leq s < 10$, we have

$$\log 1 \leq \log s < \log 10$$

or, equivalently,

$$0 \leq \log s < 1$$

That is, the mantissa is either 0 or a *positive number* between 0 and 1. To determine the value of $\log x$, we determine the value of $\log s$, where s is always between 1 and 10, and add n.

We can obtain the *approximate* value of $\log s$, where $1 \leq s < 10$, from a table of common logarithms (Table I in the Appendix). For instance, $\log 1.85 = 0.2672$ and $\log 9.40 = 0.9731$.

EXAMPLE 1 Use scientific notation and Appendix Table I to evaluate each of the following common logarithms.

(a) $\log 41{,}700$

(b) $\log 0.0024$

SOLUTION

(a) $$\begin{aligned} \log 41{,}700 &= \log(4.17 \cdot 10^4) \\ &= \log 4.17 + 4 \\ &= 0.6201 + 4 \\ &= 4.6201 \end{aligned}$$

(b) $$\begin{aligned} \log 0.0024 &= \log(2.40 \cdot 10^{-3}) \\ &= \log 2.40 + (-3) \\ &= 0.3802 + (-3) \\ &= -2.6198 \end{aligned}$$

Recall from Section 4.2 that the value x that satisfies the equation $\log x = r$ is the *antilogarithm of* r. Note that $\log x = r$ is equivalent to $x = 10^r$. To determine the antilogarithm x of a given number r by the use of Appendix Table I, we "reverse" the process of finding the logarithm of a number. For instance, the *antilogarithm of* 4.4969 or solution of $\log x = 4.4969$ to three significant digits can be found as follows:

$$\begin{aligned}\log x &= 4.4969 \\ &= 0.4969 + 4 && \text{(Standard Form)} \\ &= \log s + 4 \\ &= \log 3.14 + 4 && \text{(Table Use)} \\ &= \log(3.14 \cdot 10^4) && \text{(Scientific Notation)}\end{aligned}$$

Thus, $x = 31{,}400$, so that the antilogarithm of 4.4969 is 31,400 or $10^{4.4969} = 31{,}400$.

EXAMPLE 2 Find the value of each antilogarithm by the use of Appendix Table I.

(a) $10^{2.7210}$
(b) $10^{-2.0804}$

SOLUTION (a) If we let $x = 10^{2.7210}$, then x represents the antilogarithm of 2.7210 and we have

$$\begin{aligned}\log x &= 2.7210 \\ &= 0.7210 + 2 && \text{(Standard Form)} \\ &= \log 5.26 + 2 && \text{(Table Use)} \\ &= \log(5.26 \cdot 10^2) && \text{(Scientific Notation)}\end{aligned}$$

so that $x = 526$ or $10^{2.7210} = 526$.

(b) Here we let x represent the antilogarithm of -2.0804 so that $x = 10^{-2.0804}$ and we have

$$\log x = -2.0804$$

Because -2.0804 means $-2 - 0.0804$, and the decimal -0.0804 is negative, the antilogarithm cannot be found directly from Appendix Table I. However, to convert -2.0804 to *standard logarithm form,* we "adjust" the right-hand side of the latter equation by adding and subtracting 3 from -2.0804 to obtain a *positive* mantissa, and then proceed as follows:

$$\begin{aligned}\log x &= -2.0804 + 3 + (-3) \\ &= 0.9196 + (-3) && \text{(Standard Form)} \\ &= \log 8.31 + (-3) && \text{(Table Use)} \\ &= \log(8.31 \cdot 10^{-3}) && \text{(Scientific Notation)}\end{aligned}$$

so that $x = 0.00831$ or $10^{-2.0804} = 0.00831$.

Linear Interpolation

The logarithms and antilogarithms found thus far were special in the sense that we were able to find the necessary numbers in Appendix Table I. But what if this were not the case? Suppose, for example, that a number has four significant digits. Then it is possible to obtain an approximation of its logarithm by using a method known as **linear interpolation** or simply **interpolation.**

To illustrate the method of interpolation, let us approximate log 1.234. We find in Appendix Table I that

$$\log 1.23 = 0.0899 \qquad \text{and} \qquad \log 1.24 = 0.0934$$

We selected these numbers because

$$1.23 < 1.234 < 1.24$$

Next, let us examine that part of the graph of $y = \log x$ where $1.23 \leq x \leq 1.24$ (Figure 1a). Note that

$$\log 1.234 = 0.0899 + \bar{d}$$

where $\bar{d}$ is the number that we are to approximate. First, we "replace" the arc of the graph with a line segment (Figure 1b). Next, we assume that the length $\bar{d}$ is approximately the same as the length d in Figure 1b. (The amount of "curvature" of $y = \log x$ has been exaggerated in Figure 1b for illustrative purposes.) Finally, d can be determined by using the proportionality of the sides of the similar right triangles formed. Thus the value of d can be determined by using the ratio

$$\frac{d}{0.0035} = \frac{0.004}{0.01}$$

so that

$$d = \frac{0.0035(0.004)}{0.01} = 0.0014$$

Therefore, the approximate value of log 1.234 is given by

$$\log 1.234 = \log 1.23 + d$$
$$= 0.0899 + 0.0014 = 0.0913$$

This evaluation of log 1.234 can also be shown in a schematic arrangement to display the mechanics involved in linear interpolation. This scheme is illustrated for log 1.234 in the first of the following examples.

Figure 1

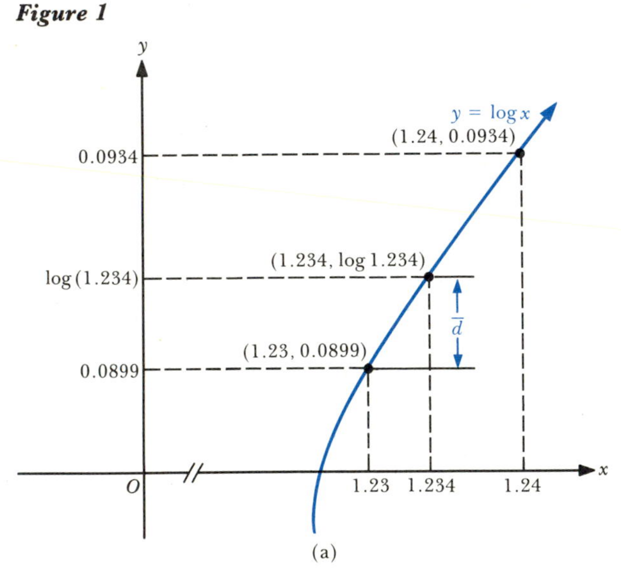

(a)

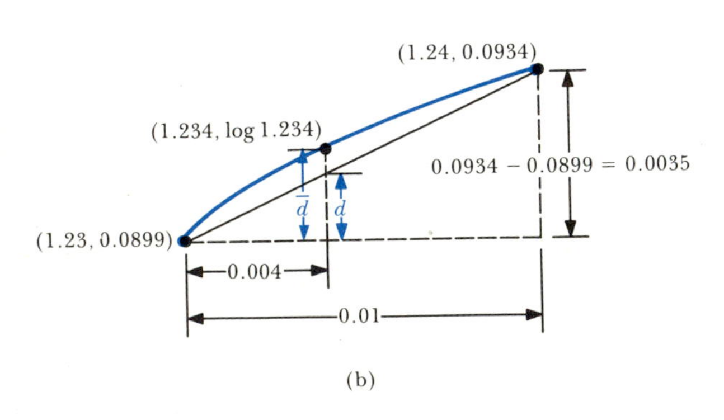

(b)

In Examples 3 and 4, use linear interpolation and Appendix Table I to approximate each value.

EXAMPLE 3 $\log 1.234$

SOLUTION We arrange the work as follows:

$$0.01\left[\begin{array}{cc} t & \log t \\ 1.24 & 0.0934 \\ 0.004\left[\begin{array}{c}1.234 \\ 1.23\end{array}\right. & \left.\begin{array}{c} m \\ 0.0899 \end{array}\right]d \end{array}\right]0.0035$$

We have used Appendix Table I to determine that

$$\log 1.24 = 0.0934 \qquad \text{and} \qquad \log 1.23 = 0.0899$$

The notation

$$0.004\left[\begin{array}{l}1.234 \\ 1.23\end{array}\right.$$

indicates that the difference between 1.234 and 1.23 (top minus bottom number) is

$$1.234 - 1.23 = 0.004$$

Other such differences are indicated similarly. For linear interpolation, we assume that corresponding differences are proportional, so that

$$\frac{d}{0.0035} = \frac{0.004}{0.01} \qquad \text{or} \qquad d = 0.0014$$

Thus, rounded off to four decimal places, we have

$$\begin{aligned} m = \log 1.234 &= 0.0899 + d \\ &= 0.0899 + 0.0014 \\ &= 0.0913 \end{aligned}$$

EXAMPLE 4 $\log 0.0007957$

SOLUTION First we write 0.0007957 in scientific notation as

$$0.0007957 = 7.957 \cdot 10^{-4}$$

so that

$$\begin{aligned} \log 0.0007957 &= \log(7.957 \cdot 10^{-4}) \\ &= \log 7.957 + (-4) \\ &= m + (-4) \end{aligned}$$

Here the mantissa

$$m = \log 7.957$$

is obtained by using linear interpolation and Appendix Table I. Arranging the work and rounding off to four decimal places, we have

$$\begin{array}{cccc} & t & \log t & \\ 0.01 & \left[\begin{array}{cc} & 7.96 \\ 0.007 & \left[\begin{array}{c} 7.957 \\ 7.95 \end{array}\right. \end{array}\right. & \left.\begin{array}{cc} 0.9009 & \\ \left.\begin{array}{c} m \\ 0.9004 \end{array}\right] & d \end{array}\right] & 0.0005 \end{array}$$

Thus,

$$\frac{d}{0.0005} = \frac{0.007}{0.01} \qquad \text{or} \qquad d = 0.0005\left(\frac{0.007}{0.01}\right) = 0.0004$$

Consequently,

$$m = \log 7.957 = 0.9004 + 0.0004 = 0.9008$$

Thus, $\log 0.0007957 = 0.9008 + (-4) = -3.0992$.

The method of interpolation can also be used to find the *antilogarithm;* that is, to find x when we are given r in the equation $\log x = r$ or $x = 10^r$. If we use Appendix Table I, then x may be found to *four significant digits*. A geometric argument similar to the one given earlier in the section can be used to justify the procedure illustrated in the next example.

EXAMPLE 5 Use Appendix Table I and linear interpolation to find the antilogarithm $10^{4.5544}$.

SOLUTION If $x = 10^{4.5544}$, then $\log x = 4.5544$. Thus,

$$\begin{aligned} \log x &= 4.5544 = 0.5544 + 4 \\ &= \log s + 4 \end{aligned}$$

To determine s, we notice that 0.5544 does not appear in the body of Appendix Table I. The numbers closest to it that do appear are 0.5551 and 0.5539. We use linear interpolation to obtain

$$\begin{array}{cccc} & s & \log s & \\ 0.01 & \left[\begin{array}{cc} & 3.59 \\ d & \left[\begin{array}{c} s \\ 3.58 \end{array}\right. \end{array}\right. & \left.\begin{array}{cc} 0.5551 & \\ \left.\begin{array}{c} 0.5544 \\ 0.5539 \end{array}\right] & 0.0005 \end{array}\right] & 0.0012 \end{array}$$

Thus,

$$\frac{d}{0.01} = \frac{0.0005}{0.0012} \qquad \text{or} \qquad d = 0.01\left(\frac{0.0005}{0.0012}\right) = 0.004$$

so that, rounded off to four significant digits,

$$\log s = 3.58 + 0.004 = 3.584$$

Therefore,

$$\begin{aligned} \log x &= \log(3.584 \cdot 10^4) \\ x &= 35{,}840 \qquad \text{or} \qquad 10^{4.5544} = 35{,}840 \end{aligned}$$

Natural Logarithm Table

Appendix Table II contains values of $\ln x$ for selected values of x where $1 \leq x < 10$. But now it is more difficult to find the logarithms of positive numbers greater than or equal to 10, because the natural logarithms of the integer powers of 10 are not themselves integers. This situation is illustrated in the next example.

EXAMPLE 6 Use Appendix Table II to determine each value.

(a) $\ln 7.47$ (b) $\ln 17.4$

SOLUTION (a) Using Appendix Table II, we have $\ln 7.47 = 2.0109$.

(b) 17.4 can be written as $(1.74)(10)$, so that

$$\begin{aligned}\ln 17.4 &= \ln 1.74 + \ln 10 = \ln 1.74 + \ln[(2)(5)] \\ &= \ln 1.74 + \ln 2 + \ln 5 = 0.5539 + 0.6931 + 1.6094 = 2.8564\end{aligned}$$

The method of linear interpolation can also be used in conjunction with Appendix Table II.

EXAMPLE 7 Use Appendix Table II, together with linear interpolation, to determine $\ln 7.115$.

SOLUTION From Appendix Table II, we have

$$0.01\left[\begin{array}{l}\\ 0.005\left[\begin{array}{l}\\ \\ \end{array}\right.\end{array}\right. \begin{array}{cc} t & \ln t \\ 7.12 & 1.9629 \\ 7.115 & ? \\ 7.11 & 1.9615 \end{array} \left.\begin{array}{l}\\ \left.\begin{array}{l}\\ \\ \end{array}\right] d\end{array}\right] 0.0014$$

and

$$\frac{d}{0.0014} = \frac{0.005}{0.01} = 0.0007$$

Thus $\ln 7.115 = 1.9615 + 0.0007 = 1.9622$.

We can use logarithms to perform numerical computations as the next example shows.

EXAMPLE 8 Use logarithms to approximate the value of $\sqrt[5]{17}$ to four significant digits.

SOLUTION Let $x = \sqrt[5]{17}$, then $x = 17^{1/5}$ so that

$$\begin{aligned}\log x &= \log 17^{1/5} = \tfrac{1}{5}\log 17 \\ &= \tfrac{1}{5}(1.2304) = 0.2461\end{aligned}$$

Using linear interpolation, we have $x = 10^{0.2461} = 1.762$. Therefore, $\sqrt[5]{17}$ is approximately 1.762.

PROBLEM SET 4.5

In problems 1–12, use Appendix Table I and the properties of logarithms to find an approximate value of each common logarithm.

1. (a) $\log 870$ (b) $\log 8.7$ (c) $\log 0.87$ (d) $\log 0.087$
2. (a) $\log 575$ (b) $\log 5.75$ (c) $\log 0.575$ (d) $\log 0.00575$
3. (a) $\log 47.3$ (b) $\log 4.73$ (c) $\log 0.473$ (d) $\log 0.000473$
4. (a) $\log 13{,}600$ (b) $\log 13.600$ (c) $\log 1.3600$ (d) $\log 0.013600$
5. (a) $\log 550{,}000$ (b) $\log 55.00$ (c) $\log 0.0055$ (d) $\log 0.55$
6. (a) $\log 34{,}300$ (b) $\log 3.430$ (c) $\log 0.3430$ (d) $\log 0.00343$
7. (a) $\log(32.5)^2$ (b) $\log(32.5)^{-2}$ (c) $\log(32.5)^{1/2}$
8. (a) $\log(1410)^4$ (b) $\log(1410)^{40}$ (c) $\log(1410)^{1/4}$
9. $\log[(365)(215)]$ **10.** $\log \dfrac{2.71}{635}$ **11.** $\log\left[\dfrac{(41.3)^3}{(31.2)^4}\right]$ **12.** $\log\left[\dfrac{(641)^3}{\sqrt{13.5}}\right]$

In problems 13–20, use Appendix Table I, together with linear interpolation, to find the value of each common logarithm.

13. $\log 6.137$ **14.** $\log 555.5$ **15.** $\log 12.35$ **16.** $\log 75.56$
17. $\log 0.5473$ **18.** $\log 1111$ **19.** $\log 472{,}800$ **20.** $\log 0.0001766$

In problems 21–30, use Appendix Table I to find the value of each antilogarithm.

21. $10^{0.9138}$ **22.** $10^{0.6021}$ **23.** $10^{1.7419}$ **24.** $10^{2.3139}$
25. $10^{2.1959}$ **26.** $10^{-0.4473}$ **27.** $10^{-1.2549}$ **28.** $10^{-9.8125}$
29. $10^{-3.3979}$ **30.** $10^{-4.9957}$

In problems 31–40, use Appendix Table I, together with linear interpolation, to find the value of each antilogarithm.

31. $10^{0.5627}$ **32.** $10^{3.8665}$ **33.** $10^{1.1979}$ **34.** $10^{2.002}$
35. $10^{-0.7777}$ **36.** $10^{-1.341}$ **37.** $10^{-2.1234}$ **38.** $10^{-5.7}$
39. $10^{0.3902+(-2)}$ **40.** $10^{0.1664+(-3)}$

In problems 41–46, use Appendix Table II to find the value of each natural logarithm.

41. $\ln 5$ **42.** $\ln 3.88$ **43.** $\ln 12$
44. $\ln 100$ **45.** $\ln 456$ **46.** $\ln 0.035$

In problems 47–50, use Appendix Table II, together with linear interpolation, to find the value of each natural logarithm.

47. $\ln 4.253$ **48.** $\ln 17.18$ **49.** $\ln 7523$ **50.** $\ln 0.09174$

In problems 51–54, use logarithms to approximate the value of each expression to four significant digits.

51. $\sqrt[3]{731}$ **52.** $\sqrt[7]{32.5}$ **53.** $(8.11)^{3/5}$ **54.** $(1.07)^5(2.72)^{-2}$

REVIEW PROBLEM SET, CHAPTER 4

1. Let $f(x) = (\sqrt{2})^x$ and $g(x) = x^x$, find
(a) $f(4)$ (b) $g(2)$ (c) $f(-6)$ (d) $g(-3)$

C **2.** Use a calculator with a $\boxed{y^x}$ key to find the value of each expression to seven significant digits.
(a) $781 \cdot 3^{4.185}$ (b) $64.38 \cdot 5^{-3.7182}$ (c) $2.781^{-2.78}$

In problems 3–7, graph the function. Indicate the domain, the range, and the intervals in which the function is increasing or decreasing. Determine the horizontal asymptote.

3. $f(x) = (\frac{1}{2})^x$ **4.** $h(x) = (\frac{3}{5})^x$ **5.** $g(x) = 4^x$
6. $g(x) = \pi^x$ **7.** $f(x) = 2 \cdot 5^x$

8. Use $f(x) = \dfrac{5^x - 5^{-x}}{5^x + 5^{-x}}$ to show that $f(a + b) = \dfrac{f(a) + f(b)}{1 + f(a)f(b)}$

C **9.** Suppose that \$1000 is put into a savings plan that yields a nominal interest rate of $6\frac{1}{2}\%$. Find how much money is accumulated after 8 years if the interest is compounded
(a) Annually (b) Semiannually (c) Quarterly
(d) Monthly (e) Continuously

C **10.** What is the effective annual interest rate of an investment that pays a nominal interest rate of 9.4% compounded quarterly?

C **11.** Money can be invested at 8.50% compounded semiannually or at 8.34% compounded continuously. Which is the better rate for an investor after one year?

C **12.** Find the present value of \$500,000 to be paid to an investor 5 years in the future, if investments during this period are earning a nominal interest rate of 13% compounded annually.

In problems 13 and 14, sketch the graph of each function. Find the domain and range.

13. $f(x) = 2 - e^x$ **14.** $g(x) = 3 + e^{-x}$

C **15.** Use a calculator with an $\boxed{e^x}$ key or $\boxed{\text{IVN}}-\boxed{\text{ln}}$ keys to find the value of each expression to seven significant digits.
(a) $e^{\sqrt{17}}$ (b) $e^{-3.11}$ (c) $e^{-5\pi}$

16. Suppose that $f(x) = a^x$. Show that

$$[1 + f(x - y)]^{-1} + [1 + f(y - x)]^{-1} = 1$$

C **17.** The owner of an oil painting estimates that the value V in dollars of the painting is approximately given by

$$V = 8000 \cdot (1.3)^{\sqrt{t}}$$

over t years. Find
(a) The present value of the painting.
(b) The value of the painting $6\frac{1}{2}$ years from now.

C **18.** The altitude x in meters is related to the atmospheric pressure y in atmospheres by the equation

$$y = e^{-0.000125x}$$

Determine the altitude of a jet if the atmospheric pressure outside the jet is 0.37 atmosphere.

In problems 19–26, find the solution set of each equation without using logarithms.

19. $2^x = 8^{x-1}$ **20.** $4^{-x} = 8^{x+2}$ **21.** $3^{x^2+x} = 9$
22. $7^{(1/2)x+3} = 1$ **23.** $(\frac{1}{2})^{x-1/3} = 8$ **24.** $(\frac{2}{3})^x = \frac{81}{16}$
25. $2^{x^2+4x} = 32$ **26.** $10^{3x^2-2x} = 100{,}000$

27. Let $f(x) = cb^x$, $f(0) = 7$, and $f(1) = 14$. Find b and c.
28. Let $f(x) = \log_{16} x$. Find each value.
(a) $f(256)$ (b) $f(64)$ (c) $f(32)$ (d) $f(\sqrt[5]{2})$

In problems 29–32, determine the domain of the function.

29. $f(x) = \log_5(2x - 1)$ **30.** $g(x) = \log|x|$
31. $h(x) = \ln|x + 2|$ **32.** $f(x) = \ln e^x$

In problems 33–38, graph the function and determine the domain and range.

33. $f(x) = \log_7 x$ **34.** $h(x) = \log_{1/5} x$ **35.** $g(x) = 2 \ln x$
36. $g(x) = \log(x - 1)$ **37.** $h(x) = \log_3|-2x|$ **38.** $F(x) = \log_5(2 + x)$

39. Find the value of each logarithm without using a calculator or a table.
(a) $\log_3 9$ (b) $\log_4 8$ (c) $\log_6 1$
(d) $\log_5 0.04$ (e) $\log_{100} 0.001$ (f) $\log_9 \frac{1}{3}$

C **40.** Determine the approximate value to four decimal places of

$$\left(1 + \frac{1}{n}\right)^n$$

for $n = 200$.

C **41.** Use a calculator with a [log] key and an [ln] key to find each value to seven significant digits.
(a) log 27,300 (b) log 74.47 (c) ln 2.95714
(d) ln 9.5127 (e) ln(ln 300) (f) ln 8.865
(g) ln(log 7.3) (h) log(ln 123.4)

C **42.** Use a calculator with a [y^x] key and an [e^x] key or [INV]–[ln] keys to find the value of x. Round off the answer to four significant digits.
(a) $\log x = 2.6471$ (b) $\log x = -1.9863$ (c) $\log x = -1.7324$
(d) $\ln x = 4.6514$ (e) $\ln x = -4.9513$ (f) $\ln x = -5.6714$

In problems 43–46, solve for x in terms of y.

43. $y = \dfrac{3^x + 3^{-x}}{2}$ **44.** $y = \dfrac{e^{2x} - e^{-2x}}{2}$

45. $y = \dfrac{e^x - e^{-x}}{e^x + e^{-x}}$ **46.** $y = \dfrac{7^x + 7^{-x}}{7^x - 7^{-x}}$

[c] In problems 47–52, use the change of base formula and a calculator with a [log] key and an [ln] key to approximate each value to four significant digits.

47. $\log_4 3$ **48.** $\log_5 1.73$ **49.** $\log_7 38$
50. $\log_5 e$ **51.** $\log_4 7$ **52.** $\log_9 382$

In problems 53–58, use the properties of logarithms to write each expression as a sum, difference, or multiple of logarithms. Assume that all variables represent positive real numbers and that $a \neq 1$.

53. $\log_2(3^6 \cdot 5^7)$ **54.** $\log_8\left(\frac{5^7}{9^3}\right)$ **55.** $\log_2(xy^5)$
56. $\log_2 \sqrt[7]{5^3 \cdot 5^6}$ **57.** $\log_a\left(\frac{x^2}{y^4}\right)$ **58.** $\log_5(2^6 \cdot 3^7 \cdot 5^2)$

In problems 59–64, use the properties of logarithms to write each expression as a single logarithm. Assume that all variables represent positive real numbers and that $a \neq 1$.

59. $\log_2 \frac{3}{7} + \log_2 \frac{14}{27}$ **60.** $\log_3 \frac{5}{13} + \log_3 \frac{4}{15}$
61. $\log_5 \frac{6}{7} - \log_5 \frac{27}{4} + \log_5 \frac{21}{16}$ **62.** $\log_9 \frac{11}{5} + \log_9 \frac{14}{5} - \log_9 \frac{22}{15}$
63. $5 \log_a x - 3 \log_a y$ **64.** $2 \log_a x^3 + \log_a\left(\frac{2}{x}\right) - \log_a\left(\frac{2}{x^4}\right)$

In problems 65–72, suppose that $\log_a 2 = 0.69$, $\log_a 3 = 1.10$, $\log_a 5 = 1.62$, and $\log_a 7 = 1.94$. Find each of the values.

65. $\log_a(3^5 \cdot 3^7)$ **66.** $\log_a \sqrt[5]{5^3 \cdot 7^4}$ **67.** $\log_a \sqrt[3]{16}$ **68.** $\sqrt[3]{\log_a 16}$
69. $\log_a\left(\frac{2^4}{3^4}\right)$ **70.** $\log_a \sqrt[3]{\frac{3}{2}}$ **71.** $\log_a\left(\frac{60}{a}\right)$ **72.** $\log_a \frac{25}{27}$

In problems 73–82, solve each equation or inequality.

73. $\log_4(x + 3) = -1$ **74.** $\log(x^2 - 4) = 0$
75. $\log_7(2x - 1) > 2$ **76.** $\log_5|3x + 7| < 0$
77. $\log_4 3x = \log_4 3 + \log_4 5$ **78.** $\log_7 2x = \log_7 8 - \log_7 2$
79. $\log_5(2x - 1) + \log_5(2x + 1) = 1$ **80.** $\log_3(x - 1) + \log_3(x - 2) = \log_3 6$
81. $\log_{1/2}(4x^2 - 1) - \log_{1/2}(2x + 1) = 1$ **82.** $\log_3(x^2 - 1) - \log_3(x - 1) = 4$

83. If $\log_b x = 3$, determine each of the following values.

(a) $\log_{1/b} x$ (b) $\log_b\left(\frac{1}{x}\right)$

84. Let $\log_b 2 = A$, $\log_b 3 = B$, and $\log_b 5 = C$. Express $\log_b 0.006$ in terms of A, B, and C.

[c] In problems 85–90, solve each equation. Round off the answer to two decimal places.

85. $(2.06)^x = 300$ **86.** $10^x = 6$ **87.** $5^{2x-1} = 2^{x+3}$
88. $e^{x^2} = 9$ **89.** $15e^x = 5$ **90.** $e^{-x} = 3^{x+1}$

c **91.** Suppose that \$10,000 is invested at a 9% nominal interest rate. Find how many years it will take for the money to double if the interest is compounded
(a) Quarterly (b) Continuously

c **92.** In the study of probability theory, the normal distribution curve is described by the equation $y = ae^{-cx^2}$. Find the value of c when $a = 2.2$ if $x = 0.5$ and $y = 0.345$.

c **93.** Find the pH of each substance to two decimal places.
(a) Vinegar: $[H^+] = 1.58 \cdot 10^{-3}$ moles per liter
(b) Milk of magnesia: $[H^+] = 3.16 \cdot 10^{-11}$ moles per liter

94. The Alaskan earthquake of 1964 had intensity $10^{8.4}$ times that of I_0 on the Richter scale. Find the magnitude of the earthquake.

c **95.** Suppose that a species of bacteria is known to triple its population every hour. If 10,000 bacteria are present initially, how long will it be before the culture contains 60,000 bacteria?

c **96.** The charge Q on a condenser (in coulombs) is described by the formula

$$Q = CE[1 - e^{-t/(C \cdot R)}]$$

where C is the capacity of the condenser in farads, E is the applied voltage, R is the resistance in ohms, and t is the time in seconds after the voltage is applied. Find Q if $C = 7 \cdot 10^{-6}$ farad, $R = 600$ ohms, $E = 130$ volts, and $t = 0.025$ second.

c **97.** The radioactive material plutonium 239 decays according to the function

$$y = y_0 \cdot 2^{-t/24,400}$$

where t is in years. How long will it take 50 grams of plutonium 239 to become 3 grams?

c **98.** The half-life of radioactive radiothorium denoted by ${}_{90}Th^{228}$ is 1.9 years. If 50 milligrams of radiothorium are present today, how much was present a year ago, assuming that $y = y_0 e^{-kt}$?

In problems 99–104, use Appendix Tables to find the approximate value of the logarithm.

99. $\log 94{,}100$ **100.** $\log 4.71$ **101.** $\log 0.342$
102. $\log(47.4)^{-3}$ **103.** $\ln (1710)^4$ **104.** $\ln 0.00416$

In problems 105–110, use Appendix Tables to find the value of the antilogarithm.

105. $10^{1.9969}$ **106.** $10^{-1.9263}$ **107.** $10^{-2.5977}$
108. $10^{3.4742}$ **109.** $e^{2.3927}$ **110.** $e^{-4.8432}$

In problems 111 and 112, use logarithms to approximate the value of each expression to four significant digits.

111. $\sqrt[3]{17.1}$ **112.** $(1.25)^{4/5}$

CHAPTER 5

Trigonometric Functions

Hipparchus

So far, our study has been limited to polynomial, exponential, and logarithmic functions, In this chapter, we consider an important class of functions called the *trigonometric functions*. The development of trigonometry started with the ancient Egyptian and Greek mathematicians and astronomers. Hipparchus (second century B.C.) is given credit for inventing trigonometry. He defined the trigonometric functions as ratios of chord lengths of a circle (a line segment that joins two points on a circle) to the radius of the circle.

Originally, trigonometry was limited to the numerical solutions of triangles and their applications, first in astronomy and later in surveying and navigation. The trigonometric functions that establish the relationships between angles and sides of triangles are still used routinely in calculations made by surveyors, engineers, and navigators. And today, in electronics, engineering, the physical sciences, and the life sciences, trigonometric functions also provide mathematical models for periodic phenomena such as sound waves, electromagnetic waves, light waves, business cycles, and biological systems.

We begin the chapter by considering trigonometric functions of numbers. Then we define trigonometric functions of angles and examine the relationships between these two types of trigonometric functions and their properties. Finally, we study the evaluations and the graphs of the trigonometric functions.

5.1 Trigonometric Functions of Numbers

Recall from Section 2.1 that the graph of the equation $x^2 + y^2 = 1$ is the *unit circle* of radius 1 whose center is at the origin O of the Cartesian coordinate system (Figure 1).

Figure 1

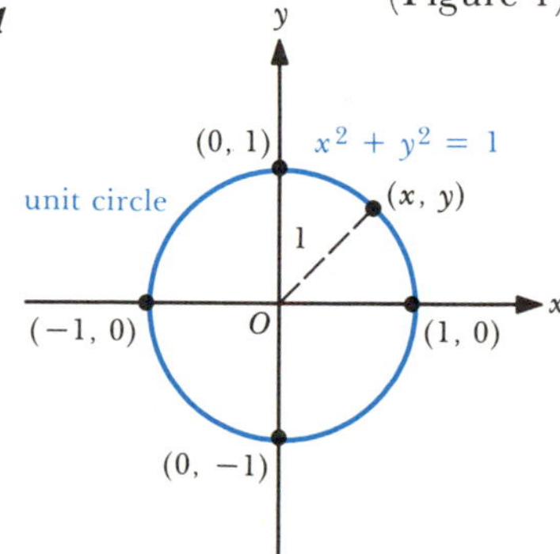

It is possible to associate each real number with the coordinates of a point on the unit circle in the following way. First, assume that a number line has the same scale unit as the one used for the unit circle. Next, place the number 0 on the real number line L so that it coincides with the point $(1, 0)$ on the unit circle. Then, the line L is "wrapped around" the circle, either in a *counterclockwise sense* (using the positive part of the real line L) or in a *clockwise sense* (using the negative part of the real line L). Thus, if the number t_1 is positive ($t_1 > 0$), it is associated with a point on the unit circle by moving $|t_1|$ units counterclockwise along the circumference of the circle, starting at the point $(1, 0)$ and ending at (x_1, y_1) (Figure 2a). If the number t_2 is negative ($t_2 < 0$), it is associated with a point on the circle by moving $|t_2|$ units clockwise along the circumference of the circle, starting at the point $(1, 0)$ and ending at (x_2, y_2) (Figure 2b).

Since the circumference C of any circle is given by $C = 2\pi r$, where r represents the length of the radius of the circle, it follows that the circumference of the unit circle is 2π. If 3.14 is used as an approximation for π, the circumference of the unit circle is approximately 6.28 units.

Figure 3 displays points along the circumference of the unit circle that are obtained by the wrapping process described in Table 1.

In Figure 4a, we observe that 0 is associated with the point $(1, 0)$; $\pi/2$ is associated with the point $(0, 1)$; π is associated with the point $(-1, 0)$; and $3\pi/2$ is associated with the point $(0, -1)$. In Figure 4b, we see that $-\pi/2$ is associated with the point $(0, -1)$; and $-\pi$ is associated with the point $(-1, 0)$.

We use the notation (or symbol) $P(t)$ to denote the point on the unit circle associated with the real number t by the wrapping process. Therefore, as a result of

Figure 2

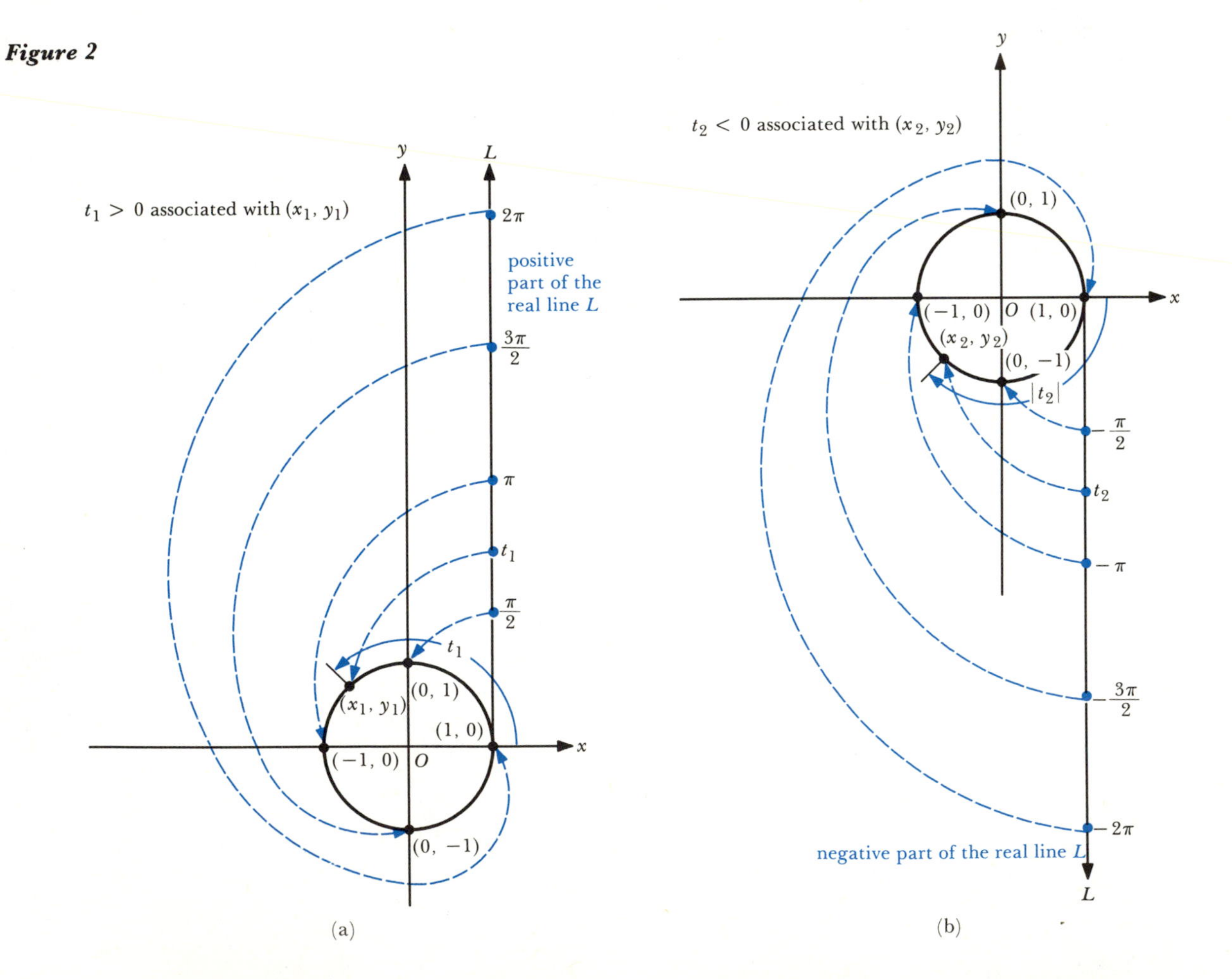

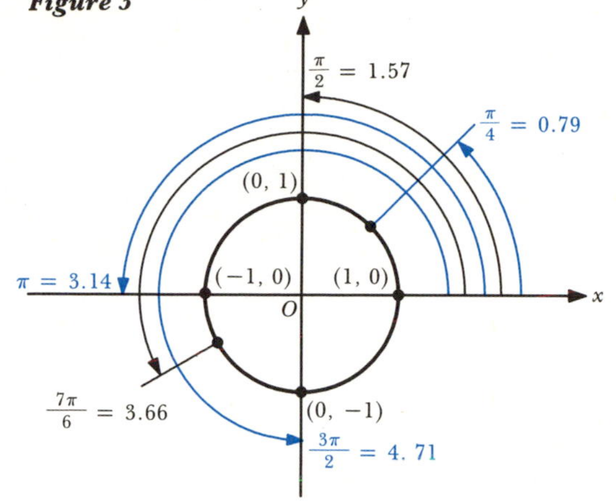

Figure 3

Table 1

Start at (1, 0) and Move Counterclockwise	Actual Distance	Approximate Distance ($\pi = 3.14$)
One-eighth of the way around the circumference	$\frac{1}{8}(2\pi) = \frac{\pi}{4}$	0.79
One-fourth of the way around the circumference	$\frac{1}{4}(2\pi) = \frac{\pi}{2}$	1.57
One-half of the way around the circumference	$\frac{1}{2}(2\pi) = \pi$	3.14
Seven-twelfths of the way around the circumference	$\frac{7}{12}(2\pi) = \frac{7\pi}{6}$	3.66
Three-fourths of the way around the circumference	$\frac{3}{4}(2\pi) = \frac{3\pi}{2}$	4.71

the wrapping process described above, every point P on the unit circle may be designated in two ways: by the Cartesian coordinates (x, y) or by a real number t such that

$$P(t) = (x, y)$$

Table 2 illustrates the designation of the points labeled in Figure 4.

Table 2

$P(t)$	$P(0)$	$P\left(\frac{\pi}{2}\right)$	$P(\pi)$	$P\left(\frac{3\pi}{2}\right)$	$P\left(-\frac{\pi}{2}\right)$	$P(-\pi)$
(x, y)	$(1, 0)$	$(0, 1)$	$(-1, 0)$	$(0, -1)$	$(0, -1)$	$(-1, 0)$

Figure 4

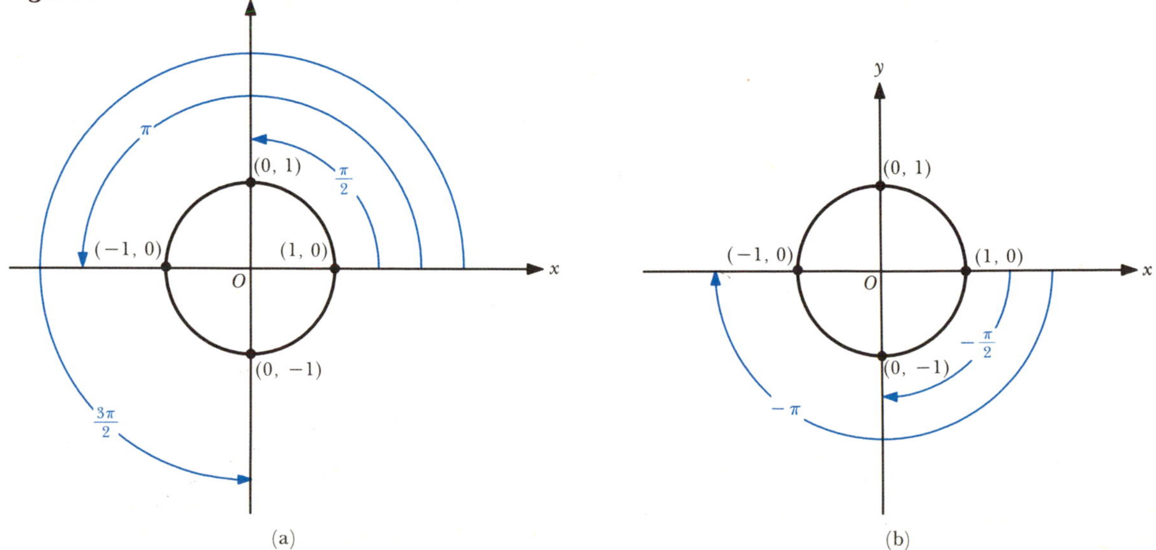

EXAMPLE 1 Find the quadrant containing the given point (use $\pi = 3.14$).

(a) $P(\pi/4)$ (b) $P(\pi/6)$ (c) $P(-3\pi/4)$
(d) $P(2)$ (e) $P(-1)$

SOLUTION (a) Since $\pi/4$ lies halfway between 0 and $\pi/2$,

$$0 < \frac{\pi}{4} < \frac{\pi}{2}$$

It follows that moving counterclockwise from the point $(1, 0)$ a distance $\pi/4$ brings us to a point in quadrant I; thus $P(\pi/4)$ lies in quadrant I (Figure 5).

(b) Since

$$0 < \frac{\pi}{6} < \frac{\pi}{2}$$

it follows that $P(\pi/6)$ lies in quadrant I, because moving counterclockwise from $(1, 0)$ a distance $\pi/6$ brings us to a point in quadrant I (Figure 5).

Figure 5

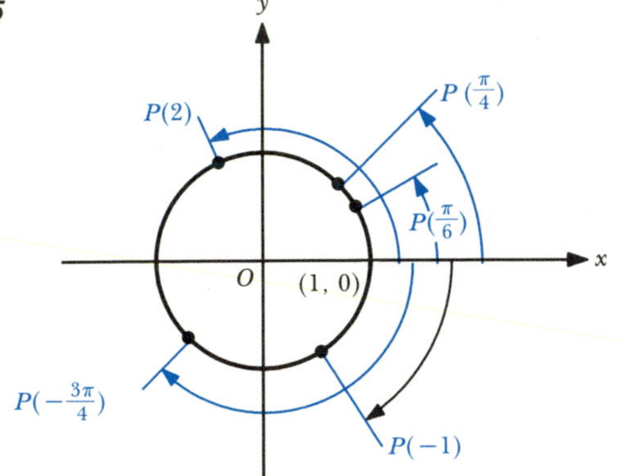

(c) Since

$$-1 < -\frac{3}{4} < -\frac{1}{2}$$

it follows that

$$-\pi < -\frac{3\pi}{4} < -\frac{\pi}{2}$$

so that moving clockwise from $(1, 0)$ a distance $|-3\pi/4|$ brings us to a point in quadrant III. So $P(-3\pi/4)$ lies in quadrant III (Figure 5).

(d) Since $\pi = 3.14$ and $\pi/2 = 1.57$,

$$\frac{\pi}{2} < 2 < \pi$$

so that $P(2)$ is in quadrant II (Figure 5).

(e) Since $-\pi/2 = -1.57$,

$$-\frac{\pi}{2} < -1 < 0$$

We locate $P(-1)$ by moving a distance $|-1| = 1$ in the clockwise direction from the point $(1, 0)$ so that $P(-1)$ lies in quadrant IV (Figure 5).

The coordinates (x, y) of point $P(t)$ on the unit circle (Figure 6) are used to define the six *trigonometric functions* or **circular functions** of t. These functions are referred to and are abbreviated as follows:

Name of Trigonometric Function	Value of the Function at t
sine	$\sin t$
cosine	$\cos t$
tangent	$\tan t$
cotangent	$\cot t$
secant	$\sec t$
cosecant	$\csc t$

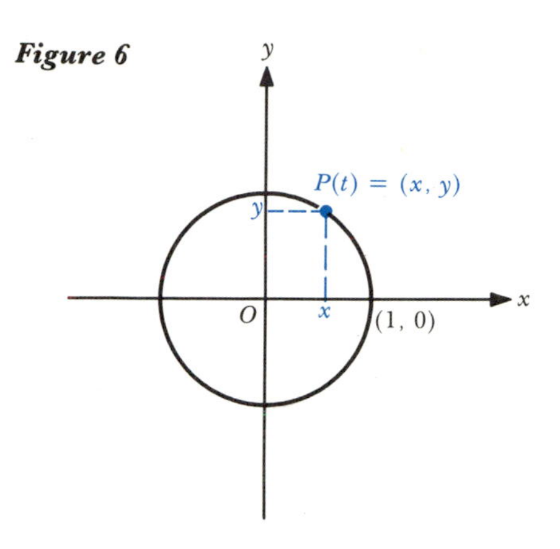

Figure 6

These six trigonometric functions have domains that are sets of real numbers and are defined as follows:

DEFINITION 1

Trigonometric Functions of a Real Number

Let t be a real number and let $P(t) = (x, y)$ be the point on the unit circle associated with t. Then

$$\sin t = y \qquad \csc t = \frac{1}{y}, \quad y \neq 0$$

$$\cos t = x \qquad \sec t = \frac{1}{x}, \quad x \neq 0$$

$$\tan t = \frac{y}{x}, \quad x \neq 0 \qquad \cot t = \frac{x}{y}, \quad y \neq 0$$

EXAMPLE 2 Use Definition 1 to determine the values of the six trigonometric functions at t if

(a) $P(t) = (-\frac{4}{5}, \frac{3}{5})$

[c] (b) $P(t) = (0.7418, y)$ and $P(t)$ is on the unit circle in quadrant IV. Round off the answers to four decimal places.

SOLUTION (a) Note that $(-\frac{4}{5})^2 + (\frac{3}{5})^2 = 1$, so that $P(t) = (-\frac{4}{5}, \frac{3}{5})$ lies on the unit circle (Figure 7). It follows from Definition 1 that

$$\sin t = y = \frac{3}{5} \qquad \csc t = \frac{1}{y} = \frac{1}{3/5} = \frac{5}{3}$$

$$\cos t = x = -\frac{4}{5} \qquad \sec t = \frac{1}{x} = \frac{1}{-4/5} = -\frac{5}{4}$$

$$\tan t = \frac{y}{x} = \frac{3/5}{-4/5} = -\frac{3}{4} \qquad \cot t = \frac{x}{y} = \frac{-4/5}{3/5} = -\frac{4}{3}$$

Figure 7

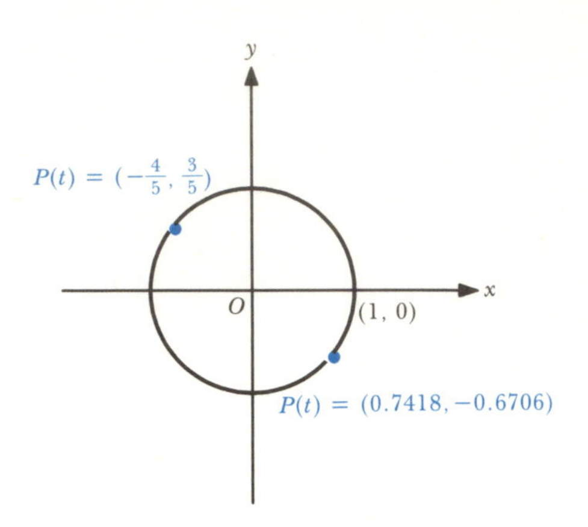

(b) Since $P(t) = (0.7418, y)$ lies on the unit circle, its coordinates satisfy the equation $x^2 + y^2 = 1$, so that

$$(0.7418)^2 + y^2 = 1 \qquad \text{or} \qquad y^2 = 1 - (0.7418)^2$$

Because the point is in quadrant IV, y must be negative. Therefore we choose

$$y = -\sqrt{1 - (0.7418)^2}$$
$$y = -0.6706$$

Thus $P(t) = (0.7418, -0.6706)$ (Figure 7). It follows from Definition 1 that

$$\sin t = y = -0.6706 \qquad \csc t = \frac{1}{y} = \frac{1}{-0.6706} = -1.4912$$

$$\cos t = x = 0.7418 \qquad \sec t = \frac{1}{x} = \frac{1}{0.7418} = 1.3481$$

$$\tan t = \frac{y}{x} = \frac{-0.6706}{0.7418} = -0.9040 \qquad \cot t = \frac{x}{y} = \frac{0.7418}{-0.6706} = -1.1062$$

EXAMPLE 3 Determine the values of the six trigonometric functions for the given value of t.

(a) $t = 0$ (b) $t = \dfrac{\pi}{2}$

SOLUTION (a) Since $P(0) = (1, 0)$ (Figure 8), it follows from Definition 1 that the values of the cosecant and the cotangent functions are undefined for $t = 0$, since division by zero is not defined. The remaining values are:

$$\cos 0 = x = 1 \qquad \sin 0 = y = 0$$

$$\tan 0 = \frac{y}{x} = \frac{0}{1} = 0 \qquad \sec 0 = \frac{1}{x} = \frac{1}{1} = 1$$

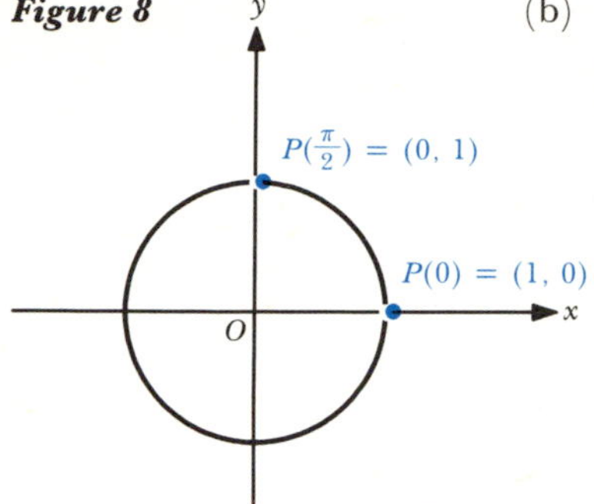

Figure 8

(b) $P(\pi/2) = (0, 1)$ (Figure 8), so that from Definition 1 it follows that the values of the tangent and secant functions are undefined for $t = \pi/2$. The remaining values are

$$\cos\frac{\pi}{2} = x = 0 \qquad \sin\frac{\pi}{2} = y = 1$$

$$\cot\frac{\pi}{2} = \frac{x}{y} = \frac{0}{1} = 0 \qquad \csc\frac{\pi}{2} = \frac{1}{y} = \frac{1}{1} = 1$$

Special Values of the Trigonometric Functions

The exact values of the trigonometric functions can be established for some "special" values of t by using a geometric argument for determining the coordinates of $P(t)$.

In Examples 4 and 5, determine the values of the six trigonometric functions of t.

EXAMPLE 4 $t = \dfrac{\pi}{4}$

SOLUTION Let $P(\pi/4) = (a, b)$. Since

$$P\left(\frac{\pi}{4}\right) = P\left(\frac{1}{2}\cdot\frac{\pi}{2}\right)$$

$P(\pi/4)$ is the midpoint of the arc joining the points $(1, 0)$ and $(0, 1)$ on the unit circle (Figure 9). Thus, $P(\pi/4)$ must lie on the line $y = x$, and we may write

$$P\left(\frac{\pi}{4}\right) = (a, a)$$

Figure 9

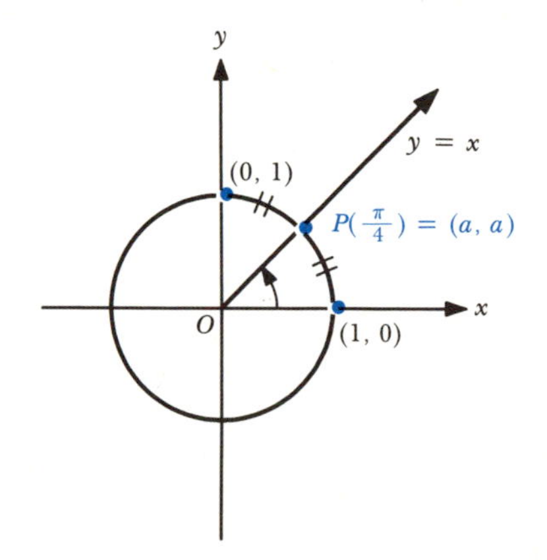

Since the coordinates of any point on the unit circle satisfy $x^2 + y^2 = 1$, we have

$$a^2 + a^2 = 1$$
$$2a^2 = 1$$
$$a^2 = \frac{1}{2}$$

Because $a > 0$ for a point in quadrant I,

$$a = \sqrt{\frac{1}{2}} = \frac{1}{\sqrt{2}} = \frac{\sqrt{2}}{2}$$

so that

$$P\left(\frac{\pi}{4}\right) = \left(\frac{\sqrt{2}}{2}, \frac{\sqrt{2}}{2}\right)$$

Using Definition 1, we have

$$\sin\frac{\pi}{4} = y = \frac{\sqrt{2}}{2} \qquad \csc\frac{\pi}{4} = \frac{1}{y} = \frac{1}{\sqrt{2}/2} = \frac{2}{\sqrt{2}} = \sqrt{2}$$

$$\cos\frac{\pi}{4} = x = \frac{\sqrt{2}}{2} \qquad \sec\frac{\pi}{4} = \frac{1}{x} = \frac{1}{\sqrt{2}/2} = \frac{2}{\sqrt{2}} = \sqrt{2}$$

$$\tan\frac{\pi}{4} = \frac{y}{x} = \frac{\sqrt{2}/2}{\sqrt{2}/2} = 1 \qquad \cot\frac{\pi}{4} = \frac{x}{y} = \frac{\sqrt{2}/2}{\sqrt{2}/2} = 1$$

EXAMPLE 5 (a) $t = \frac{\pi}{3}$ (b) $t = \frac{\pi}{6}$

SOLUTION (a) Here we let $A = (1, 0)$, $B = P(\pi/3) = (a, b)$, and $C = P(2\pi/3) = (-a, b)$. Since the lengths of arcs $\widehat{AB}$ and $\widehat{BC}$ are equal (each is of length $\pi/3$) (Figure 10), it follows from geometry that the chords $\overline{AB}$ and $\overline{BC}$ are equal in length, so that

$$|\overline{AB}| = |\overline{BC}| \quad \text{and thus} \quad |\overline{AB}|^2 = |\overline{BC}|^2$$

Figure 10

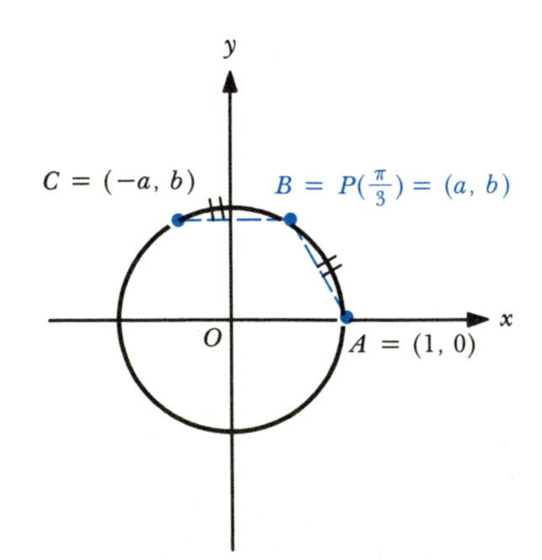

Using the distance formula, we obtain

$$(a - 1)^2 + (b - 0)^2 = (-a - a)^2 + (b - b)^2$$
$$(a - 1)^2 + b^2 = (-2a)^2$$
$$a^2 - 2a + 1 + b^2 = 4a^2$$

But $a^2 + b^2 = 1$, so that the equation becomes

$$-2a + 2 = 4a^2$$

or

$$4a^2 + 2a - 2 = 0$$
$$2(2a - 1)(a + 1) = 0$$

Thus $a = \frac{1}{2}$ or $a = -1$. But $P(\pi/3)$ is in quadrant I; therefore, the only value for a is $\frac{1}{2}$. Then, since $b > 0$,

$$b^2 = 1 - a^2 = 1 - \left(\frac{1}{2}\right)^2 = 1 - \frac{1}{4} = \frac{3}{4} \qquad \text{and} \qquad b = \frac{\sqrt{3}}{2}$$

Consequently,

$$P\left(\frac{\pi}{3}\right) = \left(\frac{1}{2}, \frac{\sqrt{3}}{2}\right)$$

Using Definition 1, we have

$$\sin\frac{\pi}{3} = y = \frac{\sqrt{3}}{2} \qquad \csc\frac{\pi}{3} = \frac{1}{y} = \frac{1}{\sqrt{3}/2} = \frac{2}{\sqrt{3}} = \frac{2\sqrt{3}}{3}$$

$$\cos\frac{\pi}{3} = x = \frac{1}{2} \qquad \sec\frac{\pi}{3} = \frac{1}{x} = \frac{1}{1/2} = 2$$

$$\tan\frac{\pi}{3} = \frac{y}{x} = \frac{\sqrt{3}/2}{1/2} = \sqrt{3} \qquad \cot\frac{\pi}{3} = \frac{x}{y} = \frac{1/2}{\sqrt{3}/2} = \frac{1}{\sqrt{3}} = \frac{\sqrt{3}}{3}$$

(b) A similar argument to the one used in part (a) (Problem 28) can be used to show that

$$P\left(\frac{\pi}{6}\right) = \left(\frac{\sqrt{3}}{2}, \frac{1}{2}\right)$$

Using Definition 1, we have

$$\sin\frac{\pi}{6} = y = \frac{1}{2} \qquad \csc\frac{\pi}{6} = \frac{1}{y} = \frac{1}{1/2} = 2$$

$$\cos\frac{\pi}{6} = x = \frac{\sqrt{3}}{2} \qquad \sec\frac{\pi}{6} = \frac{1}{x} = \frac{1}{\sqrt{3}/2} = \frac{2}{\sqrt{3}} = \frac{2\sqrt{3}}{3}$$

$$\tan\frac{\pi}{6} = \frac{y}{x} = \frac{1/2}{\sqrt{3}/2} = \frac{\sqrt{3}}{3} \qquad \cot\frac{\pi}{6} = \frac{x}{y} = \frac{\sqrt{3}/2}{1/2} = \sqrt{3}$$

Since we frequently use the values of the trigonometric functions for special real numbers, we list these numbers in Table 3.

Table 3

t	$P(t)$	$\sin t$	$\cos t$	$\tan t$	$\cot t$	$\sec t$	$\csc t$
$\frac{\pi}{6}$	$\left(\frac{\sqrt{3}}{2}, \frac{1}{2}\right)$	$\frac{1}{2}$	$\frac{\sqrt{3}}{2}$	$\frac{\sqrt{3}}{3}$	$\sqrt{3}$	$\frac{2\sqrt{3}}{3}$	2
$\frac{\pi}{4}$	$\left(\frac{\sqrt{2}}{2}, \frac{\sqrt{2}}{2}\right)$	$\frac{\sqrt{2}}{2}$	$\frac{\sqrt{2}}{2}$	1	1	$\sqrt{2}$	$\sqrt{2}$
$\frac{\pi}{3}$	$\left(\frac{1}{2}, \frac{\sqrt{3}}{2}\right)$	$\frac{\sqrt{3}}{2}$	$\frac{1}{2}$	$\sqrt{3}$	$\frac{\sqrt{3}}{3}$	2	$\frac{2\sqrt{3}}{3}$

Figure 11

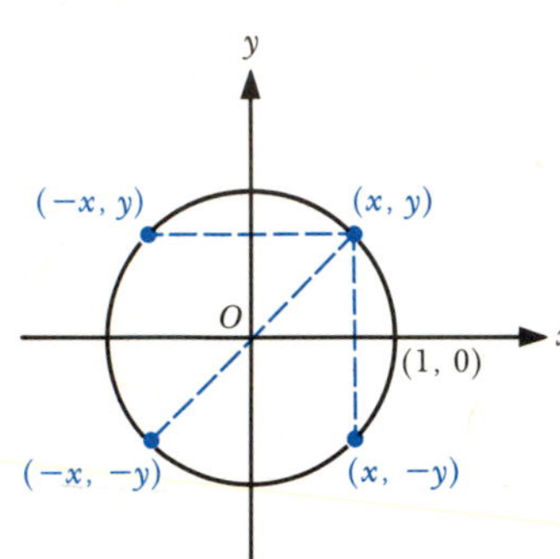

Using the symmetry of the unit circle and Table 3, we can find values of the trigonometric functions for other special values of t when $P(t)$ is in quadrants other than quadrant I. Because of the symmetry of the unit circle $x^2 + y^2 = 1$, it follows that the reflection of any point on the circle across the x axis, y axis, and the origin results in another point on the circle. Thus, if (x, y) is a point on the unit circle in quadrant I, then $(x, -y)$, $(-x, y)$, and $(-x, -y)$ are also on the unit circle (Figure 11).

We have seen in Example 5b that $P(\pi/6) = (\sqrt{3}/2, \frac{1}{2})$ lies on the unit circle, because its coordinates satisfy the equation

$$x^2 + y^2 = \left(\frac{\sqrt{3}}{2}\right)^2 + \left(\frac{1}{2}\right)^2 = \frac{3}{4} + \frac{1}{4} = 1$$

Figure 12

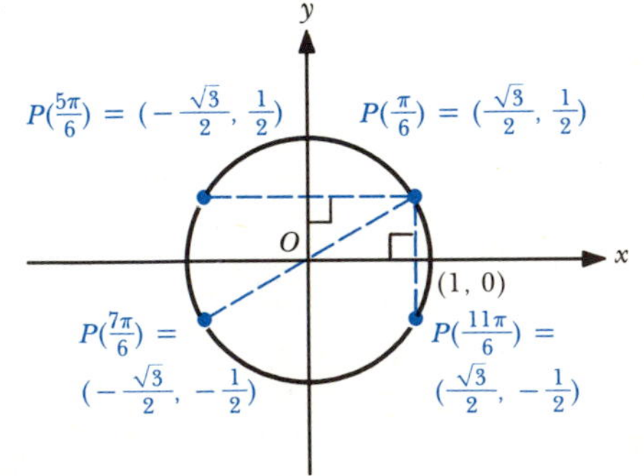

It follows that reflections of this point across the x axis, the y axis, and the origin are given by the points $P(11\pi/6) = (\sqrt{3}/2, -\frac{1}{2})$, $P(5\pi/6) = (-\sqrt{3}/2, \frac{1}{2})$, and $P(7\pi/6) = (-\sqrt{3}/2, -\frac{1}{2})$, respectively (Figure 12). With this information, the values of the trigonometric functions for $t = 5\pi/6$, $7\pi/6$, and $11\pi/6$ can be obtained by using Definition 1. For instance,

$$\sin\frac{5\pi}{6} = y = \frac{1}{2}$$

$$\cos\frac{7\pi}{6} = x = -\frac{\sqrt{3}}{2}$$

$$\tan\frac{11\pi}{6} = \frac{y}{x} = \frac{-1/2}{\sqrt{3}/2} = -\frac{1}{\sqrt{3}} = -\frac{\sqrt{3}}{3}$$

EXAMPLE 6 Use Figure 12 to find the values of the trigonometric functions at $t = -\pi/6$.

SOLUTION We locate $P(-\pi/6)$ by moving clockwise from the point $(1, 0)$ a distance of $|-\pi/6|$. We see that $P(-\pi/6)$ is in quadrant IV, and by using symmetry, we note that it coincides with $P(11\pi/6)$ (Figure 12), so that

$$P\left(-\frac{\pi}{6}\right) = \left(\frac{\sqrt{3}}{2}, -\frac{1}{2}\right)$$

Using Definition 1, we have

$$\sin\left(-\frac{\pi}{6}\right) = y = -\frac{1}{2} \qquad \csc\left(-\frac{\pi}{6}\right) = \frac{1}{y} = \frac{1}{-1/2} = -2$$

$$\cos\left(-\frac{\pi}{6}\right) = x = \frac{\sqrt{3}}{2} \qquad \sec\left(-\frac{\pi}{6}\right) = \frac{1}{x} = \frac{1}{\sqrt{3}/2} = \frac{2}{\sqrt{3}} = \frac{2\sqrt{3}}{3}$$

$$\tan\left(-\frac{\pi}{6}\right) = \frac{y}{x} = \frac{-1/2}{\sqrt{3}/2} = -\frac{1}{\sqrt{3}} = -\frac{\sqrt{3}}{3} \qquad \cot\left(-\frac{\pi}{6}\right) = \frac{x}{y} = \frac{\sqrt{3}/2}{-1/2} = -\sqrt{3}$$

We can use a similar argument to find the coordinates of the point P and the values of the trigonometric functions for $t = 3\pi/4$, $5\pi/4$, and $7\pi/4$ (Figure 13a), and for $t = 2\pi/3$, $4\pi/3$, and $5\pi/3$ (Figure 13b).

Figure 13

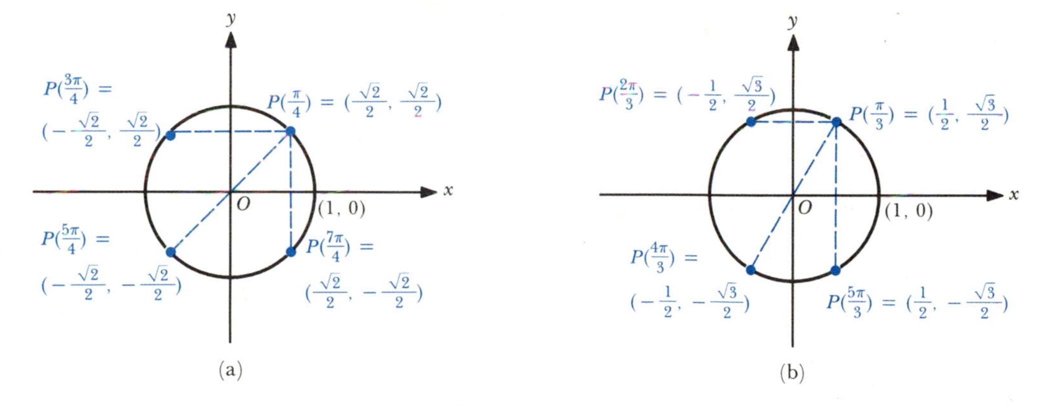

EXAMPLE 7 Use special values, the symmetry of the unit circle, and Definition 1 to find each of the following values.

(a) $\sin(3\pi/4)$ (b) $\cos(-\pi/4)$ (c) $\tan(5\pi/3)$
(d) $\cot(-2\pi/3)$ (e) $\sec(11\pi/6)$ (f) $\csc(-7\pi/6)$

SOLUTION Referring to Figures 12 and 13, we see that $P(-\pi/4) = P(7\pi/4)$, $P(-2\pi/3) = P(4\pi/3)$, and $P(-7\pi/6) = P(5\pi/6)$. Thus

(a) $P\left(\frac{3\pi}{4}\right) = \left(-\frac{\sqrt{2}}{2}, \frac{\sqrt{2}}{2}\right)$, so $\sin\frac{3\pi}{4} = y = \frac{\sqrt{2}}{2}$

(b) $P\left(-\frac{\pi}{4}\right) = P\left(\frac{7\pi}{4}\right) = \left(\frac{\sqrt{2}}{2}, -\frac{\sqrt{2}}{2}\right)$, so $\cos\left(-\frac{\pi}{4}\right) = x = \frac{\sqrt{2}}{2}$

(c) $P\left(\frac{5\pi}{3}\right) = \left(\frac{1}{2}, -\frac{\sqrt{3}}{2}\right)$, so $\tan\frac{5\pi}{3} = \frac{y}{x} = \frac{-\sqrt{3}/2}{1/2} = -\sqrt{3}$

(d) $P\left(-\frac{2\pi}{3}\right) = P\left(\frac{4\pi}{3}\right) = \left(-\frac{1}{2}, -\frac{\sqrt{3}}{2}\right)$, so $\cot\left(-\frac{2\pi}{3}\right) = \frac{x}{y} = \frac{-1/2}{-\sqrt{3}/2} = \frac{1}{\sqrt{3}} = \frac{\sqrt{3}}{3}$

(e) $P\left(\frac{11\pi}{6}\right) = \left(\frac{\sqrt{3}}{2}, -\frac{1}{2}\right)$, so $\sec\frac{11\pi}{6} = \frac{1}{x} = \frac{1}{\sqrt{3}/2} = \frac{2}{\sqrt{3}} = \frac{2\sqrt{3}}{3}$

(f) $P\left(-\frac{7\pi}{6}\right) = P\left(\frac{5\pi}{6}\right) = \left(-\frac{\sqrt{3}}{2}, \frac{1}{2}\right)$, so $\csc\left(-\frac{7\pi}{6}\right) = \frac{1}{y} = \frac{1}{1/2} = 2$

PROBLEM SET 5.1

In problems 1–4, find the value of t if the point $P(t)$ is located by starting at point $(1, 0)$ and moving around the circumference of the unit circle as specified.

1. One-third of the way, clockwise
2. Three-eighths of the way, clockwise
3. Five-fourths of the way, counterclockwise
4. One-sixth of the way, counterclockwise

In problems 5–27, display each point on the unit circle and find the quadrant (if any) containing each point (use $\pi = 3.14$).

5. $P\left(\frac{5\pi}{4}\right)$ **6.** $P\left(\frac{5\pi}{3}\right)$ **7.** $P\left(\frac{2\pi}{3}\right)$ **8.** $P\left(\frac{11\pi}{6}\right)$

9. $P(-2\pi)$ **10.** $P\left(-\frac{5\pi}{3}\right)$ **11.** $P\left(\frac{3\pi}{4}\right)$ **12.** $P\left(\frac{5\pi}{6}\right)$

13. $P\left(-\frac{7\pi}{6}\right)$ **14.** $P\left(-\frac{\pi}{3}\right)$ **15.** $P\left(-\frac{5\pi}{4}\right)$ **16.** $P\left(-\frac{11\pi}{6}\right)$

17. $P\left(-\frac{\pi}{6}\right)$ **18.** $P(0.8)$ **19.** $P(3.6)$ **20.** $P(4.5)$

21. $P(6)$ **22.** $P(1.4)$ **23.** $P(-1.3)$ **24.** $P(-2.7)$

25. $P(-3.7)$ **26.** $P(-5.7)$ **27.** $P(-4)$

28. Prove that

$$P\left(\frac{\pi}{6}\right) = \left(\frac{\sqrt{3}}{2}, \frac{1}{2}\right)$$

In problems 29–38, show that each point lies on the unit circle, and then use Definition 1 to determine the values of the trigonometric functions at t if the point $P(t)$ lies on the unit circle.

29. $P(t) = \left(-\frac{\sqrt{3}}{2}, \frac{1}{2}\right)$ **30.** $P(t) = \left(-\frac{4}{5}, \frac{3}{5}\right)$ **31.** $P(t) = \left(\frac{5}{13}, -\frac{12}{13}\right)$

32. $P(t) = \left(\frac{1}{\sqrt{10}}, \frac{3}{\sqrt{10}}\right)$ **33.** $P(t) = \left(-\frac{3}{\sqrt{13}}, \frac{2}{\sqrt{13}}\right)$ **34.** $P(t) = \left(-\frac{4}{\sqrt{17}}, -\frac{1}{\sqrt{17}}\right)$

35. $P(t) = (\frac{1}{2}, y)$ and $P(t)$ is in quadrant I. **36.** $P(t) = (x, \frac{1}{2})$ and $P(t)$ is in quadrant III.

c **37.** $P(t) = (-0.4561, y)$ and $P(t)$ is in quadrant II. Round off the answers to four decimal places.

c **38.** $P(t) = (0.75x, -0.23)$ and $P(t)$ is in quadrant IV. Round off the answers to four decimal places.

In problems 39–42, use Definition 1 to determine the values of the six trigonometric functions at t.

39. $t = \pi$ **40.** $t = -\frac{\pi}{2}$ **41.** $t = \frac{3\pi}{2}$ **42.** $t = -2\pi$

In problems 43–48, show that each point lies on the unit circle, and find the reflection of the point across (a) the x axis, (b) the y axis, and (c) the origin.

43. $\left(\frac{1}{2}, \frac{\sqrt{3}}{2}\right)$ **44.** $\left(-\frac{\sqrt{2}}{2}, -\frac{\sqrt{2}}{2}\right)$ **45.** $\left(-\frac{\sqrt{3}}{2}, -\frac{1}{2}\right)$

46. $\left(-\frac{1}{2}, \frac{\sqrt{3}}{2}\right)$ **47.** $\left(\frac{\sqrt{2}}{2}, -\frac{\sqrt{2}}{2}\right)$ **48.** $\left(\frac{1}{2}, -\frac{\sqrt{3}}{2}\right)$

In problems 49–56, use Figures 12 and 13 to determine the coordinates of the given point $P(t)$ on the unit circle. Also find $\sin t$ and $\cos t$.

49. $P\left(-\frac{3\pi}{4}\right)$ **50.** $P\left(\frac{5\pi}{4}\right)$ **51.** $P\left(\frac{7\pi}{4}\right)$ **52.** $P\left(-\frac{4\pi}{3}\right)$

53. $P\left(-\frac{5\pi}{3}\right)$ **54.** $P\left(-\frac{\pi}{3}\right)$ **55.** $P\left(\frac{11\pi}{6}\right)$ **56.** $P\left(-\frac{5\pi}{6}\right)$

In problems 57–72, use Definition 1 and the symmetry of the unit circle in Figures 12 and 13, if necessary, to find the value of the given expression.

57. $\sec(-\pi)$ **58.** $\csc(-2\pi)$ **59.** $\cot(-\pi)$ **60.** $\tan\left(-\frac{\pi}{4}\right)$

61. $\tan\frac{7\pi}{6}$ **62.** $\sec\frac{3\pi}{4}$ **63.** $\sin\frac{7\pi}{6}$ **64.** $\cos\left(-\frac{7\pi}{6}\right)$

65. $\cot\left(-\frac{11\pi}{6}\right)$ **66.** $\sec\left(-\frac{5\pi}{4}\right)$ **67.** $\csc\left(-\frac{7\pi}{4}\right)$ **68.** $\cos\frac{11\pi}{6}$

69. $\sin\frac{5\pi}{3}$ **70.** $\sin\left(-\frac{5\pi}{3}\right)$ **71.** $\cos\left(-\frac{4\pi}{3}\right)$ **72.** $\cot\left(-\frac{7\pi}{4}\right)$

5.2 Angle Measures

It is often necessary to consider trigonometric functions whose domains consist of angles rather than real numbers. This viewpoint is required in certain applications of mathematics to areas such as surveying and navigation. In this section, we review some notions regarding angles and angle measures.

In plane geometry, an **angle** is determined by rotating a ray about its endpoint (the **vertex** of the angle) from some initial position (the **initial side** of the angle) to a terminal position (the **terminal side** of the angle) (Figure 1). If the angle is formed by a counterclockwise rotation, the angle is considered to be **positive,** whereas if the angle is formed by a clockwise rotation, the angle is **negative.** In Figure 1a the angle determined by Q, P, and R, and the rotation indicated by $\angle QPR$, is a positive angle, whereas in Figure 1b, $\angle CAB$ is a negative angle. Notice in this scheme of denoting an angle that the middle letter represents the vertex, the first letter represents a point on the initial side, and the third letter represents a point on the terminal side.

Figure 1

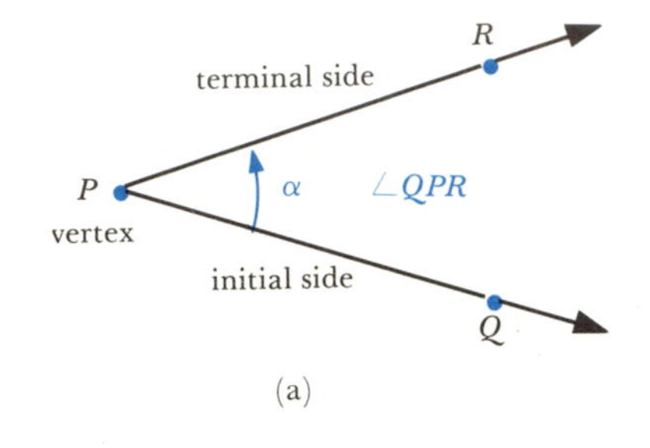

(a)

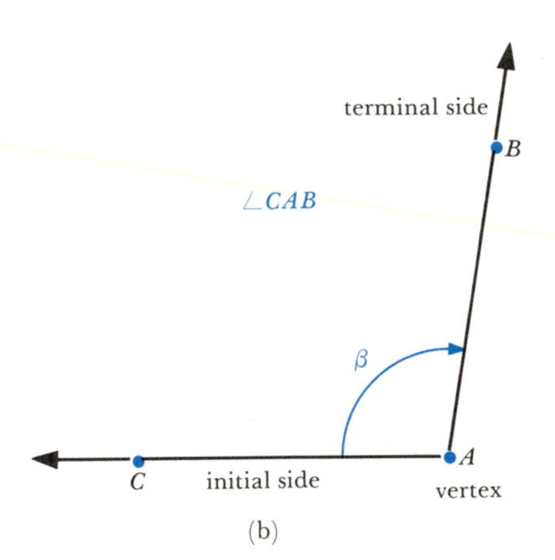

(b)

Angles are often denoted by lowercase Greek letters such as α in Figure 1a (α is the Greek letter *alpha*) or β in Figure 1b (β is the Greek letter *beta*).

An angle is in **standard position** if it is placed on a Cartesian coordinate system with its vertex at the origin O (Figure 2) and with the initial side coinciding with the positive x axis. In Figure 2a, angle α is positive; in Figure 2b, angle β is negative.

The **measure** of an angle is the amount of rotation taking the ray from its initial position to its terminal position. There are various ways to assign

Figure 2

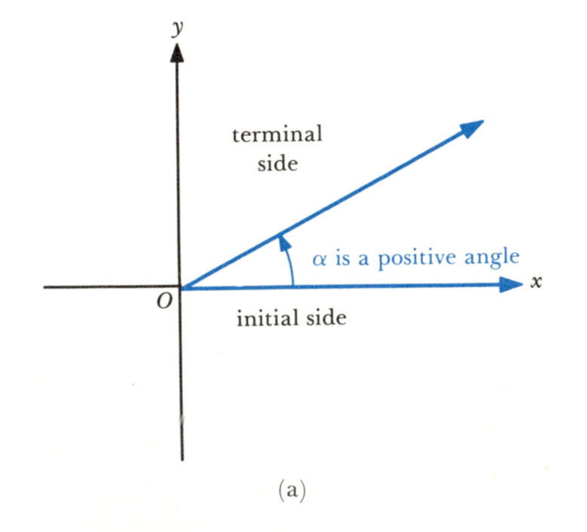

(a)

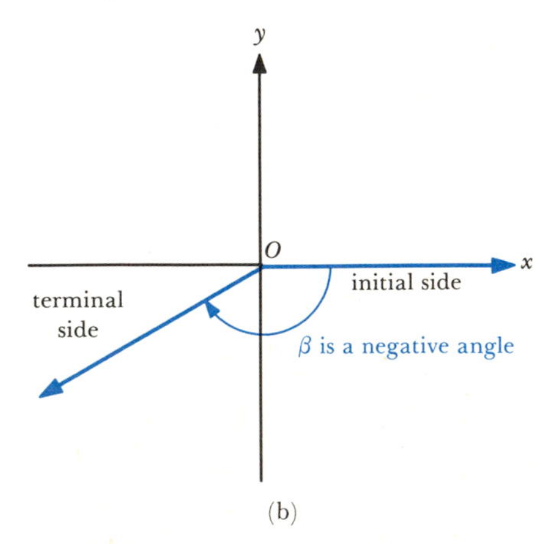

(b)

a numerical measure to an angle. One unit of measurement of an angle is called the **degree.** An angle of *one degree* (1°) is an angle formed by $\frac{1}{360}$ of a complete revolution in the counterclockwise direction. Negative angles are measured by a negative number of degrees; for instance, −360° is the measure of an angle formed by one complete clockwise revolution.

Figure 3 shows angles in standard position measured in degrees. We refer to an angle formed by one-half of a complete counterclockwise revolution as a **straight angle;** it has a measure of $\frac{1}{2}(360°) = 180°$ (Figure 3a). An angle of one-quarter of a counterclockwise revolution is a **right angle;** it has a measure of $\frac{1}{4}(360°) = 90°$ (Figure 3b). An angle is **acute** if its degree measure is between 0° and 90°; for example, 70° (Figure 3c). If an angle measures between 90° and 180°, it is **obtuse;** for example, 120° (Figure 3d).

Figure 3

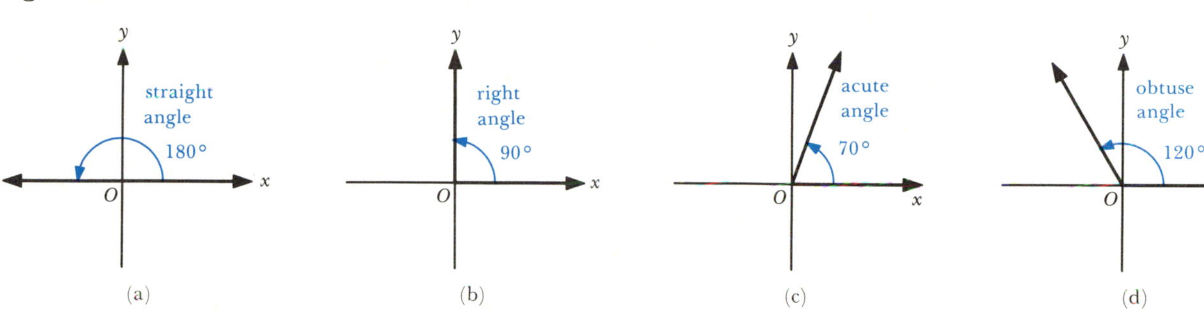

If the terminal side of an angle in standard position lies along either the x axis or the y axis, then the angle is called **quadrantal.** For example, −360°, −270°, −180°, 0°, 90°, 180°, 270°, and 360° are quadrantal angles. Figure 4 shows quadrantal angles −180°, 0°, and 90°.

Angles in standard position that have the same terminal sides are called **coterminal** angles. For example, the three angles 30°, −330°, and 750° are coterminal (Figure 5).

Although parts of a degree can be expressed as a decimal, such parts are sometimes given in *minutes* and *seconds*. A degree can be divided into 60 equal parts called **minutes** (′); a minute can be divided into 60 equal parts called **seconds** (″).

Figure 4

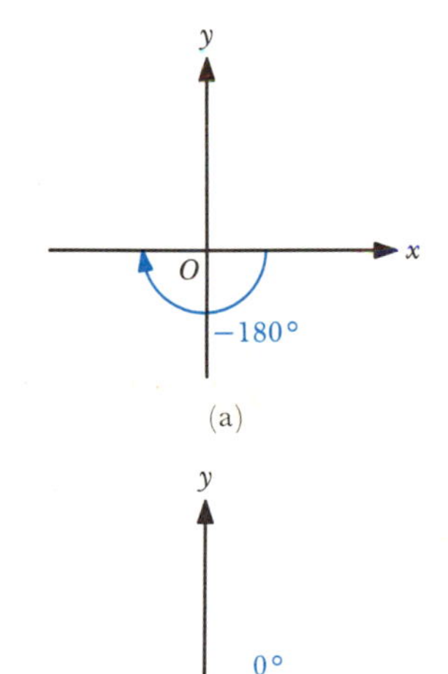

Figure 5

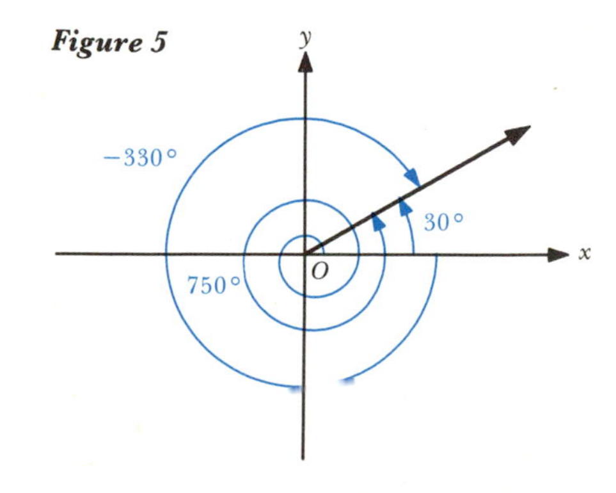

Using these relationships, we have

$$1' = \left(\frac{1}{60}\right)^\circ, \qquad 1'' = \left(\frac{1}{3600}\right)^\circ, \qquad 1^\circ = 60', \qquad \text{and} \qquad 1' = 60''$$

Some calculators have keys that automatically convert decimal degree measures to degrees, minutes, and seconds and vice versa. The next two examples show how such conversions can be made without such special keys.

EXAMPLE 1 Express the following angle measures in terms of degrees, minutes, and seconds.

(a) 37.45° (b) -84.32°

SOLUTION (a) $37.45^\circ = 37^\circ + 0.45^\circ$; and $0.45^\circ = (0.45)(60') = 27'$. Therefore, $37.45^\circ = 37^\circ 27' 0''$.

(b) $-84.32^\circ = -(84^\circ + 0.32^\circ)$; and $0.32^\circ = (0.32)(60') = 19.20'$.

Also, $0.20' = (0.20)(60'') = 12''$, so $-84.32^\circ = -84^\circ 19' 12''$.

EXAMPLE 2 [c] Express $23^\circ 17' 37''$ in decimal degree measure to four decimal places.

SOLUTION

$$23^\circ 17' 37'' = \left(23 + \frac{17}{60} + \frac{37}{3600}\right)^\circ$$

$$= 23.2936^\circ$$

Figure 6

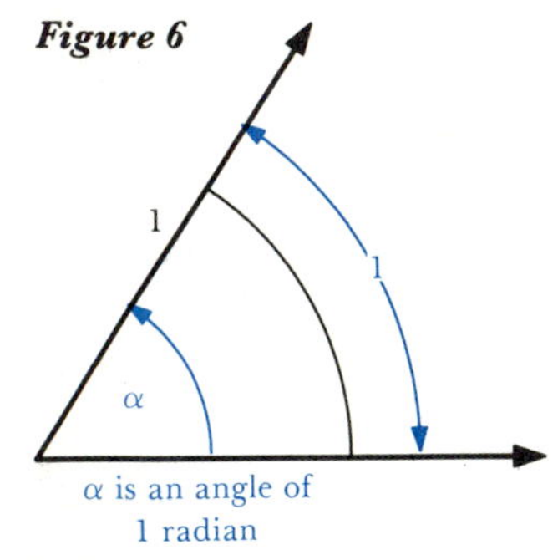

α is an angle of 1 radian

Another frequently used unit of angle measurement is called a **radian.** *One radian* is the measure of a positive angle that intercepts an arc of length 1 on a circle of radius 1 (Figure 6). An angle of radian measure t in standard position has its initial side at the positive x axis and its terminal side at the ray containing the origin and the point $P(t)$, which is located through the use of the wrapping scheme described in Section 5.1. For example, an angle θ (θ is the Greek letter *theta*) of one radian is generated by a counterclockwise rotation in which the point of intersection of the rotating ray with the unit circle travels 1 unit (Figure 7a). Similarly, an

Figure 7

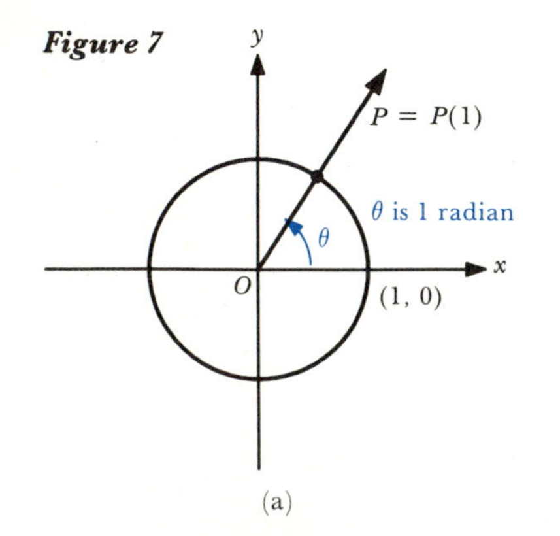

(a)

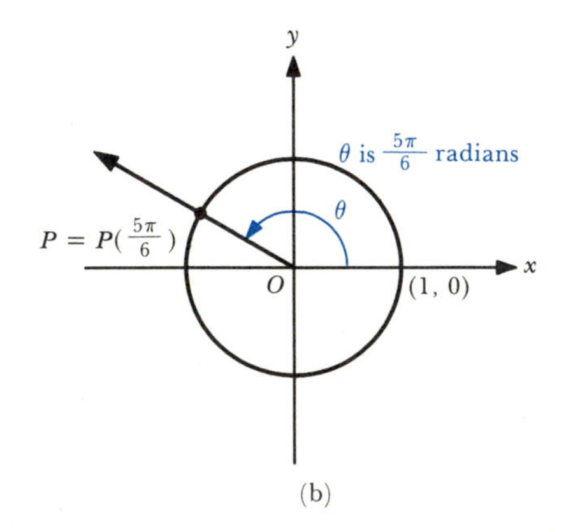

(b)

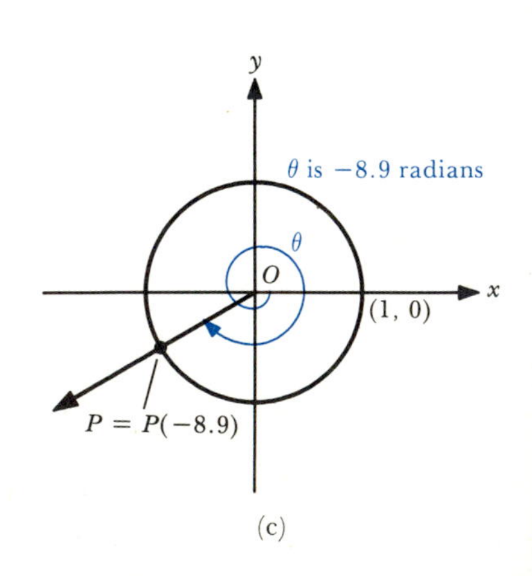

(c)

angle θ of radian measure $5\pi/6$ is generated by a counterclockwise rotation in which the point of intersection of the rotating ray with the unit circle travels $5\pi/6$ units (Figure 7b). The angle θ with radian measure -8.9 is generated by a clockwise rotation in which the point of intersection of the rotating ray with the unit circle travels 8.9 units (Figure 7c).

Figure 8

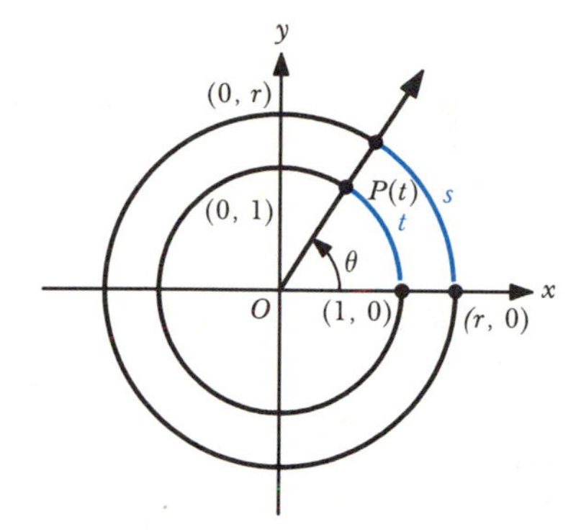

A **central angle** of a circle is an angle that has its vertex at the center of the circle. Figure 8 shows a central angle θ that subtends an arc of length t units on the unit circle and an arc of length s units on a concentric circle of radius r. We know that the radian measure of θ is t. From plane geometry, we know that the ratio of the arc lengths is the same as the ratio of the radii. That is,

$$\frac{t}{s} = \frac{1}{r}$$

so that

$$t = \frac{s}{r} \quad \text{or} \quad s = rt$$

Notice that when $s = r$, the radian measure of θ equals 1; that is, one radian is the measure of a central angle that intercepts an arc of the circle equal in length to the radius of the circle (Figure 9a). If the radian measure of θ is 2π, then the arc length $s = 2\pi r$ corresponds to the circumference of the circle (Figure 9b).

Figure 9

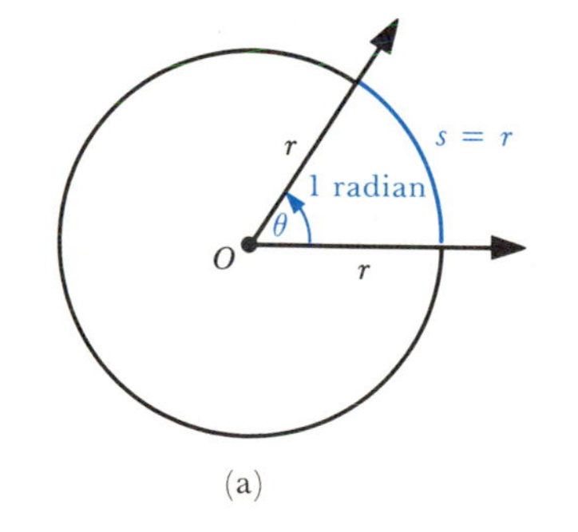

(a)

(b)

Figure 10

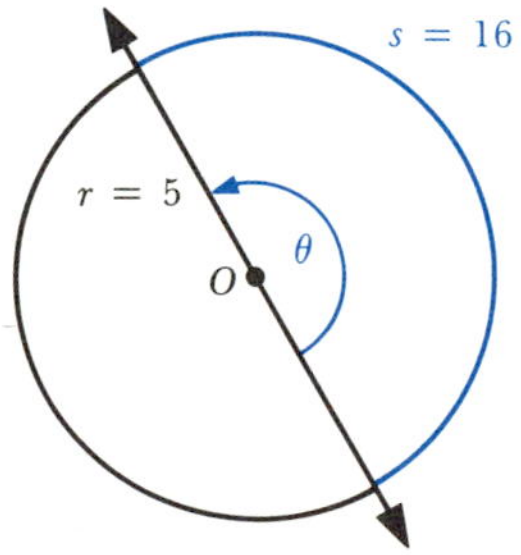

EXAMPLE 3 A central angle θ subtends an arc 16 centimeters long on a circle of radius 5 centimeters (Figure 10). Find the radian measure of θ.

SOLUTION Substituting $s = 16$ and $r = 5$ in the formula $t = s/r$, for the radian measure of θ, we have $t = \frac{16}{5} = 3.2$ radians.

Figure 11

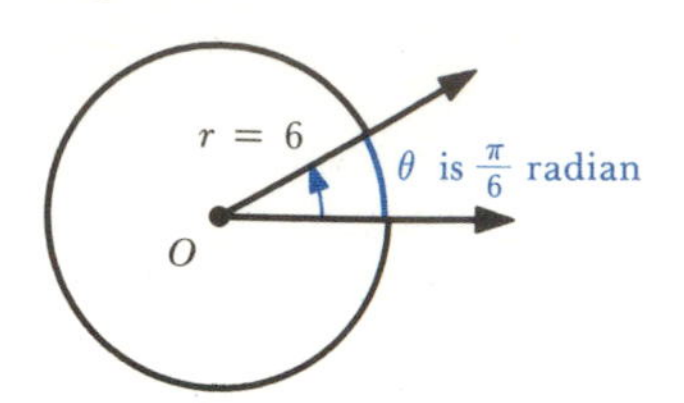

EXAMPLE 4 Find the length s of the arc subtended by a central angle of $\pi/6$ radian on a circle of radius 6 centimeters (Figure 11).

SOLUTION $s = rt = 6\left(\frac{\pi}{6}\right) = \pi$ centimeters

Conversion of Angle Measures

Often an angle measure is expressed in either degrees or radians and we need to convert its given measure to the other measure. For example, an angle of 1° subtends an arc of length $\frac{1}{360}$ of the circumference of a circle with radius 1. In determining the radian measure t of this angle, we use the fact that t is the length of the subtended arc on a circle with radius 1, together with the fact that the circumference of this circle is 2π, to get

Figure 12

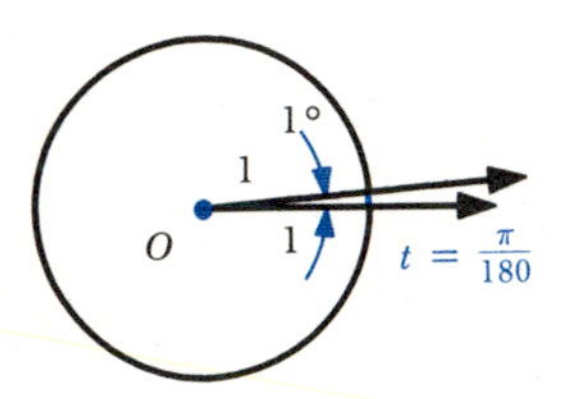

$$t = \frac{1}{360}(2\pi) = \frac{\pi}{180} \qquad \text{(Figure 12)}$$

Hence a measure of 1° corresponds to a measure of $\pi/180$ radian so that a measure of D degrees corresponds to a radian measure $R = (\pi/180)D$.

In general, if an angle has radian measure $R \neq 0$ and degree measure D, then the following ratio holds:

$$\frac{R}{D} = \frac{\pi}{180}$$

Notice that an angle of 1 radian has a degree measure given by $(180/\pi)°$, which can be approximated by using a calculator as 57.2958° or 57°17′45″.

EXAMPLE 5 Convert each degree measure to radian measure.

(a) 150°
(b) 810°
(c) −15°
[c] (d) 26.85°, to four decimal places

SOLUTION Since $R = (\pi/180)D$, there is $\pi/180$ radian in each degree, so that

(a) 150° corresponds to $\frac{\pi}{180}(150) = \frac{5\pi}{6}$ radians

(b) 810° corresponds to $\frac{\pi}{180}(810) = \frac{9\pi}{2}$ radians

(c) −15° corresponds to $\frac{\pi}{180}(-15) = -\frac{\pi}{12}$ radian

(d) 26.85° corresponds to $\frac{\pi}{180}(26.85) = 0.4686$ radian

EXAMPLE 6 Convert each radian measure to degree measure.

(a) $\frac{\pi}{3}$ radians

(b) $\frac{17\pi}{10}$ radians

(c) $-\frac{5\pi}{12}$ radians

[c] (d) -0.3152 radian, to four decimal places

SOLUTION Since $D = (180/\pi)R$, there are $(180/\pi)^\circ$ in each radian; it follows that

(a) $\frac{\pi}{3}$ radians corresponds to $\left(\frac{180}{\pi}\right)^\circ \cdot \left(\frac{\pi}{3}\right) = 60^\circ$

(b) $\frac{17\pi}{10}$ radians corresponds to $\left(\frac{180}{\pi}\right)^\circ \cdot \left(\frac{17\pi}{10}\right) = 306^\circ$

(c) $-\frac{5\pi}{12}$ radians corresponds to $\left(\frac{180}{\pi}\right)^\circ \cdot \left(-\frac{5\pi}{12}\right) = -75^\circ$

(d) -0.3152 radian corresponds to $\left(\frac{180}{\pi}\right)^\circ \cdot (-0.3152) = -18.0596^\circ$.

It is customary to omit the term "radian" when dealing with radian measure. This practice will be followed in this textbook. Thus we write $\pi/3$ to correspond to 60°, without using the word *radians* after $\pi/3$.

Figure 13

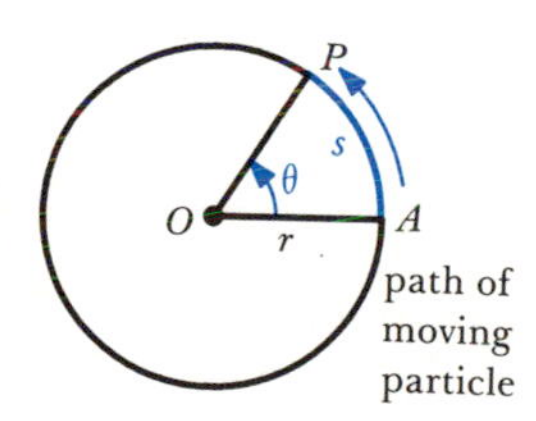

Angles measured in radians are useful in studying the *motion of a particle* moving at a constant speed around a circle of radius r with center O as displayed in Figure 13. Suppose that the moving particle starts at point A and after T units of time it is at point P. If the arc $\widehat{AP}$ has length s, then the particle has moved s units of distance in T units of time. If we assume that the speed is constant, we can use the fact that "distance divided by time equals average speed" to represent the **linear speed** v of the moving particle as

$$v = \frac{s}{T}$$

Because $s = rt$, where t is the radian measure of the subtended central angle θ, we get

$$v = \frac{s}{T} = \frac{rt}{T} = r\left(\frac{t}{T}\right)$$

The ratio that measures the rate of change of the angle (measured in radians) with respect to time is called the **angular speed** $\boldsymbol{\omega}$ (the Greek letter *omega*) and is given by

$$\omega = \frac{t}{T}$$

so that the relationship between the linear speed v and angular speed ω is given by

$$v = r\omega$$

Thus, the linear speed of a particle moving along the circumference of a circle of radius r is the product of the angular speed and the radius.

EXAMPLE 7 A belt passes over the rim of a flywheel with radius 18 centimeters.

(a) Find the angular speed of a point on the rim of the wheel if the belt drives the wheel at a linear speed of 5.76 meters per second.

(b) Find the angular speed of the wheel if the belt drives the wheel at a rate of 12 revolutions every 2 seconds.

SOLUTION (a) Since $v = r\omega$, the angular speed ω is given by

$$\omega = \frac{v}{r}$$

Changing the linear speed from meters per second to centimeters per second, we have

$$v = (5.76)(100) = 576 \text{ centimeters per second}$$

so that

$$\omega = \frac{576}{18} = 32 \text{ radians per second}$$

(b) Since the belt drives the wheel 12 times around the circle in 2 seconds, the radius of the wheel sweeps through $12(2\pi) = 24\pi$ radians in 2 seconds. Therefore, the angular speed of the belt with respect to the center of the wheel is given by

$$\omega = \frac{t}{T} = \frac{24\pi}{2} = 12\pi \text{ radians per second}$$

EXAMPLE 8 The tip of the minute hand of a clock travels $7\pi/10$ inches in 3 minutes. How long is the minute hand?

SOLUTION The angular speed for the minute hand of a clock is given by

$$\omega = \frac{t}{T} = \frac{2\pi}{60} = \frac{\pi}{30} \text{ radian per minute}$$

Also the linear speed for the tip of this minute hand is

$$v = \frac{s}{T} = \frac{7\pi/10}{3} = \frac{7\pi}{30} \text{ inch per minute}$$

Since $v = r\omega$, we get

$$r = \frac{v}{\omega} = \frac{7\pi/30}{\pi/30} = 7$$

so that $r = 7$ inches, which is the length of the minute hand.

PROBLEM SET 5.2

In problems 1–8, the measure of an angle in standard position is given. Sketch three angles in standard position that are coterminal with the given angle.

1. $40°$ **2.** $-30°$ **3.** $-\frac{\pi}{4}$ **4.** $\frac{2\pi}{3}$

5. $-220°$ **6.** $-420°$ **7.** $\frac{7\pi}{6}$ **8.** $-\frac{7\pi}{12}$

c In problems 9–14, express each angle measure in terms of degrees, minutes, and seconds.

9. $87.35°$ **10.** $-62.45°$ **11.** $-25.55°$
12. $267.32°$ **13.** $-65.37°$ **14.** $-181.41°$

c In problems 15–20, express each angle measure as a decimal to four decimal places.

15. $18°42'4''$ **16.** $-68°14'32''$ **17.** $-920°25'$
18. $-19°40'25''$ **19.** $70°35'16''$ **20.** $-51°20'$

In problems 21–26, s denotes the length of the arc intercepted on a circle of radius r by a central angle θ of t radians. Find the missing quantity.

21. $r = 7$ centimeters, $t = \frac{3\pi}{14}$, $s = ?$ **22.** $r = 1.8$ meters, $t = 4$, $s = ?$

23. $r = 6$ meters, $s = 3.6$ meters, $t = ?$ **24.** $s = 6$ feet, θ is $90°$, $r = ?$

25. $s = \frac{8\pi}{5}$ inches, θ is $72°$, $r = ?$ **26.** $s = 5\pi$ kilometers, θ is $225°$, $r = ?$

27. Convert each degree measure to radians. Do not use a calculator. Write your answer as a rational multiple of π.
(a) $40°$ (b) $75°$ (c) $240°$ (d) $330°$
(e) $-95°$ (f) $-220°$ (g) $1080°$ (h) $-3050°$
(i) $-420°$ (j) $444°$ (k) $67.5°$ (l) $-7.5°$

c **28.** Use a calculator to convert each degree measure to an approximate radian measure expressed as a decimal. Round off all answers to four significant digits.
(a) $11°$ (b) $18.33°$ (c) $-15.27°$ (d) $0.0173°$
(e) $371.2°$ (f) $-314.71°$

29. Convert each radian measure to degrees. Do not use a calculator.

(a) $\frac{2\pi}{3}$ (b) $\frac{11\pi}{6}$ (c) $\frac{7\pi}{18}$ (d) $\frac{121\pi}{360}$

(e) $\frac{7\pi}{12}$ (f) $\frac{43\pi}{6}$ (g) $-\frac{4\pi}{9}$ (h) -5π

(i) $-\frac{13\pi}{6}$ (j) $\frac{41\pi}{4}$ (k) $-\frac{3\pi}{8}$ (l) $-\frac{\pi}{14}$

C **30.** Use a calculator to convert each radian measure to an approximate decimal degree measure. Round off all answers to four significant digits.

(a) 4 (b) -2 (c) $\frac{4}{3}$ (d) 1.783

(e) 14.71 (f) -13.33 (g) -0.454 (h) π^2

31. What is the radian measure of the smaller of the angles between the hands of a clock at 11:30?

32. If θ has radian measure t, explain why the terminal side of θ contains the point $P(t)$ if θ is in standard position. Illustrate with $t = \pi/2$ and $t = -\pi$.

33. The length of each chain supporting the seat of a child's swing is 8 feet. When the swing moves from its highest forward position to its highest backward position, the radian measure of the angle swept out by one of these chains is $5\pi/6$. How far does the seat travel in one trip between these high points; that is, what is the length of the arc generated by the seat through one swing between the two highest points?

C **34.** A **nautical mile** is the arc length intercepted on the surface of the earth by a central angle of 1 minute. Assume that the radius of the earth is 3960 miles.

(a) How many feet are there in one nautical mile?

(b) How many miles are there in one nautical mile?

35. Find the length of a pendulum if the tip of the pendulum traces an arc of $11\pi/6$ units and the measure of the angle that this arc subtends is $\pi/5$ radian.

36. A sector POQ of a circle with center O is the region inside the circle bounded by the arc $\overset{\frown}{PQ}$ and the radial segments $\overline{OP}$ and $\overline{OQ}$ (Figure 14). If the central angle θ has a measure of t radians, and r is the radius of the circle, show that the **area A of the sector** is given by

Figure 14

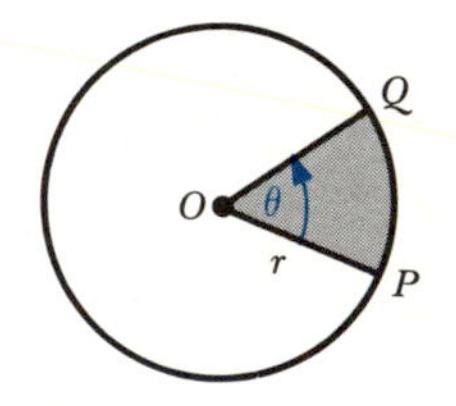

$$A = \frac{1}{2}r^2t \quad \text{or} \quad A = \frac{1}{2}rs$$

where s is the length of arc $\overset{\frown}{QP}$.

In problems 37–42, use the formula for the area in Problem 36 to find the area of a circular sector of radius r with a central angle θ of t radians.

37. $r = 7$ centimeters and $t = \frac{3\pi}{14}$

38. $r = 6$ inches and $t = \frac{5\pi}{18}$

39. $s = 10$ feet and θ is 12°

40. $s = 4$ meters and θ is 315°

41. $r = 5$ centimeters and θ is 245°

42. $s = 3$ feet and $r = 1.5$ feet

43. An object is moving in a circular path of radius 3 meters at the rate of 5 revolutions per minute. How many meters per minute is it moving on the circular path?

44. A wheel 1.5 meters in diameter rolls forward a distance of 228 meters. How many revolutions does the wheel make?

45. Assume that a circular wheel of radius 20 inches is rolled along a flat surface.
(a) Find the angular velocity ω if the linear speed v is 4 feet per second.
(b) Find the linear speed v if the angular speed ω is 7 radians per second.

46. Suppose that a vehicle with tires that are 14 inches in diameter moves at 55 miles per hour.
(a) Find the angular speed of each tire in radians per minute.
(b) How many revolutions per minute does each tire make?

47. A belt drives the rim of a flywheel 5 meters in diameter. Find the linear speed of the wheel if the belt drives the wheel at 450 revolutions per minute.

48. A satellite is orbiting the earth in a perfectly circular orbit with a radius of 6400 kilometers. If it makes three-fourths of a revolution every hour, find
(a) its angular speed ω and (b) its linear speed v.

49. Assume that an object moves in a circular path with a radius of 3 feet.
(a) Find the angular speed ω if the linear speed is 10 feet per second.
(b) Find the linear speed v if the angular speed is 2 radians per second.

5.3 Trigonometric Functions Defined on Angles

In Section 5.1, we dealt with trigonometric functions of real numbers. Now we define the trigonometric functions of angles. In the next section, we shall see how the two are related.

DEFINITION 1 **Trigonometric Functions of an Angle**

Figure 1

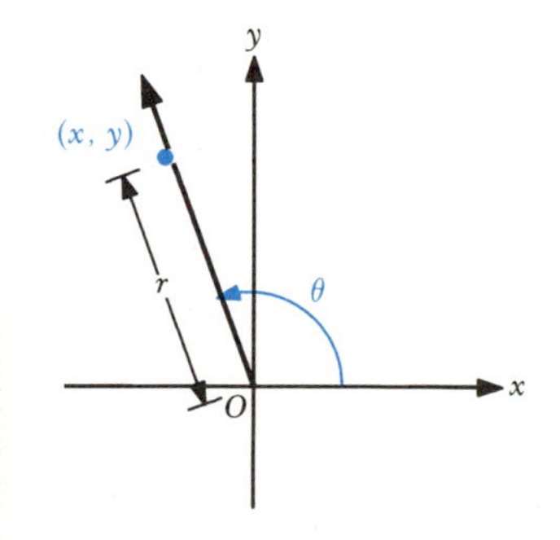

Let θ be a given angle. Place θ in standard position and let $(x, y) \neq (0, 0)$ be a point on the terminal side of θ at a distance $r = \sqrt{x^2 + y^2}$ from the origin (Figure 1). Then the **trigonometric functions of θ** are defined as follows:

$$\sin\theta = \frac{y}{r} \qquad \csc\theta = \frac{r}{y}, \quad y \neq 0$$

$$\cos\theta = \frac{x}{r} \qquad \sec\theta = \frac{r}{x}, \quad x \neq 0$$

$$\tan\theta = \frac{y}{x}, \quad x \neq 0 \qquad \cot\theta = \frac{x}{y}, \quad y \neq 0$$

EXAMPLE 1 Evaluate the six trigonometric functions of an angle θ in standard position if the terminal side of θ contains the point $(5, -12)$.

SOLUTION Consider an angle in standard position whose terminal side contains the point $(5, -12)$ (Figure 2). Since $x = 5$ and $y = -12$, the distance r from the origin to $(5, -12)$ is given by

Figure 2

$$r = \sqrt{x^2 + y^2} = \sqrt{5^2 + (-12)^2} = \sqrt{25 + 144} = \sqrt{169} = 13$$

so that

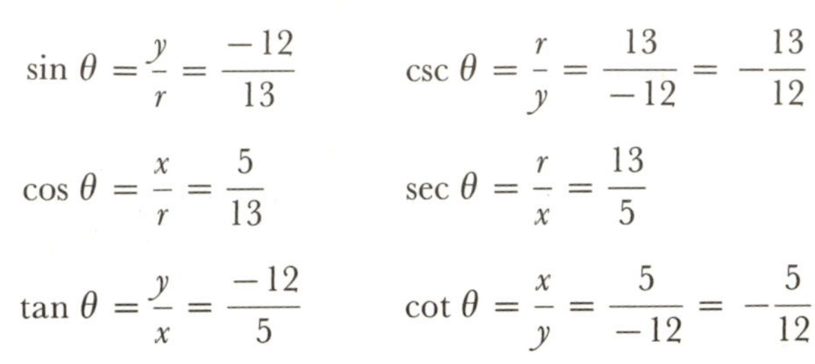

$$\sin\theta = \frac{y}{r} = \frac{-12}{13} \qquad \csc\theta = \frac{r}{y} = \frac{13}{-12} = -\frac{13}{12}$$

$$\cos\theta = \frac{x}{r} = \frac{5}{13} \qquad \sec\theta = \frac{r}{x} = \frac{13}{5}$$

$$\tan\theta = \frac{y}{x} = \frac{-12}{5} \qquad \cot\theta = \frac{x}{y} = \frac{5}{-12} = -\frac{5}{12}$$

There are many angles in standard position whose terminal side contains the same given point. All of these angles are coterminal. Because the evaluations of the trigonometric functions for these coterminal angles depend on the same point, it follows that *for coterminal angles in standard position, the values of the respective trigonometric functions are equal.*

It is important to understand that the values of the six trigonometric functions depend on the *position of the terminal side* of the angle when located in standard position. In other words, the values of the trigonometric functions are the same no matter what particular point is selected on the terminal side [other than $(0, 0)$, of course]. For instance, if (x, y) and (x_1, y_1) are two *different* points in quadrant I, both of which lie on the terminal side of θ (Figure 3), then

Figure 3

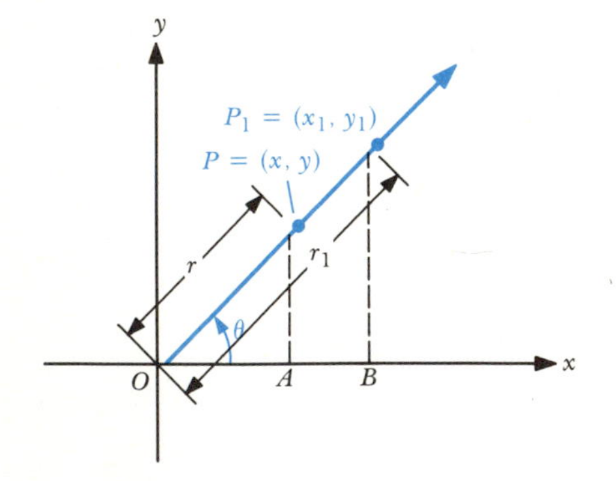

$$r = \sqrt{x^2 + y^2} \qquad \text{and} \qquad r_1 = \sqrt{{x_1}^2 + {y_1}^2}$$

Because of the similar triangles ΔOBP_1 and ΔOAP, we have the equal ratios:

$$\frac{y_1}{r_1} = \frac{y}{r} \qquad \frac{x_1}{r_1} = \frac{x}{r} \qquad \frac{y_1}{x_1} = \frac{y}{x}$$

$$\frac{x_1}{y_1} = \frac{x}{y} \qquad \frac{r_1}{x_1} = \frac{r}{x} \qquad \frac{r_1}{y_1} = \frac{r}{y}$$

Hence the values of the six trigonometric functions are the same no matter what two points are selected on the terminal side of the angle. By a similar construction, we can show that the above results are true for all angles regardless of the quadrant in which the terminal side lies (Problem 50).

EXAMPLE 2 Find the values of the six trigonometric functions of the angles with measures 180° and $-\pi$ radians.

SOLUTION First, we observe that angles with measures 180° and $-\pi$ displayed in standard position are coterminal (Figure 4). Next, we select the point $(-3, 0)$ on the terminal side of the angles so that $x = -3$, $y = 0$, and $r = 3$. Since the angles are coterminal, the values of the respective trigonometric functions are the same for each angle.

Figure 4

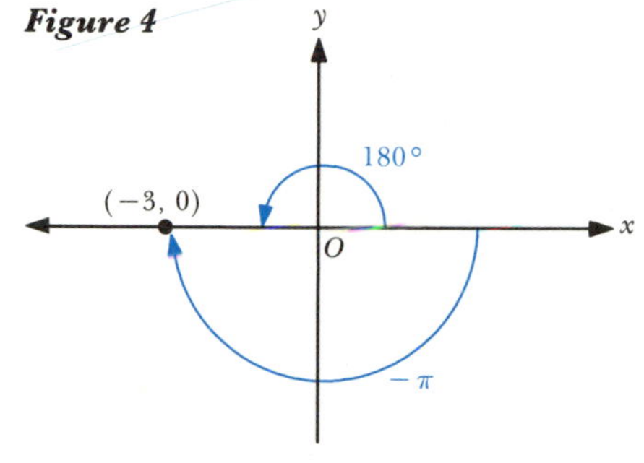

$$\sin 180^\circ = \sin(-\pi) = \frac{y}{r} = \frac{0}{3} = 0$$

$$\cos 180^\circ = \cos(-\pi) = \frac{x}{r} = \frac{-3}{3} = -1$$

$$\tan 180^\circ = \tan(-\pi) = \frac{y}{x} = \frac{0}{-3} = 0$$

$$\csc 180^\circ = \csc(-\pi) = \frac{r}{y} \text{ is undefined because } y = 0$$

$$\sec 180^\circ = \sec(-\pi) = \frac{r}{x} = \frac{3}{-3} = -1$$

$$\cot 180^\circ = \cot(-\pi) = \frac{x}{y} \text{ is undefined because } y = 0$$

Trigonometric Functions of Special Angles

The trigonometric functions enable us to establish relationships between the acute angles and ratios of the lengths of the sides of right triangles. Suppose we place a *right triangle* (Figure 5a) on a coordinate system with one of the acute angles α in standard position (Figure 5b), then the values of the trigonometric functions of α may be expressed in terms of the lengths of the sides of the right triangle. Using

Figure 5

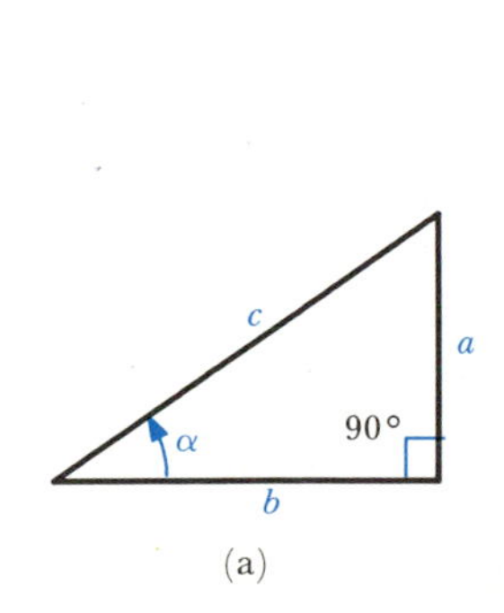

(a)

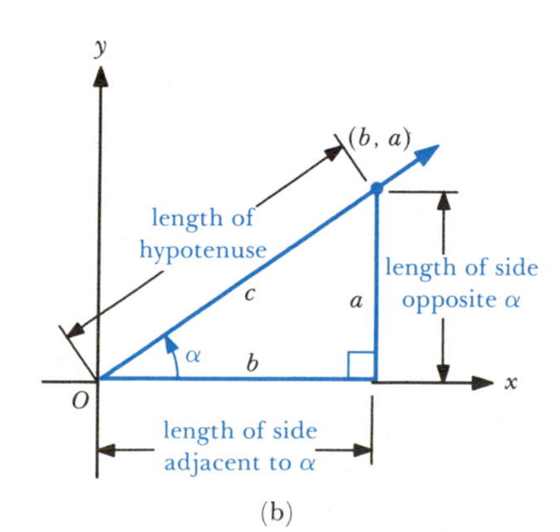

(b)

Definition 1 on page 257 with $\theta = \alpha$, $r = c$, $x = b$, and $y = a$, we express the trigonometric functions of α in terms of the lengths of the sides of the right triangle as follows:

$$\sin\alpha = \frac{a}{c} = \frac{\text{length of side opposite } \alpha}{\text{length of hypotenuse}} \qquad \csc\alpha = \frac{c}{a} = \frac{\text{length of hypotenuse}}{\text{length of side opposite } \alpha}$$

$$\cos\alpha = \frac{b}{c} = \frac{\text{length of side adjacent to } \alpha}{\text{length of hypotenuse}} \qquad \sec\alpha = \frac{c}{b} = \frac{\text{length of hypotenuse}}{\text{length of side adjacent to } \alpha}$$

$$\tan\alpha = \frac{a}{b} = \frac{\text{length of side opposite } \alpha}{\text{length of side adjacent to } \alpha} \qquad \cot\alpha = \frac{b}{a} = \frac{\text{length of side adjacent to } \alpha}{\text{length of side opposite } \alpha}$$

Figure 6

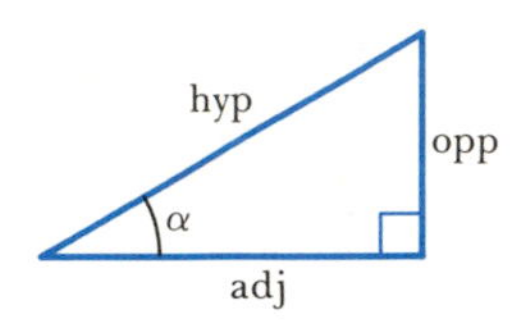

We can abbreviate the lengths of the side **opposite** α, the side **adjacent** to α, and the **hypotenuse** as **opp, adj,** and **hyp,** respectively (Figure 6). Thus the relationships between the trigonometric functions of an acute angle and the ratios of the lengths of the sides of a right triangle are summarized as follows:

Right Triangle Trigonometric Relationships

Assume α is an acute angle in a right triangle. Then

$$\sin\alpha = \frac{\text{opp}}{\text{hyp}} \qquad \csc\alpha = \frac{\text{hyp}}{\text{opp}}$$

$$\cos\alpha = \frac{\text{adj}}{\text{hyp}} \qquad \sec\alpha = \frac{\text{hyp}}{\text{adj}}$$

$$\tan\alpha = \frac{\text{opp}}{\text{adj}} \qquad \cot\alpha = \frac{\text{adj}}{\text{opp}}$$

EXAMPLE 3 Use the right triangle in Figure 7 to find the values of the six trigonometric functions of acute angle α.

SOLUTION For angle α, we have opp = 12 centimeters and hyp = 13 centimeters, but the length of the adj isn't given. Using the Pythagorean theorem, we have

$$(\text{adj})^2 + (\text{opp})^2 = (\text{hyp})^2$$

so that

$$(\text{adj})^2 = (\text{hyp})^2 - (\text{opp})^2$$
$$= 13^2 - 12^2 = 25$$

Figure 7

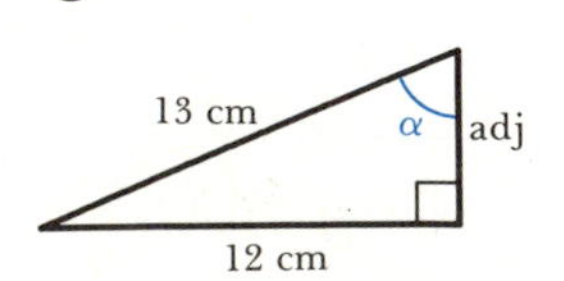

Therefore,

$$\text{adj} = \sqrt{25} = 5$$

It follows from the right triangle relationships that

$$\sin\alpha = \frac{\text{opp}}{\text{hyp}} = \frac{12}{13} \qquad \csc\alpha = \frac{\text{hyp}}{\text{opp}} = \frac{13}{12}$$

$$\cos\alpha = \frac{\text{adj}}{\text{hyp}} = \frac{5}{13} \qquad \sec\alpha = \frac{\text{hyp}}{\text{adj}} = \frac{13}{5}$$

$$\tan\alpha = \frac{\text{opp}}{\text{adj}} = \frac{12}{5} \qquad \cot\alpha = \frac{\text{adj}}{\text{opp}} = \frac{5}{12}$$

Figure 8

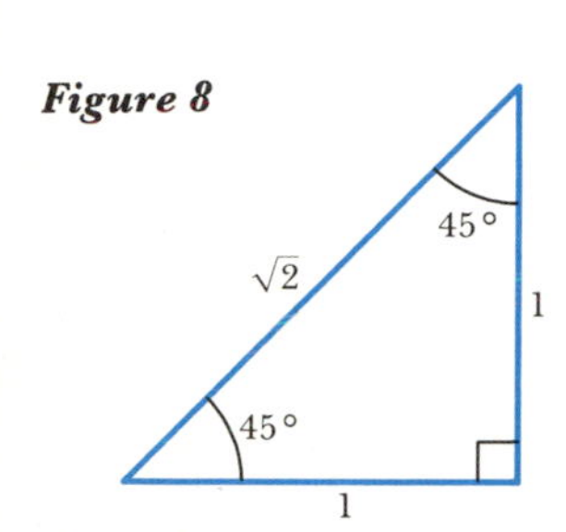

Our knowledge of right triangles enables us to determine the exact values of the trigonometric functions for **special angles 45°, 30°, and 60°.** First, we construct an *isosceles right triangle* with two equal sides of length 1 unit (Figure 8). We know from plane geometry that both of the acute angles in this right triangle measure 45°. Using the Pythagorean theorem, we have

$$1^2 + 1^2 = (\text{hyp})^2$$

so that

$$\text{hyp} = \sqrt{2}$$

From the right triangle trigonometric relationships, we obtain

$$\sin 45^\circ = \frac{1}{\sqrt{2}} = \frac{\sqrt{2}}{2} \qquad \csc 45^\circ = \frac{\sqrt{2}}{1} = \sqrt{2}$$

$$\cos 45^\circ = \frac{1}{\sqrt{2}} = \frac{\sqrt{2}}{2} \qquad \sec 45^\circ = \frac{\sqrt{2}}{1} = \sqrt{2}$$

$$\tan 45^\circ = \frac{1}{1} = 1 \qquad \cot 45^\circ = \frac{1}{1} = 1$$

Figure 9

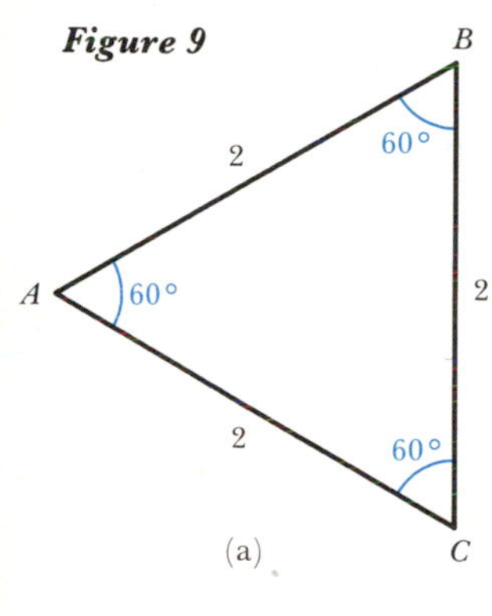

(a)

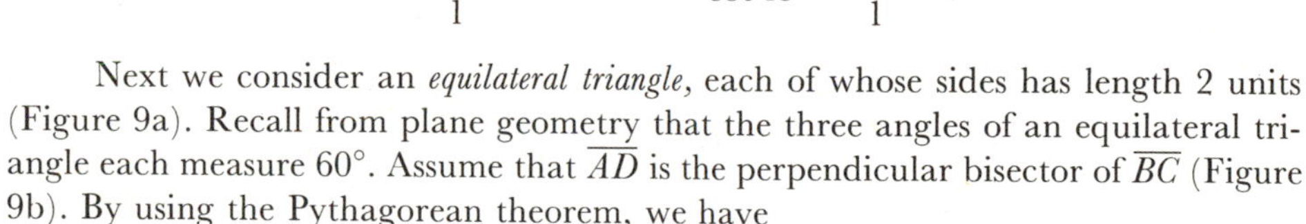

Next we consider an *equilateral triangle,* each of whose sides has length 2 units (Figure 9a). Recall from plane geometry that the three angles of an equilateral triangle each measure 60°. Assume that $\overline{AD}$ is the perpendicular bisector of $\overline{BC}$ (Figure 9b). By using the Pythagorean theorem, we have

$$|\overline{AD}|^2 = 2^2 - 1^2 = 3 \quad \text{or} \quad |\overline{AD}| = \sqrt{3}$$

Thus, from right triangle ADB in Figure 9b, we have

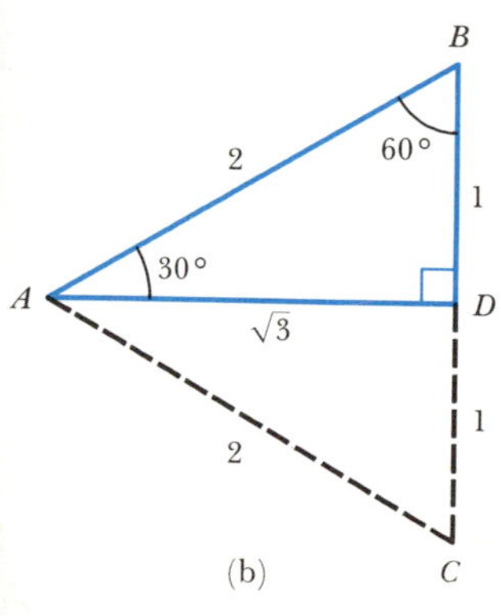

(b)

$$\sin 30^\circ = \cos 60^\circ = \frac{1}{2} \qquad \csc 30^\circ = \sec 60^\circ = \frac{2}{1} = 2$$

$$\cos 30^\circ = \sin 60^\circ = \frac{\sqrt{3}}{2} \qquad \sec 30^\circ = \csc 60^\circ = \frac{2}{\sqrt{3}} = \frac{2\sqrt{3}}{3}$$

$$\tan 30^\circ = \cot 60^\circ = \frac{1}{\sqrt{3}} = \frac{\sqrt{3}}{3} \qquad \cot 30^\circ = \tan 60^\circ = \frac{\sqrt{3}}{1} = \sqrt{3}$$

Note that when we are trying to recall the values of the trigonometric functions for 45°, 30°, or 60° (or equivalently $\pi/4$, $\pi/6$, or $\pi/3$), it is useful to construct a 45°–45° right triangle (Figure 8) or a 30°–60° right triangle (triangle ADB in Figure 9b) and then apply the right triangle trigonometric relationships. In fact, the values of the trigonometric functions for these angles are used so frequently that they should be memorized. Table 1 below summarizes these values.

Table 1 **Special Trigonometric Values**

Angle α							
Radian Measure	Degree Measure	$\sin\alpha$	$\cos\alpha$	$\tan\alpha$	$\cot\alpha$	$\sec\alpha$	$\csc\alpha$
$\frac{\pi}{6}$	30°	$\frac{1}{2}$	$\frac{\sqrt{3}}{2}$	$\frac{\sqrt{3}}{3}$	$\sqrt{3}$	$\frac{2\sqrt{3}}{3}$	2
$\frac{\pi}{4}$	45°	$\frac{\sqrt{2}}{2}$	$\frac{\sqrt{2}}{2}$	1	1	$\sqrt{2}$	$\sqrt{2}$
$\frac{\pi}{3}$	60°	$\frac{\sqrt{3}}{2}$	$\frac{1}{2}$	$\sqrt{3}$	$\frac{\sqrt{3}}{3}$	2	$\frac{2\sqrt{3}}{3}$

We can use these special values to find the values of the trigonometric functions for certain other angles. We first sketch the angle θ in standard position. Then, we either draw a 45°–45° right triangle or a 30°–60° right triangle in whichever quadrant will allow the hypotenuse to lie on the terminal side of θ and one side of the triangle to lie on the x axis. Next, we use the measurements of that particular special right triangle along with the quadrant location to obtain a point $P = (x, y)$ on the terminal side of θ. Finally, we use Definition 1 to determine the values of the trigonometric functions of θ.

In Examples 4–6, use a right triangle along with Definition 1 to determine the values of the six trigonometric functions for each angle with the given measure.

EXAMPLE 4 $\theta = 135°$

SOLUTION First we sketch the angle in standard position. Then we draw a 45°–45° right triangle in quadrant II so that the hypotenuse lies on the terminal side of θ and one leg lies on the negative x axis (Figure 10). By using the measurements of the 45°–45° right triangle along with the fact that the terminal side of θ is in quadrant II, we obtain a point (x, y) on the terminal side of θ in such a way that

$$(x, y) = (-1, 1) \qquad \text{with } r = \sqrt{2}$$

Figure 10

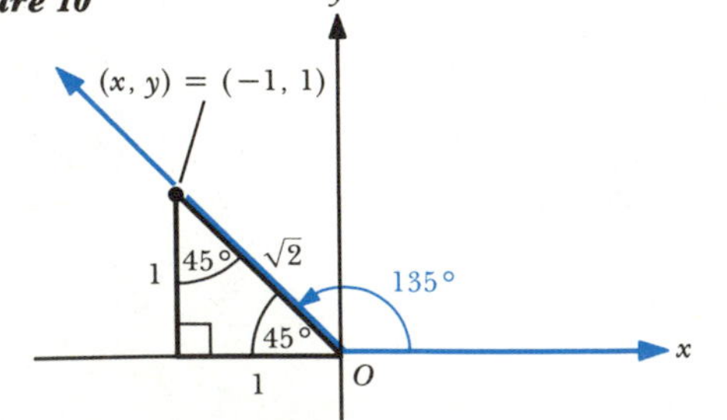

By Definition 1, we get

$$\sin 135^\circ = \frac{y}{r} = \frac{1}{\sqrt{2}} = \frac{\sqrt{2}}{2} \qquad \csc 135^\circ = \frac{r}{y} = \frac{\sqrt{2}}{1} = \sqrt{2}$$

$$\cos 135^\circ = \frac{x}{r} = \frac{-1}{\sqrt{2}} = -\frac{\sqrt{2}}{2} \qquad \sec 135^\circ = \frac{r}{x} = \frac{\sqrt{2}}{-1} = -\sqrt{2}$$

$$\tan 135^\circ = \frac{y}{x} = \frac{1}{-1} = -1 \qquad \cot 135^\circ = \frac{x}{y} = \frac{-1}{1} = -1$$

EXAMPLE 5 $\theta = \dfrac{7\pi}{6}$

SOLUTION $7\pi/6$ corresponds to 210°. First we sketch the angle θ in standard position. Then we draw a 30°–60° right triangle in quadrant III so that the hypotenuse lies on the terminal side of θ and one side lies on the negative x axis (Figure 11). Because of the quadrant location and the measurements of the 30°–60° right triangle, we obtain the point $(x, y) = (-\sqrt{3}, -1)$ on the terminal side of θ with $r = 2$. Thus by Definition 1 we get

Figure 11

$$\sin \frac{7\pi}{6} = \frac{y}{r} = \frac{-1}{2} \qquad \csc \frac{7\pi}{6} = \frac{r}{y} = \frac{2}{-1} = -2$$

$$\cos \frac{7\pi}{6} = \frac{x}{r} = \frac{-\sqrt{3}}{2} \qquad \sec \frac{7\pi}{6} = \frac{r}{x} = \frac{2}{-\sqrt{3}} = \frac{-2\sqrt{3}}{3}$$

$$\tan \frac{7\pi}{6} = \frac{y}{x} = \frac{-1}{-\sqrt{3}} = \frac{\sqrt{3}}{3} \qquad \cot \frac{7\pi}{6} = \frac{x}{y} = \frac{-\sqrt{3}}{-1} = \sqrt{3}$$

EXAMPLE 6 $\theta = -60^\circ$

SOLUTION After sketching the angle $\theta = -60^\circ$ in standard position and drawing a 30°–60° right triangle in quadrant IV, so that the hypotenuse lies on the terminal side of θ (Figure 12), we obtain point $(x, y) = (1, -\sqrt{3})$ on the terminal side of θ with $r = 2$.

Figure 12

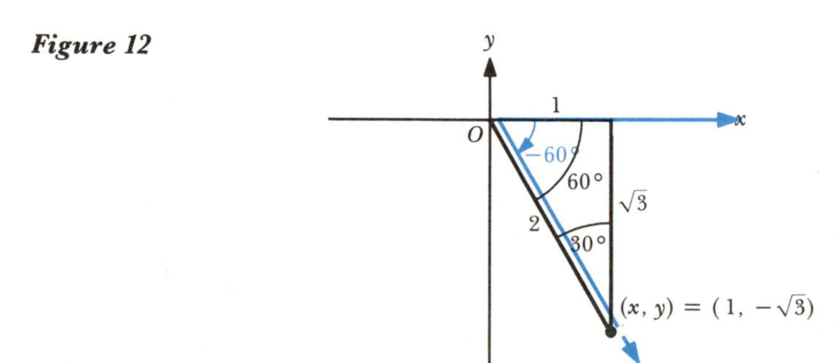

It follows from Definition 1 that

$$\sin(-60°) = \frac{y}{r} = \frac{-\sqrt{3}}{2} \qquad \csc(-60°) = \frac{r}{y} = \frac{2}{-\sqrt{3}} = \frac{-2\sqrt{3}}{3}$$

$$\cos(-60°) = \frac{x}{r} = \frac{1}{2} \qquad \sec(-60°) = \frac{r}{x} = \frac{2}{1} = 2$$

$$\tan(-60°) = \frac{y}{x} = \frac{-\sqrt{3}}{1} = -\sqrt{3} \qquad \cot(-60°) = \frac{x}{y} = \frac{1}{-\sqrt{3}} = \frac{-\sqrt{3}}{3}$$

PROBLEM SET 5.3

In problems 1–12, find the values of the six trigonometric functions of θ if θ is in standard position and the terminal side of θ contains the given point (x, y), which is r units from the origin. Sketch two possible angles that satisfy the conditions on θ. In problems 7, 8, and 12, round off the answers to two decimal places.

1. $(x, y) = (4, 3)$

2. $(x, y) = (-5, 12)$

3. $(x, y) = (-3, -4)$

4. $(x, y) = (7, -10)$

5. $(x, y) = (0, -5)$

6. $(x, y) = (-\sqrt{3}, -\sqrt{2})$

[C] **7.** $(x, y) = (1.35, -2.76)$

[C] **8.** $(-7.89, 9.39)$

9. $(x, y) = (8, b)$; $r = 17$, $b < 0$

10. $(x, y) = (a, -\frac{1}{2})$; $r = 1$, $a < 0$

11. $(x, y) = (a, a)$; $r = 2$, $a > 0$

[C] **12.** $(x, y) = (-6.83, b)$; $r = 23.69$, $b > 0$

In problems 13–16, use Definition 1 to find the values of the six trigonometric functions of the angle with the given measure.

13. $90°$ **14.** $-\frac{\pi}{2}$ **15.** $\frac{3\pi}{2}$ **16.** $-270°$

In problems 17–21, use the right triangle trigonometric relationships to find the values of the six trigonometric functions of the acute angle θ for each right triangle.

17.

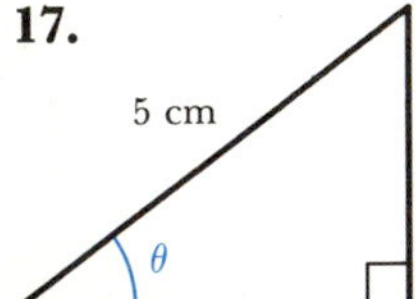
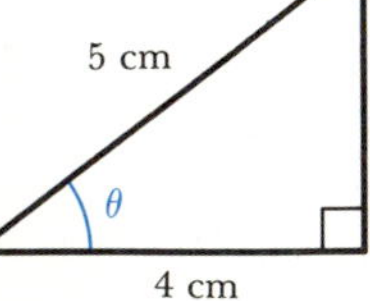

18.

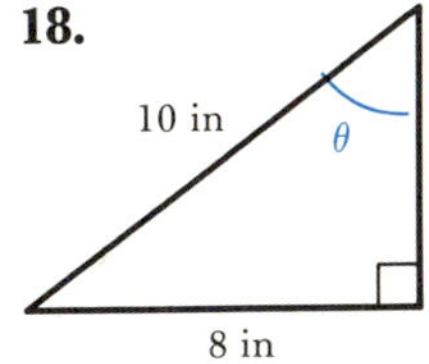
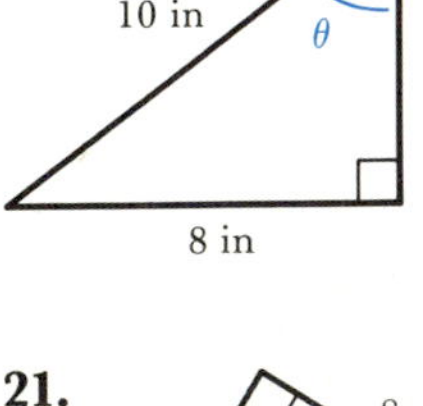

19.

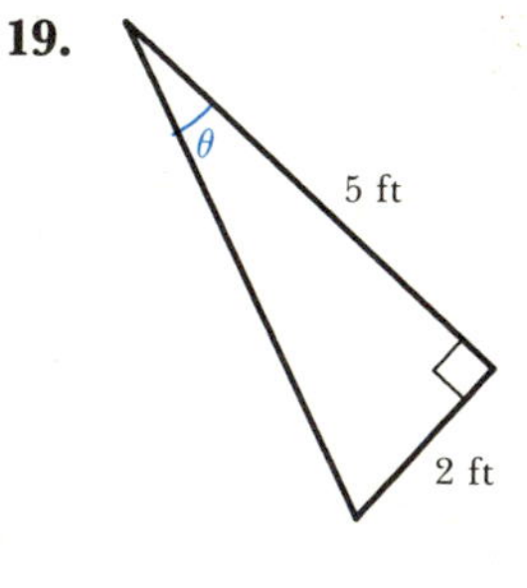

20.

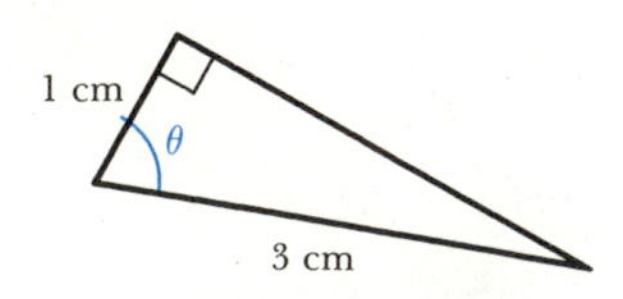

21.

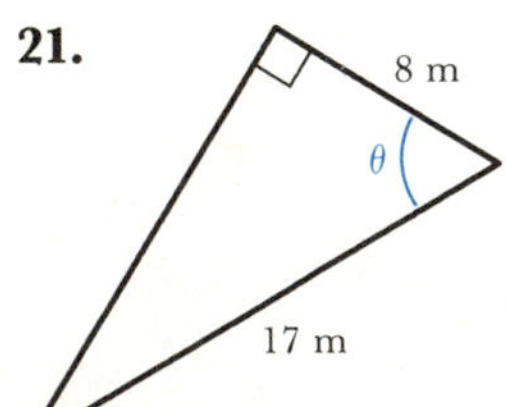

22. Use the right triangle trigonometric relationships to find the length of $\overline{AB}$ in Figure 13.

Figure 13

In problems 23–47, use either a 45°–45° right triangle or a 30°–60° right triangle along with Definition 1 to find the exact value of the given expression. Do not use a calculator.

23. $\sin 150°$
24. $\cos 225°$
25. $\tan 240°$
26. $\cot \frac{11\pi}{6}$
27. $\sec 315°$
28. $\csc 300°$
29. $\tan\left(-\frac{\pi}{4}\right)$
30. $\cot 210°$
31. $\cos\left(-\frac{7\pi}{6}\right)$
32. $\cot 330°$
33. $\sec\left(-\frac{11\pi}{6}\right)$
34. $\csc\left(-\frac{7\pi}{4}\right)$
35. $\sin 300°$
36. $\cot\left(-\frac{4\pi}{3}\right)$
37. $\cos(-315°)$
38. $\sec(-120°)$
39. $\sin \frac{3\pi}{4}$
40. $\cos\left(-\frac{\pi}{3}\right)$
41. $\sec \frac{4\pi}{3}$
42. $\cot \frac{5\pi}{6}$
43. $\csc(-330°)$
44. $\csc\left(-\frac{7\pi}{6}\right)$
45. $\tan(-330°)$
46. $\sin(-315°)$
47. $\cot 300°$
48. If angle θ in standard position has a terminal side that is in quadrant II, explain why $\sin \theta > 0$, $\cos \theta < 0$, and $\tan \theta < 0$.
49. (a) Explain why the respective six trigonometric function values are the same for 160° and −200°.
 (b) Explain why the respective six trigonometric function values are the same for 40° and 400°.
50. (a) With reference to Figure 14, suppose that (x, y) and (x_1, y_1) are two different points in quadrant II on the terminal side of θ other than $(0, 0)$. Show that

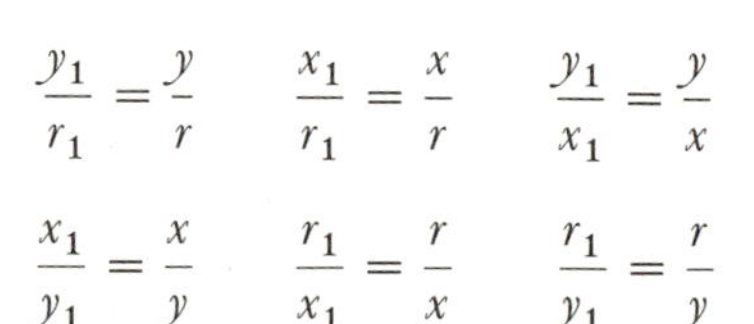

$$\frac{y_1}{r_1} = \frac{y}{r} \qquad \frac{x_1}{r_1} = \frac{x}{r} \qquad \frac{y_1}{x_1} = \frac{y}{x}$$

$$\frac{x_1}{y_1} = \frac{x}{y} \qquad \frac{r_1}{x_1} = \frac{r}{x} \qquad \frac{r_1}{y_1} = \frac{r}{y}$$

 (b) Show that the ratios in part (a) also hold if angle θ is in quadrant III.
 (c) Show that the ratios in part (a) also hold if angle θ is in quadrant IV.

Figure 14

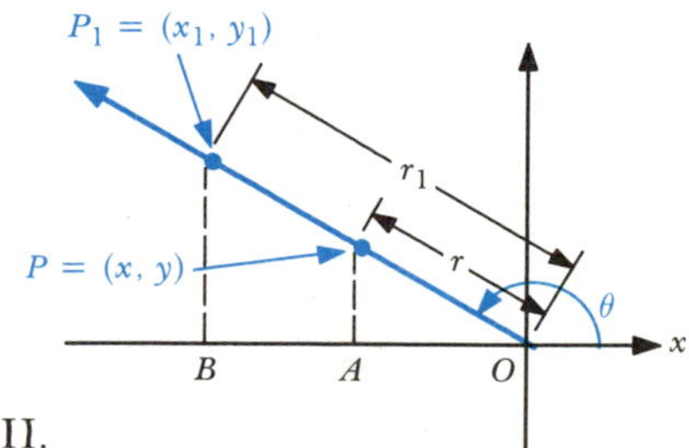

5.4 Properties of Trigonometric Functions

Now we examine the connection between the trigonometric functions of real numbers and the trigonometric functions of angles. Suppose that t is a real number and $P(t) = (x, y)$, then the trigonometric functions defined on *real numbers* (Figure 1) yield

$$\cos t = x \qquad \text{and} \qquad \sin t = y$$

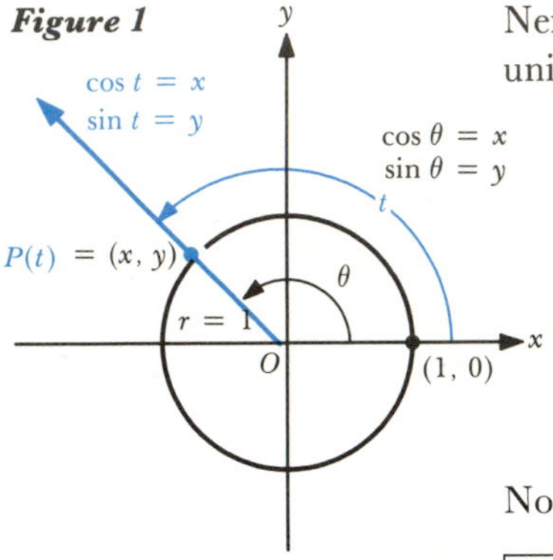

Next, we assume that θ is an angle in standard position subtended by an arc of $|t|$ units on the unit circle so that point $P(t)$ is on the terminal side of θ (Figure 1). (The argument holds regardless of the quadrant where the terminal side of θ is located.) Since (x, y) is located on the unit circle where $r = 1$ (Figure 1), then, by using the definition of trigonometric functions defined on angles, we have

Figure 1

$$\cos \theta = \frac{x}{r} = \frac{x}{1} = x \qquad \text{and} \qquad \sin \theta = \frac{y}{r} = \frac{y}{1} = y$$

Notice that the real number t is the radian measure of θ. Thus

$$\cos t = \cos \theta \qquad \text{and} \qquad \sin t = \sin \theta$$

where $\cos t$ and $\sin t$ are the values of the trigonometric functions defined on real number t, and $\cos \theta$ and $\sin \theta$ are the values of the trigonometric functions defined on angle θ.

Similarly, if angle θ has radian measure t, we can use the definition of the trigonometric functions of *real number t* and the definition of the trigonometric functions of *angle* θ to establish the following relationships:

$$\tan t = \frac{y}{x} = \tan \theta$$

$$\cot t = \frac{x}{y} = \cot \theta$$

$$\sec t = \frac{1}{x} = \sec \theta$$

$$\csc t = \frac{1}{y} = \csc \theta$$

Consequently, we consider the values of the trigonometric functions of angle θ to be equivalent to the values of the trigonometric function of a real number t, which is the radian measure of θ. Therefore, we use the terminology *trigonometric functions* regardless of whether *angles* or *real numbers* are employed. For instance, the sine of an angle of radian measure $\pi/4$ is the same as the sine of the real number $\pi/4$.

Even if an angle is given in degrees, we can associate the trigonometric functions of real numbers with the trigonometric functions of the angle by converting degrees to radians. For example,

$$\sin 30^\circ = \sin \frac{\pi}{6} = \frac{1}{2}$$

$$\cos(-45^\circ) = \cos\left(-\frac{\pi}{4}\right) = \frac{\sqrt{2}}{2}$$

$$\tan 300^\circ = \tan \frac{5\pi}{3} = -\sqrt{3}$$

Now we investigate the properties of trigonometric functions. For convenience, we use the symbol θ to represent an angle, the degree measure of the angle, or the radian measure of the angle. The context of our discussion will clarify which meaning of θ is being used.

Suppose that angle θ, which is measured in radians or degrees, is placed in standard position. Let P be the point where the terminal side of θ intersects the unit circle (Figure 2a). Angle θ is coterminal with angles of measures

$$\theta + 2\pi \qquad (\text{or } \theta + 360^\circ) \qquad (\text{Figure 2b})$$

and

$$\theta - 2\pi \qquad (\text{or } \theta - 360^\circ) \qquad (\text{Figure 2c})$$

Figure 2

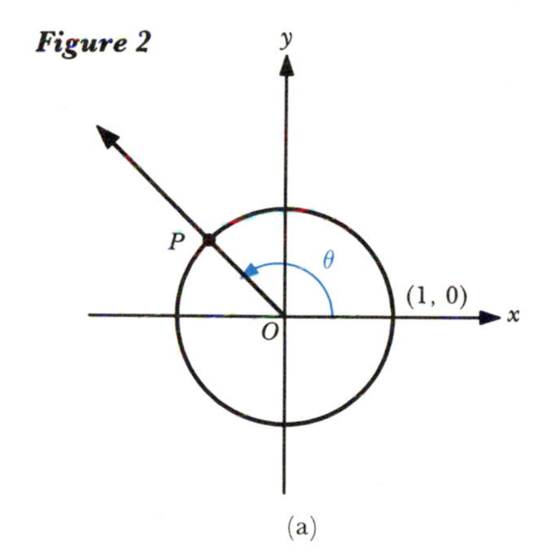

(a)

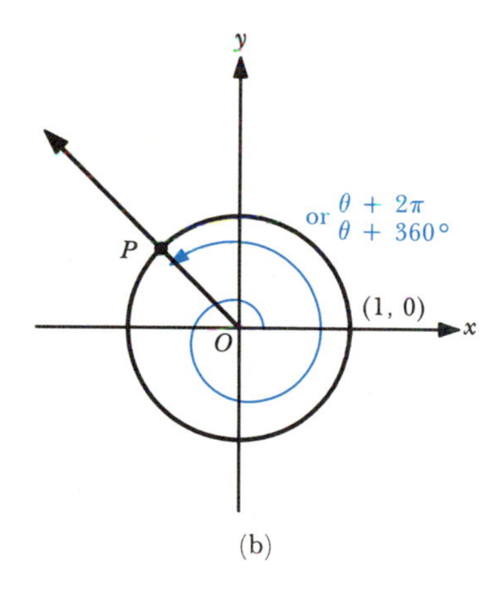

(b)

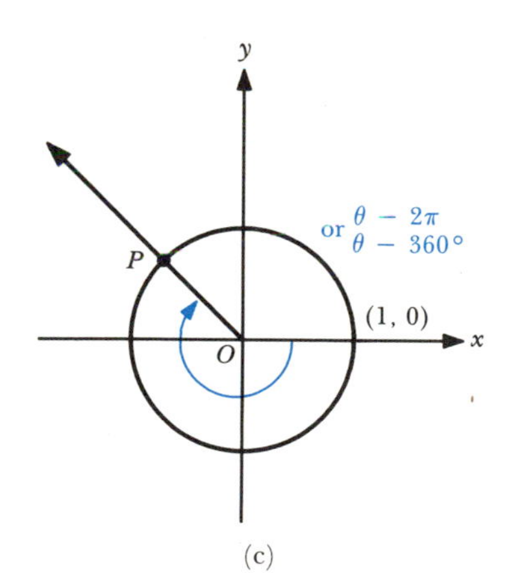

(c)

Therefore, the point P where the terminal side of each angle intersects the unit circle is the same in all three cases (Figure 2). Since the trigonometric functions for each angle can be determined by using the coordinates of the same point, it follows that the values of the trigonometric functions of

$$\theta, \qquad \theta + 2\pi \ (\text{or } \theta + 360^\circ)$$

and

$$\theta - 2\pi \ (\text{or } \theta - 360^\circ)$$

or of any angle formed by adding or subtracting multiples of 2π (or $360°$) are the same. That is:

> The values of the trigonometric functions are the same for
>
> $$\theta \quad \text{and} \quad \theta + 2\pi n \quad \text{(if radians are used)}$$
>
> or
>
> $$\theta \quad \text{and} \quad \theta + 360°n \quad \text{(if degrees are used)}$$
>
> for any integer n.

For example, since

$$\frac{5\pi}{2} = \frac{\pi}{2} + 2\pi$$

$$\frac{9\pi}{2} = \frac{\pi}{2} + 4\pi = \frac{\pi}{2} + 2 \cdot 2\pi$$

$$-\frac{3\pi}{2} = \frac{\pi}{2} - 2\pi$$

and

$$-\frac{11\pi}{2} = \frac{\pi}{2} - 6\pi = \frac{\pi}{2} - 3 \cdot 2\pi$$

it follows that the values of the trigonometric functions for

$$\frac{5\pi}{2}, \quad \frac{9\pi}{2}, \quad -\frac{3\pi}{2}, \quad \text{and} \quad -\frac{11\pi}{2}$$

are the same as the values for $\pi/2$. Similarly, the values of trigonometric functions for $450°$, $-270°$, and $-630°$ are the same as the values for $90°$, since

$$450° = 90° + 360°$$

$$-270° = 90° - 360°$$

and

$$-630° = 90° - 2(360°)$$

EXAMPLE 1 Find the value of each expression.

(a) $\sin \dfrac{9\pi}{2}$ (b) $\cos(-630°)$ (c) $\csc 450°$

SOLUTION Since $9\pi/2$ is coterminal with $\pi/2$, and $-630°$ and $450°$ are coterminal with $90°$, we have

(a) $\sin \dfrac{9\pi}{2} = \sin \dfrac{\pi}{2} = 1$ (b) $\cos(-630°) = \cos 90° = 0$

(c) $\csc 450° = \csc 90° = 1$

EXAMPLE 2 Use the special values along with the fact that the values of the trigonometric functions repeat every 2π units (or $360°$) to find the exact value of each expression.

(a) $\sin \dfrac{25\pi}{6}$

(b) $\cot\left(-\dfrac{27\pi}{4}\right)$

(c) $\sec 1140°$

SOLUTION (a) Since

$$\frac{25\pi}{6} = \frac{\pi}{6} + 4\pi$$

$$= \frac{\pi}{6} + 2 \cdot 2\pi$$

the angles $25\pi/6$ and $\pi/6$ are coterminal, so that

$$\sin \frac{25\pi}{6} = \sin \frac{\pi}{6} = \frac{1}{2}$$

(b) Since

$$-\frac{27\pi}{4} = \frac{5\pi}{4} - 8\pi$$

$$= \frac{5\pi}{4} - 4 \cdot 2\pi$$

the angles $-27\pi/4$ and $5\pi/4$ are coterminal. Therefore,

$$\cot\left(-\frac{27\pi}{4}\right) = \cot \frac{5\pi}{4} = 1$$

(c) By dividing $1140°$ by $360°$, we find the largest integer multiple of $360°$ that is less than $1140°$ to be $3(360°) = 1080°$. Thus

$$1140° = 60° + 3(360°)$$

and the angles $60°$ and $1140°$ are coterminal. Thus,

$$\sec 1140° = \sec 60° = 2$$

Figure 3

y
(x, y)
θ
O
x
(1, 0)

(a)

We say that an angle is in a certain quadrant if, when the angle is in standard position, the terminal side lies in that quadrant. For example, in Figure 3a angle θ lies in quadrant I, and in Figure 3b it lies in quadrant II. Assume that θ is an angle in standard position and θ is *not* a quadrantal angle. Figure 3 displays four possible situations, one for each quadrant, where the point (x, y) intersects the unit circle on the terminal side of angle θ. If θ is in quadrant I (Figure 3a), then x and y are positive, so that all the values of the trigonometric functions are positive. If θ

Figure 3

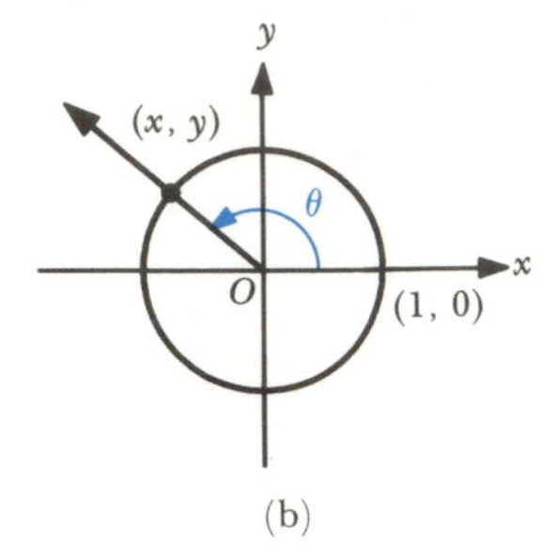

(b)

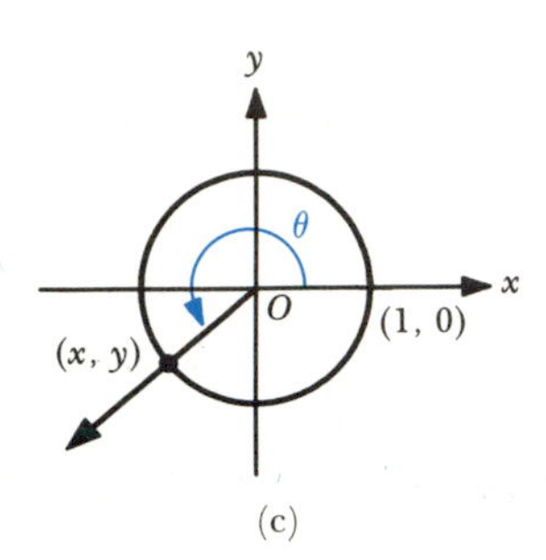

(c)

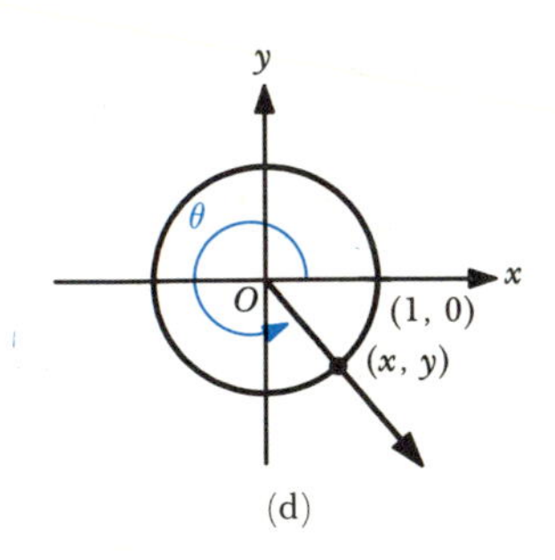

(d)

is in quadrant II (Figure 3b), then x is negative while y is positive, so that the algebraic signs of the values of the trigonometric functions of θ are:

$$\sin\theta = y \text{ (positive)} \qquad \csc\theta = \frac{1}{y \text{ (positive)}} = \text{(positive)}$$

$$\cos\theta = x \text{ (negative)} \qquad \sec\theta = \frac{1}{x \text{ (negative)}} = \text{(negative)}$$

$$\tan\theta = \frac{y \text{ (positive)}}{x \text{ (negative)}} = \text{(negative)} \qquad \cot\theta = \frac{x \text{ (negative)}}{y \text{ (positive)}} = \text{(negative)}$$

Thus, for angles in quadrant II the values of the sine and cosecant are positive, whereas the values of the other four trigonometric functions are negative.

The algebraic signs of the values of the trigonometric functions when θ is in quadrants III (Figure 3c) or IV (Figure 3d) are easily determined by observing the algebraic signs of x and y in those quadrants.

The following table summarizes the signs of the values of the trigonometric functions for angles in each of the four quadrants.

Signs of the Values of the Trigonometric Functions

Quadrant Location of Terminal Side of Angle θ	Positive Values	Negative Values
I	all	none
II	$\sin\theta$, $\csc\theta$	$\cos\theta$, $\sec\theta$, $\tan\theta$, $\cot\theta$
III	$\tan\theta$, $\cot\theta$	$\sin\theta$, $\cos\theta$, $\csc\theta$, $\sec\theta$
IV	$\cos\theta$, $\sec\theta$	$\sin\theta$, $\csc\theta$, $\tan\theta$, $\cot\theta$

EXAMPLE 3 Find the quadrant in which θ lies if $\cot\theta > 0$ and $\cos\theta < 0$.

SOLUTION Let θ be an angle in standard position. Using the above table, we see that $\cot\theta > 0$ if θ is in quadrant I or III, and $\cos\theta < 0$ if θ is in quadrant II or III. Hence, for both conditions to be satisfied, θ must be in quadrant III.

Basic Identities

Using the definitions of the six trigonometric functions, it is possible to develop special relationships that exist between these functions. These relationships are called **basic identities.** An **identity** is an equation that is true for *all* values of the variable for which both sides of the equation are defined. These identities are classified as the **reciprocal identities,** the **quotient identities,** and the **Pythagorean identities.** They are useful in simplifying expressions involving trigonometric functions.

The Reciprocal Identities

If we examine the ratios that define the six trigonometric functions, we can establish the following relationships for trigonometric functions of any angle θ.

1. $\csc\theta = \dfrac{1}{\sin\theta}$	2. $\sec\theta = \dfrac{1}{\cos\theta}$	3. $\cot\theta = \dfrac{1}{\tan\theta}$

PROOF To prove these identities, let θ be an angle in standard position whose terminal side intersects the unit circle at the point (x, y) (Figure 4). Then, provided that the denominator is not zero, we have

$$\csc\theta = \frac{1}{y} = \frac{1}{\sin\theta}$$

$$\sec\theta = \frac{1}{x} = \frac{1}{\cos\theta}$$

$$\cot\theta = \frac{x}{y} = \frac{1}{y/x} = \frac{1}{\tan\theta}$$

Figure 4

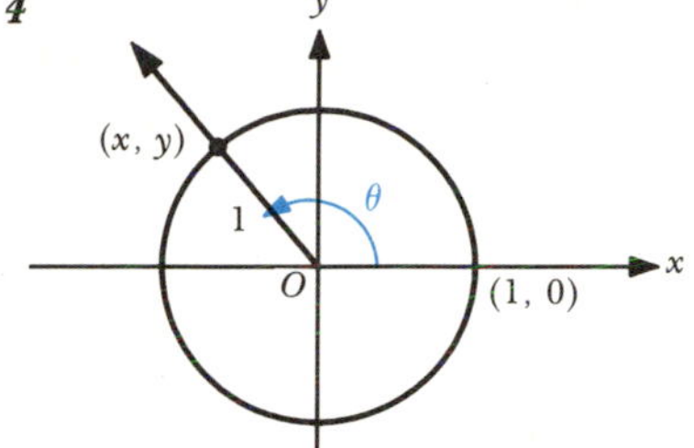

In a similar way, we can prove two more identities.

The Quotient Identities

Since

$$\sin\theta = y \qquad \text{and} \qquad \cos\theta = x \qquad \text{(Figure 4)}$$

and because $\tan\theta = y/x$ and $\cot\theta = x/y$, it follows that

4. $\tan\theta = \dfrac{\sin\theta}{\cos\theta}$	5. $\cot\theta = \dfrac{\cos\theta}{\sin\theta}$

EXAMPLE 4 If $\sin\theta = \frac{4}{5}$ and $\cos\theta = \frac{3}{5}$, use the basic identities to find the values of the other four trigonometric functions of θ.

SOLUTION Using the quotient and reciprocal identities, we obtain

$$\tan\theta = \frac{\sin\theta}{\cos\theta} = \frac{4/5}{3/5} = \frac{4}{3} \qquad \cot\theta = \frac{\cos\theta}{\sin\theta} = \frac{3/5}{4/5} = \frac{3}{4}$$

$$\sec\theta = \frac{1}{\cos\theta} = \frac{1}{3/5} = \frac{5}{3} \qquad \csc\theta = \frac{1}{\sin\theta} = \frac{1}{4/5} = \frac{5}{4}$$

It should be pointed out that although

$$\tan\theta = \frac{4}{3} \qquad \text{and} \qquad \tan\theta = \frac{\sin\theta}{\cos\theta}$$

it does *not* follow that $\sin\theta = 4$ and $\cos\theta = 3$.

Other important relationships depend on the Pythagorean theorem.

The Pythagorean Identities

Suppose that θ is an angle in standard position whose terminal side intersects the unit circle at (x, y) (Figure 4), then, since the coordinates x and y satisfy the equation of the unit circle, it follows that

$$(\cos\theta)^2 + (\sin\theta)^2 = x^2 + y^2 = 1$$

and we obtain the relationship

$$(\cos\theta)^2 + (\sin\theta)^2 = 1$$

Expressions of the form $(\sin\theta)^n$ and $(\cos\theta)^n$ occur so frequently that special notation is used. It is customary to write $(\sin\theta)^n = \sin^n\theta$ and $(\cos\theta)^n = \cos^n\theta$. Using this notation, we have

$$\sin^2\theta = (\sin\theta)^2 = (\sin\theta)(\sin\theta)$$

and

$$\cos^2\theta = (\cos\theta)^2 = (\cos\theta)(\cos\theta)$$

Thus, the above equation is written as

$$\cos^2\theta + \sin^2\theta = 1$$

The following are Pythagorean identities:

6. $\cos^2\theta + \sin^2\theta = 1$ 7. $1 + \tan^2\theta = \sec^2\theta$ 8. $1 + \cot^2\theta = \csc^2\theta$

PROOF To prove identity 7, we divide both sides of the equation

$$\cos^2\theta + \sin^2\theta = 1$$

by $\cos^2\theta$ to obtain

$$1 + \frac{\sin^2\theta}{\cos^2\theta} = \frac{1}{\cos^2\theta}$$

or

$$1 + \left(\frac{\sin\theta}{\cos\theta}\right)^2 = \left(\frac{1}{\cos\theta}\right)^2$$

Therefore, provided that $\cos\theta \neq 0$,

$$1 + \tan^2\theta = \sec^2\theta$$

Identity 8 is proved by dividing both sides of identity 6 by $\sin^2\theta$ (Problem 48).

In Examples 5–7, the value of one of the trigonometric functions of an angle θ is given. Use the basic identities along with the signs of the values of the trigonometric functions to find the values of the other five trigonometric functions of θ.

EXAMPLE 5 $\sin\theta = \frac{8}{17}$, θ in quadrant I

SOLUTION Using $\cos^2\theta + \sin^2\theta = 1$, we have

$$\cos^2\theta = 1 - \sin^2\theta \quad \text{and so} \quad \cos\theta = \pm\sqrt{1 - \sin^2\theta}$$

Since θ is in quadrant I, we know that $\cos\theta$ is positive; hence

$$\cos\theta = \sqrt{1 - \left(\frac{8}{17}\right)^2} = \sqrt{1 - \frac{64}{289}} = \sqrt{\frac{225}{289}} = \frac{15}{17}$$

It follows that

$$\tan\theta = \frac{\sin\theta}{\cos\theta} = \frac{8/17}{15/17} = \frac{8}{15} \qquad \cot\theta = \frac{\cos\theta}{\sin\theta} = \frac{15/17}{8/17} = \frac{15}{8}$$

$$\sec\theta = \frac{1}{\cos\theta} = \frac{1}{15/17} = \frac{17}{15} \qquad \csc\theta = \frac{1}{\sin\theta} = \frac{1}{8/17} = \frac{17}{8}$$

EXAMPLE 6 $\cos\theta = \frac{24}{25}$, θ in quadrant IV

SOLUTION Since $\sin^2\theta + \cos^2\theta = 1$ and $\sin\theta$ is negative in quadrant IV, we have

$$\sin\theta = \sqrt{1 - \cos^2\theta} = -\sqrt{1 - \left(\frac{24}{25}\right)^2} = -\sqrt{1 - \frac{576}{625}}$$

$$= -\sqrt{\frac{49}{625}} = -\frac{7}{25}$$

Hence

$$\tan\theta = \frac{\sin\theta}{\cos\theta} = \frac{-7/25}{24/25} = -\frac{7}{24} \qquad \cot\theta = \frac{\cos\theta}{\sin\theta} = \frac{24/25}{-7/25} = -\frac{24}{7}$$

$$\sec\theta = \frac{1}{\cos\theta} = \frac{1}{24/25} = \frac{25}{24} \qquad \csc\theta = \frac{1}{\sin\theta} = \frac{1}{-7/25} = -\frac{25}{7}$$

EXAMPLE 7 $\tan\theta = -\frac{5}{12}$ and $\cos\theta < 0$

SOLUTION Because $\tan\theta < 0$ only for θ in quadrants II and IV, and $\cos\theta < 0$ for θ in quadrants II and III, it follows that θ must be in quadrant II and $\sec\theta < 0$. Using the identity $\sec^2\theta = 1 + \tan^2\theta$, we have

$$\sec\theta = -\sqrt{1+\tan^2\theta} = -\sqrt{1+\left(\frac{-5}{12}\right)^2} = -\sqrt{\frac{169}{144}} = -\frac{13}{12}$$

Since $\sec\theta = \frac{1}{\cos\theta}$, it follows that

$$\cos\theta = \frac{1}{\sec\theta} = \frac{1}{-13/12} = -\frac{12}{13}$$

Also

$$\tan\theta = \frac{\sin\theta}{\cos\theta}$$

so that

$$\sin\theta = (\tan\theta)(\cos\theta) = \left(\frac{-5}{12}\right)\left(\frac{-12}{13}\right) = \frac{5}{13}$$

Finally,

$$\csc\theta = \frac{1}{\sin\theta} = \frac{1}{5/13} = \frac{13}{5}$$

and

$$\cot\theta = \frac{1}{\tan\theta} = \frac{1}{-5/12} = -\frac{12}{5}$$

EXAMPLE 8 Express each of the given functions in terms of u if $\sec\theta = u/2$ and θ is an angle in quadrant I.

(a) $\cos\theta$ (b) $\sin\theta$ (c) $\tan\theta$

SOLUTION Using the basic identities, we have

(a) $\cos\theta = \frac{1}{\sec\theta} = \frac{1}{u/2} = \frac{2}{u}$

(b) Since θ is in quadrant I,

$$\sin\theta = \sqrt{1-\cos^2\theta} = \sqrt{1-\left(\frac{2}{u}\right)^2} = \sqrt{\frac{u^2-4}{u^2}} = \frac{\sqrt{u^2-4}}{u}$$

(c) $\tan\theta = \frac{\sin\theta}{\cos\theta} = \frac{\frac{\sqrt{u^2-4}}{u}}{\frac{2}{u}} = \frac{\sqrt{u^2-4}}{2}$

In addition to the basic identities, there are other useful identities that relate values of the trigonometric functions of θ and $-\theta$. To establish these identities, we construct two angles in standard position, one of measure θ and the other with measure $-\theta$, where the terminal sides of the angles intersect the unit circle at points P and Q, respectively (Figure 5). (Although we illustrate the situation for θ in quadrant I, it is important to remember that the argument can be used for any quadrant.) By the symmetry of the circle, P and Q are the mirror images of each other across the x axis.

Figure 5

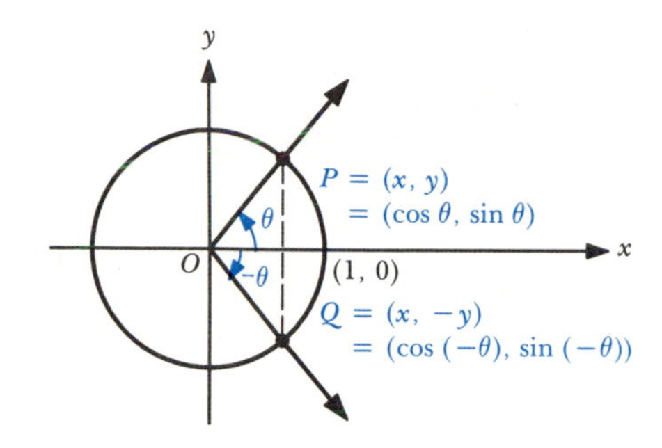

Therefore, if $P = (x, y)$, then $Q = (x, -y)$, so that by Definition 1 on page 257, we have

$$\sin\theta = y \qquad \cos\theta = x$$

and

$$\sin(-\theta) = -y \qquad \cos(-\theta) = x$$

Therefore

$$\sin(-\theta) = -y = -\sin\theta$$

and

$$\cos(-\theta) = x = \cos\theta$$

Also,

$$\tan(-\theta) = \frac{\sin(-\theta)}{\cos(-\theta)} = \frac{-\sin\theta}{\cos\theta} = -\tan\theta$$

Similar arguments apply to $\cot(-\theta)$, $\sec(-\theta)$, and $\csc(-\theta)$ (Problem 76).

Even–Odd Identities

(i) $\sin(-\theta) = -\sin\theta$	(ii) $\cos(-\theta) = \cos\theta$	(iii) $\tan(-\theta) = -\tan\theta$
(iv) $\cot(-\theta) = -\cot\theta$	(v) $\sec(-\theta) = \sec\theta$	(vi) $\csc(-\theta) = -\csc\theta$

We see from the above identities that the cosine and secant functions are even, whereas the four other trigonometric functions are odd.

EXAMPLE 9 Use the even–odd identities to evaluate each given expression.

(a) $\sin\left(-\frac{\pi}{6}\right)$ (b) $\cos\left(-\frac{\pi}{6}\right)$ (c) $\tan(-45^\circ)$

SOLUTION (a) $\sin\left(-\frac{\pi}{6}\right) = -\sin\frac{\pi}{6} = -\frac{1}{2}$ (b) $\cos\left(-\frac{\pi}{6}\right) = \cos\frac{\pi}{6} = \frac{\sqrt{3}}{2}$

(c) $\tan(-45^\circ) = -\tan 45^\circ = -1$

PROBLEM SET 5.4

In problems 1–40, use the special values along with the fact that the trigonometric function values repeat every 2π (or 360°) units to find the exact value of each expression. Do not use a calculator.

1. $\sin(315^\circ + 360^\circ)$ **2.** $\cot(60^\circ + 720^\circ)$ **3.** $\cos\left(\frac{5\pi}{6} + 2\pi\right)$ **4.** $\sec\left(\frac{5\pi}{3} + 4\pi\right)$

5. $\tan(150^\circ - 720^\circ)$ **6.** $\csc\left(\frac{5\pi}{3} - 4\pi\right)$ **7.** $\cot\frac{41\pi}{4}$ **8.** $\sin 945^\circ$

9. $\sec(2550^\circ)$ **10.** $\cos\frac{71\pi}{3}$ **11.** $\csc\frac{43\pi}{4}$ **12.** $\tan(1020^\circ)$

13. $\sin(-675^\circ)$ **14.** $\sin\left(-\frac{9\pi}{2}\right)$ **15.** $\cos(-25\pi)$ **16.** $\cos(-405^\circ)$

17. $\tan(-405^\circ)$ **18.** $\tan\left(-\frac{89\pi}{6}\right)$ **19.** $\cot\left(-\frac{91\pi}{6}\right)$ **20.** $\cot(-405^\circ)$

21. $\sec(-1230^\circ)$ **22.** $\sec\left(-\frac{129\pi}{6}\right)$ **23.** $\csc\left(-\frac{25\pi}{3}\right)$ **24.** $\csc(-1380^\circ)$

25. $\sin 390^\circ$ **26.** $\sin 7\pi$ **27.** $\cos\frac{9\pi}{4}$ **28.** $\cot 900^\circ$

29. $\tan 495^\circ$ **30.** $\tan\frac{65\pi}{6}$ **31.** $\cot\left(-\frac{11\pi}{3}\right)$ **32.** $\csc 3720^\circ$

33. $\sec 585^\circ$ **34.** $\csc\left(-\frac{9\pi}{4}\right)$ **35.** $\sin\left(-\frac{15\pi}{4}\right)$ **36.** $\cos(-750^\circ)$

37. $\tan(-1845^\circ)$ **38.** $\cot\left(-\frac{65\pi}{6}\right)$ **39.** $\csc\left(-\frac{8\pi}{3}\right)$ **40.** $\sec(-1050^\circ)$

In problems 41–47, assume that θ is an angle in standard position. Find the quadrant in which θ lies.

41. $\cot\theta > 0$ and $\sin\theta > 0$ **42.** $\sin\theta > 0$ and $\sec\theta < 0$ **43.** $\sec\theta > 0$ and $\tan\theta < 0$

44. $\cos\theta < 0$ and $\csc\theta < 0$ **45.** $\sin\theta > 0$ and $\cos\theta < 0$ **46.** $\sec\theta > 0$ and $\cot\theta < 0$

47. $\tan\theta < 0$ and $\cos\theta < 0$

48. Prove the Pythagorean identity $1 + \cot^2\theta = \csc^2\theta$. [*Hint:* Divide both sides of the identity $\cos^2\theta + \sin^2\theta = 1$ by $\sin^2\theta$.]

In problems 49 and 50, use the given information along with the quotient and reciprocal identities to find the values of the other four trigonometric functions of θ.

49. $\sin\theta = \frac{8}{17}$ and $\cos\theta = -\frac{15}{17}$

50. $\sin\theta = -\frac{1}{5}$ and $\sec\theta = -\frac{5\sqrt{6}}{12}$

In problems 51–72, the value of one of the trigonometric functions of an angle θ is given. Use the basic identities along with the signs of the values to evaluate the other five trigonometric functions of θ. In problems 69–72, round off the answers to two decimal places.

51. $\sin\theta = \frac{15}{17}$, θ in quadrant I

52. $\cos\theta = -\frac{4}{5}$, θ in quadrant II

53. $\sec\theta = -\frac{13}{5}$, θ in quadrant II

54. $\sin\theta = -\frac{24}{25}$, θ in quadrant IV

55. $\cot\theta = \frac{7}{24}$, θ in quadrant III

56. $\csc\theta = \frac{17}{8}$, θ in quadrant I

57. $\tan\theta = -\frac{7}{24}$, θ in quadrant IV

58. $\cot\theta = \sqrt{2}$, θ in quadrant III

59. $\csc\theta = 2$, θ in quadrant II

60. $\tan\theta = -3$, θ in quadrant II

61. $\cos\theta = -\frac{7}{25}$, θ in quadrant III

62. $\cot\theta = \frac{3}{4}$, θ in quadrant III

63. $\cos\theta = \frac{5}{13}$ and $\tan\theta > 0$

64. $\sin\theta = -\frac{9}{41}$ and $\cot\theta < 0$

65. $\csc\theta = \frac{\sqrt{5}}{2}$ and $\cot\theta < 0$

66. $\sec\theta = -\frac{5}{3}$ and $\sin\theta < 0$

67. $\sin\theta = -\frac{\sqrt{7}}{4}$ and $\cos\theta > 0$

68. $\tan\theta = -\frac{40}{9}$ and $\cos\theta < 0$

c **69.** $\tan\theta = -1.53$ and $\sin\theta > 0$

c **70.** $\cot\theta = 3.41$ and $\csc\theta > 0$

c **71.** $\sin\theta = 0.53$ and $\tan\theta < 0$

c **72.** $\csc\theta = -4.75$ and $\cos\theta > 0$

In problems 73–75, assume that θ is an angle in standard position in quadrant I. Express the value of each of the other five trigonometric functions of θ in terms of u.

73. $\sin\theta = u$

74. $\csc\theta = u$

75. $\tan\theta = u$

76. Prove the following even-odd identities.

(a) $\cot(-\theta) = -\cot\theta$ (b) $\sec(-\theta) = \sec\theta$ (c) $\csc(-\theta) = -\csc\theta$

In problems 77–82, use the even–odd identities to simplify each expression.

77. $\frac{\sin(-t)}{\cos(-t)}$

78. $\tan(-\theta^2)$

79. $\frac{\cos\theta}{\sin(-\theta)}$

80. $\sec|-t|$

81. $\cot\left(-\frac{1}{t}\right)$

82. $\sin(\sqrt{t^2})$, $t < 0$

In problems 83–90, use the even–odd identities to find the exact value of each expression. Do not use a calculator.

83. $\sin\left(-\frac{5\pi}{4}\right)$

84. $\cos(-300^\circ)$

85. $\tan(-690^\circ)$

86. $\cot\left(-\frac{11\pi}{6}\right)$

87. $\csc\left(-\frac{2\pi}{3}\right)$

88. $\cos\left(-\frac{5\pi}{3}\right)$

89. $\sec(-510^\circ)$

90. $\sin(-480^\circ)$

5.5 Evaluation of the Trigonometric Functions

So far, we have determined the values of the trigonometric functions only for special angles such as 0, $\pi/2$, 30°, $\pi/4$, $\pi/3$, and 180°. In order to make significant use of the trigonometric functions, we need to find their values for other angles such as 53.4°, 433°21′15″, $19\pi/45$, -4.6381, and 73.8°.

In this section, we evaluate the trigonometric functions by using scientific calculators or specially prepared tables (one for radian measures or real numbers and one for degree measures). Except in a few instances, the values of the trigonometric functions obtained from calculators or tables will be decimal approximations of the actual values. If the tables are to be used, it will be necessary to learn about *reference angles,* which are discussed in this section.

The Use of Calculators

Most calculators deal with fractions of a degree only in decimal form, and if an angle is given in degrees, minutes, and seconds, it must be converted to a decimal form before pressing a trigonometric key. When a calculator is used to evaluate trigonometric functions, correct algebraic signs are automatically provided.

[C] *In Examples 1–5, use a calculator to find each value to five decimal places.*

EXAMPLE 1 $\sin 53.4°$

SOLUTION First, we set the calculator in *degree* mode, then enter 53.4, and finally press the [SIN] key to get

$$\sin 53.4° = 0.80282$$

EXAMPLE 2 $\cos 433°21'17''$

SOLUTION We begin by using a calculator to convert 433°21′17″ to decimal form

$$433°21'17'' = 433° + \left(\frac{21}{60}\right)^{\circ} + \left(\frac{17}{3600}\right)^{\circ} = 433.3547222°$$

With the calculator set in *degree* mode, we press the [COS] key to get

$$\cos 433°21'17'' = 0.28645$$

EXAMPLE 3 $\tan \dfrac{19\pi}{45}$

SOLUTION Here we are finding the value of the tangent of a real number or angle measured in radians. First, we set the calculator in *radian* mode, press the [π] key (or enter

3.14159265), multiply by $\frac{19}{45}$, press the [=] key, and finally press the [TAN] key to obtain

$$\tan \frac{19\pi}{45} = 4.01078$$

EXAMPLE 4 $\sin(-4.6381)$

SOLUTION Because the degree symbol is not used, we know that we are evaluating the sine of a real number or an angle measured in radians. We set the calculator in *radian* mode, enter -4.6381, and then press the [SIN] key to obtain

$$\sin(-4.6381) = 0.99724$$

EXAMPLE 5 (a) $\cot 73.8°$ (b) $\sec 73.8°$ (c) $\csc 73.8°$

SOLUTION Most calculators do not have cotangent, secant, or cosecant keys. If this is the case, we use the identities

$$\cot\theta = \frac{1}{\tan\theta} \qquad \sec\theta = \frac{1}{\cos\theta} \qquad \csc\theta = \frac{1}{\sin\theta}$$

to evaluate cot 73.8°, sec 73.8°, and csc 73.8°. With the calculator in *degree* mode, we first enter 73.8° and then proceed as follows:

(a) Press the [TAN] key, then take the reciprocal of the result by pressing the [1/x] key. The result is

$$\cot 73.8° = 0.29053$$

(b) Press the [COS] key, then press the [1/x] key to obtain

$$\sec 73.8° = 3.58434$$

(c) Press the [SIN] key, then press the [1/x] key to obtain

$$\csc 73.8° = 1.04135$$

Reference Angles

Tables of values of trigonometric functions ordinarily give decimal *approximations* for the values of the functions for *selected angles* measured in radians (or degrees) between 0 and $\pi/2$ (or 0° and 90°). In order to find values of the functions for an angle outside this range, we need a way of relating the values of the functions for such an angle to the values of functions of an angle that does appear in the table. This is accomplished by associating an acute angle θ_R, called the *reference angle,* with a given angle θ.

DEFINITION 1 **Reference Angle**

> If θ is an angle in standard position whose terminal side does not lie on either coordinate axis, then the **reference angle** θ_R for θ is defined to be the positive acute angle formed by the terminal side of θ and the x axis.

Figure 1

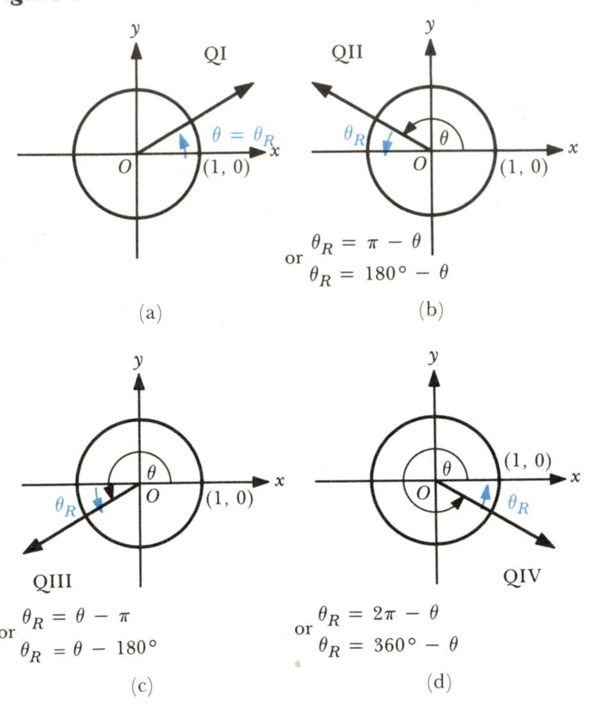

Figure 1 illustrates the reference angle θ_R associated with a positive angle θ less than 2π (or 360°) in each of the four quadrants for both radian and degree measures.

The reference angle θ_R is not always in standard position. However, if it were, it would be in the first quadrant with six positive values for the six trigonometric functions. In fact, it can be shown that the values of the trigonometric functions of a given angle θ and the values of the corresponding trigonometric functions of the reference angle θ_R are the *same in absolute value, but they may differ in sign* (according to the function being evaluated and the quadrant where the terminal side of θ lies).

We verify this assertion for the situation where the terminal side of θ lies in quadrant II (Figure 2). (See Problem 40 for the situations where the terminal side lies in quadrant III or IV.) If point (x, y) lies on the intersection of the unit circle and the terminal side of the reference angle θ_R (Figure 2 shows θ_R in standard position), then $(-x, y)$ is the point of intersection of the unit circle and the terminal side of the angle θ. By using Definition 1 on page 239, we have

Figure 2

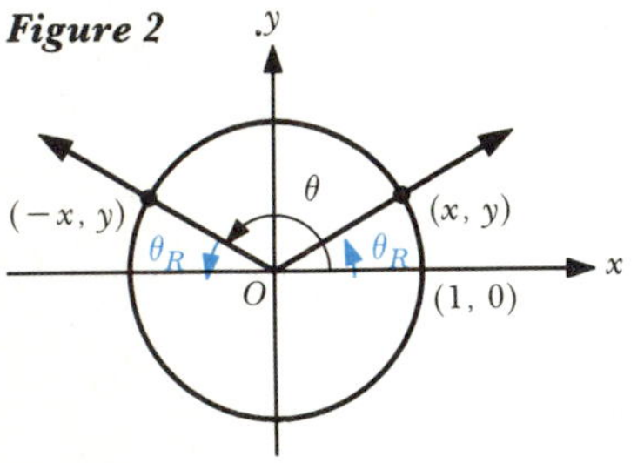

$$\sin\theta = y = \sin\theta_R \qquad \csc\theta = \frac{1}{y} = \csc\theta_R$$

$$\cos\theta = -x = -\cos\theta_R \qquad \sec\theta = \frac{1}{-x} = -\frac{1}{x} = -\csc\theta_R$$

$$\tan\theta = \frac{y}{-x} = -\frac{y}{x} = -\tan\theta_R \qquad \cot\theta = \frac{-x}{y} = -\frac{x}{y} = -\cot\theta_R$$

In general, the value of any trigonometric function of an angle θ is the same as the value of the function for the associated reference angle θ_R, except possibly for a change in algebraic sign. Thus

$$\sin\theta = \sin\theta_R \quad \text{or} \quad \sin\theta = -\sin\theta_R$$

$$\cos\theta = \cos\theta_R \quad \text{or} \quad \cos\theta = -\cos\theta_R$$

and so forth. We can always determine the correct algebraic sign by considering the quadrant in which θ lies and using the results on page 270.

EXAMPLE 6 Reduce the evaluation of each trigonometric function to an evaluation of the same function of its reference angle (use $\pi = 3.14$).

(a) $\cos 2.5$ (b) $\tan 245^\circ$ (c) $\csc 24.84$ (d) $\sin(-220^\circ)$

SOLUTION (a) The reference angle for $\theta = 2.5$ is

$$\theta_R = 3.14 - 2.50 = 0.64 \qquad \text{(Figure 3a)}$$

Since the terminal side of the angle with radian measure 2.5 is in quadrant II, where the cosine is negative, we have

$$\cos 2.5 = -\cos 0.64$$

(b) The reference angle for $\theta = 245^\circ$ is

$$\theta_R = 245^\circ - 180^\circ = 65^\circ \qquad \text{(Figure 3b)}$$

Since the terminal side of the angle with degree measure 245° lies in quadrant III where the tangent function is positive, we have

$$\tan 245^\circ = \tan 65^\circ$$

(c) Since $24.84 = 6 + 3(6.28)$, an angle of 24.84 radians is coterminal with an angle of 6 radians if both angles are in standard position. The reference angle for $\theta = 6$ is given by

$$\theta_R = 6.28 - 6 = 0.28 \qquad \text{(Figure 3c)}$$

Because the cosecant is negative for angles in quadrant IV, we get

$$\csc 24.84 = \csc 6 = -\csc 0.28$$

(d) An angle of -220° in standard position is coterminal with an angle of 140° in standard position, so that the reference angle θ_R for -220° is given by

$$\theta_R = 180^\circ - 140^\circ = 40^\circ \qquad \text{(Figure 3d)}$$

Since the sine is positive in quadrant II, it follows that

$$\sin(-220^\circ) = \sin 140^\circ = \sin 40^\circ$$

Once the evaluation of a trigonometric function has been reduced to an evaluation at a reference angle with the proper adjustment in sign, it is possible to use tables to find values of the functions at the reference angle.

Figure 3

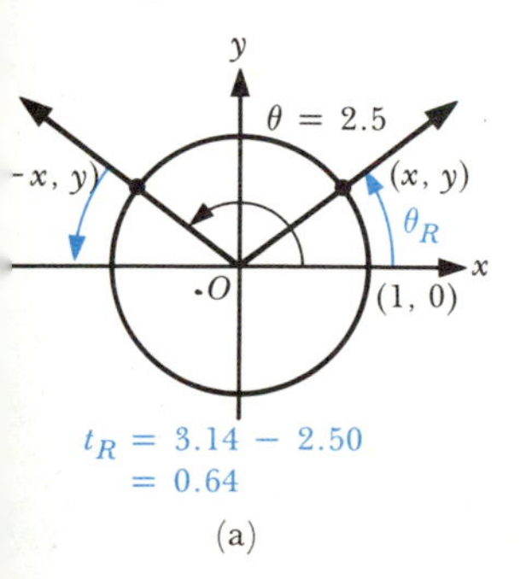

(a)

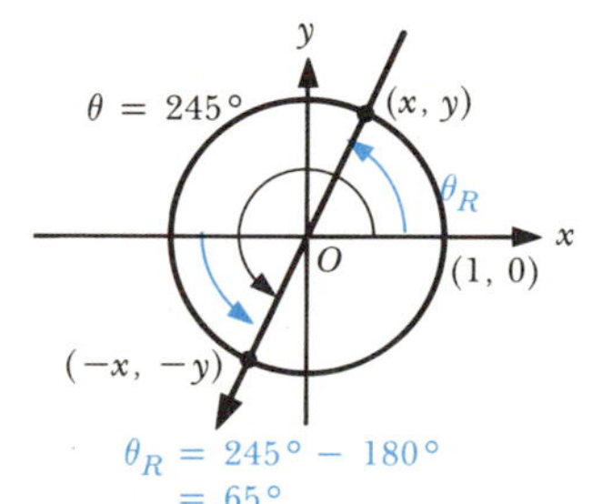

(b)

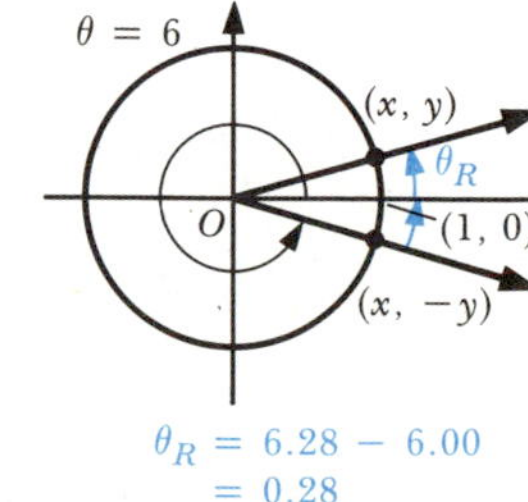

(c)

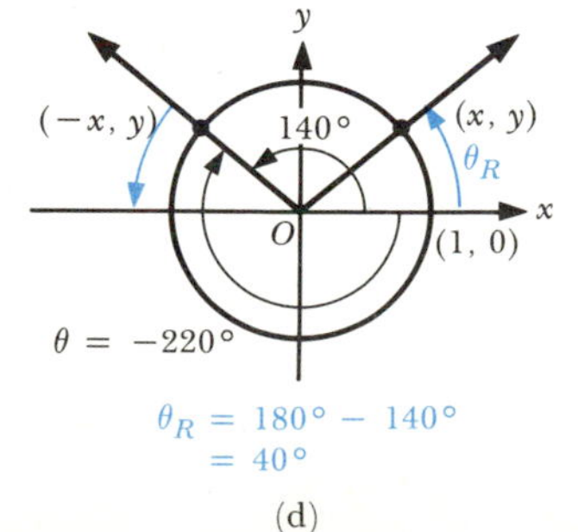

(d)

The Use of Tables

As with the values given by a calculator, the values of the trigonometric functions given by a table are usually decimal *approximations* of the actual values. In most practical situations, these approximations are sufficiently accurate.

Appendix Tables III and IV contain the trigonometric functions of angles measured in degrees and in radians, respectively. Appendix Table III gives the values rounded off to four significant digits, corresponding to angles between $0°$ and $90°$ in steps of $10'$. Angles between $0°$ and $45°$ are found in the vertical column to the left side of the tables, and the captions at the top of the tables apply to these angles. Angles between $45°$ and $90°$ are found in the vertical column to the right side of the tables, and the captions at the bottom apply to these angles. Appendix Table IV gives values of the six trigonometric functions rounded off to four significant digits, corresponding to angles between 0 and $\pi/2$ radians in steps of 0.01.

EXAMPLE 7 Use Appendix Table III or IV to find the approximate value of each trigonometric function.

(a) $\cot 43°$ (b) $\sec 75°10'$ (c) $\csc 83°40$

(d) $\cos 0.65$ (e) $\sin 0.84$ (f) $\tan 1.47$

SOLUTION

(a) Since $43° < 45°$, we begin by locating $43°$ in the vertical column on the *left side* of the degree table. Looking in the vertical column with the *top* caption "cot" we find that

$$\cot 43° = 1.072$$

(b) Because $75°10' > 45°$, we use the vertical column on the *right side* (reading upward) to locate $75°10'$ and the *bottom* caption "sec" to get

$$\sec 75°10' = 3.906$$

(c) Here again, we use the right column and bottom caption to get

$$\csc 83°40' = 1.006$$

(d) In the radian table, we locate 0.65 in the left column and line it up with the column captioned with "$\cos t$" to get

$$\cos 0.65 = 0.7961$$

(e) $\sin 0.84 = 0.7446$

(f) $\tan 1.47 = 9.887$

EXAMPLE 8 Use Appendix Table III or IV and the reference angle to determine the approximate value of each trigonometric function (use $\pi = 3.14$).

(a) $\cot 104°50'$ (b) $\sin 339°$ (c) $\cos 2.14$

(d) $\tan 41.8$ (e) $\csc 965°$

SOLUTION

(a) The angle 104°50′ lies in quadrant II, so that the cotangent value is negative. The reference angle corresponding to 104°50′ is

$$\theta_R = 180° - 104°50' = 75°10' \qquad \text{(Figure 4a)}$$

Thus,

$$\cot 104°50' = -\cot 75°10'$$

Since 75°10′ > 45°, we locate 75°10′ in the vertical column on the right side of Appendix Table III. Looking in the vertical column with the bottom caption "cot", we see that

$$\cot 75°10' = 0.2648$$

Hence, $\cot 104°50' = -\cot 75°10' = -0.2648$

(b) The terminal side of an angle with degree measure 339° is in quadrant IV; the reference angle is

$$\theta_R = 360° - 339° = 21° \qquad \text{(Figure 4b)}$$

Because the sine is negative in quadrant IV, we get

$$\sin 339° = -\sin 21° = -0.3584$$

(c) Since the terminal side of the angle with radian measure 2.14 lies in quadrant II, the reference angle is

$$\theta_R = 3.14 - 2.14 = 1.0 \qquad \text{(Figure 4c)}$$

The cosine is negative in quadrant II so that by using Appendix Table IV, we obtain

$$\cos 2.14 = -\cos 1 = -0.5403$$

(d) Here 41.8 = 4.12 + 6(6.28) so that 41.8 and 4.12 are measures of coterminal angles. Because the terminal side of the angle with radian measure 4.12 is in quadrant III, the reference angle is

$$\theta_R = 4.12 - 3.14 = 0.98 \qquad \text{(Figure 4d)}$$

The tangent is positive in quadrant III, so that

$$\tan 41.8 = \tan 4.12 = \tan 0.98 = 1.491$$

Figure 4

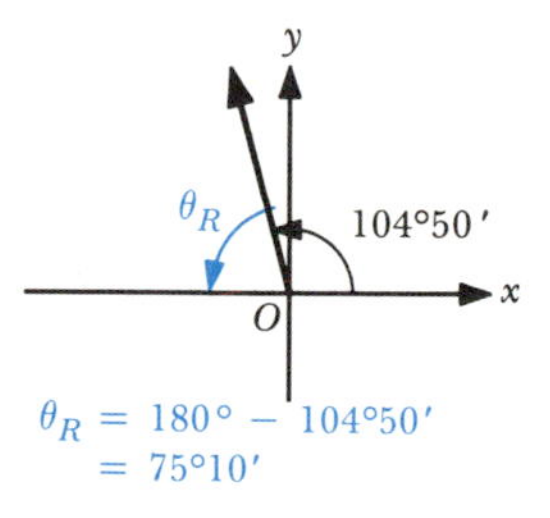

(a)

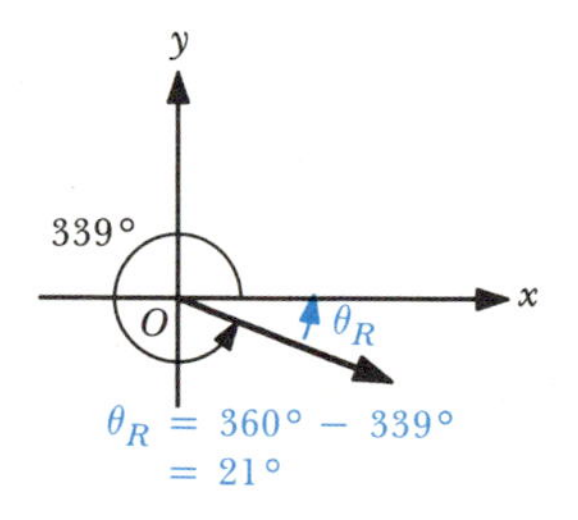

(b)

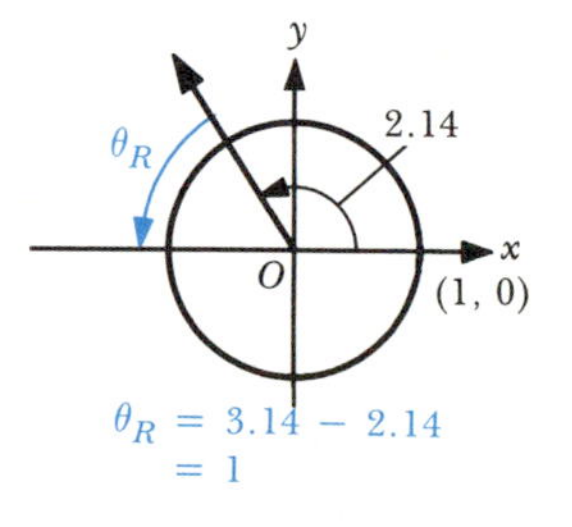

(c)

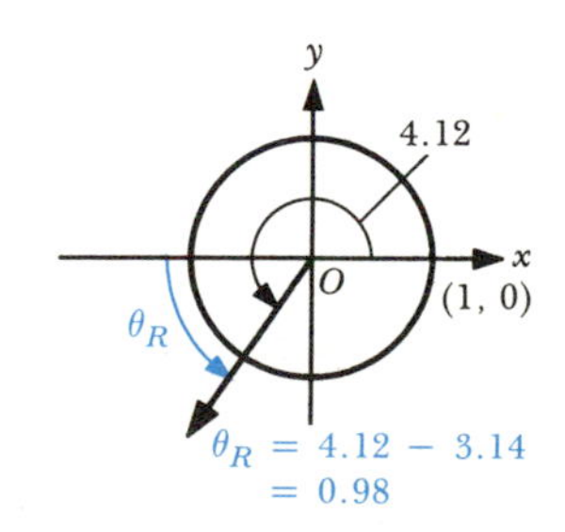

(d)

Figure 4

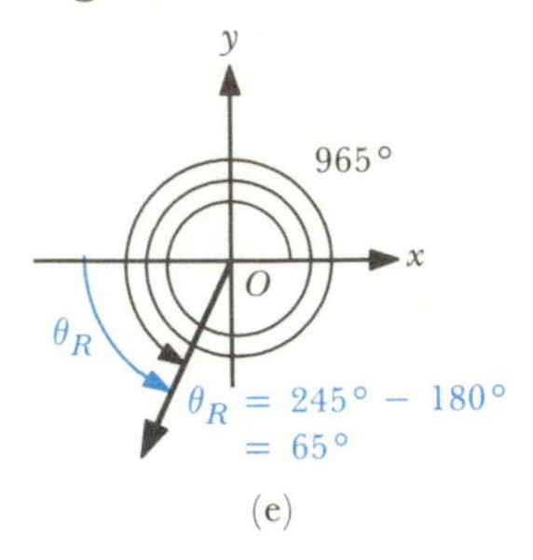

(e) Since $965° = 245° + 2(360°)$, the angle 965° is coterminal with the angle 245°. Since the terminal side of 245° is in quadrant III, the reference angle is

$$\theta_R = 245° - 180° = 65° \qquad \text{(Figure 4e)}$$

The cosecant is negative in quadrant III, so that

$$\csc 245° = -\csc 65° = -1.103$$

Hence $\csc 965° = -1.103$

It is obvious that Appendix Tables III and IV do not contain entries for all the values between 0 and $\pi/2$, or between 0° and 90°. For example, $\sin 1.514$ and $\cos 31°24'$ cannot be found in these tables. As we see in the examples below, *linear interpolation* can be used to approximate such values.

First, let us summarize the steps for finding the values of the trigonometric functions of a nonquadrantal angle when the tables are used.

Step 1. Locate the quadrant where the terminal side of the given angle lies when placed in standard position.

Step 2. Determine the reference angle, an acute angle formed by the terminal side of the given angle and the x axis.

Step 3. Use Appendix Table IV if radians are being used, or Appendix Table III if degrees are being used, and linear interpolation (if necessary) to find the value of the trigonometric function of the reference angle.

Step 4. Use the quadrant location of the angle to attach the appropriate sign (+ or −) to the number obtained in step 3, depending on whether the desired value of the trigonometric function of the given angle is positive or negative.

In Examples 9 and 10, use Appendix Table III or IV and linear interpolation to approximate each value.

EXAMPLE 9 sin 1.514

SOLUTION The terminal side of an angle of radian measure 1.514 is in quadrant I. Appendix Table IV, together with linear interpolation, can be used to find sin 1.514 as follows:

$$0.01\left[\begin{array}{cc} & \begin{array}{cc} \underline{\theta} & \underline{\sin\theta} \\ 1.51 & 0.9982 \end{array} \\ 0.006 & \left[\begin{array}{cc} 1.514 & ? \\ 1.52 & 0.9987 \end{array}\right] d \end{array}\right] 0.0005 \qquad \frac{d}{0.0005} = \frac{0.006}{0.01}$$

Hence, $d = 0.0003$ and

$$\sin 1.514 = 0.9987 - 0.0003 = 0.9984$$

Notice that we subtracted d because the unknown value is *less than* 0.9987.

EXAMPLE 10 $\sec 118°42'$

SOLUTION Since the terminal side of $118°42'$ lies in quadrant II, the secant is negative. The reference angle is $\theta_R = 180° - 118°42' = 61°18'$. Using Appendix Table III and linear interpolation, we have

		θ	$\sec\theta$		
$10'$	$8'$	$61°10'$	2.074	d	0.011
		$61°18'$	?		
		$61°20'$	2.085		

$$\frac{d}{0.011} = \frac{8}{10}$$

and d is approximately 0.009, so that

$$\sec 118°42' = -\sec 61°18' = -(2.074 + 0.009) = -2.083$$

PROBLEM SET 5.5

C In problems 1–19, use a calculator to find approximate values of the six trigonometric functions of each angle to five decimal places.

1. $43°$ **2.** $161.2°$ **3.** $119.23°$ **4.** $-121.27°$
5. $326.21°$ **6.** $-61°2'8''$ **7.** $-103°8'12''$ **8.** $-611.71°$
9. $\frac{\pi}{5}$ **10.** $\frac{\pi}{36}$ **11.** $-\frac{3\pi}{7}$ **12.** $-\frac{5\pi}{8}$
13. 1.588 **14.** -4.233 **15.** 12.407 **16.** -5.151
17. -8.764 **18.** -11.793 **19.** 16.574

20. A calculator is set in radian mode, π is entered, and the SIN key is pressed. The display shows $-4.1 \cdot 10^{-10}$, but we know that $\sin\pi = 0$. Explain.

In problems 21–39, find the reference angle θ_R for the given angle θ (use $\pi = 3.14$).

21. $\theta = 123°$ **22.** $\theta = 241.5°$ **23.** $\theta = 211°$
24. $\theta = 161.5°$ **25.** $\theta = 275°30'$ **26.** $\theta = 314.1°$
27. $\theta = -85.2°$ **28.** $\theta = -200.1°$ **29.** $\theta = 800.2°$
30. $\theta = -700.5°$ C **31.** $\theta = \frac{9\pi}{14}$ C **32.** $\theta = \frac{17\pi}{16}$
C **33.** $\theta = \frac{29\pi}{18}$ C **34.** $\theta = -\frac{11\pi}{14}$ C **35.** $\theta = 8.82$
C **36.** $\theta = 7.76$ C **37.** $\theta = -32.1$ C **38.** $\theta = -10.53$
C **39.** $\theta = 121.15$

40. (a) Let θ be an angle in quadrant III in standard position and let θ_R be its reference angle. Show that the value of any trigonometric function of θ is the same as the value of the function for θ_R except possibly for a change of algebraic sign.
(b) Repeat for θ in quadrant IV.

In problems 41–74, use Appendix Table III or IV and a reference angle, when necessary, to determine the approximate value of each trigonometric function (use $\pi = 3.14$).

41. $\sin 59°$
42. $\cos 53°$
43. $\cot 85°50'$
44. $\sec 51°20'$
45. $\csc(-19°40')$
46. $\tan(-70°30')$
47. $\sin 0.18$
48. $\cos 0.64$
49. $\tan(-1.31)$
50. $\cot(-0.44)$
51. $\sec 1.51$
52. $\csc 1.42$
53. $\sin 125°$
54. $\sin 162°$
55. $\tan 261°$
56. $\cot 245°$
57. $\sec 295°$
58. $\cot(-317°)$
59. $\sin 850°$
60. $\cot 1375°$
61. $\cos 798°$
62. $\sec(-495°)$
63. $\tan(-623°)$
64. $\cos(-625°)$
65. $\csc(-770°)$
66. $\cos(-1237°)$
67. $\cos 3$
68. $\sin 5.8$
69. $\tan 2$
70. $\cot(-15)$
71. $\sin(-14.23)$
72. $\cos 15.42$
73. $\tan 64$
74. $\csc\left(-\frac{49\pi}{11}\right)$

In problems 75–85, use Appendix Table III or IV and linear interpolation to determine each value (use $\pi = 3.142$).

75. $\cos 1.588$
76. $\csc 167°45'$
77. $\sin(-4.233)$
78. $\sec 121°27'$
79. $\tan 326°15'$
80. $\cot 5.151$
81. $\cot 760°38'$
82. $\sin(-12.468)$
83. $\sec(-326°18')$
84. $\tan 103°25'$
85. $\cos 12.407$
86. Compare the calculator value of $\tan 4.12$ to the table value of $\tan 0.98$. Explain why calculator results may vary slightly from the results of using reference angles and the tables.

5.6 Graphs of the Sine and Cosine Functions

We now consider the graphs of the sine and cosine functions. In the following section we will discuss the graphs of the other four trigonometric functions.

In Section 5.1 we introduced the wrapping scheme for associating with each real number t a unique point $P(t) = (x, y)$ on the unit circle (Figure 1). We established that $\sin t = y$ and $\cos t = x$. Since $P(t) = (x, y)$ is defined for any real number t, it follows that the cosine and sine functions exist and therefore are defined for any real number t; that is, the domain for both the cosine and the sine functions is the set of all real numbers $\mathbb{R}$.

Figure 1

Moreover, we observed in Section 5.4 that the values of the sine and cosine functions repeat every 2π units. This occurs because the coordinates of the point

$P(t) = (x, y)$ on the unit circle, associated with t by the wrapping process, will be the same for $t + 2\pi n$, where n is any integer. We now describe this repetitive behavior by saying that the sine and cosine functions are **periodic,** in the sense that their values repeat themselves whenever the independent variable t increases or decreases by 2π; that is, if t represents any real number, we have the following property:

Periodic Property of Sine and Cosine Functions

$\sin(t + 2n\pi) = \sin t$ and $\cos(t + 2n\pi) = \cos t$ for any integer n

The variation of the values of the sine and cosine functions in the interval $[0, 2\pi]$ can be determined by observing the variations of the values of the coordinates of $P(t) = (x, y) = (\cos t, \sin t)$. As t increases from 0 to 2π, the point $(\cos t, \sin t)$ (Figure 1) moves once around the unit circle in a counterclockwise direction, and the coordinates $\cos t$ and $\sin t$ behave as described in Table 1.

Table 1

As t increases from:	$\sin t$	$\cos t$
0 to $\frac{\pi}{2}$	is positive and increases from 0 to 1	is positive and decreases from 1 to 0
$\frac{\pi}{2}$ to π	is positive and decreases from 1 to 0	is negative and decreases from 0 to -1
π to $\frac{3\pi}{2}$	is negative and decreases from 0 to -1	is negative and increases from -1 to 0
$\frac{3\pi}{2}$ to 2π	is negative and increases from -1 to 0	is positive and increases from 0 to 1

We use the convention of denoting the independent variable of a function by x instead of t, and we sketch the graphs of $y = \sin x$ and $y = \cos x$ on the usual xy coordinate system. We can make a rough sketch of the graph of $y = \sin x$ for $0 \leq x \leq 2\pi$ by using the information in Table 1, if we replace t by x. The graph that results by letting x vary from 0 to 2π (corresponding to letting P make one complete counterclockwise rotation on the unit circle) is referred to as a one **cycle** graph of $y = \sin x$. We can make a more accurate sketch by using the coordinates $(x, \sin x)$ listed in Table 2. If even more accuracy is desired, additional points can be found using a calculator. By plotting these points and drawing a smooth curve through them, we obtain one cycle of the graph of $y = \sin x$ (Figure 2). Using the fact that the sine is a periodic function, we obtain a *complete* graph of $y = \sin x$ by repeating the cycle to the right and to the left every 2π units (Figure 3). The graph shows that the range of the sine function is the closed interval $[-1, 1]$; that is,

$$-1 \leq \sin x \leq 1$$

Table 2

x	$\sin x$
0	0
$\pi/6$	$1/2$
$\pi/4$	$\sqrt{2}/2$
$\pi/3$	$\sqrt{3}/2$
$\pi/2$	1
π	0
$7\pi/6$	$-1/2$
$3\pi/2$	-1
$11\pi/6$	$-1/2$
2π	0

holds for all real numbers x. Notice that the graph of $y = \sin x$ is symmetric with respect to the origin. This reflects the fact that the sine is an odd function; that is,

$$\sin(-x) = -\sin x$$

Figure 2

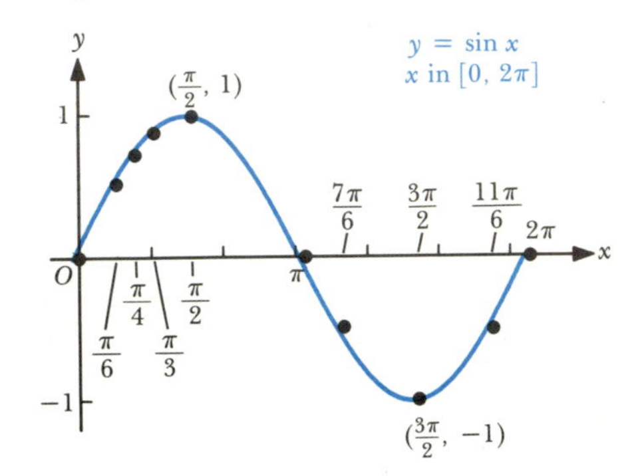

Figure 3

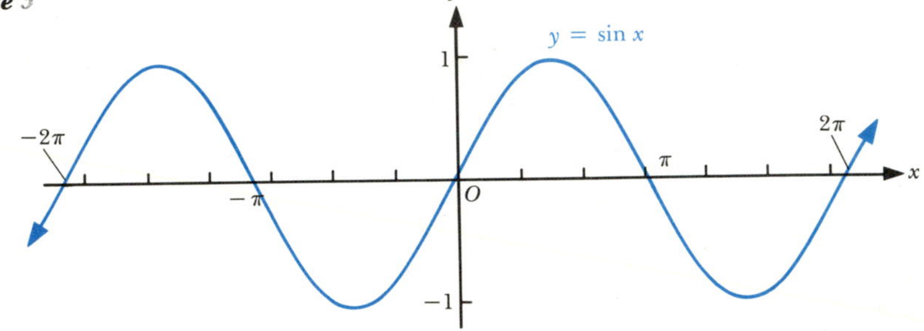

A rough sketch of the graph of the cosine function $y = \cos x$ for $0 \leq x \leq 2\pi$ can be found in a similar manner, by using the information in Table 1 on page 287. Once again, we can make a more accurate sketch by using the coordinates $(x, \cos x)$ listed in Table 3; the calculator can be used to find additional points. The graph of one cycle of $y = \cos x$ is shown in Figure 4. By repeating the cycle to the right and

Table 3

x	$\cos x$
0	1
$\pi/3$	$1/2$
$\pi/2$	0
π	-1
$4\pi/3$	$-1/2$
$3\pi/2$	0
2π	1

Figure 4

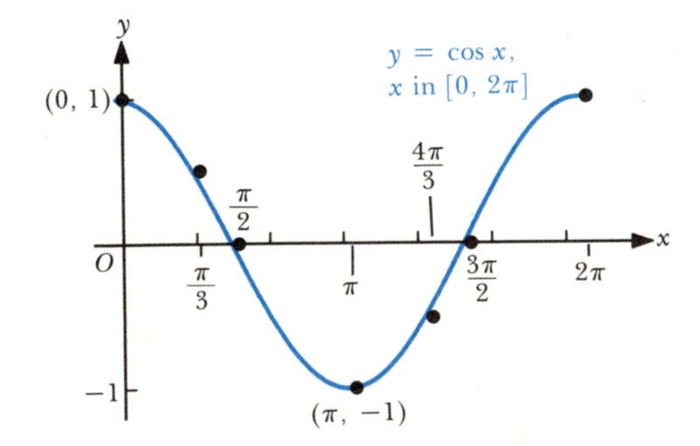

to the left every 2π units, we get the complete graph of $y = \cos x$ (Figure 5). As with the sine function, the range of the cosine function is the interval $[-1, 1]$; hence,

$$-1 \leq \cos x \leq 1$$

holds for all real numbers x. The graph is symmetric with respect to the y axis, reflecting the fact that the cosine is an even function, that is,

$$\cos(-x) = \cos x$$

Figure 5

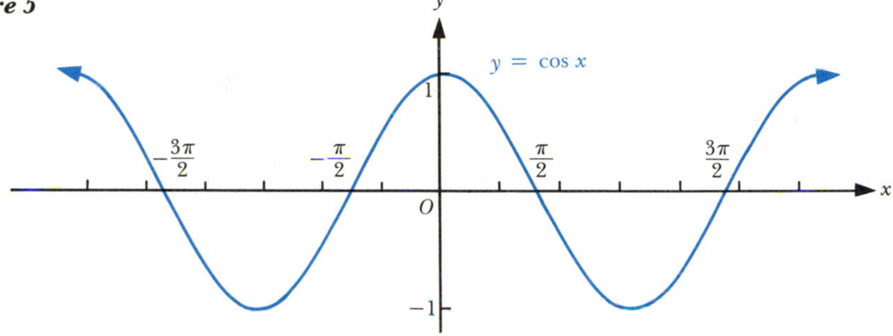

If the techniques of graphing discussed in Chapter 2 are used to stretch or shift the graph of the sine function or cosine function, the resulting curve always has the wavelike form called a *simple harmonic curve*, a *sine wave*, or a *sinusoidal curve*. This curve occurs as the graph of an equation of the form $y = a\sin(kx + b)$ or $y = a\cos(kx + b)$, where a, k, and b are constants. The graphs of these latter types of equations can be obtained from the known graphs of

$$y = \sin x \qquad \text{or} \qquad y = \cos x$$

by examining the geometrical effects of the constants a, k, and b on the sine and cosine graphs.

Consider the case in which $b = 0$ and $k = 1$, that is,

$$y = a\sin x \qquad \text{or} \qquad y = a\cos x$$

Since

$$-1 \leq \sin x \leq 1 \qquad \text{and} \qquad -1 \leq \cos x \leq 1$$

it follows that if $a > 0$, the ranges of the functions are given by $-a \leq a\sin x \leq a$ and $-a \leq a\cos x \leq a$. If $a < 0$, the ranges are given by $-a \geq a\sin x \geq a$ and $-a \geq a\cos x \geq a$. So in either case $|y| \leq |a|$, and $|a|$ is referred to as the *amplitude* of the function. In general:

> The **amplitude** of the following functions is $|a|$.
>
> $$y = a\sin(kx + b) \qquad y = a\cos(kx + b)$$

For example, the amplitude of the function $y = 7\sin 2x$ is $|7| = 7$, and the amplitude of $y = -5\cos 3x$ is $|-5| = 5$.

In Examples 1 and 2, use the sine and cosine graphs to sketch the graph of each function and find the amplitude and range of the function.

EXAMPLE 1 (a) $f(x) = 3 \sin x$ (b) $g(x) = -3 \sin x$

SOLUTION (a) The graph of $f(x) = 3 \sin x$ is obtained by multiplying each ordinate of $y = \sin x$ by 3. We first sketch the graph of $y = \sin x$, then we multiply each ordinate by 3. Geometrically, this means that the graph of $y = \sin x$ is stretched vertically by the factor 3 to obtain the graph of $f(x) = 3 \sin x$ (Figure 6a). The amplitude of $f(x) = 3 \sin x$ is 3. Notice that the maximum distance between points of the graph of $f(x) = 3 \sin x$ and the x axis is 3. Also the graph indicates that the range of the given function is the interval $[-3, 3]$.

(b) The graph of $g(x) = -3 \sin x$ is obtained by reflecting the graph of $f(x) = 3 \sin x$ across the x axis (Figure 6b). The amplitude of the reflected graph is still 3. Here again the range is $[-3, 3]$.

Figure 6

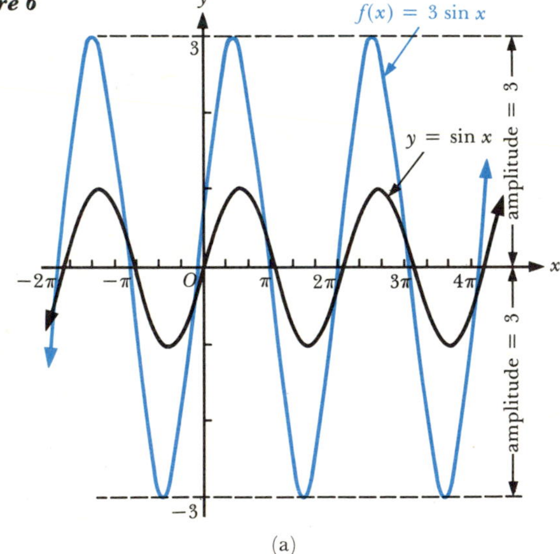

(a)

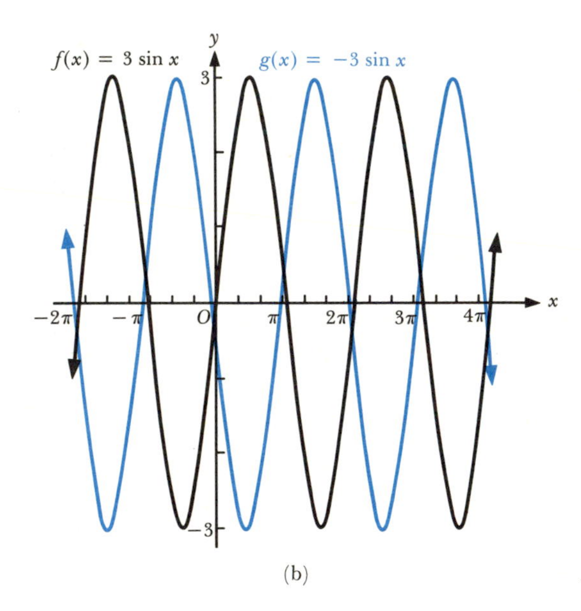

(b)

EXAMPLE 2 (a) $F(x) = \frac{1}{5} \cos x$ (b) $H(x) = 1 + \frac{1}{5} \cos x$

SOLUTION (a) The graph of $F(x) = \frac{1}{5} \cos x$ is obtained from the graph $y = \cos x$ by multiplying each ordinate by $\frac{1}{5}$. Thus, the graph of $F(x) = \frac{1}{5} \cos x$ can be obtained geometrically from the graph of $y = \cos x$ by vertically contracting the latter graph by a multiple of $\frac{1}{5}$ (Figure 7a). The amplitude is $\frac{1}{5}$ and the range is $[-\frac{1}{5}, \frac{1}{5}]$.

(b) The graph of $H(x) = 1 + \frac{1}{5} \cos x$ is obtained by vertically shifting the graph of $F(x) = \frac{1}{5} \cos x$ one unit upward. This shift does not affect the amplitude, which is still $\frac{1}{5}$ (Figure 7b). The range is $[\frac{4}{5}, \frac{6}{5}]$.

Figure 7

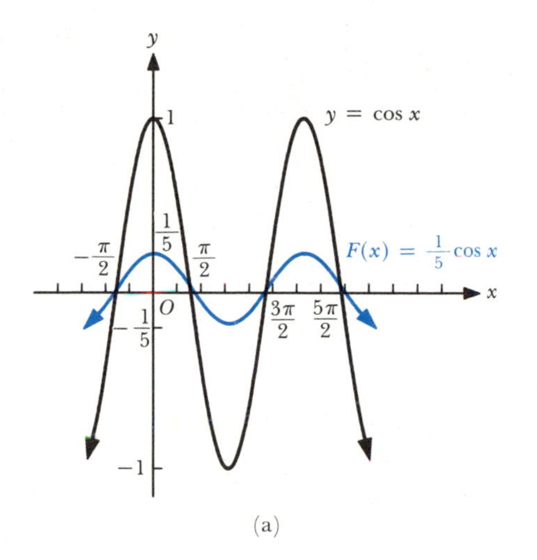

(a)

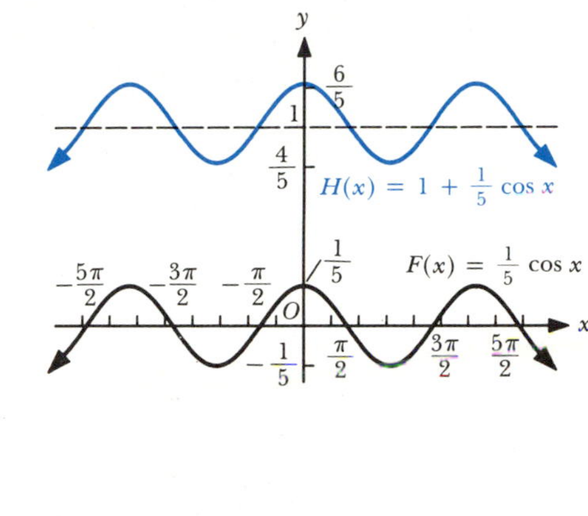

(b)

Figure 8

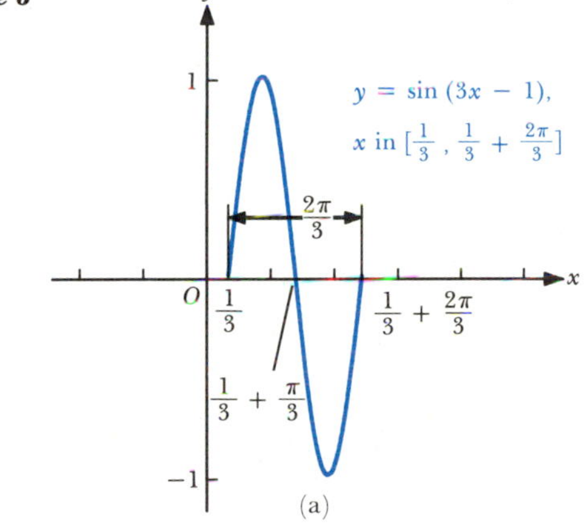

(a)

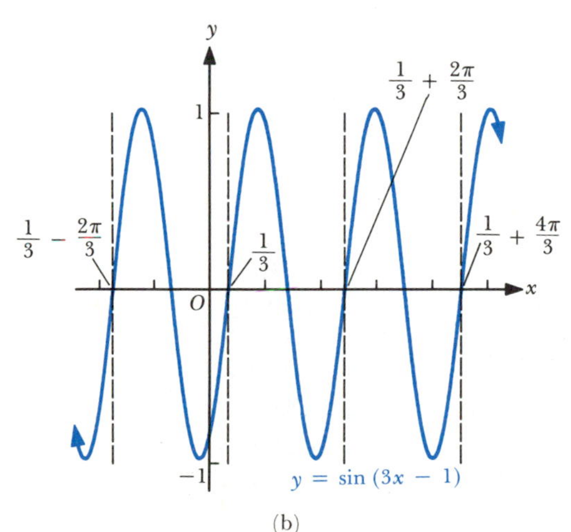

(b)

Since both the sine and the cosine are periodic functions of period 2π, it follows that $y = a\sin(kx + b)$ and $y = a\cos(kx + b)$ generate one cycle of the sine graph and the cosine graph, respectively, when

$$kx + b \text{ varies from } 0 \text{ to } 2\pi$$

that is, when $0 \leq kx + b < 2\pi$

Suppose, for example, that $y = \sin(3x - 1)$. Then the function generates one cycle of the sine graph when

$$3x - 1 \text{ varies from } 0 \text{ to } 2\pi \qquad (0 \leq 3x - 1 < 2\pi)$$

that is, when

$$3x \text{ varies from } 1 \text{ to } 2\pi + 1 \qquad (1 \leq 3x < 2\pi + 1)$$

or

$$x \text{ varies from } \frac{1}{3} \text{ to } \frac{2\pi}{3} + \frac{1}{3} \qquad \left(\frac{1}{3} \leq x < \frac{2\pi}{3} + \frac{1}{3}\right)$$

Hence, if x varies from $\frac{1}{3}$ to $[\frac{1}{3} + (2\pi/3)]$, then $3x - 1$ varies from 0 to 2π, so that, in turn, $y = \sin(3x - 1)$ generates one cycle of the sine curve (Figure 8a). The cycle *begins* at $x = \frac{1}{3}$ and covers an interval of length $2\pi/3$, the *period* of the function, to *end* at $x = (2\pi/3) + \frac{1}{3}$. A complete graph of $y = \sin(3x - 1)$ can be obtained by repeating the cycle to the right and left every $2\pi/3$ units (Figure 8b).

In general, if $k > 0$, then one cycle of the graph of a function of the form $y = a\sin(kx + b)$ or $y = a\cos(kx + b)$ can be obtained by allowing $kx + b$ to vary from 0 to 2π. Solving the equations

$$kx + b = 0$$

and

$$kx + b = 2\pi$$

we have

$$x = -\frac{b}{k}$$

and

$$x = \frac{2\pi - b}{k} = \frac{2\pi}{k} - \frac{b}{k}$$

so that one cycle of the graph of each function is obtained as x varies from $-b/k$ to $-b/k + 2\pi/k$, and the **period** of the function is

$$\frac{2\pi}{k}, \qquad \text{where } k > 0$$

The number $-b/k$ is called the **phase shift** associated with the function. One cycle of the graph of each function *begins* at the point $(-b/k, 0)$ and *ends* at the point $(-b/k + 2\pi/k, 0)$. The complete graph of each function can be obtained by repeating the cycle to the right and left every $2\pi/k$ units.

In Examples 3–6, find the amplitude, the period, and the phase shift and sketch the graph of each function.

EXAMPLE 3 (a) $f(x) = 3\sin 2x$ (b) $g(x) = 3\sin(-2x)$

SOLUTION (a) We see that the amplitude of f is 3. The graph of $f(x) = 3\sin 2x$ covers one cycle as $2x$ varies

$$\text{from} \quad 2x = 0 \qquad (x = 0)$$
$$\text{to} \quad 2x = 2\pi \qquad (x = \pi)$$

Therefore, the graph covers one cycle as x varies from 0 to π, and the period is π. The phase shift is 0. We use this information to sketch one cycle, and then repeat the cycle in both directions (Figure 9a). Geometrically, the graph of f can be obtained by stretching the graph of $y = \sin 2x$ by a multiple of 3.

(b) Since the sine function is odd, we have

$$g(x) = 3\sin(-2x) = -3\sin 2x$$

Thus we can obtain the graph of $g(x) = 3\sin(-2x)$ by reflecting the graph of $f(x) = 3\sin 2x$ across the x axis (Figure 9b). The amplitude is $|-3| = 3$, the period is $2\pi/|-2| = 2\pi/2 = \pi$, and the phase shift is 0.

Figure 9

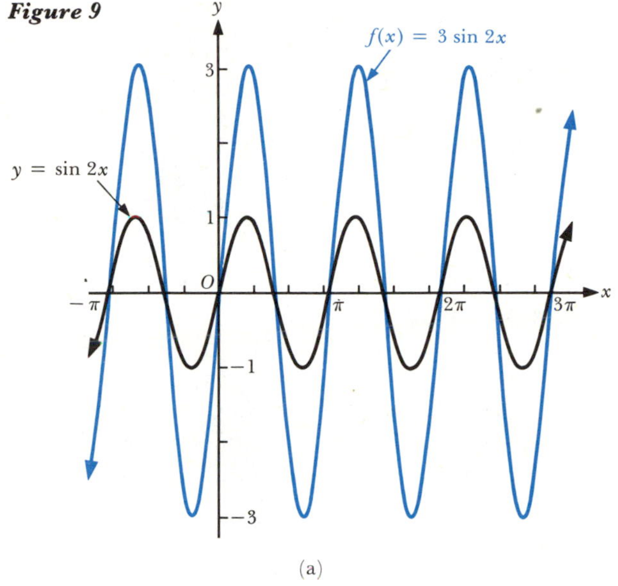

(a)

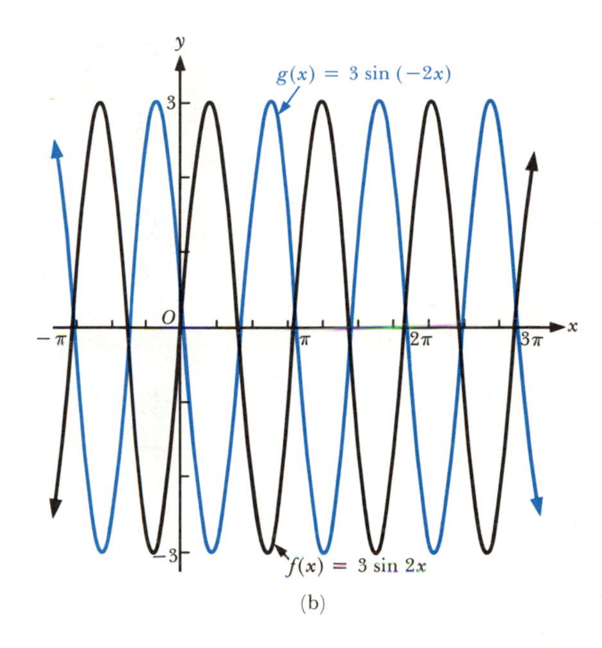

(b)

EXAMPLE 4 $h(x) = 3\cos\frac{1}{2}x$

SOLUTION The amplitude of $h(x) = 3\cos\frac{1}{2}x$ is 3. The graph of the function will cover one cycle as $\frac{1}{2}x$ varies

$$\text{from} \quad \tfrac{1}{2}x = 0 \qquad (x = 0)$$
$$\text{to} \quad \tfrac{1}{2}x = 2\pi \qquad (x = 4\pi)$$

Therefore, the graph covers one cycle as x varies

$$\text{from } 0 \text{ to } 4\pi$$

so that the phase shift is 0 and the period is 4π.

We use this information to sketch one cycle and then repeat the cycle in both directions (Figure 10). The graph of

$$y = 3\cos x$$

is also displayed in Figure 10 for contrast.

Figure 10

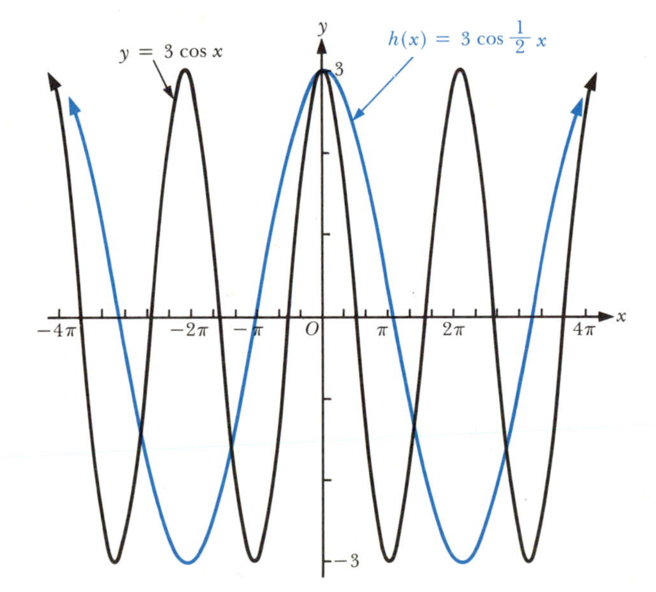

EXAMPLE 5 $y = 5\sin(3x - \pi)$

SOLUTION The amplitude is 5 and the graph will cover one cycle as $3x - \pi$ varies

Figure 11

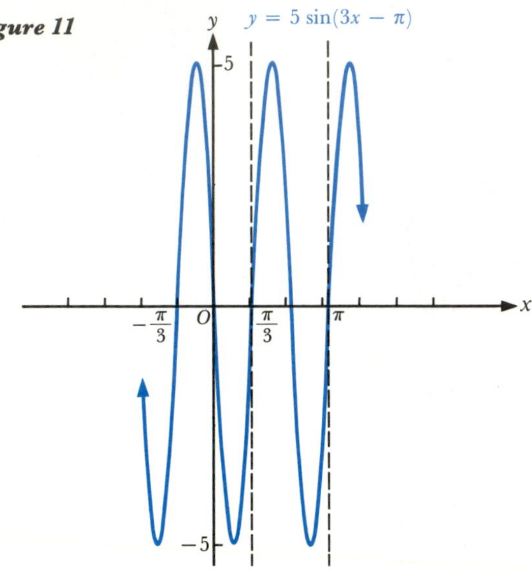

$$\text{from} \quad 3x - \pi = 0 \qquad \left(x = \frac{\pi}{3}\right)$$

$$\text{to} \quad 3x - \pi = 2\pi \qquad \left(x = \frac{\pi}{3} + \frac{2\pi}{3} = \pi\right)$$

That is, the graph covers one cycle as x varies from $\pi/3$ to π. It follows that the period is

$$\pi - \pi/3 = 2\pi/3$$

and the phase shift is $\pi/3$. The complete graph is obtained by repeating the cycle in both directions (Figure 11). Geometrically, the graph of

$$y = 5\sin(3x - \pi)$$

can also be obtained by shifting the graph of $y = 5\sin 3x$, $\pi/3$ units to the right.

EXAMPLE 6 $y = -3\cos\left(\frac{x}{8} + \frac{\pi}{16}\right)$

SOLUTION The amplitude is $|-3| = 3$ and the graph will cover one cycle as $x/8 + \pi/16$ varies

$$\text{from} \quad \frac{x}{8} + \frac{\pi}{16} = 0 \qquad \left(x = -\frac{\pi}{2}\right)$$

$$\text{to} \quad \frac{x}{8} + \frac{\pi}{16} = 2\pi \qquad \left(x = -\frac{\pi}{2} + 16\pi = \frac{31\pi}{2}\right)$$

Figure 12

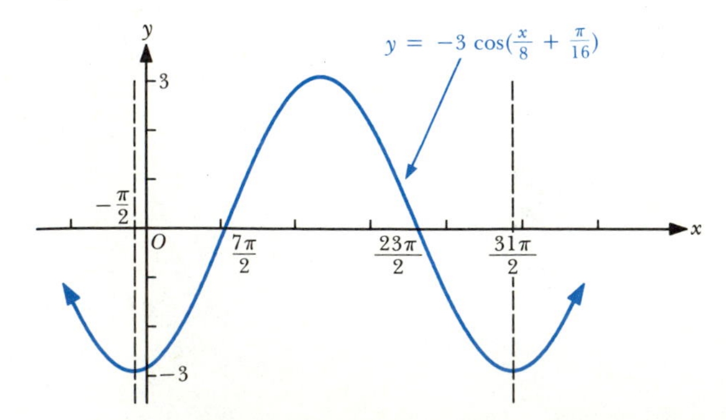

That is, the graph covers one cycle as x varies from $-\pi/2$ to $31\pi/2$. It follows that the period is given by

$$31\pi/2 - (-\pi/2) = 32\pi/2 = 16\pi$$

and the phase shift is $-\pi/2$. Geometrically, the graph of

$$y = -3\cos\left(\frac{x}{8} + \frac{\pi}{16}\right)$$

is the graph of

$$y = -3\cos(x/8)$$

shifted $\pi/2$ units to the left. The negative sign in front of 3 indicates a reflection across the x axis (Figure 12).

Simple Harmonic Motion

The phenomenon of oscillatory motion or recurring events, in which quantities periodically repeat themselves over fixed intervals of time, can often be described by using equations of the form

$$y = a\cos(\omega t - \phi) \qquad \text{or} \qquad y = a\sin(\omega t - \phi)$$

where t represents time and a, ω, and ϕ are constants for the particular motion. Examples of such phenomena include sound waves, light waves, alternating electrical currents, displacements in spring–weight systems, displacements in simple pendulum systems, human circulatory systems, and predator–prey systems. We refer to all such oscillatory behaviors as **simple harmonic motions.** In this mathematical model, a is called the *amplitude*, ω (the Greek letter *omega*) is called the **angular frequency,** and ϕ (the Greek letter *phi*) is called the **phase angle.** An examination of the graph of either equation, say,

$$y = a\cos(\omega t - \phi)$$

(Figure 13), shows that y will go through one cycle as $\omega t - \phi$ goes from 0 to 2π. When $\omega t - \phi = 0$, we have $t = \phi/\omega$; when $\omega t - \phi = 2\pi$, we have

$$t = \frac{\phi}{\omega} + \frac{2\pi}{\omega}$$

Figure 13

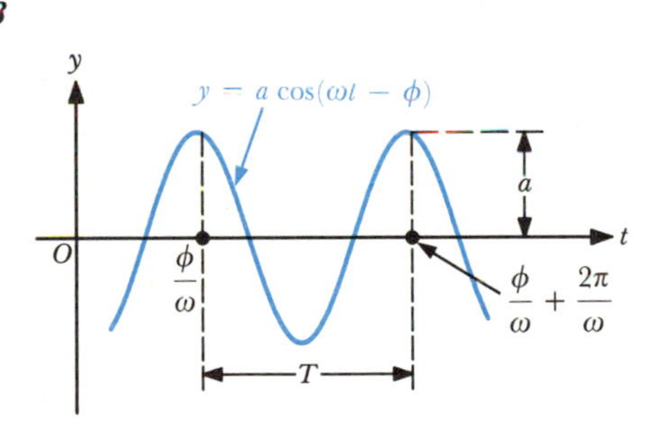

Therefore, one cycle of the graph of

$$y = a\cos(\omega t - \phi)$$

covers the interval from $t = \phi/\omega$ to $t = \dfrac{\phi}{\omega} + \dfrac{2\pi}{\omega}$ (Figure 13). The *period* of the oscillation is the amount of time T required for one cycle to occur. From Figure 13, we can see that the period

$$T = \left(\frac{\phi}{\omega} + \frac{2\pi}{\omega}\right) - \frac{\phi}{\omega} = \frac{2\pi}{\omega}$$

The equation for the period $T = \dfrac{2\pi}{\omega}$ is often written in the form $\omega = \dfrac{2\pi}{T}$ where ω is the angular frequency.

Notice that one cycle of the graph of $y = a\cos(\omega t - \phi)$ has the same general shape as one cycle of the cosine function, but it is shifted horizontally by an amount ϕ/ω. For this reason, ϕ/ω is the *phase shift* of the motion. We denote the phase shift by S, so that

$$S = \frac{\phi}{\omega} \qquad \text{or} \qquad \phi = \omega S$$

EXAMPLE 7 If y is oscillating according to the simple harmonic motion

$$y = 10\cos\left(\frac{\pi}{3}t - \frac{\pi}{2}\right)$$

find (a) the amplitude, (b) the angular frequency, (c) the phase angle ϕ, (d) the period, and (e) the phase shift. (f) Sketch the graph of one cycle.

SOLUTION The equation

$$y = 10\cos\left(\frac{\pi}{3}t - \frac{\pi}{2}\right)$$

is of the form $y = a\cos(\omega t - \phi)$, where $a = 10$, $\omega = \pi/3$, and $\phi = \pi/2$.

(a) The amplitude

$$a = 10$$

(b) The angular frequency

$$\omega = \frac{\pi}{3}$$

(c) The phase angle

$$\phi = \frac{\pi}{2}$$

(d) The period

$$T = \frac{2\pi}{\omega} = \frac{2\pi}{\pi/3} = 6$$

(e) The phase shift

$$S = \frac{\phi}{\omega} = \frac{\pi/2}{\pi/3} = \frac{3}{2}$$

(f) The graph of one cycle is obtained by sketching the cosine curve on the interval $[\frac{3}{2}, \frac{15}{2}]$ (Figure 14).

Figure 14

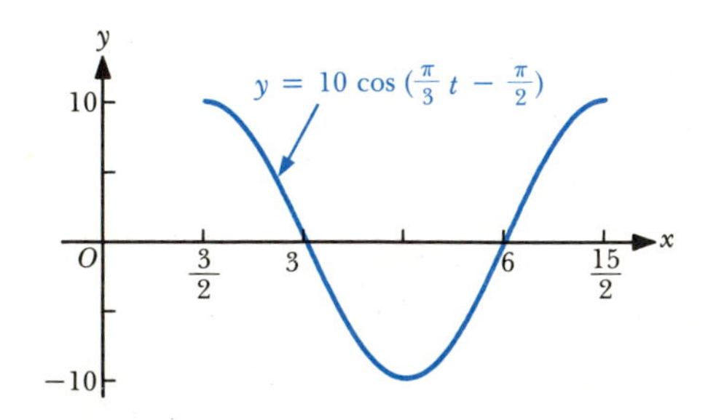

In the simple harmonic models

$$y = a\cos(\omega t - \phi) \qquad \text{or} \qquad y = a\sin(\omega t - \phi)$$

one cycle requires $T = 2\pi/\omega$ units of time, so y *oscillates* through

$$\frac{1}{T} = \frac{\omega}{2\pi}$$

cycles in one unit of time. We call $1/T$ the **frequency of oscillation.** The Greek letter ν (nu) is used to denote the frequency. Thus,

$$\nu = \frac{1}{T} = \frac{\omega}{2\pi}$$

A physical interpretation of simple harmonic oscillation is illustrated by a **simple pendulum,** which is an object suspended by a "weightless" string of length l (Figure 15). When the pendulum is pulled to one side of its vertical position and released, it oscillates periodically back and forth in a vertical plane (if resistance is neglected). Let y denote the displacement of the object from its vertical position, measured along the arc of the swing (positive to the right and negative to the left) at time t. Suppose that $y = a$ when $t = 0$, the instant of release. Then, if a is not too large and t is small, the quantity y will oscillate according to the equation

Figure 15

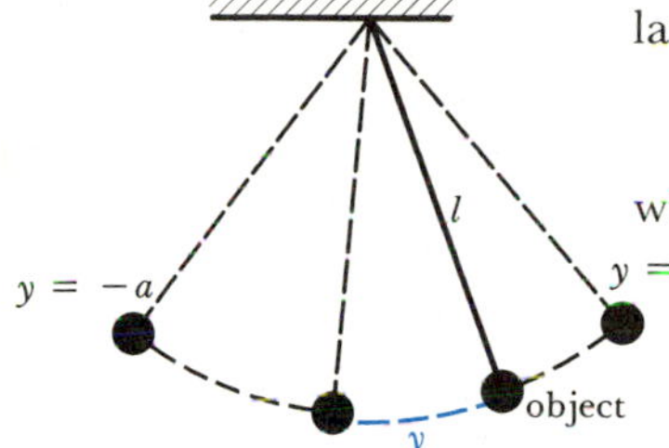

$$y = a\cos\omega t \qquad \text{or} \qquad y = a\sin\omega t$$

with amplitude a, phase angle $\phi = 0$, and period

$$T = 2\pi\sqrt{\frac{l}{g}}$$

where g is the pull of gravity (g is approximately 32 feet per second squared or 9.8 meters per second squared).

EXAMPLE 8 [c] A simple pendulum of length 15 centimeters is pulled to the right through an arc of $a = 0.18$ meter and released at $t = 0$ second. Find (a) the period T, (b) the frequency of oscillation, and (c) the angular frequency. Round off the answers to two decimal places and use $\pi = 3.14$.

SOLUTION Using the formulas derived above, we have

(a) $l = 0.15$ meter and $g = 9.8$ meters/sec^2, and so the period of oscillation

$$T = 2\pi\sqrt{\frac{l}{g}} = 2\pi\sqrt{\frac{0.15}{9.8}} = 0.78 \text{ second}$$

(b) The frequency of oscillation

$$\nu = \frac{1}{T} = \frac{1}{0.78} = 1.28$$

so that the pendulum swings through 1.28 oscillations for each second of elapsed time.

(c) The angular frequency

$$\omega = \frac{2\pi}{T} = \frac{2\pi}{0.78} = 8.05$$

EXAMPLE 9 c Consider a condenser with a capacitance of C farads and containing a charge of Q_0 coulombs. When connected in series with a coil of negligible resistance and having an inductance of L henrys, the charge Q on the condenser after t seconds is given by the function

$$Q(t) = Q_0 \sin\left(\frac{t}{\sqrt{LC}} - \frac{\pi}{2}\right)$$

where L and C are constants. If $L = 0.4$ henry and $C = 10^{-5}$ farad, find the period and the frequency of oscillation of this circuit to four significant digits (use $\pi = 3.142$).

SOLUTION Here $L = 0.4$ henry and $C = 10^{-5}$ farad, so that

$$\omega = \frac{1}{\sqrt{LC}} = \frac{1}{\sqrt{(0.4)(10^{-5})}} = \frac{1}{\sqrt{4 \cdot 10^{-6}}} = \frac{1000}{2} = 500$$

The above equation describes a simple harmonic motion, whose period

$$T = \frac{2\pi}{\omega} = \frac{2\pi}{500} = \frac{\pi}{250} = 0.004\pi = 0.01257 \text{ second}$$

The frequency of oscillation of the circuit is

$$\nu = \frac{\omega}{2\pi} = \frac{500}{2\pi} = \frac{250}{\pi} = 79.57 \text{ cycles per second}$$

PROBLEM SET 5.6

In problems 1–28, use the graphs of the sine and cosine functions to graph each function. Indicate the period, amplitude, and phase shift.

1. $y = 10 \sin x$

2. $y = \frac{2}{3} \sin x$

3. $f(x) = 2 \cos x$

4. $y = 3 \cos(-x)$

5. $y = 1 + \sin x$

6. $g(x) = -\dfrac{3}{4} + 2 \cos x$

7. $y = 2 + \cos(-x)$

8. $y = 3 - \sin(-x)$

9. $h(x) = \sin \dfrac{\pi x}{3}$

10. $y = 2 \sin 2\pi x$

11. $y = 5 \sin 10x$

12. $F(x) = 2 \cos \dfrac{\pi x}{2}$

13. $y = 3 \cos 3\pi x$

14. $y = \dfrac{1}{2} \sin \dfrac{\pi x}{4}$

15. $G(x) = \dfrac{5}{3} \sin(-6x)$

16. $y = -3 \cos 5x$

17. $y = -\pi \cos(-5x)$

18. $H(x) = -5 \sin \dfrac{2\pi x}{3}$

19. $y = 2 + 3 \cos \pi x$

20. $y = \dfrac{1}{2} - \dfrac{1}{2} \sin \dfrac{\pi x}{2}$

21. $F(x) = -\dfrac{1}{2} \sin \dfrac{3x}{2}$

22. $y = |\cos x|$

23. $y = \frac{1}{3}\sin\left(x - \frac{\pi}{6}\right)$

24. $f(x) = -5\cos\left(x + \frac{\pi}{2}\right)$

25. $y = 3\cos\left(\frac{x}{2} + \frac{\pi}{4}\right)$

26. $y = -4\sin\left(\frac{x}{3} + \frac{\pi}{5}\right)$

27. $g(x) = -2\sin(5 - 3x)$

28. $y = -2\cos(3 - 2x)$

In problems 29–32, display on the same coordinate system the graphs of the four given functions.

29. $y = \sin x,\quad y = \sin 2x,\quad y = \sin\left(2x - \frac{\pi}{2}\right),\quad y = 2\sin\left(2x - \frac{\pi}{2}\right)$

30. $y = \cos x,\quad y = \cos\frac{x}{2},\quad y = 3\cos\frac{x}{2},\quad y = 1 + 3\cos\frac{x}{2}$

31. $y = \cos x,\quad y = \cos\frac{2\pi x}{3},\quad y = \frac{1}{2}\cos\frac{2\pi x}{3},\quad y = -\frac{1}{2}\cos\frac{2\pi x}{3}$

32. $y = \sin x,\quad y = \sin 3\pi x,\quad y = -2\sin 3\pi x,\quad y = 3 - 2\sin 3\pi x$

c In problems 33–36, use a calculator to verify that the equation is true for the indicated value of x. Then display the location of the two points on the graph of the associated function f.

33. $f(-x) = -f(x)$, where $f(x) = \sin x$ and $x = 3.752$
34. $f(-x) = -f(x)$, where $f(x) = \sin x$ and $x = 4.138$
35. $f(-x) = f(x)$, where $f(x) = \cos x$ and $x = 5.384$
36. $f(-x) = f(x)$, where $f(x) = \cos x$ and $x = 2.874$

In problems 37–40, find the amplitude, period, phase angle, angular frequency, and phase shift, and graph one cycle of the simple harmonic motion defined by the given function.

37. $y = \sin\frac{5\pi}{6}t$

38. $y = \cos 0.4t$

39. $y = 4\cos\left(\frac{\pi}{4}t - \frac{\pi}{2}\right)$

c **40.** $y = 0.263\sin(3.5t - 1.4)$

41. Assume that $y = 3\sin(5t - 1)$ represents the simple harmonic motion for a spring–weight system. Find
(a) the amplitude (b) the angular frequency
(c) the phase angle (d) the period
(e) the phase shift (f) Sketch the graph of one cycle.

42. Explain why the frequency equals the reciprocal of the period.

c In problems 43–45, assume that l represents the length of the string of a simple pendulum and that the pendulum is pulled to the right through an arc of length a and released at $t = 0$ second. Find (a) the period T (b) the frequency of oscillation ν (c) the angular frequency ω. Round off the answers to two decimal places.

43. $l = 2$ feet; $a = 0.34$ foot

44. $l = 30$ centimeters; $a = 0.72$ centimeter

45. $l = 25$ centimeters; $a = 0.08$ centimeter

C **46.** A chandelier suspended from a ceiling with a cable 7 meters long is pulled to the right through an arc of length $a = 1.4$ meters and released at $t = 0$. Find
(a) the period
(b) the frequency of oscillation
(c) the angular frequency

C **47.** In Example 9 on page 298, if $L = 0.9$ henry and $C = 10^{-5}$ farad, find
(a) the period
(b) the frequency of oscillation of the circuit
(c) the time t_1 when $Q = 0$ for the first time
(d) the time t_2 when $Q = 0.5Q_0$ for the second time

C **48.** The pressure of a traveling sound wave is given by the function

$$P(t) = 1.5 \sin(x_0 - 330t)$$

where x_0 is a constant, t is elapsed time in seconds, and P is in newtons per square meter. Find the amplitude and the frequency of the sound wave.

5.7 Graphs of Other Trigonometric Functions

In Section 5.6, we discussed the graphs of the cosine and the sine functions and their basic properties. In this section we graph the other four trigonometric functions by relating them to the sine and cosine functions. We also use general graphing techniques to sketch graphs of functions consisting of combinations of trigonometric functions.

Graphs of Tangent, Cotangent, Secant, and Cosecant Functions

The domains of the tangent, cotangent, secant, and cosecant functions can be determined by using the sine and cosine functions. Since

$$\tan x = \frac{\sin x}{\cos x} \qquad \text{and} \qquad \sec x = \frac{1}{\cos x}$$

the domain for both the tangent and the secant functions consists of all real numbers x for which $\cos x \neq 0$. As we saw in Figure 5 on page 289, the cosine function $y = \cos x$ is zero at $\pm\frac{\pi}{2}, \pm\frac{3\pi}{2}, \pm\frac{5\pi}{2}$, and so on. In other words, the domain for both the tangent and the secant functions consists of all values of x for which

$$x \neq \frac{\pi}{2} + n\pi \qquad \text{or} \qquad x \neq (2n + 1)\frac{\pi}{2}$$

where n is any integer, that is, x cannot equal an odd multiple of $\pi/2$. The domain of both the cotangent and cosecant functions consists of all real numbers x for which

$\sin x \neq 0$, since

$$\cot x = \frac{\cos x}{\sin x} \quad \text{and} \quad \csc x = \frac{1}{\sin x}$$

Because $\sin x = 0$ at $x = 0, \pm\pi, \pm 2\pi$, and so on, the domain of the cotangent and the cosecant functions is the set of all values of x such that

$$x \neq n\pi$$

where n is any integer. That is, x cannot equal an integer multiple of π.

Figure 1

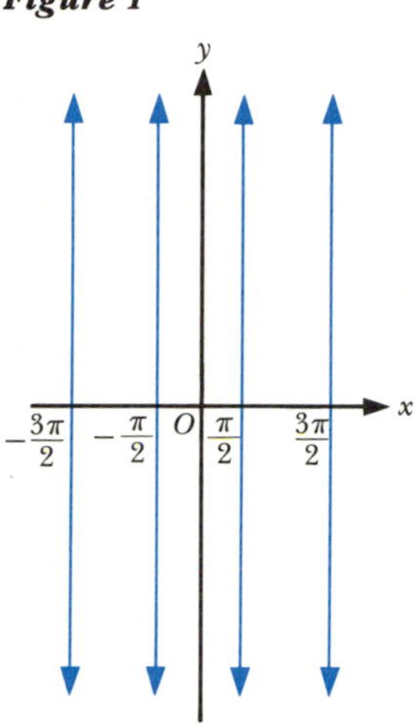

Using the domain of the tangent function, we note that its graph cannot intersect any line of the form $x = (\pi/2) + n\pi$, where n is an integer (Figure 1). Since $\tan x = \sin x/\cos x$, we can use values of $\sin x$ and $\cos x$ on $[0, \pi/2)$ to determine values of $\tan x$ as x varies from 0 to $\pi/2$. This behavior of the function $y = \tan x$ can be observed by examining the values in Table 1, which is obtained by using a calculator and rounding off the values. As x increases from 0 toward $\pi/2$, $\sin x$ increases from 0 to 1 and $\cos x$ decreases from 1 to 0. Hence, the quotient $\sin x/\cos x$ becomes increasingly large as these values of x get "closer" to $\pi/2$. That is, the values of $\tan x$ increase indefinitely as x increases from 0 toward $\pi/2$, and the line $x = \pi/2$ is a vertical asymptote of the graph of $y = \tan x$ (Figure 2a).

Since the tangent function is odd, that is, since

$$\tan(-x) = -\tan x$$

we can use the fact that the graph of the function is symmetric with respect to the origin to obtain its graph in the interval $(-\pi/2, 0)$. The line $x = -\pi/2$ is also a vertical asymptote of the graph (Figure 2b). As with the sine and cosine functions, the tangent function is periodic; however, its values repeat themselves whenever the

Table 1

x	$\sin x$	$\cos x$	$\tan x = \sin x/\cos x$
0	0	1.00	0
$\pi/6$	0.500	0.866	0.577
$\pi/4$	0.707	0.707	1.000
$\pi/3$	0.866	0.500	1.732
1.3	0.964	0.267	3.610
1.5	0.997	0.071	14.042
1.569	1.000	$1.796 \cdot 10^{-3}$	556.793
1.57	1.000	$7.963 \cdot 10^{-4}$	1255.808
$\pi/2$	1.000	0	undefined

Figure 2

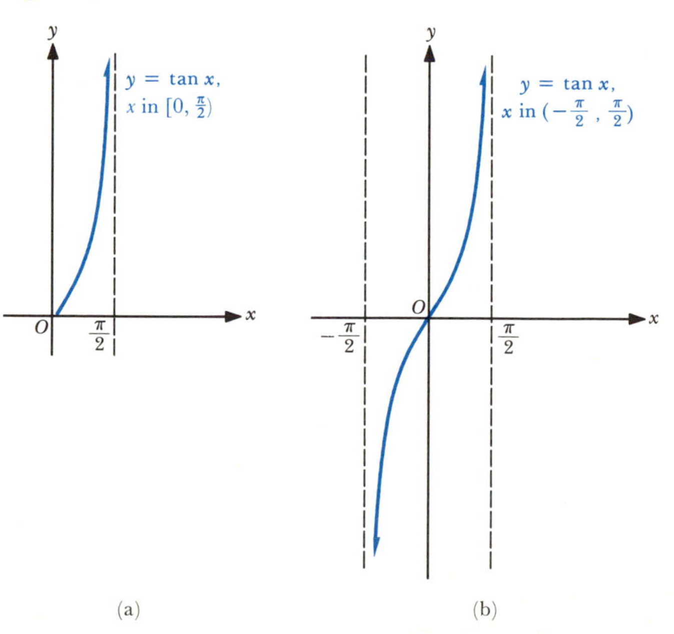

independent variable increases by π units. In other words, π is a period of the tangent function, so that

$$\tan(x + \pi) = \tan x$$

holds for every value of x in the domain of the tangent function. (This identity will be proved on page 326 in Section 6.2.) Notice that the interval from $-\pi/2$ to $\pi/2$ in Figure 2b has length π units. Therefore, since the tangent function has a period π, the graph in Figure 2b can be repeated to the right and left every π units to obtain a more complete graph of the tangent function (Figure 3). The graph shows that the range of the tangent function includes all real numbers ℝ.

Figure 3

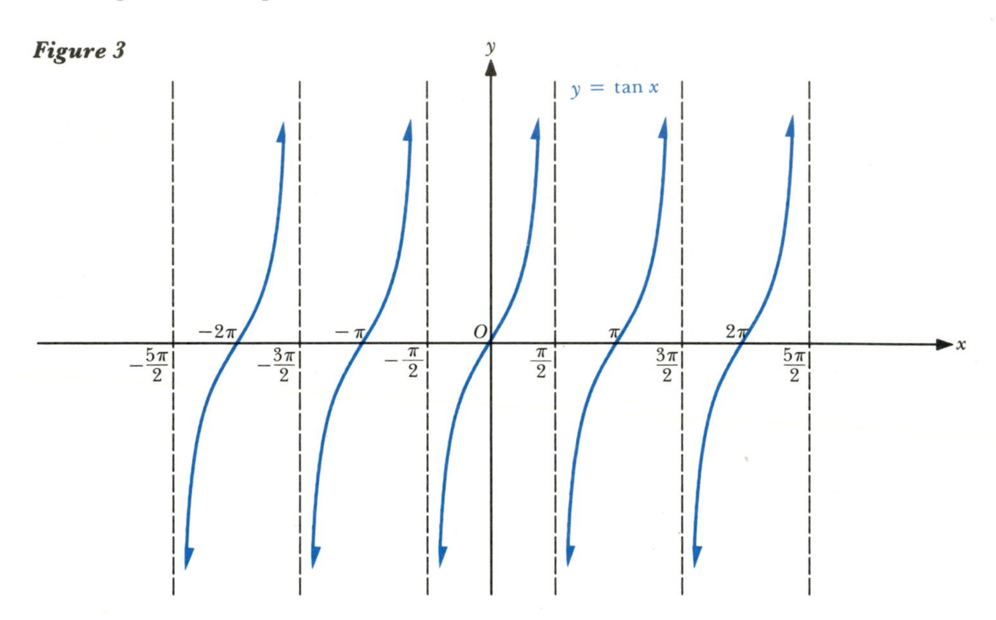

The values of the cotangent function $y = \cot x$ repeat themselves whenever the independent variable increases by π; that is,

$$\cot(x + \pi) = \cot x$$

holds for every value of x in the domain of the cotangent function (Problem 4a) and the cotangent function is periodic of period π. We have established that the cotangent function is defined for all real numbers x for which $x \neq n\pi$, where n is any integer. In fact, the graph of the cotangent function has a vertical asymptote whenever $x = n\pi$ for any integer n (Problem 6a). To sketch the graph of $y = \cot x$, we start by sketching the graph for $0 < x < \pi$ (Problem 6b), then repeat the graph to the right and to the left every π units to obtain a more complete graph of the cotangent function (Figure 4). Notice that the graph of the cotangent function can also be obtained by first reflecting the graph of the tangent about the x axis and then shifting the graph $\pi/2$ units to the right. The graph shows that the range of the cotangent function includes all real numbers ℝ. Also, the graph is symmetric

Figure 4

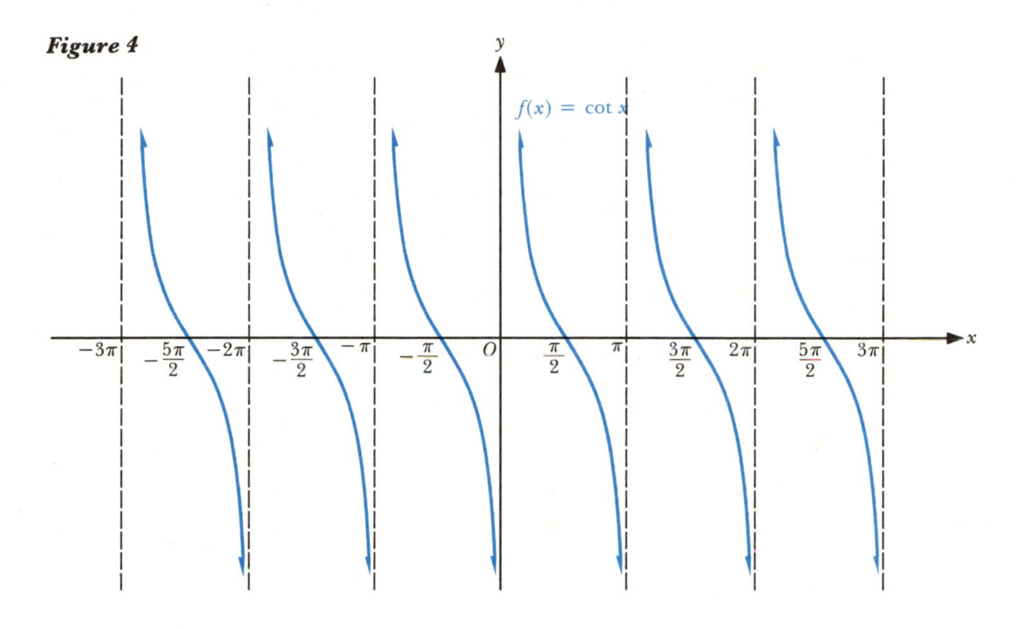

about the origin, which confirms that the cotangent is an odd function; that is,

$$\cot(-x) = -\cot x$$

for all values of x in the domain.

The graph of $g(x) = \sec x$ can be constructed by using the basic identity $\sec x = 1/\cos x$. The vertical asymptotes occur when $\cos x = 0$, that is, when $x = (\pi/2) + n\pi$, for any integer n (Figure 5). As an aid to sketching the graph of

Figure 5

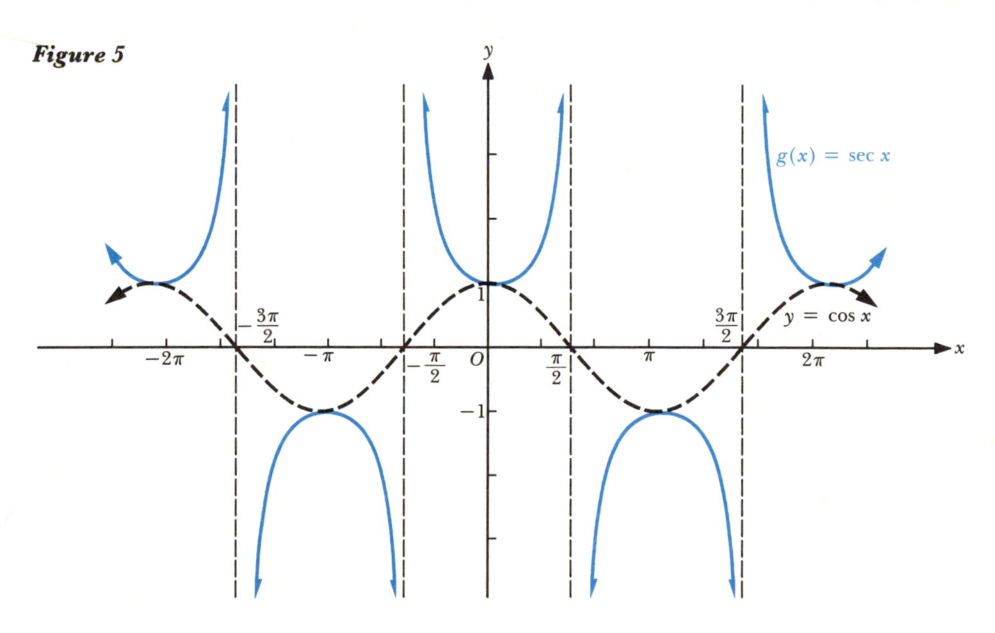

$g(x) = \sec x$, it is helpful to sketch the graph of $y = \cos x$ and then take the reciprocals of the ordinates to obtain points on the secant graph (Figure 5). The graph displays the fact that the range of the secant function includes all y in the intervals $(-\infty, -1]$ and $[1, \infty)$. Notice the manner in which the graph of the secant increases or decreases by observing the behavior of the graph of $y = \cos x$ as it increases or decreases. In addition, the graph indicates that the period of $g(x) = \sec x$ is 2π (Problem 4b).

The graph of the cosecant function $h(x) = \csc x$ may be obtained in a similar fashion by using $\csc x = 1/\sin x$. The vertical asymptotes occur when $\sin x = 0$, or when $x = n\pi$, for any integer n (Figure 6). The graph reveals that the range of the cosecant function includes all y in the intervals $(-\infty, -1]$ or $[1, \infty)$. Also, the period of $h(x) = \csc x$ is 2π (Problem 4b).

Figure 6

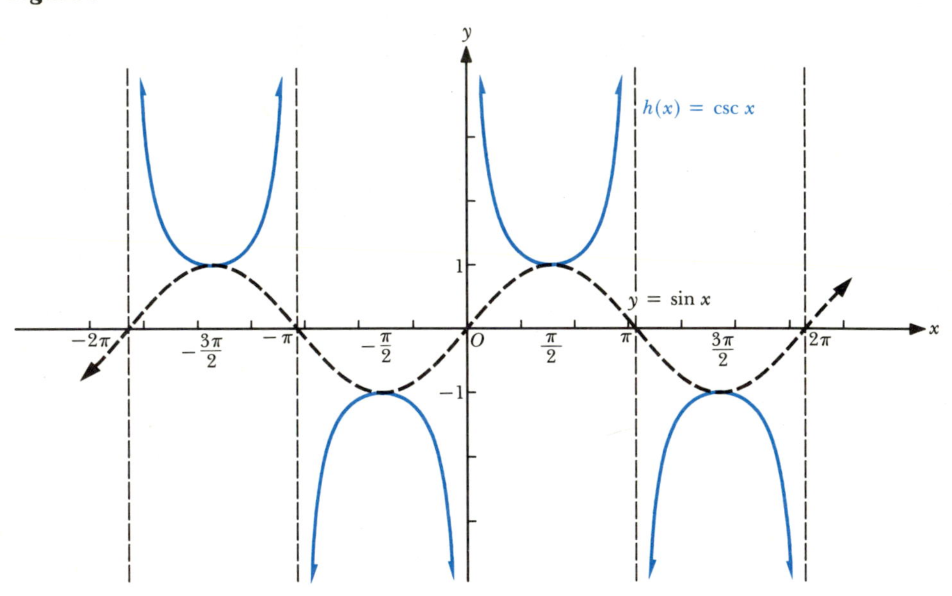

In Examples 1 and 2, sketch one cycle of the graph of each function.

EXAMPLE 1 $f(x) = \tan\left(x - \dfrac{\pi}{6}\right)$

SOLUTION One cycle of the graph of

$$f(x) = \tan\left(x - \frac{\pi}{6}\right)$$

can be obtained by shifting one cycle of the graph of $y = \tan x$ by $\pi/6$ unit to the right (Figure 7). Note that the lines $x = -\pi/3$ and $x = 2\pi/3$ are vertical asymptotes.

Figure 7

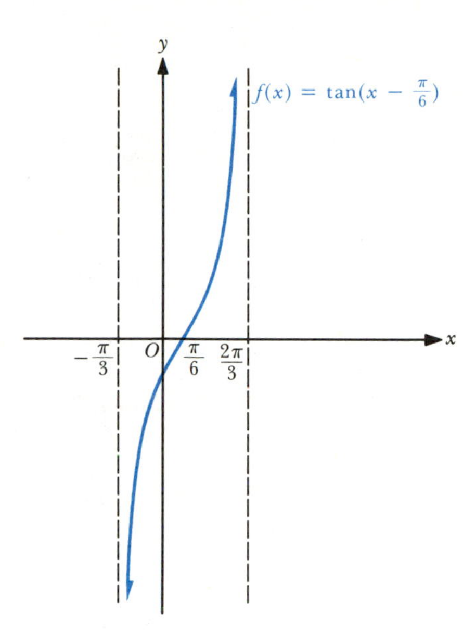

EXAMPLE 2 $g(x) = 2 \sec 5x$

SOLUTION First we graph one cycle of $y = \sec 5x$. Since the period of the secant function is 2π, this function generates one cycle of the secant graph if $5x$ varies

$$\text{from} \quad 5x = -\frac{\pi}{2} \quad \left(x = -\frac{\pi}{10}\right) \quad \text{to} \quad 5x = \frac{3\pi}{2} \quad \left(x = \frac{3\pi}{10}\right)$$

So the period of the function is $\left(\frac{3\pi}{10}\right) - \left(-\frac{\pi}{10}\right) = \frac{2\pi}{5}$

Figure 8

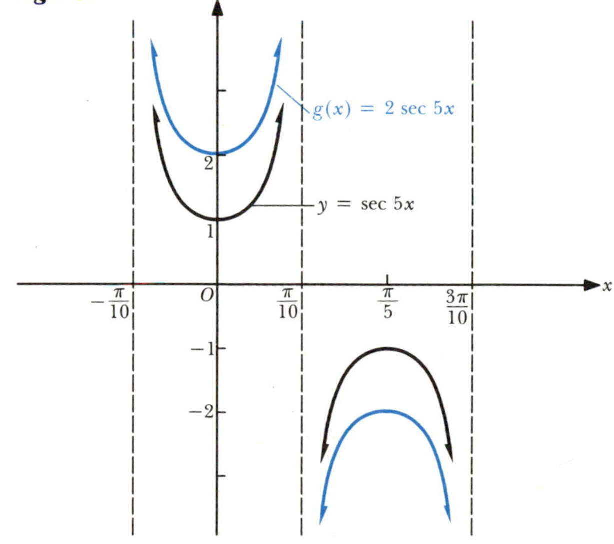

Since $\sec 5x = \dfrac{1}{\cos 5x}$ it follows that the function is not defined if

$$5x = -\frac{\pi}{2} \quad \text{or} \quad 5x = \frac{\pi}{2} \quad \text{or} \quad 5x = \frac{3\pi}{2}$$

and the lines

$$x = -\frac{\pi}{10}, \quad x = \frac{\pi}{10}, \quad \text{and} \quad x = \frac{3\pi}{10}$$

are vertical asymptotes. Finally, one cycle of the graph of $g(x) = 2 \sec 5x$ is obtained from the graph of $y = \sec 5x$ by multiplying the ordinate of each point by 2 (Figure 8). The graph of g suggests that the range includes all numbers in the intervals $(-\infty, -2]$ or $[2, \infty)$.

Graphing Techniques

The following examples show how to graph a trigonometric function that is the sum of two other functions by sketching the graphs of the two functions separately on the same coordinate axis and then adding the ordinates graphically.

In Examples 3 and 4, sketch the graph of each function by adding ordinates.

EXAMPLE 3 $y = \sin x + \cos x$

SOLUTION We begin by sketching the graphs of

$$y_1 = \sin x \qquad \text{and} \qquad y_2 = \cos x$$

on the same coordinate system (Figure 9). Since

$$y = y_1 + y_2 = \sin x + \cos x$$

we can obtain some points on the graph by adding the ordinates y_1 and y_2 for selected values of x (a calculator is useful here). After plotting the resulting points, we obtain the graph by connecting them with a smooth curve (Figure 9).

Figure 9

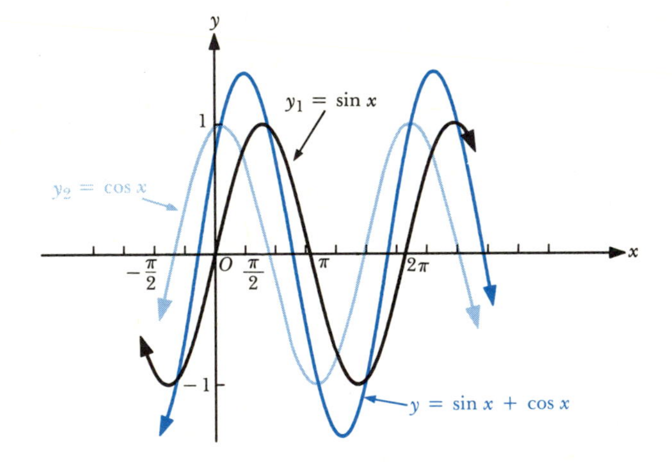

EXAMPLE 4 $y = \cos x + 2 \sin 2x$

SOLUTION We begin by sketching the graphs of

$$y_1 = \cos x \qquad \text{and} \qquad y_2 = 2 \sin 2x$$

on the same coordinate system (Figure 10). Then we add the ordinates y_1 and y_2 for selected values of x to obtain ordinates for the graph of

$$y = \cos x + 2 \sin 2x$$

By plotting the resulting points and connecting them with a smooth curve, we obtain the desired graph (Figure 10).

Figure 10

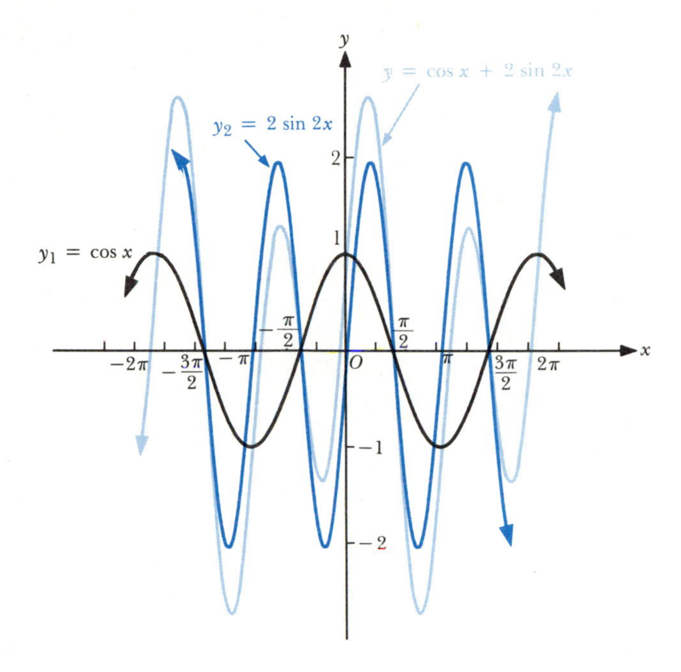

Computers are also used to generate graphs of equations involving trigonometric functions that require advanced techniques. The computer-generated graphs of $y = \sin^4 x$, where $-2 \leq x \leq 6$, and $y = \cos x^2$, where $0 \leq x \leq 6.28$, are displayed in Figures 11a and 11b, respectively.

Figure 11

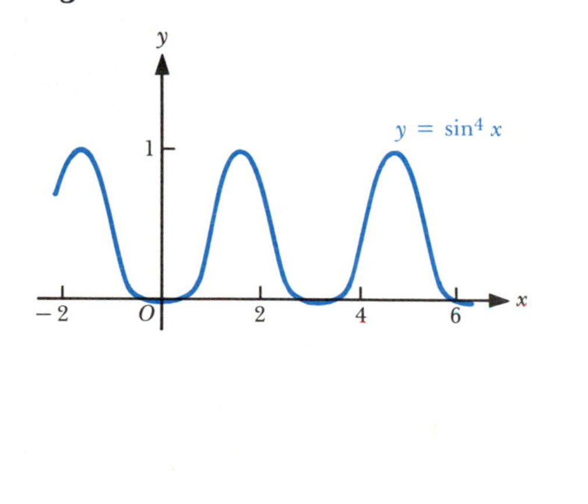

(a)

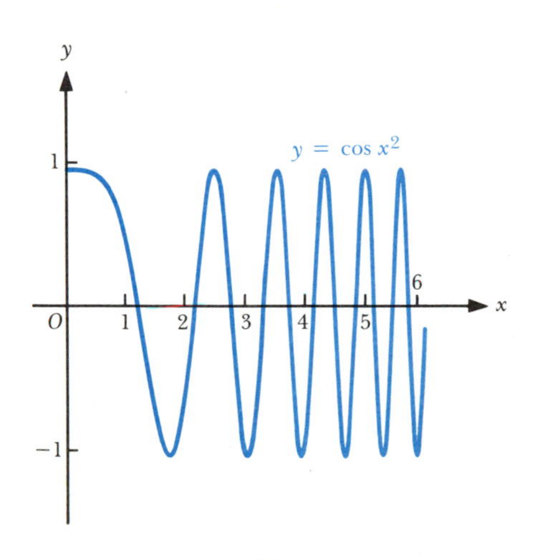

(b)

PROBLEM SET 5.7

In problems 1–3, discuss the behavior of the values of the given function in the interval $(0, 2\pi)$ by using the graph of the corresponding reciprocal function. Display both graphs on the same coordinate system.

(i) As x increases from 0 to $\pi/2$ (ii) As x increases from $\pi/2$ to π
(iii) As x increases from π to $3\pi/2$ (iv) As x increases from $3\pi/2$ to 2π

1. $y = \cot x$; $\cot x = \dfrac{1}{\tan x}$ **2.** $y = \sec x$; $\sec x = \dfrac{1}{\cos x}$ **3.** $y = \csc x$; $\csc x = \dfrac{1}{\sin x}$

4. (a) Use $\cot x = 1/\tan x$, and the fact that the tangent function has period π, to prove that the cotangent function is a periodic function of period π.
(b) Use the reciprocal identities $\sec x = 1/\cos x$ and $\csc x = 1/\sin x$, along with the fact that the cosine and sine functions are periodic functions of period 2π, to prove that the secant and cosecant functions are periodic functions of period 2π.

5. For the graph of $y = \sec x$, what is the distance from one vertical asymptote to the next?

6. (a) Sketch $y = \tan x$ and $y = \cot x$ on the same coordinate system. How are the x intercepts of each of the functions related to the locations of the asymptotes of the other function?
(b) Construct a table of values of $y = \cot x$ for $0 < x < \pi$, and then use the results to graph the function.

7. Use the graphs of the trigonometric functions to complete the following table.

	x increasing from:			
	0 to $\pi/2$	$\pi/2$ to π	π to $3\pi/2$	$3\pi/2$ to 2π
$\tan x$	increasing		increasing	
$\cot x$		decreasing		
$\sec x$				decreasing
$\csc x$				

8. How does the graph of $y = \csc x$ confirm that the cosecant function is an odd function?

In problems 9–22, use the graphs of the tangent, cotangent, secant, and cosecant functions and the techniques of graphing to sketch the graph and identify the asymptotes of each function.

9. $y = 3 \tan x$ **10.** $y = -3 \tan x$ **11.** $y = 1 - 2 \cot x$

12. $y = 2 - \sec x$ **13.** $y = 1 + 2 \sec x$ **14.** $y = 2 \tan \dfrac{\pi x}{4}$

15. $y = 2 \csc\left(x - \dfrac{\pi}{6}\right)$ **16.** $y = 3 \csc\left(x + \dfrac{\pi}{3}\right)$ **17.** $y = \sec(2x - \pi)$

18. $y = \tan\left(x + \frac{2\pi}{3}\right)$ **19.** $y = \tan\left(x - \frac{\pi}{4}\right)$ **20.** $y = -\sec\left(x + \frac{5\pi}{6}\right)$

21. $y = 1.5 \cot\left(x - \frac{\pi}{2}\right)$ **22.** $y = 3 \cot\left(x + \frac{5\pi}{6}\right)$

c In problems 23–26, use a calculator to verify that the equation is true for the indicated value of the variable.

23. $\tan(-x) = -\tan x;\ x = 1.734$ **24.** $\cot(-x) = -\cot x;\ x = 0.7532$

25. $\csc(-x) = -\csc x;\ x = 0.3544$ **26.** $\sec(-x) = \sec x;\ x = \frac{5\pi}{12}$

In problems 27–34, use the addition or subtraction of ordinates to sketch each graph, for $0 \leq x \leq 2\pi$.

27. $y = \sin x + 2\cos x$ **28.** $y = 2\sin x + 3\cos x$

29. $y = \cos x - 3\sin 2x$ **30.** $y = \sin x - 3\cos x$

31. $y = \sin x + \cos 2x$ **32.** $y = 2\cos x + \cos\frac{x}{3}$

33. $y = -2\cos x + \sin x$ **34.** $y = 2\sin x - 2\cos\frac{x}{2}$

REVIEW PROBLEM SET, CHAPTER 5

In problems 1–6, display the location of each point on the unit circle and determine the quadrant, if any, in which each point lies (use $\pi = 3.14$).

1. $P\left(-\frac{3\pi}{2}\right)$ **2.** $P\left(\frac{2\pi}{5}\right)$ **3.** $P\left(-\frac{7\pi}{5}\right)$

4. $P(3.31)$ **5.** $P(-2.32)$ **6.** $P(-1.19)$

In problems 7–10, for the given point $P(t)$ on a unit circle, use Definition 1 on page 239 to determine the values of the trigonometric functions of t.

7. $P(t) = \left(-\frac{\sqrt{3}}{2}, -\frac{1}{2}\right)$ **8.** $P(t) = \left(\frac{4}{\sqrt{17}}, \frac{1}{\sqrt{17}}\right)$ **9.** $P(t) = \left(\frac{3}{\sqrt{10}}, -\frac{1}{\sqrt{10}}\right)$

10. $P(t) = \left(\frac{3}{\sqrt{58}}, \frac{7}{\sqrt{58}}\right)$

In problems 11–18, convert each degree measure to radian measure.

11. 290° **12.** -105° **13.** 820° **14.** 36°

15. -70° **16.** -12° **17.** 190° **18.** -40°

In problems 19–26, convert each radian measure to degree measure (use $\pi = 3.14$).

19. $\frac{5\pi}{3}$ **20.** $\frac{17\pi}{5}$ **21.** $-\frac{13\pi}{4}$ **22.** $-\frac{4\pi}{5}$

C **23.** $-\frac{18\pi}{7}$ C **24.** -4.7 C **25.** 5.82 C **26.** -4.81

In problems 27–30, find the length of the circular arc and the area of the circular sector that are determined by a central angle θ in a circle of radius r. (See Problem 36 on page 256.)

27. $r = 7$ centimeters and $\theta = 75°$ **28.** $r = 2$ inches and $\theta = \frac{8\pi}{9}$

29. $r = 10$ inches and $\theta = \frac{\pi}{5}$ **30.** $r = 4$ centimeters and $\theta = 340°$

31. A circle of radius 4 inches contains a central angle θ that intercepts an arc of 10 inches. Express θ in radian measure and degree measure. Find the area of the sector whose central angle is θ.

32. A bicycle with wheels of radius 13 inches travels 52 feet. What is the radian measure of the angle through which a wheel of the bicycle has turned?

33. A point revolves in a circular path making 12 complete revolutions in 6 seconds. Find the linear speed and the angular speed of the point if the radius of the circle is 3 centimeters.

34. Suppose that two objects are moving along the circumference of a circle of radius 1 meter. One object is moving counterclockwise at a rate of 0.5 meter per second, whereas the other object is moving clockwise at a rate of 0.3 meter per second. How long will it take the two objects to collide if they both start at the same position?

In problems 35–40, evaluate the six trigonometric function values of θ if θ is in standard position and its terminal side contains the given point (x, y). In problems 39 and 40, round off the answers to two decimal places.

35. $(x, y) = (-6, -8)$ **36.** $(x, y) = (-7, 24)$ **37.** $(x, y) = (5, -6)$

38. $(x, y) = (-8, -17)$ C **39.** $(x, y) = (-3.2, 4.7)$ C **40.** $(x, y) = (2.7, -4.3)$

41. Find the six trigonometric function values of θ if $|\overline{AB}| = 25$ feet and also $|\overline{AC}| = 24$ feet in right triangle ABC in Figure 1.

42. Refer to Figure 1, and find x if $\sin\theta = \frac{7}{8}$ and $|\overline{AC}| = 4$ centimeters.

Figure 1

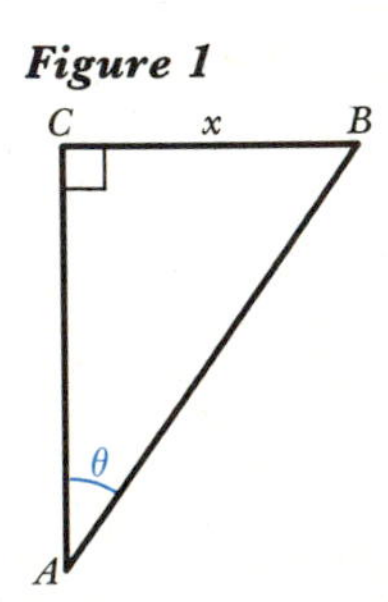

In problems 43–62, use special values to find the exact value of each expression. Do not use a calculator.

43. $\sec\left(-\frac{5\pi}{3}\right)$ **44.** $\cot\frac{37\pi}{6}$

45. $\cos\left(-\frac{47\pi}{6}\right)$ **46.** $\sin\frac{89\pi}{2}$

47. $\tan \frac{14\pi}{3}$ **48.** $\csc\left(-\frac{25\pi}{4}\right)$ **49.** $\cos\left(-\frac{125\pi}{6}\right)$ **50.** $\sec \frac{325\pi}{3}$

51. $\tan \frac{721\pi}{4}$ **52.** $\cos 780°$ **53.** $\tan(-240°)$ **54.** $\cot(-210°)$

55. $\csc(-750°)$ **56.** $\sin 1230°$ **57.** $\sec 1380°$ **58.** $\cos(-585°)$

59. $\tan(-3720°)$ **60.** $\csc(-900°)$ **61.** $\cos(-495°)$ **62.** $\csc 1845°$

In problems 63 and 64, determine the quadrant in which the terminal side of θ lies, assuming that θ is in standard position.

63. $\sin\theta < 0$ and $\cos\theta < 0$

64. $\sec\theta > 0$ and $\cot\theta < 0$

In problems 65–74, use the basic identities to find the values of the other five trigonometric functions under the given conditions. In problems 69 and 70, round off the answers to two decimal places.

65. $\sin\theta = \frac{2}{3}$, θ in quadrant II

66. $\cos\theta = \frac{3}{8}$, θ in quadrant IV

67. $\csc\theta = 5$, θ in quadrant I

68. $\tan\theta = -\frac{5}{12}$, θ in quadrant IV

c **69.** $\sec\theta = -2.61$, θ in quadrant III

c **70.** $\cot\theta = -3.71$, θ in quadrant II

71. $\tan\theta = \frac{6}{5}$, θ in quadrant III

72. $\sec\theta = \frac{5}{4}$, θ in quadrant IV

73. $\cot\theta = -\frac{25}{7}$, θ in quadrant II

74. $\sin\theta = \frac{12}{13}$, θ in quadrant I

c In problems 75–92, use a calculator to approximate the value of the given trigonometric function to five decimal places.

75. $\sin 3$ **76.** $\sin(-208°)$ **77.** $\cos \frac{3\pi}{5}$

78. $\cos(-3.92)$ **79.** $\tan(-209°40')$ **80.** $\tan 8.2$

81. $\cot 604°20'$ **82.** $\cot(-26.37)$ **83.** $\sec 18.71$

84. $\sec(-407°30')$ **85.** $\csc(-26.37)$ **86.** $\csc 248°5'$

87. $\sin 5.115$ **88.** $\cot 658°23'$ **89.** $\cos 743°56'$

90. $\sec(-19.213)$ **91.** $\tan(-241°28')$ **92.** $\tan \frac{7\pi}{9}$

In problems 93–96, find the reference angle for each angle.

93. $246°$ **94.** $148°10'$ **95.** $\frac{9\pi}{11}$ **96.** $-\frac{9\pi}{8}$

In problems 97–102, use the known graphs of the trigonometric functions to graph one cycle of each function. Indicate the amplitude, the period, and the phase shift.

97. $f(x) = \frac{1}{3}\cos x$

98. $g(x) = \sin \frac{\pi}{3}x$

99. $h(x) = 2\sin(-2x)$

100. $F(x) = \frac{1}{4}\sin\left(x - \frac{\pi}{6}\right)$

101. $G(x) = \frac{1}{5}\cos\left(2x + \frac{\pi}{8}\right)$

102. $f(x) = -2 + \sin x$

In problems 103–106, sketch the graph of each function by using the techniques of graphing.

103. $f(x) = \frac{3}{2} \cot x$

104. $g(x) = 4 \csc(x - 1)$

105. $h(x) = 3 \sec 4x$

106. $H(x) = -2 \tan\left(\frac{x}{2} - 1\right)$

In problems 107 and 108, use the addition or subtraction of ordinates to sketch each graph, for $0 \leq x \leq 2\pi$.

107. $y = 2 \cos x + 3 \cos \frac{x}{2}$

108. $y = 2 \sin x - \cos x$

109. A wheel that is 10 centimeters in diameter is driven by a belt with a linear speed of 50 centimeters per second. How many revolutions per minute is the wheel turning?

110. The 30-inch-diameter tires of a truck are turning at 300 revolutions per minute. What is the linear speed of the truck in miles per hour?

c **111.** A body moves in simple harmonic motion according to the equation given by $y = 2 \cos(0.05t - 3)$. Find

(a) the amplitude
(b) the angular frequency
(c) the phase angle
(d) the period
(e) the phase shift
(f) Sketch one cycle of the graph.

c **112.** An iron ball of mass 10 kilograms is attached to a spring that oscillates freely in a simple harmonic motion defined by the equation

$$y = 2 \sin \sqrt{\frac{k}{m}}\, t$$

Find the period and frequency of oscillation if $k = 4$ and m represents the mass.

113. A simple harmonic motion has a frequency of 0.55 oscillation per minute and an amplitude of 12 centimeters. Express the motion by means of an equation of the form $y = a \sin \omega t$.

CHAPTER 6

Analytic Trigonometry

In Chapter 5, we used the definitions of the trigonometric functions to derive the basic identities. These identities are used in this chapter to establish other useful trigonometric identities and formulas for *sums, differences,* and *multiples* of angles. They are then used in a variety of applications to simplify trigonometric expressions and to solve trigonometric equations. In this chapter, we also examine the inverse trigonometric functions and some interesting applications involving right triangles; and we use the laws of sines and cosines to solve other triangles. The chapter concludes with a discussion of polar coordinates, trigonometric forms of complex numbers, and vectors in the plane.

6.1 Fundamental Identities

The basic trigonometric identities, derived in Chapter 5, relate the trigonometric functions to one another. These basic identities, which hold for all values in the domains of the functions involved, are called the **fundamental identities.** They will be used throughout the book and should be memorized. They are also listed in the back inside cover of this book.

1. $\sec\theta = \dfrac{1}{\cos\theta}$
2. $\csc\theta = \dfrac{1}{\sin\theta}$
3. $\cot\theta = \dfrac{1}{\tan\theta}$
4. $\tan\theta = \dfrac{\sin\theta}{\cos\theta}$
5. $\cot\theta = \dfrac{\cos\theta}{\sin\theta}$
6. $\sin^2\theta + \cos^2\theta = 1$
7. $\tan^2\theta + 1 = \sec^2\theta$
8. $\cot^2\theta + 1 = \csc^2\theta$
9. $\sin(-\theta) = -\sin\theta$
10. $\cos(-\theta) = \cos\theta$
11. $\tan(-\theta) = -\tan\theta$
12. $\cot(-\theta) = -\cot\theta$
13. $\sec(-\theta) = \sec\theta$
14. $\csc(-\theta) = -\csc\theta$

Our use of the basic identities in Chapter 5 was, for the most part, restricted to the evaluations of the trigonometric functions. These identities are also often used to change trigonometric expressions from one form to another. The following examples provide us with illustrations of how we can use the identities, along with basic algebra, to manipulate trigonometric expressions and convert them to other forms. As before, the variable may be regarded as either a *real number* or an *angle* in *degree* or *radian measure.*

EXAMPLE 1 Show that each expression is equal to one.

(a) $\sin^2 31t + \cos^2 31t$ (b) $\csc^2\left(\frac{u}{3} + c\right) - \cot^2\left(\frac{u}{3} + c\right)$

SOLUTION (a) Since $\sin^2\theta + \cos^2\theta = 1$, then, by letting $\theta = 31t$, we get

$$\sin^2 31t + \cos^2 31t = 1$$

(b) Since $\cot^2\theta + 1 = \csc^2\theta$, then $\csc^2\theta - \cot^2\theta = 1$, so that

$$\csc^2\left(\frac{u}{3} + c\right) - \cot^2\left(\frac{u}{3} + c\right) = 1$$

EXAMPLE 2 Write each expression in terms of $\sin\theta$.

(a) $\dfrac{\sec\theta}{\tan\theta + \cot\theta}$ (b) $(\tan\theta + \cot\theta)\cot\theta$

SOLUTION (a)
$$\frac{\sec\theta}{\tan\theta + \cot\theta} = \frac{\frac{1}{\cos\theta}}{\frac{\sin\theta}{\cos\theta} + \frac{\cos\theta}{\sin\theta}} = \frac{\frac{1}{\cos\theta}}{\frac{\sin^2\theta + \cos^2\theta}{\sin\theta\cos\theta}} = \frac{\frac{1}{\cos\theta}}{\frac{1}{\sin\theta\cos\theta}}$$
$$= \frac{\sin\theta\cos\theta}{\cos\theta} = \sin\theta$$

(b)
$$(\tan\theta + \cot\theta)\cot\theta = \left(\frac{1}{\cot\theta} + \cot\theta\right)\cot\theta = \left(\frac{1 + \cot^2\theta}{\cot\theta}\right)\cot\theta$$
$$= 1 + \cot^2\theta = \csc^2\theta = \frac{1}{\sin^2\theta}$$

EXAMPLE 3 Write each expression in terms of $\sin t$ or $\cos t$.

(a) $(\sec t + \tan t)(1 - \sin t)$ (b) $\sin^4 t - \cos^4 t + \cos^2 t$

SOLUTION (a)
$$(\sec t + \tan t)(1 - \sin t) = \left(\frac{1}{\cos t} + \frac{\sin t}{\cos t}\right)(1 - \sin t)$$
$$= \left(\frac{1 + \sin t}{\cos t}\right)(1 - \sin t)$$
$$= \frac{1 - \sin^2 t}{\cos t} = \frac{\cos^2 t}{\cos t} = \cos t$$

(b) $$\begin{aligned}\sin^4 t - \cos^4 t + \cos^2 t &= (\sin^2 t)^2 - \cos^4 t + \cos^2 t\\ &= (1 - \cos^2 t)^2 - \cos^4 t + \cos^2 t\\ &= 1 - 2\cos^2 t + \cos^4 t - \cos^4 t + \cos^2 t\\ &= 1 - \cos^2 t = \sin^2 t\end{aligned}$$

The results in Examples 1 to 3 are considered to be simplified forms of the original expressions. Normally, to **simplify a trigonometric expression** means to write it as a constant or as a trigonometric expression containing only one of the trigonometric functions.

In calculus, algebraic expressions of the form

$$\sqrt{a^2 - u^2}, \qquad \sqrt{u^2 + a^2}, \qquad \text{and} \qquad \sqrt{u^2 - a^2}$$

where a is a positive constant, can be converted to more useful forms by making appropriate substitutions involving trigonometric functions. This technique is called **trigonometric substitution.** The idea is to write these types of radical expressions as trigonometric expressions containing no radical.

EXAMPLE 4 Write $\sqrt{25 - u^2}$ as a trigonometric expression containing no radical by using the trigonometric substitution $u = 5 \sin\theta$. Assume that θ is an acute angle.

SOLUTION We know that $\cos\theta > 0$ for θ an acute angle, so that

$$\begin{aligned}\sqrt{25 - u^2} &= \sqrt{25 - (5\sin\theta)^2} = \sqrt{25 - 25\sin^2\theta}\\ &= \sqrt{25(1 - \sin^2\theta)} = \sqrt{25\cos^2\theta}\\ &= 5\cos\theta\end{aligned}$$

Verifying Trigonometric Identities

The fundamental identities can be used to prove other identities. There is no general rule for proving that a trigonometric equation is an identity; however, we normally start with one side of the equation and try to convert it to the other side by means of a sequence of algebraic manipulations and substitutions utilizing known identities. Often it is helpful to start with the side containing more terms and try to convert it to the *simpler* side.

Here are some useful suggestions that may help in carrying out the proof.

1. Combine a sum or difference of fractions into a single fraction.

2. Reduce a fraction.

3. Factor the expression.

4. Combine like terms.

5. Multiply both the numerator and denominator by the same expression.

6. Write all trigonometric expressions in terms of sines and cosines, and then simplify.

In Examples 5–9, prove that each equation is an identity.

EXAMPLE 5 $\cos^2\theta(1 + \tan^2\theta) = 1$

SOLUTION Starting with the left side of the equation we obtain

$$\begin{aligned}\cos^2\theta\,(1 + \tan^2\theta) &= \cos^2\theta \cdot \sec^2\theta = (\cos\theta\sec\theta)^2\\ &= \left(\cancel{\cos\theta}\cdot\frac{1}{\cancel{\cos\theta}}\right)^2 = 1^2 = 1\end{aligned}$$

EXAMPLE 6 $\csc^2\theta + \sec^2\theta = \sec^2\theta\csc^2\theta$

SOLUTION Here we use the right side of the equation to get

$$\begin{aligned}\sec^2\theta\csc^2\theta &= (1 + \tan^2\theta)(1 + \cot^2\theta)\\ &= 1 + \cot^2\theta + \tan^2\theta + \tan^2\theta\cot^2\theta\\ &= 1 + \cot^2\theta + \tan^2\theta + \cancel{\tan^2\theta}\cdot\frac{1}{\cancel{\tan^2\theta}}\\ &= (1 + \cot^2\theta) + (\tan^2\theta + 1) = \csc^2\theta + \sec^2\theta\end{aligned}$$

EXAMPLE 7 $1 - \tan^4 t = 2\sec^2 t - \sec^4 t$

SOLUTION We start by factoring the left side of the equation to get

$$\begin{aligned}1 - \tan^4 t &= (1 - \tan^2 t)(1 + \tan^2 t) = (1 - \tan^2 t)\sec^2 t\\ &= [1 - (\sec^2 t - 1)]\sec^2 t = (1 - \sec^2 t + 1)\sec^2 t\\ &= (2 - \sec^2 t)\sec^2 t = 2\sec^2 t - \sec^4 t\end{aligned}$$

EXAMPLE 8 $\dfrac{1}{\sec\theta - \tan\theta} - \dfrac{1}{\sec\theta + \tan\theta} = 2\tan\theta$

SOLUTION Here we start with the left side and combine the two fractions to obtain

$$\begin{aligned}\frac{1}{\sec\theta - \tan\theta} - \frac{1}{\sec\theta + \tan\theta} &= \frac{(\sec\theta + \tan\theta) - (\sec\theta - \tan\theta)}{(\sec\theta - \tan\theta)(\sec\theta + \tan\theta)}\\ &= \frac{\sec\theta + \tan\theta - \sec\theta + \tan\theta}{\sec^2\theta - \tan^2\theta}\\ &= \frac{2\tan\theta}{1} = 2\tan\theta\end{aligned}$$

EXAMPLE 9 $\dfrac{\cos(-t)}{1 - \sin t} = \dfrac{1 - \sin(-t)}{\cos t}$

SOLUTION Since $\cos(-t) = \cos t$ and $\sin(-t) = -\sin t$, it is enough to prove that

$$\frac{\cos t}{1 - \sin t} = \frac{1 + \sin t}{\cos t}$$

We start with the left side of the latter equation and multiply the numerator and the denominator of the expression by $1 + \sin t$ to obtain

$$\begin{aligned}\frac{\cos t}{1 - \sin t} &= \frac{\cos t}{1 - \sin t} \cdot \frac{(1 + \sin t)}{(1 + \sin t)}\\ &= \frac{\cos t\,(1 + \sin t)}{1 - \sin^2 t}\\ &= \frac{\cos t\,(1 + \sin t)}{\cos^2 t}\\ &= \frac{1 + \sin t}{\cos t}\end{aligned}$$

It is important to remember that a trigonometric identity is true only for angles or real numbers in the domains of *all* of the trigonometric functions contained in the identity and for which each expression is defined. For instance, note that the identity in Example 9 is not defined when $t = \pi/2$ because this value would yield a zero in the denominator of either side of the equation.

PROBLEM SET 6.1

In problems 1–28, use the fundamental identities to simplify each expression.

1. $\sec^2 5t - \tan^2 5t$

2. $\csc^2 7\theta - \cot^2 7\theta$

3. $\sin\theta(\csc\theta - \sin\theta)$

4. $\tan\theta \sin\theta + \cos\theta$

5. $\sin^2\left(\dfrac{\theta}{7} + c\right) + \cos^2\left(\dfrac{\theta}{7} + c\right)$

6. $(\sin\theta + \cos\theta)^2 - 2\sin\theta\cos\theta$

7. $(\sin\theta - \cos\theta)^2 + 2\sin\theta\cos\theta$

8. $\csc\theta\sec\theta - \cot\theta$

9. $\sin\pi t \cot\pi t - \cos\pi t$

10. $\tan\theta + \cot\theta - \sec\theta\csc\theta$

11. $\sin^4\theta + 2\sin^2\theta\cos^2\theta + \cos^4\theta$

12. $\cos^4\theta - \sin^4\theta$

13. $(1 - \cos\theta)(1 + \cos\theta)$

14. $\cos\theta(\tan\theta + \cot\theta)$

15. $(\sin^2 100t + \cos^2 100t)^3$

16. $(\csc\theta - \cot\theta)^4(\csc\theta + \cot\theta)^4$

17. $\dfrac{1 - \cos^2(-t)}{\sin t}$

18. $\dfrac{\sec^2(-\theta) - 1}{\sec^2\theta}$

19. $\dfrac{1 + \tan^2\theta}{\sec\theta}$

20. $\dfrac{\sec t \csc t}{\tan t + \cot t}$

21. $\dfrac{\cos t + \sin^2 t \sec t}{\sec t}$

22. $\dfrac{\cos\theta \csc\theta}{\csc^2\theta - 1}$

23. $\dfrac{\csc\theta}{\tan\theta + \cot\theta}$

24. $1 + \dfrac{\tan^2\theta}{1 + \sec\theta}$

25. $\dfrac{1}{1 - \cos\theta} + \dfrac{1}{\cos\theta}$

26. $\dfrac{1}{1 + \tan\theta} - \dfrac{\cot\theta}{1 + \cot\theta}$

27. $\dfrac{\sin t}{1 + \cos t} + \dfrac{1 + \cos t}{\sin t}$

28. $\dfrac{\cos\alpha}{1 - \sin\alpha} + \dfrac{\cos\alpha}{1 + \sin\alpha}$

In problems 29–34, write each expression in terms of sines and/or cosines.

29. $\cos\theta + \tan\theta\sin\theta$

30. $\dfrac{1}{(\cot\theta + \tan\theta)\cos\theta}$

31. $\dfrac{\cos\theta + \sin^2(-\theta)\sec\theta}{\csc\theta}$

32. $(\sec t + \csc t)^2 \cot t$

33. $\dfrac{\tan t}{\sin t(1 + \tan^2 t)}$

34. $\dfrac{\sec(-t)}{\csc(-t)[\tan(-t) + \cot(-t)]}$

In problems 35–44, rewrite the given expression as a trigonometric expression containing no radical by using the given trigonometric substitution. Assume that θ is an acute angle.

35. $\sqrt{4 - u^2}$; $u = 2\sin\theta$

36. $\sqrt{u^2 + 4}$; $u = 2\tan\theta$

37. $\sqrt{u^2 + 9}$; $u = 3\tan\theta$

38. $\sqrt{64 - u^2}$; $u = 8\sin\theta$

39. $\sqrt{u^2 - 25}$; $u = 5\sec\theta$

40. $\sqrt{u^2 - 81}$; $u = 9\sec\theta$

41. $(u^2 + 9)^{3/2}$; $u = 3\tan\theta$

42. $(u^2 - 25)^{3/2}$; $u = 5\sec\theta$

43. $\sqrt{9u^2 + 4}$; $3u = 2\tan\theta$

44. $\sqrt{4 - 9u^2}$; $3u = 2\sin\theta$

In problems 45–76, show that each trigonometric equation is an identity.

45. $\sin^2\theta(1 + \cot^2\theta) = 1$

46. $\sin t \cot t \sec t = 1$

47. $\tan t \cos t = \sin t$

48. $\sin t \cot t = \cos t$

49. $\sec^2\theta(1 - \sin^2\theta) = 1$

50. $\cot^2\theta(\sec^2\theta - 1) = 1$

51. $\sin t(\csc t - \sin t) = \cos^2 t$

52. $\sin t \cos t \cot t = 1 - \sin^2 t$

53. $\cos^2\theta - \sin^2\theta = 2\cos^2\theta - 1$

54. $\cos^4\theta - \sin^4\theta = 1 - 2\sin^2\theta$

55. $\sec(-\theta)\cot(-\theta) = -\csc\theta$

56. $\csc(-t)\tan(-t) = \sec t$

57. $\cos^3\theta + \cos\theta\sin^2\theta = \cos\theta$

58. $(\tan\theta + \cot\theta)^2 = \sec^2\theta\csc^2\theta$

59. $(\cos t - \sin t)^2 + (\cos t + \sin t)^2 = 2$

60. $(a\cos t + b\sin t)^2 + (-a\sin t + b\cos t)^2 = a^2 + b^2$

61. $\dfrac{\sin\theta}{\tan\theta} + \dfrac{\cos\theta}{\cot\theta} = \cos\theta + \sin\theta$

62. $\dfrac{\sin\theta}{\cot\theta + \csc\theta} - \dfrac{\sin\theta}{\cot\theta - \csc\theta} = 2$

63. $\dfrac{\sec\theta + 1}{\tan\theta} + \dfrac{\tan\theta}{\sec\theta + 1} = 2\csc\theta$

64. $\dfrac{1 - \sin\theta}{1 - \sec\theta} - \dfrac{1 + \sin\theta}{1 + \sec\theta} = 2\cot\theta(\cos\theta - \csc\theta)$

65. $\dfrac{\sin t}{1 + \cos(-t)} + \dfrac{\sin t}{1 - \cos(-t)} = -2\csc(-t)$

66. $-2\csc(-\theta) - \dfrac{\sin\theta}{1 + \cos(-\theta)} = \dfrac{1 + \cos(-\theta)}{-\sin(-\theta)}$

67. $\dfrac{1}{\cos^2\theta} + 1 + \dfrac{\sin^2\theta}{\cos^2\theta} = 2\sec^2\theta$

68. $\dfrac{1 + \tan t}{\sin t} - \sec t = \csc t$

69. $\dfrac{\sin t}{1 - \cos t} = \csc t + \cot t$

70. $\dfrac{\sin(-t)}{1 - \cos(-t)} = \dfrac{1 + \cos(-t)}{\sin(-t)}$

71. $\csc t + \cot t = \dfrac{1}{\csc t - \cot t}$

72. $\tan^2 \theta - \sin^2 \theta = \dfrac{\sin^4 \theta}{\cos^2 \theta}$

73. $\dfrac{\cos \theta \cot \theta}{\cot \theta - \cos \theta} = \dfrac{\cot \theta + \cos \theta}{\cos \theta \cot \theta}$

74. $\dfrac{\tan \theta + \cot \theta}{\tan \theta - \cot \theta} = \dfrac{\sec^2 \theta}{\tan^2 \theta - 1}$

75. $\dfrac{\sin \theta}{\csc \theta - \cot \theta} = 1 + \cos \theta$

76. $(\sec \theta - \tan \theta)^2 = \dfrac{1 - \sin \theta}{1 + \sin \theta}$

6.2 Sum, Difference, and Related Trigonometric Formulas

In this section, we derive formulas involving trigonometric functions of a *sum* $t + s$ and a *difference* $t - s$, where s and t represent angles or real numbers.

Suppose that we want to express the value of $\cos(t + s)$ in terms of values of trigonometric functions of t and s. One *might be tempted* to say that $\cos(t + s)$ is the same as $\cos t + \cos s$. To check whether or not this is true, we compare the values of the expressions $\cos(30° + 60°)$ and $\cos 30° + \cos 60°$ (Table 1).

Table 1

$\cos(30° + 60°)$ and $\cos 30° + \cos 60°$	
$\cos(30° + 60°)$	$\cos 30° + \cos 60°$
$= \cos 90°$	$= \dfrac{\sqrt{3}}{2} + \dfrac{1}{2}$
$= 0$	$= \dfrac{\sqrt{3} + 1}{2}$

Clearly, the results are different. The next question is: Can we find *general* formulas that give $\cos(t + s)$ and $\cos(t - s)$ in terms of trigonometric functions of s and t? The answer to this question is included in the next theorem.

THEOREM 1 **Cosine of a Difference or Sum**

(i) $\cos(t - s) = \cos t \cos s + \sin t \sin s$
(ii) $\cos(t + s) = \cos t \cos s - \sin t \sin s$

PROOF

(i) Let s and t represent angles in standard position whose terminal sides intersect the unit circle at $P_1 = (\cos s, \sin s)$ and $P_2 = (\cos t, \sin t)$, respectively. For convenience, we illustrate the case where $0 < s < \pi/2 < t < \pi$ (Figure 1a). However, the results are true for all possible values of s and t.

Figure 1

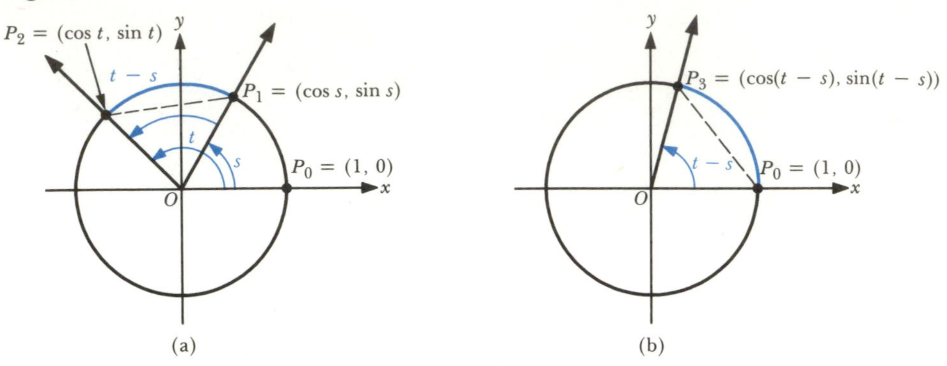

After rotating the angle $t - s$ (Figure 1a) until it is in standard position (Figure 1b), we see that the coordinates of the point of intersection of the terminal side of the angle $t - s$ and the unit circle are given by

$$P_3 = \big(\cos(t-s), \sin(t-s)\big)$$

If $P_0 = (1, 0)$, then the lengths of arcs $\widehat{P_0P_3}$ and $\widehat{P_1P_2}$ are equal. It follows from geometry that the corresponding chord lengths are equal, that is,

$$|\overline{P_0P_3}| = |\overline{P_1P_2}|$$

From the distance formula, we have

$$\begin{aligned}
|\overline{P_0P_3}|^2 &= [\cos(t-s) - 1]^2 + [\sin(t-s) - 0]^2 \\
&= \cos^2(t-s) - 2\cos(t-s) + 1 + \sin^2(t-s) \\
&= \cos^2(t-s) + \sin^2(t-s) + 1 - 2\cos(t-s) \\
&= 1 + 1 - 2\cos(t-s) \\
&= 2 - 2\cos(t-s)
\end{aligned}$$

and

$$\begin{aligned}
|\overline{P_1P_2}|^2 &= (\cos t - \cos s)^2 + (\sin t - \sin s)^2 \\
&= \cos^2 t - 2\cos t\cos s + \cos^2 s + \sin^2 t - 2\sin t\sin s + \sin^2 s \\
&= \cos^2 t + \sin^2 t + \cos^2 s + \sin^2 s - 2\cos t\cos s - 2\sin t\sin s \\
&= 1 + 1 - 2\cos t\cos s - 2\sin t\sin s \\
&= 2 - 2\cos t\cos s - 2\sin t\sin s
\end{aligned}$$

Equating $|\overline{P_0P_3}|^2$ with $|\overline{P_1P_2}|^2$, we get

$$\begin{aligned}
2 - 2\cos(t-s) &= 2 - 2\cos t\cos s - 2\sin t\sin s \\
-2\cos(t-s) &= -2\cos t\cos s - 2\sin t\sin s \\
\cos(t-s) &= \cos t\cos s + \sin t\sin s
\end{aligned}$$

(ii) $\cos(t+s) = \cos[t - (-s)]$

$= \cos t\cos(-s) + \sin t\sin(-s)$ [Theorem 1(i)]

Since $\cos(-s) = \cos s$ and $\sin(-s) = -\sin s$, we have

$$\cos(t + s) = \cos t \cos s - \sin t \sin s$$

EXAMPLE 1 Use Theorem 1 to find the exact value of each expression.

(a) $\cos 15°$ (b) $\cos \dfrac{7\pi}{12}$

SOLUTION (a) By Theorem 1(i), we have

$$\begin{aligned}\cos 15° &= \cos(45° - 30°) \\ &= \cos 45° \cos 30° + \sin 45° \sin 30° \\ &= \frac{\sqrt{2}}{2} \cdot \frac{\sqrt{3}}{2} + \frac{\sqrt{2}}{2} \cdot \frac{1}{2} = \frac{\sqrt{6}}{4} + \frac{\sqrt{2}}{4} = \frac{\sqrt{6} + \sqrt{2}}{4}\end{aligned}$$

(b) By Theorem 1(ii), we have

$$\begin{aligned}\cos \frac{7\pi}{12} &= \cos\left(\frac{\pi}{3} + \frac{\pi}{4}\right) \\ &= \cos \frac{\pi}{3} \cos \frac{\pi}{4} - \sin \frac{\pi}{3} \sin \frac{\pi}{4} \\ &= \frac{1}{2} \cdot \frac{\sqrt{2}}{2} - \frac{\sqrt{3}}{2} \cdot \frac{\sqrt{2}}{2} = \frac{\sqrt{2}}{4} - \frac{\sqrt{6}}{4} = \frac{\sqrt{2} - \sqrt{6}}{4}\end{aligned}$$

Similar sum and difference formulas can be derived for the sine function. To derive them we first establish the following identities.

THEOREM 2 **Cofunction Identities**

If t is a real number or the radian measure of an angle, then

(i) $\cos\left(\dfrac{\pi}{2} - t\right) = \sin t$ (ii) $\sin\left(\dfrac{\pi}{2} - t\right) = \cos t$ (iii) $\tan\left(\dfrac{\pi}{2} - t\right) = \cot t$

PROOF (i) By Theorem 1(i), we have

$$\cos\left(\frac{\pi}{2} - t\right) = \cos \frac{\pi}{2} \cos t + \sin \frac{\pi}{2} \sin t = 0 + \sin t = \sin t$$

(ii) In part (i), written from right to left, we replace t by $\pi/2 - t$ to obtain

$$\begin{aligned}\sin\left(\frac{\pi}{2} - t\right) &= \cos\left[\frac{\pi}{2} - \left(\frac{\pi}{2} - t\right)\right] \\ &= \cos\left(\frac{\pi}{2} - \frac{\pi}{2} + t\right) = \cos t\end{aligned}$$

(iii) By parts (i) and (ii), we have

$$\tan\left(\frac{\pi}{2} - t\right) = \frac{\sin\left(\frac{\pi}{2} - t\right)}{\cos\left(\frac{\pi}{2} - t\right)} = \frac{\cos t}{\sin t} = \cot t$$

Naturally, if we measure an angle θ in degrees instead of radians, the relationships expressed in Theorem 2 are written as

(i) $\cos(90° - \theta) = \sin\theta$ (ii) $\sin(90° - \theta) = \cos\theta$
(iii) $\tan(90° - \theta) = \cot\theta$

Thus, $\sin 40° = \cos 50°$, $\tan 80° = \cot 10°$, and so forth.

Similarly, the cotangent, secant, and cosecant are cofunctions (Problem 39) because

(iv) $\cot(90° - \theta) = \tan\theta$ (v) $\sec(90° - \theta) = \csc\theta$
(vi) $\csc(90° - \theta) = \sec\theta$

By combining Theorems 1 and 2, we can prove the following important theorem.

THEOREM 3 **Sine of a Difference or Sum**

(i) $\sin(t - s) = \sin t \cos s - \cos t \sin s$ (ii) $\sin(t + s) = \sin t \cos s + \cos t \sin s$

PROOF (i) By applying Theorem 2(i) from right to left, we obtain

$$\sin(t - s) = \cos\left[\frac{\pi}{2} - (t - s)\right] = \cos\left(\frac{\pi}{2} - t + s\right) = \cos\left[\left(\frac{\pi}{2} - t\right) + s\right]$$

so that

$$\sin(t - s) = \cos\left(\frac{\pi}{2} - t\right)\cos s - \sin\left(\frac{\pi}{2} - t\right)\sin s \qquad \text{[Theorem 1(ii)]}$$

$$= \sin t \cos s - \cos t \sin s \qquad \text{[Theorems 2(i) and 2(ii)]}$$

(ii) Clearly

$$\sin(t + s) = \sin[t - (-s)]$$

After replacing s by $-s$ in part (i) of this theorem, we obtain

$$\sin(t + s) = \sin[t - (-s)] = \sin t \cos(-s) - \cos t \sin(-s)$$

$$= \sin t \cos s + \cos t \sin s$$

EXAMPLE 2 Use Theorem 3 to find the exact value of each expression.

(a) $\sin \dfrac{11\pi}{12}$ (b) $\sin 195°$

SOLUTION (a) Since $11\pi/12 = 7\pi/6 - \pi/4$, we use Theorem 3(i) to obtain

$$\begin{aligned}\sin \frac{11\pi}{12} &= \sin\left(\frac{7\pi}{6} - \frac{\pi}{4}\right)\\ &= \sin \frac{7\pi}{6} \cos \frac{\pi}{4} - \cos \frac{7\pi}{6} \sin \frac{\pi}{4}\\ &= \left(-\frac{1}{2}\right) \cdot \frac{\sqrt{2}}{2} - \left(-\frac{\sqrt{3}}{2}\right) \cdot \frac{\sqrt{2}}{2} = -\frac{\sqrt{2}}{4} + \frac{\sqrt{6}}{4}\\ &= \frac{\sqrt{6} - \sqrt{2}}{4}\end{aligned}$$

(b) By Theorem 3(ii), we have

$$\begin{aligned}\sin 195° = \sin(150° + 45°) &= \sin 150° \cos 45° + \cos 150° \sin 45°\\ &= \frac{1}{2} \cdot \frac{\sqrt{2}}{2} + \left(-\frac{\sqrt{3}}{2}\right) \cdot \frac{\sqrt{2}}{2} = \frac{\sqrt{2}}{4} - \frac{\sqrt{6}}{4} = \frac{\sqrt{2} - \sqrt{6}}{4}\end{aligned}$$

EXAMPLE 3 Simplify each expression. Do not use a calculator.

(a) $\cos\left(\dfrac{\pi}{2} + t\right)$ (b) $\sin(90° + \theta)$

(c) $\sin 27° \cos 18° + \cos 27° \sin 18°$ (d) $\cos \dfrac{5\pi}{7} \cos \dfrac{2\pi}{7} - \sin \dfrac{5\pi}{7} \sin \dfrac{2\pi}{7}$

SOLUTION (a) Using Theorem 1(ii), we have

$$\cos\left(\frac{\pi}{2} + t\right) = \cos \frac{\pi}{2} \cos t - \sin \frac{\pi}{2} \sin t = 0 - \sin t = -\sin t$$

(b) Using Theorem 3(ii), we have

$$\sin(90° + \theta) = \sin 90° \cos \theta + \cos 90° \sin \theta = \cos \theta + 0 = \cos \theta$$

(c) Reading the formula in Theorem 3(ii) for the sine of a sum from right to left, we have

$$\sin 27° \cos 18° + \cos 27° \sin 18° = \sin(27° + 18°) = \sin 45° = \frac{\sqrt{2}}{2}$$

(d) From Theorem 1(ii) we obtain

$$\cos \frac{5\pi}{7} \cos \frac{2\pi}{7} - \sin \frac{5\pi}{7} \sin \frac{2\pi}{7} = \cos\left(\frac{5\pi}{7} + \frac{2\pi}{7}\right) = \cos \pi = -1$$

EXAMPLE 4 Suppose that s and t are angles in standard position, where s is in quadrant I, $\cos s = \frac{4}{5}$, t is in quadrant II, and $\sin t = \frac{3}{5}$. Determine each of the following values.

(a) $\sin(t + s)$ (b) $\cos(t + s)$

SOLUTION Before we can apply the sum formulas, we need to determine both $\sin s$ and $\cos t$. Since the terminal side of s is in quadrant I, $\sin s$ is positive, so that

$$\sin s = \sqrt{1 - \cos^2 s} = \sqrt{1 - \tfrac{16}{25}} = \tfrac{3}{5}$$

Also, since the terminal side of t is in quadrant II, $\cos t$ is negative and

$$\cos t = -\sqrt{1 - \sin^2 t} = -\sqrt{1 - \tfrac{9}{25}} = -\tfrac{4}{5}$$

Thus, it follows that

(a) $\sin(t + s) = \sin t \cos s + \cos t \sin s = (\frac{3}{5})(\frac{4}{5}) + (-\frac{4}{5})(\frac{3}{5}) = 0$

(b) $\cos(t + s) = \cos t \cos s - \sin t \sin s$
$= (-\frac{4}{5})(\frac{4}{5}) - (\frac{3}{5})(\frac{3}{5}) = -\frac{16}{25} - \frac{9}{25} = -\frac{25}{25} = -1$

EXAMPLE 5 Verify that each equation is an identity.

(a) $\sin(t + \pi) = -\sin t$ (b) $\cos(\theta - 180°) = -\cos\theta$

SOLUTION

(a) $\sin(t + \pi) = \sin t \cos \pi + \cos t \sin \pi = -\sin t + 0 = -\sin t$

(b) $\cos(\theta - 180°) = \cos\theta \cos 180° + \sin\theta \sin 180° = -\cos\theta + 0 = -\cos\theta$

EXAMPLE 6 Let A and B be constants and let t be an angle in standard position with the point (A, B) on the terminal side of t. Show that if $a = \sqrt{A^2 + B^2}$ and s is any angle, then

$$A = a \cos t, \qquad B = a \sin t$$

and

$$A \cos s + B \sin s = a \cos(s - t)$$

SOLUTION Notice that $a = \sqrt{A^2 + B^2}$ is the distance between $(0, 0)$ and (A, B) (Figure 2). Using Definition 1 on page 257, we have

Figure 2

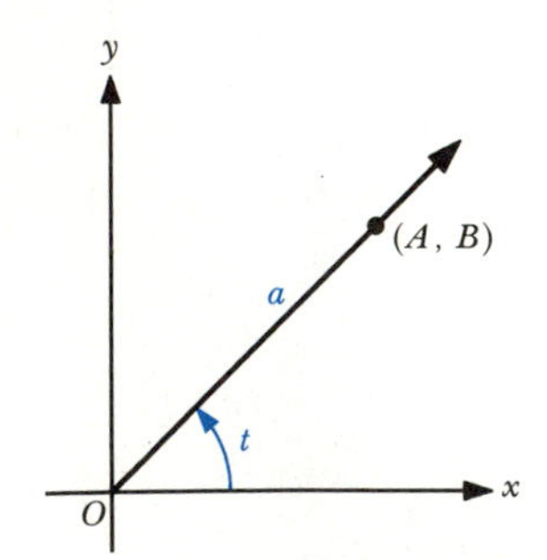

$$\cos t = \frac{A}{a} \quad \text{and} \quad \sin t = \frac{B}{a}$$

and so

$$A = a \cos t \quad \text{and} \quad B = a \sin t$$

Therefore

$$\begin{aligned} A \cos s + B \sin s &= (a \cos t)(\cos s) + (a \sin t)(\sin s) \\ &= a(\cos t \cos s + \sin t \sin s) \\ &= a \cos(s - t) \end{aligned}$$

Using the formulas for the sine and cosine of a difference and a sum, we can establish formulas for the tangent of a difference and a sum.

THEOREM 4 **Tangent of a Difference or Sum**

(i) $\tan(t - s) = \dfrac{\tan t - \tan s}{1 + \tan t \tan s}$ (ii) $\tan(t + s) = \dfrac{\tan t + \tan s}{1 - \tan t \tan s}$

PROOF (i) We use the identities for the sine and cosine of a difference to obtain

$$\tan(t - s) = \frac{\sin(t - s)}{\cos(t - s)}$$

$$= \frac{\sin t \cos s - \cos t \sin s}{\cos t \cos s + \sin t \sin s}$$

After dividing both the numerator and the denominator of the right-hand side of the above equation by $\cos t \cos s$, we get

$$\tan(t - s) = \frac{\dfrac{\sin t \cos s}{\cos t \cos s} - \dfrac{\cos t \sin s}{\cos t \cos s}}{\dfrac{\cos t \cos s}{\cos t \cos s} + \dfrac{\sin t \sin s}{\cos t \cos s}}$$

Simplifying the above expression, we have

$$\tan(t - s) = \frac{\tan t - \tan s}{1 + \tan t \tan s}$$

(ii) If we replace s with $-s$ in part (i), we get

$$\tan(t + s) = \tan[t - (-s)]$$

$$= \frac{\tan t - \tan(-s)}{1 + \tan t \tan(-s)}$$

$$= \frac{\tan t + \tan s}{1 - \tan t \tan s}$$

Care must be taken when applying the trigonometric formulas. For example, if we replaced t by $\pi/2$ in Theorem 4(i), we would get

$$\tan\left(\frac{\pi}{2} - s\right) = \frac{\tan\dfrac{\pi}{2} - \tan s}{1 + \tan\dfrac{\pi}{2}\tan s}$$

Since $\tan(\pi/2)$ is not defined, we might *erroneously* conclude that $\tan[(\pi/2) - s]$ is not defined for any value of s. However, we know that $\tan[(\pi/2) - s]$ can be written as

$$\tan\left(\frac{\pi}{2} - s\right) = \frac{\sin\left(\frac{\pi}{2} - s\right)}{\cos\left(\frac{\pi}{2} - s\right)}$$

$$= \frac{\cos s}{\sin s} = \cot s$$

This illustration emphasizes the important fact that the formulas derived in trigonometry are applicable only when the values of the variables are in the domains of *all* the functions contained in the formulas.

EXAMPLE 7 Simplify the expression. Do not use a calculator.

$$\frac{\tan 200° - \tan 65°}{1 + \tan 200° \tan 65°}$$

SOLUTION Reading Theorem 4(i) from right to left, we have

$$\frac{\tan 200° - \tan 65°}{1 + \tan 200° \tan 65°} = \tan(200° - 65°) = \tan 135° = -1$$

EXAMPLE 8 Prove that $\tan(x + \pi) = \tan x$.

SOLUTION By Theorem 4(ii), we have

$$\tan(x + \pi) = \frac{\tan x + \tan \pi}{1 - \tan x \tan \pi}$$

$$= \frac{\tan x + 0}{1 - 0} = \tan x$$

In certain applications of mathematics, it is necessary to write a product of sine and cosine functions as a sum or difference of such functions.

THEOREM 5 **Product Formulas**

(i) $\sin u \cos v = \frac{1}{2}[\sin(u + v) + \sin(u - v)]$

(ii) $\cos u \sin v = \frac{1}{2}[\sin(u + v) - \sin(u - v)]$

(iii) $\cos u \cos v = \frac{1}{2}[\cos(u - v) + \cos(u + v)]$

(iv) $\sin u \sin v = \frac{1}{2}[\cos(u - v) - \cos(u + v)]$

PROOF OF (i) Using Theorem 3, we have

$$\begin{aligned}\sin(u + v) + \sin(u - v) &= (\sin u \cos v + \cos u \sin v) + (\sin u \cos v - \cos u \sin v)\\ &= 2 \sin u \cos v\end{aligned}$$

so that

$$\sin u \cos v = \frac{1}{2}[\sin(u + v) + \sin(u - v)]$$

The proofs of parts (ii)–(iv) are left as exercises (Problem 41).

The product formulas can also be used to express a sum of sine and cosine functions as a product of such functions.

THEOREM 6 **Sum Formulas**

(i) $\sin s + \sin t = 2 \sin \frac{s + t}{2} \cos \frac{s - t}{2}$

(ii) $\sin s - \sin t = 2 \cos \frac{s + t}{2} \sin \frac{s - t}{2}$

(iii) $\cos s + \cos t = 2 \cos \frac{s + t}{2} \cos \frac{s - t}{2}$

(iv) $\cos t - \cos s = 2 \sin \frac{s + t}{2} \sin \frac{s - t}{2}$

PROOF OF (i) Let

$$s = u + v \qquad \text{and} \qquad t = u - v$$

Then

$$u = \frac{s + t}{2} \qquad \text{and} \qquad v = \frac{s - t}{2}$$

By substituting $\frac{s + t}{2}$ for u and $\frac{s - t}{2}$ for v in Theorem 5(i), we have

$$\sin\left(\frac{s + t}{2}\right) \cos\left(\frac{s - t}{2}\right) = \frac{1}{2}[\sin s + \sin t]$$

or

$$\sin s + \sin t = 2 \sin\left(\frac{s + t}{2}\right) \cos\left(\frac{s - t}{2}\right)$$

The proofs of parts (ii)–(iv) are left as exercises (Problem 42).

EXAMPLE 9 Express each product as a sum.

(a) $\sin 5x \cos 2x$ (b) $\cos 3t \cos 4t$

SOLUTION (a) Using Theorem 5(i) with $u = 5x$ and $v = 2x$, we get

$$\begin{aligned}\sin 5x \cos 2x &= \tfrac{1}{2}[\sin(5x + 2x) + \sin(5x - 2x)]\\ &= \tfrac{1}{2}[\sin 7x + \sin 3x]\\ &= \tfrac{1}{2}\sin 7x + \tfrac{1}{2}\sin 3x\end{aligned}$$

(b) Using Theorem 5(iii) with $u = 3t$ and $v = 4t$, we get

$$\begin{aligned}\cos 3t \cos 4t &= \tfrac{1}{2}[\cos(3t - 4t) + \cos(3t + 4t)]\\ &= \tfrac{1}{2}[\cos(-t) + \cos 7t]\\ &= \tfrac{1}{2}\cos t + \tfrac{1}{2}\cos 7t\end{aligned}$$

EXAMPLE 10 Express $\cos 8\theta - \cos 3\theta$ as a product.

SOLUTION Using Theorem 6(iv) with $t = 8\theta$ and $s = 3\theta$, we get

$$\begin{aligned}\cos 8\theta - \cos 3\theta &= 2\sin\left(\frac{8\theta + 3\theta}{2}\right)\sin\left(\frac{8\theta - 3\theta}{2}\right)\\ &= 2\sin\frac{11\theta}{2}\sin\frac{5\theta}{2}\end{aligned}$$

PROBLEM SET 6.2

In problems 1–10, find the *exact* value of each trigonometric function. Do not use a calculator.

1. $\sin\frac{5\pi}{12}$ $\left[\textit{Hint: } \frac{5\pi}{12} = \frac{\pi}{4} + \frac{\pi}{6}\right]$

2. $\cos\frac{13\pi}{12}$

3. $\cos 165°$ [*Hint:* $165° = 210° - 45°$]

4. $\sin\frac{19\pi}{12}$

5. $\tan\frac{19\pi}{12}$ $\left[\textit{Hint: } \frac{19\pi}{12} = \frac{11\pi}{6} - \frac{\pi}{4}\right]$

6. $\cot\frac{13\pi}{12}$

7. $\sin 105°$

8. $\cos 195°$

9. $\cos\frac{11\pi}{12}$

10. $\sin 285°$

In problems 11–30, write each expression as a single trigonometric function of one angle and simplify the result. Do not use a calculator.

11. $\sin\left(\frac{5\pi}{2} - \theta\right)$

12. $\cos\left(\frac{5\pi}{2} - \theta\right)$

13. $\tan(\pi + \theta)$

14. $\cot(90° - \theta)$

15. $\cos(270° + \theta)$

16. $\sec\left(\frac{\pi}{2} + \theta\right)$

17. $\csc(90° - \theta)$

18. $\sin(270° - \theta)$

19. $\cot(10\pi - \theta)$

20. $\tan(3\pi - \theta)$

21. $\sin 33° \cos 27° + \cos 33° \sin 27°$

22. $\cos \frac{5\pi}{7} \sin \frac{2\pi}{7} + \cos \frac{2\pi}{7} \sin \frac{5\pi}{7}$

23. $\sin 35° \cos 25° + \cos 35° \sin 25°$

24. $\cos 7t \cos t - \sin 7t \sin t$

25. $\cos 65° \cos 25° - \sin 65° \sin 25°$

26. $\cos 3\theta \cos 2\theta + \sin 3\theta \sin 2\theta$

27. $\sin(t - s) \cos s + \cos(t - s) \sin s$

28. $\sin(-5t) \cos 2t + \cos(-5t) \sin 2t$

29. $\dfrac{\tan 17° + \tan 43°}{1 - \tan 17° \tan 43°}$

30. $\dfrac{\tan 9x - \tan 8x}{1 + \tan 9x \tan 8x}$

In problems 31–34, suppose that s and t are angles in standard position. Use the given information to find the value of each expression.

(a) $\sin(t + s)$ (b) $\cos(t + s)$ (c) $\sin(t - s)$

(d) $\cos(t - s)$ (e) $\tan(t + s)$ (f) $\sec(t - s)$

31. $\sin t = \frac{12}{13}$, t is in quadrant II, $\cos s = -\frac{4}{5}$, and s is in quadrant II

32. $\sin s = \frac{3}{5}$, s is in quadrant II, $\cos t = \frac{3}{5}$, and t is in quadrant IV

33. $\sin s = -\frac{7}{25}$, s is in quadrant III, $\sin t = -\frac{15}{17}$, and t is in quadrant IV

34. $\tan s = \frac{5}{12}$, s is in quadrant I, $\tan t = \frac{8}{15}$, and t is in quadrant III

In problems 35–38, simplify each expression.

35. $\sin t \cos s - \sin\left(t + \frac{\pi}{2}\right) \sin(-s)$

36. $\cos\left(t - \frac{\pi}{2}\right) \tan s + \sin t \cot s$

37. $-\sin(\pi - t) + \cot t \sin\left(t - \frac{\pi}{2}\right)$

38. $\cos(\pi - t) - \tan t \cos\left(\frac{\pi}{2} - t\right)$

39. Show that

(a) $\cot\left(\frac{\pi}{2} - t\right) = \tan t$ (b) $\sec\left(\frac{\pi}{2} - t\right) = \csc t$ (c) $\csc\left(\frac{\pi}{2} - t\right) = \sec t$

40. Show that

(a) $\cot(s + t) = \dfrac{\cot s \cot t - 1}{\cot s + \cot t}$ (b) $\cot(s - t) = \dfrac{\cot s \cot t + 1}{\cot t - \cot s}$

41. Prove parts (ii)–(iv) of Theorem 5.

42. Prove parts (ii)–(iv) of Theorem 6.

In problems 43–53, verify that each equation is an identity.

43. $\sin(t + s) \sin(t - s) = \sin^2 t - \sin^2 s$

44. $\sin(30° + t) = \frac{1}{2}\cos t + \frac{\sqrt{3}}{2} \sin t$

45. $\tan(45° + t) = \dfrac{1 + \tan t}{1 - \tan t}$

46. $\sin(60° + t) - \cos(30° + t) = \sin t$

47. $\cos\left(\frac{\pi}{4} - t\right) = \frac{1}{\sqrt{2}} (\cos t + \sin t)$

48. $\cos(s + t) \cos(s - t) = \cos^2 s + \cos^2 t - 1$

49. $\cos\left(\frac{\pi}{6} + t\right) \cos\left(\frac{\pi}{6} - t\right) - \sin\left(\frac{\pi}{6} + t\right) \sin\left(\frac{\pi}{6} - t\right) = \frac{1}{2}$

50. $\cos(\pi + \theta)\cos(\pi - \theta) + \sin(\pi + \theta)\sin(\pi - \theta) = \cos 2\theta$

51. $\sin(\pi - s - t) = \sin s \cos t + \cos s \sin t$

52. $\dfrac{\sin(s + t)}{\sin(s - t)} = \dfrac{\tan s + \tan t}{\tan s - \tan t}$

53. $\dfrac{\tan s + \tan(t + u)}{1 - \tan s \tan(t + u)} = \dfrac{\tan t + \tan(s + u)}{1 - \tan t \tan(s + u)}$

54. Prove that each equation is an identity.

(a) $\dfrac{\sin(x + h) - \sin x}{h} = \sin x\left(\dfrac{\cos h - 1}{h}\right) + \cos x\left(\dfrac{\sin h}{h}\right)$

(b) $\dfrac{\cos(x + h) - \cos x}{h} = \cos x\left(\dfrac{\cos h - 1}{h}\right) - \sin x\left(\dfrac{\sin h}{h}\right)$

In problems 55 and 56, use Example 6 on page 324 to write each equation in the form $y = a\cos(s - t)$. Determine the radian measure of t, if t is an acute angle.

55. $y = \cos s + \sin s$

56. $y = \sqrt{3}\cos \pi s + \sin \pi s$

In problems 57–62, express each product as a sum or a difference.

57. $\sin 4w \sin w$

58. $\cos 3\theta \sin 9\theta$

59. $\cos 8v \cos 4v$

60. $\cos 7t \cos(-3t)$

61. $\cos(s + t)\cos(s - t)$

62. $\sin(s + t)\sin(s - t)$

In problems 63–66, express each sum or difference as a product.

63. $\sin 3\theta + \sin 2\theta$

64. $\cos t - \cos 10t$

65. $\cos 2t + \cos 4t$

66. $\sin 5t - \sin 7t$

In problems 67–70, verify that each equation is an identity.

67. $\dfrac{\sin 5\theta + \sin 3\theta}{\sin 5\theta - \sin 3\theta} = \dfrac{\tan 4\theta}{\tan \theta}$

68. $\dfrac{\sin 3t + \sin t}{\cos 3t + \cos t} = \tan 2t$

69. $\dfrac{\sin(t - s)}{\sin t \sin s} = \cot s - \cot t$

70. $\dfrac{\sin(t + 300^\circ) - \sin(t + 60^\circ)}{\cos(t + 30^\circ) + \cos(t - 30^\circ)} = -1$

6.3 Multiple-Angle Formulas

In this section, we study important relationships consisting of formulas that express values of trigonometric functions of *twice* an angle and *half* an angle in terms of trigonometric functions of the angle. Again these formulas also hold for real numbers.

Double-Angle Formulas

If angles s and t are equal, then the formulas for the sine [Theorem 3(ii) on page 322] and for the cosine [Theorem 1(ii) on page 319] of the sum of two angles may be transformed into formulas called the *double-angle* formulas.

THEOREM 1 **Sine of a Double-Angle**

$$\sin 2t = 2 \sin t \cos t$$

PROOF In the formula

$$\sin (s + t) = \sin s \cos t + \cos s \sin t$$

we let $s = t$ to obtain

$$\sin (t + t) = \sin t \cos t + \cos t \sin t$$

or

$$\sin 2t = 2 \sin t \cos t$$

EXAMPLE 1 Write $\sin 110°$ in terms of trigonometric functions of $55°$.

SOLUTION Using Theorem 1, we have

$$\sin 110° = \sin[2(55°)] = 2 \sin 55° \cos 55°$$

EXAMPLE 2 Write the expression $2 \sin 5\theta \cos 5\theta$ as the value of a function of 10θ.

SOLUTION Reading Theorem 1 from right to left with $t = 5\theta$, we have

$$2 \sin 5\theta \cos 5\theta = \sin[2(5\theta)] = \sin 10\theta$$

THEOREM 2 **Cosine of a Double-Angle**

$$\cos 2t = \cos^2 t - \sin^2 t = 2 \cos^2 t - 1 = 1 - 2 \sin^2 t$$

PROOF In the formula

$$\cos(s + t) = \cos s \cos t - \sin s \sin t$$

we let $s = t$, so that

$$\cos(t + t) = \cos t \cos t - \sin t \sin t$$

or

$$\cos 2t = \cos^2 t - \sin^2 t$$

However, since $\cos^2 t + \sin^2 t = 1$, we also have

$$\cos 2t = \cos^2 t - (1 - \cos^2 t) = 2 \cos^2 t - 1$$

and

$$\cos 2t = (1 - \sin^2 t) - \sin^2 t = 1 - 2 \sin^2 t$$

EXAMPLE 3 Write $\cos 50°$ in terms of trigonometric functions of $25°$.

SOLUTION Using Theorem 2, we have

$$\begin{aligned}\cos 50° &= \cos[2(25°)] = \cos^2 25° - \sin^2 25° \\ &= 2\cos^2 25° - 1 \\ &= 1 - 2\sin^2 25°\end{aligned}$$

EXAMPLE 4 Convert the given expression to a form containing a single trigonometric function of twice the given angle.

(a) $\cos^2 3t - \sin^2 3t$ (b) $2\cos^2 4t - 1$ (c) $1 - 2\sin^2 6\theta$

SOLUTION From Theorem 2 we have

(a) $\cos^2 3t - \sin^2 3t = \cos[2(3t)] = \cos 6t$

(b) $2\cos^2 4t - 1 = \cos[2(4t)] = \cos 8t$

(c) $1 - 2\sin^2 6\theta = \cos[2(6\theta)] = \cos 12\theta$

EXAMPLE 5 If $\sin\theta = \frac{12}{13}$ and θ is an angle in quadrant I, find

(a) $\sin 2\theta$ (b) $\cos 2\theta$ (c) $\tan 2\theta$

SOLUTION First, we need to find $\cos\theta$. Because θ is in quadrant I, we know that $\cos\theta$ is positive so that

$$\cos\theta = \sqrt{1 - \sin^2\theta} = \sqrt{1 - (\tfrac{12}{13})^2} = \sqrt{\tfrac{25}{169}} = \tfrac{5}{13}$$

Thus we obtain:

(a) $\sin 2\theta = 2\sin\theta\cos\theta = 2(\frac{12}{13})(\frac{5}{13}) = \frac{120}{169}$

(b) $\cos 2\theta = \cos^2\theta - \sin^2\theta = (\frac{5}{13})^2 - (\frac{12}{13})^2$

$$= \tfrac{25}{169} - \tfrac{144}{169} = -\tfrac{119}{169}$$

(c) $\tan 2\theta = \dfrac{\sin 2\theta}{\cos 2\theta} = \dfrac{\frac{120}{169}}{-\frac{119}{169}}$

$$= -\tfrac{120}{119}$$

EXAMPLE 6 Simplify the expression

$$(\sin t + \cos t)^2 - \sin 2t$$

SOLUTION

$$\begin{aligned}(\sin t + \cos t)^2 - \sin 2t &= \sin^2 t + 2\sin t\cos t + \cos^2 t - \sin 2t \\ &= 1 + 2\sin t\cos t - \sin 2t \\ &= 1 + \sin 2t - \sin 2t = 1\end{aligned}$$

EXAMPLE 7 Show that each equation is an identity.

(a) $\cos 2\theta + 2\sin^2\theta = 1$ (b) $\dfrac{2\cos 2t}{\sin 2t - 2\sin^2 t} = \cot t + 1$

SOLUTION (a) $\cos 2\theta + 2\sin^2\theta = \cos^2\theta - \sin^2\theta + 2\sin^2\theta$
$= \cos^2\theta + \sin^2\theta = 1$

(b) Here we start with the left side to obtain

$$\frac{2\cos 2t}{\sin 2t - 2\sin^2 t} = \frac{2(\cos^2 t - \sin^2 t)}{2\sin t\cos t - 2\sin^2 t}$$

$$= \frac{2(\cos t - \sin t)(\cos t + \sin t)}{2\sin t(\cos t - \sin t)}$$

$$= \frac{\cos t + \sin t}{\sin t} = \frac{\cos t}{\sin t} + 1$$

$$= \cot t + 1$$

Theorem 4(ii) on page 325 can be used to obtain the *double-angle* formula for the tangent.

THEOREM 3 **Tangent of a Double-Angle**

$$\tan 2t = \frac{2\tan t}{1 - \tan^2 t}$$

PROOF In the formula

$$\tan(s + t) = \frac{\tan s + \tan t}{1 - \tan s\tan t}$$

if we let $s = t$, then

$$\tan(t + t) = \frac{\tan t + \tan t}{1 - \tan t\tan t}$$

or

$$\tan 2t = \frac{2\tan t}{1 - \tan^2 t}$$

EXAMPLE 8 Write $\tan 80°$ in terms of $\tan 40°$.

SOLUTION By using Theorem 3, we have

$$\tan 80° = \tan[2(40°)] = \frac{2\tan 40°}{1 - \tan^2 40°}$$

Half-Angle Formulas

If we solve the equation $\cos 2t = 2\cos^2 t - 1$ in Theorem 2 for $\cos^2 t$, we obtain the formula

$$\text{(i)}\quad \cos^2 t = \frac{1 + \cos 2t}{2}$$

Similarly, the equation $\cos 2t = 1 - 2\sin^2 t$ in Theorem 2 can be solved for $\sin^2 t$ to get

$$\text{(ii)}\quad \sin^2 t = \frac{1 - \cos 2t}{2}$$

EXAMPLE 9 Use formula (i) above to express $\cos^2 7\theta$ in terms of $\cos 14\theta$.

SOLUTION If we substitute 7θ for t in formula (i), we get

$$\cos^2 7\theta = \frac{1 + \cos[2(7\theta)]}{2} = \frac{1 + \cos 14\theta}{2}$$

EXAMPLE 10 Show that $\cos^4 t = \frac{1}{8}(3 + 4\cos 2t + \cos 4t)$.

SOLUTION

$$\begin{aligned}
\cos^4 t = (\cos^2 t)^2 = \left(\frac{1 + \cos 2t}{2}\right)^2 &= \frac{1}{4}(1 + 2\cos 2t + \cos^2 2t) \\
&= \frac{1}{4}\left[1 + 2\cos 2t + \frac{1}{2}(1 + \cos 4t)\right] \\
&= \frac{1}{4}\left[1 + 2\cos 2t + \frac{1}{2} + \frac{\cos 4t}{2}\right] \\
&= \frac{1}{4}\left[\frac{3}{2} + 2\cos 2t + \frac{\cos 4t}{2}\right] \\
&= \frac{1}{8}(3 + 4\cos 2t + \cos 4t)
\end{aligned}$$

If we replace t with $s/2$ in formulas (i) and (ii) above, we obtain equivalent forms of these formulas.

THEOREM 4 **Half-Angle Formulas**

$$\text{(i)}\quad \cos^2 \frac{s}{2} = \frac{1 + \cos s}{2} \qquad \text{(ii)}\quad \sin^2 \frac{s}{2} = \frac{1 - \cos s}{2}$$

Notice that by taking the square root of both sides of the identities in Theorem 4, we have

$$\cos\frac{s}{2} = \pm\sqrt{\frac{1+\cos s}{2}} \qquad \text{and} \qquad \sin\frac{s}{2} = \pm\sqrt{\frac{1-\cos s}{2}}$$

where the sign is determined by the quadrant in which the terminal side of an angle in standard position with measure $s/2$ lies.

EXAMPLE 11 Find the exact value of each expression.

(a) $\sin 15°$ (b) $\cos\frac{\pi}{12}$

SOLUTION Here we use Theorem 4 as follows:

(a) Since $\sin 15°$ is positive and $15° = (\frac{30}{2})°$, we have

$$\sin 15° = \sin\left(\frac{30}{2}\right)° = \sqrt{\frac{1-\cos 30°}{2}} = \sqrt{\frac{1-(\sqrt{3}/2)}{2}}$$
$$= \sqrt{\frac{2-\sqrt{3}}{4}} = \frac{\sqrt{2-\sqrt{3}}}{2}$$

(b) Since $\cos(\pi/12)$ is positive,

$$\cos\frac{\pi}{12} = \cos\left[\frac{1}{2}\left(\frac{\pi}{6}\right)\right] = \sqrt{\frac{1+\cos(\pi/6)}{2}} = \sqrt{\frac{1+(\sqrt{3}/2)}{2}}$$
$$= \sqrt{\frac{2+\sqrt{3}}{4}} = \frac{\sqrt{2+\sqrt{3}}}{2}$$

EXAMPLE 12 Suppose that θ, in standard position, is in quadrant II and $\sin\theta = \frac{3}{5}$. Find

(a) $\sin\frac{\theta}{2}$ (b) $\cos\frac{\theta}{2}$ (c) $\tan\frac{\theta}{2}$

SOLUTION Since θ is in quadrant II, $\cos\theta$ is negative so that

$$\cos\theta = -\sqrt{1-\sin^2\theta} = -\sqrt{1-\left(\frac{3}{5}\right)^2}$$
$$= -\sqrt{\frac{16}{25}} = -\frac{4}{5}$$

If θ is in degrees, then $90° < \theta < 180°$ so that $45° < \theta/2 < 90°$; that is, $\theta/2$ is in quadrant I. Thus $\sin(\theta/2)$ and $\cos(\theta/2)$ are both positive and we have:

(a) $$\sin\frac{\theta}{2} = \sqrt{\frac{1-\cos\theta}{2}} = \sqrt{\frac{1-(-\frac{4}{5})}{2}} = \sqrt{\frac{9}{10}} = \frac{3}{\sqrt{10}} = \frac{3\sqrt{10}}{10}$$

(b) $\cos\frac{\theta}{2} = \sqrt{\frac{1+\cos\theta}{2}} = \sqrt{\frac{1+(-\frac{4}{5})}{2}} = \sqrt{\frac{1}{10}} = \frac{1}{\sqrt{10}} = \frac{\sqrt{10}}{10}$

(c) $\tan\frac{\theta}{2} = \dfrac{\sin\frac{\theta}{2}}{\cos\frac{\theta}{2}} = \dfrac{\frac{3\sqrt{10}}{10}}{\frac{\sqrt{10}}{10}} = 3$

EXAMPLE 13 (a) Show that $\tan^2\frac{s}{2} = \frac{1-\cos s}{1+\cos s}$.

(b) Use the result in part (a) to verify that

$$\tan^2\frac{\theta}{2}\cos\theta + \tan^2\frac{\theta}{2} = 1 - \cos\theta$$

SOLUTION (a) $\tan^2\frac{s}{2} = \dfrac{\sin^2\frac{s}{2}}{\cos^2\frac{s}{2}} = \dfrac{\frac{1-\cos s}{2}}{\frac{1+\cos s}{2}} = \dfrac{1-\cos s}{1+\cos s}$

(b) We start with the left side to obtain

$$\tan^2\frac{\theta}{2}\cos\theta + \tan^2\frac{\theta}{2} = \tan^2\frac{\theta}{2}(\cos\theta + 1)$$

$$= \frac{(1-\cos\theta)}{(1+\cos\theta)}\cdot(\cos\theta + 1) = 1 - \cos\theta$$

PROBLEM SET 6.3

In problems 1–6, use the double-angle formulas to write the given expression in terms of values of trigonometric functions of an angle half as large.

1. $\sin 80°$ **2.** $\cos 8t$ **3.** $\cos 20°$ **4.** $\sin\frac{3\pi}{14}$ **5.** $\tan\frac{2\pi}{9}$ **6.** $\tan 74°$

In problems 7–18, use the double-angle formulas to write the given expression as a single trigonometric function of an angle twice as large.

7. $2\sin 3t\cos 3t$ **8.** $1 - 2\sin^2 37°$ **9.** $1 - 2\sin^2 7t$ **10.** $2\cos^2 8t - 1$

11. $2\cos^2 4t - 1$ **12.** $\cos^2 2t - \sin^2 2t$ **13.** $2 - 4\sin^2\frac{\theta}{2}$ **14.** $4\sin^2 7\theta\cos^2 7\theta$

15. $4\cos^2\frac{\theta}{2} - 2$ **16.** $\dfrac{2\tan 2t}{1-\tan^2 2t}$ **17.** $\dfrac{2\tan 4\theta}{1-\tan^2 4\theta}$ **18.** $\dfrac{\tan\frac{\theta}{2}}{\frac{1}{2}-\frac{1}{2}\tan^2\frac{\theta}{2}}$

In problems 19–24, use the given information to find the exact value of each expression.
(a) $\sin 2t$ (b) $\cos 2t$ (c) $\tan 2t$

19. $\sin t = \frac{4}{5}$, t in quadrant I

20. $\cos t = -\frac{12}{13}$, t in quadrant III

21. $\cos t = -\frac{7}{25}$, t in quadrant III

22. $\sin t = -\frac{5}{13}$, t in quadrant IV

23. $\tan t = -\frac{5}{12}$, t in quadrant II

24. $\tan t = \frac{18}{5}$, t in quadrant I

In problems 25–32, use the half-angle formulas to find the exact value of each expression. Do not use a calculator.

25. $\cos 22.5°$

26. $\cos \dfrac{5\pi}{12}$

27. $\sin\left(-\dfrac{7\pi}{12}\right)$

28. $\cos \dfrac{3\pi}{8}$

29. $\sin 75°$

30. $\sin 105°$

31. $\tan \dfrac{\pi}{12}$

32. $\tan 67.5°$

In problems 33–36, use the given information to find the exact value of each expression.

(a) $\sin \dfrac{t}{2}$ (b) $\cos \dfrac{t}{2}$ (c) $\tan \dfrac{t}{2}$

33. $\sin t = \frac{24}{25}$, t in quadrant I

34. $\cos t = \frac{7}{25}$, t in quadrant IV

35. $\cos t = -\frac{4}{5}$, t in quadrant II

36. $\sin t = -\frac{5}{13}$, t in quadrant III

In problems 37–41, write the given expression as a single trigonometric function of an angle half as large.

37. $\sqrt{\dfrac{1 + \cos 80°}{2}}$

38. $-\sqrt{\dfrac{1 - \cos 100°}{2}}$

39. $-\sqrt{\dfrac{1 - \cos 140°}{2}}$

40. $\sqrt{\dfrac{1 - \cos 170°}{1 + \cos 170°}}$

41. $\sqrt{\dfrac{1 - \cos \dfrac{2\pi}{5}}{1 + \cos \dfrac{2\pi}{5}}}$

42. In calculus, the substitution $z = \tan(\theta/2)$ is used. Use this substitution to show that

(a) $\cos\theta = \dfrac{1 - z^2}{1 + z^2}$ (b) $\sin\theta = \dfrac{2z}{1 + z^2}$

In problems 43–62, verify that each equation is an identity.

43. $\cot\theta - \tan\theta = 2\cot 2\theta$

44. $\cos^2\theta(1 - \tan^2\theta) = \cos 2\theta$

45. $\dfrac{1 + \cos 2t}{\sin 2t} = \cot t$

46. $\csc t \sec t = 2\csc 2t$

47. $\dfrac{\sin 3\theta}{\sin\theta} - \dfrac{\cos 3\theta}{\cos\theta} = 2$

48. $4\sin^2\theta\cos^2\theta + \cos^2 2\theta = 1$

49. $\dfrac{\sin 4t}{\sin 2t} = 2\cos 2t$

50. $\dfrac{\sin 5t}{\sin t} - \dfrac{\cos 5t}{\cos t} = 4\cos 2t$

51. $\dfrac{1}{2}\sin\theta\tan\dfrac{\theta}{2}\csc^2\dfrac{\theta}{2} = 1$

52. $\cos^2 t\sin^2 t = \dfrac{\sin^2 2t}{4}$

53. $2\cos\frac{\theta}{2} = (1 + \cos\theta)\sec\frac{\theta}{2}$

54. $\tan\frac{s}{2} = \frac{\sin s}{1 + \cos s}$

55. $\frac{\cos^3 t - \sin^3 t}{\cos t - \sin t} = \frac{2 + \sin 2t}{2}$

56. $2\sin\theta\cos^3\theta + 2\sin^3\theta\cos\theta = \sin 2\theta$

57. $\sin^2\theta\cos^2\theta = \frac{1}{8}(1 - \cos 4\theta)$

58. $\sin^4\theta = \frac{1}{8}(3 - 4\cos 2\theta + \cos 4\theta)$

59. $2\tan\frac{\theta}{2}\csc\theta = \sec^2\frac{\theta}{2}$

60. $2\sin^2\frac{t}{2}\tan t = \tan t - \sin t$

61. $\cos 4t = 8\cos^4 t - 8\cos^2 t + 1$

62. $\sin 3\theta = 3\sin\theta - 4\sin^3\theta$

6.4 Inverse Trigonometric Functions

In Chapter 2, on page 123, we established the fact that a function is invertible if and only if no horizontal straight line intersects its graph more than once. An examination of the graphs of the six trigonometric functions clearly shows that because of their periodicity *none* of the trigonometric functions is invertible.

For instance, the horizontal line $y = \frac{1}{2}$ intersects the graph of $y = \sin x$ repeatedly, and therefore the sine function is not invertible (Figure 1). However, if the domain of the sine function is restricted to the interval $[-\pi/2, \pi/2]$, the resulting function is invertible. More specifically, by restricting $y = \sin x$ to the interval $[-\pi/2, \pi/2]$, we define the function

$$f(x) = \text{Sin}\, x$$

(Notice the use of the capital letter S to distinguish this function from the sine function.) By examining the graph of $f(x) = \text{Sin}\, x$ (Figure 2), we can see that the function is invertible because of the horizontal line test. When we refer to the *inverse sine function*, we mean the function that is the inverse of $f(x) = \text{Sin}\, x$, *not* of $f(x) = \sin x$.

Figure 1

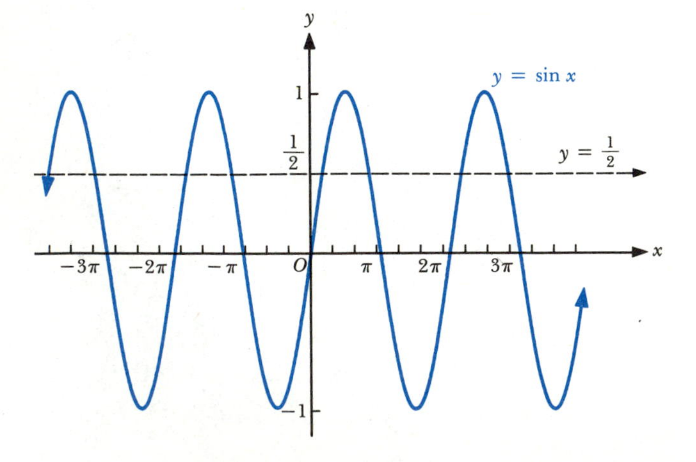

Figure 2

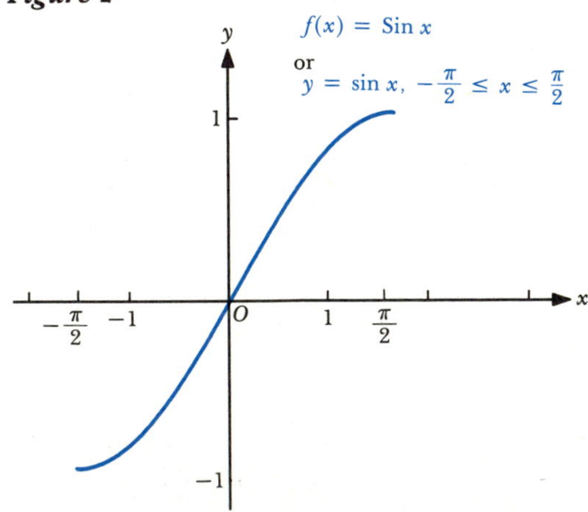

Thus we are led to the following definition.

DEFINITION 1 **The Inverse Sine or Arcsine Function**

The **inverse sine** or **Arcsine function,** denoted by $\sin^{-1}$ or Arcsin, is defined by $y = \sin^{-1} x$ (or $y = \text{Arcsin}\, x$) if and only if

$$\sin y = x \qquad \text{and} \qquad -\frac{\pi}{2} \leq y \leq \frac{\pi}{2}$$

It should be noted that even though we write $\sin^2 x$ to represent $(\sin x)^2$, we never use $\sin^{-1} x$ to represent $(\sin x)^{-1}$, since $(\sin x)^{-1} = 1/\sin x$, which is not the same as $\sin^{-1} x$. This problem arises because we use the "exponent" -1 in two different ways—to denote reciprocals and to denote inverse functions. To avoid confusion, the notation "Arcsin x" is also used. It is often helpful to think of Arcsin x, or $\sin^{-1} x$, as the number (angle) between $-\pi/2$ and $\pi/2$ ($-90°$ and $90°$) whose sine is x.

EXAMPLE 1 Find the exact value of each expression.

(a) $\sin^{-1} 0$ (b) $\sin^{-1} \dfrac{\sqrt{3}}{2}$ (c) $\text{Arcsin}\left(\dfrac{-\sqrt{2}}{2}\right)$

SOLUTION (a) If $\sin^{-1} 0 = y$, then $\sin y = 0$ and $-\dfrac{\pi}{2} \leq y \leq \dfrac{\pi}{2}$.

Since $\sin 0 = 0$ and $-\pi/2 \leq 0 \leq \pi/2$, we have $\sin^{-1} 0 = 0$ (or $0°$).

(b) If $\sin^{-1}\left(\dfrac{\sqrt{3}}{2}\right) = y$, then $\sin y = \dfrac{\sqrt{3}}{2}$ and $-\dfrac{\pi}{2} \leq y \leq \dfrac{\pi}{2}$.

Since $\sin(\pi/3) = \sqrt{3}/2$ and $-\pi/2 \leq \pi/3 \leq \pi/2$, we obtain

$$\sin^{-1} \frac{\sqrt{3}}{2} = \frac{\pi}{3} \quad (\text{or } 60°)$$

(c) If $\text{Arcsin}\left(-\dfrac{\sqrt{2}}{2}\right) = y$, then $\sin y = -\dfrac{\sqrt{2}}{2}$ and $-\dfrac{\pi}{2} \leq y \leq \dfrac{\pi}{2}$.

Because $\sin(\pi/4) = \sqrt{2}/2$, we use $\pi/4$ as the reference angle to locate y in quadrant IV so that $-\pi/2 \leq y \leq \pi/2$. Thus

$$\text{Arcsin}\left(-\frac{\sqrt{2}}{2}\right) = -\frac{\pi}{4} \quad (\text{or } -45°)$$

It is often necessary to use a scientific calculator (or Appendix Table III or IV) to find approximate values of Arcsin x or $\sin^{-1} x$. Some calculators have a $\boxed{\sin^{-1}}$ key, some have a key marked $\boxed{\text{Arcsin}}$, and others have an $\boxed{\text{INV}}$ key, which must be pressed before the [SIN] key to give the inverse sine. When a calculator is used, we must make sure it is set in radian mode if we want $\sin^{-1} x$ in radians, otherwise the answer will be the angle in degrees (between $-90°$ and $90°$) whose sine is x.

EXAMPLE 2 [C] Find each value in radians to two decimal places.

(a) $\sin^{-1} 0.8016$ (b) $\text{Arcsin}(-0.8330)$

SOLUTION Using a calculator in radian mode, we proceed as follows:

(a) We enter 0.8016 and then press the [$\sin^{-1}$] or [Arcsin] key or the [INV] and [SIN] keys in succession to obtain

$$\sin^{-1} 0.8016 = 0.93$$

(b) We enter -0.8330 and then press the [$\sin^{-1}$] or [Arcsin] key or the [INV] and [SIN] keys to obtain

$$\text{Arcsin}(-0.8330) = -0.98$$

Geometrically, the graph of the inverse sine function $y = \sin^{-1} x$ can be obtained by reflecting the graph of the function $f(x) = \text{Sin } x$ across the line $y = x$ (Figure 3). Notice that the domain of $y = \text{Arcsin } x$ is the interval $[-1, 1]$ and the range of $y = \text{Arcsin } x$ is the interval $[-\pi/2, \pi/2]$. These two intervals are the range and domain of $y = \text{Sin } x$, respectively.

x	$y = \sin^{-1} x$
-1	$-\frac{\pi}{2}$
$-\frac{1}{2}$	$-\frac{\pi}{6}$
0	0
$\frac{1}{2}$	$\frac{\pi}{6}$
1	$\frac{\pi}{2}$

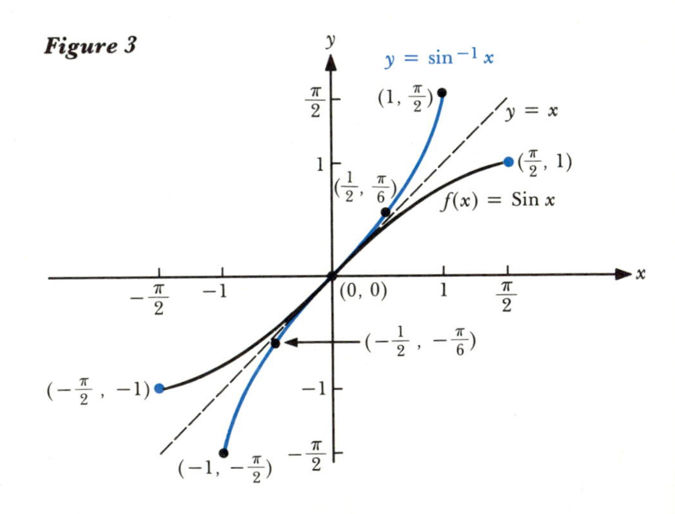

Figure 3

Similar considerations produce the inverse of the cosine function. To define the inverse cosine function, we begin by constructing the function $g(x) = \text{Cos } x$ (notice the use of capital C) from $y = \cos x$ by restricting the domain to the interval $[0, \pi]$ (Figure 4). By the horizontal line test, we see that $g(x) = \text{Cos } x$ is invertible. Its range is $[-1, 1]$. The inverse of $g(x) = \text{Cos } x$ is referred to as the inverse cosine function or Arccosine function.

Figure 4

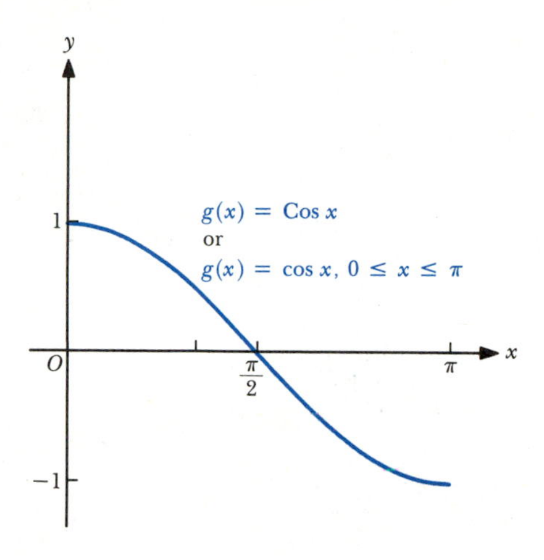

DEFINITION 2

The Inverse Cosine or Arccosine Function

The **inverse cosine** or **Arccosine function,** denoted by $\cos^{-1}$ or Arccos, is defined by $y = \cos^{-1} x$ (or $y = \text{Arccos}\, x$) if and only if

$$\cos y = x \qquad \text{and} \qquad 0 \le y \le \pi$$

The graph of $y = \cos^{-1} x$ can be obtained from the graph of $g(x) = \text{Cos}\, x$ by reflecting the latter graph across the line $y = x$ (Figure 5). The domain of the inverse cosine is $[-1, 1]$ and the range is $[0, \pi]$. Note that Arccos x or $\cos^{-1} x$ is the number (angle) between 0 and π (0° and 180°) whose cosine is x.

x	$y = \cos^{-1} x$
-1	π
$-\frac{1}{2}$	$\frac{2\pi}{3}$
0	$\frac{\pi}{2}$
$\frac{1}{2}$	$\frac{\pi}{3}$
1	0

Figure 5

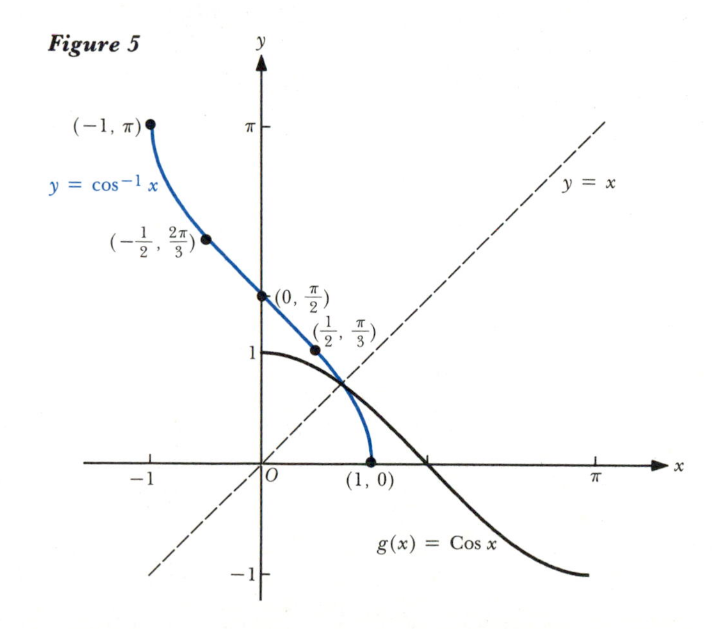

EXAMPLE 3 Find each value in radians.

(a) $\cos^{-1}(-1)$ (b) $\cos^{-1} 0$ (c) $\text{Arccos}\left(-\frac{\sqrt{2}}{2}\right)$

C (d) Arccos 0.7951 to two decimal places.

SOLUTION (a) If $\cos^{-1}(-1) = y$, then $\cos y = -1$ and $0 \le y \le \pi$. Therefore,

$$y = \pi \quad \text{and} \quad \cos^{-1}(-1) = \pi$$

(b) If $\cos^{-1} 0 = y$, then $\cos y = 0$ and $0 \le y \le \pi$. Therefore,

$$y = \frac{\pi}{2} \quad \text{and} \quad \cos^{-1} 0 = \frac{\pi}{2}$$

(c) If Arccos $(-\sqrt{2}/2) = y$, then $\cos y = -\sqrt{2}/2$ and $0 \le y \le \pi$. Because $\cos(\pi/4) = \sqrt{2}/2$, we use $\pi/4$ as the reference angle to locate y in quadrant II. Thus

$$y = \frac{3\pi}{4} \quad \text{and} \quad \text{Arccos}\left(-\frac{\sqrt{2}}{2}\right) = \frac{3\pi}{4}$$

(d) Using a calculator in radian mode, we find that Arccos 0.7951 = 0.65

EXAMPLE 4 If $y = 5 + 3\cos^{-1} 2x$, express x in terms of y.

SOLUTION $y = 5 + 3\cos^{-1} 2x$ can be written as

$$\frac{y-5}{3} = \cos^{-1} 2x$$

The latter equation is equivalent to

$$\cos\left(\frac{y-5}{3}\right) = 2x \quad \text{where} \quad 0 \le \frac{y-5}{3} \le \pi$$

Thus $$x = \tfrac{1}{2}\cos\left(\frac{y-5}{3}\right) \quad \text{where} \quad 5 \le y \le 3\pi + 5$$

To obtain the inverse tangent function, we construct $h(x) = \text{Tan}\, x$ (notice the capital T) by restricting the domain of $y = \tan x$ to the interval $(-\pi/2, \pi/2)$. The range is the set of real numbers $\mathbb{R}$ (Figure 6a). The function h is increasing, and the horizontal line test shows that it is invertible. The inverse tangent function is the inverse of $h(x) = \text{Tan}\, x$.

DEFINITION 3 **The Inverse Tangent or Arctangent Function**

The **inverse tangent** or **Arctangent function,** denoted by $\tan^{-1}$ or Arctan, is defined by $y = \tan^{-1} x$ (or $y = \text{Arctan}\, x$) if and only if

$$\tan y = x \quad \text{and} \quad -\frac{\pi}{2} < y < \frac{\pi}{2}$$

The graph of $y = \tan^{-1} x$ can be obtained as a reflection of the graph in Figure 6a (Figure 6b). Clearly the domain is the set of all real numbers $\mathbb{R}$ and the range is the interval $(-\pi/2, \pi/2)$. We can think of Arctan x, or $\tan^{-1} x$, as the number (angle) between $-\pi/2$ and $\pi/2$ ($-90°$ and $90°$) whose tangent is x.

Figure 6

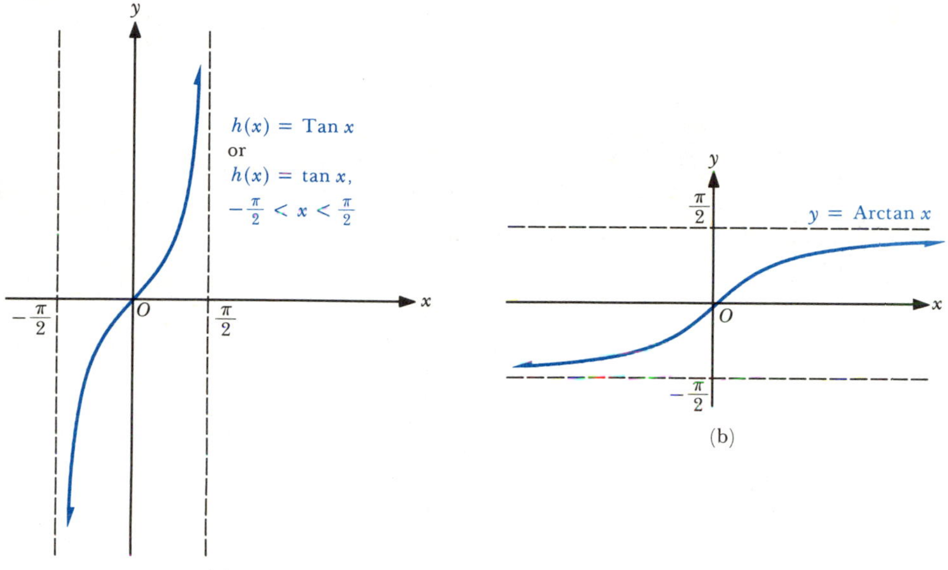

(a)

(b)

EXAMPLE 5 Find each value in radians.

(a) $\tan^{-1} 1$
(b) Arctan 0
(c) $\tan^{-1}(-\sqrt{3})$
C (d) Arctan (-0.1249) to two decimal places.

SOLUTION (a) If $\tan^{-1} 1 = y$, then $\tan y = 1$ and $-\pi/2 < y < \pi/2$. Therefore,

$$y = \frac{\pi}{4} \quad \text{and} \quad \tan^{-1} 1 = \frac{\pi}{4}$$

(b) If Arctan $0 = y$, then $\tan y = 0$ and $-\pi/2 < y < \pi/2$. Therefore,

$$y = 0 \quad \text{and} \quad \tan^{-1} 0 = 0$$

(c) If $\tan^{-1}(-\sqrt{3}) = y$, then $\tan y = -\sqrt{3}$ and $-\pi/2 < y < \pi/2$. Because $\tan(\pi/3) = \sqrt{3}$, we use $\pi/3$ as the reference angle to locate y in quadrant IV. Thus

$$y = -\frac{\pi}{3} \quad \text{and} \quad \tan^{-1}(-\sqrt{3}) = -\frac{\pi}{3}$$

(d) Using a calculator in radian mode, we find that

$$\text{Arctan}(-0.1249) = -0.12$$

In calculus, it is sometimes necessary to find exact values of expressions such as $\cos[(\tan^{-1}(3/4)]$. The following examples illustrate how to do this.

EXAMPLE 6 Without the use of a calculator, find the exact value of each expression.

(a) $\sin(\cos^{-1}\frac{3}{4})$ (b) $\tan(\cos^{-1}\frac{3}{4})$

SOLUTION Assume that $\theta = \cos^{-1}\frac{3}{4}$, so that $\cos\theta = \frac{3}{4}$ and $0° \le \theta \le 180°$. θ can be thought of as an acute angle, since $\cos\theta$ is positive and $0° < \theta < 90°$. Next we construct a right triangle with one side of length 3 adjacent to angle θ and the hypotenuse of length 4. The length of the side opposite angle θ can be found by the Pythagorean theorem to be $\sqrt{4^2 - 3^2} = \sqrt{7}$ (Figure 7). Thus, by using the trigonometric functions defined on right triangles, we obtain

Figure 7

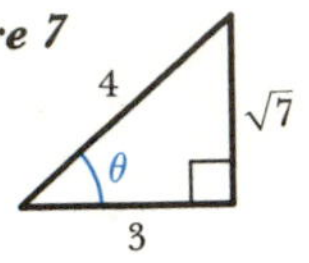

4 $\sqrt{7}$ θ 3

(a) $\sin\left(\cos^{-1}\dfrac{3}{4}\right) = \sin\theta = \dfrac{\text{opp}}{\text{hyp}} = \dfrac{\sqrt{7}}{4}$

(b) $\tan\left(\cos^{-1}\dfrac{3}{4}\right) = \tan\theta = \dfrac{\text{opp}}{\text{adj}} = \dfrac{\sqrt{7}}{3}$

EXAMPLE 7 Find the exact value of $\sin[2\cos^{-1}(-\frac{3}{5})]$. Do not use a calculator.

SOLUTION We use the double-angle formula for the sine (Theorem 1 on page 331) to get

$$\sin[2\cos^{-1}(-\tfrac{3}{5})] = 2\sin[\cos^{-1}(-\tfrac{3}{5})]\cos[\cos^{-1}(-\tfrac{3}{5})]$$

To find $\sin[\cos^{-1}(-\frac{3}{5})]$ we set $w = \cos^{-1}(-\frac{3}{5})$ so that $\cos w = -\frac{3}{5}$ and w is between 0 and π. Since $\cos w$ is negative, w must be between $\pi/2$ and π and

$$\sin w = \sqrt{1 - \cos^2 w} = \sqrt{1 - (-\tfrac{3}{5})^2} = \sqrt{\tfrac{16}{25}} = \tfrac{4}{5}$$

Also,

$$\cos[\cos^{-1}(-\tfrac{3}{5})] = -\tfrac{3}{5}.$$

Thus,

$$\sin[2\cos^{-1}(-\tfrac{3}{5})] = 2(\tfrac{4}{5})(-\tfrac{3}{5}) = \tfrac{-24}{25}$$

EXAMPLE 8 Show that $\sin^{-1}x = \cos^{-1}\sqrt{1 - x^2}$ for $0 \le x \le 1$.

SOLUTION If we let $t = \sin^{-1}x$, then $\sin t = x$, where $0 \le t \le \pi/2$ because x is nonnegative. Since $\cos t = \sqrt{1 - \sin^2 t}$ for $0 \le t \le \pi/2$, we have $\cos t = \sqrt{1 - x^2}$ or $t = \cos^{-1}\sqrt{1 - x^2}$. By substitution, we obtain

$$\sin^{-1}x = \cos^{-1}\sqrt{1 - x^2}$$

The remaining three inverse trigonometric functions for the cotangent, secant, and cosecant, respectively, are discussed in Problems 64, 65, and 66.

PROBLEM SET 6.4

In problems 1–12, determine the value of each expression without using a calculator or tables.

1. $\sin^{-1}\frac{1}{2}$
2. $\cos^{-1} 1$
3. $\cos^{-1}(-\frac{1}{2})$
4. $\sin^{-1}\left(-\frac{\sqrt{3}}{2}\right)$
5. $\text{Arcsin}(-1)$
6. $\text{Arccos}\,\frac{\sqrt{3}}{2}$
7. $\text{Arctan}\,(-1)$
8. $\text{Arctan}\,\frac{\sqrt{3}}{3}$
9. $\text{Arcsin}\,\frac{\sqrt{2}}{2}$
10. $\cos^{-1}\frac{1}{2}$
11. $\tan^{-1}\left(-\frac{\sqrt{3}}{3}\right)$
12. $\sin^{-1}\left(-\frac{1}{2}\right)$

c In problems 13–22, use a calculator (or Appendix Table III) to determine each value in radian measure to two decimal places.

13. $\sin^{-1} 0.2182$
14. $\cos^{-1} 0.8628$
15. $\text{Arccos}\, 0.2092$
16. $\text{Arctan}\, 1.072$
17. $\text{Arcsin}(-0.7771)$
18. $\tan^{-1}(-0.5334)$
19. $\cos^{-1}(-0.8473)$
20. $\sin^{-1} 0.05$
21. $\tan^{-1} 3.01$
22. $\text{Arcsin}(-0.9425)$

In problems 23–34, determine the value of each expression without using a calculator or tables.

23. $\cos^{-1}\left[\sin\frac{\pi}{6}\right]$
24. $\tan^{-1}\left[\cos\frac{\pi}{2}\right]$
25. $\sin\left[\cos^{-1}\frac{1}{2}\right]$
26. $\sin^{-1}\left[\sin\left(-\frac{3\pi}{2}\right)\right]$
27. $\sin\left[\sin^{-1}\frac{1}{2}\right]$
28. $\tan[\tan^{-1}(-1)]$
29. $\cos[\sin^{-1}\frac{4}{5}]$
30. $\sin[\cos^{-1}\frac{5}{13}]$
31. $\sin[\tan^{-1}\frac{4}{3}]$
32. $\cos\left[\tan^{-1}\left(-\frac{12}{5}\right)\right]$
33. $\sin\left[\sin^{-1}\frac{\sqrt{3}}{2} + \cos^{-1}\frac{\sqrt{2}}{2}\right]$
34. $\cos\left[\sin^{-1}\frac{1}{2} - \cos^{-1}\frac{1}{2}\right]$

In problems 35–38, express x in terms of y. Determine the restrictions on x and y.

35. $y = \cos^{-1} 2x$
36. $y = 2\sin^{-1} 3x + 7$
37. $y = \tan^{-1}(x + 1)$
38. $y = 4 - 2\sin^{-1} 4x$

In problems 39–47, find the exact value of each expression. Do not use a calculator.

39. $\sin\,[2\sin^{-1}\frac{1}{2}]$
40. $\sin\,[\frac{1}{2}\cos^{-1}\frac{\sqrt{3}}{2}]$
41. $\cos\,[2\sin^{-1} 1]$
42. $\cos\,[2\cos^{-1}\frac{3}{5}]$
43. $\tan\,[2\tan^{-1}\frac{4}{3}]$
44. $\sin\,[2\cos^{-1}(-\frac{24}{25})]$
45. $\cos t$ and $\cos 2t$, if $\sin^{-1}\frac{4}{5} = t$
46. $\sin w$ and $\sin 2w$, if $\cos^{-1}(-\frac{3}{5}) = w$
47. $\sec v$ and $\tan 2v$, if $\tan^{-1}\frac{1}{2} = v$

48. Write each value as an algebraic expression in the variable t.
(a) $\sin(\tan^{-1} t)$ (b) $\tan(\cos^{-1} t)$

In problems 49–54, show that each equation is an identity.

49. $\sin^{-1}(-x) = -\sin^{-1} x$

50. $\cos^{-1} x = \frac{\pi}{2} - \sin^{-1} x$

51. $2 \cos^{-1} x = \cos^{-1}(2x^2 - 1)$

52. $\cos [2 \sin^{-1} x] = 1 - 2x^2$

53. $\tan[2 \tan^{-1} x] = \frac{2x}{1 - x^2}$

54. $\cos\left[\frac{1}{2} \cos^{-1} x\right] = \sqrt{\frac{1 + x}{2}}$

In problems 55–58, use the graph of the inverse sine or the inverse cosine to sketch the graph of each function.

55. $y = \frac{1}{2} \sin^{-1} x$

56. $y = -3 \cos^{-1} x$

57. $y = 1 + \cos^{-1} x$

58. $y = 2 - \frac{1}{3} \sin^{-1} x$

In problems 59–62, indicate whether or not the statement is an identity on the given interval.

59. $\sin^{-1}(\sin y) = y$ for $-\frac{\pi}{2} \le y \le \frac{\pi}{2}$

60. $\sin(\sin^{-1} x) = x$ for $-1 \le x \le 1$

61. $\cos(\cos^{-1} x) = x$ for all x

62. $\cos^{-1}(\cos y) = y$ for $-\frac{\pi}{2} \le y \le \frac{\pi}{2}$

63. Explain why $\sin^{-1} \pi$ is *not* defined, whereas $\sin \pi$ is defined.

64. The **inverse cotangent** function, denoted by $\cot^{-1}$ or Arccot, is defined by $y = \cot^{-1} x$ if and only if $x = \cot y$ and $0 < y < \pi$.
(a) Sketch the graph of $y = \cot^{-1} x$ and indicate the domain and range.
(b) Show that

$$\cot^{-1} x = \frac{\pi}{2} - \tan^{-1} x$$

65. The **inverse secant** function, denoted by $\sec^{-1}$ or Arcsec, is defined by $y = \sec^{-1} x$ if and only if $x = \sec y$ and $0 \le y \le \pi$ with $y \ne \pi/2$.
(a) Sketch the graph of $y = \sec^{-1} x$ and indicate the domain and the range.
(b) Show that

$$\sec^{-1} x = \cos^{-1}\left(\frac{1}{x}\right) \quad \text{for} \quad |x| \ge 1$$

66. The **inverse cosecant** function, denoted by $\csc^{-1}$ or Arccsc, is defined by $y = \csc^{-1} x$ if and only if $x = \csc y$ and $-\pi/2 \le y \le \pi/2$ with $y \ne 0$.
(a) Sketch the graph of $y = \csc^{-1} x$ and indicate the domain and range.
(b) Show that

$$\csc^{-1} x = \sin^{-1}\left(\frac{1}{x}\right) \quad \text{for} \quad |x| \ge 1$$

67. A museum plans to hang a painting for public viewing. The painting, which is a meters high, is mounted on a wall in such a way that its lower edge is b meters above the level of a viewer's eye. The viewers pass the painting x meters from the wall on which the painting will be mounted (Figure 8). Show that the angle θ formed by the viewer's eye and the top and the bottom of the painting is given by

$$\theta = \tan^{-1}\left(\frac{ax}{x^2 + ab + b^2}\right)$$

Figure 8

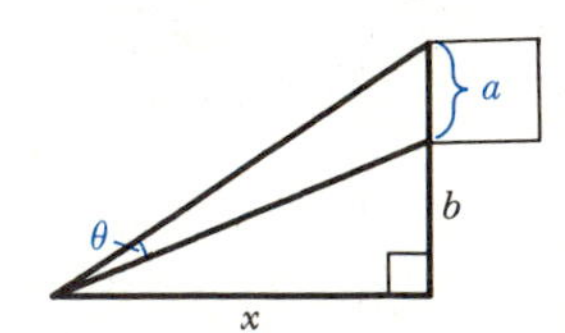

6.5 Trigonometric Equations

Early in the chapter we discussed trigonometric identities and formulas, which are true for *all* real numbers or angles at which the functions are defined. Now we use the properties of the trigonometric functions, together with the identities and algebra, to solve *conditional* trigonometric equations that can yield *more than one solution*.

For example, the equation

$$\sin t = 1$$

is true for $t = \pi/2$, since $\sin(\pi/2) = 1$. But it is also true that $\sin t = \sin(t + 2\pi n)$, where n is any integer, because the sine function is a periodic function of period 2π. Therefore,

$$\sin t = 1 \qquad \text{is true for} \qquad t = \pm\pi/2, \pm 5\pi/2, \pm 9\pi/2, \ldots$$

so that the solution consists of all real numbers t such that

$$t = \frac{\pi}{2} + 2\pi n$$

where n is an integer.

In Examples 1–9, solve each trigonometric equation. Use radian or degree measures as indicated.

EXAMPLE 1 (a) $t = \sin^{-1}\frac{1}{2}$, t is in radians (b) $\sin t = \frac{1}{2}$, $0 \le t < 2\pi$
(c) $\sin t = \frac{1}{2}$, t is in radians

SOLUTION (a) $t = \sin^{-1}\frac{1}{2}$ is equivalent to $\sin t = \frac{1}{2}$, where $-\pi/2 \le t \le \pi/2$, so that the solution is $t = \pi/6$ (Figure 1a).

(b) The restriction $0 \le t < 2\pi$ implies that there are two possible solutions to the equation $\sin t = \frac{1}{2}$, because the sine function is positive in both quadrants I and II. Since $\pi/6$ is in quadrant I and $\sin(\pi/6) = \frac{1}{2}$, the reference angle for the two solutions is $t_R = \pi/6$. Thus, the solution in quadrant I is given by $t = t_R = \pi/6$, and the solution in quadrant II is given by $t = \pi - (\pi/6) = 5\pi/6$. Therefore, the solutions are $t = \pi/6$ and $t = 5\pi/6$ (Figure 1b).

Figure 1

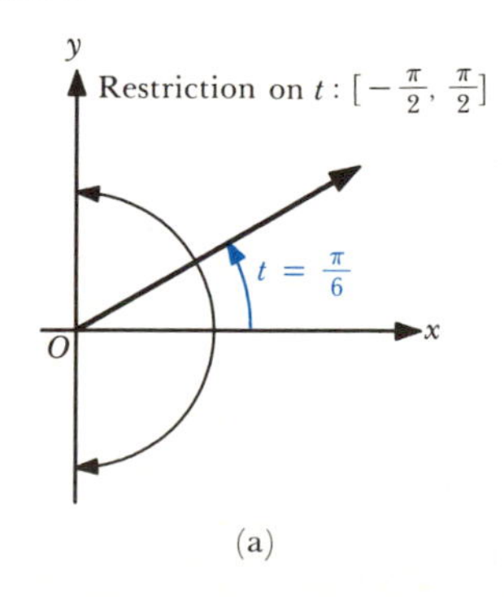

(a)

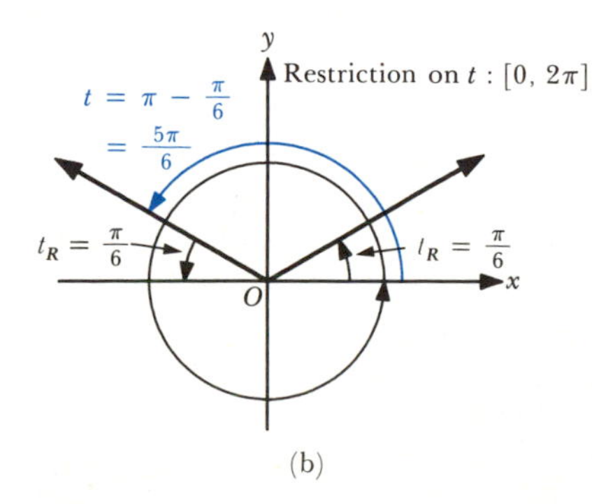

(b)

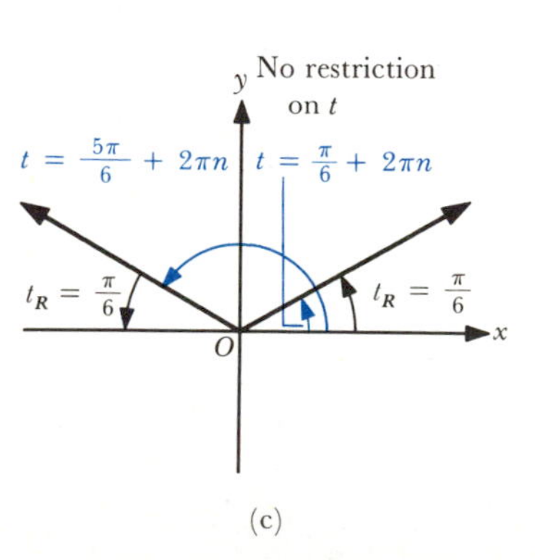

(c)

(c) We begin by solving the equation for $0 \leq t < 2\pi$. Proceeding as in part (b), we find the two solutions

$$t = \frac{\pi}{6} \quad \text{and} \quad t = \frac{5\pi}{6}$$

Using the periodicity of the sine function, we conclude that the complete solution consists of all real numbers t such that

$$t = \frac{\pi}{6} + 2\pi n \quad \text{or} \quad t = \frac{5\pi}{6} + 2\pi n$$

where n is an integer (Figure 1c).

EXAMPLE 2 $\sin \theta = -\frac{1}{2}, \quad 0^\circ \leq \theta < 360^\circ$

SOLUTION Recall that the values of $\sin \theta$ are negative for angles in quadrants III and IV, and that $\sin 30^\circ = \frac{1}{2}$. Therefore, we obtain two positive angles in quadrants III and IV with 30° as a reference angle, namely,

$$\theta = 180^\circ + 30^\circ = 210^\circ$$

and

$$\theta = 360^\circ - 30^\circ = 330^\circ \quad \text{(Figure 2)}$$

Thus the solutions of the equation are 210° and 330°.

Figure 2

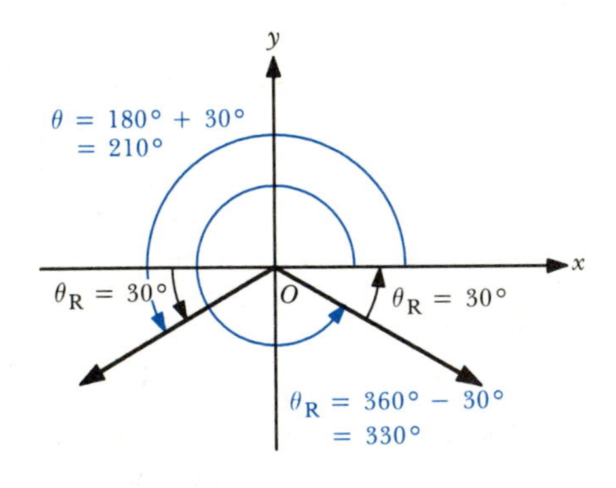

Figure 3

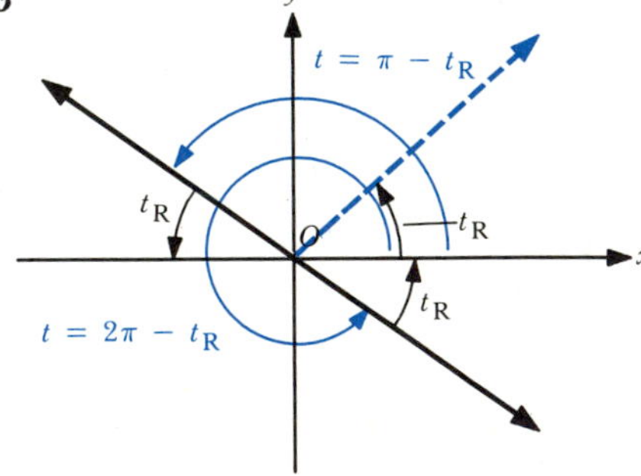

EXAMPLE 3 [c] $\tan t = -1.2137, \quad 0 \leq t < 2\pi$ (use $\pi = 3.14$)

Round off the answer to two decimal places.

SOLUTION The reference angle t_R, which satisfies $\tan t_R = 1.2137$, can be found by using a calculator in radian mode (or Appendix Table IV), rounded off to two decimal places, to get

$$t_R = \tan^{-1} 1.2137 = 0.88 \text{ (Figure 3)}$$

Recall that the tangent is negative in quadrants II and IV, so that in quadrant II

$$t = \pi - t_R = \pi - 0.88 = 2.26$$

and in quadrant IV

$$t = 2\pi - t_R = 2\pi - 0.88 = 5.40$$

Thus the solutions are 2.26 and 5.40.

EXAMPLE 4 $\cos 3\theta = \dfrac{\sqrt{3}}{2}, \qquad 0^\circ \leq \theta < 360^\circ$

SOLUTION First we let $\alpha = 3\theta$, so that the equation is $\cos \alpha = \sqrt{3}/2$. Since $0^\circ \leq \theta < 360^\circ$, then $0^\circ \leq 3\theta < 1080^\circ$. The solutions of

$$\cos \alpha = \frac{\sqrt{3}}{2}$$

for $0^\circ \leq \alpha < 360^\circ$ are $\alpha = 30^\circ$ and $\alpha = 330^\circ$. Adding 360° to these two angle measures, we obtain two more solutions

$$\alpha = 30^\circ + 360^\circ = 390^\circ \qquad \text{and} \qquad \alpha = 330^\circ + 360^\circ = 690^\circ$$

both of which satisfy the condition $0^\circ \leq \alpha < 1080^\circ$. Finally, adding 360° again, we obtain two more solutions,

$$\alpha = 390^\circ + 360^\circ = 750^\circ \qquad \text{and} \qquad \alpha = 690^\circ + 360^\circ = 1050^\circ$$

both of which satisfy the condition $0^\circ \leq \alpha < 1080^\circ$. Thus, the solutions of the equation for $0^\circ \leq 3\theta \leq 1080^\circ$ are given by

$$\alpha = 3\theta = 30^\circ, 330^\circ, 390^\circ, 690^\circ, 750^\circ, \text{and } 1050^\circ$$

and the values of θ are found by dividing each value of 3θ by 3 to obtain the following solutions of the original equation:

$$10^\circ, 110^\circ, 130^\circ, 230^\circ, 250^\circ, \text{and } 350^\circ$$

EXAMPLE 5 $\sin t \tan t = \sin t, \qquad 0 \leq t < 2\pi$

SOLUTION We rewrite the equation as

$$\sin t \tan t - \sin t = 0 \qquad \text{or} \qquad \sin t(\tan - 1) = 0$$

so that, with $0 \leq t < 2\pi$, we have

$\sin t = 0$	$\tan t - 1 = 0$
$t = 0$ or π	$\tan t = 1$
	$t = \dfrac{\pi}{4}$ or $\dfrac{5\pi}{4}$

Therefore, the solutions are $0, \dfrac{\pi}{4}, \pi$, and $\dfrac{5\pi}{4}$.

EXAMPLE 6 $4\cos^2 t - 3 = 0, \qquad 0 \le t < 2\pi$

SOLUTION By factoring the left side of the equation, we have $(2\cos t - \sqrt{3})(2\cos t + \sqrt{3}) = 0$ so that, with $0 \le t < 2\pi$, we have

$2\cos t - \sqrt{3} = 0$	$2\cos t + \sqrt{3} = 0$
$2\cos t = \sqrt{3}$	$2\cos t = -\sqrt{3}$
$\cos t = \dfrac{\sqrt{3}}{2}$	$\cos t = -\dfrac{\sqrt{3}}{2}$
$t = \dfrac{\pi}{6}$ or $\dfrac{11\pi}{6}$	$t = \dfrac{5\pi}{6}$ or $\dfrac{7\pi}{6}$

Therefore, the solutions are $\dfrac{\pi}{6}, \dfrac{5\pi}{6}, \dfrac{7\pi}{6}$, and $\dfrac{11\pi}{6}$.

EXAMPLE 7 $\tan^2\theta + \sec\theta - 1 = 0, \qquad 0° \le \theta < 360°$

SOLUTION Using the Pythagorean identity $\tan^2\theta + 1 = \sec^2\theta$, we rewrite the equation as

$$\sec^2\theta - 1 + \sec\theta - 1 = 0$$

or

$$\sec^2\theta + \sec\theta - 2 = 0$$

$$(\sec\theta + 2)(\sec\theta - 1) = 0$$

Setting each factor equal to zero and keeping in mind that $0° \le \theta < 360°$, we have

$\sec\theta - 1 = 0$	$\sec\theta + 2 = 0$
$\sec\theta = 1$ or $\cos\theta = 1$	$\sec\theta = -2$ or $\cos\theta = -\frac{1}{2}$
$\theta = 0°$	$\theta = 120°$ or $240°$

Therefore, the solutions are 0°, 120°, and 240°.

EXAMPLE 8 C $2\csc^2 t - \cot t - 5 = 0, \qquad 0 \le t < 2\pi$

Round off the answers to two decimal places.

SOLUTION Using the Pythagorean identity $\csc^2 t = \cot^2 t + 1$, we rewrite the equation as

$$2(\cot^2 t + 1) - \cot t - 5 = 0$$

$$2\cot^2 t - \cot t - 3 = 0$$

The left side of the resulting equation can be factored as

$$(2\cot t - 3)(\cot t + 1) = 0$$

Setting each factor equal to zero with $0 \le t < 2\pi$, we have

$2\cot t - 3 = 0$	$\cot t + 1 = 0$
$2\cot t = 3$	$\cot t = -1$
$\cot t = \frac{3}{2}$ or $\tan t = \frac{2}{3}$	
$t = 0.59$ or 3.73	$t = \dfrac{3\pi}{4}$ or $\dfrac{7\pi}{4}$

Therefore, the solutions are

$$0.59,\ 3.73,\ \frac{3\pi}{4} = 2.36, \text{ and } \frac{7\pi}{4} = 5.50$$

EXAMPLE 9 $\sin\theta + \cos\theta = 1, \qquad 0° \le \theta < 360°$

SOLUTION We rewrite the equation as $\sin\theta = 1 - \cos\theta$, and square both sides to get

$$\sin^2\theta = (1 - \cos\theta)^2$$

Using the Pythagorean identity $\sin^2\theta + \cos^2\theta = 1$, we have

$$1 - \cos^2\theta = (1 - \cos\theta)^2$$
$$1 - \cos^2\theta = 1 - 2\cos\theta + \cos^2\theta$$
$$0 = 2\cos^2\theta - 2\cos\theta$$

so that

$$2\cos\theta(\cos\theta - 1) = 0$$

and, since $0° \le \theta < 360°$, we have

$\cos\theta = 0$	$\cos\theta - 1 = 0$
$\theta = 90°$ or $270°$	$\cos\theta = 1$
	$\theta = 0°$

In this case, we must check for *extraneous roots* because we squared both sides of the equation. Substituting $\theta = 0°$ in the original equation, we get

$$\sin 0° + \cos 0° = 1 \qquad \text{which is } \textit{true}.$$

Substituting $\theta = 90°$, we have

$$\sin 90° + \cos 90° = 1 \qquad \text{which is } \textit{true}.$$

Finally, substituting $\theta = 270°$, we have

$$\sin 270° + \cos 270° = 1 \qquad \text{which is } \textit{false}$$

because $\sin 270° + \cos 270° = -1 + 0 = -1$. Therefore, the solutions of the original equation are $0°$ and $90°$.

EXAMPLE 10 [c] Recall the circuit problem, Example 9 on page 298, where the equation was given by

$$Q = Q_0 \sin\left(\frac{t}{\sqrt{LC}} - \frac{\pi}{2}\right)$$

Find the first time t when $Q = \frac{1}{2}Q_0$, if $L = 0.16$ henry, $C = 10^{-4}$ farad, and t represents the number of elapsed seconds (use $\pi = 3.142$). Round off the answer to three decimal places.

SOLUTION Using the given conditions, we get

$$\frac{1}{2}Q_0 = Q_0 \sin\left(\frac{t}{\sqrt{0.16 \cdot 10^{-4}}} - \frac{\pi}{2}\right)$$

which is equivalent to

$$\sin\left(\frac{t}{0.004} - \frac{\pi}{2}\right) = \frac{1}{2}$$

We know that

$$\sin\frac{\pi}{6} = \frac{1}{2}$$

Thus

$$\frac{t}{0.004} - \frac{\pi}{2} = \frac{\pi}{6} \quad \text{or} \quad \frac{t}{0.004} - 1.571 = 0.524$$

so that $t = 0.008$ second (approximately).

PROBLEM SET 6.5

In problems 1 and 2, solve each trigonometric equation and explain why the solutions are different.

1. (a) $t = \cos^{-1}\frac{1}{2}$, t is in radians (b) $\cos t = \frac{1}{2}$, $0 \le t < 2\pi$ (c) $\cos t = \frac{1}{2}$, t is in radians

2. (a) $\theta = \sin^{-1}\left(-\frac{\sqrt{3}}{2}\right)$, θ is in degrees (b) $\sin\theta = -\frac{\sqrt{3}}{2}$, $0° \le \theta < 360°$

(c) $\sin\theta = -\frac{\sqrt{3}}{2}$, θ is in degrees

In problems 3–22, solve each trigonometric equation with the condition $0 \le t < 2\pi$ or $0° \le \theta < 360°$. Do not use a calculator or tables.

3. $\sin t = \frac{\sqrt{3}}{2}$ **4.** $\sin\theta = -\frac{\sqrt{2}}{2}$ **5.** $\tan\theta = -1$

6. $\cot t = -\sqrt{3}$
7. $\sec t = \sqrt{2}$
8. $\csc \theta = \frac{-2\sqrt{3}}{3}$
9. $\cos \theta = 0$
10. $\sin t = -1$
11. $\tan \theta = \sqrt{3}$
12. $\cot \theta = -\frac{\sqrt{3}}{3}$
13. $\csc t = \sqrt{2}$
14. $\cot t = -1$
15. $2 \cos t + 1 = 0$
16. $2 \cos \theta - \sqrt{3} = 0$
17. $\sqrt{3} + 2 \sin \theta = 0$
18. $3 \tan \theta + \sqrt{3} = 0$
19. $6 \csc t - 4\sqrt{3} = 0$
20. $\sqrt{3} \tan \theta - 1 = 0$
21. $2 \sec t + 4 = 0$
22. $4 \csc \theta - 8 = 0$

c In problems 23–30, use a calculator or tables to solve each trigonometric equation for $0 \leq t < 2\pi$ or $0° \leq \theta < 360°$. Round off all answers to two decimal places.

23. $\sin t = 0.8134$
24. $\cos t = \frac{2}{3}$
25. $\cos \theta = -0.4176$
26. $\tan \theta = 0.6696$
27. $\tan t = \frac{5}{6}$
28. $\sec \theta = 10.1561$
29. $\cot \theta = 6.6173$
30. $\csc t = 1.5763$

In problems 31–66, solve each trigonometric equation with the condition $0 \leq t < 2\pi$ or $0° \leq \theta < 360°$. Do not use a calculator.

31. $\cos 3\theta = -\frac{\sqrt{2}}{2}$
32. $\sin 3\theta = 1$
33. $\sec 2t = \sqrt{2}$
34. $\tan 5t = \sqrt{3}$
35. $\csc \frac{t}{3} = 2$
36. $\sin^2 t = \frac{1}{2} \sin t$
37. $4 \sin^2 \theta - 3 = 0$
38. $3 - \tan^2 t = 0$
39. $\sec^2 \theta - 4 = 0$
40. $3 \cot^2 t - 1 = 0$
41. $\sin t \cos t = \sin t$
42. $(\sin \theta - 1) \cos \theta = 0$
43. $(2 \sin \theta + \sqrt{3}) \cos \theta = 0$
44. $\sec^2 \theta + \sec \theta = 0$
45. $2 \sin \theta \cos \theta = \cos \theta$
46. $\sin 2t = \sqrt{2} \cos t$
47. $\sin^2 t = \cos^2 t$
48. $\cos 2t = 2 \sin^2 t$
49. $\cos 2\theta = \cos \theta$
50. $\cos 2t = 1 - \sin t$
51. $2 \cos^2 \theta - \cos \theta = 0$
52. $\sqrt{3} \csc^2 \theta + 2 \csc \theta = 0$
53. $2 \tan \theta - \sec^2 \theta = 0$
54. $\sin^2 t - 2 \sin t + 1 = 0$
55. $2 - \sin t = 2 \cos^2 t$
56. $\cot^2 t + \csc^2 t = 3$
57. $2 \sin^2 \theta - \sin \theta - 1 = 0$
58. $\tan^2 \theta - \sec \theta + 1 = 0$
59. $\tan^2 t - 2 \tan t + 1 = 0$
60. $1 - \sin t = \sqrt{3} \cos t$
61. $2 \cos^2 \theta - \sin \theta = 1$
62. $\cos 2\theta + \sin 2\theta = 0$
63. $\tan \theta - 3 \cot \theta = 0$
64. $\tan \theta + \sec \theta = 1$
65. $\cos \theta - \sin \theta = 1$
66. $\sec^5 t = 4 \sec t$

c In problems 67–72, solve each equation for $0 \leq t < 2\pi$ or $0° \leq \theta < 360°$. Use a calculator or tables and round off all answers to two decimal places.

67. $15 \sin^2 t - 8 \sin t + 1 = 0$
68. $\tan^2 \theta - 3 \tan \theta + 2 = 0$
69. $9(\cos^2 \theta + \sin \theta) = 11$
70. $\cot^2 t - 5 \cot t + 4 = 0$
71. $5 \sin^2 t + 2 \sin t - 3 = 0$
72. $4 \cos^2 \theta - 5 \sin \theta \cot \theta - 6 = 0$

c In problems 73–76, use the following information to solve for the unknown angle to the nearest hundredth of a degree. In physics there is a formula, called **Snell's**

law, which deals with the change in direction of a ray of light as it passes through a medium (Figure 4). In Figure 4 the dashed line is perpendicular to the surface of the medium. The angle α is called the **angle of incidence** and the angle β is called the **angle of refraction.** Snell's law states that, for a given medium,

$$\frac{\sin \alpha}{\sin \beta} = C$$

where C is a constant. The constant C is called the **index of refraction** of the medium.

Figure 4

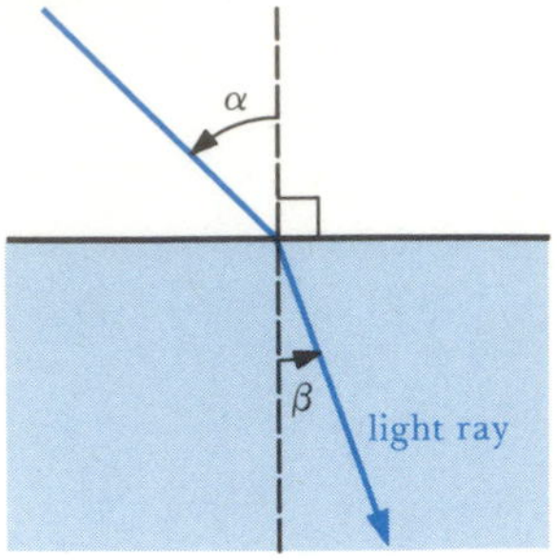

73. Determine the angle of refraction of a light ray traveling through a diamond if the angle of incidence is 30° and the index of refraction is 2.42.

74. Find the angle of refraction of a light ray traveling through ice if the angle of incidence is 45° and the index of refraction is 1.31.

75. Find the angle of incidence of a light ray striking a rock-salt crystal if the angle of refraction of the ray is 35°. The index of refraction of rock salt is 1.54.

76. Find the angle of incidence of a light ray striking the surface of turpentine in a container if the angle of refraction of the ray is 10°40′. The index of refraction of turpentine is 1.47.

6.6 Applications Involving Right Triangle Trigonometry

The right triangle trigonometric relationships that were established in Section 5.3 enable us to answer questions about right triangles that arise in such applications as surveying, engineering, and navigation.

In this section we use some standard notation to describe the sides and angles of any triangle. The triangle determined by points A, B, and C is denoted by triangle ABC. The angle at vertex A is denoted by the Greek letter *alpha*, α; the angle at vertex B is denoted by *beta*, β; and the angle at vertex C is denoted by *gamma*, γ. The length of the side opposite α is denoted by a; the length of the side opposite β by b; and the length of the side opposite γ by c (Figure 1). To **solve a triangle** means to find all of its parts; that is, to find the lengths of the three sides and the measures of the three angles.

Figure 1

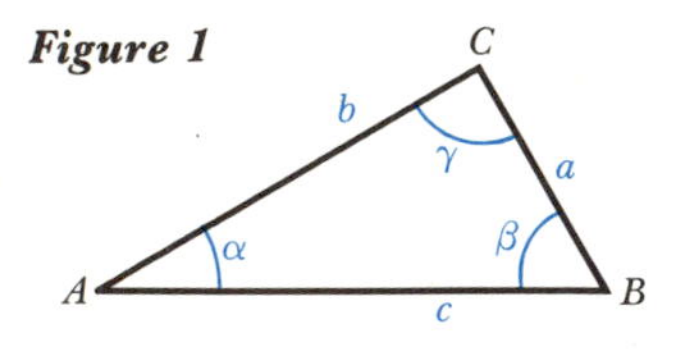

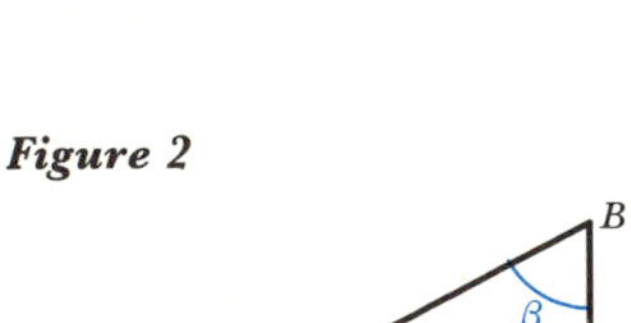

Recall that in right triangle ABC (Figure 2), if the angle at vertex C is the right angle and c is the hypotenuse, then

Figure 2

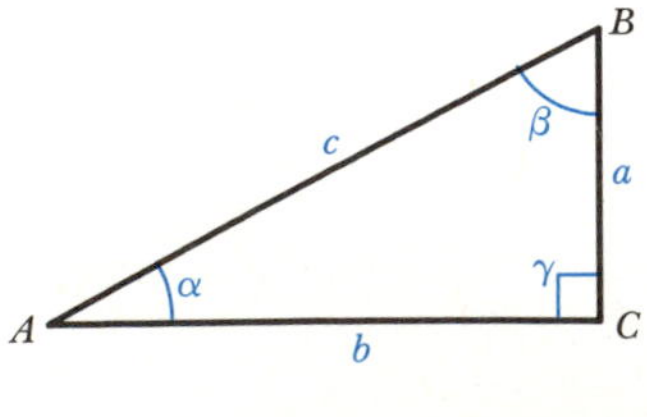

$$\sin \alpha = \cos \beta = \frac{a}{c} \qquad \csc \alpha = \sec \beta = \frac{c}{a}$$

$$\cos \alpha = \sin \beta = \frac{b}{c} \qquad \sec \alpha = \csc \beta = \frac{c}{b}$$

$$\tan \alpha = \cot \beta = \frac{a}{b} \qquad \cot \alpha = \tan \beta = \frac{b}{a}$$

We should always keep in mind that the solutions obtained using either a calculator or tables are often *approximations*. In the solutions of triangles, we round off all angles to the nearest hundredth of a degree, and all side lengths to four significant digits.

C *In Examples 1 and 2, assume that ACB is a right triangle labeled as in Figure 2 and solve the triangle, with the given parts.*

EXAMPLE 1 $b = 4$ and $\alpha = 40°$

SOLUTION We see from Figure 3 that

$$\beta = 90° - \alpha = 90° - 40° = 50°$$

Figure 3

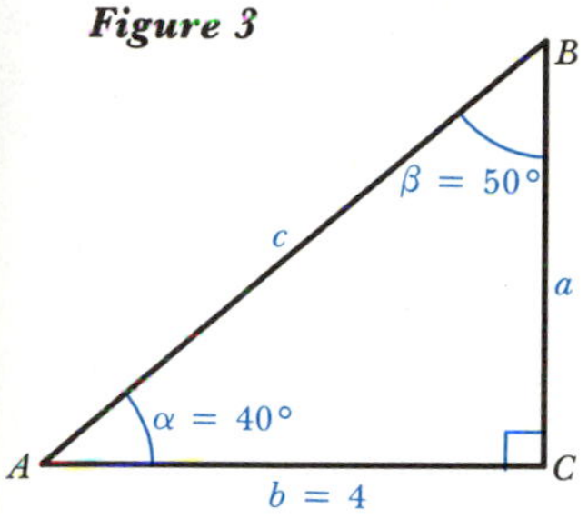

To find a, we notice that

$$\tan 40° = \frac{a}{b} = \frac{a}{4}$$

or

$$a = 4 \tan 40° = 3.356$$

To find c, we notice that

$$\cos 40° = \frac{4}{c}$$

so that

$$c = \frac{4}{\cos 40°} = 5.222$$

EXAMPLE 2 $a = 4$ and $b = 3$

SOLUTION By the Pythagorean theorem,

$$c = \sqrt{a^2 + b^2} = \sqrt{16 + 9} = 5$$

Figure 4

From Figure 4, we see that

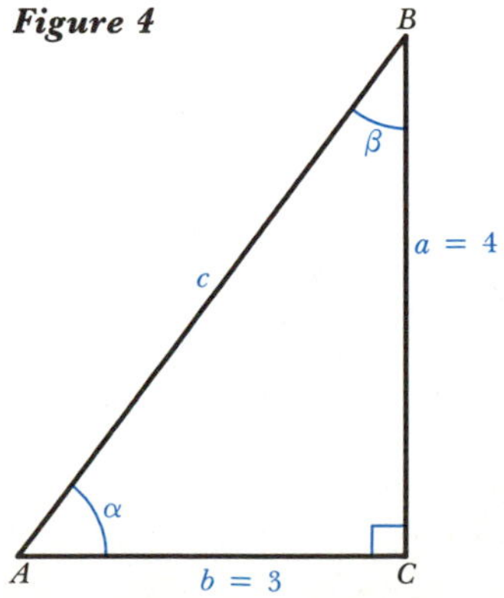

$$\sin \alpha = \frac{a}{c} = \frac{4}{5} = 0.8$$

so that

$$\alpha = \sin^{-1} 0.8 = 53.13°$$

It follows that

$$\beta = 90° - \alpha = 90° - 53.13° = 36.87°$$

Frequently, distances and angles that arise in applications may be calculated by using right triangle trigonometry.

EXAMPLE 3 [C] A roadway rises 10 feet for every 200 feet along the horizontal. Find the angle of inclination of the roadway (Figure 5).

SOLUTION From Figure 5 it is clear that α represents the angle of inclination. Thus

Figure 5

$$\tan \alpha = \frac{10}{200} = 0.05$$

so that $\alpha = \tan^{-1}(0.05) = 2.86°$.

The technical terms *angle of elevation* and *angle of depression* occur frequently in practical applications. In Figure 6a an observer at A views an object at B. The acute angle CAB formed by a horizontal line $\overrightarrow{AC}$ and an observer's line of sight $\overline{AB}$ to any object B above the horizontal is called the **angle of elevation.** Similarly, the acute angle EAF formed by a horizontal line $\overline{AE}$ and an observer's line of sight $\overline{AF}$ to an object F below the horizontal is called the **angle of depression** (Figure 6b).

Figure 6

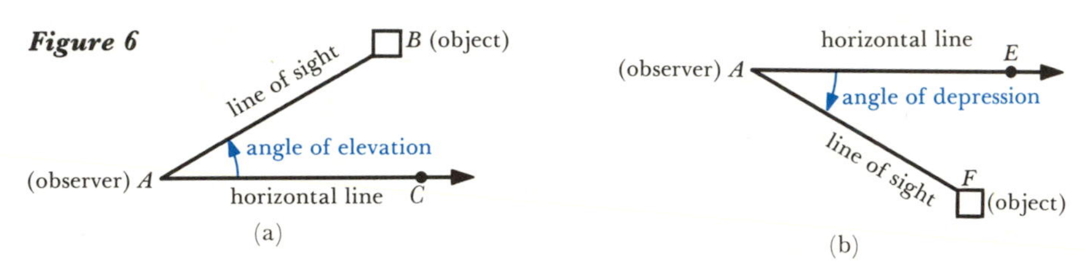

EXAMPLE 4 [C] A rocket rises vertically upward. A camera on the ground is located at a point 1000 meters horizontally from the base of the launching pad. At a certain instant, the angle of elevation of the camera, focusing on the bottom of the rocket, is 53° (Figure 7). How high above the ground is the rocket at that instant?

SOLUTION Let h represent the altitude of the rocket. In Figure 7,

$$\tan 53° = \frac{h}{1000}$$

so that $h = 1000 \tan 53° = 1327$

Therefore, the rocket is approximately 1327 meters above the ground.

Figure 7

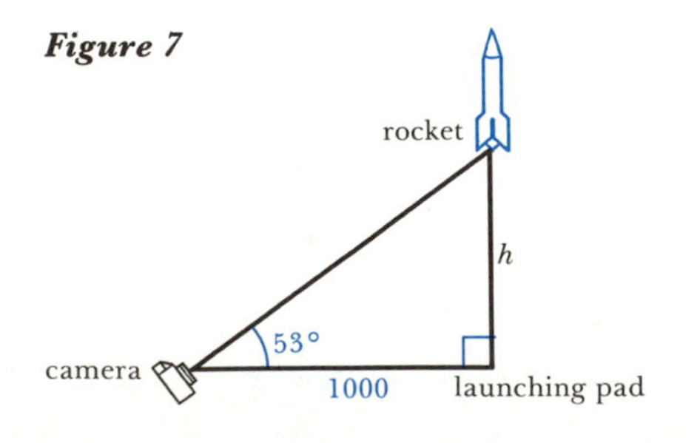

EXAMPLE 5 [c] A Coast Guard tower 150 feet high is situated on the shore of a lake. At the top of the tower, a Coast Guard observer at A determines that the angles of depression of two sailboats located at C and D in the lake, are 25° and 12°, respectively. Assuming that the Coast Guard tower and the two sailboats lie in the same vertical plane, find the distance $|\overline{CD}|$ between the two sailboats (Figure 8).

SOLUTION In Figure 8, A represents the position of the Coast Guard observer, C the position of the first sailboat, and D the position of the second sailboat. We know from plane geometry that if two parallel lines are cut by a transversal, the alternate interior angles are equal, so that angle $BCA = 25°$ and angle $BDA = 12°$. In right triangle ABC

Figure 8

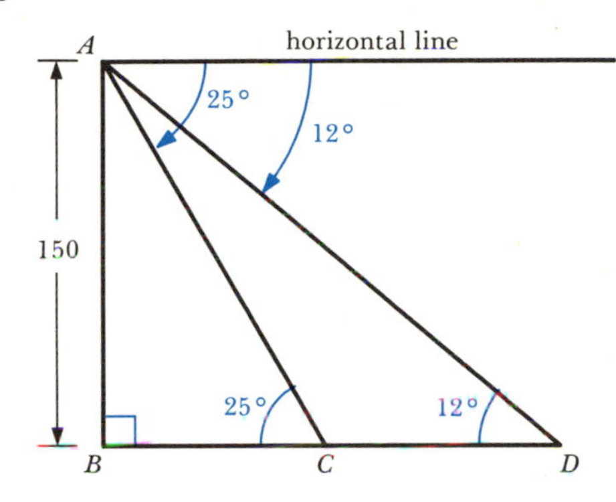

$$\tan 25° = \frac{|\overline{AB}|}{|\overline{BC}|} = \frac{150}{|\overline{BC}|}$$

so that

$$|\overline{BC}| = \frac{150}{\tan 25°} = 321.7$$

In right triangle ABD

$$\tan 12° = \frac{|\overline{AB}|}{|\overline{BD}|} = \frac{150}{|\overline{BD}|}$$

so

$$|\overline{BD}| = \frac{150}{\tan 12°} = 705.7$$

Hence,

$$|\overline{CD}| = |\overline{BD}| - |\overline{BC}| = 705.7 - 321.7 = 384$$

Therefore, the sailboats are approximately 384 feet apart.

Figure 9

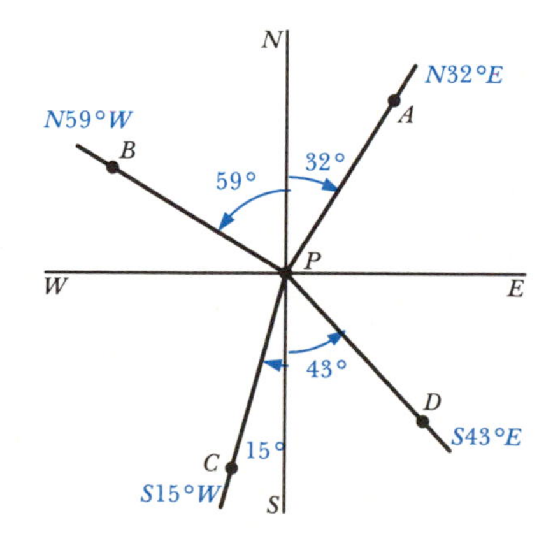

In some applications of trigonometry, especially in surveying and navigation, the concept of **direction** or **bearing** of a point A as viewed from point P is defined to be the positive acute angle that the ray $\overrightarrow{PA}$ makes with a north–south line through P. Such an angle is specified as being measured east or west from north or south. For example, in Figure 9, the bearing from P to A is 32° east of north or N32°E. Similarly, the bearing from P to B is N59°W; the bearing from P to C is S15°W; and the bearing from P to D is S43°E.

EXAMPLE 6 [C] A lifeguard at station A sights a swimmer C directly south of her. Another lifeguard at station B, 70 meters directly east of A, sights the same swimmer at a bearing of S43°W. How far is the swimmer from station A?

SOLUTION In Figure 10, the measure of angle CBA is given by $90° - 43° = 47°$. Also,

$$\tan 47° = \frac{d}{70}$$

so

$$d = 70 \tan 47° = 75.07$$

Therefore, the swimmer is approximately 75.07 meters due south of lifeguard A.

Figure 10

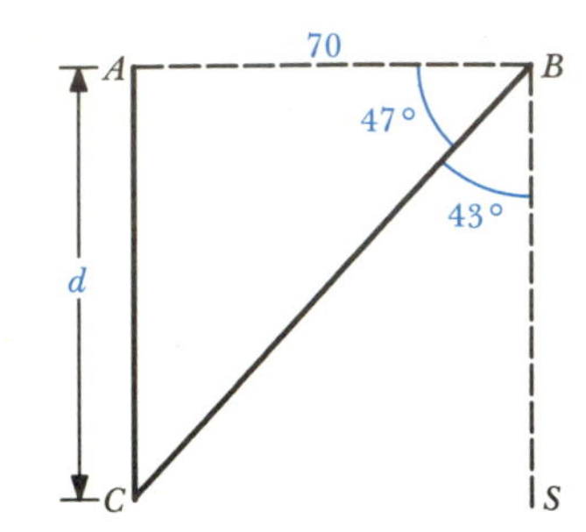

PROBLEM SET 6.6

[C] In all problems, round off all angles to the nearest hundredth of a degree and side lengths to four significant digits.

In problems 1–14, assume that ACB is a right triangle with $\gamma = 90°$ (Figure 11). In each case, solve the triangle.

Figure 11

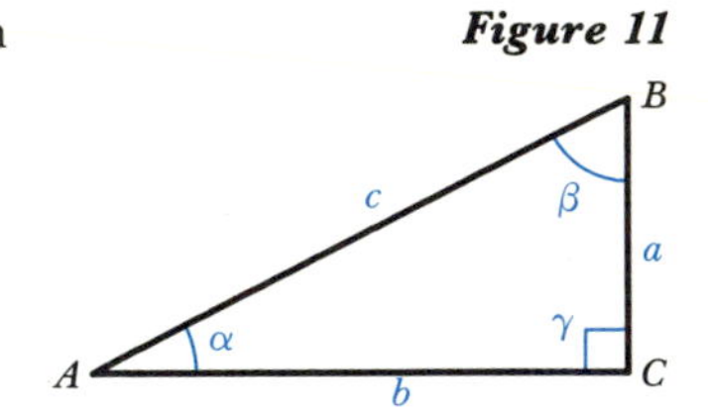

1. $a = 5, b = 3$
2. $a = 10, \alpha = 30°$
3. $c = 10, \beta = 50°$
4. $a = 7, c = 12$
5. $a = 17, \beta = 23°$
6. $c = 100, \beta = 14.6°$
7. $a = 22, \alpha = 36°$
8. $a = 25, \beta = 42.5°$
9. $a = 73.96, c = 123.5$
10. $a = 8.14, \alpha = 3.14°$
11. $c = 46.13, \alpha = 17.7°$
12. $b = 8.27, \alpha = 6.3°$
13. $c = 100, \alpha = 14.6°$
14. $c = 200, \alpha = 34.4°$

15. The lot for a new home site is uniformly pitched from the horizontal at an angle of 0.51° (Figure 12). How many inches will the ground drop if one walks 100 feet down the slope?

Figure 12

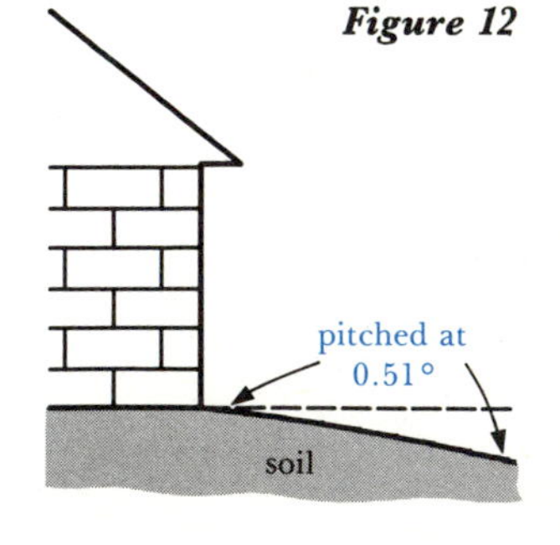

16. A straight sidewalk is inclined to the horizontal at an angle of 5.3°. How far must one walk along the sidewalk to change elevation by 2 meters?

17. A ladder 5 meters long is leaning against a building. The angle formed by the ladder and the ground is 65°. How far from the building is the foot of the ladder?

18. A monument is 180 meters high. What is the length of the shadow cast by the monument if the angle of elevation of the sun is 58.4°?

19. A guy wire attached to a utility pole makes an angle of 71.4° with the ground. If the end of the wire attached to the ground is 15.5 feet from the pole, how high up the pole is the other end of the wire attached?

20. A kite is flying 8.3 meters above the ground and 10 meters of string is let out. What is the angle of elevation of the string if we assume that the string is straight?

21. A rocket rises vertically upward at a speed of 800 feet per second. A camera is mounted at a point 2000 feet from the base of the rocket launching pad (see Figure 7 on page 356). What is the angle of elevation of the bottom of the rocket from the camera 10 seconds after the rocket is fired?

22. A military aircraft is climbing at an angle of 9.2° at a speed of 800 kilometers per hour. What is its gain in altitude in 8 minutes?

23. A missile is launched from sea level and climbs at a constant angle of 71° for a distance of 180 kilometers. What is the altitude of the missile at that moment?

24. A lamp is suspended above the center of a round table of radius 4 feet. How high above the table is the lamp if the angle of elevation from the edge of the table to the lamp is 34°?

25. A cable television company has its master antenna located at a point A on the bank of a straight river 1 kilometer wide (Figure 13). It is going to run a cable from A to a point C on the opposite side of the river. If point B is on the same side as C and located directly opposite A, and if the angle BCA is 48.3°, what is the length of the cable?

Figure 13

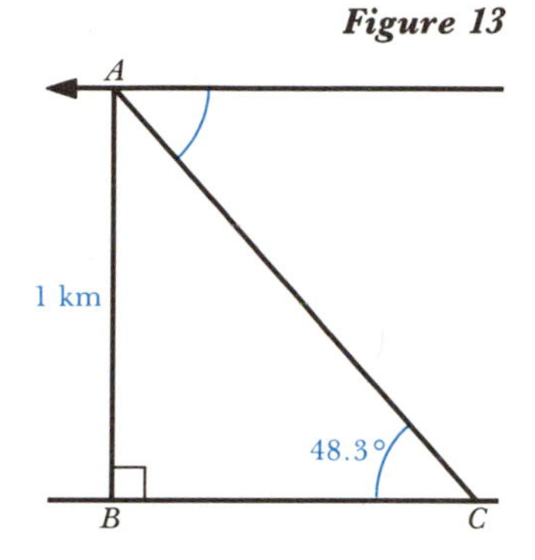

26. An offshore oil well is located at point A, which is 6 miles from the nearest point B on a straight shoreline. The oil is to be piped from A to a terminal at a point C on the same shoreline as B. If angle BCA is 35°, find the distance from A to C.

27. If the angle of elevation of the top of a tower from a distance 200 feet away on the ground level is 60°, find the height of the tower.

28. What is the angle of elevation of the sun when a woman 2 meters tall casts a 3.1-meter shadow?

29. A railroad bridge runs perpendicular to a roadway and is 25 feet above the road level. Find the angle of elevation of the bottom of the bridge from a point P on the roadway that is 189.33 feet away from the bridge (Figure 14).

Figure 14

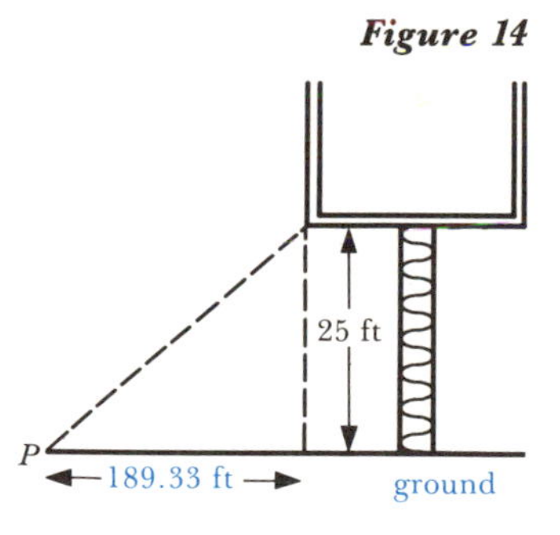

30. From the top of a lighthouse the angle of depression of a ship is 30°. The base of the lighthouse is on the same plane as the ship. If the distance between the top of the lighthouse and the ship is 4500 feet, find the height of the lighthouse.

31. From the top of a 100 meter tower, the angles of depression of the top and the bottom of a flagpole standing on a plane level with the base of the tower are observed to be 45° and 60°, respectively. Find the height of the flagpole.

32. From the top of a cliff, an observer views two boats sailing on a lake. The observer and the two boats lie in the same vertical plane and the base of the cliff is on the lake shore. If the observer is 2800 feet above the lake level and the angles of depression of the two boats are 18° and 14°, respectively, find the distance between the boats.

33. From an altitude of 5000 meters, a balloonist observes the angle of depression of the base of a building to be 20°. How far is the building from a point on the ground directly beneath the balloon?

34. A television antenna is situated at the edge of a flat roof on top of a house that is located on level ground. From a point 14 feet from the base of the house on the side where the antenna is placed, the angles of elevation of the top and bottom of the antenna measure 70° and 66°, respectively. How high is the house? How tall is the television antenna?

35. A mountainside hotel is located on a shoreline 3000 meters above a lake. From the hotel it is observed that the angles of depression of two ships in line with the hotel are 24.3° and 11.2°, respectively. Find the distance between the two ships.

36. A high-altitude reconnaissance jet photographs a missile silo under construction near a small town. The jet is at an altitude of 16 kilometers, and the angles of depression of the town and the silo are 58° and 27°, respectively. Assuming that the jet, the silo, and the town lie in the same vertical plane, find the distance between the town and the silo.

37. Two tourists, standing on the same side of and in line with the Washington Monument, are looking at its top. The angle of elevation from the first tourist is 33.8°, and from the second is 59.4°. If the two tourists stand on level ground and are 170 meters apart, approximately how tall is the monument? (Neglect the tourists' heights.)

38. The *pitch* of a roof is defined to be the ratio h/w where h is the "rise" of the roof rafter and w is the width of the house (Figure 15). Assume that a house of width 32 feet has a roof with a pitch of $\frac{1}{6}$. Find the rise of the roof, the length of a rafter, and the angle α formed by a rafter and the horizontal support that goes across the width of the house.

Figure 15

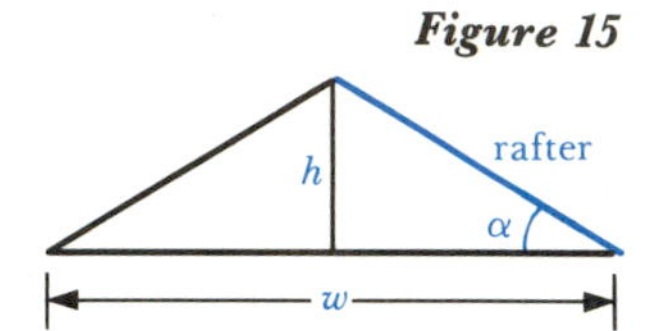

39. A Coast Guard observer at lookout A sights a ship directly north of him. Another observer at lookout B, 2 kilometers directly west of A, sights the same ship at a bearing of N41.2°E. How far is the ship from lookout A?

40. Two observation stations are located 10 kilometers apart on a straight north–south coastline. The bearings of a ship as measured from these two stations are S52°E and N73°E, respectively. Find the distance from the ship to the shore.

41. A police helicopter flying at a constant altitude spots a car directly below it. Another police helicopter flying at the same altitude, 4 miles directly east of the first helicopter, spots the same car at a bearing of S79.3°W. What is the altitude of the two helicopters in feet?

6.7 Law of Sines and Law of Cosines

In this section, we study relationships among the three angles α, β, and γ and the lengths of the opposite sides a, b, and c of *any* triangle ABC (Figure 1), by using two formulas called the *law of sines* and the *law of cosines*.

Figure 1

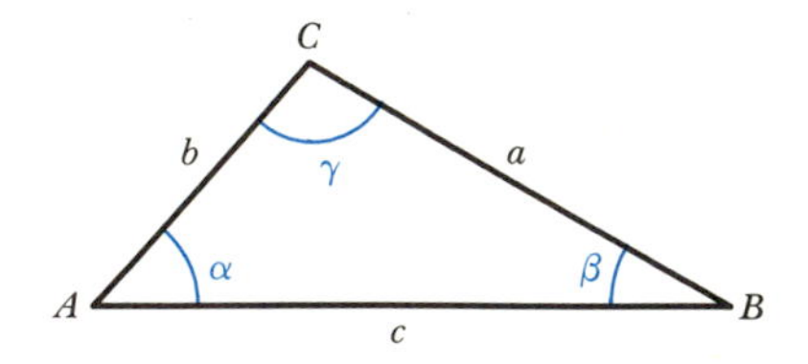

The Law of Sines

The *law of sines,* which relates the lengths of the three sides of *any* triangle to the three angles, is derived as follows.

THEOREM 1 **Law of Sines**

> In any triangle *ABC*,
>
> $$\frac{\sin\alpha}{a} = \frac{\sin\beta}{b} = \frac{\sin\gamma}{c}$$

PROOF (See Figure 2.) Let h_1 be the altitude of triangle ABC from vertex C to side $\overline{AB}$ and let h_2 be the altitude from vertex B to side $\overline{AC}$.

Figure 2

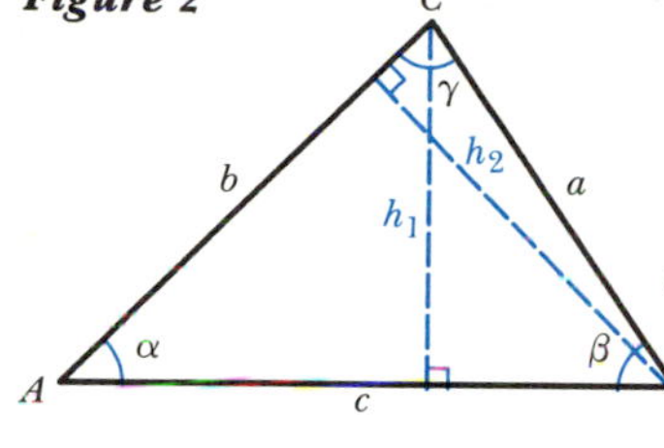

By using right triangle trigonometry, we have

$$\sin\alpha = \frac{h_1}{b} \quad \text{and} \quad \sin\beta = \frac{h_1}{a}$$

or $h_1 = b\sin\alpha$ and $h_1 = a\sin\beta$, so that $b\sin\alpha = a\sin\beta$. That is,

$$\frac{\sin\alpha}{a} = \frac{\sin\beta}{b}$$

Again by using right triangle trigonometry, we obtain

$$\sin\alpha = \frac{h_2}{c} \quad \text{and} \quad \sin\gamma = \frac{h_2}{a}$$

so that $h_2 = c\sin\alpha = a\sin\gamma$ or $$\frac{\sin\alpha}{a} = \frac{\sin\gamma}{c}$$

Thus $$\frac{\sin\alpha}{a} = \frac{\sin\beta}{b} = \frac{\sin\gamma}{c}$$

It is important to note that the law of sines can also be written in the equivalent reciprocal form

$$\frac{a}{\sin\alpha} = \frac{b}{\sin\beta} = \frac{c}{\sin\gamma}$$

The following two examples illustrate the procedure for solving triangles by using the law of sines when we know the measures of two angles and a side. Unless otherwise stated, we round off all angles to the nearest hundredth of a degree and all side lengths to four significant digits.

EXAMPLE 1 In triangle ABC (Figure 3) suppose that $a = 20$, $\gamma = 51°$, and $\beta = 42°$. Solve the triangle for α, b, and c.

SOLUTION Since $\alpha + \beta + \gamma = 180°$, then

$$\alpha = 180° - (\beta + \gamma)$$
$$= 180° - 93°$$
$$= 87°$$

Figure 3

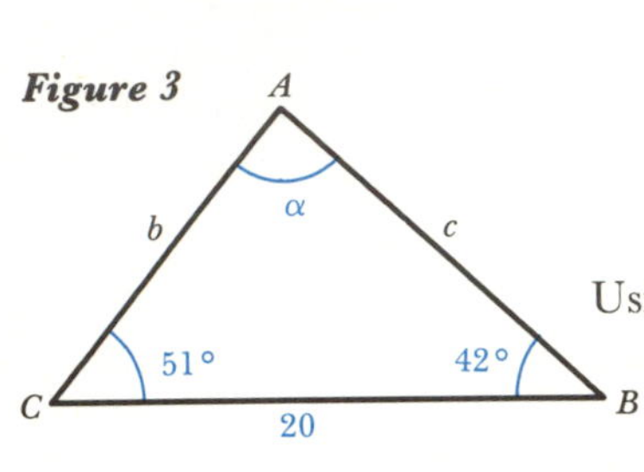

Using the law of sines, we have $\dfrac{b}{\sin 42°} = \dfrac{20}{\sin 87°}$

so that $$b = \frac{20 \sin 42°}{\sin 87°} = 13.40$$

To find c, we use the law of sines to get

$$\frac{c}{\sin 51°} = \frac{20}{\sin 87°}$$

so that

$$c = \frac{20 \sin 51°}{\sin 87°} = 15.56$$

EXAMPLE 2 A surveyor wishes to find the distance between two cities situated on opposite banks of a river at points D and W (Figure 4). He measures the length of line segment $\overline{DC}$ to be 550 feet. Also, he determines that angle $CDW = 125.67°$ and angle $DCW = 48.83°$. Find the distance $|\overline{DW}|$ between the two cities.

SOLUTION Angle DWC is easily found, since angle $DWC = 180° - (125.67° + 48.83°) = 5.50°$. In triangle DWC, we can use the law of sines to find $c = |\overline{DW}|$ as follows:

$$\frac{c}{\sin 48.83°} = \frac{550}{\sin 5.50°}$$

Figure 4

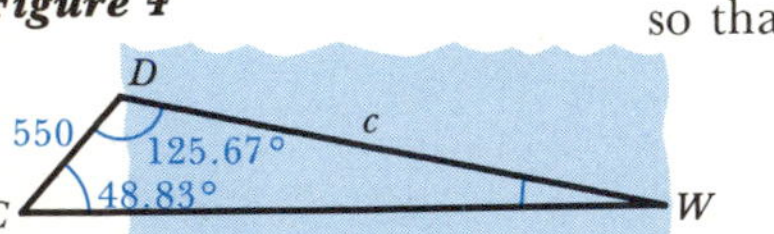

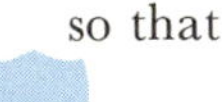

so that

$$c = \frac{550 \sin 48.83°}{\sin 5.50°} = 4320$$

Therefore, the distance $|\overline{DW}|$ between the two cities is approximately 4320 feet.

As illustrated in the above examples, *the law of sines can be applied to find unknown parts of a triangle if we are given the measurements of any two angles and any side*. This situation always yields a unique triangle. *The law of sines is also applicable if we are given the measurements of two sides and an angle opposite one of them*. This situation does not always determine a unique triangle.

The Ambiguous Case

Let us examine the various cases that may occur if we are given two sides and an angle opposite one of them. In Figure 5, consider a, b, and α in each situation and let h represent the length of the perpendicular line segment. It may be that there is no possible triangle with these parts (Figure 5a); there may be two different triangles with these parts (Figure 5b); or there may be only one triangle with these parts (Figure 5c). Figure 5d shows that we may obtain a right triangle.

Figure 5

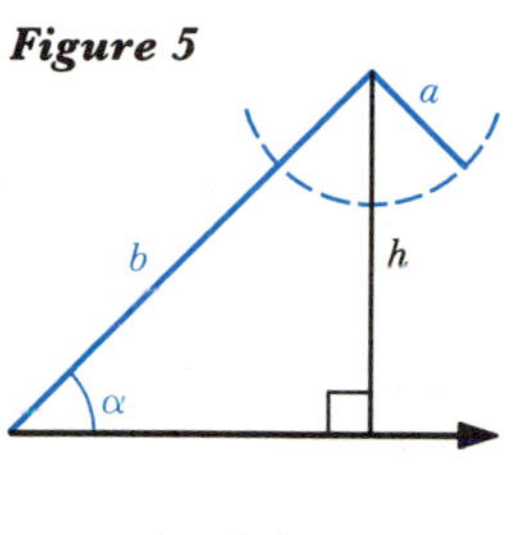

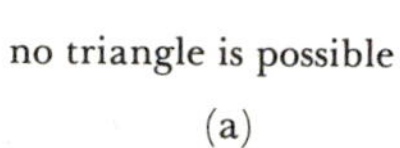

no triangle is possible

(a)

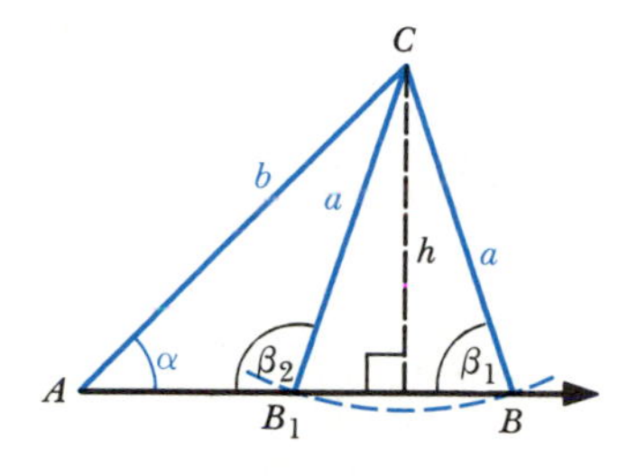

both $\triangle ABC$ and $\triangle AB_1C$ are possible

(b)

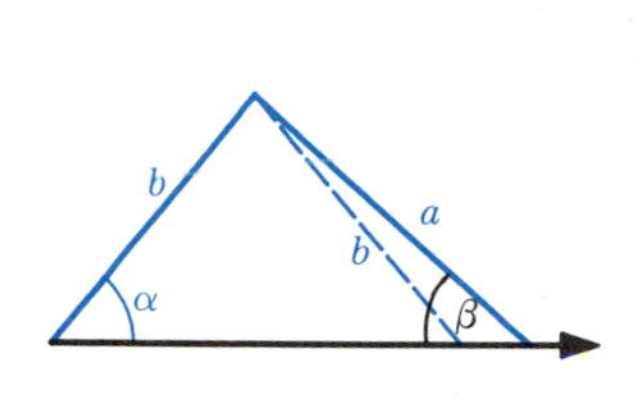

only one triangle is possible

(c)

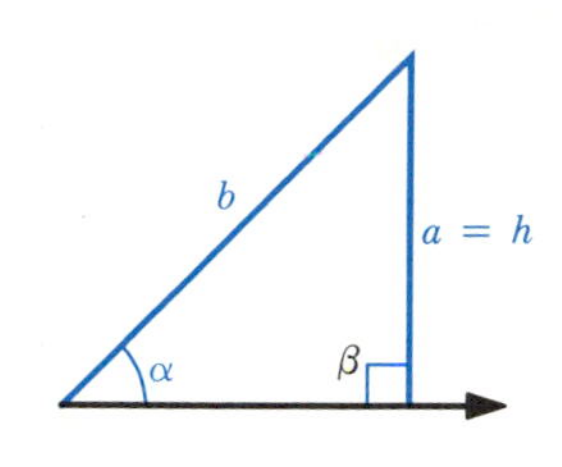

a right triangle is possible

(d)

Because there are different possibilities here, we sometimes refer to this situation as the **ambiguous case.** Thus, to solve a triangle where a, b, and α are given, we begin by using the law of sines,

$$\frac{\sin \alpha}{a} = \frac{\sin \beta}{b}$$

to evaluate $\sin \beta$ as

$$\sin \beta = \frac{b \sin \alpha}{a}$$

Note that β must satisfy $0 < \beta < 180°$, since β is an angle of a triangle.

The following results correspond to the possibilities illustrated in Figure 5.

1. $\mathbf{\sin \beta > 1}$**:** There is no triangle that satisfies the given condition (Figure 5a).
2. $\mathbf{0 < \sin \beta < 1}$**:** There are two possible choices for β in the equation

$$\sin \beta = \frac{b \sin \alpha}{a}$$

In one solution β_1 is an acute angle, and in the other, $\beta_2 = 180° - \beta_1$ is an obtuse angle. If $\alpha + \beta_2 < 180°$, the situation results in two possible triangles (Figure 5b). If $\alpha + \beta_2 > 180°$, there is only one triangle (Figure 5c).

3. $\mathbf{\sin \beta = 1}$**:** Here $\beta = 90°$ and there is one right triangle (Figure 5d).

Memorizing special rules to handle each of these possibilities is confusing and unnecessary.

C *In Examples 3–6, solve each ambiguous case triangle by using the law of sines.*

EXAMPLE 3 $a = 5, \quad b = 20, \quad \text{and} \quad \alpha = 30^\circ$

SOLUTION Using the law of sines, we have

$$\sin \beta = \frac{b \sin \alpha}{a} = \frac{20 \sin 30^\circ}{5} = 2 > 1$$

so that there is *no possible triangle* satisfying the given conditions.

EXAMPLE 4 $a = 5, \quad b = 9, \quad \text{and} \quad \alpha = 33^\circ$

SOLUTION Using the law of sines, we have

$$\sin \beta = \frac{b \sin \alpha}{a} = \frac{9 \sin 33^\circ}{5} = 0.9804$$

Therefore, an acute angle

$$\beta_1 = \sin^{-1} 0.9804 = 78.64^\circ$$

and an obtuse angle

$$\beta_2 = 180^\circ - 78.64^\circ = 101.36^\circ$$

both satisfy the trigonometric equation $\sin \beta = 0.9804$. The solution $\beta_1 = 78.64^\circ$ leads to a triangle AB_1C with

$$\gamma_1 = 180^\circ - (33^\circ + 78.64^\circ) = 68.36^\circ \quad \text{(Figure 6)}$$

so that

$$\frac{c_1}{\sin \gamma_1} = \frac{a}{\sin \alpha}$$

or

$$c_1 = \frac{a \sin \gamma_1}{\sin \alpha} = \frac{5 \sin 68.36^\circ}{\sin 33^\circ} = 8.533 \quad \text{(Figure 6)}$$

Figure 6

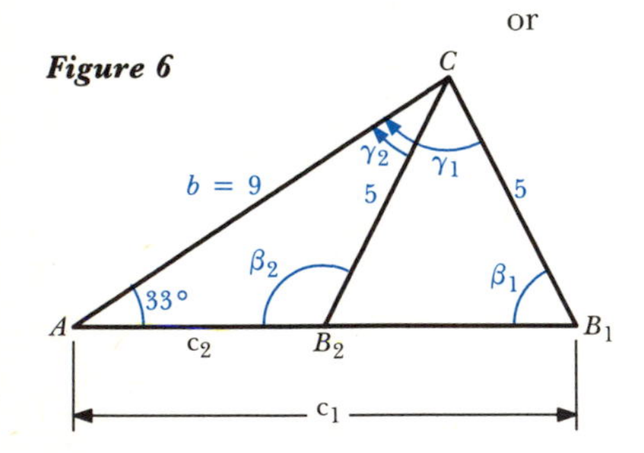

The solution $\beta_2 = 101.36^\circ$ leads to a second triangle AB_2C with

$$\gamma_2 = 180^\circ - (33^\circ + 101.36^\circ) = 45.64^\circ \quad \text{(Figure 6)}$$

so that

$$c_2 = \frac{a \sin \gamma_2}{\sin \alpha} = \frac{5 \sin 45.64^\circ}{\sin 33^\circ} = 6.564$$

EXAMPLE 5 $a = 20, \quad b = 15, \quad \text{and} \quad \alpha = 29^\circ$

SOLUTION Applying the law of sines, we have

$$\sin \beta = \frac{b \sin \alpha}{a} = \frac{15 \sin 29^\circ}{20} = 0.3636$$

Thus an acute angle

$$\beta_1 = \sin^{-1} 0.3636 = 21.32°$$

and an obtuse angle

$$\beta_2 = 180° - 21.32° = 158.68°$$

both satisfy $\sin\beta = 0.3636$. Since it is impossible to have an angle $\beta_2 = 158.68°$ in a triangle known to have an angle of 29° ($\alpha + \beta_2 = 29° + 158.68° = 187.68° > 180°$), the latter solution β_2 must be rejected, and there is only one possible triangle (Figure 7). The solution $\beta_1 = 21.32°$ leads to

Figure 7

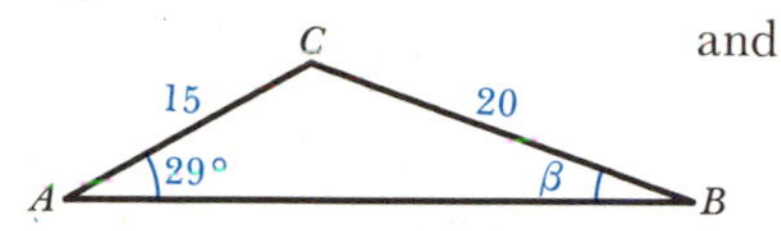

$$\gamma = 180° - (29° + 21.32°) = 129.68°$$

and

$$c = \frac{a \sin\gamma}{\sin\alpha} = \frac{20 \sin 129.68°}{\sin 29°} = 31.75$$

EXAMPLE 6 $a = 10, \quad b = 20, \quad \text{and} \quad \alpha = 30°$

SOLUTION Using the law of sines, we have

Figure 8

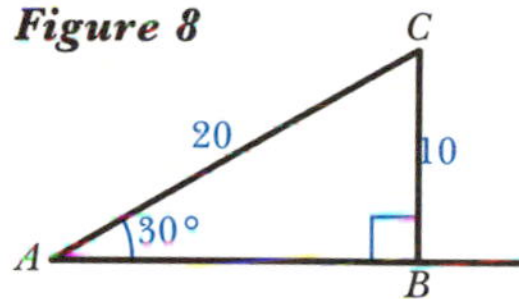

$$\sin\beta = \frac{b \sin\alpha}{a} = \frac{20 \sin 30°}{10} = 1$$

Therefore, $\beta = 90°$, and there is only one possible triangle, a right triangle (Figure 8). Thus, $\gamma = 90° - 30° = 60°$, and by using the Pythagorean theorem, $c = \sqrt{20^2 - 10^2} = \sqrt{300} = 10\sqrt{3} = 17.32$.

The Law of Cosines

If we are given the measurements of two sides and the included angle of a triangle or if we are given the lengths of the three sides of a triangle, then we use the law of cosines to solve the triangle.

THEOREM 2 **Law of Cosines**

In any triangle ABC,

(i) $c^2 = a^2 + b^2 - 2ab\cos\gamma$
(ii) $b^2 = a^2 + c^2 - 2ac\cos\beta$
(iii) $a^2 = b^2 + c^2 - 2bc\cos\alpha$

Figure 9

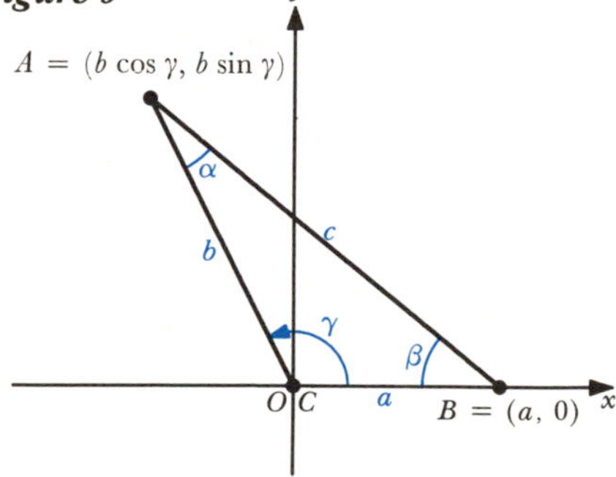

PROOF First, consider triangle ABC on a Cartesian coordinate system with γ in standard position (Figure 9). By the definition of the trigonometric functions on page 257, point

A has coordinates $(b\cos\gamma, b\sin\gamma)$. Also point B has coordinates $(a, 0)$. So, by the distance formula, we have

$$\begin{aligned} c^2 &= (b\cos\gamma - a)^2 + (b\sin\gamma - 0)^2 \\ &= b^2\cos^2\gamma - 2ab\cos\gamma + a^2 + b^2\sin^2\gamma \\ &= b^2\cos^2\gamma + b^2\sin^2\gamma + a^2 - 2ab\cos\gamma \\ &= a^2 + b^2(\cos^2\gamma + \sin^2\gamma) - 2ab\cos\gamma \\ &= a^2 + b^2 - 2ab\cos\gamma \end{aligned}$$

In view of the fact that the location of the coordinate axes in the plane is merely a matter of convenience, the other two formulas in Theorem 2 are obtained in a similar manner, by placing the other two angles in standard position and repeating the steps in the above proof.

Only one of the above formulas need be memorized; the other two can be obtained by appropriately changing the letters. The law of cosines can be stated as follows:

The square of the length of any side of a triangle is equal to the sum of the squares of the lengths of the other two sides minus twice their product times the cosine of the angle included between these other two sides.

C *In Examples 7–9, use the law of cosines to find the specified unknown part of the triangle.*

EXAMPLE 7 In triangle ABC, $a = 8$, $b = 6$, and $\gamma = 60°$ (Figure 10). Find c.

SOLUTION Using the law of cosines, we have

$$\begin{aligned} c^2 &= a^2 + b^2 - 2ab\cos\gamma \\ &= 64 + 36 - 2(8)(6)(\tfrac{1}{2}) \\ &= 100 - 48 = 52 \end{aligned}$$

so $c = \sqrt{52} = 7.211$.

Figure 10

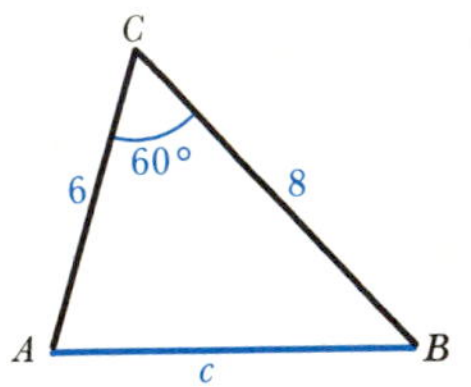

EXAMPLE 8 In triangle ABC, $a = 7$, $b = 4$, and $c = 5$ (Figure 11). Find α.

SOLUTION Using the law of cosines, we have

$$a^2 = b^2 + c^2 - 2bc\cos\alpha$$

or

$$7^2 = 4^2 + 5^2 - 2(4)(5)\cos\alpha$$

so that

$$\cos\alpha = \frac{4^2 + 5^2 - 7^2}{(2)(4)(5)} = -0.2000$$

Thus $\alpha = \cos^{-1}(-0.2000) = 101.54°$.

Figure 11

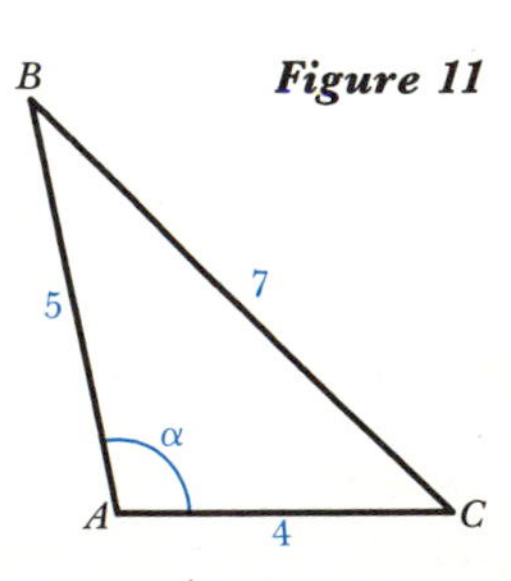

EXAMPLE 9 A highway engineer plans to construct a straight tunnel from point A on one side of a hill to point B on the other side (Figure 12). From a third point C, the distances $|\overline{CA}| = 573$ meters and $|\overline{CB}| = 819$ meters are measured. If angle $ACB = 67°$, find the length of the tunnel $|\overline{AB}|$ when it is completed.

SOLUTION (See Figure 12.) Using the law of cosines, we have

$$|\overline{AB}|^2 = |\overline{CB}|^2 + |\overline{CA}|^2 - 2|\overline{CB}| \cdot |\overline{CA}| \cos \gamma$$
$$= 819^2 + 573^2 - 2(819)(573) \cos 67°$$

Figure 12

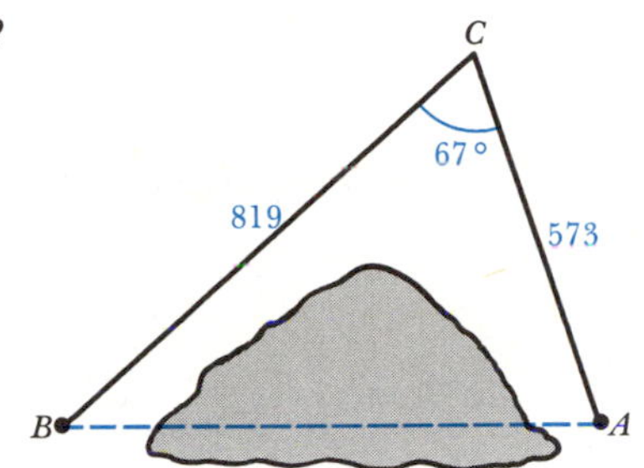

Hence

$$|\overline{AB}| = \sqrt{819^2 + 573^2 - 2(819)(573) \cos 67°}$$
$$= 795.2$$

Therefore, the length of the tunnel is approximately 795.2 meters.

PROBLEM SET 6.7

[c] In all problems, round off all angle measures to the nearest hundredth of a degree and all lengths of sides to four significant digits.

In problems 1–4, use the law of sines to solve for the specified side of triangle ABC satisfying the given conditions.

1. a if $\alpha = 60°$, $\beta = 45°$, and $b = 10$
2. b if $\gamma = 75°$, $\beta = 25°$, and $c = 8$
3. c if $\alpha = 50°$, $\beta = 100°$, and $a = 5$
4. b if $\alpha = 21°$, $\beta = 35°$, and $a = 14$

In problems 5–10, use the law of sines and the ambiguous case to find all parts of all the possible triangles.

5. $\alpha = 50°$, $a = 12$, and $b = 17$
6. $\alpha = 30°$, $a = 4$, and $b = 10$
7. $\alpha = 22°$, $a = 5$, and $b = 8$
8. $\alpha = 60°$, $a = 9$, and $b = 10$
9. $\alpha = 35°$, $a = 9$, and $b = 6$
10. $\alpha = 22°$, $a = 10$, and $b = 50$

In problems 11–14, use the law of cosines to find the specified part of triangle ABC satisfying the given conditions.

11. a if $\alpha = 60°$, $b = 20$, and $c = 6$
12. b if $\beta = 28.50°$, $a = 25$, and $c = 30$
13. α if $a = 3$, $b = 4$, and $c = 6$
14. β if $a = 144$, $b = 180$, and $c = 108$

In problems 15–22, use the law of sines or the law of cosines to find the specified part of the triangle satisfying the given conditions.

15. c if $\beta = 42°$, $\gamma = 31°$, and $a = 29$
16. b if $\alpha = 12°$, $\beta = 97°$, and $a = 14$
17. a if $\alpha = 100.30°$, $\beta = 5.50°$, and $b = 3.71$
18. a if $\beta = 70°$ and $b = c = 17.13$
19. γ if $a = 10$, $b = 15$, and $c = 20$
20. c if $\gamma = 108°$, $a = 10$, and $b = 12$
21. a if $\alpha = 17°$, $c = 39$, and $b = 98$
22. β if $a = 5$ and $b = c = 10$

23. Two forest rangers are 1.2 kilometers apart. One ranger sights a fire at an angle of 41.43° from the line between the two observation points. The other ranger sights the same fire at an angle of 61.4° from the same line between the two observation points. How far is the fire from each observation point?

24. An airplane is sighted simultaneously from two towns that are 3 miles apart. The angle of elevation from one of the towns is 40.83°, and the angle of elevation from the other town is 75°. If the airplane is directly over a line from one town to the other, how far is the airplane from each town?

25. Highway engineers determine that a tunnel is to be dug between points A and B on opposite sides of a hill. A point C is chosen that is 473 meters from A and 367 meters from B. If angle ABC is 74.2°, find the length of the tunnel.

26. A vertical tower 125 feet high is on a hill on the bank of a river. It is observed that the angle of depression from the top of the tower to a point on the opposite shore is 28.67°, and the angle of depression from the base of the tower to the same point on the shore is 18.33°. How wide is the river and how high is the hill?

27. A surveyor marks points A and B 400 meters apart on the same bank of a river. He sights a point C on the opposite bank and determines that angle CAB is 79.4° and angle CBA is 35.8°. Find the distance $|\overline{AC}|$.

28. A communication satellite A traveling in a circular orbit 1600 kilometers above the earth is located by a tracking station B on earth at a certain time (Figure 13). If the tracking antenna is aimed 35° above the horizon and if the radius of the earth is 6400 kilometers, what is the distance $|\overline{AB}|$ from the antenna to the satellite?

Figure 13

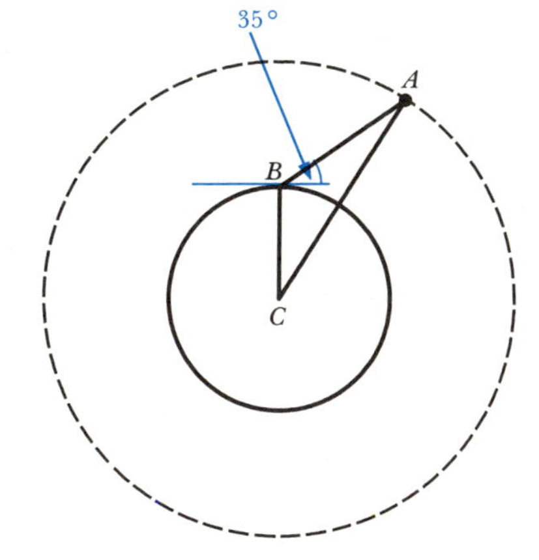

29. Two straight roads cross at an angle of 65°. A jogger on one road is 5 miles from the intersection and moving away from it at the rate of 7 miles per hour. At the same instant, another jogger on the other road is 4 miles from the intersection and moving away from it at the rate of 10 miles per hour. If the two joggers maintain constant speeds, and if they travel in directions that differ by 65°, what is the distance between them after 2 hours?

30. Two jets leave an airbase at the same time and fly, at the same speed, along straight courses forming an angle of 112.45° with each other. After the jets have each flown 504 kilometers, how far apart are they?

31. Suppose that two trains depart from a railroad station (where the tracks meet) at the same time and travel along tracks that differ in direction by 75.5°. Find how far apart the two trains are after $2\frac{1}{2}$ hours if their average speeds are 55 and 70 miles per hour, respectively.

32. Assume that one diagonal of a parallelogram is 80 inches long and that at one end it forms angles of 35° and 27°, respectively, with the two sides. Find the lengths of the sides of the parallelogram.

33. In order to find the horizontal distance between two towers A and B that are separated by a group of trees, a point C that is readily accessible from both A and B (Figure 14) is used to make the following measurements: $|\overline{CA}| = 361$ meters, $|\overline{CB}| = 575$ meters, angle $ACB = 105.5°$. Find $|\overline{AB}|$.

34. A sign is to be mounted on a building using two pairs of steel brackets. Figure 15 displays a side view of one pair of brackets, $\overline{AB}$ and $\overline{BC}$, and the angles they make with the building. How long are $\overline{AB}$ and $\overline{BC}$ if the points where they are attached to the building are 7 feet apart?

35. A ship sails 19 nautical miles in a direction of S29.3°W, and then turns onto a course of S51.7°W and sails 24 nautical miles. How far is the ship from the starting point?

36. Assume a, b, and c are the lengths of the three sides of a triangle ABC and

$$s = \frac{a + b + c}{2} \quad \text{(Figure 16)}$$

(a) Use the law of sines to show that the area A of the triangle is given by

$$A = \frac{c^2 \sin\alpha \sin\beta}{2 \sin\gamma} = \frac{a^2 \sin\beta \sin\gamma}{2 \sin\alpha} = \frac{b^2 \sin\gamma \sin\alpha}{2 \sin\beta}$$

(b) Show that

(i) $\cos\dfrac{\alpha}{2} = \sqrt{\dfrac{s(s-a)}{bc}}$ (ii) $\sin\dfrac{\alpha}{2} = \sqrt{\dfrac{(s-b)(s-c)}{bc}}$

(c) Show that

$$\frac{\cos\alpha}{a} + \frac{\cos\beta}{b} + \frac{\cos\gamma}{c} = \frac{a^2 + b^2 + c^2}{2abc}$$

(d) Combine parts (a), (b), and (c) to show that the area A is given by

$$A = \sqrt{s(s-a)(s-b)(s-c)}$$

This formula is known as **Hero's** (or **Heron's**) **formula.**

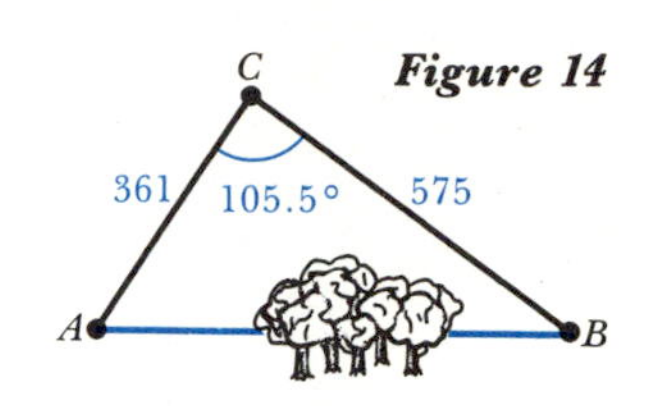

Figure 14

Figure 15

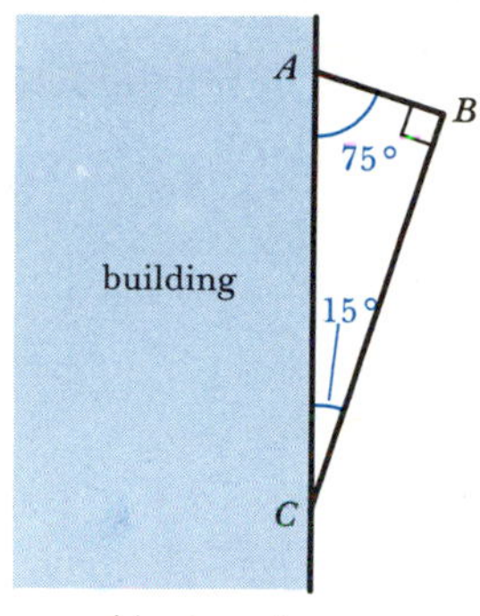

side view of one pair of brackets

Figure 16

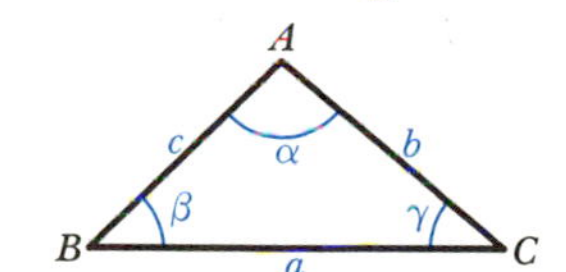

In problems 37–40, use Hero's formula from problem 36 to find the area of the triangle with the given side lengths a, b, and c.

37. $a = 5$ centimeters, $b = 7$ centimeters, and $c = 10$ centimeters

38. $a = 5$ feet, $b = 8$ feet, and $c = 7$ feet

39. $a = 1.2$ kilometers, $b = 2.3$ kilometers, and $c = 3.4$ kilometers

40. $a = 976$ yards, $b = 725$ yards, and $c = 543$ yards

6.8 Polar Coordinates

We have seen that points in the plane can be associated with pairs of real numbers by using a Cartesian coordinate system, which is also called a *rectangular coordinate system*. Another way of associating pairs of numbers with points in the plane is based on a "grid" composed of concentric circles and rays emanating from the common center of the circles (Figure la). Such a system is called a *polar coordinate system*. We will see how the trigonometric functions are used to develop this system.

Figure 1

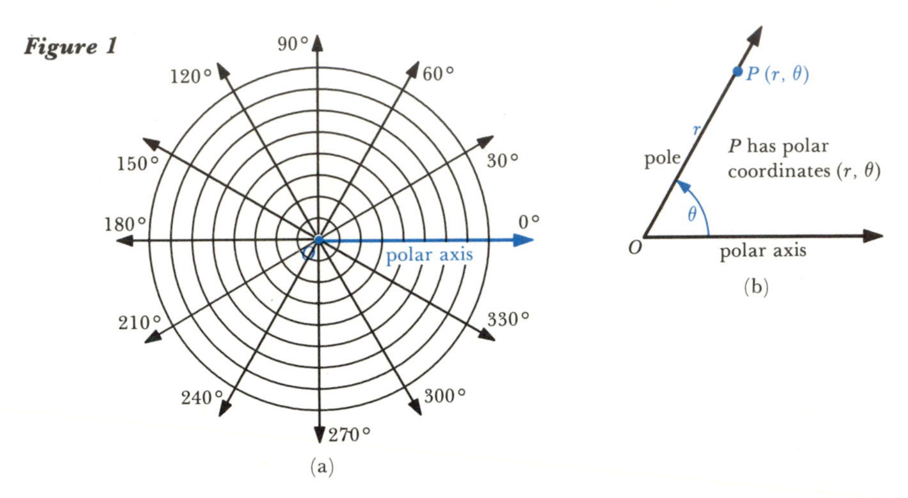

The frame of reference for the **polar coordinate system** consists of a fixed point O, called the **pole,** and a fixed ray called the **polar axis** (Figure lb). The position of a point P is uniquely determined by r and θ, where θ is any angle having the polar axis as its initial side and the ray from the pole through point P as its terminal side, and r as the *directed* distance along the terminal side of θ from the pole O to P. The pair (r, θ) are called the **polar coordinates** of P (Figure lb). Each point in a polar coordinate system can be represented by infinitely many ordered pairs of numbers. For example, $(2, 30°)$, $(2, 390°)$, and $(2, -330°)$ each represent the same point (Figure 2). Also, r need not be positive. If $r < 0$, the point (r, θ) is determined by plotting

$$(|r|, \theta + 180°) \qquad \text{or} \qquad (|r|, \theta + \pi)$$

depending on whether θ is measured in degrees or radians. For example, $(-2, 30°)$ is the same as $(2, 210°)$ (Figure 3).

Figure 2

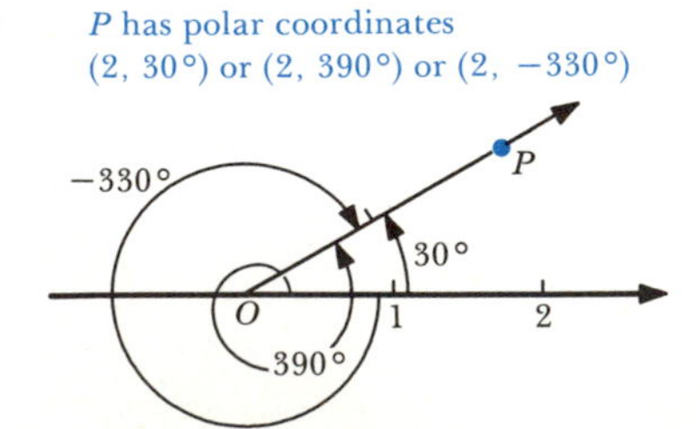

Figure 3

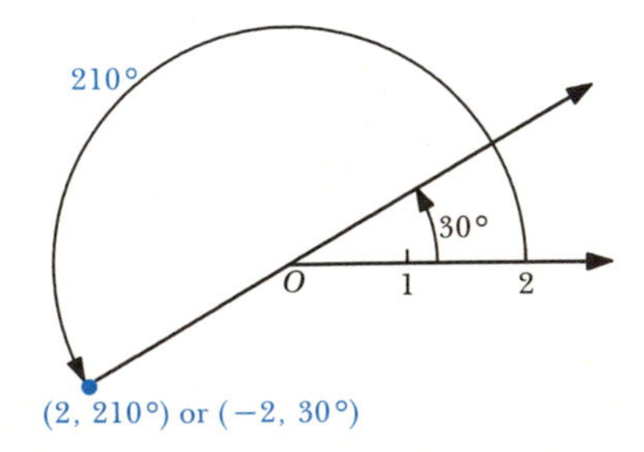

EXAMPLE 1 Locate the points that have the given polar coordinates.

(a) $(3, 70°)$ (b) $(0, 0)$ (c) $\left(7, \frac{7\pi}{5}\right)$

(d) $(3, 100°)$ (e) (π, π) (f) $(3, 5)$

(g) $(5, 0)$ (h) $(-5, 15°)$

SOLUTION The points are plotted in Figure 4.

Figure 4

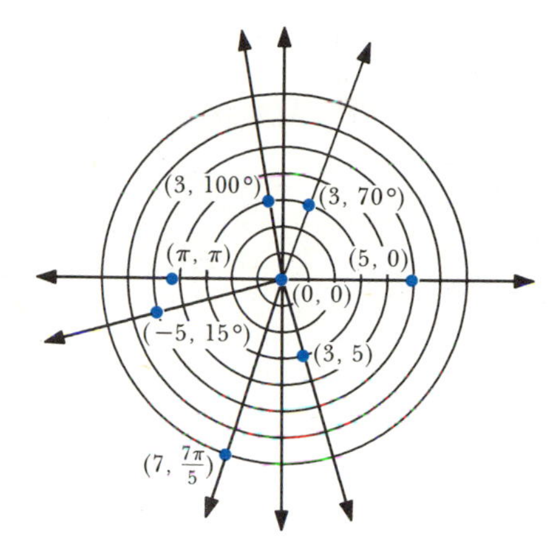

Conversion of Coordinates

Figure 5

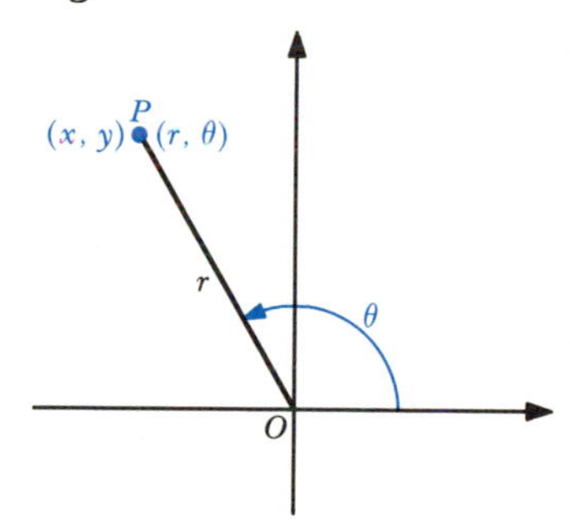

Suppose that the polar axis of the polar coordinate system coincides with the positive x axis of the rectangular coordinate system. Any point P in the plane can be located either by polar coordinates (r, θ) or by rectangular coordinates (x, y) (Figure 5). Assuming that $r > 0$, we know from trigonometry that

$$\cos\theta = \frac{x}{r}, \qquad \sin\theta = \frac{y}{r},$$

and

$$\tan\theta = \frac{y}{x}$$

Also, $x^2 + y^2 = r^2$. In general, we have the following relationships.

Transformation or Conversion Formulas

(i) $x = r\cos\theta$	(ii) $y = r\sin\theta$
(iii) $\tan\theta = \frac{y}{x}$	(iv) $r^2 = x^2 + y^2$

These formulas enable us to convert polar coordinates of a given point to rectangular coordinates of the same point and vice versa.

EXAMPLE 2 Convert the given polar coordinates to rectangular coordinates.

(a) $(3, 60°)$ (b) $(-2, 180°)$ (c) $\left(4, -\frac{5\pi}{6}\right)$

SOLUTION From transformation formulas (i) and (ii) above, we obtain

(a) $x = r\cos\theta = 3\cos 60° = \frac{3}{2}$ and $y = r\sin\theta = 3\sin 60° = \frac{3\sqrt{3}}{2}$

Hence the rectangular coordinates are $\left(\frac{3}{2}, \frac{3\sqrt{3}}{2}\right)$.

(b) $x = r\cos\theta = -2\cos 180° = 2$ and $y = r\sin\theta = -2\sin 180° = 0$

Hence the rectangular coordinates are $(2, 0)$.

(c) $x = r\cos\theta = 4\cos\left(-\frac{5\pi}{6}\right) = 4\left(-\frac{\sqrt{3}}{2}\right) = -2\sqrt{3}$ and

$y = r\sin\theta = 4\sin\left(-\frac{5\pi}{6}\right) = 4\left(-\frac{1}{2}\right) = -2$

Hence the rectangular coordinates are $(-2\sqrt{3}, -2)$.

EXAMPLE 3 Convert the given rectangular coordinates to polar coordinates.

(a) $(-1, 1)$ (b) $(3, -\sqrt{3})$ C (c) $(-3, 4)$

SOLUTION From transformation formulas (iii) and (iv) above, we get

(a) $r = \sqrt{x^2 + y^2} = \sqrt{(-1)^2 + 1^2} = \sqrt{2}$ and $\tan\theta = -1$

Since the point is in quadrant II, $\tan\theta = -1$ implies that one angle is $\theta = 135°$. Thus, one pair of polar coordinates is $(\sqrt{2}, 135°)$, while another is $(\sqrt{2}, -225°)$.

(b) $r = \sqrt{3^2 + (-\sqrt{3})^2} = \sqrt{9 + 3} = \sqrt{12} = 2\sqrt{3}$ and $\tan\theta = -\frac{\sqrt{3}}{3}$

so $\theta = -30°$ is one possible angle, since the point is in quadrant IV. Hence one possible pair of polar coordinates is given by $(2\sqrt{3}, -30°)$.

(c) $r = \sqrt{(-3)^2 + 4^2} = 5$ and $\tan\theta = -\frac{4}{3}$

The reference angle θ_R for the angle θ is determined by

$$\begin{aligned}\theta_R &= \tan^{-1}\left(\frac{4}{3}\right)\\ &= 53.13°\end{aligned}$$

The point is located in quadrant II. Therefore, it follows that one possible value for θ is approximately equal to $126.87°$. Thus, $(5, 126.87°)$ is one pair of polar coordinates.

The transformation formulas can also be used to convert an equation of a graph given in one coordinate system to an equation of the same graph in another coordinate system. The examples below demonstrate how this can be done.

EXAMPLE 4 Convert the given equation in the rectangular system to a corresponding polar equation in r and θ.

(a) $x^2 + y^2 = 9$ (b) $y = x^2$

SOLUTION (a) Since $x = r\cos\theta$ and $y = r\sin\theta$, if we assume that $r > 0$, we get

$$
\begin{aligned}
x^2 + y^2 &= 9 \\
(r\cos\theta)^2 + (r\sin\theta)^2 &= 9 \\
r^2\cos^2\theta + r^2\sin^2\theta &= 9 \\
r^2(\cos^2\theta + \sin^2\theta) &= 9 \\
r^2 &= 9 \\
r &= 3
\end{aligned}
$$

(b) Similarly, we have

$$
\begin{aligned}
y &= x^2 \\
r\sin\theta &= (r\cos\theta)^2 \\
r\sin\theta &= r^2\cos^2\theta \\
\sin\theta &= r\cos^2\theta
\end{aligned}
$$

Thus

$$r = \frac{\sin\theta}{\cos^2\theta} = \frac{\sin\theta}{\cos\theta}\cdot\frac{1}{\cos\theta}$$

so that

$$r = \tan\theta\sec\theta$$

EXAMPLE 5 Convert each of the given polar equations to corresponding equations in the rectangular system in terms of x and y. Sketch the graph.

(a) $r = 2\cos\theta$ (b) $r = \dfrac{5}{\sin\theta - 3\cos\theta}$

SOLUTION (a) Since $r^2 = x^2 + y^2$ and $x = r\cos\theta$, we have

$$
\begin{aligned}
r &= 2\cos\theta \\
r^2 &= 2r\cos\theta \\
x^2 + y^2 &= 2x \\
x^2 - 2x + y^2 &= 0 \\
(x^2 - 2x + 1) + y^2 &= 1 \\
(x - 1)^2 + y^2 &= 1
\end{aligned}
$$

Thus $r = 2\cos\theta$ and $(x-1)^2 + y^2 = 1$ are equivalent in the sense that both yield the graph of a circle with center $(1, 0)$ and radius 1 (Figure 6).

Figure 7

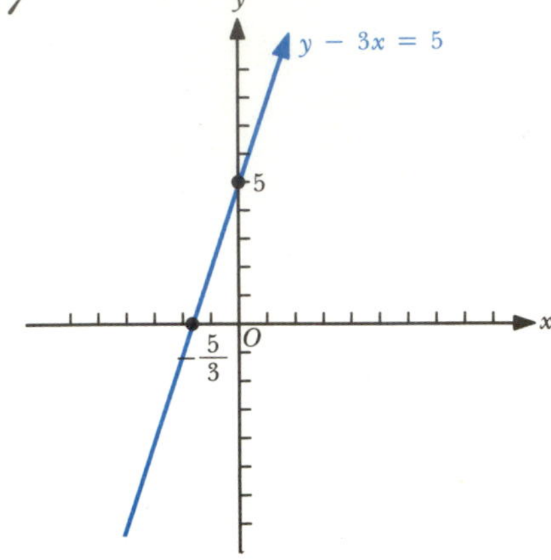

Figure 6

$(x-1)^2 + y^2 = 1$

(b) We use $x = r\cos\theta$ and $y = r\sin\theta$ to obtain

$$r = \frac{5}{\sin\theta - 3\cos\theta}$$

$$r\sin\theta - 3r\cos\theta = 5$$

$$y - 3x = 5$$

Thus, both

$$r = \frac{5}{\sin\theta - 3\cos\theta} \quad \text{and} \quad y - 3x = 5$$

represent the graph of a line (Figure 7).

Graphs of Polar Equations

In Example 5, we graphed each polar equation by converting the given equation to rectangular form and then using the latter form to obtain the graph. It is possible to graph polar equations *directly* without such conversions. The **graph of a polar equation** in r and θ consists of all points with polar coordinates (r, θ) that satisfy the equation. Although polar equation graphs are studied more extensively in calculus, we can survey the topic here by looking at some examples.

EXAMPLE 6 Graph each of the following polar equations.

(a) $r = 3$ (b) $\theta = \dfrac{\pi}{3}$

SOLUTION (a) Since the equation $r = 3$ places no restriction on the angle, θ can take on any value, and the graph consists of all points of the form $(3, \theta)$; that is, all points that are three units from the pole. The graph is a circle of radius 3 with center at the pole (Figure 8a). Note that this situation is confirmed in Example 4(a) on page 374.

(b) The graph of $\theta = \pi/3$ consists of all points of the form $(r, \pi/3)$. Since r can be positive, negative, or zero, the graph is the line displayed in Figure 8b.

Figure 8

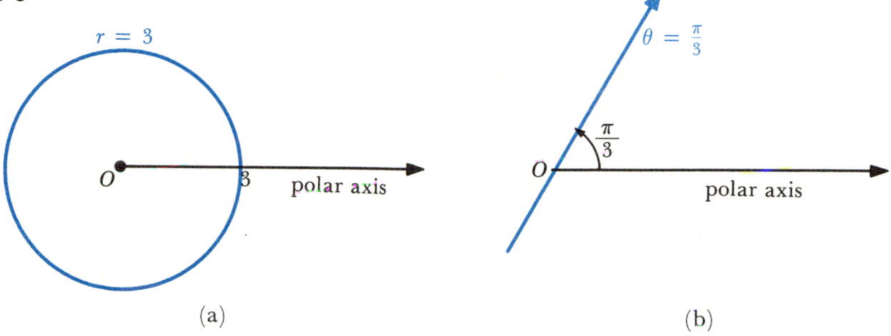

(a) (b)

EXAMPLE 7 Graph $r = 2 \cos \theta$.

SOLUTION After plotting the points contained in the table below, the graph appears to be a circle (Figure 9).

θ	0	$\frac{\pi}{6}$	$\frac{\pi}{4}$	$\frac{\pi}{3}$	$\frac{\pi}{2}$	$\frac{2\pi}{3}$	$\frac{3\pi}{4}$	$\frac{5\pi}{6}$	π	$\frac{4\pi}{3}$
r	2	$\sqrt{3}$	$\sqrt{2}$	1	0	-1	$-\sqrt{2}$	$-\sqrt{3}$	-2	-1

Figure 9

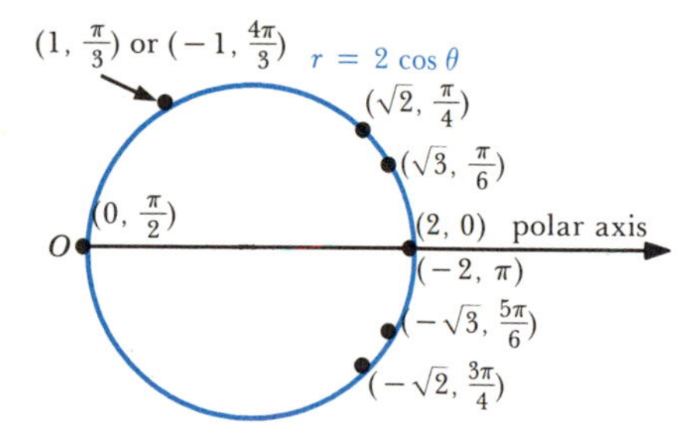

The fact that the graph of $r = 2\cos\theta$ is a circle is confirmed in Example 5(a) on page 373. The observation that the complete graph is generated by allowing θ to vary from 0 to π is noteworthy. θ values equal to or larger than π will yield overlapping points. For example, $(1, \pi/3)$ and $(-1, 4\pi/3)$ represent the same point. Note also that the graph of $r = 2 \cos \theta$ is symmetric with respect to the polar axis.

EXAMPLE 8 Graph $r = 1 + \sin\theta$.

SOLUTION The table below can be used to sketch the graph (Figure 10). This graph is called a **cardioid** because of its heart shape.

θ	0	$\frac{\pi}{6}$	$\frac{\pi}{4}$	$\frac{\pi}{3}$	$\frac{\pi}{2}$	$\frac{2\pi}{3}$	$\frac{3\pi}{4}$	$\frac{5\pi}{6}$	π	$\frac{5\pi}{4}$	$\frac{3\pi}{2}$	$\frac{7\pi}{4}$	2π
r	1	$\frac{3}{2}$	$\frac{2+\sqrt{2}}{2}$	$\frac{2-\sqrt{3}}{2}$	2	$\frac{2+\sqrt{3}}{2}$	$\frac{2+\sqrt{2}}{2}$	$\frac{3}{2}$	1	$\frac{2-\sqrt{2}}{2}$	0	$\frac{2-\sqrt{2}}{2}$	1

Figure 10

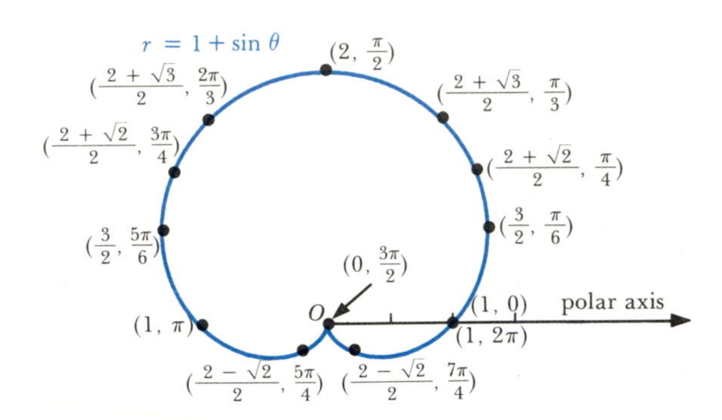

The graph is symmetric with respect to a line perpendicular to the polar axis at the pole and values of θ larger than 2π will yield overlapping points.

PROBLEM SET 6.8

In problems 1–12, locate each point with the given polar coordinates, then convert the given representation to rectangular coordinates. Round off the answers in problems 7 and 11 to two decimal places.

1. $(6, 30^\circ)$ **2.** $(10, \pi/3)$ **3.** $(7, 2\pi/3)$ **4.** $(4, -\pi/6)$
5. $(-4, 3\pi/4)$ **6.** $(-8, 45^\circ)$ c **7.** $(3, 27^\circ)$ **8.** $(5, \pi/2)$
9. $(4, 210^\circ)$ **10.** $(4, -330^\circ)$ c **11.** $(2, 5.1)$ **12.** $(-3, -3)$

In problems 13–24, locate each point with the given rectangular coordinates, then convert the given representation to polar coordinates (r, θ), where $r > 0$ and also $0^\circ \le \theta < 360^\circ$. In problems 15, 22, 23, and 24, round off both r and θ to two decimal places.

13. $(-1, \sqrt{3})$ **14.** $(-3, 0)$ c **15.** $(4, 3)$ **16.** $(-6, 6\sqrt{3})$
17. $(5, 5)$ **18.** $(0, -2)$ **19.** $(-3, 3\sqrt{3})$ **20.** $(2\sqrt{3}, -2)$
21. $(-5, -5)$ c **22.** $(-2, 5)$ c **23.** $(-7, -1)$ c **24.** $(8.3, -3.4)$

In problems 25–30, convert each polar equation to a corresponding equation in the rectangular system. Sketch the graph.

25. $\theta = \dfrac{\pi}{4}$ [c] **26.** $\theta = 29°$ **27.** $r = 3 \sin \theta$ **28.** $r = \dfrac{1}{\sin \theta + \cos \theta}$

29. $r = 3 \sec \theta$ **30.** $r = 2 \cot \theta \sec \theta$

In problems 31–36, convert each equation given in the rectangular system to a corresponding equation in the polar system.

31. $x^2 + y^2 = 25$ **32.** $x - y = 5$ **33.** $y = 3x$ **34.** $x = 1$

35. $x^2 - 6x + y^2 = 0$ **36.** $x^2 - y^2 = 4$

In problems 37–48, graph each polar equation.

37. $r = 5$ **38.** $r = -2$ **39.** $\theta = \dfrac{\pi}{9}$ **40.** $\theta = -50°$

41. $r = 3 \sin \theta$ **42.** $r = \theta$ **43.** $r = 3 + 3 \cos \theta$ **44.** $r = \dfrac{1}{\theta}$

45. $r = 1 - \sin \theta$ **46.** $r \cos \theta = 5$ **47.** $r = 1 + 2 \cos \theta$ **48.** $r = -3 \cos \theta$

6.9 Trigonometric Forms of Complex Numbers

In Section 1.4 we introduced complex numbers and their operations. The geometric representation of a complex number as a point in the plane leads to a method that links complex numbers to trigonometry. Each ordered pair of real numbers (a, b) can be associated with the complex number $z = a + bi$, and each complex number $z = a + bi$ can be associated with the ordered pair of real numbers (a, b). Because of this one-to-one correspondence between the complex numbers and the ordered pairs of real numbers, we use the points in the plane associated with the ordered pairs of real numbers to represent the complex numbers. For example, the ordered pairs $(2, -3)$, $(5, 2)$, and (e, π) are used to represent complex numbers $z_1 = 2 - 3i$, $z_2 = 5 + 2i$, and $z_3 = e + \pi i$, respectively, as points in the plane (Figure 1). The plane on which the complex numbers are represented is called the **complex plane;** the horizontal axis (x axis) is called the **real axis** and the vertical axis (y axis) is called the **imaginary axis.** Thus, complex numbers of the form $z = bi$ are represented by points of the form $(0, b)$, that is, as points on the imaginary axis; whereas complex numbers of the form $z = a$ are represented by points of the form $(a, 0)$, that is, as points on the real axis.

Figure 1

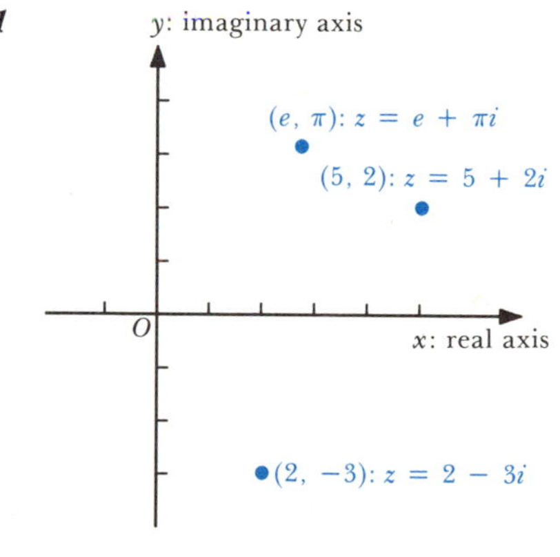

Geometrically, we interpret the *absolute value* $|a + bi|$ of a complex number as the distance $\sqrt{a^2 + b^2}$ between the origin and the point (a, b) corresponding to $a + bi$ (Figure 2). Thus, we have the following definition.

DEFINITION 1 **Absolute Value**

If $z = a + bi$, then the **absolute value** of z, written $|z|$, is defined by

$$|z| = \sqrt{a^2 + b^2}$$

Figure 2

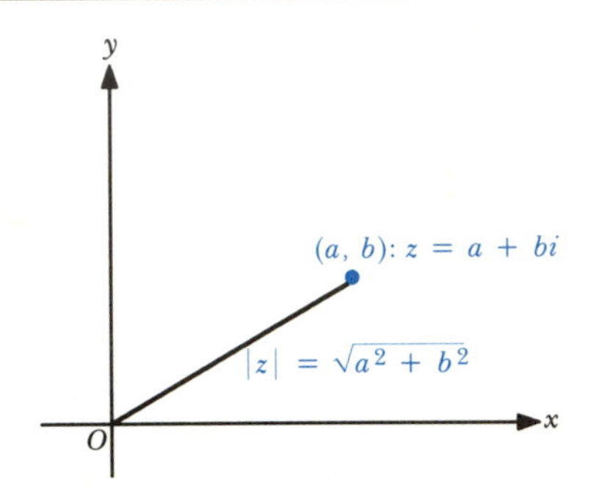

Notice that $|z| = \sqrt{z \cdot \bar{z}}$ (Problem 9).

EXAMPLE 1 Let $z_1 = 4 + 3i$ and $z_2 = \sqrt{3} - i$. Find

(a) $|z_1|$ (b) $|z_2|$ (c) $|z_1 z_2|$ and $|z_1| \cdot |z_2|$ (d) $\left|\dfrac{z_1}{z_2}\right|$ and $\dfrac{|z_1|}{|z_2|}$

SOLUTION

(a) $|z_1| = \sqrt{4^2 + 3^2} = \sqrt{16 + 9} = \sqrt{25} = 5$

(b) $|z_2| = \sqrt{(\sqrt{3})^2 + (-1)^2} = \sqrt{4} = 2$

(c)
$$\begin{aligned} |z_1 z_2| &= |(4 + 3i)(\sqrt{3} - i)| \\ &= |(4\sqrt{3} + 3) + (3\sqrt{3} - 4)i| \\ &= \sqrt{(4\sqrt{3} + 3)^2 + (3\sqrt{3} - 4)^2} = \sqrt{100} = 10 \end{aligned}$$

Also, $|z_1||z_2| = 5 \cdot 2 = 10$.

(d)
$$\begin{aligned} \left|\frac{z_1}{z_2}\right| &= \left|\frac{4 + 3i}{\sqrt{3} - i}\right| = \left|\frac{(4 + 3i)(\sqrt{3} + i)}{3 + 1}\right| = \left|\frac{(4\sqrt{3} - 3) + (3\sqrt{3} + 4)i}{4}\right| \\ &= \sqrt{\left(\frac{4\sqrt{3} - 3}{4}\right)^2 + \left(\frac{3\sqrt{3} + 4}{4}\right)^2} = \sqrt{\frac{25}{4}} = \frac{5}{2} \end{aligned}$$

Also, $\dfrac{|z_1|}{|z_2|} = \dfrac{5}{2}$.

Parts (c) and (d) of Example 1 illustrate the general properties which are proved in Problem 10.

(i) $|z_1 z_2| = |z_1||z_2|$ (ii) $\left|\dfrac{z_1}{z_2}\right| = \dfrac{|z_1|}{|z_2|}$

It is worth noting that the points corresponding to all complex numbers that have a fixed absolute value lie on a circle with center at the origin in the plane. For example, the points corresponding to the complex number $z = x + iy$, with $|z| = 1$, satisfy

$$|z| = \sqrt{x^2 + y^2} = 1 \qquad \text{or} \qquad x^2 + y^2 = 1$$

so that they lie on the unit circle.

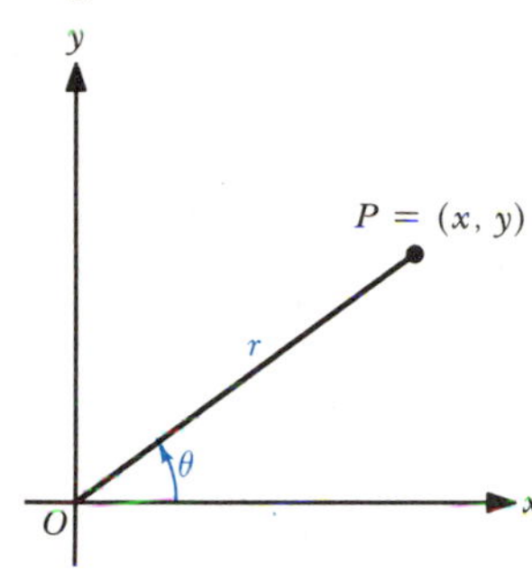

Now consider a nonzero complex number $z = x + iy$ and its geometric representation $P = (x, y)$. Let θ be any angle in standard position whose terminal side lies on the segment $\overline{OP}$ and $r = |z| = \sqrt{x^2 + y^2}$ (Figure 3). Since $x = r\cos\theta$ and $y = r\sin\theta$, then $z = x + yi$ is written as

$$z = r\cos\theta + ri\sin\theta$$

which is a **trigonometric form** or **polar form** of the complex number z; that is,

$$z = x + iy = r(\cos\theta + i\sin\theta)$$

where (r, θ) are polar coordinates for the point with rectangular coordinates (x, y). We call r the **modulus** of z and θ, the angle associated with z, the **argument** of z.

Notice that θ is not unique, since $r(\cos\theta + i\sin\theta) = r(\cos\theta_1 + i\sin\theta_1)$ holds whenever $\theta - \theta_1$ is an integer multiple of 2π. Hence *two complex numbers are equal if and only if their moduli are equal and their arguments differ by an integer multiple of 2π*. Thus,

$$z = x + iy = r[\cos(\theta + 2\pi k) + i\sin(\theta + 2\pi k)]$$

where k is an integer.

EXAMPLE 2 Express each complex number in trigonometric form.

(a) $z = -\sqrt{3} - i$
(b) $z = 2i$
(c) $z = 2$

SOLUTION (a) $z = -\sqrt{3} + (-1)i = r(\cos\theta + i\sin\theta)$, where $r = \sqrt{(-\sqrt{3})^2 + (-1)^2} = 2$, and θ satisfies $\tan\theta = 1/\sqrt{3}$ with the terminal side of θ in quadrant III, so that one value of θ is $7\pi/6$. Hence

$$z = 2\left(\cos\frac{7\pi}{6} + i\sin\frac{7\pi}{6}\right)$$

(b) $z = 2i = 0 + 2i = r(\cos\theta + i\sin\theta)$, where $r = \sqrt{0^2 + 2^2} = 2$. One value of θ is $\pi/2$. Hence

$$z = 2\left(\cos\frac{\pi}{2} + i\sin\frac{\pi}{2}\right)$$

(c) $z = 2 = 2 + 0i = r(\cos\theta + i\sin\theta)$, where $r = \sqrt{2^2 + 0^2} = 2$ and an argument is $\theta = 0$, so that $z = 2(\cos 0 + i\sin 0)$.

EXAMPLE 3 Express

$$z = 2\left(\cos\frac{\pi}{3} + i\sin\frac{\pi}{3}\right)$$

in the form of $x + iy$.

SOLUTION

$$\begin{aligned} z = x + iy &= 2\cos\frac{\pi}{3} + i\cdot 2\sin\frac{\pi}{3} \\ &= 2\cdot\frac{1}{2} + i\cdot 2\cdot\frac{\sqrt{3}}{2} = 1 + \sqrt{3}i \end{aligned}$$

If complex numbers are expressed in trigonometric form, then multiplication and division may be carried out as follows.

THEOREM 1 **Multiplication and Division of Complex Numbers**

Suppose that trigonometric forms of the complex numbers z_1 and z_2 are given by $z_1 = r_1(\cos\theta_1 + i\sin\theta_1)$ and $z_2 = r_2(\cos\theta_2 + i\sin\theta_2)$. Then

(i) $z_1 z_2 = r_1 r_2[\cos(\theta_1 + \theta_2) + i\sin(\theta_1 + \theta_2)]$

(ii) $\dfrac{z_1}{z_2} = \dfrac{r_1}{r_2}[\cos(\theta_1 - \theta_2) + i\sin(\theta_1 - \theta_2)], \quad z_2 \neq 0$

PROOF OF (i)

$$\begin{aligned} z_1 z_2 &= r_1(\cos\theta_1 + i\sin\theta_1)\cdot r_2(\cos\theta_2 + i\sin\theta_2) \\ &= r_1 r_2[(\cos\theta_1\cos\theta_2 - \sin\theta_1\sin\theta_2) + i(\cos\theta_1\sin\theta_2 + \cos\theta_2\sin\theta_1)] \\ &= r_1 r_2[\cos(\theta_1 + \theta_2) + i\sin(\theta_1 + \theta_2)] \end{aligned}$$

The proof of part (ii) is left as an exercise (Problem 34).

EXAMPLE 4 Express $z_1 = -1 + i$ and $z_2 = -4i$ in trigonometric form; then find

(a) $z_1 z_2$ (b) $\dfrac{z_1}{z_2}$

SOLUTION First we find $|z_1| = \sqrt{(-1)^2 + 1^2} = \sqrt{2}$ and $|z_2| = \sqrt{(-4)^2 + 0^2} = 4$; so that an angle in standard position with $(-1, 1)$ on its terminal side has measure $3\pi/4$; and one measure of an angle in standard position with $(0, -4)$ on its terminal side is $3\pi/2$. Thus, trigonometric forms of z_1 and z_2 are, respectively,

$$z_1 = -1 + i = \sqrt{2}\left(\cos\frac{3\pi}{4} + i\sin\frac{3\pi}{4}\right)$$

and

$$z_2 = 0 - 4i = 4\left(\cos\frac{3\pi}{2} + i\sin\frac{3\pi}{2}\right)$$

Applying Theorem 1, we have

$$\text{(a)}\quad z_1 z_2 = 4\sqrt{2}\left[\cos\left(\frac{3\pi}{4} + \frac{3\pi}{2}\right) + i\sin\left(\frac{3\pi}{4} + \frac{3\pi}{2}\right)\right]$$

$$= 4\sqrt{2}\left[\cos\frac{9\pi}{4} + i\sin\frac{9\pi}{4}\right] = 4\sqrt{2}\left[\frac{\sqrt{2}}{2} + i\frac{\sqrt{2}}{2}\right] = 4 + 4i$$

$$\text{(b)}\quad \frac{z_1}{z_2} = \frac{\sqrt{2}}{4}\left[\cos\left(\frac{3\pi}{4} - \frac{3\pi}{2}\right) + i\sin\left(\frac{3\pi}{4} - \frac{3\pi}{2}\right)\right]$$

$$= \frac{\sqrt{2}}{4}\left[\cos\left(-\frac{3\pi}{4}\right) + i\sin\left(-\frac{3\pi}{4}\right)\right]$$

$$= \frac{\sqrt{2}}{4}\left(\cos\frac{3\pi}{4} - i\sin\frac{3\pi}{4}\right) = \frac{\sqrt{2}}{4}\left(-\frac{\sqrt{2}}{2} - i\frac{\sqrt{2}}{2}\right) = -\frac{1}{4} - \frac{1}{4}i$$

Powers and Roots of Complex Numbers

If $z = r(\cos\theta + i\sin\theta)$, then

$$z^2 = r \cdot r[\cos(\theta + \theta) + i\sin(\theta + \theta)] = r^2(\cos 2\theta + i\sin 2\theta)$$

But $z^3 = z^2 \cdot z$, so that

$$z^3 = r^2 \cdot r[\cos(2\theta + \theta) + i\sin(2\theta + \theta)] = r^3(\cos 3\theta + i\sin 3\theta)$$

If we repeat the process one more time, we get

$$z^4 = z^3 \cdot z = r^4(\cos 4\theta + i\sin 4\theta)$$

This scheme for repeated multiplication of a complex number in trigonometric form is generalized in the next theorem. The proof of Theorem 2 entails the use of the principle of mathematical induction (Problem 10 on page 523).

THEOREM 2 **DeMoivre's Theorem**

Let $z = r(\cos\theta + i\sin\theta)$. Then, for n a positive integer,

$$z^n = [r(\cos\theta + i\sin\theta)]^n = r^n(\cos n\theta + i\sin n\theta)$$

EXAMPLE 5 Use DeMoivre's theorem to determine each of the given powers. Express the answer in the form $a + bi$.

(a) $[3(\cos 60° + i \sin 60°)]^4$

(b) $(1 + i)^{20}$

SOLUTION (a) By DeMoivre's theorem,

$$[3(\cos 60° + i \sin 60°)]^4 = 3^4(\cos 240° + i \sin 240°)$$
$$= 81\left(-\frac{1}{2} - \frac{i\sqrt{3}}{2}\right) = -\frac{81}{2} - \frac{81\sqrt{3}}{2}i$$

(b) The complex number $1 + i$ can be expressed in trigonometric form as

$$\sqrt{2}\left(\cos\frac{\pi}{4} + i \sin\frac{\pi}{4}\right)$$

where the modulus is $\sqrt{2}$ and an argument is $\pi/4$. Using DeMoivre's theorem, we have

$$(1 + i)^{20} = \left[\sqrt{2}\left(\cos\frac{\pi}{4} + i \sin\frac{\pi}{4}\right)\right]^{20} = 2^{10}(\cos 5\pi + i \sin 5\pi)$$
$$= 1024(-1 + 0i) = -1024 + 0i$$

EXAMPLE 6 Use DeMoivre's theorem to express $\cos 2\theta$ and $\sin 2\theta$ in terms of $\sin\theta$ and $\cos\theta$.

SOLUTION By using DeMoivre's theorem with $n = 2$ and $r = 1$, we have

$$\cos 2\theta + i \sin 2\theta = (\cos\theta + i \sin\theta)^2$$
$$= \cos^2\theta + i \cdot 2 \sin\theta\cos\theta - \sin^2\theta$$
$$= (\cos^2\theta - \sin^2\theta) + i \cdot 2 \sin\theta\cos\theta$$

Since the two complex numbers are equal, the real parts are equal; that is,

$$\cos 2\theta = \cos^2\theta - \sin^2\theta$$

and the imaginary parts are also equal; that is,

$$\sin 2\theta = 2 \sin\theta\cos\theta$$

DeMoivre's theorem is useful in finding the nth roots of a complex number. If $w = R(\cos\phi + i \sin\phi)$, and $z = r(\cos\theta + i \sin\theta)$ is any root of $z^n = w$, where n is a positive integer, then, by DeMoivre's theorem, it follows that

$$[r(\cos\theta + i \sin\theta)]^n = r^n(\cos n\theta + i \sin n\theta) = R(\cos\phi + i \sin\phi)$$

so that

$$r^n = R \qquad \text{and} \qquad n\theta = \phi + 2k\pi \qquad (n\theta = \phi + 360°k \text{ if degrees are used})$$

Hence

$$z = r(\cos\theta + i\sin\theta)$$

is a root of $z^n = w$ whenever

$$r = \sqrt[n]{R} \qquad (\text{note that } R \geq 0)$$

and

$$\theta = \frac{\phi}{n} + \frac{2k\pi}{n} \qquad \left(\theta = \frac{\phi}{n} + \frac{360°k}{n} \qquad \text{if degrees are used}\right)$$

where $k = 0, \pm 1, \pm 2, \pm 3, \ldots$.

Thus it would appear that there are an infinite number of roots for the equation $z^n = w$. However, for n a positive integer, there are only *n distinct roots*. These n roots can be determined by letting k take on the values $0, 1, 2, 3, 4, \ldots, n - 1$. This is true because if we let $k = n$, the argument θ will be

$$\frac{\phi}{n} + \frac{2n\pi}{n} = \frac{\phi}{n} + 2\pi$$

and this angle has the same terminal side as ϕ/n, the value of θ where $k = 0$; similarly, the value of θ obtained by letting $k = n + 1$ gives an angle with the same terminal side as $(\phi/n) + (2\pi/n)$, the value of θ when $k = 1$; and so on. Thus, we have proved the next theorem.

THEOREM 3 **nth Roots of a Complex Number**

If $w = R(\cos\phi + i\sin\phi)$ is any nonzero complex number, and if n is any positive integer, then w has precisely n distinct nth roots that are given by

$$\sqrt[n]{R}\left[\cos\left(\frac{\phi}{n} + \frac{2\pi k}{n}\right) + i\sin\left(\frac{\phi}{n} + \frac{2\pi k}{n}\right)\right], \qquad k = 0, 1, 2, \ldots, n - 1$$

Since each root has modulus $\sqrt[n]{R}$, the points representing these n roots are equally spaced around the circle with center at $(0, 0)$ and with radius $\sqrt[n]{R}$. Sometimes we use degree measures for ϕ in which case the nth roots of w are given by

$$\sqrt[n]{R}\left[\cos\left(\frac{\phi}{n} + \frac{360°k}{n}\right) + i\sin\left(\frac{\phi}{n} + \frac{360°k}{n}\right)\right], \qquad k = 0, 1, 2, \ldots, n - 1$$

EXAMPLE 7 Find the four fourth roots of $1 + i$ in trigonometric form and represent them geometrically.

SOLUTION First, we determine $R(\cos\phi + i\sin\phi)$, a trigonometric representation of $1 + i$. Here $R = \sqrt{1 + 1} = \sqrt{2}$ and $\phi = \pi/4$, so that

$$1 + i = \sqrt{2}\left(\cos\frac{\pi}{4} + i\sin\frac{\pi}{4}\right)$$

Since $n = 4$, the fourth roots are given by

$$\sqrt[4]{\sqrt{2}}\left[\cos\left(\frac{\pi/4}{4} + \frac{2\pi k}{4}\right) + i\sin\left(\frac{\pi/4}{4} + \frac{2\pi k}{4}\right)\right] \qquad \text{where } k = 0, 1, 2, 3$$

Since $\sqrt[4]{\sqrt{2}} = \sqrt[8]{2}$, after substituting these values for k, we obtain

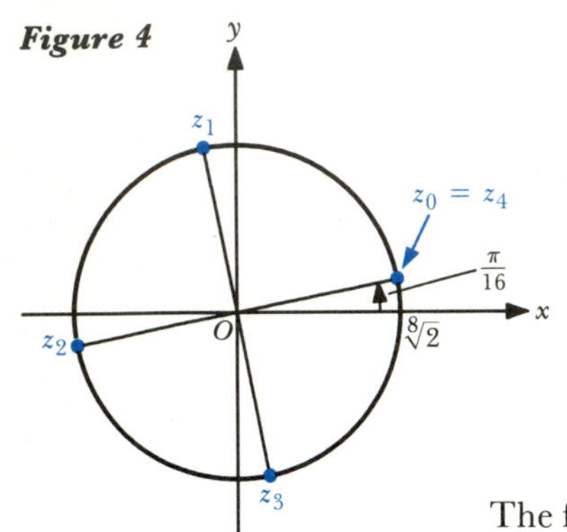

Figure 4

$$\sqrt[8]{2}\left(\cos\frac{\pi}{16} + i\sin\frac{\pi}{16}\right) \qquad \text{for } k = 0$$

$$\sqrt[8]{2}\left(\cos\frac{9\pi}{16} + i\sin\frac{9\pi}{16}\right) \qquad \text{for } k = 1$$

$$\sqrt[8]{2}\left(\cos\frac{17\pi}{16} + i\sin\frac{17\pi}{16}\right) \qquad \text{for } k = 2$$

$$\sqrt[8]{2}\left(\cos\frac{25\pi}{16} + i\sin\frac{25\pi}{16}\right) \qquad \text{for } k = 3$$

The fourth roots of $1 + i$ are equally spaced on the circumference of a circle of radius $\sqrt[8]{2}$ and their arguments differ by $\pi/2$ (Figure 4). Notice that if we were to set $k = 4$, we would get

$$\sqrt[8]{2}\left(\cos\frac{33\pi}{16} + i\sin\frac{33\pi}{16}\right) = \sqrt[8]{2}\left(\cos\frac{\pi}{16} + i\sin\frac{\pi}{16}\right)$$

which is the same number as the root we obtained when $k = 0$. In general, if $k = 4, 5, 6, \ldots$, we would get a repetition of the roots that we have already found.

The solutions of $z^n = 1$ are called the ***n*th roots of unity.**

EXAMPLE 8 Find the three third roots of unity in the form $a + bi$.

SOLUTION The number 1 can be written as $1 = 1(\cos 0° + i\sin 0°)$, so that $R = 1$, $\phi = 0°$, and $n = 3$ in Theorem 3. The three roots are given by

$$\sqrt[3]{1}\left[\cos\left(\frac{0° + 360°k}{3}\right) + i\sin\left(\frac{0° + 360°k}{3}\right)\right]$$

where $k = 0$, 1, and 2. Substituting these values for k yields the three roots:

$$1(\cos 0° + i\sin 0°) = 1 + 0i \qquad \text{for } k = 0$$

$$1(\cos 120° + i\sin 120°) = -\frac{1}{2} + \frac{\sqrt{3}}{2}i \qquad \text{for } k = 1$$

$$1(\cos 240° + i\sin 240°) = -\frac{1}{2} - \frac{\sqrt{3}}{2}i \qquad \text{for } k = 2$$

PROBLEM SET 6.9

In problems 1–8, find the value of each expression if $z_1 = 2 + 7i$ and $z_2 = 5 - 2i$.

1. $|z_1|$ **2.** $|z_2|$ **3.** $|z_1 z_2|$ **4.** $|z_1||z_2|$

5. $\left|\dfrac{z_1}{z_2}\right|$ **6.** $\dfrac{|z_1|}{|z_2|}$ **7.** $|z_1 \bar{z}_1|$ **8.** $|3z_1 - 2z_2|$

9. Let $z = a + bi$ and verify that $z\bar{z} = |z|^2$.

10. Use $z_1 = a + bi$ and $z_2 = c + di$ to verify that

(a) $|z_1 z_2| = |z_1||z_2|$ (b) $\left|\dfrac{z_1}{z_2}\right| = \dfrac{|z_1|}{|z_2|}$

In problems 11–17, express each complex number in trigonometric form, and then represent each of them graphically.

11. $-1 - i$ **12.** 7 **13.** $-2i$ **14.** $\dfrac{\sqrt{3}}{2} + \dfrac{1}{2}i$

15. $-\sqrt{3} - i$ C **16.** $3 + 4i$ **17.** $-\dfrac{1}{2} + \dfrac{\sqrt{3}}{2}i$

18. Describe geometrically the set of complex numbers $z = x + yi$ that satisfy each equation.
(a) $|z| = 2$ (b) $|z - i| = 1$

In problems 19–24, express each complex number in the form of $a + bi$. In problems 19 and 21, round off each answer to four decimal places.

C **19.** $z = 2(\cos 10° + i \sin 10°)$

20. $z = 10\left(\cos \dfrac{3\pi}{4} + i \sin \dfrac{3\pi}{4}\right)$

C **21.** $z = 3[\cos(-75°) + i \sin(-75°)]$

22. $z = 2\left(\cos \dfrac{\pi}{2} + i \sin \dfrac{\pi}{2}\right)$

23. $z = 4(\cos 0° + i \sin 0°)$

24. $z = 7\left[\cos\left(-\dfrac{3\pi}{2}\right) + i \sin\left(-\dfrac{3\pi}{2}\right)\right]$

In problems 25–33, find (a) $z_1 z_2$ and (b) z_1/z_2 and express the answers in the form $a + bi$. In problems 25, 27, and 33, round off each answer to four decimal places.

C **25.** $z_1 = 5(\cos 170° + i \sin 170°)$ and $z_2 = \cos 55° + i \sin 55°$

26. $z_1 = 4\left(\cos \dfrac{3\pi}{4} + i \sin \dfrac{3\pi}{4}\right)$ and $z_2 = 2(\cos \pi + i \sin \pi)$

C **27.** $z_1 = 2(\cos 50° + i \sin 50°)$ and $z_2 = 3(\cos 40° + i \sin 40°)$

28. $z_1 = 5(\cos 30° + i \sin 30°)$ and $z_2 = 6(\cos 240° + i \sin 240°)$

29. $z_1 = 4\left(\cos \dfrac{5\pi}{6} + i \sin \dfrac{5\pi}{6}\right)$ and $z_2 = 2\left(\cos \dfrac{\pi}{3} + i \sin \dfrac{\pi}{3}\right)$

30. $z_1 = \cos 30° + i \sin 30°$ and $z_2 = \cos 60° + i \sin 60°$

31. $z_1 = \sqrt{2}(\cos 45° + i \sin 45°)$ and $z_2 = \sqrt{2}\left(\cos \dfrac{3\pi}{4} + i \sin \dfrac{3\pi}{4}\right)$

32. $z_1 = 2\left[\cos\left(-\frac{\pi}{6}\right) + i\sin\left(-\frac{\pi}{6}\right)\right]$ and $z_2 = 2\left[\cos\left(-\frac{5\pi}{6}\right) + i\sin\left(-\frac{5\pi}{6}\right)\right]$

c **33.** $z_1 = 4[\cos(-240°) + i\sin(-240°)]$ and $z_2 = 3\sqrt{2}(\cos 45° + i\sin 45°)$

34. Prove Theorem 1(ii).

In problems 35–48, use DeMoivre's theorem to compute each of the powers. Express the answer in the form $a + bi$.

35. $(\cos 30° + i\sin 30°)^7$

36. $(\cos 15° + i\sin 15°)^8$

37. $\left[2\left(\cos\frac{\pi}{6} + i\sin\frac{\pi}{6}\right)\right]^{10}$

38. $\left[3\left(\cos\frac{\pi}{18} + i\sin\frac{\pi}{18}\right)\right]^6$

39. $\left[2\left(\cos\frac{5\pi}{4} + i\sin\frac{5\pi}{4}\right)\right]^8$

40. $[4(\cos 36° + i\sin 36°)]^5$

41. $(5 + 5i)^6$

42. $(1 + i\sqrt{3})^5$

43. $(\sqrt{3} - i)^4$

44. $\left(-\frac{1}{2} - i\frac{\sqrt{3}}{2}\right)^8$

45. $(\sqrt{3} + i)^{30}$

46. $(1 + i)^{50}$

47. $\left(\frac{1}{\sqrt{2}} + i\frac{1}{\sqrt{2}}\right)^{100}$

48. $\left(\frac{1}{2} + i\frac{\sqrt{3}}{2}\right)^{30}$

In problems 49–56, find the indicated roots of each complex number in the form $a + bi$. In problem 53, round off the answers to four decimal places.

49. The square roots of i

50. The square roots of $3 - 3i$

51. The cube roots of 8

52. The cube roots of i

c **53.** The fifth roots of $-\sqrt{3} - i$

54. The fifth roots of unity

55. The fourth roots of $-8 - 8\sqrt{3}i$

56. The fourth roots of 16

In problems 57–61, find all the roots of each equation.

57. $z^3 + 8 = 0$

58. $z^3 + 8i = 0$

59. $z^4 + 81 = 0$

60. $z^5 + 1 = 0$

61. $z^6 + 64 = 0$

62. Use DeMoivre's theorem to derive formulas for $\cos 3\theta$ and $\sin 3\theta$. [*Hint:* Use the identity $\cos 3\theta + i\sin 3\theta = (\cos\theta + i\sin\theta)^3$.]

63. Let

$$z_1 = \left[2\left(\cos\frac{\pi}{8} + i\sin\frac{\pi}{8}\right)\right]^4 \quad \text{and} \quad z_2 = \left[4\left(\cos\frac{\pi}{12} + i\sin\frac{\pi}{12}\right)\right]^4$$

Find $z_1 z_2$.

64. Use DeMoivre's theorem, together with Theorem 1(ii) on page 380, to show that if $z = r(\cos\theta + i\sin\theta)$, then

$$z^{-n} = r^{-n}[\cos(-n\theta) + i\sin(-n\theta)]$$

65. Use the result of problem 64 to write each expression in the form $a + bi$.

(a) $\left(\frac{\sqrt{3}}{2} + \frac{1}{2}i\right)^{-5}$ (b) $(-2 + 2i)^{-3}$

6.10 Plane Vectors and Their Applications

Until now, we have dealt exclusively with quantities that can be measured or represented by single real numbers, such as area, volume, angle, time, temperature, and speed. Since real numbers can be represented by points on a number scale, these quantities are often called **scalars.** In this section, we deal with quantities that have both magnitude and direction, and thus cannot be described or represented by a *single* real number. Such quantities, which are called *vectors*, are used to represent force, velocity, acceleration, and displacement. We will consider a few applications of vectors that utilize the tools of trigonometry.

Geometry of Vectors

Figure 1

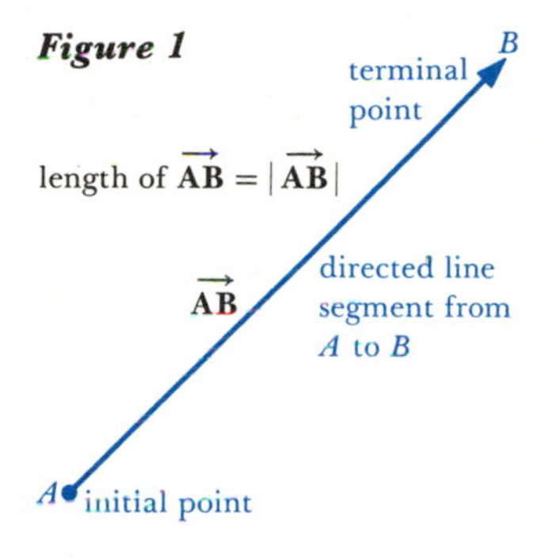

A **vector** in the plane is a line segment, with a direction usually denoted by an arrowhead at the end of the segment (Figure 1). The length of the line segment is called the **magnitude** of the vector, and the orientation of the line segment is called the **direction** of the vector. The endpoint of the vector containing the arrowhead is called the **terminal point** of the vector, and the other endpoint is called the **initial point** of the vector. If a vector is determined by a directed line segment with the initial point A and the terminal point B, then it can be denoted by $\overrightarrow{AB}$; its magnitude is denoted by $|\overrightarrow{AB}|$ (Figure 1).

The **zero vector,** denoted by **0**, is a vector whose initial and terminal points are the same. Note that $|\mathbf{0}| = 0$.

Two vectors are considered to be **equal** if they agree both in magnitude and in direction. Note that $\overrightarrow{AB}$ and $\overrightarrow{BA}$ (Figure 2) are not equal (even though their lengths are the same) because they are opposite in direction; that is, $\overrightarrow{AB} \neq \overrightarrow{BA}$, even though $|\overrightarrow{AB}| = |\overrightarrow{BA}|$.

Figure 2

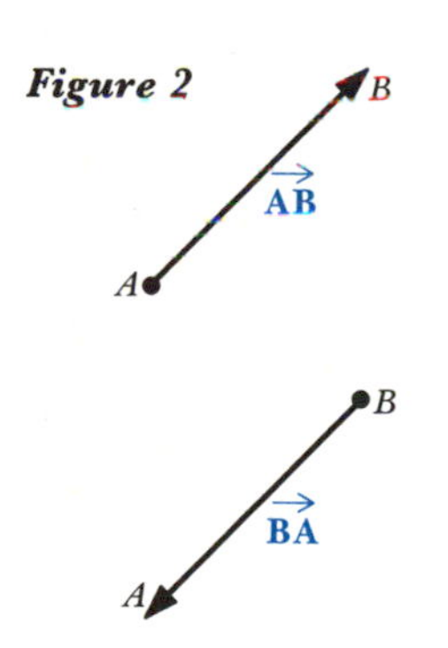

We use *lowercase boldface* letters to denote vectors. Hence we speak of "vector **u**" and write it as **u**.

Suppose that we are given two vectors, **u** and **v** (Figure 3a). First, we "shift" **v** so that its initial point coincides with the terminal point of **u**. The vector having the same initial point as **u** and the same terminal point as **v** is defined to be the **sum** (**resultant vector**) **u** + **v** of **u** and **v** (Figure 3b).

Figure 3b indicates that the resultant vector is, in a manner of speaking, a *diagonal vector* of the parallelogram "determined" by vectors **u** and **v**. Figure 3b also displays the fact that **u** + **v** = **v** + **u**; that is, *vector addition is commutative.*

Figure 3

u v

(a)

v u + v u u v

(b)

Figure 4

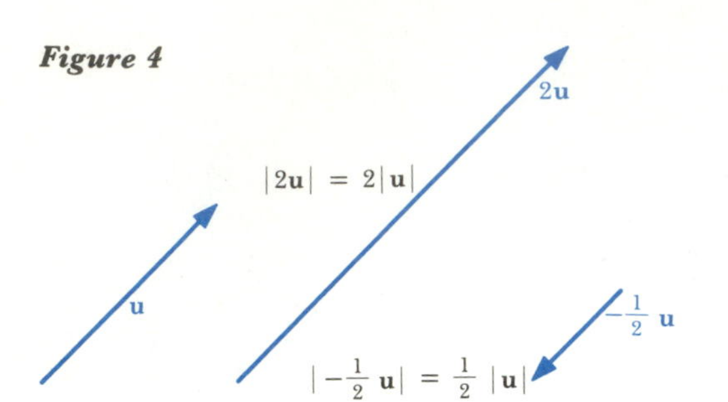

Suppose that we are given a real number a and a vector **u**. The real number a is a scalar; a**u** is defined to be the vector with magnitude $|a|\,|\mathbf{u}|$, with the *same* direction as **u** if $a > 0$ and with the *opposite* direction if $a < 0$. Combining a and **u** to form the vector a**u** is called **scalar multiplication.** Thus, in Figure 4, 2**u** is a vector in the same direction as **u**, with magnitude two times that of **u**, and $-\frac{1}{2}\mathbf{u}$ is a vector opposite in direction to **u**, with a length of $\frac{1}{2}$ that of **u**.

Now we can use vector addition and scalar multiplication to define **vector subtraction** as

$$\mathbf{u} - \mathbf{v} = \mathbf{u} + (-\mathbf{v}) \qquad \text{(Figure 5)}$$

Figure 5

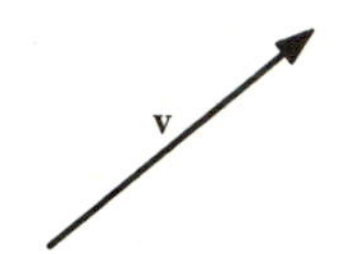

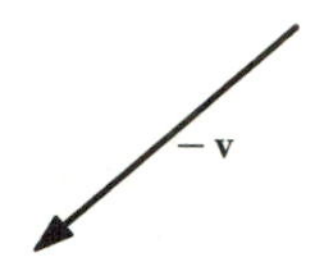

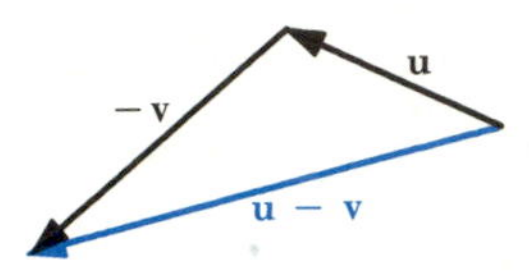

Figure 6

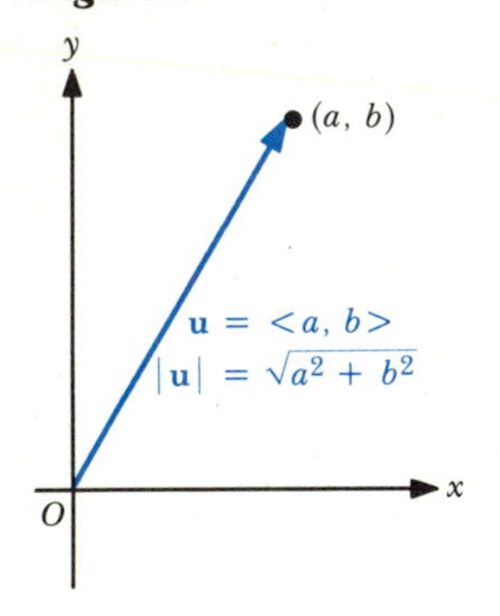

Analytic Representation of Vectors in a Plane

If vector **u** is positioned in a plane with a Cartesian coordinate system so that the initial point of **u** is the origin and the terminal point is at point (a, b) (Figure 6), then **u** is called a **radius vector** or **position vector.** We identify such a vector as $\mathbf{u} = \langle a, b\rangle$, and a and b are called the ***x* component** and the ***y* component** of **u**, respectively. Notice that

$$|\mathbf{u}| = \sqrt{a^2 + b^2}$$

Figure 7

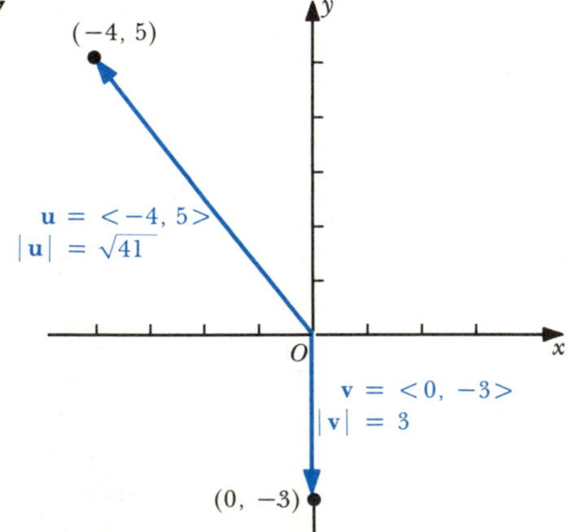

For example, $\mathbf{u} = \langle -4, 5\rangle$ has x component -4 and y component 5, so

$$|\mathbf{u}| = \sqrt{(-4)^2 + 5^2} = \sqrt{41} \qquad \text{(Figure 7)}$$

All vectors with the x component 0 have terminal points on the y axis. For example, $\mathbf{v} = \langle 0, -3\rangle$ has its terminal point on the negative y axis and

$$|\mathbf{v}| = \sqrt{0^2 + (-3)^2} = \sqrt{9} = 3 \qquad \text{(Figure 7)}$$

If $\mathbf{u}_1$ and $\mathbf{u}_2$ are *equal* position vectors, they have the same initial point $(0, 0)$, so the components must also be equal; that is,

> If $\mathbf{u}_1 = \langle a_1, b_1 \rangle$ and $\mathbf{u}_2 = \langle a_2, b_2 \rangle$, then $\mathbf{u}_1 = \mathbf{u}_2$ whenever $a_1 = a_2$ and $b_1 = b_2$.

Suppose that $\mathbf{w}$ is a vector in a plane, and furthermore that it is not a radius vector. To be specific, we assume that both the initial point (x_1, y_1) and the terminal point (x_2, y_2) of $\mathbf{w}$ are in quadrant I (Figure 8). Now if $\mathbf{w}$ is shifted or "translated" from its original location to become a radius vector, observe that we can represent $\mathbf{w}$ as $\langle x_2 - x_1, y_2 - y_1 \rangle$ (Figure 8). In general:

> If $\mathbf{w}$ is a vector with initial point (x_1, y_1) and terminal point (x_2, y_2), then we have $\mathbf{w} = \langle x_2 - x_1, y_2 - y_1 \rangle$.

Figure 8

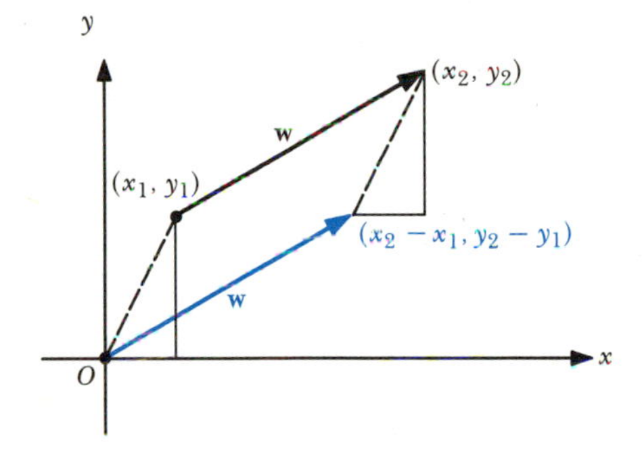

If θ represents an angle formed by $\mathbf{u}$ and the positive x axis, θ is sometimes called a **direction angle** for $\mathbf{u}$.

EXAMPLE 1 Find the x and y components of the radius vector $\mathbf{u} = \langle x, y \rangle$ for each of the following situations. Assume that θ is a direction angle.

(a) $|\mathbf{u}| = 5$, $\theta = 30°$ (b) $|\mathbf{u}| = 1$, $\theta = \frac{5\pi}{4}$ c (c) $|\mathbf{u}| = 2$, $\theta = -50°$

SOLUTION Figure 9 shows the locations of $\mathbf{u}$ for each situation. From trigonometry, we have

$$x = |\mathbf{u}| \cos \theta \quad \text{and} \quad y = |\mathbf{u}| \sin \theta$$

Thus we have the following results.

(a) $x = 5 \cos 30° = 5\left(\frac{\sqrt{3}}{2}\right) = \frac{5\sqrt{3}}{2}$

$y = 5 \sin 30° = 5\left(\frac{1}{2}\right) = \frac{5}{2}$ (Figure 9a)

(b) $x = 1\cos\frac{5\pi}{4} = -\frac{\sqrt{2}}{2}$

$y = 1\sin\frac{5\pi}{4} = -\frac{\sqrt{2}}{2}$ (Figure 9b)

(c) $x = 2\cos(-50^\circ) = 1.2856$ $y = 2\sin(-50^\circ) = -1.5321$ (Figure 9c)

Figure 9

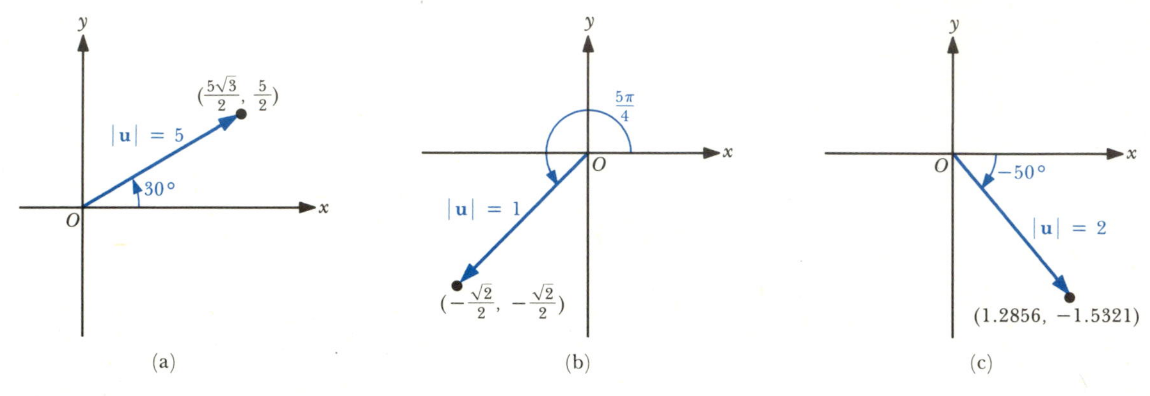

EXAMPLE 2 Assume that **u** is a vector whose initial point is $(-2, 3)$ and whose terminal point is $(5, 10)$. Represent **u** as a radius vector and then find $|\mathbf{u}|$ and θ, where θ is a direction angle for **u**.

SOLUTION Since $\mathbf{u} = \langle 5 - (-2), 10 - 3\rangle = \langle 7, 7\rangle$,

$$|\mathbf{u}| = \sqrt{7^2 + 7^2} = \sqrt{98} = 7\sqrt{2}$$

and

$$\tan\theta = \frac{7}{7} = 1$$

so $\theta = 45^\circ$ (Figure 10).

Figure 10

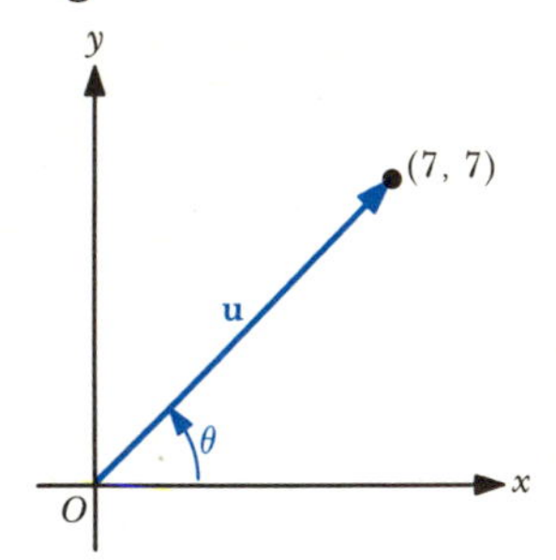

The components of position vectors completely characterize the operations of vectors in the sense that the addition, scalar multiplication, and subtraction of position vectors can be performed by using components. Addition and scalar multiplication of position vectors can be accomplished by using the components as follows.

> Assume that $\mathbf{u}_1 = \langle a_1, b_1\rangle$, $\mathbf{u}_2 = \langle a_2, b_2\rangle$, and c is a scalar. Then
> (i) $\mathbf{u}_1 + \mathbf{u}_2 = \langle a_1, b_1\rangle + \langle a_2, b_2\rangle = \langle a_1 + a_2, b_1 + b_2\rangle$
> and
> (ii) $c\mathbf{u}_1 = c\langle a_1, b_1\rangle = \langle ca_1, cb_1\rangle$

Geometric representation of $\mathbf{u}_1 + \mathbf{u}_2$ and $c\mathbf{u}_1$ are illustrated in Figures 11a and 11b, respectively.

Figure 11

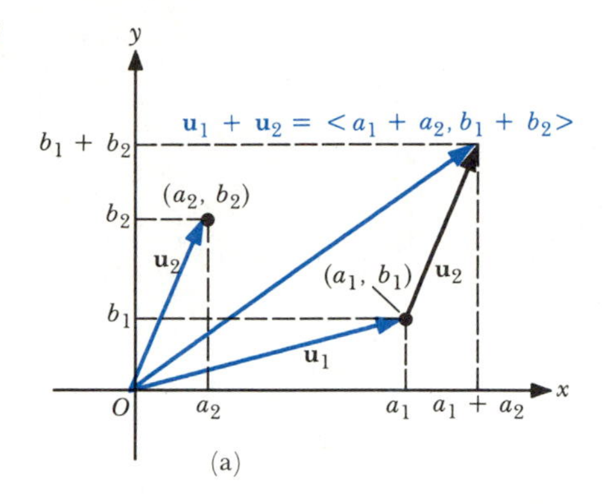

(a)

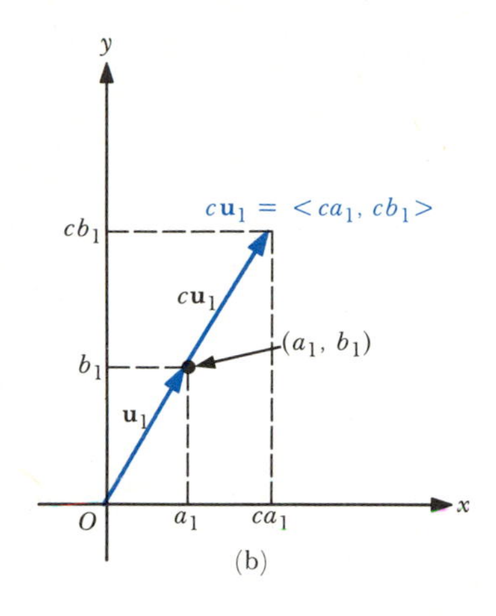

(b)

Figure 12

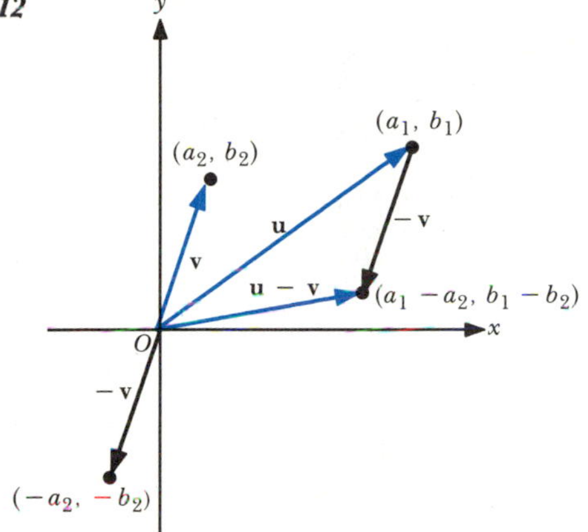

If $\mathbf{u} = \langle a_1, b_1 \rangle$ and $\mathbf{v} = \langle a_2, b_2 \rangle$, then

$$\begin{aligned}\mathbf{u} - \mathbf{v} &= \mathbf{u} + (-\mathbf{v}) \\ &= \langle a_1, b_1 \rangle + (-\langle a_2, b_2 \rangle) \\ &= \langle a_1, b_1 \rangle + \langle -a_2, -b_2 \rangle \\ &= \langle a_1 - a_2, b_1 - b_2 \rangle\end{aligned}$$

Thus (Figure 12),

$$\mathbf{u} - \mathbf{v} = \langle a_1 - a_2, b_1 - b_2 \rangle$$

EXAMPLE 3 Let $\mathbf{u} = \langle 3, 4 \rangle$ and $\mathbf{v} = \langle -5, 6 \rangle$. Find

(a) $\mathbf{u} + \mathbf{v}$ (b) $\mathbf{u} - \mathbf{v}$ (c) $4\mathbf{u} - 3\mathbf{v}$ (d) $|4\mathbf{u} - 3\mathbf{v}|$

SOLUTION

(a) $\mathbf{u} + \mathbf{v} = \langle 3, 4 \rangle + \langle -5, 6 \rangle = \langle -2, 10 \rangle$

(b) $\mathbf{u} - \mathbf{v} = \langle 3, 4 \rangle - \langle -5, 6 \rangle = \langle 8, -2 \rangle$

(c) $\begin{aligned}4\mathbf{u} - 3\mathbf{v} &= 4\langle 3, 4 \rangle - 3\langle -5, 6 \rangle \\ &= \langle 12, 16 \rangle + \langle 15, -18 \rangle = \langle 27, -2 \rangle\end{aligned}$

(d) $\begin{aligned}|4\mathbf{u} - 3\mathbf{v}| &= \sqrt{(27)^2 + (-2)^2} \\ &= \sqrt{729 + 4} = \sqrt{733}\end{aligned}$

Applications of Vectors

The following examples illustrate how vectors, together with the tools of trigonometry, are used to solve force problems.

EXAMPLE 4 [c] Two forces, one of 10 pounds and the other of 15 pounds, are applied to the same object, yielding a resultant force of 18 pounds. To the nearest hundredth of a degree, what angle does the resultant vector make with the vector representation of the 15-pound force?

SOLUTION Assume that the force vectors are **u**, **v**, and **r**, as shown in Figure 13. Then, by the law of cosines, we obtain

$$10^2 = 18^2 + 15^2 - 2(18)(15)\cos\theta$$

so that

$$\theta = \cos^{-1}\left[\frac{18^2 + 15^2 - 10^2}{2(18)(15)}\right] = 33.75°$$

Figure 13

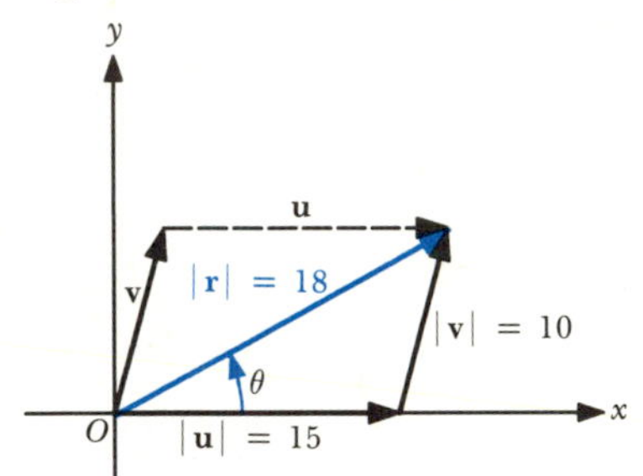

Figure 14

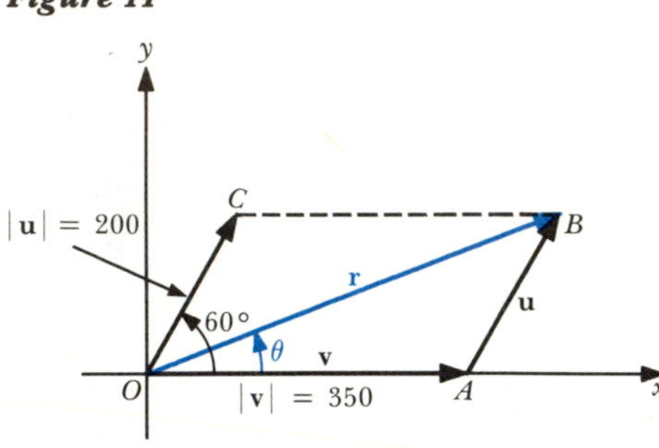

EXAMPLE 5 [c] A raft is pulled along a canal by two ropes on opposite sides of the canal. The two ropes form an angle of 60° with each other; one pulls with a force of 200 pounds, while the force of the other is 350 pounds. Assume that the force vector for 350 pounds is horizontal. What is the magnitude of the resultant vector, to four significant digits, and the angle that the resultant vector makes with the 350-pound force vector, to the nearest hundredth of a degree?

SOLUTION Suppose that the 350-pound force is exerted in the direction of the positive x axis and that it is represented by vector **v** so that the 200-pound force has a 60° direction angle and is represented by vector **u** in Figure 14. Therefore, by using the definition of the trigonometric functions, we get

$$\mathbf{u} = \langle 200\cos 60°, 200\sin 60°\rangle = \langle 100, 100\sqrt{3}\rangle$$

and

$$\mathbf{v} = \langle 350, 0\rangle$$

The resultant vector is

$$\mathbf{r} = \mathbf{u} + \mathbf{v} = \langle 100, 100\sqrt{3}\rangle + \langle 350, 0\rangle = \langle 450, 100\sqrt{3}\rangle$$

and

$$|\mathbf{r}| = \sqrt{450^2 + (100\sqrt{3})^2} = 482.2 \text{ pounds}$$

The angle θ between $\mathbf{v}$ and $\mathbf{r}$ is given by

$$\tan\theta = \frac{100\sqrt{3}}{450}$$

so that $\theta = 21.05°$.

Another important application of vectors is in navigation, where the following terminology is commonly used. The **heading** of an airplane is the direction in which it is pointed and its **air speed** is its speed relative to the air. The vector $\mathbf{v}_1$, whose magnitude is the air speed and whose direction is the heading, represents the **velocity of the airplane relative to the air** (Figure 15). The **course** or **track** of an airplane is the direction in which it is actually moving over the ground, and its **ground speed** is its speed relative to the ground. Assume that vector $\mathbf{v}_2$ is the **velocity vector** for the wind, that is, $|\mathbf{v}_2|$ is the speed of the wind, and the direction angle of $\mathbf{v}_2$ is the direction of the wind. Then the resultant vector $\mathbf{v}$, whose magnitude is the ground speed and whose direction is the course, represents the actual **velocity of the airplane relative to the ground** (Figure 15). The angle α between the vectors $\mathbf{v}$ and $\mathbf{v}_1$ is called the **drift angle.**

Figure 15

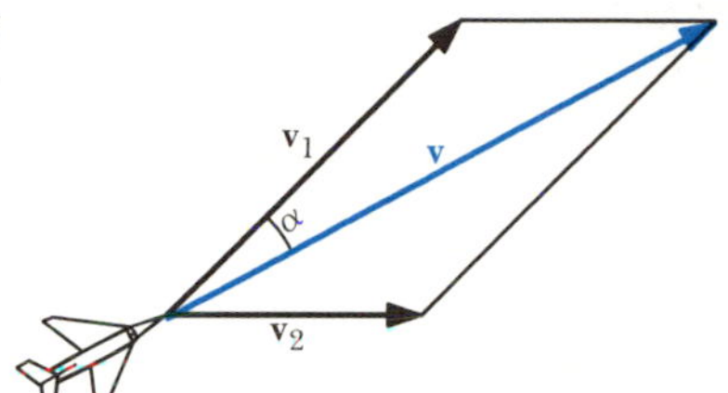

EXAMPLE 6 [c] An airplane is headed N30°E with an air speed of 500 miles per hour. The wind is blowing S29°E at a speed of 50 miles per hour. Find (a) the ground speed to four significant digits, (b) the drift angle to the nearest hundredth of a degree, and (c) the course of the airplane, that is, the direction angle of the velocity vector relative to the ground.

SOLUTION Let $\mathbf{v}_1$ represent the velocity of the jet relative to the air; let $\mathbf{v}_2$ represent the velocity of the wind relative to the ground; and let $\mathbf{v}$ represent the velocity of the jet relative to the ground (Figure 16). Then

$$\mathbf{v}_1 = \overrightarrow{AB} = \langle 500\cos 60°, 500\sin 60°\rangle$$
$$= \langle 250, 433.0\rangle$$

and

$$\mathbf{v}_2 = \overrightarrow{AD} = \langle 50\cos(-61°), 50\sin(-61°)\rangle = \langle 24.24, -43.73\rangle$$

(a) Thus $\mathbf{v} = \mathbf{v}_1 + \mathbf{v}_2 = \langle 274.2, 389.3\rangle$

so that

$$|\mathbf{v}| = \sqrt{(274.2)^2 + (389.3)^2}$$
$$= 476.2$$

Therefore, the ground speed is 476.2 miles per hour.

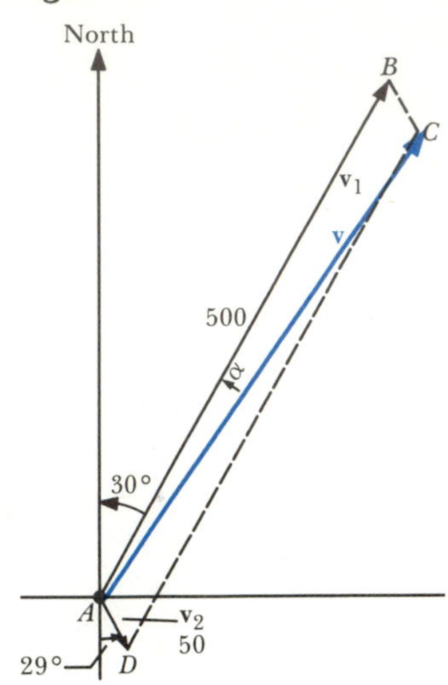

Figure 16

(b) From the geometry of parallelograms, we know that $|\overrightarrow{BC}| = |\overrightarrow{AD}| = |\mathbf{v}_2| = 50$. Also, the sum of two consecutive angles in a parallelogram is 180°, so that

$$\begin{aligned}\text{angle } ABC &= 180° - (\text{angle } DAB)\\ &= 180° - (180° - 30° - 29°)\\ &= 180° - 121° = 59°\end{aligned}$$

Using the law of sines in triangle ABC, we have

$$\frac{\sin \alpha}{50} = \frac{\sin 59°}{|\mathbf{v}|} = \frac{\sin 59°}{476.2}$$

or

$$\sin \alpha = \frac{50 \sin 59°}{476.2} = 0.0900$$

so that

$$\alpha = \sin^{-1}(0.0900) = 5.16°$$

Therefore, the drift angle is 5.16°.

(c) From Figure 16, a direction angle of **v** is

$$90° - (30° + \alpha) = 90° - 35.16° = 54.84°$$

PROBLEM SET 6.10

In problems 1–4, let **u** be a vector in the positive direction of the x axis and let **v** be a vector with a direction angle of 30°. In addition, assume that $|\mathbf{u}| = 3$ and $|\mathbf{v}| = 2$. Make a sketch of each of the following vectors.

1. $2\mathbf{u} + \mathbf{v}$ **2.** $2\mathbf{u} + \frac{1}{2}\mathbf{v}$ **3.** $\frac{1}{2}\mathbf{u} - 2\mathbf{v}$ **4.** $-\mathbf{u} - 3\mathbf{v}$

In problems 5–12, find the x and y components of $\mathbf{u} = \langle x, y \rangle$ if θ represents a direction angle of **u**. In problems 10–12, round off each answer to four decimal places.

5. $|\mathbf{u}| = 3, \theta = 0°$ **6.** $|\mathbf{u}| = 2, \theta = 180°$
7. $|\mathbf{u}| = 5, \theta = 45°$ **8.** $|\mathbf{u}| = 1, \theta = 300°$
9. $|\mathbf{u}| = 1, \theta = -150°$ c **10.** $|\mathbf{u}| = 4, \theta = -75°$
c **11.** $|\mathbf{u}| = \frac{1}{2}, \theta = 100°10'$ c **12.** $|\mathbf{u}| = 7, \theta = 253°20'$

c In problems 13–20, find the magnitude and a direction angle of each given vector. Round off each angle to the nearest hundredth of a degree.

13. $\langle -2, 2 \rangle$ **14.** $\langle \sqrt{3}, -1 \rangle$ **15.** $\langle -5, 0 \rangle$
16. $\langle 1, 3 \rangle$ **17.** $\langle -8, -6 \rangle$ **18.** $\langle 3, 4 \rangle$
19. $\langle 6, -5 \rangle$ **20.** $\langle -5, -1 \rangle$

In problems 21–26, if $\mathbf{u}$ is a vector whose initial point is the first point given and whose terminal point is the second point given, represent $\mathbf{u}$ as a position vector. Also find $|\mathbf{u}|$.

21. $(8, 6), (3, 4)$
22. $(-7, -6), (2, 3)$
23. $(-2, 6)\ (3, -5)$
24. $(-3, 2), (1, -3)$
25. $(3, 7), (-3, 1)$
26. $(1, -3), (5, -1)$

In problems 27–36, use $\mathbf{u}_1 = \langle 3, 4\rangle$, $\mathbf{u}_2 = \langle -2, 4\rangle$, and $\mathbf{u}_3 = \langle 7, 8\rangle$ to find the value of each expression.

27. $2\mathbf{u}_1 + 3\mathbf{u}_2$
28. $3\mathbf{u}_2 - \mathbf{u}_1$
29. $|\mathbf{u}_1| + |\mathbf{u}_2|$
30. $|\mathbf{u}_1 + \mathbf{u}_2|$
31. $\mathbf{u}_1 + (\mathbf{u}_2 + \mathbf{u}_3)$
32. $(\mathbf{u}_1 + \mathbf{u}_2) + \mathbf{u}_3$
33. $2\mathbf{u}_1 - 3\mathbf{u}_2$
34. $\mathbf{u}_2 - \mathbf{u}_1 - \mathbf{u}_3$
35. $3|\mathbf{u}_1 - \mathbf{u}_2|$
36. $3|\mathbf{u}_1| - 3|\mathbf{u}_2|$

37. Let $P_1 = (-6, -10)$ and $P_2 = (8, 8)$ be two points in a plane.
(a) Use vectors to find the point that is two-thirds of the way from P_1 to P_2.
(b) Use vectors to find the point that is $\sqrt{130}$ units from P_1 along the directed line from P_1 to P_2.

38. A **unit vector** is a vector of magnitude 1. Find k so that $k\mathbf{u}$ is a unit vector for each of the following vectors.
(a) $\mathbf{u} = \langle 3, 4\rangle$ (b) $\mathbf{u} = \langle -2, 5\rangle$

39. The **inner product** or **dot product** of $\mathbf{u}$ and $\mathbf{v}$, denoted by $\mathbf{u}\cdot\mathbf{v}$, is defined as $\mathbf{u}\cdot\mathbf{v} = |\mathbf{u}||\mathbf{v}|\cos\theta$, where θ is the angle between $\mathbf{u}$ and $\mathbf{v}$ and $0° \le \theta \le 180°$. Find $\mathbf{u}\cdot\mathbf{v}$ for each of the following situations.
(a) $|\mathbf{u}| = 3$, $|\mathbf{v}| = 2$, and $\theta = 60°$ (b) $|\mathbf{u}| = 5$, $|\mathbf{v}| = 1$, and $\theta = 150°$

40. Use the definition in problem 39 to prove that two nonzero vectors $\mathbf{u}$ and $\mathbf{v}$ are perpendicular (*orthogonal*) if and only if $\mathbf{u}\cdot\mathbf{v} = 0$.

41. Assume that $\mathbf{u} = \langle a, b\rangle$ and $\mathbf{v} = \langle c, d\rangle$. It can be proved that

$$\mathbf{u}\cdot\mathbf{v} = ac + bd$$

(see problem 39). Use this result to determine $\mathbf{u}\cdot\mathbf{v}$ for each of the following situations.
(a) $\mathbf{u} = \langle 1, 1\rangle$ and $\mathbf{v} = \langle 1, -1\rangle$ (b) $\mathbf{u} = \langle 2, -3\rangle$ and $\mathbf{v} = \langle -4, -1\rangle$
(c) $\mathbf{u} = \langle 4, 2\rangle$ and $\mathbf{v} = \langle -3, 5\rangle$ (d) $\mathbf{u} = \langle 0, 3\rangle$ and $\mathbf{v} = \langle 5, 0\rangle$

c **42.** Use the definition in problem 39 and the formula in problem 41 to find the angle to the nearest hundredth of a degree between the following pairs of vectors.
(a) $\langle 1, 1\rangle$ and $\langle 1, -1\rangle$ (b) $\langle 3, 1\rangle$ and $\langle 0, 5\rangle$
(c) $\langle 1, 2\rangle$ and $\langle -1, 1\rangle$ (d) $\langle -3, -4\rangle$ and $\langle 2, -3\rangle$

c In problems 43–50, compute each scalar value to four significant digits and each angle to the nearest hundredth of a degree.

43. Two forces $\mathbf{F}_1$ and $\mathbf{F}_2$ act on a point. If $|\mathbf{F}_1| = 40$ pounds, $|\mathbf{F}_2| = 23$ pounds, and the angle between $\mathbf{F}_1$ and $\mathbf{F}_2$ is 58°, find (a) the magnitude $|\mathbf{F}|$ of the resultant force $\mathbf{F}$ and (b) the angle between $\mathbf{F}_1$ and $\mathbf{F}$.

44. Two forces acting on the same object produce a resultant force of 17 pounds with an angle of 20° relative to a horizontal axis. Assume that one of the two forces is 20 pounds and has an angle of 50° relative to the horizontal axis. Find the other force and a direction angle relative to the horizontal axis.

45. Two children are pulling a third child across the ice on a sled. The first child pulls on his rope with a force of 5 newtons and the second child pulls on her rope with a force of 8 newtons. If the angle between the ropes is 28°, find (a) the magnitude of the resultant force and (b) the angle the resultant force vector makes with the first child's rope.

46. A long-distance swimmer swims across a river at the rate of 3.2 kilometers per hour perpendicular to the current. The river is flowing at 4.8 kilometers per hour. Find the speed and the angle at which the swimmer is moving in the water.

47. A commercial airliner with an air speed of 510 miles per hour is headed N85°E. The wind is blowing N35°E at a speed of 45 miles per hour. Find (a) the ground speed, (b) the drift angle, and (c) the course of the airliner.

48. An airplane is flying in the direction N25°E with an air speed of 450 kilometers per hour. Its ground speed is 500 kilometers per hour and its course is N40°E. Find (a) the speed of the wind and (b) the direction of the wind.

49. A boat heading S40°E at a still water speed of 40 kilometers per hour is pushed off course by a current of 32 kilometers per hour flowing in the direction S50°W. Find (a) the speed of the boat, (b) the drift angle, and (c) the course which the boat is traveling.

50. Suppose that a motor boat, which has a speed in still water of 12 miles per hour, heads west across a river that is flowing south at a speed of 3 miles per hour. (a) What is the speed of the boat? (b) What is the angle that the boat's path makes with the vector representing the current?

REVIEW PROBLEM SET, CHAPTER 6

In problems 1–8, simplify each expression.

1. $\cos\theta \csc\theta$

2. $\dfrac{\sec\theta}{\csc\theta}$

3. $\sin(-\theta)\sec(-\theta)$

4. $\dfrac{\cos(-\theta)}{\cot(-\theta)}$

5. $\csc^2\theta \tan^2\theta - \tan^2\theta$

6. $(\sin\theta + \cos\theta)^2 - 2\sin\theta\cos\theta$

7. $\dfrac{\tan^2\theta}{\sec^2\theta} + \dfrac{\cot^2\theta}{\csc^2\theta}$

8. $\dfrac{1}{\csc\theta - \cot\theta} - \dfrac{1}{\csc\theta + \cot\theta}$

In problems 9 and 10, write each radical expression as a trigonometric expression containing no radical by making the specified trigonometric substitution.

9. $\sqrt{(36 - u^2)^3}$; $u = 6\sin\theta$, $-\dfrac{\pi}{2} \le \theta \le \dfrac{\pi}{2}$

10. $\dfrac{x}{\sqrt{x^2 - 9}}$; $x = 3\sec\theta$, $0 \le \theta \le \dfrac{\pi}{2}$

In problems 11–18, prove that each equation is an identity.

11. $\cos t \sin t \csc t \sec t = 1$

12. $\dfrac{\sin^4 t - \cos^4 t}{\sin^2 t - \cos^2 t} = 1$

13. $\sec\theta - \cos\theta = \sin\theta \tan\theta$

14. $\cot\theta \cos\theta + \sin\theta = \csc\theta$

15. $\cos^2 t - \sin^2 t = \dfrac{1 - \tan^2 t}{1 + \tan^2 t}$

16. $\dfrac{\sin^2 t \cos t + \cos^3 t}{\cot t} = \sin t$

17. $1 + \sin\theta = \dfrac{\cos\theta}{\sec\theta - \tan\theta}$

18. $\dfrac{\sec^2\theta + 2\tan\theta}{\sin^2\theta} = (\csc\theta + \sec\theta)^2$

In problems 19–30, write each expression as a single term involving only one angle.

19. $\sin 7t \cos 2t - \cos 7t \sin 2t$

20. $\cos\left(\dfrac{\pi}{2} - t\right) \tan\left(\dfrac{\pi}{2} - t\right)$

21. $\sec^2 3t - 1$

22. $\csc(90° - \theta) \cos\theta - \cot\theta$

23. $\dfrac{\tan 5t + \tan t}{1 - \tan 5t \tan t}$

24. $\sin(3\pi - t)$

25. $2\sin^2 4\theta - 1$

26. $\tan(270° + \theta)$

27. $\sin 7t \cos 7t$

28. $2\cos^2 3t - 1$

29. $\sin(2\pi - t)$

30. $\cos(720° - t)$

In problems 31–36, use the given information, together with an appropriate trigonometric formula, to find the exact value of each expression.

31. $\sin 22.5°$; $22.5° = \frac{1}{2}(45°)$

32. $\cos 75°$; $75° = 45° + 30°$

33. $\cos\dfrac{11\pi}{12}$; $\dfrac{11\pi}{12} = \dfrac{1}{2}\left(\dfrac{11\pi}{6}\right)$

34. $\cos 75°$; $75° = \frac{1}{2}(150°)$

35. $\tan 255°$; $255° = 210° + 45°$

36. $\sin\dfrac{\pi}{12}$; $\dfrac{\pi}{12} = \dfrac{\pi}{3} - \dfrac{\pi}{4}$

In problems 37–46, let $\sin t = \frac{3}{5}$, for $0 < t < \pi/2$, and $\cos s = -\frac{8}{17}$, for $\pi/2 < s < \pi$. Evaluate each given expression.

37. $\sin 2s$

38. $\tan 2t$

39. $\cos(t + s)$

40. $\cos\left(\dfrac{s}{2}\right)$

41. $\sin(t - s)$

42. $\cos 2t$

43. $\sin\left(\dfrac{t}{2}\right)$

44. $\sin(s + t)$

45. $\tan(t + s)$

46. $\sec(t - s)$

In problems 47–59, prove that the given equation is an identity.

47. $\sin\left(\theta + \dfrac{\pi}{6}\right) + \cos\left(\theta + \dfrac{\pi}{3}\right) = \cos\theta$

48. $\tan\left(t + \dfrac{3\pi}{4}\right) = \dfrac{\tan t - 1}{\tan t + 1}$

49. $\sin(s - t)\cos t + \cos(s - t)\sin t = \sin s$

50. $\dfrac{\cos(\alpha - \beta)}{\cos\alpha \sin\beta} = \tan\alpha + \cot\beta$

51. $\dfrac{1 - \cos 2\theta}{\sin 2\theta} = \tan\theta$

52. $\sin\dfrac{\theta}{2}\cos\dfrac{\theta}{2} = \dfrac{\sin\theta}{2}$

53. $\dfrac{\sin t}{1 + \cos t} = \csc t - \cot t$

54. $\csc 2t + \cot 2t = \cot t$

55. $2 \csc 2\theta \cot \theta = 1 + \cot^2 \theta$

56. $\dfrac{\cos \theta}{\sec \theta + \tan \theta} = 1 - \sin \theta$

57. $\csc \theta \sin 2\theta = 2 \cos \theta$

58. $\dfrac{1 - 3 \tan^2 \theta}{3 \tan \theta - \tan^3 \theta} = \cot 3\theta$

59. $\dfrac{\sec t}{2 \cos t - \sec t} = \sec 2t$

60. Express $\cos 4\theta$ in terms of $\cos \theta$.

In problems 61 and 62, express each product as a sum or difference of sine or cosine functions.

61. $\sin \dfrac{5\theta}{2} \cos \dfrac{\theta}{2}$

62. $\sin 7t \sin 3t$

In problems 63 and 64, rewrite each expression as a product of sines and cosines.

63. $\sin 4t - \sin t$

64. $\cos 7\theta + \cos 3\theta$

In problems 65–70, evaluate the given expression without the use of a calculator or table.

65. $\tan^{-1} 1$

66. $\cos^{-1} \frac{1}{2}$

67. $\text{Arcsin}(-\frac{1}{2})$

68. $\sin\left(\tan^{-1} \dfrac{\sqrt{3}}{3}\right)$

69. $\tan^{-1}\left(\sin \dfrac{\pi}{2}\right)$

70. $\cos[\text{Arctan}(-1)]$

c In problems 71–76, use a calculator to evaluate each expression. Give the answer in radians and round off to four decimal places.

71. $\sec(\sin^{-1} \frac{2}{3})$

72. $\sin^{-1} 0.4969$

73. $\text{Arctan } 1.006$

74. $\cos^{-1}(-0.6294)$

75. $\text{Arcsin}(-0.9959)$

76. $\tan^{-1} 41.32$

77. Let ABC be a right triangle and let D be a point on the side $\overline{BC}$ such that $|\overline{AB}| = |\overline{DC}| = 1$ and $|\overline{BD}| = 2$. If angle $ADB = \alpha$ and angle $ACD = \beta$ (Figure 1), find (a) $\cos(\alpha + \beta)$ and (b) $\alpha + \beta$.

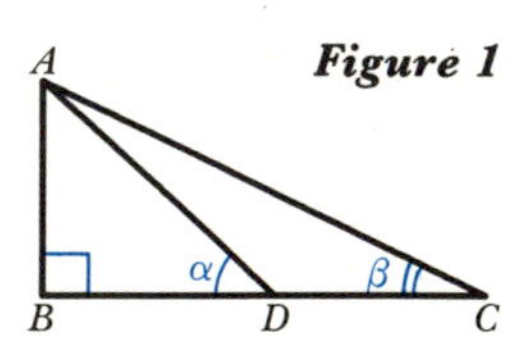

Figure 1

78. Show that $4 \tan^{-1} \frac{1}{7} + 8 \tan^{-1} \frac{1}{3} = \pi$.

In problems 79–84, find the exact value of each expression without using a calculator or table.

79. $\tan[\sin^{-1} \frac{3}{5}]$

80. $\sec[\tan^{-1}(-\frac{5}{3})]$

81. $\cot[\cos^{-1}(-\frac{1}{2})]$

82. $\cos[\tan^{-1}(-\sqrt{3})]$

83. $\sin[2 \text{ Arcsin } \frac{2}{3}]$

84. $\cos[2 \sin^{-1} \frac{1}{4}]$

85. Show that $\sin[2 \sin^{-1} x] = 2x\sqrt{1 - x^2}$, $-1 \le x \le 1$.

86. Show that $\sin^{-1} x = \tan^{-1}\left(\dfrac{x}{\sqrt{1 - x^2}}\right)$, $-1 < x < 1$.

In problems 87–90, solve for x in terms of y. Determine the restrictions on x and y.

87. $y = \sin^{-1}\left(\dfrac{x}{2}\right)$

88. $y = 2 \text{ Arctan}(x + 3)$

89. $y = 3 + \sec^{-1} 4x$

90. $y = \pi - \cos^{-1}(2x - 1)$

In problems 91–104, solve each trigonometric equation with the condition $0° \leq \theta < 360°$ or $0 \leq t < 2\pi$. Do not use a calculator or table.

91. $\sin\theta = -1$
92. $\cos\theta = \dfrac{\sqrt{3}}{2}$
93. $\tan\theta = \sqrt{3}$
94. $\cos\theta = -\dfrac{\sqrt{2}}{2}$
95. $\csc\theta = -2$
96. $\sin\theta = \cos 2\theta$
97. $\csc t = \dfrac{2\sqrt{3}}{3}$
98. $\sec t = 2$
99. $\sin t = \dfrac{\sqrt{2}}{2}$
100. $4\sin^2 t - 3 = 0$
101. $2\cos^2 t - \cos t - 1 = 0$
102. $2\sin^2 t + \sqrt{2}\sin t = 0$
103. $3\tan^2 t - \sqrt{3}\tan t = 0$
104. $2\cos^2 t + 5\cos t + 2 = 0$

c In problems 105–112, solve for the unknown parts of triangle ABC in Figure 2 if $\gamma = 90°$. Round off each angle to the nearest hundredth of a degree and each side length to four significant digits.

Figure 2

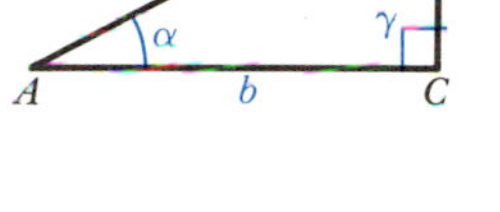

105. $a = 5$ and $b = 12$
106. $c = 15$ and $\alpha = 37°$
107. $b = 25$ and $\beta = 65°$
108. $a = 5$ and $c = 14$
109. $a = 17$ and $\beta = 51°$
110. $\sin\alpha = \frac{4}{5}$ and $c = 25$
111. $\tan\alpha = \frac{3}{4}$ and $a = 12$
112. $a = 6$ and $b = 8$

c In problems 113–122, use the given parts to find the unknown parts of triangle ABC in Figure 3. Round off each angle to the nearest hundredth of a degree and each side length to four significant digits.

Figure 3

113. $a = 5$, $b = 7$, and $\gamma = 30°$
114. $\alpha = 120°$, $c = 8$, and $b = 3$
115. $c = 10$, $\alpha = 45°$, and $\beta = 75°$
116. $a = 162$, $b = 215$, and $\beta = 110°$
117. $b = 4$, $c = 6$, and $\beta = 30°$
118. $a = 13.6$, $b = 7.82$, and $\alpha = 60°$
119. $a = 4.8$, $c = 4.3$, and $\alpha = 115°$
120. $b = 66.2$, $c = 42.3$, and $\alpha = 30°$
121. $a = 10$, $\beta = 42°$, and $\gamma = 51°$
122. $a = 4$, $c = 10$, and $\beta = 150°$

c In problems 123–130, round off the answers to two decimal places.

123. An observer is 635 meters from a launching pad when a rocket is launched vertically upward. If the angle of elevation of the bottom of the rocket at its highest point is 64.3°, how high does the rocket rise?

124. The side of a regular pentagon (a five-sided polygon with sides of equal length) is 12 centimeters. Find the radius of the circumscribed circle.

125. To find the distance d between two points A and B on opposite shores of a lake, a surveyor locates a point C that is 150 feet from A such that $\overline{CA}$ is perpendicular to $\overline{AB}$. The surveyor measures the angle ACB with a transit and finds it to be 72.67°. What is the distance d?

126. In alternating current circuits, the **impedance** Z (in ohms), the **resistance** R (in ohms), the **reactance** X (in ohms), and the **phase angle** ϕ obey the right triangle relationship as shown in Figure 4. If $R = 30$ ohms and $Z = 60$ ohms, find X and ϕ.

Figure 4

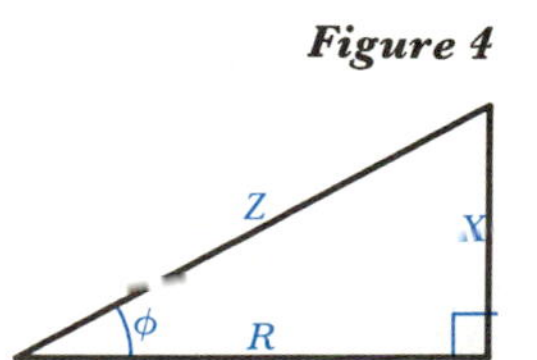

127. Two men standing 600 feet apart on level ground observe a balloon in the sky between them. The balloon is in the same vertical plane with the two men. The angles of elevation of the balloon are observed by the men to measure 75° and 48°, respectively. Find how high the balloon is above the ground.

128. Two points A and B are 50 feet apart on one side of a river. A point C across the river is located so that angle CAB is 70° and angle ABC is 80°. How wide is the river?

129. A guy wire attached to the top of a pole is 40 feet long and forms a 50° angle with the ground. How tall is the pole if it is tilted 15° from the vertical directly away from the guy line?

130. A diagonal of a parallelogram is 16 inches long and forms angles of 43° and 15°, respectively, with the two sides. How long are the sides of the parallelogram?

In problems 131–136, locate the point with the given polar coordinates, then find the corresponding rectangular coordinates of the point.

131. $\left(5, \frac{\pi}{4}\right)$ **132.** $\left(2, -\frac{\pi}{6}\right)$ **133.** $(\sqrt{2}, -135°)$

134. $(3, \pi)$ **135.** $(3, 270°)$ **136.** $\left(-2, \frac{4\pi}{3}\right)$

In problems 137–142, locate the point with the given rectangular coordinates, then find polar coordinates that correspond to the point, where $r > 0$ and $0° \leq \theta < 360°$.

137. $(-3, 0)$ **138.** $(-2, 2\sqrt{3})$ **139.** $(-5\sqrt{3}, -5)$

140. $(10, 10)$ **141.** $(0, -14)$ **142.** $(-15, 0)$

In problems 143–146, convert each polar equation to a corresponding equation in rectangular coordinates.

143. $r + 4 \sin \theta = 0$ **144.** $r = 5 \sec \theta$ **145.** $r = 4 \tan \theta$ **146.** $r = -3 \cos \theta$

In problems 147–150, convert each equation to a corresponding equation in polar coordinates.

147. $3x^2 + 3y^2 = 48$ **148.** $x^2 + \frac{y^2}{4} = 1$ **149.** $y^2 = 4x$ **150.** $x = -5$

In problems 151–154, sketch the graph of each polar equation.

151. $\theta = \frac{2\pi}{3}$ **152.** $r = -4 \cos \theta$ **153.** $r = 1 + \cos \theta$ **154.** $r\theta = 1$

In problems 155–162, express each complex number in a trigonometric form. In problem 161, round off the angle measure to two decimal places.

155. $z = 5 + 5i$ **156.** $z = \sqrt{3} + i$ **157.** $z = 6\sqrt{3} + 6i$ **158.** $z = 8i$

159. $z = 2 + 2i$ **160.** $z = -1 + \sqrt{3}i$ c **161.** $z = 3 + 4i$ **162.** $z = (2 + 2\sqrt{3}i)^{10}$

In problems 163–166 find (a) $z_1 z_2$ and (b) z_1/z_2 for each pair of complex numbers. Write the results in trigonometric form and in rectangular form. In problems 164–166, round off each answer to four decimal places.

163. $z_1 = 2(\cos \pi + i \sin \pi)$ and $z_2 = 3[\cos(\pi/2) + i \sin(\pi/2)]$

c **164.** $z_1 = 6(\cos 230° + i \sin 230°)$ and $z_2 = 3(\cos 75° + i \sin 75°)$

[c] **165.** $z_1 = 6(\cos 110° + i \sin 110°)$ and $z_2 = 2(\cos 212° + i \sin 212°)$
[c] **166.** $z_1 = 14(\cos 305° + i \sin 305°)$ and $z_2 = 7(\cos 65° + i \sin 65°)$

In problems 167–171, use DeMoivre's theorem to write each expression in trigonometric form and rectangular form.

167. $(\cos 60° + i \sin 60°)^5$ **168.** $(1 + i)^{40}$ **169.** $[\sqrt{2}/2 + i(\sqrt{2}/2)]^{100}$
170. $(\cos 0° + i \sin 0°)^{150}$ **171.** $(\sqrt{3} + i)^{30}$

172. Use DeMoivre's theorem to find expressions for $\cos 5\theta$ and $\sin 5\theta$. [*Hint:* $\cos 5\theta + i \sin 5\theta = (\cos \theta + i \sin \theta)^5$.]

[c] **173.** Let

$$z = \frac{10(\cos 17° + i \sin 17°)^{10}}{(1 + i)^2}$$

Express z in the form $a + bi$ and round off the answer to four decimal places.

174. Let $z = \cos(2\pi/5) + i \sin(2\pi/5)$. Show that

$$\left|\frac{z^2 - z^3}{z^4 - z^5}\right| = 1$$

In problems 175–178, find all the roots of each equation.

175. $z^3 = -64$ **176.** $z^4 = -8i$ **177.** $z^4 = 1 + i$ **178.** $z^4 = 8 - 8i$

In problems 179–182, let $\mathbf{u} = \langle 3, 4 \rangle$ and $\mathbf{v} = \langle 4, 3 \rangle$. Find each of the following vectors and represent them graphically.

179. $3\mathbf{u}$ **180.** $\mathbf{v} - \mathbf{u}$ **181.** $-2\mathbf{u} + \mathbf{v}$ **182.** $3\mathbf{u} + 2\mathbf{v}$

In problems 183–185, determine the components of $\mathbf{u}$ if θ is a direction angle of $\mathbf{u}$.

183. $|\mathbf{u}| = 5$ and $\theta = 30°$ **184.** $|\mathbf{u}| = 6$ and $\theta = 45°$ **185.** $|\mathbf{u}| = 8$ and $\theta = 150°$

186. Let $\mathbf{u} = \langle 2, 3 \rangle$, $\mathbf{v} = \langle -1, 4 \rangle$, and $\mathbf{w} = \langle 4, 5 \rangle$. Determine $\mathbf{z}$ such that $\mathbf{u} + \mathbf{v} = \mathbf{w} + \mathbf{z}$.

187. If $\mathbf{u}$ is a vector whose initial point is the first point given and whose terminal point is the second point given, write $\mathbf{u}$ in the form $\mathbf{u} = \langle a, b \rangle$ and find $|\mathbf{u}|$.
(a) $(-6, 8)$, $(4, 3)$ (b) $(-7, 6)$, $(3, -1)$

188. Let $\mathbf{u} = \langle -1, 4 \rangle$ and $\mathbf{v} = \langle 3, 5 \rangle$. Find
(a) $\mathbf{u} \cdot \mathbf{v}$ (b) $\mathbf{v} \cdot \mathbf{u}$
[c] (c) the angle between $\mathbf{u}$ and $\mathbf{v}$ to the nearest hundredth of a degree

[c] **189.** Two forces, one of 63 pounds and one of 45 pounds, yield a resultant force of 75 pounds. Find the angle between the two forces to the nearest hundredth of a degree.

[c] **190.** An airplane, which has an air speed of 550 miles per hour, heading N30°W, is affected by a wind blowing at 45 miles per hour from S45°W. Find the velocity vector of the airplane's path. What is the speed of the airplane to four significant digits, and in what direction is it traveling, relative to the direction of the wind, to the nearest hundredth of a degree?

CHAPTER 7

Systems of Equations and Inequalities

In science, business, and economics we often encounter mathematical statements or models of problems that involve more than one equation and contain two or more variables. Such a set of equations is referred to as a **system of equations.** If all the equations in a system are linear, it is called a **linear system.** In this chapter we explore several techniques for solving linear systems. We restrict our attention to systems with either two equations and two variables or three equations and three variables. We also discuss the decomposition of rational expressions into partial fractions. The chapter includes a brief treatment of systems of nonlinear equations and a discussion of systems of linear inequalities and *linear programming*.

7.1 Linear Systems of Equations

The set of equations

$$\begin{cases} 3x + 4y = 12 \\ 3x - 8y = 0 \end{cases}$$

is a system of two linear equations in the two variables x and y. A **solution** of such a system is a pair of numbers that satisfy each of the equations in the system, simultaneously. For instance, the pair $x = \frac{8}{3}$ and $y = 1$ is a solution of the above system because both equations are satisfied by $x = \frac{8}{3}$ and $y = 1$. Such a solution is often written as an *ordered pair* of numbers, $(\frac{8}{3}, 1)$ in this instance, with the understanding that the first value is to be used for one variable (here x) and the second value is to be used for the other variable (here y).

A *solution* of a system of three equations with three variables is a set of three numbers (a triple) that satisfy each of the equations in the system, simultaneously. Such a solution is often written as an *ordered triple* of numbers in the form (x_0, y_0, z_0). For instance, the triple of numbers $x = -1$, $y = 1$, and $z = 2$, or $(-1, 1, 2)$, is a solution of the linear system

$$\begin{cases} 3x + 4y - 2z = -3 \\ 5x - 7y + 9z = 6 \\ x + 2y - z = -1 \end{cases}$$

because the three equations are satisfied by the triple of numbers.

Figure 1

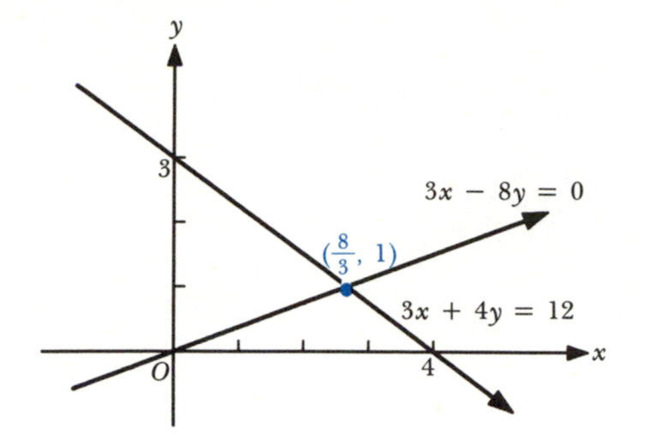

The solution $(\frac{8}{3}, 1)$ of the linear system

$$\begin{cases} 3x + 4y = 12 \\ 3x - 8y = 0 \end{cases}$$

can be interpreted geometrically as the point where the graph of $3x + 4y = 12$ intersects the graph of $3x - 8y = 0$ (Figure 1).

If we graph two linear equations with two variables on the same coordinate axes, one of the following cases occurs:

Case 1 The graphs of the two linear equations intersect at one point; we say that the system is **consistent.**

Case 2 The graphs of the two linear equations are parallel lines; there is no solution, and we say the system is **inconsistent.**

Case 3 The graphs of the two linear equations coincide so that the intersection is every point of each line. In this case, the system is said to be **dependent.**

EXAMPLE 1 Use graphs to determine whether the given linear system is consistent, inconsistent, or dependent. Indicate the solution of the system (if any) on the graphs.

(a) $\begin{cases} x - y = 1 \\ 3x - y = -1 \end{cases}$ (b) $\begin{cases} x + 3y = 4 \\ 2x + 6y = -6 \end{cases}$ (c) $\begin{cases} 3x - y = -1 \\ 6x - 2y = -2 \end{cases}$

SOLUTION The graphs of the two equations in systems (a), (b), and (c) are shown in Figures 2a, 2b, and 2c, respectively.

(a) In Figure 2a, the graphs of the two equations show a point of intersection. By writing each of the two equations in the slope-intercept form

$$\begin{cases} y = x - 1 \\ y = 3x + 1 \end{cases}$$

we see that the slopes are not equal (they are 1 and 3, respectively), so that the lines are not parallel. This means that there is one solution, in this case $(-1, -2)$, and the system is consistent.

(b) In Figure 2b, the graphs of the two equations suggest that the two lines do *not* intersect. By writing each of the two equations in the slope-intercept form, we have

$$\begin{cases} y = -\frac{1}{3}x + \frac{4}{3} \\ y = -\frac{1}{3}x - 1 \end{cases}$$

Figure 2

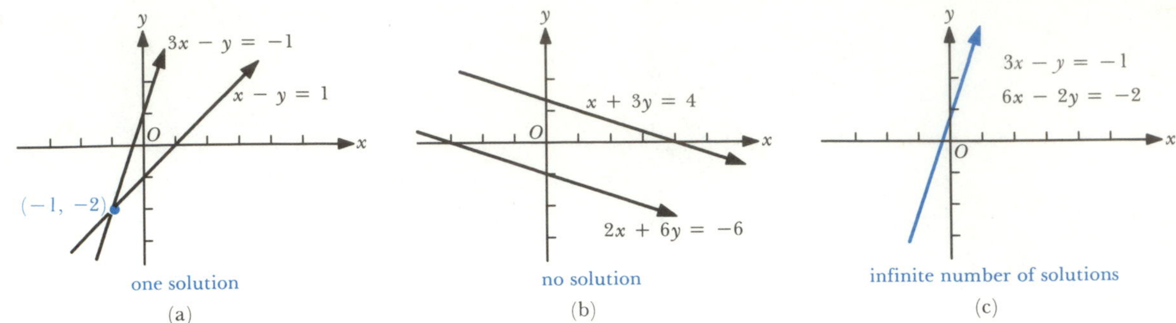

Since the slopes are equal but the y intercepts are different, the two lines are parallel and there is no solution. The system is inconsistent.

(c) Figure 2c suggests that the two equations have the same graph. By writing each of the two equations in the slope-intercept form, we obtain

$$\begin{cases} y = 3x + 1 \\ y = 3x + 1 \end{cases}$$

The slopes are equal and the y intercepts are the same. Thus, there are infinitely many solutions; namely any pair of numbers that are coordinates of a point on the line, and the system is dependent.

As illustrated in Example 1, whether a linear system of two equations in two variables is consistent, inconsistent, or dependent can be determined by investigating its graphs. However, it is difficult to find the exact values of a solution by using the graphs. For this reason, we introduce two *algebraic* methods for solving linear systems—the *substitution* method and the *elimination* method.

The Substitution Method

To solve a linear system of two equations in *two* variables by the **substitution method,** we use one of the equations to express one of the variables in terms of the other, and then we substitute this expression for the variable in the other equation to obtain an equation in *one* variable. This technique is illustrated in the next example.

EXAMPLE 2 Use the substitution method to solve the system

$$\begin{cases} 4x - y = 2 \\ 3x + y = 5 \end{cases}$$

SOLUTION Solving the first equation for y in terms of x yields

$$y = 4x - 2$$

Replacing y with $4x - 2$ in the second equation of the system gives us

$$3x + (4x - 2) = 5$$

so that

$$7x = 7 \qquad \text{or} \qquad x = 1$$

The corresponding value of y is obtained by substituting 1 for x in $y = 4x - 2$, so that

$$y = 4(1) - 2 = 2$$

Thus, the solution is the pair $x = 1$ and $y = 2$, or $(1, 2)$.

Check To check the solution, replace x by 1 and y by 2 in the original system to get

$$\begin{cases} 4(1) - 2 = 2 \\ 3(1) + 2 = 5 \end{cases}$$

The next example illustrates how the method of substitution can be used to solve a linear system of three equations and three variables.

EXAMPLE 3 Use the substitution method to solve the system

$$\begin{cases} x + y + z = 3 \\ x - 2y + 4z = 6 \\ 2x - y + z = 11 \end{cases}$$

SOLUTION First we solve the first equation for x in terms of y and z to get

$$x = 3 - y - z$$

Replacing x by $3 - y - z$ in the second and third equations yields the following linear system in two variables:

$$\begin{cases} (3 - y - z) - 2y + 4z = 6 \\ 2(3 - y - z) - y + z = 11 \end{cases}$$

or

$$\begin{cases} -y + z = 1 \\ -3y - z = 5 \end{cases}$$

Applying the substitution method to the latter linear system we have from the first equation

$$z = y + 1$$

so that by substituting $y + 1$ for z in the second equation, we obtain

$$-3y - (y + 1) = 5$$
$$y = -\tfrac{3}{2}$$

and

$$z = (-\tfrac{3}{2}) + 1 = -\tfrac{1}{2}$$

Thus

$$x = 3 - (-\tfrac{3}{2}) - (-\tfrac{1}{2}) = 5$$

The solution of the original linear system is the triple of numbers

$$x = 5,\ y = -\frac{3}{2},\ \text{and } z = -\frac{1}{2} \qquad \text{or} \qquad \left(5, -\frac{3}{2}, -\frac{1}{2}\right)$$

The Elimination Method

An alternative method for solving a linear system, known as the **elimination method,** is based on the idea of *equivalent* systems of linear equations, and on the strategy of *eliminating* one of the variables from equations, by adding or subtracting equations in a system. Two systems of equations are said to be **equivalent** if they have the same solutions. To solve a system of linear equations by the elimination method, we perform one or more of the following operations on the system to obtain an equivalent system.

1. Interchange the positions of two equations in the system.
2. Replace an equation in the system by a nonzero multiple of itself. (A *multiple* of an equation is the equation that results when we multiply both sides of the equation by the same number.)
3. Replace an equation in the system by the sum of a nonzero multiple of that equation and a nonzero multiple of another equation of the system.

In Examples 4 and 5, use the elimination method to solve each system.

EXAMPLE 4

$$\begin{cases} 3x - 2y = 3 \\ 2x + 5y = -1 \end{cases}$$

SOLUTION

We begin by eliminating x from both equations. In order to do so, we multiply the first equation by 2 and the second equation by -3 to obtain the equivalent system

$$\begin{cases} 6x - 4y = 6 \\ -6x - 15y = 3 \end{cases}$$

Next, we replace the second equation by the sum of the first and second equations to obtain the equivalent system

$$\begin{cases} 6x - 4y = 6 \\ \quad - 19y = 9 \end{cases}$$

We see from the last equation that the only possible value for y is $-\frac{9}{19}$. The corresponding value for x may be found by substituting $-\frac{9}{19}$ for y in the first equation. This gives us

$$3x - 2(-\tfrac{9}{19}) = 3$$
$$3x = 3 - \tfrac{18}{19}$$
$$3x = \tfrac{39}{19}$$
$$x = \tfrac{13}{19}$$

Thus, the solution is the pair

$$x = \tfrac{13}{19} \text{ and } y = -\tfrac{9}{19} \qquad \text{or} \qquad (\tfrac{13}{19}, -\tfrac{9}{19})$$

EXAMPLE 5

$$\begin{cases} 2x_1 + x_2 - 2x_3 = 10 \\ 3x_1 + 2x_2 + 2x_3 = 1 \\ 5x_1 + 4x_2 + 3x_3 = 4 \end{cases}$$

SOLUTION

Let us refer to the given system as system (A). We begin by replacing the second equation in system (A) by the sum of -3 times the first equation and 2 times the second equation to obtain system (B):

$$\text{(B)} \quad \begin{cases} 2x_1 + x_2 - 2x_3 = 10 \\ x_2 + 10x_3 = -28 \\ 5x_1 + 4x_2 + 3x_3 = 4 \end{cases}$$

System (B) is *equivalent* to system (A), and x_1 has been eliminated from the second equation. In the next step, replace the third equation of (B) by the sum of -5 times the first equation and 2 times the third equation to obtain the equivalent system (C), in which x_1 has been eliminated from the third equation:

$$\text{(C)} \quad \begin{cases} 2x_1 + x_2 - 2x_3 = 10 \\ x_2 + 10x_3 = -28 \\ 3x_2 + 16x_3 = -42 \end{cases}$$

Now, replace the third equation of (C) by the sum of the third equation and -3 times the second equation to get the equivalent system (D), in which x_2 has been removed from the third equation:

$$\text{(D)} \quad \begin{cases} 2x_1 + x_2 - 2x_3 = 10 \\ x_2 + 10x_3 = -28 \\ -14x_3 = 42 \end{cases}$$

Next, multiply the third equation by $-\frac{1}{14}$, and add -1 times the second equation to the first equation, to obtain the equivalent system (E), in which x_2 has been eliminated from the first equation:

$$\text{(E)} \quad \begin{cases} 2x_1 - 12x_3 = 38 \\ x_2 + 10x_3 = -28 \\ x_3 = -3 \end{cases}$$

Finally, in order to reduce the coefficient of x_1 in the first equation to 1, multiply the first equation by $\frac{1}{2}$ to obtain the equivalent system (F):

$$\text{(F)} \quad \begin{cases} x_1 - 6x_3 = 19 \\ x_2 + 10x_3 = -28 \\ x_3 = -3 \end{cases}$$

After substituting $x_3 = -3$ into the second equation, we get $x_2 = 2$; then, after substituting $x_3 = -3$ into the first equation, we obtain $x_1 = 1$. Thus, the solution is the triple of numbers

$$x_1 = 1, x_2 = 2, \text{ and } x_3 = -3 \qquad \text{or} \qquad (1, 2, -3)$$

Our strategy in performing operations on system (A) of Example 5 was to replace the system, after a number of steps, with an equivalent system in which two of the variables were eliminated in one equation and one of the variables was eliminated in the other two equations. This was accomplished in the equivalent system (F), where the coefficient of x_1 is 1 in the first equation, the coefficient of x_2 is 1 in the second equation, and the coefficient of x_3 is 1 in the third equation.

Many applied problems can be solved by using systems of linear equations.

EXAMPLE 6 A special diet has been prescribed. It is prepared from two kinds of food. The first kind contains 40% of nutrient A and 16% of nutrient B; the second contains 35% of nutrient A and 20% of nutrient B. The diet is to contain exactly 5.75 ounces of nutrient A and 2.60 ounces of nutrient B. How many ounces of each food should be mixed for the result to contain the specified diet?

SOLUTION Let x equal the number of ounces that will be required of the first kind of food and let y equal the number of ounces of the second kind of food. The quantities involved here are summarized in the following table:

Kind of Food	Percent of Nutrient A in Food	Percent of Nutrient B in Food	Required Number of Ounces of Food	Number of Ounces of Nutrient A	Number of Ounces of Nutrient B
First	40%	16%	x	$0.40x$	$0.16x$
Second	35%	20%	y	$0.35y$	$0.20y$

Then,

$$\begin{cases} 0.40x + 0.35y = 5.75 \\ 0.16x + 0.20y = 2.60 \end{cases}$$

We "clear" the system of decimals by multiplying each side of each equation by 100, so that

$$\begin{cases} 40x + 35y = 575 \\ 16x + 20y = 260 \end{cases} \quad \text{or} \quad \begin{cases} 8x + 7y = 115 \\ 16x + 20y = 260 \end{cases}$$

After multiplying the first equation by -2, we get

$$\begin{cases} -16x - 14y = -230 \\ 16x + 20y = 260 \end{cases}$$

Now we eliminate x from both equations by adding the two equations, so that

$$6y = 30 \quad \text{or} \quad y = 5$$

and

$$16x + 20(5) = 260 \quad \text{or} \quad x = 10$$

Thus, 10 ounces of the first kind of food should be mixed with 5 ounces of the second kind of food.

PROBLEM SET 7.1

In problems 1–6, sketch the graph of both equations in the given linear system on the same coordinate axes. Use the graphs to determine whether the system is consistent, inconsistent, or dependent.

1. $\begin{cases} 3x - 2y = 1 \\ x + y = 5 \end{cases}$

2. $\begin{cases} y = 2x - 3 \\ 4x - 2y = 6 \end{cases}$

3. $\begin{cases} 2x + 6y = 7 \\ x + 3y = 1 \end{cases}$

4. $\begin{cases} 2x = y + 3 \\ 4x - 2y = 5 \end{cases}$

5. $\begin{cases} x + 2y = 3 \\ 2x - 3y = 1 \end{cases}$

6. $\begin{cases} 3x + 2y = 11 \\ -2x + y = 2 \end{cases}$

In problems 7–16, use the substitution method to solve each linear system.

7. $\begin{cases} 2x - y = 5 \\ x + 3y = 13 \end{cases}$

8. $\begin{cases} 3x + y = 4 \\ 7x - y = 6 \end{cases}$

9. $\begin{cases} y = x - 2 \\ 2y = x - 3 \end{cases}$

10. $\begin{cases} 3x + 5y = 1 \\ x - 4y = -6 \end{cases}$

11. $\begin{cases} x + y = 1 \\ 5x - y = 13 \end{cases}$

12. $\begin{cases} 3x - 4y = -5 \\ 2x + 2y = 3 \end{cases}$

13. $\begin{cases} x + y = 5 \\ x + z = 1 \\ y + z = 2 \end{cases}$

14. $\begin{cases} x - 3y = -11 \\ 2y - 5z = 26 \\ 7x - 3z = -2 \end{cases}$

15. $\begin{cases} x + y + 2z = 11 \\ x - y + z = 3 \\ 2x + y + 3z = 17 \end{cases}$

16. $\begin{cases} x + 3y - z = 4 \\ 3x - 2y + 4z = 11 \\ 2x + y + 3z = 13 \end{cases}$

In problems 17–30, use the elimination method to solve each linear system.

17. $\begin{cases} 2x - y = 1 \\ x + y = 2 \end{cases}$

18. $\begin{cases} x + y = 0 \\ x - y = 1 \end{cases}$

19. $\begin{cases} x + 3y = 9 \\ x - y = 1 \end{cases}$

20. $\begin{cases} x + y = 1 \\ y - 2x = 3 \end{cases}$

21. $\begin{cases} 2x + 4y = 3 \\ -x + y = 3 \end{cases}$

22. $\begin{cases} 3y + 2z = 5 \\ 2y = -3z + 1 \end{cases}$

23. $\begin{cases} 3x + 2y = 4 \\ 5x + 3y = 7 \end{cases}$

24. $\begin{cases} \frac{1}{2}x + \frac{1}{3}y = 13 \\ \frac{1}{5}x + \frac{1}{8}y = 5 \end{cases}$

25. $\begin{cases} x + y + 2z = 4 \\ x + y - 2z = 0 \\ x - y = 0 \end{cases}$

26. $\begin{cases} x + y + z = 2 \\ x + 2y - z = 4 \\ 2x - y + z = 0 \end{cases}$

27. $\begin{cases} x + y + z = 6 \\ x - y + 2z = 12 \\ 2x + y + z = 1 \end{cases}$

28. $\begin{cases} x + y + 2z = 4 \\ x - 5y + z = 5 \\ 3x - 4y + 7z = 24 \end{cases}$

29. $\begin{cases} 2x + y - 3z = 9 \\ x - 2y + 4z = 5 \\ 3x + y - 2z = 15 \end{cases}$

30. $\begin{cases} x + 3y - 2z = -21 \\ 7x - 5y + 4z = 31 \\ 2x + y + 3z = 17 \end{cases}$

In problems 31–47, use a system of linear equations to solve each problem.

31. The sum of two numbers is 12. If one of the numbers is multiplied by 5 and the other is multiplied by 8, the sum of the products is 75. Find the numbers.

32. In a given two-digit number, the tens digit is 5 more than the units digit, and the number is 63 more than the sum of its digits. Find the number.

33. A farmer sold two grades of wheat, one grade at $4.25 per bushel and another grade at $5.50 per bushel. How many bushels of each kind did he sell if he received $15,078.75 for 3000 bushels?

34. A rowing team practices on a river that flows at a relatively constant speed. They can row downstream at 7 meters per second and upstream at 3 meters per second. Find the team's rate in still water and the rate of the river's current.

35. A man invested a total of $50,000 in securities. Part of the money was invested at 10%, the rest at 8%. His annual income from both investments was $4560. Find the amount of money invested at each rate.

36. A woman has 11 coins, consisting of nickels and dimes. She has three more dimes than nickels. How many of each type of coin does she have if the total value of the coins is 90 cents?

37. Two cars start from the same point, at the same time, and travel in opposite directions. One car travels at a rate that is 10 miles per hour faster than the other. After 3 hours, they are 330 miles apart. What is the speed of each car?

38. The perimeter of an isosceles triangle is 26 centimeters. The base is 2 centimeters longer than one of the equal sides of the triangle. Find the length of each side of the triangle.

39. A car that traveled at an average speed of 40 miles per hour went 35 miles farther than one that traveled at 55 miles per hour. The slower car traveled 2 hours longer than the faster one. What was the length of time each car traveled?

40. A chemist prepares two acid solutions. The first solution is 20% acid and the second solution is 50% acid. How many milliliters of each solution should be mixed to obtain 12 milliliters of a 30% acid solution?

41. A diet provides 1200 milligrams of protein and 1000 milligrams of iron. These nutrients will be obtained by eating meat and beans. Each pound of meat contains 500 milligrams of protein and 100 milligrams of iron. Each pound of beans contains 200 milligrams of protein and 800 milligrams of iron. How many pounds of meat and beans should be eaten to provide exactly the specified amounts of nutrients?

42. One salt solution is 10% salt and another is 35% salt. How many milliliters of each solution should be mixed to obtain 200 milliliters of a 25% salt solution?

43. A chemist at work in a laboratory has containers of the same acid in two strengths. Six parts of the first mixed with four parts of the second gives a mixture that is 86% acid, and four parts of the first mixed with six parts of the second gives a mixture that is 84% acid. What is the percentage of purity of each container of acid?

44. Three men bought a grocery store for $100,000. If the second man invested twice as much as the first, and the third invested $20,000 more than the second, how much did each invest?

45. A woman bought three different kinds of common stock for \$20,000, one paying a 6% annual dividend, one paying a 7% annual dividend, and the other paying an 8% annual dividend. At the end of the first year, the sum of the dividends from the 6% and the 7% stocks was \$940, and the sum of the dividends from the 6% and the 8% stocks was \$720. How much did she invest in each kind of stock?

46. The perimeter of a triangle is 45 inches. One side is twice as long as the shortest side, and the third side is 5 inches longer than the shortest side. Find the lengths of the three sides of the triangle.

47. A manufacturer of fertilizer wishes to mix 20 tons of three brands of artificial fertilizer. One brand contains 5% phosphate and 18% nitrogen, another brand contains 6% phosphate and 10% nitrogen, while the third brand contains 9% phosphate and 7% nitrogen. How much of each of the three brands should be mixed to obtain a mixture containing 6.8% phosphate and 12% nitrogen?

7.2 Matrices and Row Reduction

In the elimination method, it is not necessary to continue writing the variables. All we need to do is maintain a record of the numerical coefficients and constants in the equations. This suggests abbreviating a system of linear equations by writing *only* the numerical coefficients and constants. For example, we can abbreviate the system

$$\begin{cases} 3x - y = 1 \\ x + 2y = 0 \end{cases}$$

by "lining up" the coefficients of the corresponding variables and the constants vertically in the following manner:

$$\left[\begin{array}{cc:c} 3 & -1 & 1 \\ 1 & 2 & 0 \end{array}\right]$$

If a variable does not appear in one of the equations, a zero is inserted in the appropriate position in the array of coefficients and constants. The vertical dashed line separating the coefficients and constants is optional.

A rectangular array of numbers of this type is called a **matrix.** The **rows** of the matrix are the numbers that appear next to one another horizontally. The **columns** of the matrix are the vertical lines of numbers. The numbers that appear in the matrix are called the **entries** or **elements** of the matrix. The **size** of the matrix is described by specifying the number of rows and columns. We use capital letters A, B, C, and so on to denote matrices.

For example, the matrix

$$A = \begin{bmatrix} 1 & 2 \\ 3 & -1 \end{bmatrix}$$

has two rows and two columns; we say that A is a 2×2 matrix and that 1, 2, 3, and -1 are the entries of A.

The matrix

$$B = \begin{bmatrix} 1 & 3 & 5 \\ -2 & 5 & 6 \end{bmatrix}$$

has two rows and three columns, so that B is a 2×3 matrix and its entries are 1, 3, 5, -2, 5, and 6.

The matrix

$$C = \begin{bmatrix} 3 \\ 1 \\ 2 \end{bmatrix}$$

is a 3×1 matrix with entries 3, 1, and 2. Notice that when the size of a matrix is specified, the number of rows is given first and the number of columns is given second.

An entry of a matrix can also be identified by using subscripts to indicate its row position and its column position, as in matrix A displayed below:

$$A = \begin{bmatrix} a_{11} & a_{12} & a_{13} & a_{14} \\ a_{21} & a_{22} & a_{23} & a_{24} \\ a_{31} & a_{32} & a_{33} & a_{34} \\ a_{41} & a_{42} & a_{43} & a_{44} \end{bmatrix}$$

Thus, the entry in the second row and the second column is denoted by a_{22}, while the entry in the first row and the fourth column is a_{14}.

For example, if

$$B = \begin{bmatrix} 3 & 2 & -1 \\ 4 & 1 & 5 \end{bmatrix}$$

we could write $b_{11} = 3$, $b_{12} = 2$, $b_{13} = -1$, $b_{21} = 4$, $b_{22} = 1$, and $b_{23} = 5$.

If we only consider the coefficients of the variables of a system of linear equations, the resulting matrix is called the **coefficient matrix.** The coefficient matrix of the system

$$\text{(A)} \quad \begin{cases} 3x - y = 1 \\ x + 2y = 0 \end{cases} \qquad \text{is} \qquad \begin{bmatrix} 3 & -1 \\ 1 & 2 \end{bmatrix}$$

If the third column, consisting of the constants of the equations, is appended to the coefficient matrix, the new matrix is called the **augmented matrix** of the system of equations. The augmented matrix of the above system (A) is

$$A = \left[\begin{array}{cc|c} 3 & -1 & 1 \\ 1 & 2 & 0 \end{array}\right]$$

An augmented matrix can be used to "maintain a record" of the coefficients of the linear equations when the elimination method is applied to solve the linear system. (A vertical arrow $\downarrow$ will be used to indicate that the systems are equivalent.)

System

(A) $\begin{cases} 3x - y = 1 \\ x + 2y = 0 \end{cases}$

↓ First, multiply the second equation by -3.

(B) $\begin{cases} 3x - y = 1 \\ -3x - 6y = 0 \end{cases}$

↓ Next, replace equation two with the sum of equations one and two.

(C) $\begin{cases} 3x - y = 1 \\ \quad -7y = 1 \end{cases}$

↓ Multiply equation two by $-\frac{1}{7}$.

(D) $\begin{cases} 3x - y = 1 \\ \quad y = -\frac{1}{7} \end{cases}$

↓ Replace equation one with the sum of equations one and two.

(E) $\begin{cases} 3x = \frac{6}{7} \\ \quad y = -\frac{1}{7} \end{cases}$

↓ Multiply equation one by $\frac{1}{3}$.

(F) $\begin{cases} x = \frac{2}{7} \\ \quad y = -\frac{1}{7} \end{cases}$

↓ Hence, the solution of the linear system (A) is $(\frac{2}{7}, -\frac{1}{7})$.

Augmented Matrix of System

$$A = \left[\begin{array}{cc|c} 3 & -1 & 1 \\ 1 & 2 & 0 \end{array}\right]$$

↓ First, multiply the second row by -3.

$$B = \left[\begin{array}{cc|c} 3 & -1 & 1 \\ -3 & -6 & 0 \end{array}\right]$$

↓ Next, replace row two with the sum of rows one and two.

$$C = \left[\begin{array}{cc|c} 3 & -1 & 1 \\ 0 & -7 & 1 \end{array}\right]$$

↓ Multiply row two by $-\frac{1}{7}$.

$$D = \left[\begin{array}{cc|c} 3 & -1 & 1 \\ 0 & 1 & -\frac{1}{7} \end{array}\right]$$

↓ Replace row one with the sum of row one and row two.

$$E = \left[\begin{array}{cc|c} 3 & 0 & \frac{6}{7} \\ 0 & 1 & -\frac{1}{7} \end{array}\right]$$

↓ Multiply row one by $\frac{1}{3}$.

$$F = \left[\begin{array}{cc|c} 1 & 0 & \frac{2}{7} \\ 0 & 1 & -\frac{1}{7} \end{array}\right]$$

↓ The numbers to the left of the dashed line represent the coefficients of x and y, so the solution is $(\frac{2}{7}, -\frac{1}{7})$.

EXAMPLE 1 Solve the system of linear equations

$$\text{(A)} \quad \begin{cases} x_1 - 2x_2 + 3x_3 = -1 \\ 2x_1 - x_2 + 2x_3 = 2 \\ 3x_1 + x_2 + 2x_3 = 3 \end{cases}$$

by using the elimination method and displaying the corresponding augmented matrix.

SOLUTION The augmented matrix form of system (A) is written as

$$A = \left[\begin{array}{ccc|c} 1 & -2 & 3 & -1 \\ 2 & -1 & 2 & 2 \\ 3 & 1 & 2 & 3 \end{array}\right]$$

The elimination method proceeds as follows:

System	*Augmented Matrix of System*
(A) $\begin{cases} x_1 - 2x_2 + 3x_3 = -1 \\ 2x_1 - x_2 + 2x_3 = 2 \\ 3x_1 + x_2 + 2x_3 = 3 \end{cases}$	$A = \left[\begin{array}{ccc\|c} 1 & -2 & 3 & -1 \\ 2 & -1 & 2 & 2 \\ 3 & 1 & 2 & 3 \end{array}\right]$
↓ First, replace the second equation by the sum of the second equation and -2 times the first equation.	↓ First, replace the second row by the sum of the second row and -2 times the first row.
(B) $\begin{cases} x_1 - 2x_2 + 3x_3 = -1 \\ 3x_2 - 4x_3 = 4 \\ 3x_1 + x_2 + 2x_3 = 3 \end{cases}$	$B = \left[\begin{array}{ccc\|c} 1 & -2 & 3 & -1 \\ 0 & 3 & -4 & 4 \\ 3 & 1 & 2 & 3 \end{array}\right]$
↓ Next, replace equation three with the sum of the third equation and -3 times equation one.	↓ Next, replace row three with the sum of the third row and -3 times row one.
(C) $\begin{cases} x_1 - 2x_2 + 3x_3 = -1 \\ 3x_2 - 4x_3 = 4 \\ 7x_2 - 7x_3 = 6 \end{cases}$	$C = \left[\begin{array}{ccc\|c} 1 & -2 & 3 & -1 \\ 0 & 3 & -4 & 4 \\ 0 & 7 & -7 & 6 \end{array}\right]$
↓ In the next step, we multiply the second equation by $\frac{1}{3}$.	↓ In the next step, we multiply the second row by $\frac{1}{3}$.
(D) $\begin{cases} x_1 - 2x_2 + 3x_3 = -1 \\ x_2 - \frac{4}{3}x_3 = \frac{4}{3} \\ 7x_2 - 7x_3 = 6 \end{cases}$	$D = \left[\begin{array}{ccc\|c} 1 & -2 & 3 & -1 \\ 0 & 1 & -\frac{4}{3} & \frac{4}{3} \\ 0 & 7 & -7 & 6 \end{array}\right]$
↓ Now, replace the third equation with the sum of equation three and -7 times equation two.	↓ Now, replace the third row with the sum of row three and -7 times row two.
(E) $\begin{cases} x_1 - 2x_2 + 3x_3 = -1 \\ x_2 - \frac{4}{3}x_3 = \frac{4}{3} \\ \frac{7}{3}x_3 = -\frac{10}{3} \end{cases}$	$E = \left[\begin{array}{ccc\|c} 1 & -2 & 3 & -1 \\ 0 & 1 & -\frac{4}{3} & \frac{4}{3} \\ 0 & 0 & \frac{7}{3} & -\frac{10}{3} \end{array}\right]$
↓ Next, replace equation one with equation one plus 2 times equation two.	↓ Next, replace row one with row one plus 2 times row two.
(F) $\begin{cases} x_1 + \frac{1}{3}x_3 = \frac{5}{3} \\ x_2 - \frac{4}{3}x_3 = \frac{4}{3} \\ \frac{7}{3}x_3 = -\frac{10}{3} \end{cases}$	$F = \left[\begin{array}{ccc\|c} 1 & 0 & \frac{1}{3} & \frac{5}{3} \\ 0 & 1 & -\frac{4}{3} & \frac{4}{3} \\ 0 & 0 & \frac{7}{3} & -\frac{10}{3} \end{array}\right]$
↓ In the next step, we multiply the third equation by $\frac{3}{7}$.	↓ In the next step, we multiply the third row by $\frac{3}{7}$.

(G) $\begin{cases} x_1 \quad\quad + \frac{1}{3}x_3 = \frac{5}{3} \\ \quad x_2 - \frac{4}{3}x_3 = \frac{4}{3} \\ \quad\quad\quad x_3 = -\frac{10}{7} \end{cases}$ $\qquad G = \left[\begin{array}{ccc|c} 1 & 0 & \frac{1}{3} & \frac{5}{3} \\ 0 & 1 & -\frac{4}{3} & \frac{4}{3} \\ 0 & 0 & 1 & -\frac{10}{7} \end{array}\right]$

↓ Replace equation one with the sum of equation one and $-\frac{1}{3}$ times equation three. ↓ Replace row one with the sum of row one and $-\frac{1}{3}$ times row three.

(H) $\begin{cases} x_1 \quad\quad = \frac{15}{7} \\ \quad x_2 - \frac{4}{3}x_3 = \frac{4}{3} \\ \quad\quad\quad x_3 = -\frac{10}{7} \end{cases}$ $\qquad H = \left[\begin{array}{ccc|c} 1 & 0 & 0 & \frac{15}{7} \\ 0 & 1 & -\frac{4}{3} & \frac{4}{3} \\ 0 & 0 & 1 & -\frac{10}{7} \end{array}\right]$

↓ Finally, replace equation two with the sum of equation two and $\frac{4}{3}$ times equation three. ↓ Finally, replace row two with the sum of row two and $\frac{4}{3}$ times row three.

(I) $\begin{cases} x_1 \quad\quad = \frac{15}{7} \\ \quad x_2 \quad = -\frac{4}{7} \\ \quad\quad x_3 = -\frac{10}{7} \end{cases}$ $\qquad I = \left[\begin{array}{ccc|c} 1 & 0 & 0 & \frac{15}{7} \\ 0 & 1 & 0 & -\frac{4}{7} \\ 0 & 0 & 1 & -\frac{10}{7} \end{array}\right]$

↓ Hence, the solution of system (A) is $(\frac{15}{7}, -\frac{4}{7}, -\frac{10}{7})$. ↓ Hence, the solution is read from the augmented matrix as $(\frac{15}{7}, -\frac{4}{7}, -\frac{10}{7})$.

This example shows that we can work more efficiently with the augmented matrix than with the actual equations. In our work we have performed the following row operations on the augmented matrix. [Compare these with boxed equation operations (1) through (3) on page 406]:

1′ Interchange two rows of the matrix ($R_i \leftrightarrow R_j$).

2′ Multiply the elements in a row of the matrix by a nonzero number ($R_i \to kR_i$, where $k \neq 0$).

3′ Replace a row with the sum of a nonzero multiple of itself and a nonzero multiple of another row ($R_i \to kR_j + cR_i$, where $c \neq 0$ and $k \neq 0$).

Operations (1′), (2′), and (3′) are called the **elementary row operations.** The resulting matrix I in the above example is called a *row-reduced echelon matrix*. In general, a matrix is a **row-reduced echelon matrix** if all of the following conditions hold:

(i) The first nonzero entry in each row is 1; all other entries in that column are zeros. (That is, if a column contains 1 as a leading entry of some row, then the remaining entries in that column must be zero.)

(ii) Any rows that consist entirely of zeros are below all rows that do not consist entirely of zeros.

(iii) The first nonzero entry in each row is to the right of the first nonzero entry in the preceding row.

In Examples 2 and 3, solve the given linear system by performing elementary row operations.

EXAMPLE 2

$$\begin{cases} 4x + y = 2 \\ x + 4y = 1 \end{cases}$$

SOLUTION The augmented matrix of the system is

$$\left[\begin{array}{cc|c} 4 & 1 & 2 \\ 1 & 4 & 1 \end{array}\right]$$

First, we multiply row one by $\frac{1}{4}$, that is, $R_1 \to \frac{1}{4}R_1$, to get

$$\left[\begin{array}{cc|c} 1 & \frac{1}{4} & \frac{1}{2} \\ 1 & 4 & 1 \end{array}\right]$$

Next, we replace row two in the matrix with the sum of row two and -1 times row one $[R_2 \to R_2 + (-1)R_1]$ to get

$$\left[\begin{array}{cc|c} 1 & \frac{1}{4} & \frac{1}{2} \\ 0 & \frac{15}{4} & \frac{1}{2} \end{array}\right]$$

Then we multiply row two by $\frac{4}{15}$, that is, $R_2 \to \frac{4}{15}R_2$, to get

$$\left[\begin{array}{cc|c} 1 & \frac{1}{4} & \frac{1}{2} \\ 0 & 1 & \frac{2}{15} \end{array}\right]$$

Finally, we replace row one in the resulting matrix with the sum of row one and $-\frac{1}{4}$ times row two $[R_1 \to R_1 + (-\frac{1}{4})R_2]$ to get

$$\left[\begin{array}{cc|c} 1 & 0 & \frac{7}{15} \\ 0 & 1 & \frac{2}{15} \end{array}\right]$$

which is equivalent to the system

$$\begin{cases} x = \frac{7}{15} \\ y = \frac{2}{15} \end{cases}$$

Therefore, the solution is the pair of numbers $x = \frac{7}{15}$ and $y = \frac{2}{15}$ or $(\frac{7}{15}, \frac{2}{15})$.

EXAMPLE 3

$$\begin{cases} 3x_1 - x_2 + x_3 = 1 \\ 7x_1 + x_2 - x_3 = 6 \\ 2x_1 + x_2 - x_3 = 2 \end{cases}$$

SOLUTION The augmented matrix that corresponds to this linear system is

$$\left[\begin{array}{ccc|c} 3 & -1 & 1 & 1 \\ 7 & 1 & -1 & 6 \\ 2 & 1 & -1 & 2 \end{array}\right]$$

which can be reduced by performing the elementary row operations that are specified below each arrow:

$$\left[\begin{array}{ccc|c} 3 & -1 & 1 & 1 \\ 7 & 1 & -1 & 6 \\ 2 & 1 & -1 & 2 \end{array}\right] \xrightarrow[R_3 \to R_1 + R_3]{} \left[\begin{array}{ccc|c} 3 & -1 & 1 & 1 \\ 7 & 1 & -1 & 6 \\ 5 & 0 & 0 & 3 \end{array}\right]$$

$$\xrightarrow[R_2 \to R_1 + R_2]{} \left[\begin{array}{ccc|c} 3 & -1 & 1 & 1 \\ 10 & 0 & 0 & 7 \\ 5 & 0 & 0 & 3 \end{array}\right]$$

$$\xrightarrow[R_2 \to R_2 + (-2)R_3]{} \left[\begin{array}{ccc|c} 3 & -1 & 1 & 1 \\ 0 & 0 & 0 & 1 \\ 5 & 0 & 0 & 3 \end{array}\right]$$

But the second row, R_2, in the latter matrix implies that $0 \cdot x_1 + 0 \cdot x_2 + 0 \cdot x_3 = 1$ or $0 = 1$, which, of course, is false. Hence, there is no solution possible, and the system is inconsistent.

In summary, one procedure for solving a linear system is to form an augmented matrix and proceed to reduce it to echelon form. If the resulting augmented echelon matrix has no row with its first nonzero entry in the last column, then the system of linear equations either has exactly one solution, in which case the system is *consistent*, or the system has infinitely many solutions, in which case the system is *dependent*. If the resulting augmented echelon matrix has a row with its first nonzero entry appearing in the last column, then the system of linear equations has no solution, in which case the system is *inconsistent*.

Operations on Matrices

Matrices are not only used to solve linear systems. They have other important applications in mathematics. We will now examine some fundamental operations of matrices.

To **add** (or **subtract**) matrices, we add (or subtract) the corresponding elements. These operations can be performed *only* if the matrices are the same size; that is, in order to add (or subtract) matrices A and B, matrix A must have the same number of rows and the same number of columns as matrix B.

The following example illustrates these operations.

EXAMPLE 4 Let $A = \begin{bmatrix} 3 & 5 & 1 \\ 6 & 2 & -3 \end{bmatrix}$ and $B = \begin{bmatrix} 1 & 3 & 7 \\ 5 & -2 & 4 \end{bmatrix}$.

Determine (a) $A + B$ (b) $A - B$

SOLUTION

(a)
$$A + B = \begin{bmatrix} 3 & 5 & 1 \\ 6 & 2 & -3 \end{bmatrix} + \begin{bmatrix} 1 & 3 & 7 \\ 5 & -2 & 4 \end{bmatrix}$$
$$= \begin{bmatrix} 3+1 & 5+3 & 1+7 \\ 6+5 & 2+(-2) & -3+4 \end{bmatrix}$$
$$= \begin{bmatrix} 4 & 8 & 8 \\ 11 & 0 & 1 \end{bmatrix}$$

(b)
$$A - B = \begin{bmatrix} 3 & 5 & 1 \\ 6 & 2 & -3 \end{bmatrix} - \begin{bmatrix} 1 & 3 & 7 \\ 5 & -2 & 4 \end{bmatrix}$$
$$= \begin{bmatrix} 3-1 & 5-3 & 1-7 \\ 6-5 & 2-(-2) & -3-4 \end{bmatrix}$$
$$= \begin{bmatrix} 2 & 2 & -6 \\ 1 & 4 & -7 \end{bmatrix}$$

A matrix having zeros for all of its elements is called a **zero matrix,** and is usually denoted by O. When a zero matrix is added to (or subtracted from) another matrix A of the same size, we get

$$A + \mathrm{O} = A \qquad \text{and} \qquad A - \mathrm{O} = A$$

For example,

$$\begin{bmatrix} 3 & 5 & 2 \\ 6 & 3 & -7 \end{bmatrix} + \begin{bmatrix} 0 & 0 & 0 \\ 0 & 0 & 0 \end{bmatrix} = \begin{bmatrix} 3 & 5 & 2 \\ 6 & 3 & -7 \end{bmatrix}$$

Note that the matrix difference $A - A = \mathrm{O}$. For instance,

$$\begin{bmatrix} 3 & 5 & 2 \\ 6 & 3 & -7 \end{bmatrix} - \begin{bmatrix} 3 & 5 & 2 \\ 6 & 3 & -7 \end{bmatrix} = \begin{bmatrix} 0 & 0 & 0 \\ 0 & 0 & 0 \end{bmatrix}$$

If A is a matrix and k is a real number, we define kA to be the matrix obtained by multiplying each element of A by k.

EXAMPLE 5

Let $A = \begin{bmatrix} -5 & 1 \\ 3 & 2 \end{bmatrix}$. Find the value of each matrix.

(a) $4A$ (b) $-2A$

SOLUTION

(a) $4A = 4\begin{bmatrix} -5 & 1 \\ 3 & 2 \end{bmatrix} = \begin{bmatrix} 4(-5) & 4(1) \\ 4(3) & 4(2) \end{bmatrix} = \begin{bmatrix} -20 & 4 \\ 12 & 8 \end{bmatrix}$

(b) $-2A = -2\begin{bmatrix} -5 & 1 \\ 3 & 2 \end{bmatrix} = \begin{bmatrix} (-2)(-5) & (-2)(1) \\ (-2)(3) & (-2)(2) \end{bmatrix} = \begin{bmatrix} 10 & -2 \\ -6 & -4 \end{bmatrix}$

To find the **product** AB of the two matrices A and B, the number of columns of A must be the same as the number of rows of B. The product AB will have as many rows as A and as many columns as B. In particular, if A has m rows and n columns and B has n rows and p columns, then AB will have m rows and p columns.

For example, if

$$A = \begin{bmatrix} 1 & 3 & 2 \\ -2 & 4 & 5 \end{bmatrix} \quad \text{and} \quad B = \begin{bmatrix} -4 & 7 \\ 3 & -5 \\ 1 & 6 \end{bmatrix}$$

then AB has two rows and two columns. Specifically, the product AB is found as follows:

$$AB = \begin{bmatrix} 1 & 3 & 2 \\ -2 & 4 & 5 \end{bmatrix} \begin{bmatrix} -4 & 7 \\ 3 & -5 \\ 1 & 6 \end{bmatrix}$$

The numbers in row 1 of A are multiplied by the numbers in column 1 of B, and the numbers in row 2 of A are multiplied by the numbers in column 1 of B . The products obtained are then added.

The numbers in row 1 of A are multiplied by the numbers in column 2 of B, and the numbers in row 2 of A are multiplied by the numbers in column 2 of B. The products obtained are then added.

$$= \begin{bmatrix} 1(-4) + 3(3) + 2(1) & 1(7) + 3(-5) + 2(6) \\ -2(-4) + 4(3) + 5(1) & -2(7) + 4(-5) + 5(6) \end{bmatrix} = \begin{bmatrix} 7 & 4 \\ 25 & -4 \end{bmatrix}$$

EXAMPLE 6 Find AB if

$$A = \begin{bmatrix} 3 & -2 \\ 1 & 5 \\ 4 & 6 \end{bmatrix} \quad \text{and} \quad B = \begin{bmatrix} 7 & -2 \\ 3 & 0 \end{bmatrix}$$

SOLUTION

$$AB = \begin{bmatrix} 3 & -2 \\ 1 & 5 \\ 4 & 6 \end{bmatrix} \begin{bmatrix} 7 & -2 \\ 3 & 0 \end{bmatrix}$$

$$= \begin{bmatrix} 3(7) + (-2)(3) & 3(-2) + (-2)(0) \\ 1(7) + 5(3) & 1(-2) + 5(0) \\ 4(7) + 6(3) & 4(-2) + 6(0) \end{bmatrix} = \begin{bmatrix} 15 & -6 \\ 22 & -2 \\ 46 & -8 \end{bmatrix}$$

The 2×2 matrix $\begin{bmatrix} 1 & 0 \\ 0 & 1 \end{bmatrix}$ and the 3×3 matrix $\begin{bmatrix} 1 & 0 & 0 \\ 0 & 1 & 0 \\ 0 & 0 & 1 \end{bmatrix}$ are referred to as **identity matrices.** If I is an identity matrix and A is an $n \times n$ matrix so that AI and IA are defined, then

$$AI = IA = A$$

For example,

$$\begin{bmatrix} 1 & -3 \\ 2 & 4 \end{bmatrix}\begin{bmatrix} 1 & 0 \\ 0 & 1 \end{bmatrix} = \begin{bmatrix} 1 & -3 \\ 2 & 4 \end{bmatrix} \quad \text{and} \quad \begin{bmatrix} 1 & 0 \\ 0 & 1 \end{bmatrix}\begin{bmatrix} 1 & -3 \\ 2 & 4 \end{bmatrix} = \begin{bmatrix} 1 & -3 \\ 2 & 4 \end{bmatrix}$$

It should be noted that the product AB may be defined even when BA is not defined. For instance, the product BA is not defined in Example 6 above because the number of *columns* of B is not the same as the number of *rows* of A. If both products are defined, AB may not be equal to BA. For instance, if $A = \begin{bmatrix} 2 & 1 \\ 4 & 2 \end{bmatrix}$ and $B = \begin{bmatrix} -1 & 2 \\ 3 & 0 \end{bmatrix}$, then

$$AB = \begin{bmatrix} 2 & 1 \\ 4 & 2 \end{bmatrix}\begin{bmatrix} -1 & 2 \\ 3 & 0 \end{bmatrix} = \begin{bmatrix} 1 & 4 \\ 2 & 8 \end{bmatrix}$$

whereas

$$BA = \begin{bmatrix} -1 & 2 \\ 3 & 0 \end{bmatrix}\begin{bmatrix} 2 & 1 \\ 4 & 2 \end{bmatrix} = \begin{bmatrix} 6 & 3 \\ 6 & 3 \end{bmatrix}$$

Thus $AB \neq BA$, so that matrix multiplication is *not* commutative.

We can use matrix multiplication to represent a linear system of equations. Suppose that we are given a system consisting of two linear equations and two variables:

$$\begin{cases} a_1x + a_2y = k_1 \\ b_1x + b_2y = k_2 \end{cases}$$

Since

$$\begin{bmatrix} a_1 & a_2 \\ b_1 & b_2 \end{bmatrix}\begin{bmatrix} x \\ y \end{bmatrix} = \begin{bmatrix} a_1x + a_2y \\ b_1x + b_2y \end{bmatrix}$$

this system can be written in the multiplicative form

$$\begin{bmatrix} a_1 & a_2 \\ b_1 & b_2 \end{bmatrix}\begin{bmatrix} x \\ y \end{bmatrix} = \begin{bmatrix} k_1 \\ k_2 \end{bmatrix}$$

or $AX = K$, where

$$A = \begin{bmatrix} a_1 & a_2 \\ b_1 & b_2 \end{bmatrix}$$

is the coefficient matrix,

$$X = \begin{bmatrix} x \\ y \end{bmatrix} \quad \text{and} \quad K = \begin{bmatrix} k_1 \\ k_2 \end{bmatrix}$$

We refer to the equation

$$AX = K$$

as the **matrix equation form** of the system. For example, the linear system

$$\begin{cases} 2x - y = 1 \\ x + 3y = 7 \end{cases}$$

can be written as the matrix equation

$$\begin{bmatrix} 2 & -1 \\ 1 & 3 \end{bmatrix}\begin{bmatrix} x \\ y \end{bmatrix} = \begin{bmatrix} 1 \\ 7 \end{bmatrix}$$

Similar representations are possible for linear systems with more equations and variables.

There is an important similarity in the algebra of real numbers and the algebra of matrices. If $a \neq 0$ and k are given constants, then the equation

$$ax = k$$

can be solved in the real number system as follows:

$$\begin{aligned} ax &= k \\ a^{-1}ax &= a^{-1}k \\ 1 \cdot x &= a^{-1}k \\ x &= a^{-1}k \end{aligned}$$

Thus, the solution is $a^{-1}k$.

Analogously, if we are given the matrix equation $AX = K$, where A is an $n \times n$ matrix, and *if* there is a matrix B such that $BA = I$, an identity matrix, then

$$\begin{aligned} AX &= K \\ BAX &= BK \\ IX &= BK \\ X &= BK \end{aligned}$$

so that the solution is BK.

If we can find a matrix B such that

$$AB = BA = I$$

for a given $n \times n$ matrix A, then A is said to be **invertible** and matrix B is called the **inverse** of A and is written as A^{-1}. Thus,

$$AA^{-1} = A^{-1}A = I$$

For example, if $A = \begin{bmatrix} 2 & -1 \\ 1 & 3 \end{bmatrix}$ and $A^{-1} = \begin{bmatrix} \frac{3}{7} & \frac{1}{7} \\ -\frac{1}{7} & \frac{2}{7} \end{bmatrix}$, then

$$AA^{-1} = \begin{bmatrix} 2 & -1 \\ 1 & 3 \end{bmatrix}\begin{bmatrix} \frac{3}{7} & \frac{1}{7} \\ -\frac{1}{7} & \frac{2}{7} \end{bmatrix} = \begin{bmatrix} 1 & 0 \\ 0 & 1 \end{bmatrix}$$

and

$$A^{-1}A = \begin{bmatrix} \frac{3}{7} & \frac{1}{7} \\ -\frac{1}{7} & \frac{2}{7} \end{bmatrix}\begin{bmatrix} 2 & -1 \\ 1 & 3 \end{bmatrix} = \begin{bmatrix} 1 & 0 \\ 0 & 1 \end{bmatrix}$$

Thus, to solve the system

$$\begin{cases} 2x - y = 1 \\ x + 3y = 7 \end{cases}$$

we can proceed by first writing the system in matrix form as

$$\begin{bmatrix} 2 & -1 \\ 1 & 3 \end{bmatrix}\begin{bmatrix} x \\ y \end{bmatrix} = \begin{bmatrix} 1 \\ 7 \end{bmatrix}$$

Next we multiply each side of the equation *on the left* by the inverse of the coefficient matrix to get

$$\begin{bmatrix} \frac{3}{7} & \frac{1}{7} \\ -\frac{1}{7} & \frac{2}{7} \end{bmatrix}\begin{bmatrix} 2 & -1 \\ 1 & 3 \end{bmatrix}\begin{bmatrix} x \\ y \end{bmatrix} = \begin{bmatrix} \frac{3}{7} & \frac{1}{7} \\ -\frac{1}{7} & \frac{2}{7} \end{bmatrix}\begin{bmatrix} 1 \\ 7 \end{bmatrix}$$

$$\begin{bmatrix} 1 & 0 \\ 0 & 1 \end{bmatrix}\begin{bmatrix} x \\ y \end{bmatrix} = \begin{bmatrix} \frac{10}{7} \\ \frac{13}{7} \end{bmatrix}$$

$$\begin{bmatrix} x \\ y \end{bmatrix} = \begin{bmatrix} \frac{10}{7} \\ \frac{13}{7} \end{bmatrix}$$

so that the solution of the system is the pair of numbers $x = \frac{10}{7}$ and $y = \frac{13}{7}$ or $(\frac{10}{7}, \frac{13}{7})$.

EXAMPLE 7 Use

$$\begin{bmatrix} 1 & 2 & 3 \\ 1 & 1 & 2 \\ 0 & 1 & 2 \end{bmatrix}^{-1} = \begin{bmatrix} 0 & 1 & -1 \\ 2 & -2 & -1 \\ -1 & 1 & 1 \end{bmatrix}$$

to solve

$$\begin{cases} x + 2y + 3z = 4 \\ x + y + 2z = 5 \\ y + 2z = 4 \end{cases}$$

SOLUTION If we let

$$A = \begin{bmatrix} 1 & 2 & 3 \\ 1 & 1 & 2 \\ 0 & 1 & 2 \end{bmatrix}, \quad X = \begin{bmatrix} x \\ y \\ z \end{bmatrix}, \quad \text{and} \quad K = \begin{bmatrix} 4 \\ 5 \\ 4 \end{bmatrix}$$

we can write the system as

$$AX = K$$

which implies that

$$X = A^{-1}K$$

so that

$$\begin{bmatrix} x \\ y \\ z \end{bmatrix} = X = A^{-1}K = \begin{bmatrix} 0 & 1 & -1 \\ 2 & -2 & -1 \\ -1 & 1 & 1 \end{bmatrix}\begin{bmatrix} 4 \\ 5 \\ 4 \end{bmatrix} = \begin{bmatrix} 1 \\ -6 \\ 5 \end{bmatrix}$$

Thus, $x = 1, y = -6$, and $z = 5$, and the solution is $(1, -6, 5)$.

Since *not every matrix has an inverse* (see Problems 23–28 on page 435), the technique illustrated above is not always applicable.

PROBLEM SET 7.2

In problems 1–11, solve the linear system by using the elimination method. Show the corresponding augmented matrix of the system in each step of the process.

1. $\begin{cases} 3x + y = 14 \\ 2x - y = 1 \end{cases}$

2. $\begin{cases} 4x + 3y = 15 \\ 3x + 5y = 14 \end{cases}$

3. $\begin{cases} -2x + 3y = 8 \\ 2x - y = 5 \end{cases}$

4. $\begin{cases} x - 2y = 5 \\ 3x - 6y = 4 \end{cases}$

5. $\begin{cases} \frac{1}{2}x + \frac{1}{6}y = \frac{2}{3} \\ 3x + y = 4 \end{cases}$

6. $\begin{cases} x - 2y - 4 = 0 \\ x + y + 3 = 0 \end{cases}$

7. $\begin{cases} x + y + z = 6 \\ 3x - y + 2z = 7 \\ 2x + 3y - z = 5 \end{cases}$

8. $\begin{cases} 2x + 3y + z = 6 \\ x - 2y + 3z = -3 \\ 3x + y - z = 8 \end{cases}$

9. $\begin{cases} x + y + 2z = 4 \\ x + y - 2z = 0 \\ x - y = 0 \end{cases}$

10. $\begin{cases} x + y + z = 4 \\ x - y + 2z = 8 \\ 2x + y - z = 3 \end{cases}$

11. $\begin{cases} 2x + y - z = 7 \\ y - x = 1 \\ z - y = 1 \end{cases}$

12. Suppose that

$$A = \begin{bmatrix} -4 & 0 & 1 \\ 2 & 3 & -1 \\ 5 & 2 & 8 \end{bmatrix}$$

(a) What is the size of matrix A?
(b) Use subscript notation to identify each of the entries of matrix A.

In problems 13–20, each row-reduced echelon matrix is the augmented matrix form of a corresponding linear system of equations. Solve each system.

13. $\left[\begin{array}{cc|c} 1 & 0 & 1 \\ 0 & 1 & 1 \end{array}\right]$

14. $\left[\begin{array}{ccc|c} 1 & 0 & 0 & 0 \\ 0 & 1 & 0 & 0 \\ 0 & 0 & 1 & 2 \end{array}\right]$

15. $\left[\begin{array}{ccc|c} 1 & 0 & 0 & 3 \\ 0 & 1 & 0 & 0 \\ 0 & 0 & 1 & 5 \end{array}\right]$

16. $\left[\begin{array}{cccc|c} 1 & 0 & -1 & 2 & 3 \\ 0 & 1 & -2 & 1 & 4 \end{array}\right]$

17. $\left[\begin{array}{ccc|c} 1 & 0 & 1 & 3 \\ 0 & 1 & 0 & 4 \\ 0 & 0 & 0 & 0 \end{array}\right]$

18. $\left[\begin{array}{cccc|c} 1 & 0 & 0 & 0 & 3 \\ 0 & 1 & 0 & 0 & -2 \\ 0 & 0 & 1 & 0 & 5 \\ 0 & 0 & 0 & 1 & -2 \end{array}\right]$

19. $\left[\begin{array}{ccccc|c} 1 & -1 & 0 & -2 & 0 & 1 \\ 0 & 0 & 1 & 3 & 0 & 1 \\ 0 & 0 & 0 & 0 & 1 & 1 \end{array}\right]$

20. $\left[\begin{array}{ccc|c} 0 & 0 & 0 & 0 \\ 0 & 0 & 0 & 0 \\ 0 & 0 & 0 & 1 \end{array}\right]$

In problems 21–30, form the augmented matrix, reduce it to a row-reduced echelon matrix, and determine the solution if it is unique. If the solution is not unique, indicate whether the system is dependent or inconsistent.

21. $\begin{cases} 4x - y = 5 \\ -3x + 2y = 0 \end{cases}$ **22.** $\begin{cases} 4x + 3y = 240 \\ -x + y = 10 \end{cases}$ **23.** $\begin{cases} x + y = 14 \\ 3x - y = 6 \end{cases}$ **24.** $\begin{cases} 2x - 8y = 7 \\ 3x + 2y = 9 \end{cases}$

25. $\begin{cases} x - 2y + z = -1 \\ 3x + y - 2z = 4 \\ y - z = 1 \end{cases}$ **26.** $\begin{cases} x + y - 2z = 3 \\ 3x - y + z = 5 \\ 3x + 3y - 6z = 9 \end{cases}$ **27.** $\begin{cases} 2x + y + z = 1 \\ 4x + 2y + 3z = 1 \\ -2x - y + z = 2 \end{cases}$

28. $\begin{cases} x + y + z = 0 \\ 2x - y - 4z = 15 \\ x - 2y - z = 7 \end{cases}$ **29.** $\begin{cases} 2x - 3y + z = 4 \\ x - 4y - z = 3 \\ x - 9y - 4z = 5 \end{cases}$ **30.** $\begin{cases} 2x + 3y - z = -2 \\ x - y + 2z = 4 \\ 3x + y + z = 7 \end{cases}$

In problems 31–36, perform each operation on the given matrices A and B.

(a) $A + B$ (b) $A - B$ (c) $4A$ (d) $-5B$ (e) $4A - 5B$

31. $A = \begin{bmatrix} 3 & 5 \\ -1 & 4 \end{bmatrix}$ and $B = \begin{bmatrix} 5 & 6 \\ -1 & 3 \end{bmatrix}$ **32.** $A = \begin{bmatrix} -1 & 2 \\ 3 & 0 \end{bmatrix}$ and $B = \begin{bmatrix} 2 & -1 \\ 3 & -4 \end{bmatrix}$

33. $A = \begin{bmatrix} 3 & 1 & 2 \\ -1 & 3 & 1 \end{bmatrix}$ and $B = \begin{bmatrix} 5 & 11 & 6 \\ 3 & 0 & -1 \end{bmatrix}$

34. $A = \begin{bmatrix} 0 & 1 & 4 & 2 \\ -5 & 6 & 1 & 3 \end{bmatrix}$ and $B = \begin{bmatrix} 6 & -1 & 3 & 1 \\ -2 & 0 & 1 & 4 \end{bmatrix}$

35. $A = \begin{bmatrix} -5 & 2 & 1 \\ -2 & 1 & 2 \\ 3 & -1 & 2 \end{bmatrix}$ and $B = \begin{bmatrix} -4 & 6 & 2 \\ -1 & 2 & 5 \\ 3 & -1 & 2 \end{bmatrix}$

36. $A = \begin{bmatrix} 1 & 0 & 0 & 1 & -1 \\ 2 & 1 & 3 & 5 & 7 \\ 3 & 6 & 2 & -1 & 4 \end{bmatrix}$ and $B = \begin{bmatrix} 6 & 1 & -1 & 3 & 5 \\ 7 & 1 & 2 & 8 & -1 \\ 0 & 1 & 2 & 0 & 1 \end{bmatrix}$

37. Let $A = \begin{bmatrix} 3 & 2 \\ -1 & 4 \end{bmatrix}$, $B = \begin{bmatrix} 0 & 1 \\ -1 & 2 \end{bmatrix}$, and $C = \begin{bmatrix} 2 & -3 \\ 1 & -4 \end{bmatrix}$.

(a) Verify the commutative property: $A + B = B + A$.
(b) Verify the associative property: $A + (B + C) = (A + B) + C$.
(c) Find $A + (-A)$ and $C + (-C)$.
(d) Can a matrix D be found such that $A + D = C$?

38. Find the value of x and y if

(a) $\begin{bmatrix} 3 & 1 \\ 5 & 7 \end{bmatrix} = \begin{bmatrix} 3 & x \\ -y & 7 \end{bmatrix}$ (b) $\begin{bmatrix} 1 & -1 \\ 2 & 6 \end{bmatrix} + \begin{bmatrix} x & 3 \\ y & 2 \end{bmatrix} = \begin{bmatrix} 8 & 2 \\ 0 & 8 \end{bmatrix}$

In problems 39–44, find (a) AB and (b) BA, if possible.

39. $A = \begin{bmatrix} 1 & 2 \\ -1 & 1 \end{bmatrix}$ and $B = \begin{bmatrix} 5 & -1 \\ 7 & 0 \end{bmatrix}$ **40.** $A = \begin{bmatrix} 1 & -1 \\ 0 & 2 \end{bmatrix}$ and $B = \begin{bmatrix} 3 & 1 & -1 \\ -2 & 0 & 1 \end{bmatrix}$

41. $A = \begin{bmatrix} 2 & -3 & 5 \\ -1 & 1 & 3 \end{bmatrix}$ and $B = \begin{bmatrix} -3 & 1 \\ 1 & 2 \\ 0 & -5 \end{bmatrix}$

42. $A = \begin{bmatrix} 3 & 2 & 1 \\ 1 & 4 & -1 \\ 2 & 1 & -3 \end{bmatrix}$ and $B = \begin{bmatrix} -3 & 0 & 0 \\ 0 & -3 & 0 \\ 0 & 0 & -3 \end{bmatrix}$

43. $A = \begin{bmatrix} 4 & 3 & -1 \\ 8 & -2 & 3 \\ 6 & 5 & 2 \end{bmatrix}$ and $B = \begin{bmatrix} 1 & 0 & 0 \\ 0 & 1 & 0 \\ 0 & 0 & 1 \end{bmatrix}$

44. $A = \begin{bmatrix} -1 & 1 & 2 & 0 \\ -2 & 3 & 5 & 1 \\ 2 & -1 & 3 & 2 \end{bmatrix}$ and $B = \begin{bmatrix} 1 & -3 & 4 \\ 2 & 1 & 0 \\ -1 & 0 & 2 \\ 0 & -1 & 3 \end{bmatrix}$

In problems 45–50, verify that $AA^{-1} = A^{-1}A = I$ for each pair of matrices.

45. $A = \begin{bmatrix} 1 & -1 \\ 2 & 1 \end{bmatrix}$ and $A^{-1} = \begin{bmatrix} \frac{1}{3} & \frac{1}{3} \\ -\frac{2}{3} & \frac{1}{3} \end{bmatrix}$

46. $A = \begin{bmatrix} 2 & 3 \\ 5 & 1 \end{bmatrix}$ and $A^{-1} = \begin{bmatrix} -\frac{1}{13} & \frac{3}{13} \\ \frac{5}{13} & -\frac{2}{13} \end{bmatrix}$

47. $A = \begin{bmatrix} 3 & 2 \\ 2 & -3 \end{bmatrix}$ and $A^{-1} = \begin{bmatrix} \frac{3}{13} & \frac{2}{13} \\ \frac{2}{13} & -\frac{3}{13} \end{bmatrix}$

48. $A = \begin{bmatrix} 1 & 1 & 1 \\ 2 & -1 & -1 \\ 1 & -1 & 2 \end{bmatrix}$ and $A^{-1} = \begin{bmatrix} \frac{1}{3} & \frac{1}{3} & 0 \\ \frac{5}{9} & -\frac{1}{9} & -\frac{1}{3} \\ \frac{1}{9} & -\frac{2}{9} & \frac{1}{3} \end{bmatrix}$

49. $A = \begin{bmatrix} 1 & 2 & 4 \\ 2 & -3 & 1 \\ 3 & -1 & -2 \end{bmatrix}$ and $A^{-1} = \begin{bmatrix} \frac{1}{7} & 0 & \frac{2}{7} \\ \frac{1}{7} & -\frac{2}{7} & \frac{1}{7} \\ \frac{1}{7} & \frac{1}{7} & -\frac{1}{7} \end{bmatrix}$

50. $A = \begin{bmatrix} 1 & 1 & 1 \\ 2 & 3 & -1 \\ 3 & 5 & 1 \end{bmatrix}$ and $A^{-1} = \begin{bmatrix} 2 & 1 & -1 \\ -\frac{5}{4} & -\frac{1}{2} & \frac{3}{4} \\ \frac{1}{4} & -\frac{1}{2} & \frac{1}{4} \end{bmatrix}$

In problems 51–56, use the inverse of the matrix of coefficients to solve each system. (Refer to the inverses given in problems 45–50.)

51. $\begin{cases} x - y = 1 \\ 2x + y = 5 \end{cases}$ (see problem 45)

52. $\begin{cases} 2x_1 + 3x_2 = 7 \\ 5x_1 + x_2 = -2 \end{cases}$ (see problem 46)

53. $\begin{cases} 3x + 2y = 8 \\ 2x - 3y = 14 \end{cases}$ (see problem 47)

54. $\begin{cases} x + y + z = 6 \\ 2x - y - z = 0 \\ x - y + 2z = 7 \end{cases}$ (see problem 48)

55. $\begin{cases} x_1 + 2x_2 + 4x_3 = 12 \\ 2x_1 - 3x_2 + x_3 = 10 \\ 3x_1 - x_2 - 2x_3 = 1 \end{cases}$ (see problem 49)

56. $\begin{cases} x + y + z = 2 \\ 2x + 3y - z = 3 \\ 3x + 5y + z = 8 \end{cases}$ (see problem 50)

7.3 Determinants

In this section, we assume that all matrices under consideration are *square* matrices, that is, matrices with the same number of rows and columns. A real number called the *determinant* of A is associated with each square matrix A. The determinant of a square matrix A will be denoted by **det A** or $|A|$. We shall see how determinants are used to solve certain systems of linear equations.

The **determinant of the 2 × 2 matrix**

$$A = \begin{bmatrix} a_{11} & a_{12} \\ a_{21} & a_{22} \end{bmatrix}$$

is defined to be the number $a_{11}a_{22} - a_{21}a_{12}$, and we write

$$\det A = \det \begin{bmatrix} a_{11} & a_{12} \\ a_{21} & a_{22} \end{bmatrix} = a_{11}a_{22} - a_{21}a_{12}$$

Using $|A|$ notation, we have

$$|A| = \begin{vmatrix} a_{11} & a_{12} \\ a_{21} & a_{22} \end{vmatrix} = a_{11}a_{22} - a_{21}a_{12}$$

For example,

$$\begin{vmatrix} 5 & 3 \\ -3 & -6 \end{vmatrix} = 5(-6) - (-3)(3) = -30 + 9 = -21$$

Next, we consider the **determinant of the 3 × 3 matrix**

$$A = \begin{bmatrix} a_{11} & a_{12} & a_{13} \\ a_{21} & a_{22} & a_{23} \\ a_{31} & a_{32} & a_{33} \end{bmatrix}$$

We define $\det A$ or $|A|$ by

$$\det A = |A| = \begin{vmatrix} a_{11} & a_{12} & a_{13} \\ a_{21} & a_{22} & a_{23} \\ a_{31} & a_{32} & a_{33} \end{vmatrix} = a_{11}\begin{vmatrix} a_{22} & a_{23} \\ a_{32} & a_{33} \end{vmatrix} - a_{12}\begin{vmatrix} a_{21} & a_{23} \\ a_{31} & a_{33} \end{vmatrix} + a_{13}\begin{vmatrix} a_{21} & a_{22} \\ a_{31} & a_{32} \end{vmatrix}$$

For example,

$$\begin{vmatrix} 3 & 2 & 7 \\ -1 & 5 & 3 \\ 2 & -3 & -6 \end{vmatrix} = 3\begin{vmatrix} 5 & 3 \\ -3 & -6 \end{vmatrix} - 2\begin{vmatrix} -1 & 3 \\ 2 & -6 \end{vmatrix} + 7\begin{vmatrix} -1 & 5 \\ 2 & -3 \end{vmatrix}$$

$$= 3(-30 + 9) - 2(6 - 6) + 7(3 - 10)$$

$$= 3(-21) - 2(0) + 7(-7)$$

$$= -63 - 49$$

$$= -112$$

Notice that a determinant is actually a function that assigns a real number to each square matrix. A determinant has the following important properties that can be used to simplify the task of its evaluation.

THEOREM 1

A common factor that appears in all entries in a row of a matrix can be factored out of the row when evaluating the determinant of the matrix.

PROOF We prove the theorem for $n = 2$.

$$\begin{vmatrix} ka_{11} & ka_{12} \\ a_{21} & a_{22} \end{vmatrix} = ka_{11}a_{22} - ka_{21}a_{12}$$

$$= k(a_{11}a_{22} - a_{21}a_{12})$$

$$= k\begin{vmatrix} a_{11} & a_{12} \\ a_{21} & a_{22} \end{vmatrix}$$

For example,

$$\begin{vmatrix} 6 & 9 \\ 1 & 4 \end{vmatrix} = 3\begin{vmatrix} 2 & 3 \\ 1 & 4 \end{vmatrix}$$

THEOREM 2

If two (not necessarily adjacent) rows of a square matrix are interchanged, the values of the determinants of the two matrices differ only in the algebraic sign.

PROOF We prove the theorem for $n = 2$.

$$\begin{vmatrix} a_{11} & a_{12} \\ a_{21} & a_{22} \end{vmatrix} = a_{11}a_{22} - a_{21}a_{12}$$

$$= -(a_{21}a_{12} - a_{11}a_{22})$$

$$= -\begin{vmatrix} a_{21} & a_{22} \\ a_{11} & a_{12} \end{vmatrix}$$

For example,

$$\begin{vmatrix} 1 & -1 & 0 \\ 3 & 0 & 4 \\ 2 & 1 & 5 \end{vmatrix} = 3 \qquad \text{whereas} \qquad \begin{vmatrix} 2 & 1 & 5 \\ 3 & 0 & 4 \\ 1 & -1 & 0 \end{vmatrix} = -3$$

THEOREM 3

> If any nonzero multiple of one row is added to any other row of a square matrix, the value of the determinant is unaltered.

[For the proof of this theorem for $n = 2$, see Problem 16(a).]

For example, consider

$$\begin{vmatrix} 1 & 0 & 2 \\ 4 & 6 & 1 \\ -1 & 0 & -1 \end{vmatrix}$$

If we multiply the third row by 2 and add the result to the first row, we get

$$\begin{vmatrix} -1 & 0 & 0 \\ 4 & 6 & 1 \\ -1 & 0 & -1 \end{vmatrix}$$

and we are assured by Theorem 3 that the latter determinant has the same value as the original. Note that this operation affects only the first row, whereas the other two rows remain the same. But we need not stop here; indeed, we can add the third row to the second row in the latter determinant to obtain

$$\begin{vmatrix} -1 & 0 & 0 \\ 3 & 6 & 0 \\ -1 & 0 & -1 \end{vmatrix}$$

Again, this does not change the value of the determinant. Determinants such as the last one, which contain many zero entries, are relatively easy to evaluate. Theorem 3 can be used to simplify the task of evaluating determinants.

In Examples 1 and 2, evaluate each of the given determinants.

EXAMPLE 1

$$\begin{vmatrix} 1 & 0 & 2 \\ 4 & 6 & 1 \\ -1 & 0 & -1 \end{vmatrix}$$

SOLUTION

$$\begin{vmatrix} 1 & 0 & 2 \\ 4 & 6 & 1 \\ -1 & 0 & -1 \end{vmatrix} = \begin{vmatrix} -1 & 0 & 0 \\ 4 & 6 & 1 \\ -1 & 0 & -1 \end{vmatrix}$$

(Theorem 3)
($R_1 \to R_1 + 2R_3$)

$$= \begin{vmatrix} -1 & 0 & 0 \\ 3 & 6 & 0 \\ -1 & 0 & -1 \end{vmatrix} \qquad \begin{array}{l} \text{(Theorem 3)} \\ (R_2 \to R_2 + R_3) \end{array}$$

$$= \begin{vmatrix} -1 & 0 & 0 \\ 3 & 6 & 0 \\ 0 & 0 & -1 \end{vmatrix} \qquad \begin{array}{l} \text{(Theorem 3)} \\ [R_3 \to R_3 + (-1)R_1] \end{array}$$

$$= 3 \begin{vmatrix} -1 & 0 & 0 \\ 1 & 2 & 0 \\ 0 & 0 & -1 \end{vmatrix} \qquad \text{(Theorem 1)}$$

$$= (3)(-1) \begin{vmatrix} 1 & 0 & 0 \\ 1 & 2 & 0 \\ 0 & 0 & -1 \end{vmatrix} \qquad \text{(Theorem 1)}$$

$$= (3)(-1)(-1) \begin{vmatrix} 1 & 0 & 0 \\ 1 & 2 & 0 \\ 0 & 0 & 1 \end{vmatrix} \qquad \text{(Theorem 1)}$$

$$= 3 \begin{vmatrix} 1 & 0 & 0 \\ 0 & 2 & 0 \\ 0 & 0 & 1 \end{vmatrix} \qquad \begin{array}{l} \text{(Theorem 3)} \\ [R_2 \to R_2 + (-1)R_1] \end{array}$$

$$= (3)(2) \begin{vmatrix} 1 & 0 & 0 \\ 0 & 1 & 0 \\ 0 & 0 & 1 \end{vmatrix} \qquad \text{(Theorem 1)}$$

$$= 6 \cdot 1 = 6$$

EXAMPLE 2

$$\begin{vmatrix} 3 & 1 & -1 \\ 0 & 2 & 4 \\ -1 & 4 & 2 \end{vmatrix}$$

SOLUTION

$$\begin{vmatrix} 3 & 1 & -1 \\ 0 & 2 & 4 \\ -1 & 4 & 2 \end{vmatrix} = 2 \begin{vmatrix} 3 & 1 & -1 \\ 0 & 1 & 2 \\ -1 & 4 & 2 \end{vmatrix} \qquad \text{(Theorem 1)}$$

$$= 2 \begin{vmatrix} 0 & 13 & 5 \\ 0 & 1 & 2 \\ -1 & 4 & 2 \end{vmatrix} \qquad \begin{array}{l} \text{(Theorem 3)} \\ (R_1 \to R_1 + 3R_3) \end{array}$$

$$= 2 \begin{vmatrix} 0 & 13 & 5 \\ 0 & 1 & 2 \\ -1 & 3 & 0 \end{vmatrix} \qquad \begin{array}{l} \text{(Theorem 3)} \\ [R_3 \to R_3 + (-1)R_2] \end{array}$$

$$= (2)(-1)\begin{vmatrix} 0 & 13 & 5 \\ 0 & 1 & 2 \\ 1 & -3 & 0 \end{vmatrix} \qquad \text{(Theorem 1)}$$

$$= (2)(-1)(-1)\begin{vmatrix} 1 & -3 & 0 \\ 0 & 1 & 2 \\ 0 & 13 & 5 \end{vmatrix} \qquad \text{(Theorem 2)}$$

$$= 2\begin{vmatrix} 1 & -3 & 0 \\ 0 & 1 & 2 \\ 0 & 0 & -21 \end{vmatrix} \qquad \begin{matrix}\text{(Theorem 3)} \\ [R_3 \to R_3 + (-13)R_2]\end{matrix}$$

$$= (2)(-21)\begin{vmatrix} 1 & -3 & 0 \\ 0 & 1 & 2 \\ 0 & 0 & 1 \end{vmatrix} \qquad \text{(Theorem 1)}$$

$$= (-42)\left(1\begin{vmatrix} 1 & 2 \\ 0 & 1 \end{vmatrix} + 3\begin{vmatrix} 0 & 2 \\ 0 & 1 \end{vmatrix} + 0\begin{vmatrix} 0 & 1 \\ 0 & 0 \end{vmatrix}\right)$$

$$= -42$$

Cramer's Rule

We have investigated methods for computing determinants of 2×2 and 3×3 matrices. **Cramer's rule** provides us with a technique for using determinants to solve systems of linear equations that have *unique solutions*. Although Cramer's rule is not always the most practical way to solve linear systems, we shall use the rule as an example of an application of determinants.

Before stating Cramer's rule, let us establish some useful notation. Suppose that we are given a linear system (S) containing the same number of equations as unknowns. The system (S) can be written in the form:

$$\text{(S)} \begin{cases} a_{11}x_1 + a_{12}x_2 + \cdots + a_{1n}x_n = c_1 \\ a_{21}x_1 + a_{22}x_2 + \cdots + a_{2n}x_n = c_2 \\ \cdots\cdots\cdots\cdots\cdots\cdots\cdots\cdots \\ a_{n1}x_1 + a_{n2}x_2 + \cdots + a_{nn}x_n = c_n \end{cases}$$

The determinant of the matrix of coefficients occurring in the system is called the *determinant of the coefficient matrix* and is denoted by D. Hence

$$D = \begin{vmatrix} a_{11} & a_{12} & \cdots & a_{1n} \\ a_{21} & a_{22} & \cdots & a_{2n} \\ \cdots & \cdots & \cdots & \cdots \\ a_{n1} & a_{n2} & \cdots & a_{nn} \end{vmatrix}$$

D_j will be used to denote the determinant of the matrix obtained by replacing the jth column in D, by the column of constant terms in the system, the column on the right of the linear system (S), so that

$$D_j = \begin{vmatrix} a_{11} & a_{12} & \cdots & \overset{\overset{j\text{th column}}{\downarrow}}{c_1} & \cdots & a_{1n} \\ a_{21} & a_{22} & \cdots & c_2 & \cdots & a_{2n} \\ \cdots & \cdots & \cdots & \cdots & \cdots & \cdots \\ a_{n1} & a_{n2} & \cdots & c_n & \cdots & a_{nn} \end{vmatrix}$$

For example, for the linear system

$$\begin{cases} 3x - y + 3z = 1 \\ x + y = 4 \\ -5x + 7y - 2z = -2 \end{cases}$$

$$D = \begin{vmatrix} 3 & -1 & 3 \\ 1 & 1 & 0 \\ -5 & 7 & -2 \end{vmatrix} = 28 \qquad D_1 = \begin{vmatrix} 1 & -1 & 3 \\ 4 & 1 & 0 \\ -2 & 7 & -2 \end{vmatrix} = 80$$

$$D_2 = \begin{vmatrix} 3 & 1 & 3 \\ 1 & 4 & 0 \\ -5 & -2 & -2 \end{vmatrix} = 32 \quad \text{and} \quad D_3 = \begin{vmatrix} 3 & -1 & 1 \\ 1 & 1 & 4 \\ -5 & 7 & -2 \end{vmatrix} = -60$$

We use the above notation to state the next theorem.

THEOREM 4 **Cramer's Rule**

> Let (S) be a system of n linear equations in n unknowns and let D be the determinant of the coefficient matrix of (S). If $D \neq 0$, then the system (S) has exactly one solution, $(x_1, x_2, \ldots, x_n)$, where
>
> $$x_j = \frac{D_j}{D} \qquad \text{for } j = 1, 2, \ldots, n$$
>
> and D_j is the determinant defined above.

We shall consider application of Cramer's rule to cases in which $n = 2$ and $n = 3$.

For $n = 2$, Cramer's rule indicates that the solution of the system

$$\text{(A)} \quad \begin{cases} a_{11}x_1 + a_{12}x_2 = c_1 \\ a_{21}x_1 + a_{22}x_2 = c_2 \end{cases}$$

is given by (x_1, x_2), where

$$x_1 = \frac{\begin{vmatrix} c_1 & a_{12} \\ c_2 & a_{22} \end{vmatrix}}{\begin{vmatrix} a_{11} & a_{12} \\ a_{21} & a_{22} \end{vmatrix}} \quad \text{and} \quad x_2 = \frac{\begin{vmatrix} a_{11} & c_1 \\ a_{21} & c_2 \end{vmatrix}}{\begin{vmatrix} a_{11} & a_{12} \\ a_{21} & a_{22} \end{vmatrix}} \quad \text{if} \quad \begin{vmatrix} a_{11} & a_{12} \\ a_{21} & a_{22} \end{vmatrix} \neq 0$$

For $n = 3$, Cramer's rule can be applied to the system

$$\text{(B)} \quad \begin{cases} a_{11}x_1 + a_{12}x_2 + a_{13}x_3 = c_1 \\ a_{21}x_1 + a_{22}x_2 + a_{23}x_3 = c_2 \\ a_{31}x_1 + a_{32}x_2 + a_{33}x_3 = c_3 \end{cases}$$

First, we determine

$$D = \begin{vmatrix} a_{11} & a_{12} & a_{13} \\ a_{21} & a_{22} & a_{23} \\ a_{31} & a_{32} & a_{33} \end{vmatrix} \qquad D_1 = \begin{vmatrix} c_1 & a_{12} & a_{13} \\ c_2 & a_{22} & a_{23} \\ c_3 & a_{32} & a_{33} \end{vmatrix}$$

$$D_2 = \begin{vmatrix} a_{11} & c_1 & a_{13} \\ a_{21} & c_2 & a_{23} \\ a_{31} & c_3 & a_{33} \end{vmatrix} \quad \text{and} \quad D_3 = \begin{vmatrix} a_{11} & a_{12} & c_1 \\ a_{21} & a_{22} & c_2 \\ a_{31} & a_{32} & c_3 \end{vmatrix}$$

If $D \neq 0$, then

$$x_1 = \frac{D_1}{D}, \qquad x_2 = \frac{D_2}{D}, \qquad \text{and} \qquad x_3 = \frac{D_3}{D}$$

and (x_1, x_2, x_3) is the solution of system (B).

Notice that Cramer's rule has nothing to say about the *existence* of solutions in the case in which $D = 0$. Actually, if $D = 0$, either the system has no solution (inconsistent) or the system has an infinite number of different solutions (dependent).

In Examples 3–5, use Cramer's rule to solve each linear system, if possible.

EXAMPLE 3

$$\begin{cases} 3x + 4y = 12 \\ 3x - 8y = 0 \end{cases}$$

SOLUTION

Here we use the notation of Theorem 4 to obtain

$$D = \begin{vmatrix} 3 & 4 \\ 3 & -8 \end{vmatrix} = -24 - 12 = -36$$

$$D_1 = \begin{vmatrix} 12 & 4 \\ 0 & -8 \end{vmatrix} = -96 \quad \text{and} \quad D_2 = \begin{vmatrix} 3 & 12 \\ 3 & 0 \end{vmatrix} = -36$$

Since $D \neq 0$, then

$$x = \frac{D_1}{D} = \frac{-96}{-36} = \frac{8}{3} \quad \text{and} \quad y = \frac{D_2}{D} = \frac{-36}{-36} = 1$$

The solution is $(\frac{8}{3}, 1)$.

EXAMPLE 4

$$\begin{cases} x + 2y = 4 \\ 3x + 6y = -3 \end{cases}$$

SOLUTION Here

$$D = \begin{vmatrix} 1 & 2 \\ 3 & 6 \end{vmatrix} = 0$$

Since $D = 0$, Cramer's rule is not applicable. Notice that if we graph the lines represented by these equations, we find that they are parallel.

EXAMPLE 5

$$\begin{cases} x + y + z = 4 \\ 1.4x + 1.3y + 1.5z = 5.5 \\ y = 2x \end{cases}$$

SOLUTION By multiplying the second equation by 10 and rearranging the equations, we have the equivalent system

$$\begin{cases} x + y + z = 4 \\ 14x + 13y + 15z = 55 \\ 2x - y = 0 \end{cases}$$

Using the notation in Theorem 4, we have

$$D = \begin{vmatrix} 1 & 1 & 1 \\ 14 & 13 & 15 \\ 2 & -1 & 0 \end{vmatrix} = 5 \qquad D_1 = \begin{vmatrix} 4 & 1 & 1 \\ 55 & 13 & 15 \\ 0 & -1 & 0 \end{vmatrix} = 5$$

$$D_2 = \begin{vmatrix} 1 & 4 & 1 \\ 14 & 55 & 15 \\ 2 & 0 & 0 \end{vmatrix} = 10 \qquad D_3 = \begin{vmatrix} 1 & 1 & 4 \\ 14 & 13 & 55 \\ 2 & -1 & 0 \end{vmatrix} = 5$$

so that

$$x = \frac{D_1}{D} = \frac{5}{5} = 1, \qquad y = \frac{D_2}{D} = \frac{10}{5} = 2, \qquad \text{and} \qquad z = \frac{D_3}{D} = \frac{5}{5} = 1$$

The solution is $(1, 2, 1)$.

PROBLEM SET 7.3

In problems 1–6, evaluate each determinant.

1. $\begin{vmatrix} -1 & 3 \\ -7 & 4 \end{vmatrix}$

2. $\begin{vmatrix} 2 & 3 \\ 9 & 4 \end{vmatrix}$

3. $\begin{vmatrix} 2 & -1 & 3 \\ 9 & -7 & 4 \\ 11 & -6 & 2 \end{vmatrix}$

4. $\begin{vmatrix} 3 & -1 & 2 \\ 0 & 1 & -5 \\ 6 & 7 & 4 \end{vmatrix}$

5. $\begin{vmatrix} 2 & 2 & 2 \\ 3 & 3 & 3 \\ 4 & 4 & 4 \end{vmatrix}$

6. $\begin{vmatrix} \frac{1}{2} & 4 & 7 \\ 1 & -1 & 2 \\ 3 & 2 & 5 \end{vmatrix}$

In problems 7–10, solve for x.

7. $\begin{vmatrix} x & -x \\ 5 & 3 \end{vmatrix} = 2$

8. $\begin{vmatrix} x & 4 & 5 \\ 0 & 1 & x \\ 5 & 2 & 1 \end{vmatrix} = 7$

9. $\begin{vmatrix} x & 0 & 0 \\ 3 & 1 & 2 \\ 0 & 4 & 1 \end{vmatrix} = 5$

10. $\begin{vmatrix} 5x & 0 & 1 \\ 2x & 1 & 2 \\ 3x & 2 & 3 \end{vmatrix} = 0$

In problems 11–15, show why each statement is true, not by evaluating each side, but by citing which theorems have been used.

11. $\begin{vmatrix} 4 & 5 \\ 3 & -2 \end{vmatrix} = -\begin{vmatrix} 3 & -2 \\ 4 & 5 \end{vmatrix}$

12. $\begin{vmatrix} 3 & 0 & 1 \\ 1 & 1 & 2 \\ 3 & 0 & 1 \end{vmatrix} = \begin{vmatrix} 0 & 0 & 0 \\ 1 & 1 & 2 \\ 3 & 0 & 1 \end{vmatrix}$

13. $\begin{vmatrix} 3 & -6 & 2 \\ 5 & -3 & 0 \\ 0 & 9 & 18 \end{vmatrix} = 9\begin{vmatrix} 3 & -6 & 2 \\ 5 & -3 & 0 \\ 0 & 1 & 2 \end{vmatrix}$

14. $\begin{vmatrix} 2 & 4 & 12 \\ -1 & 0 & 3 \\ 1 & 0 & 6 \end{vmatrix} = 18\begin{vmatrix} 1 & 2 & 6 \\ -1 & 0 & 3 \\ 0 & 0 & 1 \end{vmatrix}$

15. $\begin{vmatrix} 1 & 1 & 1 \\ 3 & 3 & 3 \\ 2 & 2 & 2 \end{vmatrix} = 6\begin{vmatrix} 0 & 0 & 0 \\ 0 & 0 & 0 \\ 1 & 1 & 1 \end{vmatrix}$

16. (a) Prove Theorem 3 for $n = 2$. *Hint:* Compare

$$\begin{vmatrix} a_{11} & a_{12} \\ a_{21} & a_{22} \end{vmatrix} \quad \text{and} \quad \begin{vmatrix} a_{11} + ca_{21} & a_{12} + ca_{22} \\ a_{21} & a_{22} \end{vmatrix}$$

(b) Prove, for $n = 2$, that if two rows of a matrix are the same, the determinant is zero.
(c) Prove, for $n = 2$, that if all the entries in one row of a matrix are zeros, the determinant is zero.

In problems 17–21, use Theorems 1–3 to evaluate each of the determinants.

17. $\begin{vmatrix} -1 & 0 & 2 \\ 0 & 0 & 0 \\ -1 & 5 & 1 \end{vmatrix}$

18. $\begin{vmatrix} 3 & 1 & 1 \\ -1 & 0 & 3 \\ 2 & 1 & 1 \end{vmatrix}$

19. $\begin{vmatrix} 2 & 1 & 3 \\ 1 & 2 & 1 \\ 4 & 0 & 0 \end{vmatrix}$

20. $\begin{vmatrix} 20 & 12 & 8 \\ 5 & 3 & 2 \\ 5 & 7 & 2 \end{vmatrix}$

21. $\begin{vmatrix} 1 & 0 & 2 \\ 1 & -3 & 0 \\ 0 & 3 & 1 \end{vmatrix}$

22. (a) *Principle of duality.* Consider Theorems 1, 2, and 3. If we replace the word "row" with the word "column," then the theorems are still true. Rewrite the three theorems with this substitution.
(b) Use the theorems of part (a) to evaluate the determinants in problems 17–21.

In problems 23–28, use the fact that a square matrix A has an inverse if and only if $\det A \neq 0$ to determine whether or not the given matrix has an inverse.

23. $A = \begin{bmatrix} 3 & 1 \\ 2 & 1 \end{bmatrix}$

24. $A = \begin{bmatrix} -1 & 3 \\ 5 & 8 \end{bmatrix}$

25. $A = \begin{bmatrix} 2 & 17 \\ -4 & -34 \end{bmatrix}$

26. $A = \begin{bmatrix} 2 & 1 & 1 \\ 4 & 2 & 3 \\ 1 & 3 & 0 \end{bmatrix}$

27. $A = \begin{bmatrix} 3 & -1 & 2 \\ 6 & -2 & 4 \\ 4 & 7 & 3 \end{bmatrix}$

28. $A = \begin{bmatrix} 2 & 6 & 1 \\ 3 & -6 & 9 \\ 0 & 4 & 3 \end{bmatrix}$

In problems 29–38, use Cramer's rule to solve each system, if possible.

29. $\begin{cases} 2x - y = 0 \\ x + y = 1 \end{cases}$

30. $\begin{cases} -3x + y = 3 \\ -2x - y = -5 \end{cases}$

31. $\begin{cases} x + y = 0 \\ x - y = 0 \end{cases}$

32. $\begin{cases} 3x + y = 1 \\ 9x + 3y = -4 \end{cases}$

33. $\begin{cases} 2x_1 - x_2 + x_3 = 3 \\ -x_1 + 2x_2 - x_3 = 1 \\ 3x_1 + x_2 + 2x_3 = -1 \end{cases}$

34. $\begin{cases} 3x + 2z = 8 - 2y \\ x - 5y + 6z = 8 \\ 6x - 8z = 4 \end{cases}$

35. $\begin{cases} x + y + 2z = 4 \\ x + y - 2z = 0 \\ x - y = 0 \end{cases}$

36. $\begin{cases} 2x_1 - 3x_2 = 4 \\ x_1 + x_2 - 2x_3 = 1 \\ x_1 - x_2 - x_3 = 5 \end{cases}$

37. $\begin{cases} x + y + z = 4 \\ x - y + 2z = 8 \\ 2x + y - z = 3 \end{cases}$

38. $\begin{cases} 2x + 3y + z = 6 \\ x - 2y + 3z = -3 \\ 3x + y - z = 8 \end{cases}$

39. Let triangle PQR be a triangle in the xy plane with vertices $P_1 = (x_1, y_1)$, $Q = (x_2, y_2)$, and $R = (x_3, y_3)$. Then the *area A of the triangle PQR* is given by

$$A = \text{absolute value of } \tfrac{1}{2} \begin{vmatrix} x_1 & y_1 & 1 \\ x_2 & y_2 & 1 \\ x_3 & y_3 & 1 \end{vmatrix}$$

Find the area of triangle PQR if $P = (1, -5)$, $Q = (-3, -4)$, and $R = (6, 2)$.

40. The equation

$$\begin{vmatrix} x & y & 1 \\ x_1 & y_1 & 1 \\ x_2 & y_2 & 1 \end{vmatrix} = 0$$

represents the line in the xy plane containing the two points (x_1, y_1) and (x_2, y_2). Use this result to find an equation of the line containing the points $(3, 6)$ and $(-5, 7)$.

7.4 Partial Fractions

We obtain the sum of the fractions $2/(x - 2)$ and $3/(x + 1)$ as follows:

$$\begin{aligned}\frac{2}{x-2} + \frac{3}{x+1} &= \frac{2(x+1) + 3(x-2)}{(x-2)(x+1)} \\ &= \frac{2x + 2 + 3x - 6}{(x-2)(x+1)} \\ &= \frac{5x - 4}{(x-2)(x+1)}\end{aligned}$$

The *reverse process* of writing $(5x - 4)/[(x - 2)(x + 1)]$ as a sum or difference of *simple fractions* (fractions with numerators of lower degree than their denominators) is frequently important in the study of calculus. Each such simple fraction is called a **partial fraction,** and the process itself is called **decomposition into partial fractions.**

Fractions Whose Denominators Contain Linear Factors

Decomposition of a fraction into partial fractions is easy to accomplish if the factors of the denominator are distinct linear factors. In this case, it is only necessary to provide a partial fraction of the form $A/(ax + b)$, where A is a constant number, for each of the linear factors in the denominator.

For instance, to express $(5x - 4)/[(x - 2)(x + 1)]$ in terms of partial fractions, we write

$$\frac{5x-4}{(x-2)(x+1)} = \frac{A}{x-2} + \frac{B}{x+1}$$

where A and B are constants that can be determined as follows. First, we multiply both sides of the above equation by the denominator $(x - 2)(x + 1)$ to get

$$5x - 4 = A(x + 1) + B(x - 2) \qquad \text{or} \qquad 5x - 4 = (A + B)x + (A - 2B)$$

Next, we equate the coefficients of like powers of x on the two sides of the latter equation to obtain

$$\begin{cases} 5 = A + B \\ -4 = A - 2B \end{cases}$$

Solving this system of linear equations yields $A = 2$ and $B = 3$, so that

$$\frac{5x-4}{(x-2)(x+1)} = \frac{2}{x-2} + \frac{3}{x+1}$$

EXAMPLE 1 Find the partial fractions decomposition of $\dfrac{2x-2}{x^2 + 7x + 10}$.

SOLUTION We have

$$\frac{2x - 2}{x^2 + 7x + 10} = \frac{2x - 2}{(x + 2)(x + 5)} = \frac{A}{x + 2} + \frac{B}{x + 5}$$

where the constants A and B must be determined. After multiplying the two sides of the above equation by $(x + 2)(x + 5)$, we get

$$2x - 2 = A(x + 5) + B(x + 2) \qquad \text{or} \qquad 2x - 2 = (A + B)x + (5A + 2B)$$

Equating the coefficients of like powers of x on the two sides of the latter equation results in

$$\begin{cases} 2 = A + B \\ -2 = 5A + 2B \end{cases}$$

Solving this system of linear equations, we obtain $A = -2$ and $B = 4$. Therefore

$$\frac{2x - 2}{x^2 + 7x + 10} = \frac{-2}{x + 2} + \frac{4}{x + 5}$$

If the factors of the denominator of the fraction are all linear, but one of the linear factors is *repeated*, we proceed in a somewhat similar manner, as shown in the next example.

EXAMPLE 2 Find the partial fractions decomposition of $\dfrac{3x^2 + 4x + 2}{x(x + 1)^2}$.

SOLUTION We write

$$\frac{3x^2 + 4x + 2}{x(x + 1)^2} = \frac{A}{x} + \frac{B}{x + 1} + \frac{C}{(x + 1)^2}$$

where A, B, and C are constant numbers that must be determined. The fraction $C/(x + 1)^2$ is needed because of the repeated linear factor $(x + 1)^2$. After multiplying both sides of the equation by $x(x + 1)^2$, we obtain

$$3x^2 + 4x + 2 = A(x + 1)^2 + B(x + 1)x + Cx$$

or

$$3x^2 + 4x + 2 = (A + B)x^2 + (2A + B + C)x + A$$

Equating the coefficients of like powers of x on the two sides of the latter equation, we obtain the linear system

$$\begin{cases} 3 = A + B \\ 4 = 2A + B + C \\ 2 = A \end{cases}$$

Solving this system of linear equations yields $A = 2$, $B = 1$, and $C = -1$, so that

$$\frac{3x^2 + 4x + 2}{x(x + 1)^2} = \frac{2}{x} + \frac{1}{x + 1} + \frac{-1}{(x + 1)^2}$$

If the degree of the numerator is greater than or equal to the degree of the denominator, then we employ long division before finding the partial fractions decomposition, as the next example shows.

EXAMPLE 3 Find the partial fractions decomposition of

$$\frac{3x^3 + 15x^2 - 38x + 18}{x^3 + 5x^2 - 13x + 7}$$

SOLUTION Using long division, we have

$$\frac{3x^3 + 15x^2 - 38x + 18}{x^3 + 5x^2 - 13x + 7} = 3 + \frac{x - 3}{x^3 + 5x^2 - 13x + 7}$$

By factoring the denominator, we obtain

$$\frac{3x^3 + 15x^2 - 38x + 18}{x^3 + 5x^2 - 13x + 7} = 3 + \frac{x - 3}{(x + 7)(x - 1)^2}$$

Now we decompose the last fraction as follows:

$$\frac{x - 3}{(x + 7)(x - 1)^2} = \frac{A}{x + 7} + \frac{B}{x - 1} + \frac{C}{(x - 1)^2}$$

where the constants A, B, and C must be determined. Multiplying both sides of the above equation by $(x + 7)(x - 1)^2$, we obtain

$$x - 3 = A(x - 1)^2 + B(x - 1)(x + 7) + C(x + 7)$$

or

$$x - 3 = (A + B)x^2 + (-2A + 6B + C)x + (A - 7B + 7C)$$

After equating the coefficients of like powers of x, we are led to the linear system

$$\begin{cases} 0 = A + B \\ 1 = -2A + 6B + C \\ -3 = A - 7B + 7C \end{cases}$$

Solving this system of linear equations, we obtain $A = -\frac{5}{32}$, $B = \frac{5}{32}$, and $C = -\frac{1}{4}$. Therefore,

$$\frac{x - 3}{(x + 7)(x - 1)^2} = \frac{-\frac{5}{32}}{x + 7} + \frac{\frac{5}{32}}{x - 1} + \frac{-\frac{1}{4}}{(x - 1)^2}$$

and

$$\frac{3x^3 + 15x^2 - 38x + 18}{x^3 + 5x^2 - 13x + 7} = 3 + \frac{x - 3}{(x + 7)(x - 1)^2}$$

$$= 3 + \frac{-\frac{5}{32}}{x + 7} + \frac{\frac{5}{32}}{x - 1} + \frac{-\frac{1}{4}}{(x - 1)^2}$$

Fractions Whose Denominators Contain Quadratic Factors

If some of the *prime* factors (in the real number system) of the denominator of a fraction are quadratic, that is, of the form $ax^2 + bx + c$, with $a \neq 0$, $b^2 - 4ac < 0$, and none of the factors is repeated, then the numerators of the partial fractions corresponding to these factors are of the form $Ax + B$. Again, it is necessary to determine the numerical values of A and B. This procedure is illustrated in the following example.

EXAMPLE 4 Find the partial fractions decomposition of $\dfrac{8x^2 + 3x + 20}{(x + 1)(x^2 + 4)}$.

SOLUTION We write

$$\frac{8x^2 + 3x + 20}{(x + 1)(x^2 + 4)} = \frac{A}{x + 1} + \frac{Bx + C}{x^2 + 4}$$

where A, B, and C must be determined. Multiplying both sides of this equation by $(x + 1)(x^2 + 4)$, we obtain

$$8x^2 + 3x + 20 = A(x^2 + 4) + (Bx + C)(x + 1)$$

or

$$8x^2 + 3x + 20 = (A + B)x^2 + (B + C)x + (4A + C)$$

so that

$$\begin{cases} 8 = A + B \\ 3 = B + C \\ 20 = 4A + C \end{cases}$$

Solving this system yields $A = 5$, $B = 3$, and $C = 0$. Therefore,

$$\frac{8x^2 + 3x + 20}{(x + 1)(x^2 + 4)} = \frac{5}{x + 1} + \frac{3x}{x^2 + 4}$$

If the denominator of the fraction contains prime quadratic factors (and perhaps linear factors) and if one or more of the quadratic factors are repeated, then we use a procedure analogous to that used in the case of the repeated linear factors. The next example illustrates this procedure.

EXAMPLE 5 Find the partial fractions decomposition of $\dfrac{x^3 + x + 2}{x(x^2 + 1)^2}$.

SOLUTION First, we write

$$\frac{x^3 + x + 2}{x(x^2 + 1)^2} = \frac{A}{x} + \frac{Bx + C}{x^2 + 1} + \frac{Dx + E}{(x^2 + 1)^2}$$

where the constants A, B, C, D, and E must be determined. Multiplying both sides of the equation by $x(x^2+1)^2$, we have

$$\begin{aligned} x^3 + x + 2 &= A(x^2+1)^2 + (Bx + C)x(x^2+1) + (Dx + E)x \\ &= A(x^4 + 2x^2 + 1) + (Bx^4 + Cx^3 + Bx^2 + Cx) + (Dx^2 + Ex) \end{aligned}$$

or

$$x^3 + x + 2 = (A + B)x^4 + Cx^3 + (2A + B + D)x^2 + (C + E)x + A$$

so that

$$\begin{cases} 0 = A + B \\ 1 = C \\ 0 = 2A + B + D \\ 1 = C + E \\ 2 = A \end{cases}$$

Solving this system of linear equations yields $A = 2$, $B = -2$, $C = 1$, $D = -2$, and $E = 0$. Hence,

$$\frac{x^3 + x + 2}{x(x^2+1)^2} = \frac{2}{x} - \frac{2x - 1}{x^2 + 1} - \frac{2x}{(x^2+1)^2}$$

PROBLEM SET 7.4

In problems 1–30, find the partial fractions decomposition of each given fraction.

1. $\dfrac{3x - 5}{(x-1)(x-3)}$

2. $\dfrac{x + 14}{(x+4)(x+12)}$

3. $\dfrac{8x - 5}{x^2 - x - 6}$

4. $\dfrac{2x + 19}{(2x-5)(2x-1)}$

5. $\dfrac{-9x - 7}{6x^2 + 7x + 2}$

6. $\dfrac{13x - 24}{3x^2 - 11x + 10}$

7. $\dfrac{x^2 - 3x + 4}{(x-1)(x+1)(x+2)}$

8. $\dfrac{-21x + 11}{(x-1)(x-2)(x-3)}$

9. $\dfrac{2x^2 - x + 8}{x(x-2)^2}$

10. $\dfrac{x^2 + x + 29}{(x-4)(x+3)^2}$

11. $\dfrac{5x^2 - 21x + 13}{(x-3)^2(x+2)}$

12. $\dfrac{-x^2 + 13x - 26}{(x+1)^2(x-4)}$

13. $\dfrac{2y^3 - 8y^2 + 9y + 1}{y^2 - 4y + 4}$

14. $\dfrac{t^2 + 3t + 3}{t(1+t)}$

15. $\dfrac{6}{x(x^2+3)}$

16. $\dfrac{30x}{(x-6)(x^2+4)}$

17. $\dfrac{8}{(2x-1)(4x^2+1)}$

18. $\dfrac{4x^2 + 5}{(x-2)(x^2+x+1)}$

19. $\dfrac{2p^3 + 9p^2 + 3p + 2}{2p^3 - p^2 + 2p - 1}$

20. $\dfrac{x^2 + 1}{(2x-1)(2x^2+x+4)}$

21. $\dfrac{x^2 - x - 21}{2x^3 - x^2 + 8x - 4}$

22. $\dfrac{u^4 - u^3 + 3u^2 - 10u + 8}{u^3 + u^2 - 4u - 4}$

23. $\dfrac{4x^2}{(x + 1)(x^2 + 1)^2}$

24. $\dfrac{2x^3 + 6x^2 + 13x + 22}{(x - 2)(x^2 + 4)^2}$

25. $\dfrac{2x^4 - 2x^3 + 3x^2 - 6x + 4}{x(x^2 + 1)^2}$

26. $\dfrac{x^3 - x^2}{(x^2 + 3)^2}$

27. $\dfrac{2x^3 + 5x^2 + 16x}{x^5 + 8x^3 + 16x}$

28. $\dfrac{2x^4 + x^3 + 7x^2 + 2}{x^5 + 2x^3 + x}$

29. $\dfrac{-5x^4 + 3x^3 - 27x^2 + 16x - 12}{(x - 1)(x^2 + 4)^2}$

30. $\dfrac{4x^2 - 2}{(x + 1)^2(x^2 + 1)^2}$

7.5 Systems Containing Nonlinear Equations

The substitution and elimination methods that were introduced in Section 7.1 can be used to solve systems containing *nonlinear* equations. However, these methods *alone* are generally not sufficient to find a solution to these types of systems. Here we study only the relatively simple case of two equations in two variables. The solution of such nonlinear systems is greatly facilitated by sketching graphs of the two equations on the same coordinate system to determine the number of points where the two graphs intersect. Then algebraic methods can be used to find the exact values of the solutions.

In Examples 1 and 2, sketch the graphs of both equations on the same coordinate axes to determine the number of solutions of each system, and then solve the system.

EXAMPLE 1

$$\begin{cases} x^2 - y = 0 \\ x + y = 2 \end{cases}$$

SOLUTION

The graph of $x^2 - y = 0$ or $y = x^2$ is a parabola opening upward with its vertex at the origin, and the graph of $x + y = 2$ is a line that intersects the parabola at two points (Figure 1). Thus, there are two solutions of the system. To find the solutions algebraically, we use the elimination method. If we add the second equation to the first equation, we eliminate y and obtain

$$x^2 + x = 2$$

so that

$$x^2 + x - 2 = 0 \qquad \text{or} \qquad (x + 2)(x - 1) = 0$$

and

$$x = -2 \qquad \text{or} \qquad x = 1$$

After substituting each of these values of x in the first equation, we obtain $y = 4$ and $y = 1$, respectively. Hence, the two solutions of the system are $(1, 1)$ and $(-2, 4)$.

Figure 1

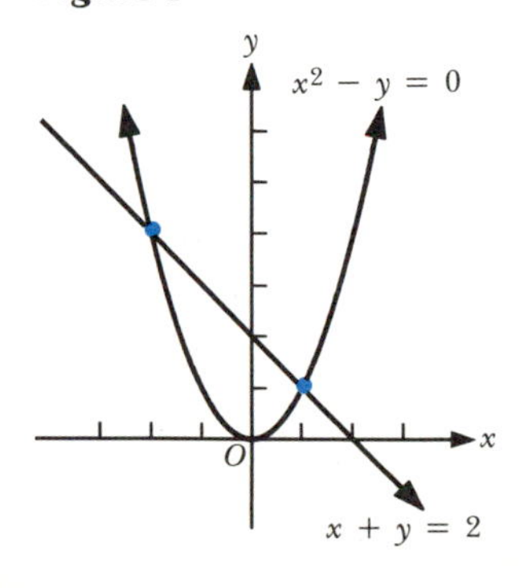

EXAMPLE 2

$$\begin{cases} 3x - 2y = -6 \\ \quad\;\; xy = 12 \end{cases}$$

SOLUTION First $xy = 12$ is written as $y = 12/x$. The graph of $3x - 2y = -6$ intersects the graph of $y = 12/x$ at two points (Figure 2). Thus, there are two solutions of the system. To find the solutions, we use the substitution method. After writing the first equation as

$$y = \tfrac{3}{2}x + 3$$

Figure 2

we substitute this equation into the second equation to obtain

$$x(\tfrac{3}{2}x + 3) = 12 \qquad \text{or} \qquad 3x^2 + 6x = 24$$

so that

$$3x^2 + 6x - 24 = 0$$

$$x^2 + 2x - 8 = 0$$

$$(x + 4)(x - 2) = 0$$

or

$$x = -4 \qquad \text{and} \qquad x = 2$$

Substituting these two values of x in the equation $y = \frac{3}{2}x + 3$, we obtain

$$y = -3 \qquad \text{and} \qquad y = 6$$

respectively. Hence, the two solutions are $(-4, -3)$ and $(2, 6)$.

EXAMPLE 3 Solve the system

$$\begin{cases} 4x^2 + 7y^2 = 32 \\ -3x^2 + 11y^2 = 41 \end{cases}$$

by using the elimination method.

SOLUTION After multiplying the first equation by 3 and the second by 4, we get

$$\begin{cases} 12x^2 + 21y^2 = 96 \\ -12x^2 + 44y^2 = 164 \end{cases}$$

We then add the two equations to get $65y^2 = 260$, so that $y^2 = 4$; that is, $y = -2$ or $y = 2$. When $y = -2$, we have

$$4x^2 + 7(-2)^2 = 32$$

$$4x^2 + 28 = 32$$

$$x^2 = 1$$

so

$$x = -1 \qquad \text{or} \qquad x = 1$$

and when $y = 2$ we have

$$4x^2 + 7(2)^2 = 32$$

so that $x = -1$ or $x = 1$. Hence, the solutions are $(1,2)$, $(-1,2)$, $(1,-2)$, and $(-1,-2)$.

Systems involving exponential and logarithmic functions may sometimes be solved by the elimination method.

EXAMPLE 4 Solve the system

$$\begin{cases} \log(x - 21) + y = 5 \\ \log x - y = -3 \end{cases}$$

SOLUTION After adding the first equation to the second equation, we obtain

$$\log(x - 21) + \log x = 2$$

so that

$$\log x(x - 21) = 2 \qquad \text{or} \qquad x(x - 21) = 10^2$$

that is,

$$x^2 - 21x - 100 = 0$$

$$(x - 25)(x + 4) = 0$$

so that

$$x = 25 \qquad \text{or} \qquad x = -4$$

Because $\log x$ is undefined when $x = -4$, we reject -4. We substitute 25 into the second equation in the system and obtain

$$y = \log 25 + 3$$

Hence the solution is $(25, 3 + \log 25)$.

Note that when solving a nonlinear system one should check for extraneous roots.

PROBLEM SET 7.5

In problems 1–6, sketch the graphs of both equations on the same coordinate axes to determine the number of solutions and then solve the system.

1. $\begin{cases} x - y = 1 \\ x^2 + y^2 = 5 \end{cases}$

2. $\begin{cases} x - 2y = 3 \\ x^2 - y^2 = 24 \end{cases}$

3. $\begin{cases} x + 2y = 6 \\ x^2 - 2y = 0 \end{cases}$

4. $\begin{cases} x^2 + y = 13 \\ x^2 + y^2 = 25 \end{cases}$

5. $\begin{cases} 2x + y = 10 \\ xy = 12 \end{cases}$

6. $\begin{cases} 6x - y = 5 \\ xy = 4 \end{cases}$

In problems 7–40, solve each system of equations by the elimination or substitution method. In problem 39, round off the answers to three decimal places.

7. $\begin{cases} 3x - y = 2 \\ x^2 + y^2 = 20 \end{cases}$

8. $\begin{cases} x + y = 3 \\ 3x^2 - y^2 = \frac{9}{2} \end{cases}$

9. $\begin{cases} 3x + 2y = 1 \\ 3x^2 - y^2 = -4 \end{cases}$

10. $\begin{cases} x + y = 6 \\ x^2 + y^2 = 20 \end{cases}$

11. $\begin{cases} 5x - 3y = 10 \\ x^2 - y^2 = 6 \end{cases}$

12. $\begin{cases} uv = 5 \\ 2u^2 - v^2 = 5 \end{cases}$

13. $\begin{cases} 2x + 3y = 7 \\ x^2 + y^2 + 4y + 4 = 0 \end{cases}$

14. $\begin{cases} x - y + 4 = 0 \\ x^2 + 3y^2 = 12 \end{cases}$

15. $\begin{cases} 4h - k^2 = 0 \\ h^2 - 4k = 0 \end{cases}$

16. $\begin{cases} x^2 - 25y^2 = 20 \\ 2x^2 + 25y^2 = 88 \end{cases}$

17. $\begin{cases} x - y^2 = 0 \\ x^2 + 2y^2 = 24 \end{cases}$

18. $\begin{cases} 3x^2 - 8y^2 = 40 \\ 5x^2 + y^2 = 81 \end{cases}$

19. $\begin{cases} 2x^2 - 3y^2 = 6 \\ 3x^2 + 2y^2 = 35 \end{cases}$

20. $\begin{cases} x^2 - y^2 = 7 \\ x^2 + y^2 = 25 \end{cases}$

21. $\begin{cases} u^2 + 9v^2 = 33 \\ u^2 + v^2 = 25 \end{cases}$

22. $\begin{cases} x^2 + 5y^2 = 70 \\ 3x^2 - 5y^2 = 30 \end{cases}$

23. $\begin{cases} 4x^2 - y^2 = 4 \\ 4x^2 + \frac{5}{3}y^2 = 36 \end{cases}$

24. $\begin{cases} r^2 - 2s^2 = 17 \\ 2r^2 + s^2 = 54 \end{cases}$

25. $\begin{cases} 2x^2 - 3y^2 = 20 \\ x^2 + 2y = 20 \end{cases}$

26. $\begin{cases} 4x^2 + 3y^2 = 43 \\ 3x^2 - y^2 = 3 \end{cases}$

27. $\begin{cases} x^2 - 2y^2 = 1 \\ x^2 + 4y^2 = 25 \end{cases}$

28. $\begin{cases} 2x^2 - 5y^2 + 8 = 0 \\ x^2 - 7y^2 + 4 = 0 \end{cases}$

29. $\begin{cases} x^2 + 4y = 8 \\ x^2 + y^2 = 5 \end{cases}$

30. $\begin{cases} \sqrt{s} - \sqrt{t} = 2 \\ s - 4t = 0 \end{cases}$

31. $\begin{cases} x^2 + y^2 = 16 \\ x^2 - y^2 = -34 \end{cases}$

32. $\begin{cases} x^2 - 4y^2 = -15 \\ -x^2 + 3y^2 = 11 \end{cases}$

33. $\begin{cases} x^2 + y^2 = 25 \\ (x - 5)^2 + y^2 = 9 \end{cases}$

34. $\begin{cases} x^2 - y = 0 \\ x^2 + (y - 6)^2 = 36 \end{cases}$

35. $\begin{cases} b - \sqrt[4]{a} = 0 \\ b^2 - \sqrt[4]{a} = 2 \end{cases}$

36. $\begin{cases} \sin(u + v) = 1 \\ \tan(u - v) = 1 \end{cases}$

37. $\begin{cases} \log_3(r^2 + s^2) = 2 \\ r^2 - s = 3 \end{cases}$

38. $\begin{cases} c + 3(2^d) = 2^{2d} \\ c - 2^{d+1} = -6 \end{cases}$

[c] **39.** $\begin{cases} u + 1.923v = 4.693 \\ \log_{10} u + \log_{10} v = \log_{10} 2.548 \end{cases}$

40. $\begin{cases} u^2 - uv = 0 \\ u - v - w^2 = -1 \\ 2u - 3v - w^2 = 0 \end{cases}$

41. The sum of the squares of two numbers is 117 and the difference of the squares of the same two numbers is 45. Find the numbers.

42. The area of a rectangle is 96 square centimeters and its perimeter is 40 centimeters. Find the length and the width of the rectangle.

43. A template in the shape of a right triangle has a perimeter 60 centimeters and an area 150 square centimeters. Find the lengths of the sides of the triangle.

44. The simple interest received from a loan after one year is \$170. If the annual interest rate had been 1% higher, the amount of interest after one year would be \$238. How much is the loan and what is its annual interest rate?

45. Environmentalists determine that two chemicals are needed to treat sewage discharged into a lake. One chemical is to purify the water in the lake and the other is to provide nourishment for fish. The sum of the required amounts of the two chemicals is 310 kilograms and the sum of the squares of the required amounts is 48,500. How much of each chemical is required?

7.6 Systems of Linear Inequalities and Linear Programming

Linear programming is the name given to a field of mathematics that deals, in general, with the problem of finding the maximum or minimum value of a given linear expression, where the variables are subject to certain conditions that are expressed as linear inequalities. The basic concepts and techniques of linear programming were invented by George B. Dantzig, who developed them while working on applied problems for the U.S. Air Force in 1947. Since that time linear programming has proven to be of great importance in solving problems in the fields of health, transportation, economics, engineering, and agriculture.

In this section, we restrict our attention to linear expressions with *two variables,* although it is possible to use linear programming for expressions containing more than two variables. In order to understand the basic notions of linear programming, it is necessary to investigate linear inequalities by extending the idea of linear systems covered in the preceding sections.

Systems of Linear Inequalities

The graph of a linear equation in two variables divides the plane into two disjoint **half planes.** For example, the graph of $y + 2x = 3$ divides the plane into the two regions R_1 and R_2 shown in Figure 1. *All* points that satisfy either $y + 2x > 3$ or $y + 2x < 3$ lie in either R_1 or R_2. By testing *one point* in either region, we can determine which inequality is satisfied by all the points in that region. For example, the point $(4, 0)$ lies in R_1, and since the coordinates satisfy $y + 2x > 3$, we conclude that R_1 contains *all* points that satisfy $y + 2x > 3$. Similarly, R_2 contains all points whose coordinates satisfy $y + 2x < 3$.

In summary, we have the following:

Condition	Location of Points
$y + 2x = 3$	on the line
$y + 2x > 3$	R_1
$y + 2x < 3$	R_2

Figure 1

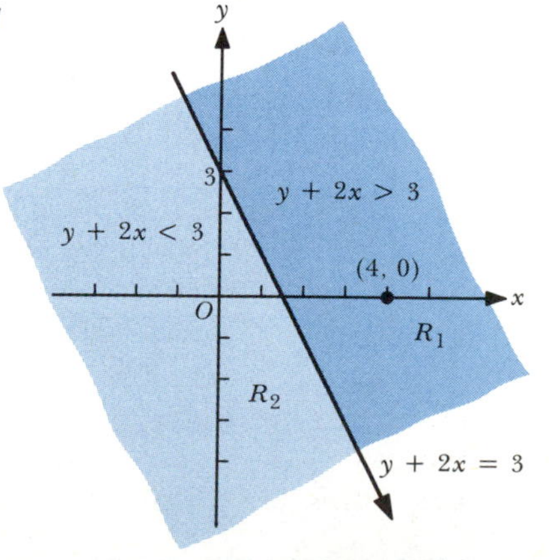

Figure 2

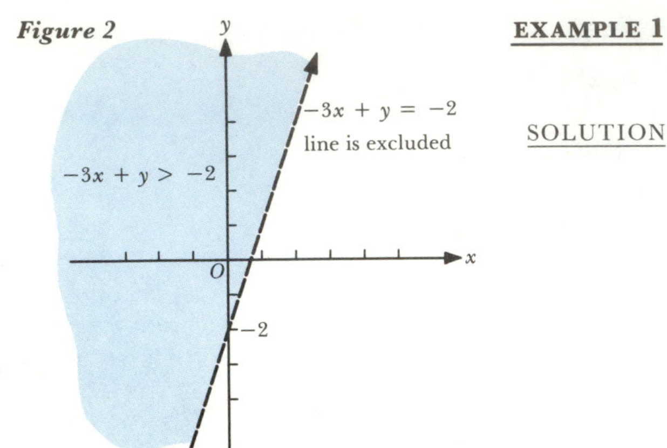

EXAMPLE 1 Sketch the region containing all the points whose coordinates satisfy the inequality $-3x + y > -2$.

SOLUTION First, we sketch the graph of the associated linear equation $-3x + y = -2$ (Figure 2). Next, we select point $(0, 0)$ to test for the region. Since $x = 0$ and $y = 0$ satisfy $-3x + y > -2$, we conclude that *all* the points above the dashed line satisfy the inequality $-3x + y > -2$. The region in question is shaded in Figure 2. Notice that a dashed line is used to represent the graph of the equation $-3x + y = -2$; this emphasizes that the line is *not* included in the region satisfying the inequality.

The region consisting of points whose coordinates satisfy a **system of linear inequalities** can be determined by finding the regions where the graphs of all the respective inequalities in the system overlap. This technique is illustrated in the following examples.

In Examples 2 and 3, sketch the region containing all points whose coordinates satisfy the given system of linear inequalities. Also, find the coordinates of the corner points of the boundaries of the region.

EXAMPLE 2

$$\begin{cases} x \geq 0 \\ y \geq 0 \\ x + y \leq 1 \end{cases}$$

SOLUTION The region for $x \geq 0$ contains all points on or to the right of the y axis; the region for $y \geq 0$ contains all points on or above the x axis; and the region for $x + y \leq 1$ contains all points on or below the line $x + y = 1$ (Figure 3). Thus the desired region is the set of all points where these three regions overlap. The corner points of the boundary of this triangular region are $(0, 0)$, $(1, 0)$, and $(0, 1)$.

EXAMPLE 3

$$\begin{cases} y < x + 2 \\ x + y \leq 4 \end{cases}$$

SOLUTION The graph of $y < x + 2$ contains all points below the line $y = x + 2$. The graph of $x + y \leq 4$ contains all points on or below the line $x + y = 4$ (Figure 4). Thus the solution of the given system contains all points contained in both regions. The coordinates of the corner point of the boundary can be found by solving the linear system

$$\begin{cases} y = x + 2 \\ x + y = 4 \end{cases}$$

to get $(1, 3)$.

Figure 3

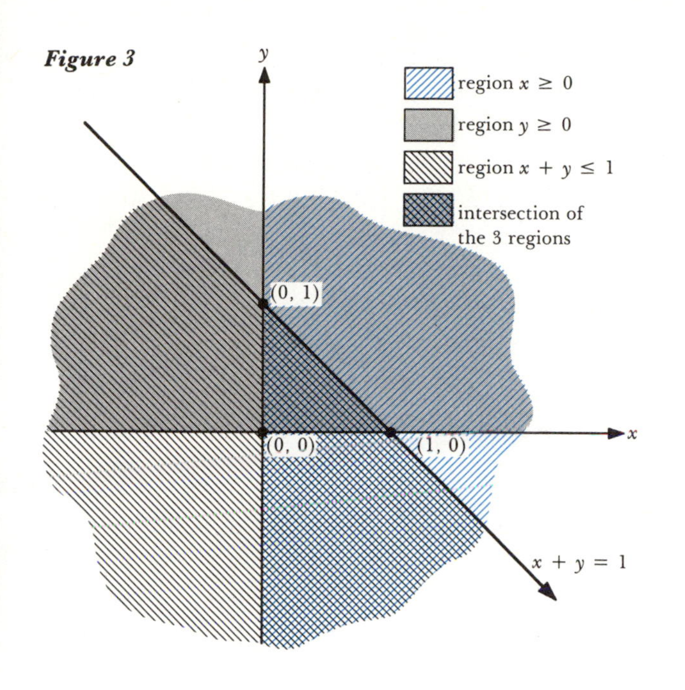

Figure 4

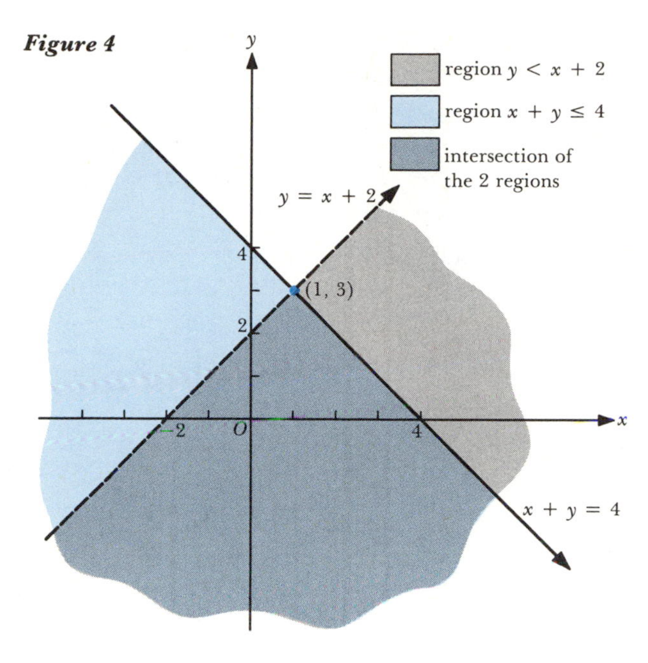

Linear Programming

Figure 5

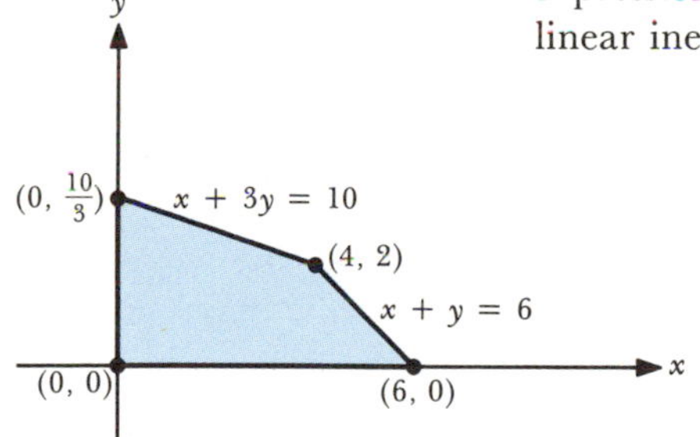

Suppose that we are given the problem of finding the maximum value of the linear expression $3x + 2y$, where x and y are subject to the conditions of the system of linear inequalities:

$$\begin{cases} x \geq 0 \\ y \geq 0 \\ x + y \leq 6 \\ x + 3y \leq 10 \end{cases}$$

These four inequalities are referred to as the **constraints** or **constraint set** of the problem.

The constraint set is graphed and the corner points are labeled as shown in Figure 5. This region represents *all* possible values for (x, y) that satisfy the system of inequalities (the constraint set).

The theory of **linear programming** establishes the fact that the **maximum** value and the **minimum** value of a linear expression occur at a corner point of the boundary of the constraint set.

Table 1 indicates the value of $3x + 2y$ for each corner point.

Table 1

Corner Point	Value of $3x + 2y$
$(0, 0)$	0
$(6, 0)$	18
$(4, 2)$	16
$(0, \frac{10}{3})$	$\frac{20}{3}$

Thus, by using linear programming, the maximum value of the expression $3x + 2y$ under the given constraints is 18; this occurs when $x = 6$ and $y = 0$. The minimum value of $3x + y$ is 0 and this occurs when $x = 0$ and $y = 0$.

The next three examples illustrate the use of linear programming techniques and their practical applications.

EXAMPLE 4 Find the maximum and minimum values of $2x + 3y$ under the constraints:

$$\begin{cases} x \geq 0 \\ y \geq 0 \\ 2y + x \leq 16 \\ x - y \leq 10 \end{cases}$$

SOLUTION The constraint set is graphed and the corner points are labeled in Figure 6. Table 2 gives the values of $2x + 3y$ at the corner points.

Figure 6

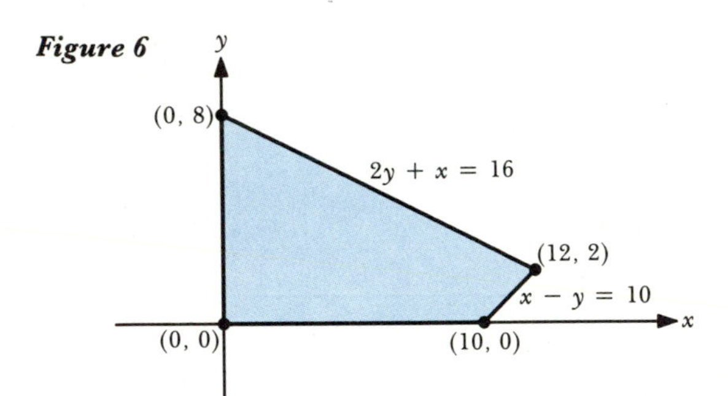

Table 2

Corner Point	Value of $2x + 3y$
(0, 0)	0
(10, 0)	20
(12, 2)	30
(0, 8)	24

Thus, $2x + 3y$ has a maximum value of 30 when $x = 12$ and $y = 2$, and a minimum value of 0 when $x = y = 0$.

EXAMPLE 5 Suppose that we are given the information in Table 3 regarding the vitamin content and cost of two kinds of pills, as well as the minimum units of each vitamin required per day.

Table 3

	White Pill	Red Pill	Minimum Units Required per Day
Vitamin A	4 units	1 unit	8
Vitamin B	1 unit	1 unit	5
Vitamin C	2 units	7 units	20
Cost per pill	7¢	14¢	

How many of each kind of pill should be taken in order to fulfill the minimum daily requirements at the least cost?

SOLUTION

If we let x represent the number of white pills and y represent the number of red pills, then we must minimize $7x + 14y$ under the following constraints:

$$\begin{cases} x \geq 0 \\ y \geq 0 \\ 4x + y \geq 8 \\ x + y \geq 5 \\ 2x + 7y \geq 20 \end{cases}$$

Figure 7 shows the graph of the constraint set, together with the corner points found by solving the system of linear equations. The corner points yield the values for the expression $7x + 14y$ given in Table 4.

Figure 7

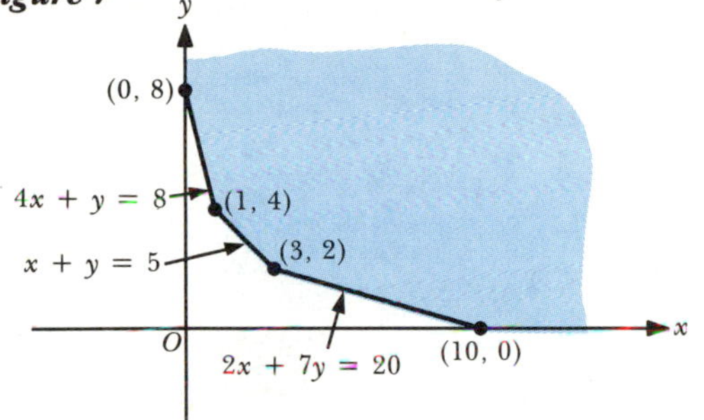

Table 4

Corner Point	Value of $7x + 14y$
(0, 8)	\$1.12
(1, 4)	\$0.63
(3, 2)	\$0.49
(10, 0)	\$0.70

Thus, three white and two red pills should be taken to fulfill the minimum daily requirement at the minimum cost.

EXAMPLE 6

A company produces automobile moldings on two different production lines. The total work force provides 500 hours of production per week and the production budget is restricted to \$3000 per week. The newer production line takes $\frac{1}{2}$ hour to produce a single part at \$5 per part, whereas the older line requires 2 hours at \$4 per part. How many moldings should be produced on each line in order to maximize productivity each week?

SOLUTION

Let x represent the number of moldings produced by the new line each week and y the number of moldings produced by the older line each week. Thus we need to maximize $x + y$ under the constraints

$$\begin{cases} x \geq 0 \\ y \geq 0 \\ \frac{1}{2}x + 2y \leq 500 \\ 5x + 4y \leq 3000 \end{cases}$$

The constraints are graphed in Figure 8 and the corner points are labeled. Table 5 gives the values of $x + y$ at the corner points.

Figure 8

Table 5

Corner Point	Value of $x + y$
$(0, 0)$	0
$(600, 0)$	600
$(500, 125)$	625
$(0, 250)$	250

Thus production is maximized at 625 moldings per week if the newer line produces 500 parts and the other line produces 125 parts per week.

PROBLEM SET 7.6

In problems 1–10, sketch the region in the plane determined by the given inequality.

1. $x \geq 2$ **2.** $y < -1$ **3.** $y \geq -\frac{1}{2}x$ **4.** $3y \leq 2x + 1$

5. $2x - 3y < 0$ **6.** $y > 2 - x$ **7.** $x - 2y \leq 1$

8. $y + 2x \geq 5$ **9.** $2x + 3y < 4$ **10.** $-3x - 4y > 12$

In problems 11–20, sketch the region defined by the given system and find the coordinates of the corner points of the boundary of the region.

11. $\begin{cases} x < 3 \\ y \geq 2 \end{cases}$ **12.** $\begin{cases} 1 \leq x < 2 \\ -1 < y \leq 2 \end{cases}$ **13.** $\begin{cases} y \geq x \\ y \leq 2 \end{cases}$ **14.** $\begin{cases} y \leq x \\ y \geq -1 \end{cases}$

15. $\begin{cases} x + y \leq 2 \\ y - 1 > 2x \end{cases}$ **16.** $\begin{cases} y - x \leq -1 \\ 3y - x > 4 \end{cases}$ **17.** $\begin{cases} x + y \leq 3 \\ x - y \geq 3 \end{cases}$ **18.** $\begin{cases} x + 2y > 5 \\ 5x - y < 9 \end{cases}$

19. $\begin{cases} 2x - 3y \leq -3 \\ 5x - 2y \geq 9 \end{cases}$ **20.** $\begin{cases} y - 2x \leq 4 \\ y + 2x \geq 3 \end{cases}$

In problems 21–30, graph the region defined by each constraint system, label each corner point, and then find the maximum and minimum values of the given linear expression over the region.

21. $3x + 5y$
$\begin{cases} x \geq 0 \\ y \geq 0 \\ 2x + y \leq 6 \end{cases}$

22. $2x - y$
$\begin{cases} 0 \leq x \leq 4 \\ 0 \leq y \leq 3 \end{cases}$

23. $2x + y$
$\begin{cases} x \geq 0 \\ y \geq 0 \\ 4x + y \leq 36 \\ 4x + 3y \leq 60 \end{cases}$

24. $15x + 25y$
$$\begin{cases} x \geq 0 \\ y \geq 0 \\ x + y \leq 50 \\ 2x - y \leq 40 \\ -3x + y \leq 10 \end{cases}$$

25. $7x - 3y$
$$\begin{cases} x \geq 0 \\ y \leq 4 \\ x + y \geq 1 \\ x - y \leq 1 \end{cases}$$

26. $4x + 2y$
$$\begin{cases} x \geq 0 \\ y \geq 0 \\ x + 3y \leq 15 \\ 2x + y \leq 10 \end{cases}$$

27. $5x + 4y$
$$\begin{cases} x \geq 0 \\ y \geq 0 \\ x + 2y \geq 3 \\ 2y \leq 5 - x \end{cases}$$

28. $x + 2y$
$$\begin{cases} y \geq 0 \\ x + y \geq 1 \\ x + 2y \leq 2 \end{cases}$$

29. $3x + 4y$
$$\begin{cases} 2x + y \geq 2 \\ x + 2y \geq 2 \\ x + y \leq 2 \end{cases}$$

30. $x + 5y$
$$\begin{cases} x \geq 0 \\ y \geq 0 \\ 5x + 4y \leq 2000 \\ 5x + 12y \leq 3000 \end{cases}$$

In problems 31–35, use linear programming to solve each problem.

31. Suppose that because of limited storage capacity, a restaurant owner can order no more than 200 pounds of ground beef per week for making hamburgers and tacos. Each hamburger contains $\frac{1}{3}$ pound of ground beef, while each taco contains $\frac{1}{4}$ pound. His profit is 15 cents on each hamburger and 20 cents on each taco. Labor cost is 8 cents for each hamburger and 12 cents for each taco. If he is willing to pay at most \$60 for labor costs, how many tacos and hamburgers should he sell to maximize his profits?

32. A refinery produces a combined maximum of 25,000 barrels of gasoline and diesel oil per day, of which no less than 5000 barrels must be diesel oil. If the profit is \$31.50 per barrel of gasoline and \$24 per barrel of diesel oil, find the maximum profit and how many barrels of each product must be made to yield this maximum.

33. A boat builder has available 100 units of wood, 160 units of plastic, and 400 units of fiberglass. Each regular boat produced requires 1 unit of wood, 2 units of plastic, and 2 units of fiberglass, and yields a profit of \$1000. By comparison, each deluxe model produced requires 1 unit of wood, 1 unit of plastic, and 5 units of fiberglass, and yields a \$1500 profit. How many boats of each type should be produced in order to maximize the profit?

34. Suppose that each serving of a particular food contains 2 units of vitamin B and 5 units of iron and that each glass of a special drink contains 4 units of vitamin B and 2 units of iron. A minimum of 80 units of vitamin B and 60 units of iron must be provided each day. How much of the food and drink must be consumed in order to meet the daily requirements and at the same time minimize costs if each serving of the food is \$1.00 and each drink is \$0.80?

35. A manufacturer makes two types of fertilizer. One type is 80% nitrogen and 20% potassium. The other type is 60% nitrogen and 40% potassium. He is obliged to make at least 30 tons of the first type and at least 50 tons of the second type. In attempting to make as much fertilizer as possible, how many tons of each type should he produce if he has 100 tons of nitrogen and 50 tons of potassium?

REVIEW PROBLEM SET, CHAPTER 7

In problems 1–6, use the graphs of the equations in each system to determine whether the system is dependent, inconsistent, or consistent. Solve each system, if possible.

1. $\begin{cases} y = -2x + 2 \\ y = x - 4 \end{cases}$

2. $\begin{cases} y = 5x + 2 \\ 10x - 2y + 4 = 0 \end{cases}$

3. $\begin{cases} 3x + 2y = 1 \\ 3x - 2y = 2 \end{cases}$

4. $\begin{cases} x + y = 4 \\ 2x - y = 8 \end{cases}$

5. $\begin{cases} \frac{1}{2}x + \frac{1}{3}y = \frac{1}{5} \\ x - \frac{1}{4}y = 7 \end{cases}$

6. $\begin{cases} 0.3x - 0.7y = 1 \\ 2.3x + y = -2 \end{cases}$

In problems 7–10, use the substitution method to solve each linear system.

7. $\begin{cases} x - 2y = 3 \\ x + y = -3 \end{cases}$

8. $\begin{cases} 8r - 7s = 28 \\ 5r + 2s = -8 \end{cases}$

9. $\begin{cases} x + y = 1 \\ 3x - y + z = 16 \\ 2y + 3z = 11 \end{cases}$

10. $\begin{cases} x + y + z = 1 \\ x - y + z = 1 \\ x - y - z = 1 \end{cases}$

In problems 11–14, use the elimination method to solve each linear system.

11. $\begin{cases} x - y = 3 \\ 2x + y = 3 \end{cases}$

12. $\begin{cases} 5x + 2y = 3 \\ 2x - 3y = 5 \end{cases}$

13. $\begin{cases} x - y + 2z = 0 \\ 3x + y + z = 2 \\ 2x - y + 5z = 5 \end{cases}$

14. $\begin{cases} 3x + 2y - z = -4 \\ x - y + 2z = 13 \\ 5x + 3y - 4z = -15 \end{cases}$

In problems 15–18, use row reduction to solve each linear system.

15. $\begin{cases} 4x - y = -4 \\ x + 2y = 6 \end{cases}$

16. $\begin{cases} 2x + 3y = 9 \\ 5x - 2y = 5 \end{cases}$

17. $\begin{cases} 3x - y + 2z = 1 \\ x - 2y + 4z = 2 \\ x - y + z = 0 \end{cases}$

18. $\begin{cases} x + 2y + 3z = 4 \\ -x - 4y + z = 1 \\ x + y + z = 0 \end{cases}$

In problems 19–22, each row-reduced echelon matrix is the augmented matrix form of a corresponding system of linear equations. Solve each system.

19. $\left[\begin{array}{cc|c} 1 & 0 & -9 \\ 0 & 1 & 5 \end{array}\right]$

20. $\left[\begin{array}{ccc|c} 1 & 0 & 0 & -2 \\ 0 & 1 & 3 & 7 \\ 0 & 0 & 0 & 0 \end{array}\right]$

21. $\left[\begin{array}{ccc|c} 1 & 0 & 0 & -7 \\ 0 & 1 & 0 & 4 \\ 0 & 0 & 1 & 3 \end{array}\right]$

22. $\left[\begin{array}{cc|c} 1 & 1 & 2 \\ 0 & 0 & 0 \end{array}\right]$

In problems 23–30, perform the indicated matrix operations, if possible.

23. (a) $A + 2B$ and (b) $-3A + 4B$ if $A = \begin{bmatrix} 2 & 1 \\ 1 & -1 \end{bmatrix}$ and $B = \begin{bmatrix} 5 & -2 \\ 3 & 4 \end{bmatrix}$

24. (a) $3A - 2B$ and (b) $4A + 2B$ if $A = \begin{bmatrix} 1 & -4 & -8 \\ 5 & 20 & 3 \end{bmatrix}$ and $B = \begin{bmatrix} 7 & -2 & 7 \\ 3 & 5 & 3 \end{bmatrix}$

25. (a) $3A - B$ and (b) $-5A + 2B$ if $A = \begin{bmatrix} 3 & 4 \\ 2 & -1 \\ 4 & 3 \end{bmatrix}$ and $B = \begin{bmatrix} 2 & -3 \\ 4 & 1 \\ 5 & 7 \end{bmatrix}$

26. (a) $7A - \frac{1}{2}B$ and (b) $\frac{3}{2}A + B$ if $A = \begin{bmatrix} 1 & 5 \\ 2 & -1 \\ -6 & 3 \end{bmatrix}$ and $B = \begin{bmatrix} 6 & 1 \\ -1 & 7 \\ 4 & 2 \end{bmatrix}$

27. (a) AB and (b) BA if $A = \begin{bmatrix} 3 & -2 \\ 1 & 3 \end{bmatrix}$ and $B = \begin{bmatrix} 2 & 3 \\ 5 & 2 \end{bmatrix}$

28. (a) AB and (b) BA if $A = \begin{bmatrix} 3 & -1 & 7 \\ 4 & 1 & 2 \end{bmatrix}$ and $B = \begin{bmatrix} 2 & 6 \\ -3 & 2 \\ 5 & -9 \end{bmatrix}$

29. (a) AB and (b) BA if $A = \begin{bmatrix} 1 & -2 & 1 \\ 3 & 1 & -2 \\ 0 & 1 & -1 \end{bmatrix}$ and $B = \begin{bmatrix} 4 & -1 \\ 3 & 1 \\ 2 & 8 \end{bmatrix}$

30. $(A - 2B)B$ if $A = \begin{bmatrix} 4 & -1 \\ -3 & 2 \end{bmatrix}$ and $B = \begin{bmatrix} 4 & 3 \\ -1 & 1 \end{bmatrix}$

In problems 31–34, verify that $AA^{-1} = A^{-1}A = I$.

31. $A = \begin{bmatrix} 3 & -2 \\ 8 & -5 \end{bmatrix}$ and $A^{-1} = \begin{bmatrix} -5 & 2 \\ -8 & 3 \end{bmatrix}$

32. $A = \begin{bmatrix} 4 & 5 \\ -3 & -1 \end{bmatrix}$ and $A^{-1} = \begin{bmatrix} -\frac{1}{11} & -\frac{5}{11} \\ \frac{3}{11} & \frac{4}{11} \end{bmatrix}$

33. $A = \begin{bmatrix} 1 & -1 & 1 \\ 0 & 2 & -1 \\ 2 & 3 & 0 \end{bmatrix}$ and $A^{-1} = \begin{bmatrix} 3 & 3 & -1 \\ -2 & -2 & 1 \\ -4 & -5 & 2 \end{bmatrix}$

34. $A = \begin{bmatrix} 2 & 3 & 1 \\ 4 & 6 & 0 \\ -1 & 1 & 4 \end{bmatrix}$ and $A^{-1} = \begin{bmatrix} \frac{24}{10} & -\frac{11}{10} & -\frac{6}{10} \\ -\frac{16}{10} & \frac{9}{10} & \frac{4}{10} \\ \frac{10}{10} & -\frac{5}{10} & 0 \end{bmatrix}$

In problems 35–38, use the inverse of the matrix of coefficients to solve each system. (Refer to the inverses given in problems 31–34.)

35. $\begin{cases} 3x - 2y = -7 \\ 8x - 5y = -18 \end{cases}$ (see problem 31)

36. $\begin{cases} 4x + 5y = 2 \\ -3x - y = -7 \end{cases}$ (see problem 32)

37. $\begin{cases} x - y + z = 7 \\ 2y - z = -7 \\ 2x + 3y = -1 \end{cases}$ (see problem 33)

38. $\begin{cases} 2x + 3y + z = 3 \\ 4x + 6y = 16 \\ -x + y + 4z = 2 \end{cases}$ (see problem 34)

In problems 39–44, evaluate each determinant. In problems 41 and 44, round off each answer to three decimal places.

39. $\begin{vmatrix} 1 & -1 \\ 3 & 2 \end{vmatrix}$

40. $\begin{vmatrix} \sin 25^\circ & \cos 25^\circ \\ -\cos 25^\circ & \sin 25^\circ \end{vmatrix}$

c **41.** $\begin{vmatrix} 3.041 & 14.037 \\ -2.815 & 6.385 \end{vmatrix}$

42. $\begin{vmatrix} 4 & 2 & 1 \\ 5 & 7 & 1 \\ 6 & 2 & 3 \end{vmatrix}$

43. $\begin{vmatrix} 1 & 2 & 1 \\ 1 & 3 & 4 \\ 1 & 4 & 9 \end{vmatrix}$

c **44.** $\begin{vmatrix} 2.41 & 7.31 & -1.07 \\ -1.42 & 3.04 & -7.51 \\ 2.57 & 4.13 & 3.02 \end{vmatrix}$

In problems 45 and 46, solve for x.

45. $\begin{vmatrix} x & 0 & 0 \\ 3 & x & 2 \\ 0 & 4 & 1 \end{vmatrix} = 5$

46. $\begin{vmatrix} -2 & 1 & x \\ 1 & x+1 & -2 \\ x-6 & 3 & -1 \end{vmatrix} = 16$

In problems 47–52, use the properties of determinants to evaluate each determinant.

47. $\begin{vmatrix} -1 & 5 \\ -2 & 10 \end{vmatrix}$

48. $\begin{vmatrix} 2 & 7 & 4 \\ 3 & -1 & 5 \\ 4 & 14 & 8 \end{vmatrix}$

49. $\begin{vmatrix} 2 & -4 & 7 \\ 0 & 1 & 2 \\ 0 & 0 & 5 \end{vmatrix}$

50. $\begin{vmatrix} 3 & 2 & 4 \\ 6 & 7 & -1 \\ 3 & 2 & 4 \end{vmatrix}$

51. $\begin{vmatrix} 0 & 1 & 4 \\ 3 & -6 & 1 \\ -9 & 18 & -3 \end{vmatrix}$

52. $\begin{vmatrix} 2 & 1 & 1 \\ 4 & 2 & 2 \\ 1 & 3 & 0 \end{vmatrix}$

In problems 53–58, use Cramer's rule to solve each system. In problem 55, round off the answers to three decimal places.

53. $\begin{cases} 3x - y = 7 \\ 2x + 3y = 12 \end{cases}$

54. $\begin{cases} 5x + 3y = 13 \\ 7x - 5y = 18 \end{cases}$

c **55.** $\begin{cases} 3.742x + 1.573y = 8.621 \\ 1.602x + 2.237y = 7.431 \end{cases}$

56. $\begin{cases} 2x - 3y = 2a \\ 5x + y = 3b \end{cases}$

57. $\begin{cases} 3x - y + 2z = 5 \\ 2x + 3y + z = 1 \\ 5x + y + 4z = 8 \end{cases}$

58. $\begin{cases} x_1 + x_2 - x_3 = -2 \\ 2x_1 + x_3 = 7 \\ x_1 + x_2 + 3x_3 = 10 \end{cases}$

In problems 59–64, decompose each fraction into partial fractions.

59. $\dfrac{x+2}{x^2 - 6x - 7}$

60. $\dfrac{3x+1}{x^2 + x}$

61. $\dfrac{1}{x(x-1)^2}$

62. $\dfrac{1}{x^4 + x^2}$

63. $\dfrac{-7x^3 + 19x^2 - 14x + 23}{x^4 + 5x^2 + 6}$

64. $\dfrac{x}{x^4 - 1}$

In problems 65–72, solve each system of equations.

65. $\begin{cases} 3x - 4y = 25 \\ x^2 + y^2 = 25 \end{cases}$

66. $\begin{cases} 2x - y = 2 \\ x^2 + 2y^2 = 12 \end{cases}$

67. $\begin{cases} x + y^2 = 6 \\ x^2 + y^2 = 36 \end{cases}$

68. $\begin{cases} 3x^2 - 2y^2 = 35 \\ 7x^2 + 5y^2 = 43 \end{cases}$

69. $\begin{cases} p^2 + 2q^2 = 22 \\ 2p^2 + q^2 = 17 \end{cases}$

70. $\begin{cases} 5r^2 + s^2 = 23 \\ r^2 - s^2 = 1 \end{cases}$

71. $\begin{cases} \log_6 x + \log_6 y = 1 \\ x^2 + y^2 = 13 \end{cases}$

72. $\begin{cases} \log y - \log x = \log 3 \\ x^2 + y^2 = 40 \end{cases}$

In problems 73–78, graph the region defined by each constraint system, locate each corner point, and then find the maximum and minimum values of the given linear expression over the region specified.

73. $7x + 3y$

$\begin{cases} x \geq 0 \\ y \geq 0 \\ 7x + 2y \leq 14 \end{cases}$

74. $2x + y$

$\begin{cases} x \geq 1 \\ y \geq 2 \\ x + 2y \leq 10 \end{cases}$

75. $x + 5y$

$\begin{cases} x \geq 0 \\ y \geq 0 \\ x + y \leq 4 \\ 4x + y \leq 7 \end{cases}$

76. $x + y$

$$\begin{cases} x \geq 0 \\ y \geq 0 \\ 3y - 2x \leq 6 \\ 3y + 4x \leq 24 \end{cases}$$

77. $10x + 33y$

$$\begin{cases} x \geq 0 \\ y \geq 0 \\ x + 3y \leq 90 \\ 2x + y \leq 80 \end{cases}$$

78. $x + 2y$

$$\begin{cases} x \geq 0 \\ y \geq 0 \\ 2x + 5y \leq 20 \\ x + y \geq 6 \end{cases}$$

79. A coin collection containing nickels, dimes, and quarters consists of 35 coins altogether. If there are twice as many nickels as quarters, and one-fourth as many dimes as nickels, find the number of each type of coin in the collection.

80. The specific gravity of an object is defined to be its weight in air divided by its loss of weight when submerged in water. An object made partly of gold (specific gravity 19.3) and partly of silver (specific gravity 10.5) weighs 8 grams in air and 7.3 grams when submerged in water. How many grams of gold and how many grams of silver are in the object?

c **81.** An income tax service developed the following relationships among taxes for a corporation:

$$\begin{cases} x + 0.095y = 2{,}584.42 \\ 0.01x + 1.01y + 0.01z = 1{,}529.70 \\ 0.48x + 0.48y + z = 12{,}253.24 \end{cases}$$

where x (in dollars) denotes the state income tax owed, y (in dollars) denotes the city income tax owed, and z (in dollars) denotes the federal income tax owed. Solve the system to determine the taxes owed to the nearest dollar.

82. Suppose that \$10,000 is invested in two funds paying 13% and 14% annual simple interest rate. If the annual return from both investments is \$1400, what amount was invested at each rate?

83. The sum of the squares of two numbers is 113. When 5 times the square of one is added to the square of the other, the sum is 309. What are the numbers?

84. The perimeter of a rectangular garden is 46 meters and the area of the garden is 60 square meters. Find the length and the width of the garden.

CHAPTER 8

Analytic Geometry and the Conics

Early Greek mathematicians are credited with developing the geometric properties of *circles, ellipses, parabolas,* and *hyperbolas.* These *plane curves* are called **conics** because they are formed by the intersections of planes and right circular cones with two nappes (Figure 1).

Figure 1

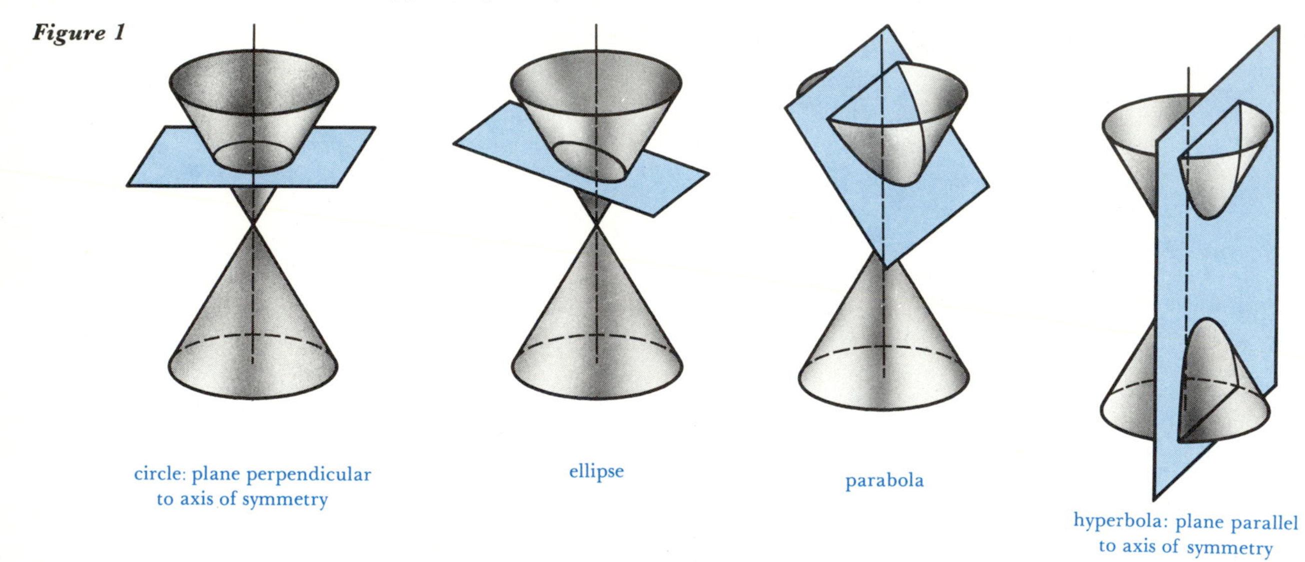

Conics are used to study the paths of orbiting planets and satellites, and the trajectories of projectiles. They are important tools in optics and atomic physics. In this chapter, we use the techniques of *analytic geometry* in the plane to develop standard equation forms for ellipses, hyperbolas, and parabolas in both the Cartesian and polar systems. Also, we consider equations of the conics in conjunction with translations and rotations of the coordinate axes.

8.1 Circles and Ellipses

In Section 2.1, we saw how the geometric definition of a circle led to the **standard form** for an equation of a circle with center at (h, k) and radius r:

$$(x - h)^2 + (y - k)^2 = r^2$$

For example, the circle with center $(-1, 3)$ and radius 7 consists of all points whose coordinates satisfy the equation

$$(x + 1)^2 + (y - 3)^2 = 49$$

EXAMPLE 1 Find an equation for the circle that satisfies the given conditions.

(a) Center at $(4, -1)$ and tangent to the line $y = 2$

(b) Center at $(-1, -3)$ and tangent to the line $3x + 4y = 10$

SOLUTION (a) Figure 2 shows the graph of the circle that has center $(4, -1)$ and is tangent to the line $y = 2$. Clearly, $(4, 2)$ is the point of tangency. The distance between the points $(4, 2)$ and $(4, -1)$ is 3, which is the radius of the circle. Thus, a standard form of the equation of this circle is

$$(x - 4)^2 + (y + 1)^2 = 9$$

(b) We begin by sketching the graph of the circle that is centered at $(-1, -3)$ and tangent to the line $3x + 4y = 10$; also, we label the point of tangency with the coordinates (a, b) (Figure 3). Since (a, b) lies on the line, the coordinates must satisfy the equation

$$3a + 4b = 10$$

In addition, the line containing the center of the circle $(-1, -3)$ and the point (a, b) has slope $(b + 3)/(a + 1)$ (see Definition 1, page 75). We also know that this line is perpendicular to the line $3x + 4y = 10$, which has slope $-\frac{3}{4}$. Thus,

$$\frac{b + 3}{a + 1} = \frac{4}{3} \qquad \text{or} \qquad 4a - 3b = 5$$

since the slope of a line perpendicular to another line is equal to the negative of the reciprocal of the slope of the other line. Hence, the coordinates a and b must satisfy the linear system

$$\begin{cases} 3a + 4b = 10 \\ 4a - 3b = 5 \end{cases}$$

Solving this system results in $(a, b) = (2, 1)$. Now we can use the distance formula

Figure 2

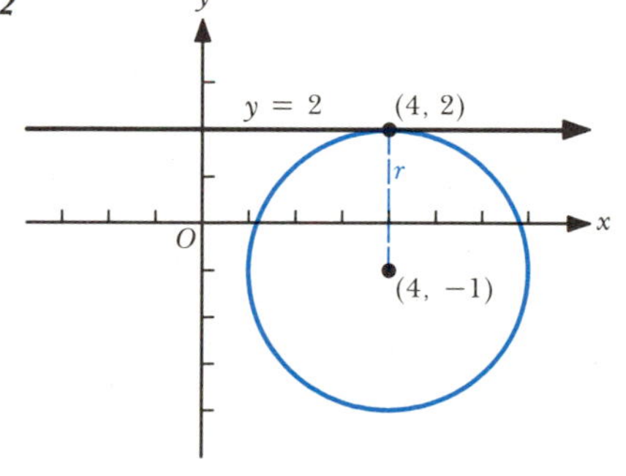

Figure 3

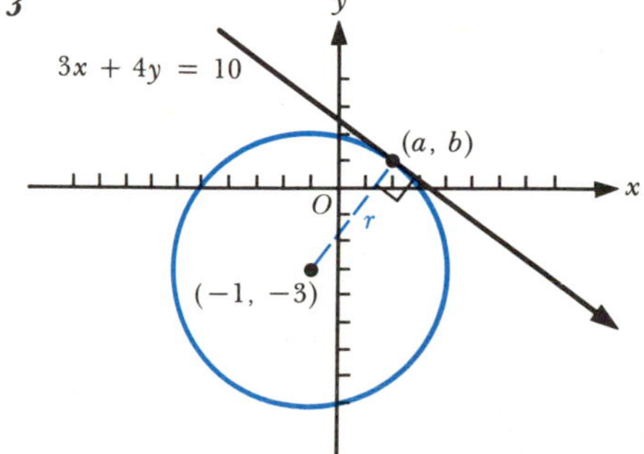

to determine the radius r, which is the distance between the center $(h,k) = (-1,-3)$ and the point $(a,b) = (2,1)$, that is,

$$r = \sqrt{(-1-2)^2 + (-3-1)^2} = \sqrt{9+16} = \sqrt{25} = 5$$

Hence, an equation of the desired circle is given by

$$(x+1)^2 + (y+3)^2 = 25$$

An alternate form of an equation of a circle is obtained by multiplying out the squared terms and simplifying the equation

$$(x-h)^2 + (y-k)^2 = r^2$$

to get an equation of the form

$$x^2 + y^2 + Ax + By + C = 0$$

where A, B, and C are constants. This latter equation is referred to as the **general form** of an equation of a circle.

EXAMPLE 2 Find both the general form and the standard form equations for the circle that contains the three points $(2,6)$, $(3,-1)$, and $(-5,5)$. Also find the center and radius of the circle and sketch the graph.

SOLUTION Since each of the three points lies on the circle, it follows that the coordinates of each point must satisfy the general equation of the circle

$$x^2 + y^2 + Ax + By + C = 0$$

After substituting the coordinates of each point into this equation, we obtain the linear system

$$\begin{cases} 2^2 + 6^2 + 2A + 6B + C = 0 \\ 3^2 + (-1)^2 + 3A - B + C = 0 \\ (-5)^2 + 5^2 - 5A + 5B + C = 0 \end{cases} \quad \text{or} \quad \begin{cases} 2A + 6B + C = -40 \\ 3A - B + C = -10 \\ -5A + 5B + C = -50 \end{cases}$$

Solving this system of linear equations, we obtain $A = 2$, $B = -4$, and $C = -20$. Thus, the general equation of the circle is

$$x^2 + y^2 + 2x - 4y - 20 = 0$$

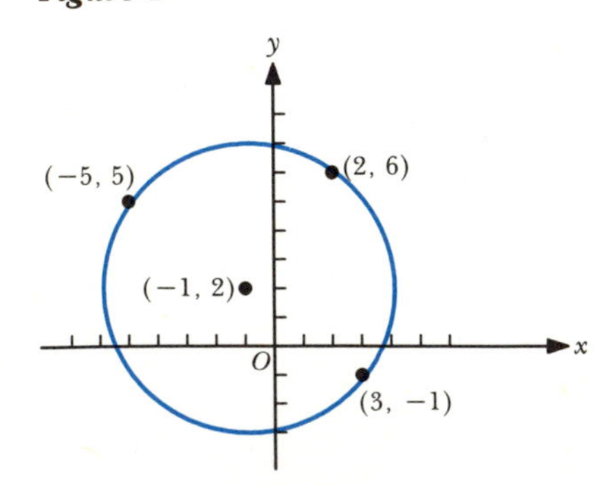

By completing the squares, we have

$$(x^2 + 2x + 1) + (y^2 - 4y + 4) = 20 + 1 + 4$$

or

$$(x+1)^2 + (y-2)^2 = 25$$

which is the standard equation of the circle. Thus, the desired circle has radius 5 units and center $(-1,2)$ (Figure 4).

Most of the nine planets in our solar system revolve around the sun in a "nearly circular" path, which is described as an *ellipse*. Geometrically, an ellipse is defined as follows.

DEFINITION 1

The Ellipse

> An **ellipse** is the set of all points P in a plane such that the sum of the distances from P to two fixed points is a positive constant. The fixed points are called the **foci** (plural for **focus**) of the ellipse.

In Figure 5a, F_1 and F_2 are the foci; P_1, P_2, and P_3 are points on the ellipse, and

$$d_1 + c_1 = d_2 + c_2 = d_3 + c_3 = k$$

where k is the constant sum referred to in the definition of the ellipse. In order to derive a standard equation form for an ellipse, we choose the x axis as the line containing the foci $F_1 = (-c, 0)$ and $F_2 = (c, 0)$, where $c > 0$. The origin is the midpoint of the line segment $\overline{F_1F_2}$. The midpoint of the segment $\overline{F_1F_2}$ is called the **center of the ellipse.**

Figure 5

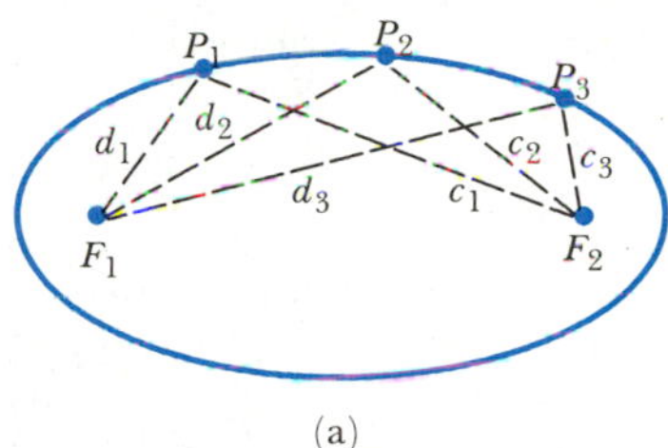

(a)

y

$|\overline{PF_1}| = \sqrt{(x + c)^2 + y^2}$

$|\overline{PF_2}| = \sqrt{(x - c)^2 + y^2}$

$P = (x, y)$

O

x

$F_1 = (-c, 0)$

$F_2 = (c, 0)$

(b)

Also, we assume that the constant sum of the distances between any point P on the ellipse and the foci is $2a$. (This constant is written in the form $2a$ so that the equation of the ellipse will have a simple form.) If we let $P = (x, y)$ be a point on the ellipse (Figure 5b), then in triangle F_1PF_2, the sum of the lengths of the two sides of the triangle $|\overline{PF_1}| + |\overline{PF_2}| = 2a$ is *greater than* the length of the third side $|\overline{F_1F_2}| = 2c$. Thus

$$2a > 2c \qquad \text{or} \qquad a > c$$

since

$$|\overline{PF_1}| + |\overline{PF_2}| = 2a$$

we can use the distance formula to obtain

$$\sqrt{(x + c)^2 + y^2} + \sqrt{(x - c)^2 + y^2} = 2a$$

or

$$\sqrt{(x + c)^2 + y^2} = 2a - \sqrt{(x - c)^2 + y^2}$$

Squaring both sides of this equation, we get

$$x^2 + 2xc + c^2 + y^2 = 4a^2 - 4a\sqrt{(x - c)^2 + y^2} + x^2 - 2cx + c^2 + y^2$$

so that

$$4cx - 4a^2 = -4a\sqrt{(x - c)^2 + y^2} \qquad \text{or} \qquad cx - a^2 = -a\sqrt{(x - c)^2 + y^2}$$

After squaring both sides of the latter equation, we obtain

$$c^2x^2 - 2a^2cx + a^4 = a^2(x^2 - 2cx + c^2 + y^2)$$

so that

$$a^4 - a^2c^2 = (a^2 - c^2)x^2 + a^2y^2$$

and so

$$(a^2 - c^2)x^2 + a^2y^2 = a^2(a^2 - c^2)$$

Since $a > c > 0$, $\sqrt{a^2 - c^2}$ is always a *positive* real number and we denote it by b. After substituting $b^2 = a^2 - c^2$ into the latter equation, we obtain

$$b^2x^2 + a^2y^2 = a^2b^2$$

Finally, we divide both sides of this equation by a^2b^2 to get

$$\boxed{\frac{x^2}{a^2} + \frac{y^2}{b^2} = 1}$$

Note that since $b^2 = a^2 - c^2$, it follows that

$$a^2 = b^2 + c^2$$

so that

$$a^2 > b^2 \qquad \text{and} \qquad a > b$$

because both a and b are positive numbers.

The boxed equation is called the **standard form** for an equation of an ellipse with *center* at the origin and foci on the x axis. Every point (x, y) on the ellipse in Figure 5b has coordinates that satisfy the last equation; and, conversely, any point with coordinates that satisfy the equation is on the ellipse.

From the equation, we can derive the following characteristics of the graph of an ellipse. (Refer to Figure 6.)

1. After substituting $y = 0$ into the equation, we find that the x intercepts of the ellipse are $-a$ and a. The y intercepts are obtained by letting $x = 0$ to get $-b$ and b. The corresponding points $V_1 = (-a, 0)$, $V_2 = (a, 0)$, $V_3 = (0, -b)$, and $V_4 = (0, b)$ on the graph are called the **vertices** of the ellipse.

2. The line segment $\overline{V_1V_2}$ is referred to as the **major axis,** and it has length $2a$; the segment $\overline{V_3V_4}$ is called the **minor axis** of the ellipse, and it has length $2b$. The major axis is longer than the minor axis, since $a > b$.

3. The graph is symmetric with respect to both the x axis (or major axis) and the y axis (or minor axis). Also, $c^2 = a^2 - b^2$, and the foci lie on the major axis c units from the center at points $(-c, 0)$ and $(c, 0)$.

Assume that the major axis of an ellipse is located on the y axis, the center of the ellipse is at the origin, the foci F_1 and F_2 are at $(0, c)$ and $(0, -c)$, and the constant sum is $2a$. Then an argument similar to the one used in the derivation of $(x^2/a^2) + (y^2/b^2) = 1$ shows that the *standard equation* of such an ellipse is

$$\boxed{\frac{x^2}{b^2} + \frac{y^2}{a^2} = 1}$$

where $c^2 = a^2 - b^2$ and $a > b$ (Figure 7).

Figure 6

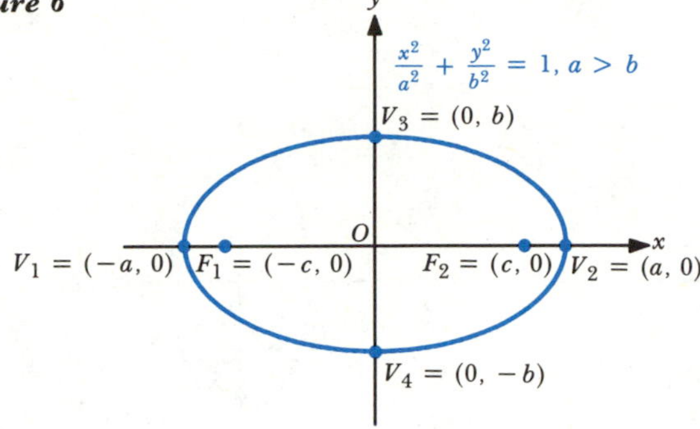

Figure 7

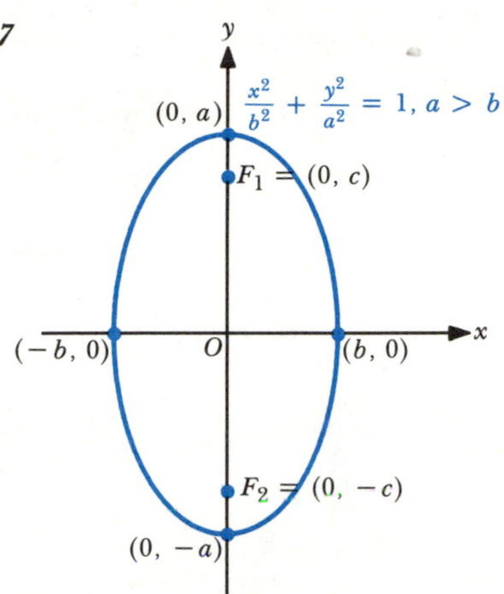

Table 1 summarizes the characteristics and standard form equations of the ellipses that we have discussed so far. We assume that $a > b$ and $c^2 = a^2 - b^2$.

Table 1 **Ellipses**

Major Axis	Center	Foci	Vertices: On Major Axis	Vertices: On Minor Axis	Standard Form Equation
Horizontal (on the x axis)	$(0,0)$	$(-c,0)$, $(c,0)$	$(-a,0)$, $(a,0)$	$(0,-b)$, $(0,b)$	$\frac{x^2}{a^2} + \frac{y^2}{b^2} = 1$
Vertical (on the y axis)	$(0,0)$	$(0,-c)$, $(0,c)$	$(0,-a)$, $(0,a)$	$(-b,0)$, $(b,0)$	$\frac{x^2}{b^2} + \frac{y^2}{a^2} = 1$

EXAMPLE 3 Show that each of the following equations represents an ellipse. Then find the vertices and the foci, and sketch the graph.

(a) $x^2 + 9y^2 = 9$

(b) $4x^2 + y^2 = 4$

SOLUTION (a) First we divide both sides of the equation by 9 to obtain

$$\frac{x^2}{9} + \frac{y^2}{1} = 1$$

which is the standard equation of an ellipse with

$$a^2 = 9, \qquad b^2 = 1, \qquad \text{and} \qquad c^2 = a^2 - b^2 = 9 - 1 = 8$$

From the equation, we see that the center is $(0,0)$, the major axis lies on the x axis, and the vertices are $(0,1)$, $(0,-1)$, $(3,0)$, and $(-3,0)$. Since $c^2 = 8$, it follows that $c = 2\sqrt{2}$, so that the foci are $(2\sqrt{2},0)$ and $(-2\sqrt{2},0)$ (Figure 8a).

Figure 8

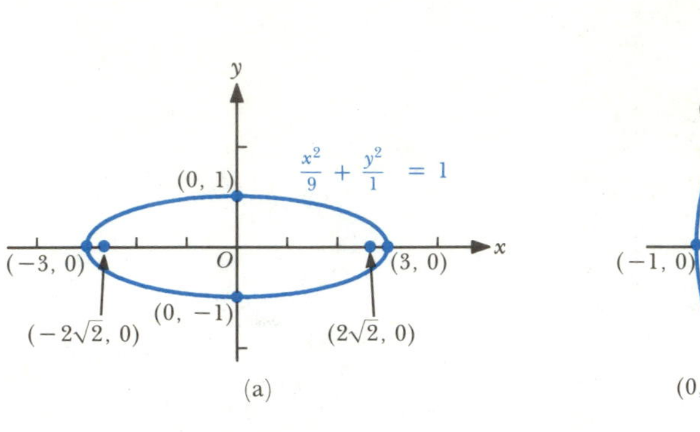

(b) After dividing both sides of the equation by 4, we obtain the equation

$$\frac{x^2}{1} + \frac{y^2}{4} = 1$$

which is the standard equation of an ellipse with

$$b^2 = 1, \qquad a^2 = 4, \qquad \text{and} \qquad c^2 = a^2 - b^2 = 4 - 1 = 3$$

The graph is an ellipse centered at the origin with major axis on the y axis and vertices $(0, 2)$, $(0, -2)$, $(1, 0)$, and $(-1, 0)$. The foci are $(0, -\sqrt{3})$ and $(0, \sqrt{3})$ (Figure 8b).

EXAMPLE 4 Find the standard equation of an ellipse with foci $(2, 0)$ and $(-2, 0)$, and vertices $(3, 0)$ and $(-3, 0)$. Also, sketch the graph.

SOLUTION Since the center of the ellipse is at the origin and its foci lie on the x axis, its equation has the standard form

$$\frac{x^2}{a^2} + \frac{y^2}{b^2} = 1$$

Figure 9

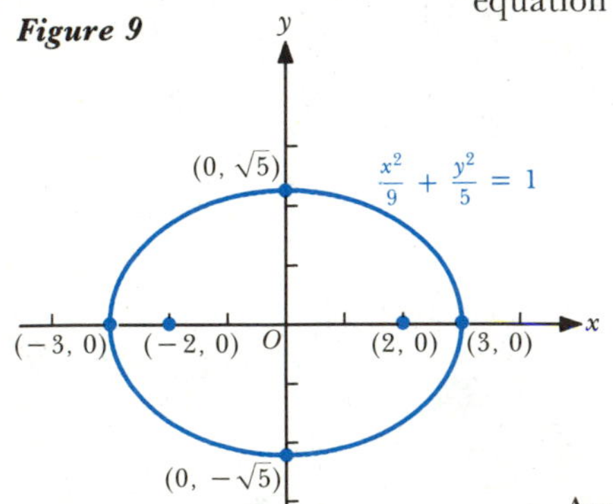

Here, $a = 3$ and $c = 2$ so that

$$b^2 = a^2 - c^2 = 9 - 4 = 5$$

and the equation is

$$\frac{x^2}{9} + \frac{y^2}{5} = 1 \quad \text{(Figure 9)}$$

Analogous to the situation for circles, if an ellipse is located on a coordinate system so that its *center is at the point* (h, k), and if its two axes of symmetry are parallel to the coordinate axes, then the equation of the ellipse can be expressed in one of the following standard forms. Here again

$$a > b \qquad \text{and} \qquad c^2 = a^2 - b^2$$

Standard Forms of Ellipses Centered at (h, k)

(i) $\dfrac{(x-h)^2}{a^2} + \dfrac{(y-k)^2}{b^2} = 1$, if the major axis is horizontal (Figure 10a)

(ii) $\dfrac{(x-h)^2}{b^2} + \dfrac{(y-k)^2}{a^2} = 1$, if the major axis is vertical (Figure 10b)

Figure 10

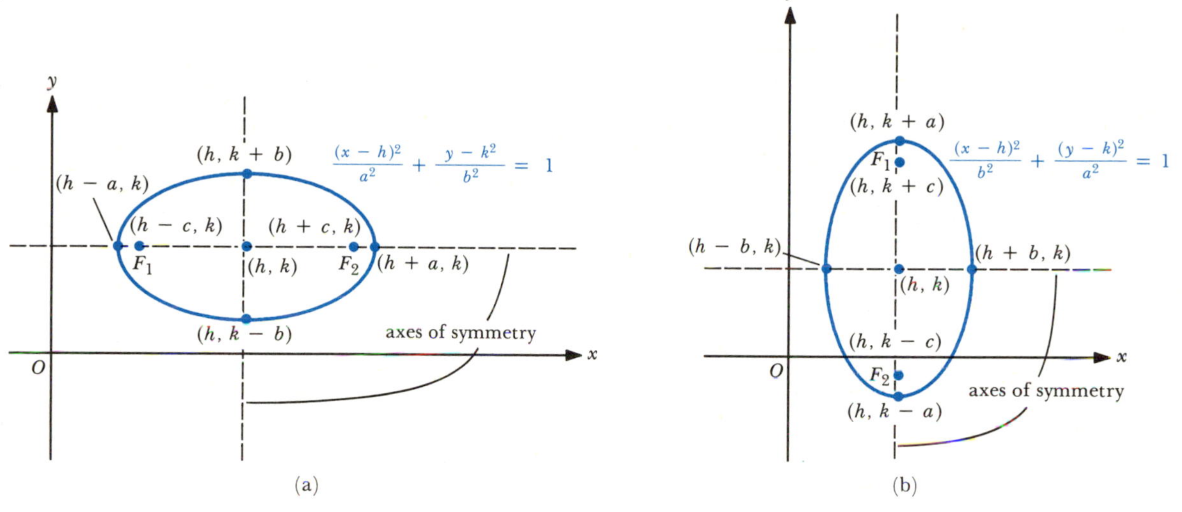

EXAMPLE 5 Given that $9x^2 + 4y^2 - 18x + 16y - 11 = 0$ is an equation of an ellipse, find the coordinates of the center, vertices, and foci. Also, sketch the graph.

SOLUTION First, we associate the x and y terms, and then we complete the square in both x and y. Since the coefficients of the squared terms must be 1, we obtain

$$9(x^2 - 2x + 1) + 4(y^2 + 4y + 4) = 11 + 9 + 16$$

Thus

$$9(x-1)^2 + 4(y+2)^2 = 36$$

That is,

$$\frac{(x-1)^2}{4} + \frac{(y+2)^2}{9} = 1$$

so that the center is $(1, -2)$.

Here, $a^2 = 9$ and $b^2 = 4$, so that $a = 3$ and $b = 2$, and the major axis is parallel to the y axis. The coordinates of the vertices are given by

$(1, -2 + 3) = (1, 1)$, $\quad (1, -2 - 3) = (1, -5)$, $\quad (1 - 2, -2) = (-1, -2)$,

and $\quad (1 + 2, -2) = (3, -2)$

From the equation $c^2 = a^2 - b^2 = 9 - 4 = 5$, we have $c = \sqrt{5}$ and the coordinates of the foci are

$$F_1 = (1, -2 + \sqrt{5}) \quad \text{and} \quad F_2 = (1, -2 - \sqrt{5}) \quad \text{(Figure 11)}$$

Figure 11

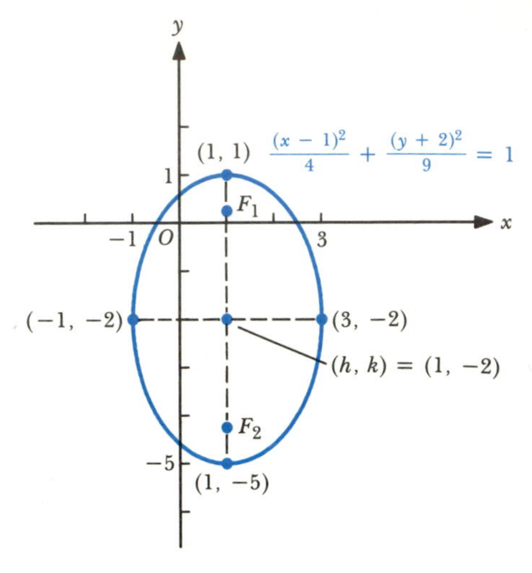

EXAMPLE 6 Write an equation of the ellipse whose vertices are $(-7, 1)$, $(3, 1)$, $(-2, -3)$, and $(-2, 5)$, and sketch its graph.

SOLUTION First, we locate the given vertices and then sketch the graph (Figure 12). These four points are the ends of the major axis and the minor axis. The major axis is parallel to the x axis and is 10 units long, which is the distance between $(-7, 1)$ and $(3, 1)$. The minor axis is parallel to the y axis and is 8 units long, the distance between $(-2, 5)$ and $(-2, -3)$. Thus $a = 5$ and $b = 4$. In this case, the center is at $(-2, 1)$, which is the midpoint of the segment with endpoints $(-7, 1)$ and $(3, 1)$. Since the major axis is horizontal, we obtain the standard equation

$$\frac{(x + 2)^2}{25} + \frac{(y - 1)^2}{16} = 1$$

Figure 12

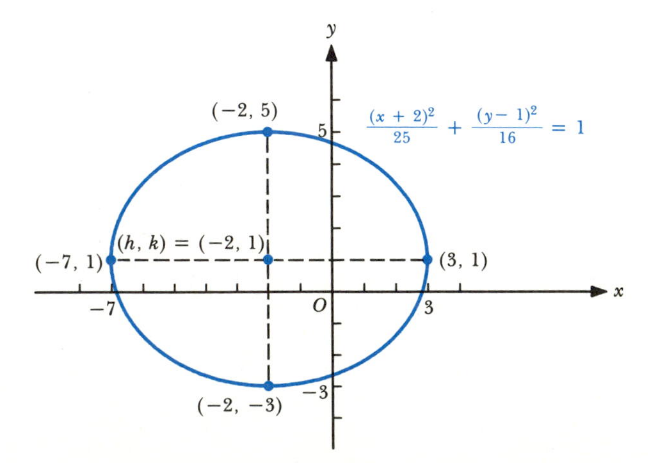

Ellipses have many applications. Some of them are shown in these photographs.

(a)

(a) *Ellipses are often used in architecture and in the construction of gardens. The* ***Ellipse*** *in Washington, D.C., is a well-known park located between the White House and the Washington Monument. It is bounded by an elliptical path.*

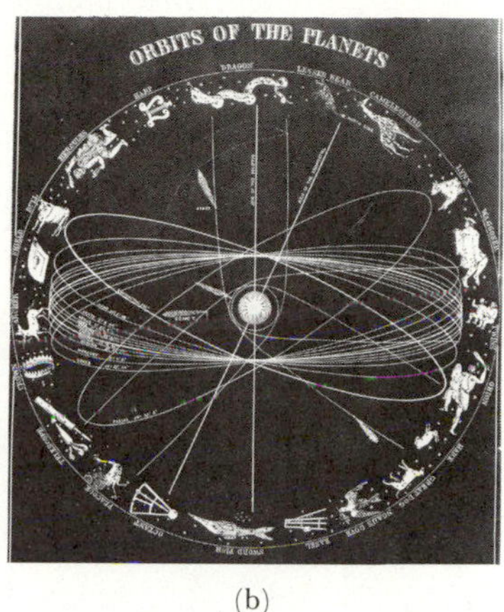

(b)

(b) *In this pre-Kepler depiction of the solar system, the orbits are circular with the sun at the center. Because we are viewing them at an angle, however, the circular orbits are seen as ellipses.*

(c)

(c) *The dome of a* ***whispering gallery*** *has an elliptical inner surface. Sound emanating from one focus is reflected from any point within the dome to the other focus, where it will be heard clearly. Two famous whispering galleries are the National Statuary Hall of the Capitol in Washington, D.C., and the dome in St. Paul's Cathedral in London.*

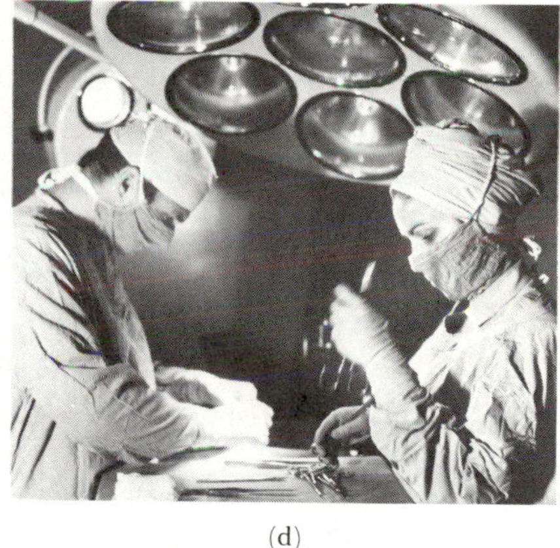

(d)

(d) *Ellipses are used in the design of dental and surgical lights. The reflecting surface has a shape that is elliptical, with the light source at one of the foci.*

PROBLEM SET 8.1

In problems 1–13, find the standard equation of the circle (or circles) that satisfies the given conditions.

1. Center at $(2, -5)$ and tangent to the x axis

2. Center at $(-3, 10)$ and tangent to the line $x = 5$

3. $(3, 7)$ and $(-3, -1)$ are the endpoints of a diameter

4. Contains points $(3, 7)$ and $(5, 5)$ and center on the line $x - 4y = 1$

5. Radius of $\sqrt{10}$ units and tangent to the line $3x + y = 6$ at $(3, -3)$ (two circles)

6. Radius of 2 units, contains the point $(3, 4)$, and tangent to the circle whose equation is $x^2 + y^2 = 25$

7. Center 1 unit from the y axis, 2 units from the x axis, and tangent to the y axis (four circles)

8. Radius of 5 units, contains the point $(3, -2)$, and has center on the line $2x - y + 2 = 0$

9. Contains the origin, center on the line $y = x$, and tangent to the line $x + y = 1$
10. Tangent to the x axis, tangent to the line $y = 3$, and tangent to the line $x = 5$
11. Contains the three points $(3, 0)$, $(-1, 2\sqrt{2})$, and $(1, -2\sqrt{2})$
12. Contains the three points $(8, -2)$, $(3, -7)$, and $(6, 2)$
13. Contains the three points $(7, 1)$, $(-7, 5)$, and $(-3, 1)$
14. Determine the points of intersection of the two circles $x^2 + y^2 = 1$ and $(x - 1)^2 + y^2 = 1$.

In problems 15–20, find the coordinates of the vertices and of the foci, and sketch the graph of each ellipse.

15. $\dfrac{x^2}{16} + \dfrac{y^2}{9} = 1$
16. $\dfrac{y^2}{25} + \dfrac{x^2}{16} = 1$
17. $\dfrac{y^2}{16} + \dfrac{x^2}{4} = 1$
18. $4x^2 + 9y^2 = 36$
19. $4x^2 + 16y^2 = 64$
20. $25x^2 + 9y^2 = 1$

In problems 21–28, find the coordinates of the center, of the vertices, and of the foci, then sketch the graph of each ellipse.

21. $3(x - 1)^2 + 4(y + 2)^2 = 192$
22. $x^2 + 4y^2 - 2x - 16y + 13 = 0$
23. $x^2 + 4y^2 + 2x - 8y + 1 = 0$
24. $25(x - 3)^2 + 4(y - 1)^2 = 100$
25. $9x^2 + 4y^2 + 18x - 16y - 11 = 0$
26. $9x^2 + y^2 - 18x + 2y + 9 = 0$
27. $16(x + 2)^2 + 25(y - 1)^2 = 400$
28. $4x^2 + 9y^2 - 24x + 36y + 36 = 0$

In problems 29–35, find the standard equation of the ellipse for each situation and sketch the graph.

29. Vertices at $(1, -2)$, $(5, -2)$, $(3, -7)$, and $(3, 3)$
30. Vertices at $(0, -1)$, $(12, -1)$, $(6, -4)$, and $(6, 2)$
31. Vertices at $(1, 1)$, $(5, 1)$, $(3, 6)$, and $(3, -4)$
32. Vertices at $(-5, 0)$ and $(5, 0)$, and containing the point $(4, \frac{12}{5})$
33. Foci at $(1, 4)$ and $(3, 4)$, and major axis 4 units long
34. Center at $(0, 0)$, axes parallel to coordinate axes, and containing the points $(3\sqrt{3}/2, 1)$ and $(2, 2\sqrt{5/3})$
35. Center at $(-3, 1)$, major axis parallel to the y axis and 10 units long, and minor axis 2 units long
36. Determine the equations of the lines of symmetry (that is, the lines containing the major and minor axes) of an ellipse of the form

$$\frac{(x - h)^2}{a^2} + \frac{(y - k)^2}{b^2} = 1$$

37. Describe the graph of $\dfrac{x^2}{a^2} + \dfrac{y^2}{b^2} = 1$, if $a = b$.

38. (a) The segment cut by an ellipse from a line that contains a focus and is perpendicular to the major axis is called a **focal chord** or **latus rectum** of the ellipse. Show that the length of the focal chord of the ellipse with the equation $(x^2/a^2) + (y^2/b^2) = 1$ is $2b^2/a$.
(b) Find the length of a focal chord of an ellipse whose equation is given by $9x^2 + 16y^2 = 144$.

39. An arch in the shape of the upper half of an ellipse with a horizontal major axis supports a railroad bridge 100 feet long over a river. The center of the arch is to be 25 feet above the railroad tracks, which lie on the major axis. Find an equation of the ellipse.

40. A mathematician has accepted a position at a university situated 6 miles from the straight shoreline of a large lake (Figure 13). The professor wishes to build a home that is half as far from the university as it is from the shore of the lake. The possible homesites satisfying this condition lie along a curve. Describe this curve and find its equation with respect to a coordinate system having the shoreline as the x axis and the university at the point $(0, 6)$ on the y axis.

Figure 13

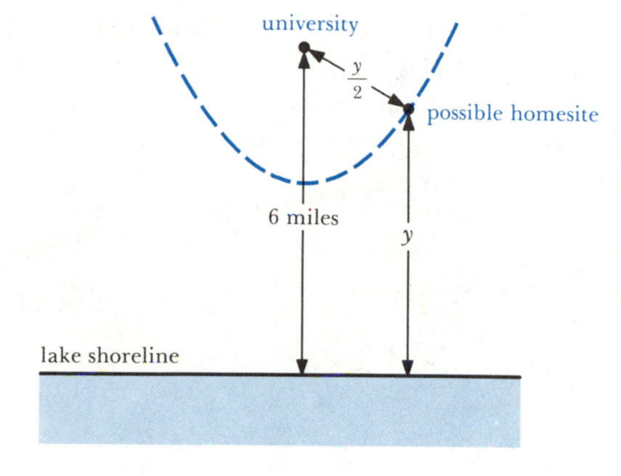

c **41.** The Ellipse in Washington, D.C., has an equation that is approximately given by

$$193{,}000x^2 + 288{,}000y^2 = 55{,}584$$

where x and y are measured in kilometers. Determine the distance between the two foci to two decimal places.

42. A communications satellite is to be placed in an elliptical orbit having a minimum altitude of 640 kilometers and a maximum altitude of 3520 kilometers above the surface of the earth. Assuming that the earth is a sphere with a radius of 6400 kilometers, determine an equation that describes the path followed by the satellite if the major axis lies on the y axis.

8.2 Parabolas

In Section 3.1 we discussed parabolas that were the graphs of quadratic functions and had vertical axes of symmetry. In this section, we derive equations for parabolas that have either vertical or horizontal axes of symmetry.

DEFINITION 1 **The Parabola**

A **parabola** is the set of all points P in a plane such that the distance from P to a fixed point is equal to the distance from P to a fixed line. The fixed point is called the **focus** and the fixed line is called the **directrix** of the parabola.

Figure 1

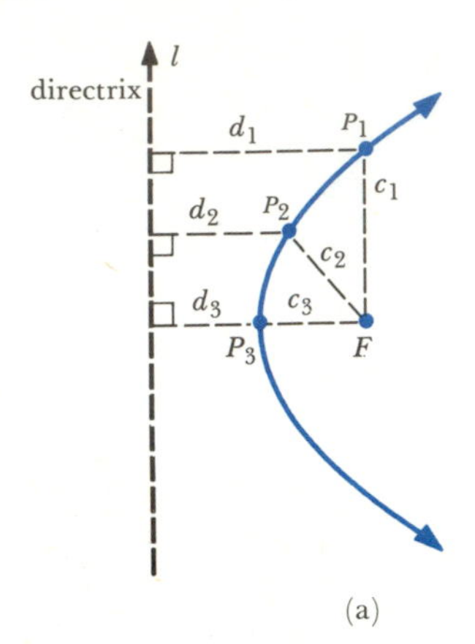

(a)

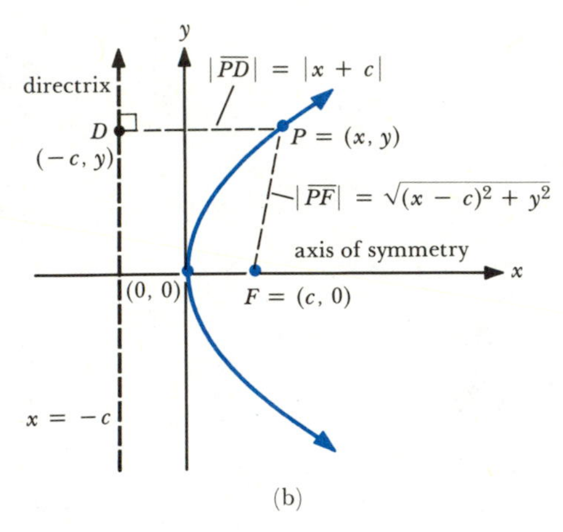

(b)

In Figure 1a, the points P_1, P_2, and P_3 are on a parabola with focus at point F and a *vertical directrix* l to the *left* of the focus; also, $d_1 = c_1$, $d_2 = c_2$, and $d_3 = c_3$. To derive a standard equation form for a parabola, we choose the coordinate system so that the focus F is at point $(c, 0)$ and the directrix is the line $x = -c$, where $c > 0$, and we let $P = (x, y)$ be any point on the parabola (Figure 1b). Assume that line segment $\overline{PD}$ is perpendicular to the directrix and point D is on the directrix. Since the point $P = (x, y)$ is equidistant from the focus and the directrix, we have $|\overline{PF}| = |\overline{PD}|$, so that, by the distance formula, we get

$$\sqrt{(x - c)^2 + y^2} = |x + c|$$

After squaring both sides of the equation, we get

$$(x - c)^2 + y^2 = (x + c)^2$$

or

$$x^2 - 2cx + c^2 + y^2 = x^2 + 2cx + c^2$$

so that

$$y^2 = 4cx, \qquad c > 0$$

We refer to the above equation as the **standard form** equation for a parabola.

We have shown that the coordinates of every point (x, y) on the parabola satisfy the equation $y^2 = 4cx$. Conversely, if (x, y) is a point satisfying the equation, then by reversing the steps above, we find that the point (x, y) is on the parabola.

Given the equation $y^2 = 4cx$, with $c > 0$, we note the following general characteristics of the graph of the parabola (Figure 1b).

1. The curve has an **axis of symmetry,** which is the line containing the focus and is perpendicular to the directrix. In this situation, the axis of symmetry is the x axis because of the y^2 term.

2. By setting $y = 0$, we find that $x = 0$ so that $(0, 0)$ is the point of intersection of the axis of symmetry and the graph. This point is called the **vertex** of the parabola.

3. Since $c > 0$, then x must be nonnegative so that the graph "opens to the right." Note that c is the distance between the vertex and the focus, and between the vertex and the directrix.

Similar derivations can be given for three other *standard form equations* for parabolas with vertices at the origin, and with vertical or horizontal axes of symmetry.

They are summarized in Table 1 below, along with the equation that was derived earlier. We assume that $c > 0$ for each situation.

Table 1 **Standard Forms of Parabolas with Vertex at the Origin**

Parabola Opens		Axis of Symmetry	Vertex	Focus	Directrix	Standard Form Equation
Right	(Figure 1b)	x axis	$(0, 0)$	$(c, 0)$	$x = -c$	$y^2 = 4cx$
Left	(Figure 2a)	x axis	$(0, 0)$	$(-c, 0)$	$x = c$	$y^2 = -4cx$
Upward	(Figure 2b)	y axis	$(0, 0)$	$(0, c)$	$y = -c$	$x^2 = 4cy$
Downward	(Figure 2c)	y axis	$(0, 0)$	$(0, -c)$	$y = c$	$x^2 = -4cy$

Figure 2

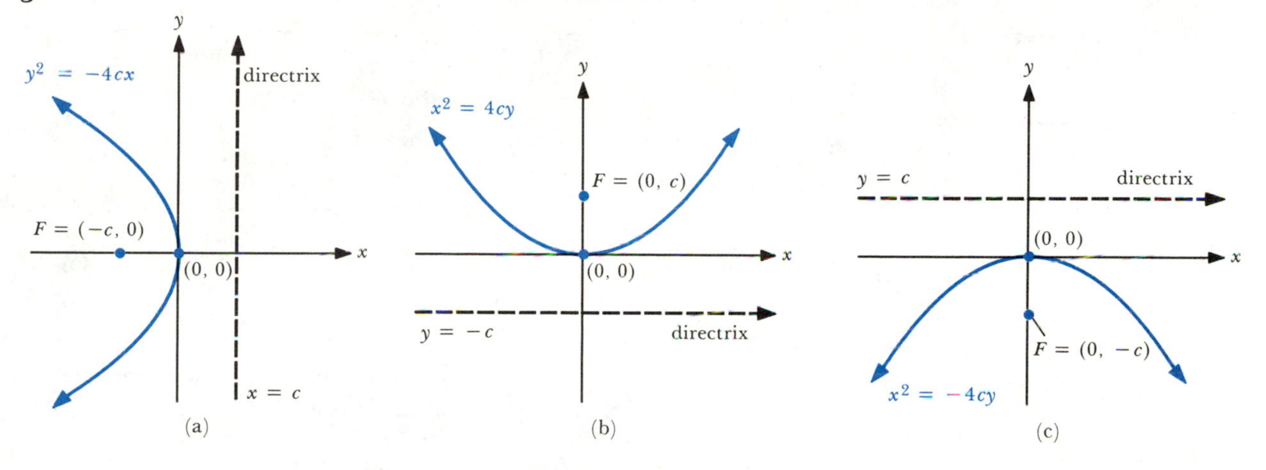

EXAMPLE 1 Find the focus and directrix of the parabola $y^2 = -24x$ and sketch the graph.

SOLUTION The equation is in standard form $y^2 = -4cx$. Since $4c = 24$, it follows that $c = 6$, so that the focus is at $(-6, 0)$ and the directrix is the line $x = 6$. The x axis is the axis of symmetry and the graph opens to the left (Figure 3).

Figure 3

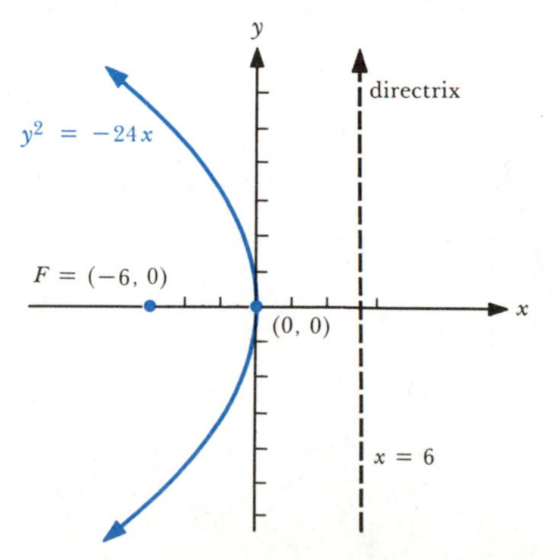

EXAMPLE 2 Find the standard equation of a parabola with vertex at the origin, focus at $(0, 3)$, and the y axis as its axis of symmetry. Sketch the graph.

SOLUTION The focus $(0, 3)$ is above the vertex and the y axis is the axis of symmetry, so that the parabola opens upward and has the standard form $x^2 = 4cy$. Since c is the distance between the focus and vertex, we have $c = 3$ and $x^2 = 12y$ (Figure 4).

Figure 4

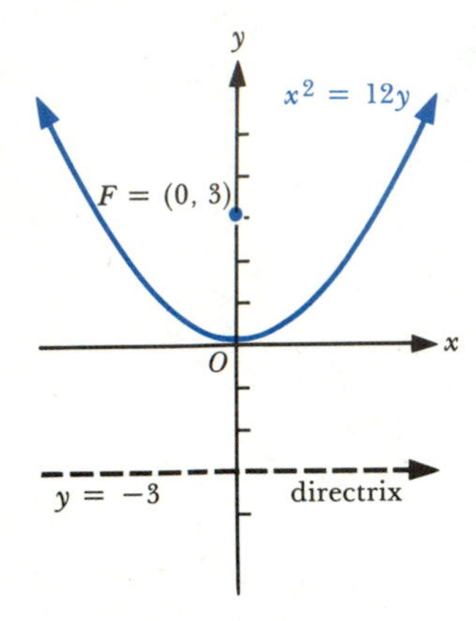

If a parabola is located in a coordinate system so that the *vertex is the point* (h, k) and the directrix is parallel to one of the coordinate axes, then the equation of the parabola has one of the following standard forms, where $c > 0$.

Standard Forms of Parabolas with Vertex at (h, k)

(i)	$(y - k)^2 = 4c(x - h)$	if the parabola opens to the right
(ii)	$(y - k)^2 = -4c(x - h)$	if the parabola opens to the left
(iii)	$(x - h)^2 = 4c(y - k)$	if the parabola opens upward
(iv)	$(x - h)^2 = -4c(y - k)$	if the parabola opens downward

Note that the axis of symmetry is the horizontal line $y = k$ for (i) and (ii), and for (iii) and (iv) it is the vertical line $x = h$.

EXAMPLE 3 Show that the graph of $y^2 + 2y - 8x - 3 = 0$ is a parabola. Find the vertex, the focus, the directrix, the axis of symmetry, and sketch the graph.

SOLUTION First, we associate the x and y terms and then complete the square in the equation $y^2 + 2y - 8x - 3 = 0$ to get

$$y^2 + 2y + 1 = 8x + 3 + 1$$
$$(y + 1)^2 = 8x + 4$$
$$(y + 1)^2 = 8(x + \tfrac{1}{2})$$

Figure 5

This is the standard form for an equation of a parabola with the vertex at the point $(-\frac{1}{2}, -1)$. The standard form implies that the parabola opens to the right and the axis of symmetry is the horizontal line $y = -1$. Since $4c = 8$, $c = 2$ and the focus is located at

$$(-\tfrac{1}{2} + 2, -1) = (\tfrac{3}{2}, -1)$$

Also, the directrix is the vertical line

$$x = -\tfrac{1}{2} - 2 = -\tfrac{5}{2} \quad \text{(Figure 5)}$$

EXAMPLE 4 Find the standard equation of a parabola with vertex $(-1, 2)$ and focus $(-1, 1)$. Also identify the axis of symmetry and sketch the graph.

SOLUTION Since the focus $(-1, 1)$ lies below the vertex $(-1, 2)$, it follows that the axis of symmetry is the vertical line $x = -1$ and the parabola opens downward (Figure 6). Thus, the standard equation has the form

$$(x + 1)^2 = -4c(y - 2)$$

Note that c, which is the distance between the vertex and focus, is given by $c = 1$, so that the equation is

$$(x + 1)^2 = -4(y - 2)$$

Figure 6

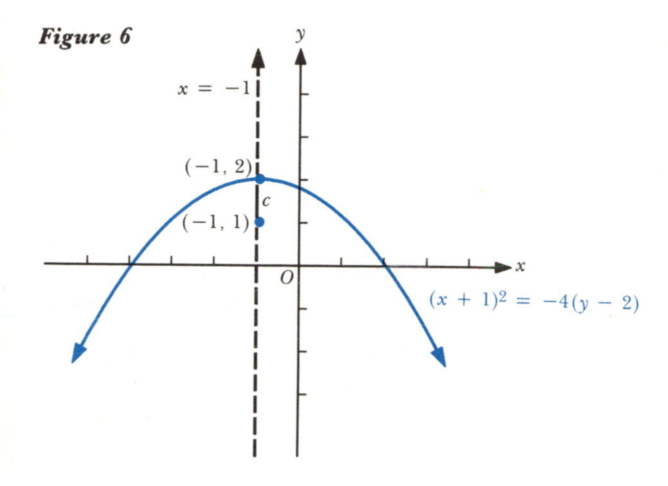

Parabolas have the following practical applications.

(a)

(a) A ***parabolic reflector mirror*** *is obtained by revolving a parabola about its axis of symmetry. Such shapes are used in automobile headlights, reflecting telescopes, searchlights, and in the construction of radar antennas.*

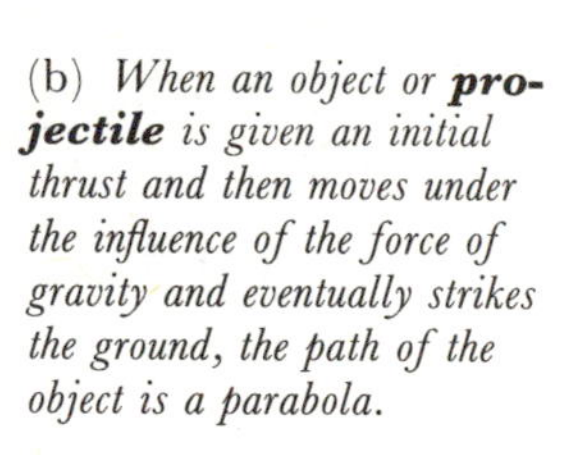

(b) *When an object or* ***projectile*** *is given an initial thrust and then moves under the influence of the force of gravity and eventually strikes the ground, the path of the object is a parabola.*

(c) ***Suspension bridges*** *are supported by cables suspended from two or more towers. These cables have the shapes of parabolas.*

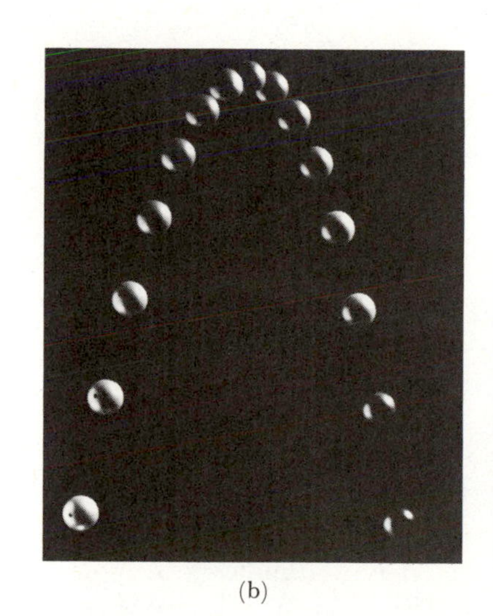

(b)

(c)

PROBLEM SET 8.2

In problems 1–9, find the vertex, the focus, the axis of symmetry, the directrix, and sketch the graph of each parabola.

1. $y^2 = 8x$
2. $x^2 = 2y$
3. $x^2 + 4y = 0$
4. $y^2 + 4x = 0$
5. $y^2 - 5x = 0$
6. $3x^2 - 2y = 0$
7. $2y - 7x^2 = 0$
8. $\dfrac{x^2}{3} - \dfrac{y}{2} = 0$
9. $-y^2 - 3x = 0$

10. The line segment with endpoints on the parabola that is perpendicular to the axis of symmetry and passes through the focus is called the **focal chord** or **latus rectum** of the parabola (Figure 7).
(a) Show that if c is the distance between the vertex V and the focus F of a parabola, then its focal chord has length $4c$.
(b) Find the length of the focal chord for each of the following parabolas.
(i) $y^2 = 8x$ (ii) $x^2 + 4y = 0$
(iii) $y^2 - 5x = 0$ (iv) $2y - 7x^2 = 0$

Figure 7

11. Sketch the graph of each parabola if the vertex $(h, k) = (3, -1)$ and $c = 2$.
(a) $(y - k)^2 = 4c(x - h)$ (b) $(y - k)^2 = -4c(x - h)$
(c) $(x - h)^2 = 4c(y - k)$ (d) $(x - h)^2 = -4c(y - k)$

12. (a) Graph $y^2 = x$ and determine the values of x for which the equation is defined in the real number system.
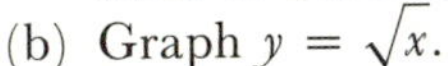
(b) Graph $y = \sqrt{x}$.
(c) Which of the two equations in parts (a) and (b) represents a function?

In problems 13–22, find the vertex, the focus, the axis of symmetry, the directrix, and sketch the graph of each parabola.

13. $(x - 2)^2 = -6(y + 1)$
14. $(y + 3)^2 = 4(x - 1)$
15. $(y - 2)^2 + 2x = 2$
16. $5y - (x - 3)^2 = 1$
17. $y^2 + 2x + 2y + 7 = 0$
18. $2y^2 + 2x + 6y + 9 = 0$
19. $3x^2 + 12x + 6y + 13 = 0$
20. $x^2 + 4x + 5y - 11 = 0$
21. $y^2 - 6x = 4y - 13$
22. $x^2 = \frac{5}{3}y + \frac{2}{3} + 3x$

In problems 23–32, find the standard equation of the parabola that satisfies the given conditions and sketch the graph.

23. Vertex at $(-1, 2)$ and focus at $(3, 2)$
24. Vertex at $(3, -4)$ and directrix $x = 1$
25. Focus at $(0, 4)$ and directrix $y = 2$
26. Vertex at $(2, -5)$ and directrix $y = 3$
27. Vertex at $(1, 3)$, contains point $(5, 7)$, and the axis of symmetry is parallel to the y axis
28. Focus at $(-3, -5)$ and vertex at $(-3, 0)$
29. Contains the three points $(0, 0)$, $(2, 2)$, and $(4, 8)$ and opens upward
30. Contains the three points $(-2, 0)$, $(0, 3)$, and $(2, 0)$ and opens downward
31. Contains the three points $(0, 0)$, $(-1, 2)$, and $(-3, -2\sqrt{3})$ and opens to the left
32. Focal chord with endpoints $(3, 5)$ and $(3, -3)$, and opens to the right
33. Find the points of intersection of the ellipse $(x^2/18) + (y^2/8) = 1$ and the parabola $4x - 3y^2 = 0$. Sketch the graphs on the same coordinate axes.
34. Find an equation of the line through the points on the parabola $y^2 = 3x$ whose ordinates are 2 and 3.

35. A suspension bridge 400 meters long is held up by a parabolic main cable (Figure 8). This cable is 100 meters above the roadway at the ends and 4 meters above the roadway at the center. Vertical supporting cables run at 50-meter intervals along the roadway. Find the lengths of these vertical cables. (*Hint:* Set up an xy coordinate system so that the x axis coincides with the roadway and the vertex of the parabola is 4 units above the origin.)

Figure 8

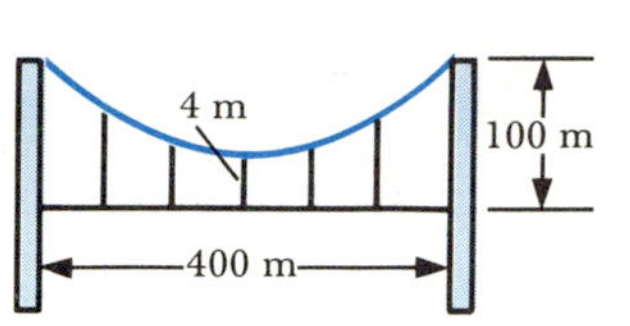

36. The surface of a roadway over a stone bridge follows a parabolic curve with the vertex in the middle of the bridge. The span of the bridge is 60 feet and the road surface is 1 foot higher in the middle than at the ends. How much higher than the ends is a point on the roadway 15 feet from an end?

c 37. A searchlight reflector is designed in such a way that a cross section through its axis is a parabola and the light source is at the focus. If the reflector is 92 centimeters in diameter across the opening and 30 centimeters deep, find the location of the light source by determining how many centimeters it is from the opening, to two decimal places.

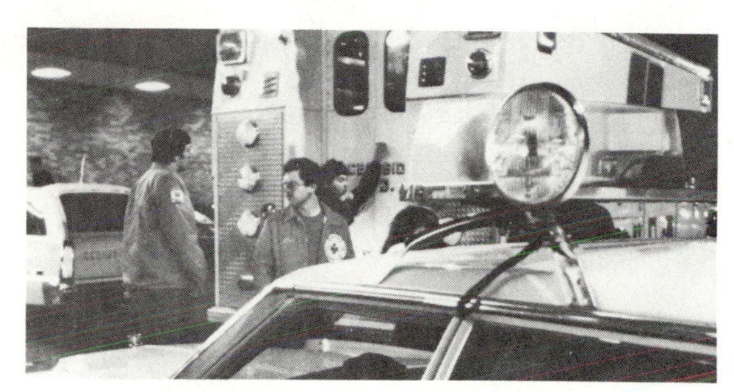

38. A baseball is hit so that it travels in a parabolic path a distance of 350 feet from home plate as measured along the ground and reaches an altitude of 150 feet. If the origin represents the home plate, and if the ball travels in the positive direction over the ground, then find an equation of the parabolic path of the baseball.

c 39. A radar antenna is constructed in such a way that any cross section through its axis is a parabola. If the receiver is located at the focus, and if the antenna is 1.5 meters in diameter across at the opening and 0.5 meter deep, find the location of the receiver by determining how far the receiver is from the opening, to two decimal places.

8.3 Hyperbolas

Geometrically, a hyperbola is defined as follows.

DEFINITION 1 **The Hyperbola**

> A **hyperbola** is the set of all points P in a plane such that the absolute value of the difference of the two distances from P to two fixed points is a constant. The fixed points are called the **foci** of the hyperbola.

In Figure 1a, the points P_1 and P_2 are on a hyperbola with foci F_1 and F_2. Also,

$$|d_2 - c_2| = |d_1 - c_1| = k$$

where k is the constant difference referred to in the definition of a hyperbola. We derive an equation of the hyperbola by choosing the coordinate system so that the foci are $F_1 = (-c, 0)$ and $F_2 = (c, 0)$, with $c > 0$, and we assume that the constant difference k equals $2a$. If $P = (x, y)$ is a point on the hyperbola (Figure 1b), then in triangle F_1PF_2,

$$|\overline{PF_2}| + |\overline{F_1F_2}| > |\overline{PF_1}| \quad \text{and} \quad |\overline{PF_1}| + |\overline{F_1F_2}| > |\overline{PF_2}|$$

so that

$$|\overline{F_1F_2}| > |\overline{PF_1}| - |\overline{PF_2}| \quad \text{and} \quad |\overline{F_1F_2}| > |\overline{PF_2}| - |\overline{PF_1}|$$

Thus

$$|\overline{F_1F_2}| > \big||\overline{PF_1}| - |\overline{PF_2}|\big|$$

that is,

$$2c > 2a \quad \text{or} \quad c > a$$

Since

$$\big||\overline{PF_1}| - |\overline{PF_2}|\big| = 2a$$

by using the distance formula, we get

$$\left|\sqrt{(x+c)^2 + y^2} - \sqrt{(x-c)^2 + y^2}\right| = 2a$$

Hence

$$\sqrt{(x+c)^2 + y^2} - \sqrt{(x-c)^2 + y^2} = \pm 2a$$

or

$$\sqrt{(x+c)^2 + y^2} = \pm 2a + \sqrt{(x-c)^2 + y^2}$$

After squaring both sides of the equation and simplifying we get

$$cx - a^2 = \pm a\sqrt{(x-c)^2 + y^2}$$

Figure 1

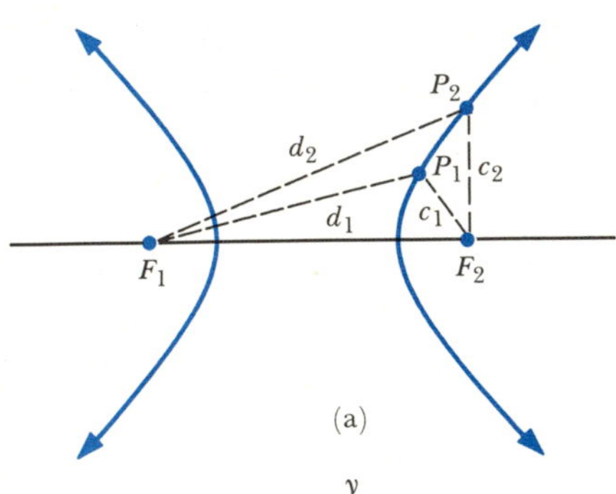

(a)

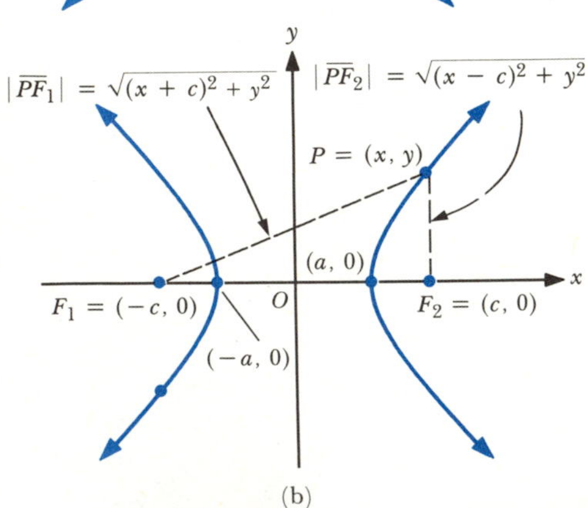

(b)

Again, we square both sides and simplify to obtain

$$(c^2 - a^2)x^2 - a^2y^2 = a^2(c^2 - a^2)$$

Since $c > a > 0$, then $c^2 - a^2 > 0$, so that we can divide both sides of the latter equation by $a^2(c^2 - a^2)$ to obtain

$$\frac{x^2}{a^2} - \frac{y^2}{c^2 - a^2} = 1$$

Since $c^2 - a^2$ is positive, we can write $b = \sqrt{c^2 - a^2}$, so that

$$b^2 = c^2 - a^2$$

and so

$$\boxed{\frac{x^2}{a^2} - \frac{y^2}{b^2} = 1}$$

This last equation is called the **standard form** of an equation of a hyperbola. The hyperbola has its **center** at the origin, which is the midpoint of the line segment determined by the foci $(-c, 0)$ and $(c, 0)$, where $c^2 = a^2 + b^2$. In contrast to an equation of an ellipse (where a was *always* greater than b) a may be greater than, equal to, or less than b in an equation of a hyperbola. The coordinates of every point (x, y) on the hyperbola in Figure 1b satisfy the boxed equation; and conversely, if the coordinates (x, y) satisfy the boxed equation, then by reversing the preceding steps, we can show that the point (x, y) is on the hyperbola.

We can derive the following characteristics of a hyperbola from this equation. (Refer to Figure 2a.)

Figure 2

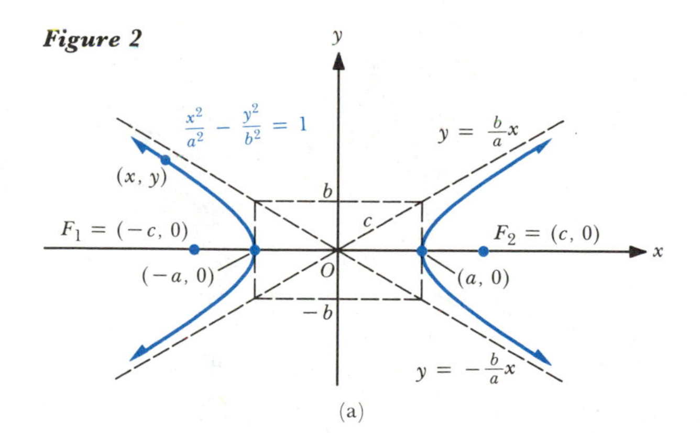

(a)

1. We can determine the x intercepts by setting $y = 0$ to obtain $-a$ and a. The corresponding points on the graph, $(-a, 0)$ and $(a, 0)$, are referred to as the **vertices** of the hyperbola.

2. The **transverse axis** is the line segment that has the vertices as its endpoints. Note that the length of the transverse axis is $2a$, and a is the distance between the center of the hyperbola and each of its two vertices. The line segment of length $2b$ that contains the center of the hyperbola as its midpoint and is perpendicular to the transverse axis is called the **conjugate axis.** The hyperbola does not intersect the conjugate axis.

3. Because of the y^2 and x^2 terms, the graph is symmetric with respect to both the x axis (or transverse axis) and the y axis (or conjugate axis). The line determined by the transverse axis is called the **axis of symmetry** of the hyperbola. In this case, the axis of symmetry is the x axis. Also, the foci lie on the axis of symmetry c units from the center at points $(-c, 0)$ and $(c, 0)$, where $c^2 = a^2 + b^2$.

4. If we write

$$\frac{x^2}{a^2} - \frac{y^2}{b^2} = 1$$

as

$$y^2 = \frac{b^2}{a^2}(x^2 - a^2)$$

then

$$y = \pm\frac{b}{a}x\sqrt{1 - \frac{a^2}{x^2}}$$

Note that as $|x|$ gets very large, a^2/x^2 approaches 0 so that, in turn, the expression $1 - (a^2/x^2)$ approaches 1. In other words, the larger x is in absolute value, the closer the graph of the hyperbola is to the lines whose equations are given by

$$y = \frac{b}{a}x \qquad \text{and} \qquad y = -\frac{b}{a}x$$

These lines are **asymptotes** for the hyperbola.

If we repeat the argument for a hyperbola with foci at $(0, -c)$ and $(0, c)$, we obtain a *standard form equation* for a hyperbola with *center* at the origin and a *vertical transverse axis* (Figure 2b) which is given by:

Figure 2

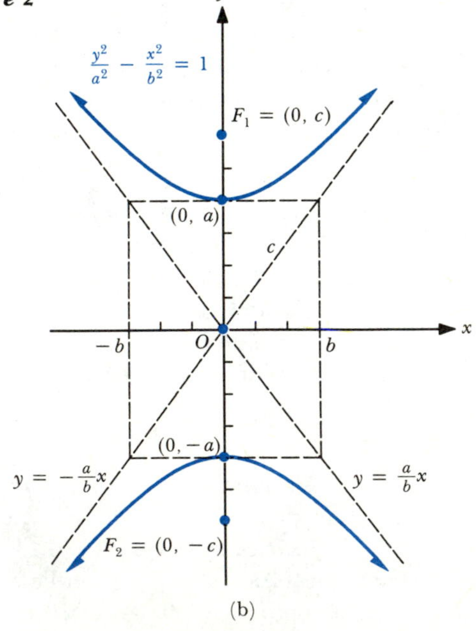

(b)

$$\frac{y^2}{a^2} - \frac{x^2}{b^2} = 1$$

Here again $c^2 = a^2 + b^2$. The equations of the asymptotes are

$$y = \frac{a}{b}x \qquad \text{and} \qquad y = -\frac{a}{b}x$$

Note that the slopes of these lines are $\pm a/b$ in contrast to the slopes of the asymptotes for a hyperbola with equation

$$\frac{x^2}{a^2} - \frac{y^2}{b^2} = 1$$

which are $\pm b/a$.

In both cases discussed, the asymptotes can be easily drawn by noting that they are the two lines determined by the diagonals of a rectangle of dimensions $2a$ and $2b$ with its center at the center of the hyperbola, as displayed in Figure 2.

Table 1 summarizes the characteristics and standard form equations of the hyperbolas we have discussed. We assume that $c^2 = a^2 + b^2$.

Table 1 **Hyperbolas**

Axis of Symmetry (or transverse axis)	Center	Foci	Vertices (endpoints of the transverse axis)	Standard Form Equation	Equations of Asymptotes
Horizontal (on the x axis)	$(0, 0)$	$(-c, 0)$, $(c, 0)$	$(-a, 0)$, $(a, 0)$	$\frac{x^2}{a^2} - \frac{y^2}{b^2} = 1$	$y = \pm\frac{b}{a}x$
Vertical (on the y axis)	$(0, 0)$	$(0, -c)$, $(0, c)$	$(0, -a)$, $(0, a)$	$\frac{y^2}{a^2} - \frac{x^2}{b^2} = 1$	$y = \pm\frac{a}{b}x$

EXAMPLE 1 Show that the graph of $25x^2 - 16y^2 = 400$ is a hyperbola. Find the coordinates of the vertices, of the foci, and equations of the asymptotes. Also sketch the graph.

SOLUTION First, we divide both sides of the equation by 400 to get

$$\frac{x^2}{16} - \frac{y^2}{25} = 1$$

which is the standard form of an equation of a hyperbola. The vertices are determined by setting $y = 0$ to get $(4, 0)$ and $(-4, 0)$. Since

$$c^2 = a^2 + b^2 = 16 + 25 = 41$$

Then

$$c = \sqrt{41}$$

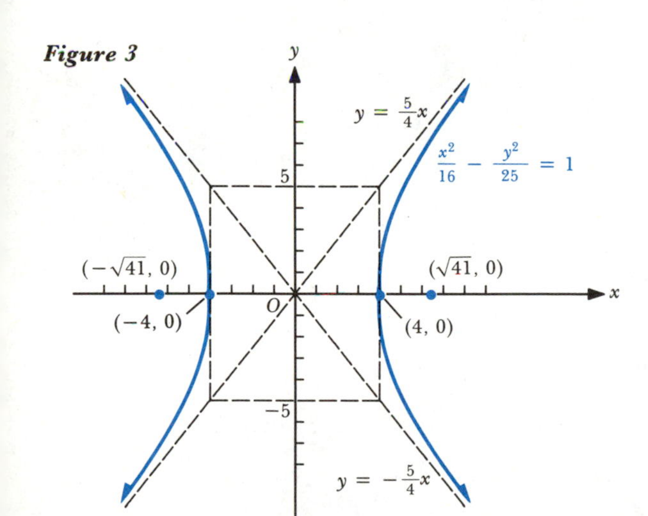

Figure 3

so that the foci are $(-\sqrt{41}, 0)$ and $(\sqrt{41}, 0)$. To sketch the graph we first locate the vertices and then draw a rectangle, centered at the origin, that contains the vertices of the hyperbola and has dimensions 8 by 10. After sketching the asymptotes as extensions of the diagonals of the rectangle, we obtain the graph (Figure 3). Equations of the asymptotes are given by

$$y = \pm\frac{b}{a}x = \pm\frac{5}{4}x$$

As with the ellipse, if a hyperbola is located on a coordinate system so that its *center is at the point* (h, k) and the transverse axis is either horizontal or vertical, then the standard equation is one of the following forms.

Standard Forms of Hyperbolas Centered at (h, k)

(i) $\dfrac{(x-h)^2}{a^2} - \dfrac{(y-k)^2}{b^2} = 1$, if the axis of symmetry is the horizontal line $y = k$ (Figure 4a)

(ii) $\dfrac{(y-k)^2}{a^2} - \dfrac{(x-h)^2}{b^2} = 1$, if the axis of symmetry is the vertical line $x = h$ (Figure 4b)

Figure 4

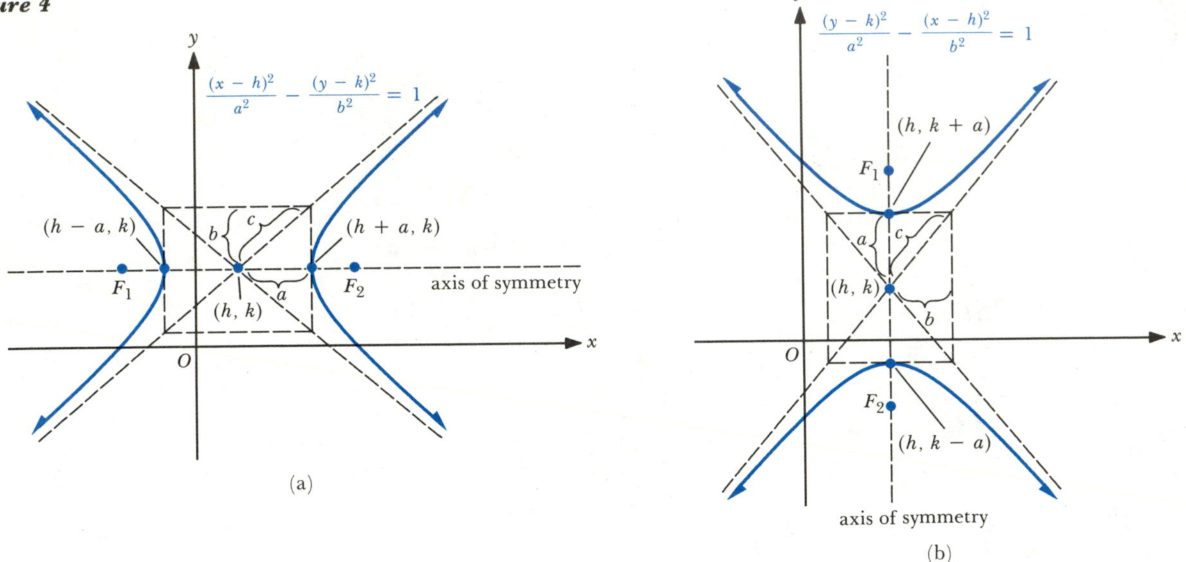

Equations of the asymptotes are, respectively,

$$y - k = \pm\frac{b}{a}(x - h) \qquad \text{and} \qquad y - k = \pm\frac{a}{b}(x - h)$$

The equations of the asymptotes in both cases can be obtained when the right-hand side of the standard form equation is replaced by 0.

EXAMPLE 2 Show that the graph of $4x^2 - y^2 - 8x + 2y + 7 = 0$ is a hyperbola. Find the coordinates of the center, the foci, and the vertices. Also, sketch the graph and determine equations for the asymptotes.

SOLUTION First, we associate the x and y terms and then complete the square to get

$$4(x^2 - 2x + 1) - (y^2 - 2y + 1) = -7 + 4 - 1$$
$$4(x - 1)^2 - (y - 1)^2 = -4$$

Figure 5

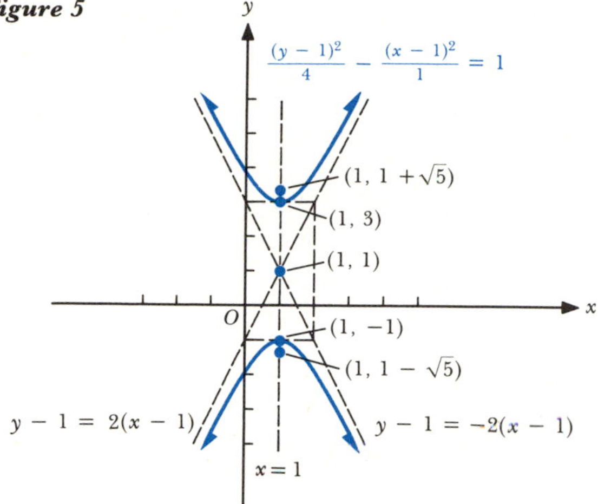

or

$$\frac{(y-1)^2}{4} - \frac{(x-1)^2}{1} = 1$$

This is the standard form for an equation of a hyperbola with center $(1, 1)$ and with axis of symmetry the vertical line $x = 1$. Since $a = 2$ and $b = 1$,

$$c^2 = a^2 + b^2 = 4 + 1 = 5$$

so that $c = \sqrt{5}$ and the foci are $(1, 1 + \sqrt{5})$ and $(1, 1 - \sqrt{5})$. Since $a = 2$ is the distance between the center $(1, 1)$ and each of the two vertices, we can locate the vertices by vertically moving from $(1, 1)$ 2 units up and 2 units down along the axis of symmetry $x = 1$ to obtain the points $(1, 3)$ and $(1, -1)$. After locating the vertices and drawing a rectangle centered at $(1, 1)$ containing the vertices, and with dimensions 2 and 4, we obtain the asymptotes which, in turn, lead to the graph (Figure 5). The asymptotes have equations

$$y - 1 = \pm 2(x - 1)$$

EXAMPLE 3 Find the standard equation for a hyperbola with vertices $(-6, 1)$ and $(2, 1)$, and a focus at $(-2 + \sqrt{41}, 1)$. Also sketch the graph.

SOLUTION First we plot the given points (Figure 6). Since the center (h, k) is at the midpoint of the line segment with endpoints on the vertices $(-6, 1)$ and $(2, 1)$, we obtain

$$(h, k) = \left(\frac{-6 + 2}{2}, \frac{1 + 1}{2}\right) = (-2, 1)$$

Figure 6

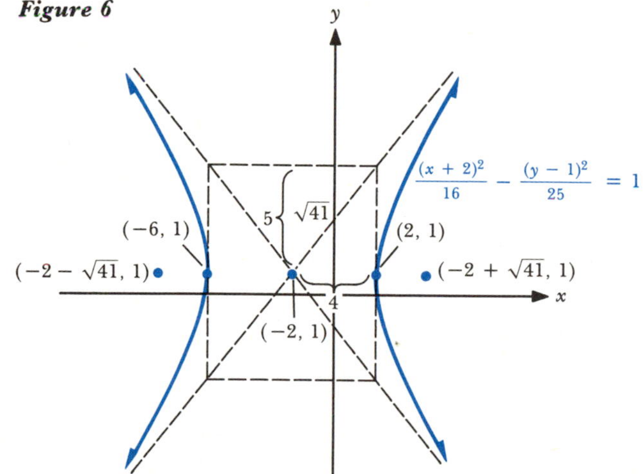

Also, since a is the distance between the center $(-2, 1)$ and vertex $(2, 1)$, we have $a = 4$. In addition, because c is the distance between the center $(-2, 1)$ and focus $(-2 + \sqrt{41}, 1)$, we conclude that $c = \sqrt{41}$. Thus, $b^2 = c^2 - a^2 = 41 - 16 = 25$, that is, $b = 5$. Since the equation is of the form

$$\frac{(x-h)^2}{a^2} - \frac{(y-k)^2}{b^2} = 1$$

we substitute for h, k, a, and b to obtain the equation

$$\frac{(x+2)^2}{16} - \frac{(y-1)^2}{25} = 1$$

As before, we locate the vertices, the asymptotes, and then the graph (Figure 6).

Applications of hyperbolas include the following.

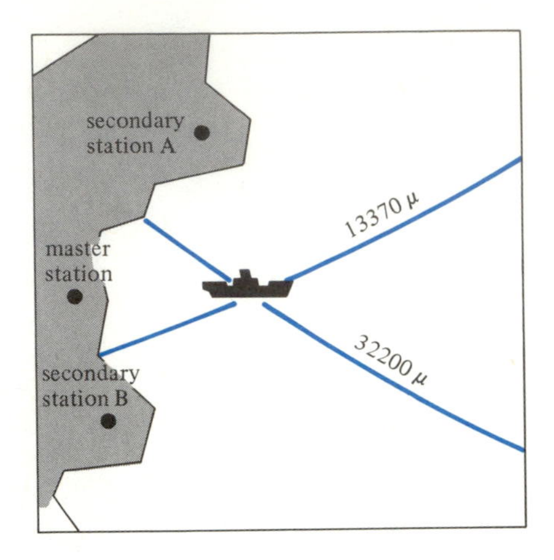

*The **LORAN** (long range navigation) **SYSTEM** of ship or aircraft navigation utilizes hyperbolas.*

PROBLEM SET 8.3

In problems 1–9, the graph of each equation is a hyperbola. Find the coordinates of the vertices and foci. Also, sketch the graph and determine equations for the asymptotes.

1. $y^2 - x^2 = 9$

2. $9x^2 - y^2 = 9$

3. $x^2 - 9y^2 = 9$

4. $36x^2 - 49y^2 = 1764$

5. $16y^2 - 4x^2 = 48$

6. $36x^2 - 9y^2 = 1$

7. $4y^2 + 20 - 5x^2 = 0$

8. $3y^2 = x^2 + 1$

9. $\dfrac{x^2}{3} - \dfrac{y^2}{4} = \dfrac{1}{6}$

10. (a) Graph the **unit hyperbola** $x^2 - y^2 = 1$ and determine equations for the asymptotes.

(b) Verify that any point of the form $\left(\dfrac{e^u + e^{-u}}{2}, \dfrac{e^u - e^{-u}}{2}\right)$, where u represents any real number, lies on the graph of $x^2 - y^2 = 1$.

In problems 11–21, find the coordinates of the center, vertices, and foci. Sketch the graph and find equations of the asymptotes for each hyperbola.

11. $\dfrac{(x-5)^2}{25} - \dfrac{(y-4)^2}{4} = 1$

12. $\dfrac{(y-2)^2}{16} - \dfrac{(x+3)^2}{25} = 1$

13. $\dfrac{(y-3)^2}{16} - \dfrac{(x+2)^2}{16} = 1$

14. $5x^2 - (y-1)^2 = 20$

15. $4x^2 - 3y^2 - 32x + 6y + 73 = 0$

16. $4x^2 - 9y^2 - 32x + 36y + 27 = 0$

17. $9y^2 - 25x^2 + 72y - 100x + 269 = 0$
18. $9x^2 - 16y^2 - 90x - 256y = 223$
19. $4x^2 - y^2 + 8x - 2y + 6 = 0$
20. $x^2 - 2y^2 + x + 8y = 8$
21. $\frac{x^2}{4} + \frac{x}{4} = \frac{y^2}{3} - \frac{4y}{3} + \frac{3}{2}$

22. (a) The line segment cut by a hyperbola from a line that contains a focus and is perpendicular to the transverse axis is called a **focal chord** or **latus rectum** of the hyperbola. Show that the length of the focal chord of the hyperbola

$$\frac{x^2}{a^2} - \frac{y^2}{b^2} = 1 \text{ is } \frac{2b^2}{a}.$$

(b) Find the length of the focal chord for each of the given hyperbolas.

(i) $\frac{x^2}{9} - \frac{y^2}{4} = 1$ (ii) $\frac{y^2}{25} - \frac{x^2}{10} = 1$

(iii) $\frac{(y+2)^2}{16} - \frac{x^2}{9} = 1$ (iv) $7(x-3)^2 - 5(y+1)^2 = 11$

In problems 23–30, find the standard equation for the hyperbola that satisfies the given conditions and sketch the graph.

23. Vertices at $(-16, 0)$ and $(16, 0)$, and asymptotes $y = \pm\frac{5}{4}x$
24. Center at $(2, 3)$, a focus at $(2, 5)$, and a focal chord of length 6 units
25. Vertices at $(-1, 4)$ and $(-1, 6)$, and foci at $(-1, 3)$ and $(-1, 7)$
26. Transverse axis of length 8 units and foci at $(0, 5)$ and $(0, -5)$
27. Center at the origin, a vertex at $(3, 0)$, and a focus at $(4, 0)$
28. Center at $(-2, 1)$, a focus at $(-2, 6)$, and a focal chord of length $\frac{32}{3}$ units
29. Center at $(0, 0)$, horizontal axis of symmetry, and contains the points $(2, 5)$ and $(3, -10)$
30. Center at $(0, 0)$, horizontal axis of symmetry, and contains points $(4, 3)$ and $(-7, 6)$
31. Find the points of intersection of the ellipse with equation $4x^2 + 7y^2 = 32$ and the hyperbola with equation $11y^2 - 3x^2 = 41$.
32. Two microphones are located at the points $(-c, 0)$ and $(c, 0)$ on the x axis (Figure 7). An explosion occurs at an unknown location P to the right of the y axis. The sound of the explosion is detected by the microphone at $(c, 0)$ exactly t seconds before it is detected by the microphone at $(-c, 0)$. Assuming that sound travels in air at the constant speed of v feet per second, show that P must be located on the right-hand branch of the hyperbola whose equation is

$$\frac{x^2}{a^2} - \frac{y^2}{b^2} = 1 \qquad \text{where } a = \frac{vt}{2} \text{ and } b = \frac{\sqrt{4c^2 - v^2t^2}}{2}$$

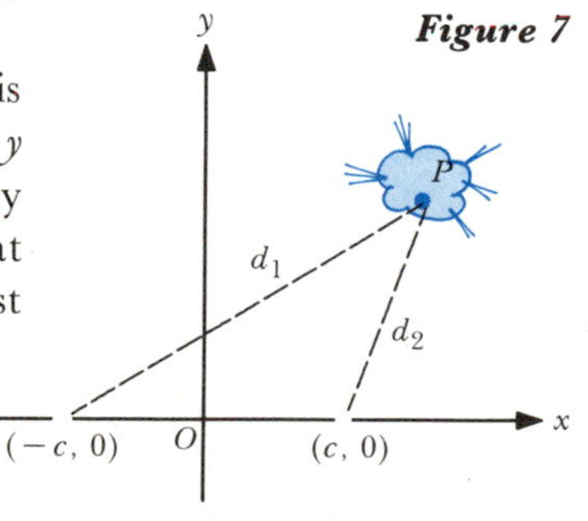

Figure 7

8.4 Translations and Rotations of Axes

An equation for a graph depends not only on the shape of the curve but also on the orientation of the coordinate axes. For instance, Figure 1 contrasts the different equations of the *same elliptical shape* for three *different* coordinate axes. In Figure 1b, the graph of the ellipse $(x^2/36) + (y^2/16) = 1$ is shifted horizontally 3 units to the right and vertically 2 units up, or "translated"; in Figure 1c, the graph is rotated through an angle θ.

Figure 1

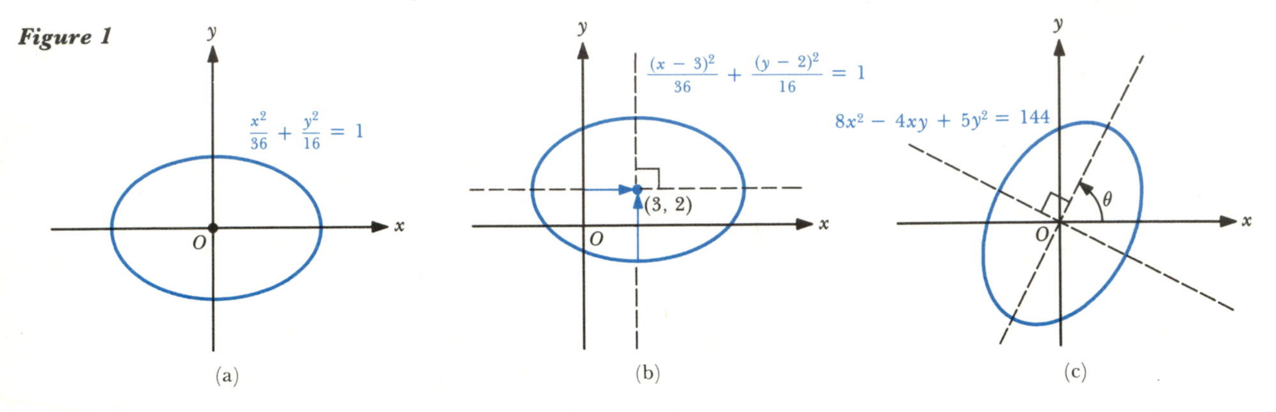

Translations of Axes

If a "new" set of coordinate axes were constructed in Figure 1b by horizontally shifting the y axis 3 units to the right and vertically shifting the x axis 2 units up, then the equation of the curve relative to the "new" coordinate axes would take on the simpler form of Figure 1a. Such a relocation of coordinate axes is called a *translation* of axes. In general, if two Cartesian coordinate systems have corresponding axes that are parallel and have the same positive directions, then we say that these systems are obtained from one another by **translation.**

Figure 2 shows a translation of the "old" xy coordinate system to a "new" $\bar{x}\bar{y}$ system whose origin $\bar{O}$ has the "old" coordinates (h, k). If P is a point in the plane having "old" coordinates (x, y), and "new" coordinates $(\bar{x}, \bar{y})$, then we have the following rule for the translation of Cartesian coordinates that enables us to relate the two pairs of coordinates to each other.

Figure 2

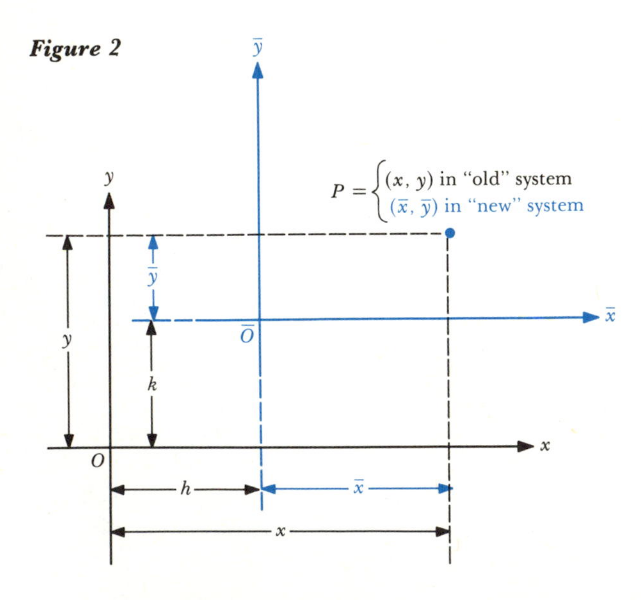

Translation Formulas

$$\begin{cases} x = \bar{x} + h \\ y = \bar{y} + k \end{cases} \quad \text{or} \quad \begin{cases} \bar{x} = x - h \\ \bar{y} = y - k \end{cases}$$

EXAMPLE 1 Translate the xy axes to form the $\bar{x}\bar{y}$ axes so that the origin $\bar{O}$ in the $\bar{x}\bar{y}$ system corrresponds to $(4, -3)$ in the xy system.

(a) If $P_1 = (2, 1)$ with respect to the xy coordinate system, find the coordinates of P_1 in the $\bar{x}\bar{y}$ system.

(b) If $P_2 = (-6, 6)$ in the $\bar{x}\bar{y}$ system, find the coordinates of P_2 in the xy coordinate system.

SOLUTION Here $(h, k) = (4, -3)$ so that by the translation formulas, we have

$$\begin{cases} x = \bar{x} + 4 \\ y = \bar{y} - 3 \end{cases} \quad \text{or} \quad \begin{cases} \bar{x} = x - 4 \\ \bar{y} = y + 3 \end{cases}$$

Figure 3

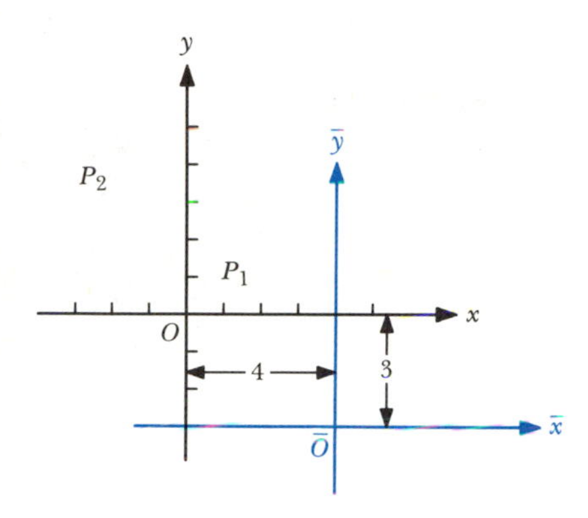

Thus, we have the following results (Figure 3):

(a) $P_1 = (x, y) = (2, 1)$ so that

$$\bar{x} = 2 - 4 = -2 \qquad \text{and} \qquad \bar{y} = 1 + 3 = 4$$

Thus $(2, 1)$ in the xy system is represented by $(-2, 4)$ in the $\bar{x}\bar{y}$ system.

(b) For $P_2 = (\bar{x}, \bar{y}) = (-6, 6)$ we have

$$x = -6 + 4 = -2 \qquad \text{and} \qquad y = 6 - 3 = 3$$

so that $(-6, 6)$ in the $\bar{x}\bar{y}$ system has coordinates $(-2, 3)$ in the xy system.

EXAMPLE 2 Find a translation of axes that transforms the equation of a circle given by the equation $x^2 + y^2 - 4x + 2y - 4 = 0$ into the form $\bar{x}^2 + \bar{y}^2 = r^2$.

SOLUTION Completing the square in both x and y in

$$x^2 + y^2 - 4x + 2y - 4 = 0$$

we get the equation

$$(x - 2)^2 + (y + 1)^2 = 9$$

Figure 4

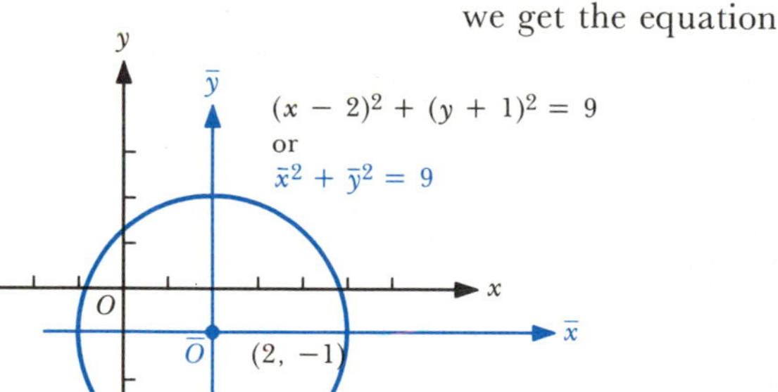

so that if we let

$$\bar{x} = x - 2 \qquad \text{and} \qquad \bar{y} = y + 1$$

we get an equation of the circle in the form (Figure 4)

$$\bar{x}^2 + \bar{y}^2 = 9$$

Notice that the origin of the $\bar{x}\bar{y}$ system is represented by $(2, -1)$ in the xy system.

A translation of coordinate axes shifts the axes horizontally or vertically or both, but it does *not change* the *position* of a curve in the plane *or* the *shape* of the curve– it only changes the *equation* of the curve. Clearly, in Example 2, the translated axes yield a "simpler" equation form than the original axes.

Rotation of Axes

If the coordinate axes in Figure 1c are rotated by keeping the origin fixed, it is possible to determine an *angle of rotation* that would yield a new coordinate system within which the equation of the ellipse would take the form of Figure 1a. The process of constructing a new coordinate system by rotating the coordinate axes of a given system is called a **rotation of axes.**

Figure 5

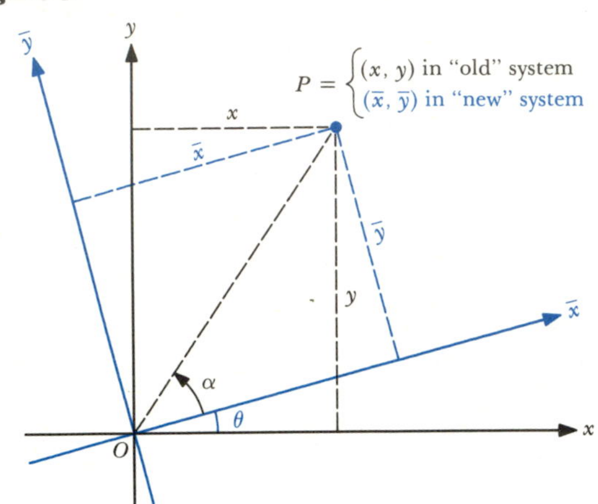

Assume that a "new" $\bar{x}\bar{y}$ coordinate system is formed from an "old" xy coordinate system by rotating the positive x axis counterclockwise through an angle θ. Let P be a given point with coordinates $(\bar{x}, \bar{y})$ relative to the new system and coordinates (x, y) relative to the old system. We can derive formulas that relate the pair of coordinates to each other. From Figure 5, we see that

$$\begin{cases}\bar{x} = |\overline{OP}| \cos \alpha \\ \bar{y} = |\overline{OP}| \sin \alpha\end{cases} \quad \text{and} \quad \begin{cases}x = |\overline{OP}| \cos(\alpha + \theta) \\ y = |\overline{OP}| \sin(\alpha + \theta)\end{cases}$$

so that

$$\begin{aligned} x = |\overline{OP}| \cos(\alpha + \theta) &= |\overline{OP}| \cos \alpha \cos \theta - |\overline{OP}| \sin \alpha \sin \theta \\ &= \bar{x} \cos \theta - \bar{y} \sin \theta \end{aligned}$$

and

$$\begin{aligned} y = |\overline{OP}| \sin(\alpha + \theta) &= |\overline{OP}| \sin \alpha \cos \theta + |\overline{OP}| \cos \alpha \sin \theta \\ &= \bar{y} \cos \theta + \bar{x} \sin \theta \end{aligned}$$

Thus, we have the following result.

Rotation of Axes Formulas

$$x = \bar{x} \cos \theta - \bar{y} \sin \theta$$
$$y = \bar{x} \sin \theta + \bar{y} \cos \theta$$

These formulas express the *old* coordinates (x, y) in terms of the *new* coordinates $(\bar{x}, \bar{y})$. By treating the two equations above as a linear system of two equations and two unknowns $\bar{x}$ and $\bar{y}$, we can solve for $\bar{x}$ and $\bar{y}$ (Problem 12) to obtain:

Rotation Formulas

$$\bar{x} = x \cos \theta + y \sin \theta$$
$$\bar{y} = -x \sin \theta + y \cos \theta$$

EXAMPLE 3 Assume that a new $\bar{x}\bar{y}$ coordinate system is formed by rotating the xy coordinate system through an angle of 60°.

(a) Find the coordinates of P with respect to the $\bar{x}\bar{y}$ coordinate system if $P = (-1, 7)$ with respect to the xy system.

(b) Find the coordinates Q with respect to the xy coordinate system if $Q = (8, 2)$ with respect to the $\bar{x}\bar{y}$ system.

SOLUTION (a) Here we can use the second set of rotation formulas with $x = -1$, $y = 7$, and $\theta = 60°$ to get

$$\begin{cases} \bar{x} = -1 \cos 60° + 7 \sin 60° = -\frac{1}{2} + \frac{7\sqrt{3}}{2} = \frac{-1 + 7\sqrt{3}}{2} \\ \bar{y} = 1 \sin 60° + 7 \cos 60° = \frac{\sqrt{3}}{2} + \frac{7}{2} = \frac{\sqrt{3} + 7}{2} \end{cases}$$

Thus,

$$P = \left(\frac{-1 + 7\sqrt{3}}{2}, \frac{\sqrt{3} + 7}{2}\right)$$

in the $\bar{x}\bar{y}$ system.

(b) The first set of rotation formulas with $\bar{x} = 8$, $\bar{y} = 2$, and $\theta = 60°$ are used to obtain

$$\begin{cases} x = 8 \cos 60° - 2 \sin 60° = 4 - \sqrt{3} \\ y = 8 \sin 60° + 2 \cos 60° = 4\sqrt{3} + 1 \end{cases}$$

Thus, $Q = (4 - \sqrt{3}, 4\sqrt{3} + 1)$ in the xy coordinate system.

EXAMPLE 4 Assume an xy coordinate system is rotated through a 30° angle to form an $\bar{x}\bar{y}$ coordinate system. Write the equation $2x^2 + \sqrt{3}xy + y^2 = 10$ in terms of the $\bar{x}\bar{y}$ coordinate system and then sketch the graph.

SOLUTION For $\theta = 30°$, the rotation formulas for x and y yield

$$\begin{cases} x = \bar{x} \cos 30° - \bar{y} \sin 30° = \frac{\sqrt{3}}{2}\bar{x} - \frac{1}{2}\bar{y} \\ y = \bar{x} \sin 30° + \bar{y} \cos 30° = \frac{1}{2}\bar{x} + \frac{\sqrt{3}}{2}\bar{y} \end{cases}$$

Next, we substitute these expressions for x and y in the given equation to get

$$2\left(\frac{\sqrt{3}}{2}\bar{x} - \frac{1}{2}\bar{y}\right)^2 + \sqrt{3}\left(\frac{\sqrt{3}}{2}\bar{x} - \frac{1}{2}\bar{y}\right)\left(\frac{1}{2}\bar{x} + \frac{\sqrt{3}}{2}\bar{y}\right) + \left(\frac{1}{2}\bar{x} + \frac{\sqrt{3}}{2}\bar{y}\right)^2 = 10$$

After simplifying the left side of the equation, we obtain

$$\frac{5}{2}\bar{x}^2 + \frac{\bar{y}^2}{2} = 10 \qquad \text{or} \qquad \frac{\bar{x}^2}{4} + \frac{\bar{y}^2}{20} = 1$$

Figure 6

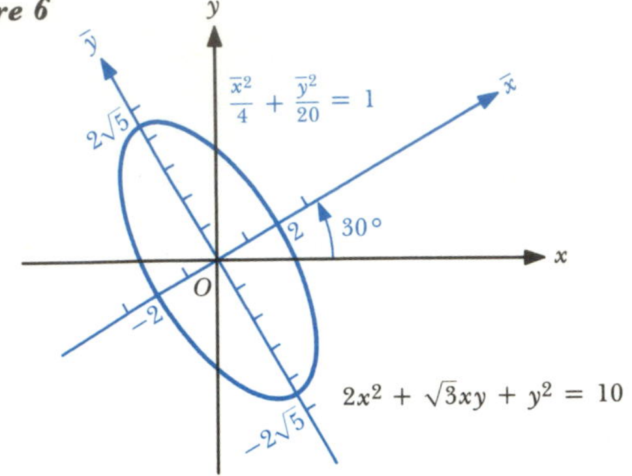

This is the standard equation of an ellipse. After drawing the $\bar{x}\bar{y}$ coordinate axes by rotating the xy coordinate axes 30°, we sketch the ellipse in the $\bar{x}\bar{y}$ system by using $a = 2\sqrt{5}$ and $b = 2$, where the major axis lies on the $\bar{y}$ axis (Figure 6). The graph we obtain is also the graph of the original equation in the xy system.

Example 4 illustrates a technique in which a rotation of axes is used to convert an equation in x and y in the **general quadratic form**

$$Ax^2 + Bxy + Cy^2 + Dx + Ey + F = 0, \qquad B \neq 0$$

to the form

$$\bar{A}\bar{x}^2 + \bar{C}\bar{y}^2 + \bar{D}\bar{x} + \bar{E}\bar{y} + \bar{F} = 0$$

This latter equation, in turn, can be written in one of the standard equation forms for conics (by completing the square if necessary) in the $\bar{x}\bar{y}$ system obtained by the rotation. It can be shown (Problem 24) that if θ satisfies

$$\cot 2\theta = \frac{A - C}{B}$$

then a rotation of axes through θ, $0° < \theta < 90°$, will accomplish the conversion or "elimination of the xy term" in the original equation.

EXAMPLE 5 Determine the angle θ for a rotation of axes that will eliminate the xy term from the quadratic equation $x^2 - 4xy + y^2 - 25 = 0$. Express x and y in terms of $\bar{x}$ and $\bar{y}$ in the new coordinate system.

SOLUTION For $x^2 - 4xy + y^2 - 25 = 0$, $A = 1$, $B = -4$, and $C = 1$, so that

$$\cot 2\theta = \frac{1 - 1}{-4} = 0$$

Thus, $2\theta = 90°$ or $\theta = 45°$ and

$$\begin{cases} x = \bar{x}\cos 45° - \bar{y}\sin 45° = \frac{\sqrt{2}}{2}\bar{x} - \frac{\sqrt{2}}{2}\bar{y} \\ y = \bar{x}\sin 45° + \bar{y}\cos 45° = \frac{\sqrt{2}}{2}\bar{x} + \frac{\sqrt{2}}{2}\bar{y} \end{cases}$$

The process of converting a given equation with a rotation of axes through angle θ is facilitated by using the following identities for $0° < \theta < 90°$ (Problem 26):

$$\cos 2\theta = \frac{\cot 2\theta}{\sqrt{\cot^2 2\theta + 1}}, \qquad \cos\theta = \sqrt{\frac{1 + \cos 2\theta}{2}}, \qquad \text{and} \qquad \sin\theta = \sqrt{\frac{1 - \cos 2\theta}{2}}$$

This technique is illustrated in the next example.

EXAMPLE 6 [C] Use a rotation of axes to eliminate the xy term in $x^2 + 4xy - 2y^2 = 12$. Write the equation in standard form relative to the rotated $\bar{x}\bar{y}$ system. Graph the equation and determine the angle of rotation to the nearest degree.

SOLUTION In this case, $A = 1$, $B = 4$, and $C = -2$, so that

$$\cot 2\theta = \frac{A - C}{B} = \frac{1 - (-2)}{4} = \frac{3}{4}$$

Thus,

$$\cos 2\theta = \frac{\cot 2\theta}{\sqrt{\cot^2 2\theta + 1}} = \frac{\frac{3}{4}}{\sqrt{\frac{9}{16} + 1}} = \frac{3}{5}$$

$$\cos\theta = \sqrt{\frac{1 + \cos 2\theta}{2}} = \sqrt{\frac{1 + \frac{3}{5}}{2}} = \frac{2}{\sqrt{5}}$$

and

$$\sin\theta = \sqrt{\frac{1 - \cos 2\theta}{2}} = \sqrt{\frac{1 - \frac{3}{5}}{2}} = \frac{1}{\sqrt{5}}$$

Consequently, from the rotation formulas, we obtain

$$\begin{cases} x = \bar{x}\cos\theta - \bar{y}\sin\theta = \dfrac{2}{\sqrt{5}}\bar{x} - \dfrac{1}{\sqrt{5}}\bar{y} \\ y = \bar{x}\sin\theta + \bar{y}\cos\theta = \dfrac{1}{\sqrt{5}}\bar{x} + \dfrac{2}{\sqrt{5}}\bar{y} \end{cases}$$

Next, we substitute these expressions into the given equation to get

$$\left(\frac{2}{\sqrt{5}}\bar{x} - \frac{1}{\sqrt{5}}\bar{y}\right)^2 + 4\left(\frac{2}{\sqrt{5}}\bar{x} - \frac{1}{\sqrt{5}}\bar{y}\right)\left(\frac{1}{\sqrt{5}}\bar{x} + \frac{2}{\sqrt{5}}\bar{y}\right) - 2\left(\frac{1}{\sqrt{5}}\bar{x} + \frac{2}{\sqrt{5}}\bar{y}\right)^2 = 12$$

After simplifying the left side, we obtain

$$2\bar{x}^2 - 3\bar{y}^2 = 12$$

or

$$\frac{\bar{x}^2}{6} - \frac{\bar{y}^2}{4} = 1$$

which is the standard equation of a hyperbola. Since $\cos\theta = 2/\sqrt{5}$, then

$$\theta = \cos^{-1}\left(\frac{2}{\sqrt{5}}\right) = 27^\circ \quad \text{(approximately)}$$

After rotating the xy system, we graph the hyperbola in the $\bar{x}\bar{y}$ system in the usual way (Figure 7). Thus, we also obtain the graph in the xy system.

Figure 7

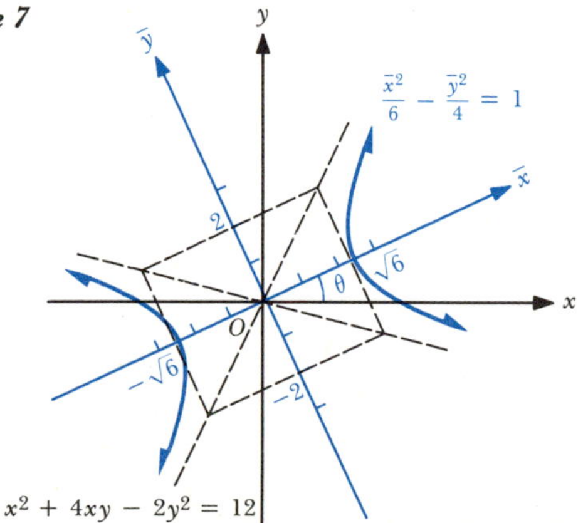

PROBLEM SET 8.4

1. Indicate the xy coordinates of the point whose $\bar{x}\bar{y}$ coordinates are
(a) $(3, -4)$ (b) $(2, 2)$ (c) $(5, 7)$
(d) $(3, -5)$ (e) $(0, 0)$ (f) $(1, -3)$
(i) if $\bar{x} = x + 1$ and $\bar{y} = y - 3$
(ii) if the translation of axes is such that $(5, 6)$ in the xy system is represented by $(-1, 3)$ in the $\bar{x}\bar{y}$ system

2. We say that a quantity is **invariant** under an operation if the quantity remains the same after the operation has been carried out. Prove that each of the following is invariant under a *translation.*
(a) The distance between $P_1 = (x_1, y_1)$ and $P_2 = (x_2, y_2)$
(b) The slope of a line joining $P_1 = (x_1, y_1)$ and $P_2 = (x_2, y_2)$

3. Indicate the $\bar{x}\bar{y}$ coordinates of the point whose xy coordinates are
(a) $(2, -7)$ (b) $(-5, 3)$ (c) $(-3, 4)$
(d) $(\frac{7}{4}, -\frac{13}{4})$ (e) $(-7, -5)$ (f) $(3, -7)$
(i) if $\bar{x} = x + 1$ and $\bar{y} = y - 3$
(ii) if the translation of axes is such that $(3, 4)$ in the xy system is represented by $(0, 0)$ in the $\bar{x}\bar{y}$ system

4. Given an equation of a line $x + y = 1$, determine an equation of the line in the $\bar{x}\bar{y}$ coordinate system that is obtained by translating the xy system so that the line contains the origin of the $\bar{x}\bar{y}$ system. Sketch a graph of the situation.

In problems 5–11, if the conic is a circle, ellipse, or hyperbola, find a translation of axes that converts the equation to a form in which the center is located at the origin. If the conic is a parabola, find a translation that converts the equation to a form in which the vertex is located at the origin.

5. $(x + 5)^2 + (y - 3)^2 = 16$
6. $x^2 - (y + 3)^2 = 1$
7. $y^2 - 4y - 8x - 20 = 0$
8. $3x^2 + 3y^2 - 9x - 7y - 36 = 0$
9. $x^2 - 2y^2 + x + 8y - 8 = 0$
10. $2x^2 - 5y + 20x + 55 = 0$
11. $2x^2 + 2y^2 + 16x - 7y = 0$

12. Use the rotation formulas that express x and y in terms of $\bar{x}$ and $\bar{y}$ on page 484 to derive the rotation formulas that express $\bar{x}$ and $\bar{y}$ in terms of x and y. (*Hint:* Treat the first set of rotation formulas on page 484 as a system of linear equations containing two unknowns $\bar{x}$ and $\bar{y}$.)

In problems 13–17, assume that new coordinate axes are obtained by rotating old coordinate axes through angle θ. Let P be a point with coordinates (x, y) in the old system and coordinates $(\bar{x}, \bar{y})$ in the new system. In problems 16 and 17, round off each answer to two decimal places.

13. If $(x, y) = (-2, -5)$ and $\theta = 30°$, find $(\bar{x}, \bar{y})$.
14. If $(\bar{x}, \bar{y}) = (0, 3\sqrt{2})$ and $\theta = 60°$, find (x, y).
15. If $(\bar{x}, \bar{y}) = (1, -10)$ and $\theta = 45°$, find (x, y).
c **16.** If $(x, y) = (3.52, 5.73)$ and $\theta = 55°$, find $(\bar{x}, \bar{y})$.
c **17.** If $(\bar{x}, \bar{y}) = (-4.71, 2.13)$ and $\theta = 22°$, find (x, y).

18. Matrix Form of a Rotation of Axes

(a) Multiply $\begin{bmatrix} \cos\theta & -\sin\theta \\ \sin\theta & \cos\theta \end{bmatrix}\begin{bmatrix} \bar{x} \\ \bar{y} \end{bmatrix}$ and compare the result to the first set of rotation formulas on page 484.

(b) Multiply $\begin{bmatrix} \cos\theta & \sin\theta \\ -\sin\theta & \cos\theta \end{bmatrix}\begin{bmatrix} x \\ y \end{bmatrix}$ and compare the result to the second set of rotation formulas on page 484.

(c) Use the matrix in part (b) to determine a matrix that will yield the $\bar{x}\bar{y}$ coordinates of a given point (x, y) if the coordinate axes are rotated 45°.

In problems 19–23, assume an xy coordinate system is rotated through angle θ to form an $\bar{x}\bar{y}$ coordinate system. Write the given equation in terms of the $\bar{x}\bar{y}$ system and sketch the graph. Display both sets of coordinate axes.

19. $xy = 1;\ \theta = 45°$

20. $2x + 3y = 6;\ \theta = 30°$

21. $2y^2 - \sqrt{3}xy + x^2 = 4;\ \theta = \dfrac{\pi}{6}$

22. $x^2 - 4xy + y^2 - 6 = 0;\ \theta = \dfrac{\pi}{4}$

23. $4x^2 + 9y^2 = 36;\ \theta = 60°$

24. (a) Substitute the rotation formulas that express x and y (old coordinates) in terms of $\bar{x}$ and $\bar{y}$ (new coordinates) into the general quadratic form

$$Ax^2 + Bxy + Cy^2 + Dx + Ey + F = 0$$

to determine that $\bar{B}$ in the transformed equation

$$\bar{A}\bar{x}^2 + \bar{B}\bar{x}\bar{y} + \bar{C}\bar{y}^2 + \bar{D}\bar{x} + \bar{E}\bar{y} + \bar{F} = 0$$

is given by

$$\bar{B} = 2(C - A)\sin\theta\cos\theta + B(\cos^2\theta - \sin^2\theta)$$

(b) Set $\bar{B} = 0$ and verify that

$$\cot 2\theta = \frac{A - C}{B}$$

25. Assume that the xy coordinate axes are rotated 45°. Find the equation of the line $x + y = 1$ in the rotated $\bar{x}\bar{y}$ system. Sketch the graph and display the $\bar{x}\bar{y}$ coordinate system.

26. Prove the identity $\cos 2\theta = \dfrac{\cot 2\theta}{\sqrt{\cot^2 2\theta + 1}}$.

In problems 27–32, use a rotation of axes to eliminate the xy term in the given equation. Write the transformed equation in standard form, graph the equation, and determine the angle of rotation to the nearest tenth of a degree.

27. $5x^2 - 4xy + 5y^2 = 9$

28. $xy - 4y - 2x = 0$

c **29.** $8x^2 - 4xy + 5y^2 = 144$

c **30.** $5x^2 - 4xy + 8y^2 = 144$

c **31.** $9x^2 + 4xy + 6y^2 - 10 = 0$

c **32.** $4x^2 - 6xy + \frac{9}{4}y^2 + 25 = 0$

8.5 Polar Forms of Conics and Parametric Equations

Figure 1

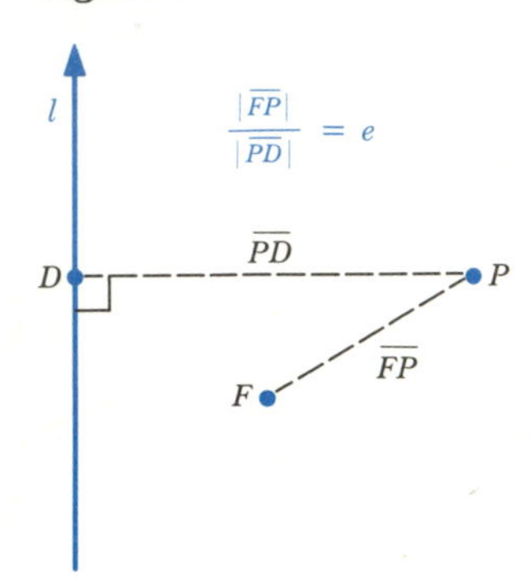

In our earlier work, we derived standard equations for the ellipse, parabola, and hyperbola by using their geometric properties and the distance formula. In this section, we give a general approach that simultaneously applies to all three of these conics. Here we describe all conics as follows.

A **conic** is determined by a given fixed point F called the **focus,** a given fixed line l called the **directrix,** and a positive constant e called the **eccentricity.** The conic consists of all points P that satisfy the ratio

$$\frac{|\overline{FP}|}{|\overline{PD}|} = e$$

where D is the foot of the perpendicular line segment from P to l (Figure 1).

EXAMPLE 1 Find the standard equation of the conic with focus $(2, 0)$, directrix $x = 0$, and eccentricity $e = 2$.

SOLUTION In reference to Figure 2, assume that $P = (x, y)$ is a point on the graph of the conic so that

$$\frac{|\overline{FP}|}{|\overline{PD}|} = \frac{r_1}{r_2} = 2$$

Figure 2

It follows that

$$|\overline{FP}|^2 = 4|\overline{PD}|^2$$

$$(x - 2)^2 + y^2 = 4x^2$$

Thus

$$3x^2 + 4x - y^2 = 4$$

After completing the square, we obtain

$$3\left(x + \frac{2}{3}\right)^2 - y^2 = \frac{16}{3}$$

or equivalently,

$$\frac{\left(x + \frac{2}{3}\right)^2}{\frac{16}{9}} - \frac{y^2}{\frac{16}{3}} = 1$$

which is a standard equation for a hyperbola (Figure 2).

Using the geometric definition of a *parabola* in Section 8.2, it can be seen that the *eccentricity* $e = 1$. Also, it can be shown that if *$e < 1$, the conic is an ellipse;* and if *$e > 1$, the conic is a hyperbola.*

The eccentricity of an ellipse or a hyperbola can be determined from its standard form equation.

THEOREM 1

Consider ellipses and hyperbolas whose equations have the forms

$$\frac{x^2}{a^2} + \frac{y^2}{b^2} = 1 \quad \text{and} \quad \frac{x^2}{a^2} - \frac{y^2}{b^2} = 1$$

where $c^2 = a^2 - b^2$ for the ellipse and $c^2 = a^2 + b^2$ for the hyperbola. Then the eccentricity $e = c/a$.

EXAMPLE 2 Find the eccentricity of the ellipse whose equation is $4x^2 + 9y^2 = 36$.

SOLUTION The equation $4x^2 + 9y^2 = 36$ can be written in the standard form

$$\frac{x^2}{9} + \frac{y^2}{4} = 1$$

This is an equation of an ellipse where $a = 3$ and $b = 2$, so that we have

$$c = \sqrt{a^2 - b^2} = \sqrt{9 - 4} = \sqrt{5}$$

Thus, by Theorem 1, the eccentricity is

$$e = \frac{c}{a} = \frac{\sqrt{5}}{3}$$

Polar Equations of Conics

The eccentricity of a conic can be used to derive standard equations in the polar coordinate system if the focus is located at the pole. Assume that $P = (r, \theta)$ is a point on the graph of a conic with focus O, eccentricity e, and directrix l to the left of the focus, as displayed in Figure 3.

Figure 3

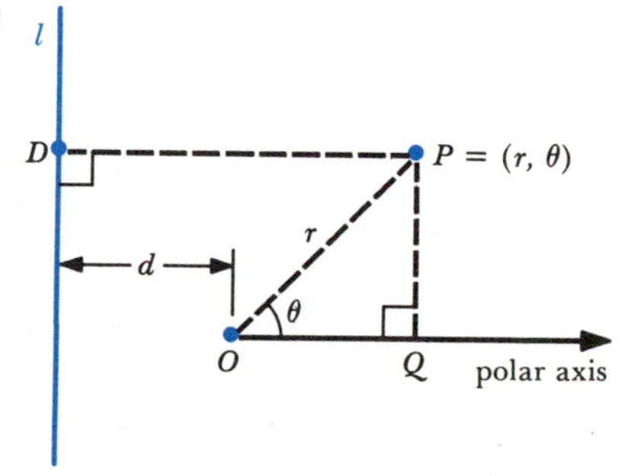

Then

$$\frac{|\overline{OP}|}{|\overline{PD}|} = e$$

or

$$|\overline{OP}| = e|\overline{PD}|$$

so that

$$r = e(d + |\overline{OQ}|)$$

$$r = e(d + r\cos\theta)$$

Solving for r yields the *standard polar equation*

$$r = \frac{ed}{1 - e\cos\theta}$$

Similar derivations lead to the following standard equations in the polar coordinate system for conics with vertical or horizontal directrices and a focus at the pole.

Standard Polar Equations for Conics with a Focus at the Pole and Eccentricity *e*

Equation	Direction and Location of Directrix
(i) $r = \dfrac{ed}{1 - e\cos\theta}$	Vertical, d units left of the pole
(ii) $r = \dfrac{ed}{1 + e\cos\theta}$	Vertical, d units right of the pole
(iii) $r = \dfrac{ed}{1 + e\sin\theta}$	Horizontal, d units above the pole
(iv) $r = \dfrac{ed}{1 - e\sin\theta}$	Horizontal, d units below the pole

In Examples 3 and 4, identify and then graph the given conic.

EXAMPLE 3 $r = \dfrac{1}{1 + 3\sin\theta}$

SOLUTION For

$$r = \frac{1}{1 + 3\sin\theta}$$

$e = 3$ so that the conic is a hyperbola. Because of the equation form, the directrix is horizontal and above the pole so that the axis of symmetry is a vertical line that contains the pole. Thus, the vertices can be found by substituting $\theta = \pi/2$ and also $\theta = 3\pi/2$ into the equation to get

$$\left(\frac{1}{4}, \frac{\pi}{2}\right) \quad \text{and} \quad \left(-\frac{1}{2}, \frac{3\pi}{2}\right) \quad \text{or} \quad \left(\frac{1}{2}, \frac{\pi}{2}\right)$$

The center is located at the midpoint of the line segment determined by the vertices, that is, $(\frac{3}{8}, \pi/2)$. The distance between the center and a vertex is given by

$$a = \frac{3}{8} - \frac{1}{4} = \frac{1}{8}$$

Figure 4

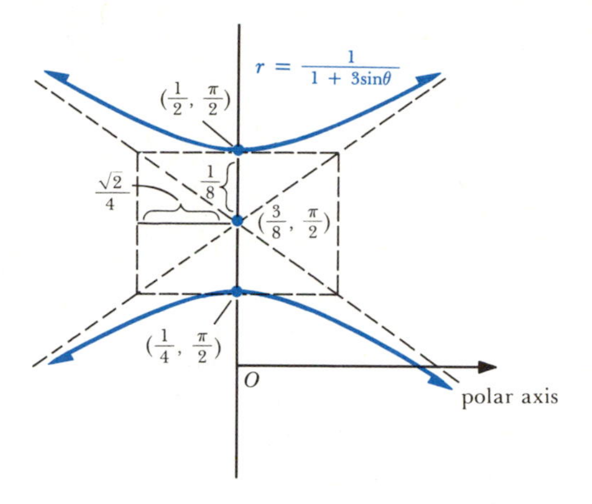

Because the pole is a focus and the center is at $(\frac{3}{8}, \pi/2)$, it follows that $c = \frac{3}{8}$. Thus,

$$b^2 = c^2 - a^2 = \left(\frac{3}{8}\right)^2 - \left(\frac{1}{8}\right)^2 = \frac{1}{8}$$

or

$$b = \sqrt{\frac{1}{8}} = \frac{\sqrt{2}}{4}$$

The graph is obtained by locating the center and the vertices, utilizing the values of a and b to draw the asymptotes, and then sketching the curves (Figure 4).

EXAMPLE 4 $r = \dfrac{5}{2 + 2\cos\theta}$

SOLUTION First we divide the numerator and denominator by 2 to obtain the standard form

$$r = \frac{\frac{5}{2}}{1 + \cos\theta}$$

Figure 5

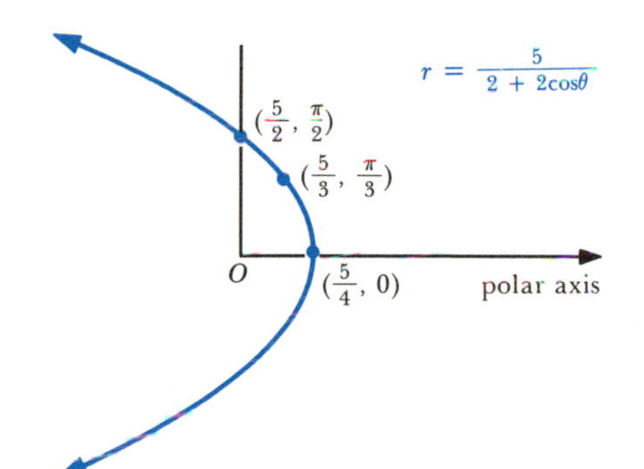

The conic is a parabola because $e = 1$. Since the directrix is to the right of the pole, the parabola opens to the left. The focus is at the pole and the vertex is found by setting $\theta = 0$ to get $(\frac{5}{4}, 0)$. In order to sketch the graph, we locate two points on the graph, in addition to the vertex, namely, $(\frac{5}{2}, \pi/2)$ and $(\frac{5}{3}, \pi/3)$ (Figure 5).

Parametric Equations

Although conics do not necessarily represent functions, they (and many other curves) can be defined by functions. In this case, we describe the location of a point (x, y) on a curve by using functions of a third variable t. This sort of representation is especially important when describing movement along a curve in terms of elapsed time. For instance, assume all points (x, y) on a curve satisfy the equations

$$x = t + 1 \qquad \text{and} \qquad y = t^2$$

where $0 \leq t \leq 5$. These equations are examples of *parametric equations* for x and y. The variable t is called the *parameter*.

Table 1 contains some values of t and the associated x and y values. Each value of t in the interval $[0, 5]$ defines a point (x, y) on the curve. To obtain the graph defined by these parametric equations, we could plot the points in Table 1 and then connect them with a smooth curve. However, it is possible, in this situation, to determine an equation of the curve in terms of x and y alone. This process of "eliminating the parameter" proceeds as follows.

We use the first parametric equation $x = t + 1$ to write t in terms of x as

$$t = x - 1$$

Table 1

t	x	y
0	1	0
1	2	1
$\frac{3}{2}$	$\frac{5}{2}$	$\frac{9}{4}$
3	4	9
$\frac{7}{2}$	$\frac{9}{2}$	$\frac{49}{4}$
4	5	16
5	6	25

Figure 6

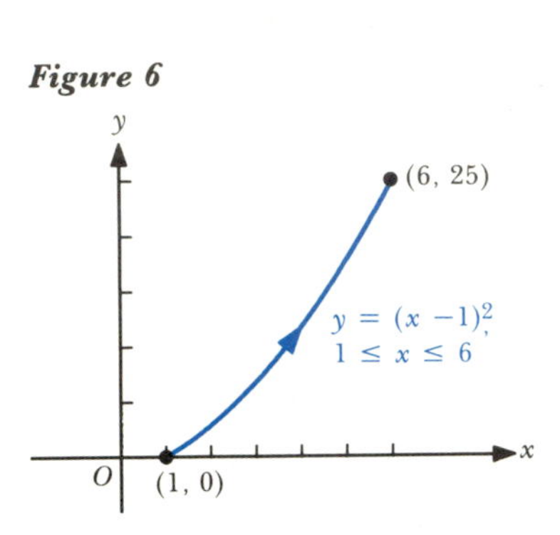

Then we substitute this representation of t into the second parametric equation $y = t^2$ to obtain

$$y = (x - 1)^2$$

which is an equation of a parabola with vertex $(1, 0)$ and axis of symmetry the line $x = 1$. This equation *alone* cannot be used to determine the graph because we have not yet used the restriction on the parameter t that $0 \le t \le 5$. This restriction on t implies that $0 \le x - 1 \le 5$ so that $1 \le x \le 6$. Thus, the graph of the parametric equations is the portion of the graph of $y = (x - 1)^2$ with endpoints $(1, 0)$ and $(6, 25)$ (see Table 1) (Figure 6). Note that an arrow is used to suggest that the curve is "generated" by *increasing* values of t from 0 to 5. If t represents time, we say that the point moves along the path defined by the curve *from* location $(1, 0)$ *to* location $(6, 25)$.

In general, if we define a **curve** C as a set of ordered pairs of the form $(x, y) = (f(t), g(t))$, where f and g are functions defined on an interval I, then the equations

$$x = f(t) \qquad \text{and} \qquad y = g(t)$$

for t in I, are called **parametric equations** for the curve C and t is called the **parameter.** We can often use the parametric equations to derive a *nonparametric* equation for C in terms of x and y. This process is referred to as *eliminating the parameter.*

EXAMPLE 5 Each given pair of parametric equations represents a curve and t represents elapsed time. Eliminate the parameter to write a nonparametric equation for the curve and sketch the graph. Display on the graph the direction in which a point moves along the curve as t increases.

(a) $x = 2\cos t$ and $y = 3\sin t$, $0 \le t \le 2\pi$

(b) $x = 2\cos t$ and $y = 3\sin t$, $0 \le t \le \dfrac{\pi}{2}$

SOLUTION We can eliminate the parameter in both situations as follows. First we write the equations as

$$\frac{x}{2} = \cos t \qquad \text{and} \qquad \frac{y}{3} = \sin t$$

Next we square both sides of each equation and then add the results to each other:

$$\left(\frac{x}{2}\right)^2 = \cos^2 t \qquad \text{and} \qquad \left(\frac{y}{3}\right)^2 = \sin^2 t$$

so that

$$\left(\frac{x}{2}\right)^2 + \left(\frac{y}{3}\right)^2 = \cos^2 t + \sin^2 t$$

or

$$\frac{x^2}{4} + \frac{y^2}{9} = 1$$

which is an equation of an ellipse with center $(0, 0)$, major axis on the y axis, and minor axis on the x axis. However, because of the different restrictions on the parameter in each part, the graphs of the parametric equations will be different portions of the graph of the ellipse, as explained below.

Figure 7

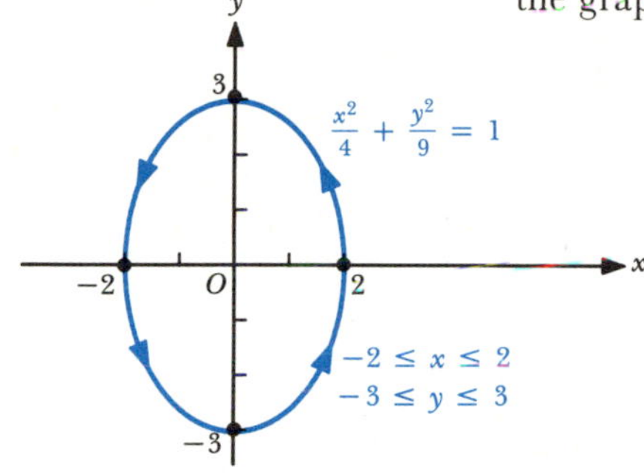

(a)

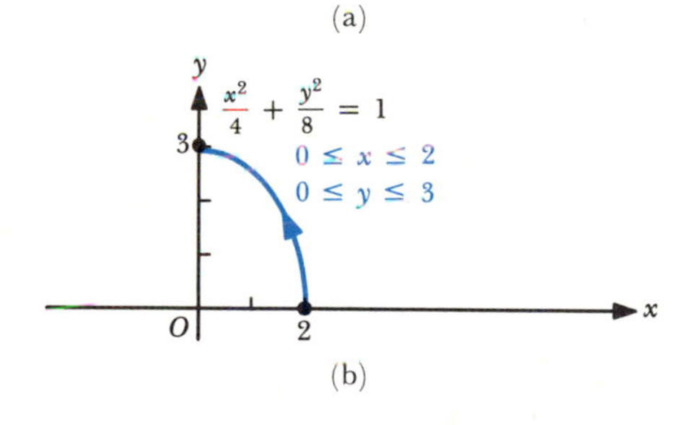

(b)

(a) Here the condition $0 \le t \le 2\pi$ suggests that a point moving along the curve, as t increases from 0 to 2π, "starts" at point $(2\cos 0, 3\sin 0) = (2, 0)$, passes through points $(2\cos(\pi/2), 3\sin(\pi/2)) = (0, 3)$ and $(2\cos\pi, 3\sin\pi) = (-2, 0)$, and continues around the ellipse to end at

$$(2\cos 2\pi, 3\sin 2\pi) = (2, 0)$$

In other words, a point moves along the ellipse in a counterclockwise direction and traverses the ellipse once as t increases from 0 to 2π (Figure 7a).

(b) In contrast, the condition on the parameter t, namely, $0 \le t \le \pi/2$, suggests that a point moving along the curve starts at $(2, 0)$ and continues counterclockwise through the part of the ellipse in quadrant I to end at point $(2\cos(\pi/2), 3\sin(\pi/2)) = (0, 3)$ (Figure 7b).

PROBLEM SET 8.5

In problems 1–6, determine the standard equation for the conic that satisfies the given conditions and sketch the graph.

1. Focus at $(3, 0)$, directrix $y = \frac{5}{3}$, and eccentricity $e = 3/\sqrt{5}$
2. Focus at $(5, 0)$, directrix $x = \frac{16}{5}$, and eccentricity $e = \frac{5}{4}$
3. Focus at $(3, 0)$, directrix $x = \frac{25}{3}$, and eccentricity $e = \frac{3}{5}$
4. Focus at $(1, 2)$, directrix $y = -2$, and eccentricity $e = \frac{1}{2}$
5. Focus at $(2, 0)$, directrix $x = 4$, and eccentricity $e = 1$
6. Directrix has the polar equation $r = 2\csc\theta$, and eccentricity $e = 3$

In problems 7–14, find the eccentricity for the given conic.

7. $25x^2 + 16y^2 = 400$

8. $x^2 + 2y^2 = 1$

9. $x^2 + 3y^2 = 4$

10. $y^2 - 8y + 3x + 5 = 0$

11. $9x^2 - 16y^2 + 144 = 0$

12. $4x^2 - 4y^2 + 1 = 0$

13. $3x^2 - 5x + y^2 + 22y = 1$

14. $x^2 + 16x - y + 7 = 0$

15. Given the equation $25x^2 - 4y^2 + 50x - 12y + 116 = 0$, find the center, the foci, the vertices, equations of the asymptotes, and the eccentricity. Also, sketch the graph.

16. We know that the eccentricity $e = \frac{c}{a} < 1$ for the ellipse $\frac{x^2}{a^2} + \frac{y^2}{b^2} = 1$, where $c^2 = a^2 - b^2$. Now, as e becomes closer and closer to zero in value, $c/a = e$ approaches zero; hence c^2 approaches zero or a approaches b in value (why?). What is the shape of an ellipse in which a and b are close in value? What if $a = b$? Does this give an indication of how to define a conic with eccentricity $e = 0$? Use sketches and examples to answer these questions.

17. Suppose that the foci of a particular ellipse lie midway between the center and the vertices.

(a) Find the eccentricity.

(b) If the length of the major axis is $2a$, find the length of the minor axis and the distance from the center to the directrices.

(c) Sketch the graph.

18. Except for minor perturbations, the orbit of the earth is an ellipse with the sun at one focus. The least and greatest distances from the earth to the sun have a ratio of $\frac{29}{30}$. Find the eccentricity of this elliptical orbit.

In problems 19–24, find the eccentricity, and identify and graph the conic.

19. $r = \dfrac{5}{1 + \sin\theta}$

20. $r = \dfrac{2}{-1 + 3\sin\theta}$

21. $r = \dfrac{2}{2 - \cos\theta}$

22. $r = \dfrac{3}{4 + 2\sin\theta}$

23. $r = \dfrac{5}{3 + 9\cos\theta}$

24. $r = \dfrac{10}{2 - 5\cos\theta}$

In problems 25–34, eliminate the parameter to write a nonparametric equation for the curve defined by the given parametric equations. Also sketch the graph and display the direction in which a point moves along the curve as t increases.

25. $x = 4\cos t, y = 4\sin t, 0 \le t \le 2\pi$

26. $x = \sec t, y = \tan t$

27. $x = 2t + 1, y = 4t^2 - 1, -1 \le t \le 1$

28. $x = 2t, y = t^2, -2 \le t \le 2$

29. $x = t, y = \sqrt{1 - t^2}, -1 \le t \le 1$

30. $x = \sin t + 1, y = 4\cos t - 1$

31. $x = 3 + 2\cos t, y = 5 + 2\sin t$

32. $x = e^t, y = e^{-2t}$

33. $x = \csc t, y = \cot t$

34. $x = \sin 2t, y = \cos 2t$

REVIEW PROBLEM SET, CHAPTER 8

In problems 1–5, find the standard equation of the circle (or circles) that satisfies the given conditions.

1. Center at $(4, -3)$ and tangent to the y axis
2. Center at $(-5, 0)$ and tangent to the line $y = -3$
3. Center at $(1, 2)$ and contains the point $(3, 1)$
4. Center at the origin and tangent to the circle $(x - 3)^2 + y^2 = 8$ (two possible circles)
5. Contains the three points $(-2, -1)$, $(0, 2)$, and $(4, -1)$
6. The endpoints of the base of a triangle are $(-2, 0)$ and $(2, 0)$, and the sum of the lengths of the other two sides is 6. Find an equation of the set of all points that are possible vertices.

In problems 7–11, find the standard equation of the ellipse satisfying the given conditions and sketch the graph.

7. Vertices at $(-5, 0)$ and $(5, 0)$, and foci at $(-3, 0)$ and $(3, 0)$
8. Foci at $(-1, 0)$ and $(1, 0)$, and minor axis of length $2\sqrt{2}$
9. Vertices at $(0, 2)$, $(1, 5)$, $(2, 2)$, and $(1, -1)$
10. Foci at $(1, 0)$ and $(9, 0)$, and the sum of the distances from any point on the ellipse to the foci is 10
11. Contains the point $(3, 1)$, center at the origin, symmetric with respect to the coordinate axes, and the length of the horizontal major axis is 3 times the length of the minor axis
12. Except for minor perturbations, a satellite orbiting the earth moves in an elliptical orbit with the center of the earth at one focus. Suppose that a satellite at *perigee* (the point of its orbit nearest to the center of the earth) is 400 kilometers from the surface of the earth and at *apogee* (the point in its orbit farthest from the center of the earth) is 600 kilometers from the surface of the earth. Assume that the earth is a sphere of radius 6371 kilometers. Find the semiminor axis b (one-half the minor axis) of the elliptical orbit.

In problems 13–17, find the coordinates of the center, the vertices, and the foci of the given ellipse, then sketch the graph.

13. $\dfrac{x^2}{8} + \dfrac{y^2}{12} = 1$
14. $144x^2 + 169y^2 = 24{,}336$
15. $9x^2 + 25y^2 + 18x - 50y - 191 = 0$
16. $3x^2 + 4y^2 - 28x - 16y + 48 = 0$
17. $9x^2 + 4y^2 + 72x - 48y + 144 = 0$
18. Find an equation of the line containing the points with y coordinates of 2 and 8, respectively, that lie on the parabola $y^2 = 8x$.

In problems 19–23, find the standard equation of the parabola satisfying the given conditions and sketch the graph.

19. Focus at $(-4, 0)$ and directrix $y = 3$
20. Focus at $(0, 3)$ and directrix $x = \frac{5}{2}$

21. Focus at $(1, -\frac{5}{2})$ and directrix $y = -4$

22. Focus at $(3, 0)$, containing point $(2, 2\sqrt{2})$, the x axis as the axis of symmetry, and opening to the right

23. Vertex at $(-2, -3)$ and focus at $(-5, -3)$

24. Find the points of intersection of the line with equation $x - 2y - 3 = 0$ and the parabola with equation $y^2 = x$.

In problems 25–29, find the vertex, the focus, the axis of symmetry, and sketch the graph of each parabola.

25. $x^2 = -8y$ **26.** $(y - 3)^2 = 4x$ **27.** $y^2 = -5(x + 2)$

28. $y^2 - 2y - 6x = 53$ **29.** $x^2 - 16y + 2x + 49 = 0$

30. Show that if a point P is equidistant from the y axis and the point $(6, 0)$, then its coordinates (x, y) must satisfy the equation $y^2 = 12(x - 3)$. Is the converse true?

In problems 31–34, find the standard equation of the hyperbola that satisfies the given conditions and sketch the graph.

31. Vertices at $(-2, 0)$ and $(2, 0)$, and foci at $(-3, 0)$ and $(3, 0)$

32. Contains the point $(5, 9)$ and with asymptotes $y = \pm x$

33. Center at $(0, 0)$, focus at $(0, 5)$, and vertex at $(0, 4)$

34. Vertices at $(-15, 0)$ and $(15, 0)$, and asymptotes $y = \pm\frac{4}{5}x$

In problems 35–38, find the coordinates of the center, the vertices, and the foci of the given hyperbola. Also find equations of the asymptotes and sketch the graph.

35. $x^2 - 9y^2 = 72$ **36.** $y^2 - 9x^2 = 54$

37. $x^2 - 4y^2 + 4x + 24y - 48 = 0$ **38.** $4y^2 - x^2 - 24y + 2x + 34 = 0$

39. Show that the hyperbolas whose equations are

$$\frac{x^2}{4} - \frac{y^2}{9} = 1 \quad \text{and} \quad \frac{y^2}{9} - \frac{x^2}{4} = 1$$

have the same set of asymptotes, and sketch them on the same coordinate system.

40. Let d_1 be the distance from point $P = (x, y)$ to point $A = (-10, 0)$ and d_2 be the distance from point $P = (x, y)$ to point $B = (10, 0)$. Show that if $d_1 - d_2 = 12$, then the coordinates of P must satisfy the equation $\frac{x^2}{36} - \frac{y^2}{64} = 1$. Is it true that every point P whose coordinates satisfy this equation also satisfies $d_1 - d_2 = 12$?

41. Two points are 2000 feet apart. At one of these points the report of a cannon is heard 1 second later than at the other. By means of the definition of a hyperbola, show that the cannon is somewhere on a particular hyperbola, then after making a suitable choice of axes, write its equation. (Consider the velocity of sound to be 1100 feet per second.)

42. Consider the equation $(b^2 - k)x^2 + (a^2 - k)y^2 = 1$, where $a \geq b$.

(a) For what choices of k is the graph an ellipse?

(b) For what choices of k is the graph a hyperbola?

(c) For what choices of k is the graph a circle?

43. If $\bar{x} = x - 2$ and $\bar{y} = y + 4$, find the xy coordinates of the point whose $\bar{x}\bar{y}$ coordinates are
(a) $(-5, 2)$ (b) $(5, -3)$ (c) $(-7, -1)$

44. If $\bar{x} = x + 3$ and $\bar{y} = y - 1$, find the $\bar{x}\bar{y}$ coordinates of the point whose xy coordinates are
(a) $(6, 1)$ (b) $(-2, 9)$ (c) $(-1, -4)$

In problems 45 and 46, find a translation of axes that transforms the given equation of a conic to a form in which the center is located at the origin. Sketch the graph of the conic, displaying both sets of coordinate axes.

45. $x^2 - 2x + y^2 + 6y - 37 = 0$

46. $9y^2 - 16x^2 - 90y - 256x = 233$

In problems 47 and 48, assume that a new $\bar{x}\bar{y}$ coordinate system is formed by rotating the xy coordinate system through angle θ.

47. Find the coordinates of each of the given points with respect to the $\bar{x}\bar{y}$ coordinate system if $\theta = 45°$ and the points have xy coordinates given by
(a) $(-1, 3)$ (b) $(4, 5)$ (c) $(-4, -1)$
(d) $(2, -4)$ (e) $(2, 0)$

48. Find the coordinates with respect to the xy system if $\theta = 30°$ and the $\bar{x}\bar{y}$ coordinates are given by
(a) $(2, 3)$ (b) $(1, -5)$ (c) $(-1, -3)$
(d) $(0, -2)$ (e) $(-4, 2)$

In problems 49 and 50, assume an xy coordinate system is rotated through angle θ to form an $\bar{x}\bar{y}$ coordinate system. Write the given equation in terms of the $\bar{x}\bar{y}$ system, sketch the graph, and display both sets of coordinate axes.

49. $x^2 - 4xy + y^2 - 6 = 0$; $\theta = 45°$

50. $3x + y = 1$; $\theta = 60°$

c In problems 51 and 52, use a rotation of axes to eliminate the xy term in the given equation. Write the transformed equation in standard form, graph the equation, and determine the angle of rotation to the nearest tenth of a degree.

51. $6x^2 - 6xy + 14y^2 = 45$

52. $x^2 + 4xy - 2y^2 = 12$

In problems 53–56, determine the standard equation for the conic in Cartesian form that satisfies the given conditions and sketch the graph.

53. Focus at $(2, 0)$, directrix $x = \frac{9}{2}$, and eccentricity $e = \frac{2}{3}$

54. Vertices at $(-5, 0)$ and $(5, 0)$, and eccentricity $e = \frac{3}{5}$

55. Foci at $(-10, 0)$ and $(10, 0)$, and eccentricity $e = \frac{5}{4}$

56. Focus at $(0, 2)$, directrix $y = -3$, and eccentricity $e = 1$

In problems 57–59, find the eccentricity for the given conic and identify the conic.

57. $5x^2 - y^2 = 10$

58. $3x^2 + 2y^2 = 6$

59. $x^2 - 2x + 2y^2 + 6y - 8 = 0$

60. The earth moves in an elliptical orbit with eccentricity 0.017, major axis 185.8 million miles, and the sun at one focus. When the earth is positioned on the major axis, how close is it to the sun?

In problems 61–64, find the eccentricity, and identify and graph the conic.

61. $r = \dfrac{6}{3 + \cos\theta}$

62. $r = \dfrac{1}{1 - \sin\theta}$

63. $r = \dfrac{1}{2 - 4\cos\theta}$

64. $r = \dfrac{3}{1 + 3\sin\theta}$

In problems 65–68, eliminate the parameter to write a nonparametric equation for the curve defined by the given parametric equations. Also sketch the graph and display the direction in which a point moves along the curve as t increases.

65. $x = 5 + \cos t, y = \sin t - 2$

66. $x = -t, y = \sqrt{1 - t^2}, -1 \leq t \leq 1$

67. $x = e^t, y = e^{-t}$

68. $x = \sqrt{t}, y = 2t + 7, t \geq 0$

CHAPTER 9

Sequences, Series, and Mathematical Induction

We conclude our study of functions with a brief discussion of sequences and series. In addition, we discuss summation notation, mathematical induction, factorial notation, and the binomial theorem.

9.1 Sequences

An **infinite sequence** is a function whose domain is the set of positive integers. The range values of a sequence are referred to as the **terms** of the sequence. For instance,

$$f(n) = \frac{1}{n}$$

where n is a positive integer, is an infinite sequence with terms

$$1, \frac{1}{2}, \frac{1}{3}, \frac{1}{4}, \ldots, \frac{1}{n}, \ldots$$

It is conventional to use **subscript notation** rather than regular function notation to denote terms of a sequence. Thus $\boldsymbol{a_n}$ denotes that a sequence has terms

$$a_1, a_2, a_3, \ldots, a_n, \ldots$$

For example, the first five terms of the sequence defined by $a_n = n^2 - n$ are:

$$\begin{aligned} a_1 &= 1^2 - 1 = 0 \\ a_2 &= 2^2 - 2 = 2 \\ a_3 &= 3^2 - 3 = 6 \\ a_4 &= 4^2 - 4 = 12 \\ a_5 &= 5^2 - 5 = 20 \end{aligned}$$

Also, the notation $\{a_n\}$ is used as a shorthand for the sequence whose ***n*th term** is a_n. For instance, $\{2n\}$ has terms

$$2, 4, 6, 8, \ldots, 2n, \ldots$$

and $\{1 - 1/n\}$ has terms

$$0, \frac{1}{2}, \frac{2}{3}, \frac{3}{4}, \ldots, 1 - \frac{1}{n}, \ldots$$

EXAMPLE 1 Find the first five terms of each sequence.

(a) $\{a_n\} = \{(-1)^n\}$ (b) $\{a_n\} = \left\{3 - \frac{1}{n}\right\}$

SOLUTION To find the first five terms of each sequence, we substitute the positive integers 1, 2, 3, 4, and 5 in turn for n in the formula for the general term. Thus, we obtain:

(a) $a_1 = (-1)^1 = -1, \quad a_2 = (-1)^2 = 1, \quad a_3 = (-1)^3 = -1$

$$a_4 = (-1)^4 = 1, \quad \text{and} \quad a_5 = (-1)^5 = -1$$

so that the first five terms of this sequence are $-1, 1, -1, 1$, and -1.

(b) $a_1 = 3 - \frac{1}{1} = 2, \quad a_2 = 3 - \frac{1}{2} = \frac{5}{2}, \quad a_3 = 3 - \frac{1}{3} = \frac{8}{3}$

$$a_4 = 3 - \frac{1}{4} = \frac{11}{4}, \quad \text{and} \quad a_5 = 3 - \frac{1}{5} = \frac{14}{5}$$

Therefore, the first five terms of the sequence $\{a_n\}$ are 2, $\frac{5}{2}$, $\frac{8}{3}$, $\frac{11}{4}$, and $\frac{14}{5}$.

Arithmetic and Geometric Sequences

We now consider special types of sequences. The first type we consider, an *arithmetic sequence,* is defined as follows.

DEFINITION 1 **Arithmetic Sequence**

A sequence $a_1, a_2, a_3, \ldots, a_n, \ldots$ is called an **arithmetic sequence** (or an **arithmetic progression**) if each term (after the first term) differs from the preceding term by a fixed number.

For example, the sequence $\{a_n\}$ whose general term a_n is given by

$$a_n = 1 + 2n$$

is an arithmetic sequence. The terms of $\{a_n\}$ are

$$a_1 = 3, a_2 = 5, a_3 = 7, a_4 = 9, \ldots, a_n = 1 + 2n, \ldots$$

Note that after the first term a_1, each term in the sequence is always 2 more than the preceding term:

$$\begin{aligned}
a_1 &= 3 && = 3\\
a_2 &= 3 + 2 && = 3 + 1 \cdot 2\\
a_3 &= (3 + 2) + 2 && = 3 + 2 \cdot 2\\
a_4 &= [(3 + 2) + 2] + 2 && = 3 + 3 \cdot 2\\
&\vdots\\
a_n &= [(3 + 2) + 2] + 2 + \cdots + 2 && = 3 + (n - 1) \cdot 2
\end{aligned}$$

In general, a sequence

$$a_1, a_2, a_3, \ldots, a_n, \ldots$$

is an arithmetic sequence (or an arithmetic progression) if it can be expressed in the form

$$a_1, a_1 + d, a_1 + 2d, a_1 + 3d, \ldots, a_1 + (n - 1)d, \ldots$$

for every positive integer n. The number d is called the **common difference** associated with the arithmetic sequence. The nth term a_n of such a sequence is given by

$$a_n = a_1 + (n - 1)d$$

EXAMPLE 2 Find the tenth term of the arithmetic sequence in which the first four terms are 2, -1, -4, and -7.

SOLUTION Because $-1 - 2 = -3$, the common difference is $d = -3$. We substitute $a_1 = 2$, $d = -3$, and $n = 10$ in the formula $a_n = a_1 + (n - 1)d$ to obtain

$$\begin{aligned} a_{10} &= 2 + (10 - 1)(-3) \\ &= 2 - 27 \\ &= -25 \end{aligned}$$

EXAMPLE 3 If the third term of an arithmetic sequence is 7 and the seventh term is 15, find the fifth term.

SOLUTION We substitute $n = 3$ and $a_3 = 7$, and then $n = 7$ and $a_7 = 15$ in $a_n = a_1 + (n - 1)d$ to obtain the following system of linear equations in the unknowns a_1 and d:

$$\begin{cases} 7 = a_1 + (3 - 1)d \\ 15 = a_1 + (7 - 1)d \end{cases} \quad \text{or} \quad \begin{cases} a_1 + 2d = 7 \\ a_1 + 6d = 15 \end{cases}$$

After solving for d and a_1, we get $a_1 = 3$ and $d = 2$. Therefore,

$$\begin{aligned} a_5 &= 3 + (5 - 1)(2) \\ &= 3 + 8 \\ &= 11 \end{aligned}$$

Another important type of sequence is called a *geometric sequence*.

DEFINITION 2 **Geometric Sequence**

A sequence $a_1, a_2, a_3, \ldots, a_n, \ldots$ is called a **geometric sequence** (or a **geometric progression**) if each term (after the first) is obtained by multiplying the preceding term by a fixed number.

For example, the sequence $\{a_n\}$ whose general term a_n is given by

$$a_n = 3(2)^{n-1}$$

is a geometric sequence. The terms of $\{a_n\}$ are

$$3, 3(2), 3(2^2), 3(2^3), \ldots$$

Note that each term in this geometric sequence (after the first term) is obtained by multiplying the preceding term by 2.

In general, a sequence of the form

$$a_1, a_1r, a_1r^2, \ldots, a_1r^{n-1}, \ldots$$

for every positive integer n is a geometric sequence or a geometric progression. The number r is called the **common ratio** associated with the geometric progression, and the nth term a_n is given by

$$\boxed{a_n = a_1r^{n-1}}$$

EXAMPLE 4 Find the tenth term of the geometric sequence having the first term $a_1 = \frac{1}{2}$ and common ratio $r = \frac{1}{2}$.

SOLUTION Substituting $a_1 = \frac{1}{2}$, $r = \frac{1}{2}$, and $n = 10$ in the formula

$$a_n = a_1r^{n-1}$$

we have

$$\begin{aligned} a_{10} &= \frac{1}{2}\left(\frac{1}{2}\right)^{10-1} \\ &= \frac{1}{2}\left(\frac{1}{2}\right)^9 = \frac{1}{1024} \end{aligned}$$

EXAMPLE 5 If the first two terms of a geometric sequence are 2 and 4, respectively, which term of the sequence is equal to 512?

SOLUTION Since $a_1 = 2$ and $a_2 = 4$, we conclude that $r = a_2/a_1 = \frac{4}{2} = 2$. If $a_n = 512$, then

$$512 = 2(2^{n-1})$$

or

$$2^9 = 2^{1+n-1} = 2^n$$

so that

$$n = 9$$

Therefore, the ninth term of the sequence is 512.

PROBLEM SET 9.1

In problems 1–6, find the first five terms of each sequence.

1. $\{a_n\} = \left\{\frac{n(n+2)}{2}\right\}$

2. $\{b_n\} = \left\{\frac{n+4}{n}\right\}$

3. $\{c_n\} = \left\{\frac{n(n-3)}{2}\right\}$

4. $\{a_n\} = \left\{\frac{3}{n(n+1)}\right\}$

5. $\{a_n\} = \{(-1)^n + 3\}$

6. $\{c_n\} = \left\{\frac{n^2-2}{2}\right\}$

In problems 7–14, determine which sequences are arithmetic and find the common difference d for each arithmetic sequence.

7. 2, 5, 8, 11, . . .

8. 3, 5, 7, 9, . . .

9. 7, 12, 17, 22, . . .

10. $11a + 7b, 7a + 2b, 3a - 3b, \ldots$

11. 67, 54, 41, 28, . . .

12. $9a^2, 16a^2, 23a^2, 30a^2, \ldots$

13. 5.7, 6.9, 8.1, 9.3, . . .

14. 1.4, 4.5, 7.6, 10.7, . . .

15. Find the tenth and fifteenth terms of the following arithmetic progression: $-13, -6, 1, 8, \ldots$.

16. Find the twelfth and thirty-fifth terms of the following arithmetic progression: 19, 17, 15, 13, ...

17. Find the sixth and ninth terms of the following arithmetic sequence: $a + 24b, 4a + 20b, 7a + 16b, \ldots$.

18. Find the third and sixteenth terms of the following arithmetic sequence: $7a^2 - 4b, 2a^2 + 7b, -3a^2 + 18b, \ldots$.

In problems 19–26, determine which sequences are geometric and give the value of the common ratio r for each geometric sequence.

19. 2, 6, 18, . . .

20. $1, \frac{1}{5}, \frac{1}{25}, \ldots$

21. $1, -2, 4, \ldots$

22. $\frac{4}{9}, \frac{1}{6}, \frac{1}{16}, \ldots$

23. 81, 54, 36, . . .

24. $147, -21, 3, \ldots$

25. $9, -6, 4, \ldots$

26. $64, -32, 16, \ldots$

In problems 27–32, find the indicated term of each geometric sequence.

27. The tenth term of $-4, 2, -1, \frac{1}{2}, \ldots$

28. The eighth term of $\frac{1}{8}, \frac{1}{4}, \frac{1}{2}, \ldots$

29. The fifth term of 32, 16, 8, . . .

30. The eleventh term of $1, 1.03, (1.03)^2, \ldots$

31. The nth term of $1, 1 + a, (1 + a)^2, \ldots$

32. The twelfth term of $10^{-5}, 10^{-7}, 10^{-9}, \ldots$

33. Find the sixth and tenth terms of the geometric sequence 6, 12, 24, 48, ...

34. Find the sixth and eighth terms of the geometric sequence 2, 6, 18, ...

35. Find the fifth term of the geometric progression 3, 6, 12, ...

36. Find the eleventh term of the geometric progression $10, 10^2, 10^3, \ldots$.

37. If the first two terms of a geometric sequence are 2 and 1, respectively, which term of the sequence is equal to $\frac{1}{16}$?

C **38.** Given the sequence $\{a_n\}$ where $a_n = \left(1 + \frac{1}{n}\right)^n$

(a) Evaluate a_{50}, a_{100}, a_{1000}, and a_{10000} to 4 significant digits.

(b) Compare the results from part(a) to an approximation of e to 4 significant digits. Note that for increasing values of n, a_n approaches the value of e.

9.2 Summation Notation and Series

Some applications of mathematics involve finding the *sum* of the terms of an infinite sequence.

Although the sum of the first n terms of a sequence

$$a_1, a_2, a_3, \ldots, a_n, \ldots$$

can be written as

$$a_1 + a_2 + a_3 + \cdots + a_n$$

a more compact notation is useful. The Greek capital letter $\sum$ (sigma) is used for this purpose. We write the sum in **sigma notation** as

$$\sum_{k=1}^{n} a_k = a_1 + a_2 + a_3 + \cdots + a_n$$

Here $\sum$ indicates a sum, and the symbols above and below $\sum$ indicate that k takes on integer values from 1 to n inclusive. k is called the **index of summation.** There is no particular reason to use k for the index of summation, any letter will do; however, i, j, k, and n are the most commonly used indices. For instance,

$$\sum_{k=1}^{n} 5^k = \sum_{i=1}^{n} 5^i = \sum_{j=1}^{n} 5^j = 5^1 + 5^2 + 5^3 + \cdots + 5^n$$

In Examples 1 and 2, evaluate each sum.

EXAMPLE 1 $\sum_{k=1}^{3} (4k^2 - 3k)$

SOLUTION Here we have $a_k = 4k^2 - 3k$. To find the indicated sum, we substitute the integers 1, 2, and 3 for k in succession, and then add the resulting numbers. Thus,

$$\sum_{k=1}^{3} (4k^2 - 3k) = [4(1^2) - 3(1)] + [4(2^2) - 3(2)] + [4(3^2) - 3(3)]$$

$$= 1 + 10 + 27 = 38$$

EXAMPLE 2 $\sum_{k=2}^{5} \frac{k-1}{k+1}$

SOLUTION Here we have $a_k = (k - 1)/(k + 1)$, where the index starts at $k = 2$. Thus,

$$\sum_{k=2}^{5} \frac{k-1}{k+1} = \frac{2-1}{2+1} + \frac{3-1}{3+1} + \frac{4-1}{4+1} + \frac{5-1}{5+1}$$

$$= \frac{1}{3} + \frac{2}{4} + \frac{3}{5} + \frac{4}{6}$$

$$= \frac{21}{10}$$

Next we state some basic properties of summation that are used in calculus, statistics, probability, and other advanced courses.

Assume $\{a_n\}$ and $\{b_n\}$ are given sequences and c represents a constant, then the following properties hold.

Basic Properties of Summation

1. **Constant Property:** $\sum_{k=1}^{n} c = nc$
2. **Homogeneous Property:** $\sum_{k=1}^{n} ca_k = c \sum_{k=1}^{n} a_k$
3. **Additive Property:** $\sum_{k=1}^{n} (a_k + b_k) = \sum_{k=1}^{n} a_k + \sum_{k=1}^{n} b_k$
4. **Sum of Successive Integers:** $\sum_{k=1}^{n} k = \frac{n(n+1)}{2}$
5. **Sum of Successive Squares:** $\sum_{k=1}^{n} k^2 = \frac{n(n+1)(2n+1)}{6}$

The first three properties can be verified by first expanding the summation and then using basic properties of algebra. Property 2 can be confirmed as follows:

$$\begin{aligned}\sum_{k=1}^{n} ca_k &= ca_1 + ca_2 + \cdots + ca_n \\ &= c(a_1 + a_2 + \cdots + a_n) \\ &= c \sum_{k=1}^{n} a_k\end{aligned}$$

(See Problem 22 for Properties 1 and 3. Properties 4 and 5 will be proved later in Section 9.3 on pages 515 and 517.)

EXAMPLE 3 Use the basic summation properties to evaluate

$$\sum_{k=1}^{20} (2k^2 - 3k + 4)$$

SOLUTION We evaluate the given sum by applying the basic summation properties:

$$\begin{aligned}\sum_{k=1}^{20} (2k^2 - 3k + 4) &= \sum_{k=1}^{20} 2k^2 + \sum_{k=1}^{20} (-3k) + \sum_{k=1}^{20} 4 && \text{(Property 3)} \\ &= 2 \sum_{k=1}^{20} k^2 + (-3) \sum_{k=1}^{20} k + 80 && \text{(Properties 1 and 2)} \\ &= 2\left[\frac{(20)(21)(41)}{6}\right] - 3\left[\frac{(20)(21)}{2}\right] + 80 && \text{(Properties 4 and 5)} \\ &= 5190\end{aligned}$$

The Sum of the First n Terms of Arithmetic and Geometric Sequences

We begin by deriving formulas for the sum of the first n terms of an arithmetic sequence. Assume that

$$a_1, a_2, a_3, \ldots, a_n, \ldots$$

is an arithmetic sequence with a first term a_1 and a common difference d. Let S_n represent the sum of the first n terms, that is,

$$S_n = a_1 + a_2 + \cdots + a_n = \sum_{k=1}^{n} a_k$$

From Section 9.1 on page 503, we know that

$$a_k = a_1 + (k - 1)d$$

so that, by using the basic properties of summation, we get

$$\begin{aligned} S_n &= \sum_{k=1}^{n} a_k \\ &= \sum_{k=1}^{n} [a_1 + (k - 1)d] \\ &= \sum_{k=1}^{n} a_1 + \sum_{k=1}^{n} (k - 1)d = na_1 + d \sum_{k=1}^{n} k - \sum_{k=1}^{n} d \\ &= na_1 + d\left[\frac{n(n + 1)}{2}\right] - nd = \frac{n}{2}[2a_1 + (n - 1)d] \end{aligned}$$

Therefore, the formula for the sum S_n of the first n terms of an arithmetic sequence $\{a_1 + (n - 1)d\}$ is:

$$S_n = \frac{n}{2}[2a_1 + (n - 1)d]$$

Using the fact that $a_n = a_1 + (n - 1)d$, we can rewrite this formula as follows:

$$S_n = \frac{n}{2}[2a_1 + (n - 1)d] = \frac{n}{2}[a_1 + a_1 + (n - 1)d] = \frac{n}{2}(a_1 + a_n)$$

Therefore,

$$S_n = \frac{n}{2}(a_1 + a_n)$$

EXAMPLE 4 Find the sum of the first twenty terms of an arithmetic sequence whose first term is 2 and whose common difference is 4.

SOLUTION We substitute $a_1 = 2$, $d = 4$, and $n = 20$ in $S_n = (n/2)[2a_1 + (n-1)d]$ to obtain

$$S_{20} = \frac{20}{2}[2(2) + (20 - 1)4]$$
$$= 10(4 + 76) = 800$$

EXAMPLE 5 How many terms are there in the arithmetic sequence for which $a_1 = 3$, $d = 5$, and $S_n = 255$?

SOLUTION Since $S_n = 255$, $a_1 = 3$, and $d = 5$, we use the formula $S_n = (n/2)[2a_1 + (n-1)d]$ to get

$$255 = \frac{n}{2}[6 + 5(n - 1)]$$
$$510 = n[6 + 5n - 5]$$
$$510 = n(1 + 5n)$$
$$510 = n + 5n^2$$
$$5n^2 + n - 510 = 0$$
$$(5n + 51)(n - 10) = 0$$

so that

$$n = 10 \qquad \text{or} \qquad n = -\tfrac{51}{5}$$

Since n must be a positive integer, the sequence has ten terms.

We now consider a way of finding the sum of the first n terms of a geometric sequence. Let us assume that the geometric sequence is given by

$$a_1, a_1r, a_1r^2, \ldots, a_1r^{n-1}, \ldots$$

To find a formula for the sum

$$S_n = a_1 + a_1r + a_1r^2 + \cdots + a_1r^{n-1} = \sum_{k=1}^{n} a_1r^{k-1}$$

we start with the expression for S_n in expanded form, that is,

$$S_n = a_1 + a_1r + a_1r^2 + \cdots + a_1r^{n-1}$$

We then multiply both sides of the equation by r to obtain

$$rS_n = a_1r + a_1r^2 + a_1r^3 + \cdots + a_1r^n$$

Next we subtract rS_n from S_n to get

$$S_n - rS_n = a_1 - a_1r^n$$

so that the equation becomes

$$(1 - r)S_n = a_1(1 - r^n) \qquad \text{or} \qquad S_n = \frac{a_1(1 - r^n)}{1 - r}, \qquad \text{if } r \neq 1$$

Therefore, the formula for the sum S_n of the first n terms of a geometric sequence $\{a_1 r^{n-1}\}$ is

$$S_n = \sum_{k=1}^{n} a_1 r^{k-1} = \frac{a_1(1 - r^n)}{1 - r}, \qquad r \neq 1$$

EXAMPLE 6 Find the sum of the first ten terms of the geometric sequence whose first term is $\frac{1}{2}$ and whose common ratio is 2.

SOLUTION We substitute $n = 10$, $a_1 = \frac{1}{2}$, and $r = 2$ in the formula

$$S_n = \frac{a_1(1 - r^n)}{1 - r}$$

to obtain

$$S_{10} = \frac{\frac{1}{2}(1 - 2^{10})}{1 - 2} = \frac{\frac{1}{2}(-1023)}{-1} = \frac{1023}{2}$$

EXAMPLE 7 The sum of the first five terms of a geometric sequence is $\frac{61}{27}$ and the common ratio is $-\frac{1}{3}$. Find the first four terms of the sequence.

SOLUTION Using the formula for S_n, we have

$$\frac{61}{27} = \frac{a_1[1 - (-\frac{1}{3})^5]}{1 - (-\frac{1}{3})}$$

Thus

$$\frac{61}{27} = \left(\frac{\frac{244}{243}}{\frac{4}{3}}\right) a_1 \qquad \text{or} \qquad \frac{61}{27} = \frac{61}{81} a_1 \qquad \text{so that} \qquad a_1 = \frac{81}{61} \cdot \frac{61}{27} = 3$$

Hence, the first four terms of the sequence are 3, $3(-\frac{1}{3})$, $3(-\frac{1}{3})^2$, and $3(-\frac{1}{3})^3$, or 3, -1, $\frac{1}{3}$, and $-\frac{1}{9}$.

Geometric Series

We can write the sum of the terms of a sequence, such as

$$a_1 + a_2 + a_3 + \cdots + a_n + \cdots$$

more compactly, using sigma notation $\sum_{k=1}^{\infty} a_k$. Such a sum represents an **infinite series** or a **series.** The numbers $a_1, a_2, a_3, \ldots$ are called the terms of the series, and a_n is called the ***n*th term,** or the **general term,** of the series.

Although we cannot add an infinite number of terms, it is sometimes useful to assign a numerical value to an infinite series by means of a special definition and to refer to this value as "the sum" of the series. This is accomplished by using the *partial sums* of the series. The sum s_n of the first n terms of the series

$$\sum_{k=1}^{\infty} a_k = a_1 + a_2 + a_3 + \cdots a_n + \cdots$$

is called the ***n*th partial sum** of the series; thus

$$s_1 = a_1$$

$$s_2 = a_1 + a_2$$

$$s_3 = a_1 + a_2 + a_3$$

$$\cdots\cdots\cdots\cdots\cdots\cdots$$

$$s_n = a_1 + a_2 + \cdots + a_n = \sum_{k=1}^{n} a_k$$

Then the sum S, $\sum_{k=1}^{\infty} a_k$, which is written as $S = \sum_{k=1}^{\infty} a_k$, is defined to be the "limit value" that s_n approaches as "n approaches infinity," if there is a (finite) limit value.

Let us apply this notion to determine the formula for the **sum of a geometric series,** that is, a series of the form

$$\sum_{k=1}^{\infty} ar^{k-1} = a + ar + ar^2 + \cdots + ar^{n-1} + \cdots$$

where a and r are constants and $|r| < 1$.

Here, $a_k = ar^{k-1}$, so that the partial sums are given by

$$s_1 = a$$

$$s_2 = a + ar$$

$$s_3 = a + ar + ar^2$$

$$\cdots\cdots\cdots\cdots\cdots\cdots$$

$$s_n = a + ar + ar^2 + ar^3 + \cdots + ar^{n-1}$$

However, we have already established (see page 510) that

$$s_n = \frac{a(1 - r^n)}{1 - r} = \frac{a - ar^n}{1 - r}$$

The latter equation can be written as

$$s_n = \frac{a}{1 - r} - \frac{ar^n}{1 - r}$$

Since $|r| < 1$, we see, intuitively, that as n becomes increasingly large (as n approaches infinity), r^n approaches 0. [Consider what happens, for example to the values of $(\frac{1}{3})^n$ as n becomes larger and larger by examining the graph of the function $f(x) = (\frac{1}{3})^x$ (Section 4.1, page 188).]

Consequently, s_n approaches $a/(1-r)$ as n approaches infinity, so that the sum S is

$$S = \sum_{k=1}^{\infty} ar^{k-1} = \frac{a}{1-r} \qquad \text{where } |r| < 1$$

For example,

$$\sum_{k=1}^{\infty} \left(\frac{1}{2}\right)^{k-1} = 1 + \frac{1}{2} + \frac{1}{4} + \frac{1}{8} + \cdots + \left(\frac{1}{2}\right)^{n-1} + \cdots$$

is a geometric series in which $a = 1$ and $r = \frac{1}{2}$, so that

$$\sum_{k=1}^{\infty} \left(\frac{1}{2}\right)^{k-1} = \frac{1}{1-\frac{1}{2}} = 2$$

The next example shows that geometric series have an interesting application in connection with the repeating decimals in Section 1.1.

EXAMPLE 8 Use geometric series to find the rational number that corresponds to each of the following decimals.

(a) $0.\overline{3}$ (b) $0.\overline{24}$

SOLUTION (a) From the expression $0.\overline{3}$ we obtain the geometric series

$$\begin{aligned} 0.\overline{3} &= \frac{3}{10} + \frac{3}{100} + \frac{3}{1000} + \cdots + \frac{3}{10^n} + \cdots \\ &= \frac{3}{10} + \frac{3}{10}\left(\frac{1}{10}\right) + \frac{3}{10}\left(\frac{1}{10}\right)^2 + \cdots + \frac{3}{10}\left(\frac{1}{10}\right)^{n-1} + \cdots \\ &= \sum_{k=1}^{\infty} \frac{3}{10}\left(\frac{1}{10}\right)^{k-1} \end{aligned}$$

Here $a = \frac{3}{10}$ and $r = \frac{1}{10}$, so the sum is given by

$$0.\overline{3} = \frac{\frac{3}{10}}{1-\frac{1}{10}} = \frac{1}{3}$$

(b) $0.\overline{24}$ can be written as

$$\begin{aligned} 0.\overline{24} &= \frac{24}{100} + \frac{24}{10{,}000} + \frac{24}{1{,}000{,}000} + \cdots + \frac{24}{(100)^n} + \cdots \\ &= \sum_{k=1}^{\infty} \left(\frac{24}{100}\right)\left(\frac{1}{100}\right)^{k-1} \end{aligned}$$

We have a geometric series in which $a = \frac{24}{100}$ and $r = \frac{1}{100}$, so that

$$0.\overline{24} = \frac{\frac{24}{100}}{1-\frac{1}{100}} = \frac{24}{99} = \frac{8}{33}$$

PROBLEM SET 9.2

In problems 1–12, evaluate each sum. Use the properties of summation if applicable.

1. $\sum_{k=1}^{15} k$
2. $\sum_{k=0}^{4} \frac{2^k}{k+1}$
3. $\sum_{i=1}^{10} 2i(i-1)$
4. $\sum_{k=0}^{4} 3^{2k}$
5. $\sum_{k=2}^{5} 2^{k-2}$
6. $\sum_{i=2}^{6} \frac{1}{i(i+1)}$
7. $\sum_{k=1}^{3} (2k+1)^2$
8. $\sum_{k=1}^{10} (3k^2 - 5k + 1)$
9. $\sum_{i=1}^{5} 2 \cdot 5^{i-1}$
10. $\sum_{k=1}^{4} k^k$
11. $\sum_{k=1}^{100} (5k^2 - 3)$
12. $\sum_{i=3}^{7} (i+2)^2$

In problems 13–16, express each finite sum in sigma notation.

13. $1 + 4 + 7 + 10 + 13$
14. $\frac{1}{2} + \frac{1}{4} + \frac{1}{8} + \frac{1}{16} + \frac{1}{32}$
15. $\frac{3}{5} + \frac{9}{25} + \frac{27}{125} + \frac{81}{625}$
16. $\frac{1}{6} + \frac{2}{11} + \frac{3}{16} + \frac{4}{21}$

In problems 17–21, determine whether the statement is true or false. Give a reason.

17. $\sum_{k=0}^{100} k^3 = \sum_{k=1}^{100} k^3$
18. $\sum_{k=0}^{100} 2 = 200$
19. $\sum_{k=0}^{100} (k+2) = \left(\sum_{k=0}^{100} k\right) + 2$
20. $\sum_{k=1}^{20} (8k)^2 = 64 \sum_{k=2}^{21} (k-1)^2$
21. $\sum_{k=0}^{99} (k+1)^2 = \sum_{k=1}^{100} k^2$

22. (a) Verify the constant property: $\sum_{k=1}^{n} c = nc$, where c is a constant.

    (b) Verify the additive property: $\sum_{k=1}^{n} (a_k + b_k) = \sum_{k=1}^{n} a_k + \sum_{k=1}^{n} b_k$.
23. Find the sum of the first ten terms of an arithmetic sequence whose first term is 1 and whose common difference is 3.
24. Find the sum of the first fifteen terms of an arithmetic sequence whose first term is $\frac{1}{2}$ and whose common difference is $\frac{1}{2}$.
25. Find the sum of the first eight terms of an arithmetic sequence whose first term is -5 and whose common difference is $\frac{3}{7}$.
26. Find the sum of the first twelve terms of an arithmetic sequence whose first term is 11 and whose common difference is -2.
27. Find S_7 for the arithmetic sequence $6, 3b + 1, 6b - 4, \ldots$.
28. Find S_{10} for the arithmetic sequence $x + 2y, 3y, -x + 4y, \ldots$.

In problems 29–34, certain information about an arithmetic sequence is given. Find the indicated unknowns.

29. $a_1 = 6$; $d = 3$; a_{10} and S_{10}
30. $a_1 = 38$; $d = -2$; $n = 25$; S_n
31. $a_1 = 17$; $S_{18} = 2310$; d and a_{18}
32. $d = 3$; $S_{25} = 400$; a_1 and a_{25}

33. $a_1 = 27$; $a_n = 48$; $S_n = 1200$; n and d

34. $a_1 = -3$; $d = 2$; $S_n = 140$; n

35. Find the sum of the first six terms of the geometric sequence whose first term is $\frac{3}{2}$ and whose common ratio is 2.

36. Find the sum of the first ten terms of the geometric sequence whose first term is 6 and whose common ratio is $\frac{1}{2}$.

37. Find the sum of the first twelve terms of the geometric sequence whose first term is -4 and whose common ratio is -2.

38. Find the sum of the first eight terms of the geometric sequence whose first term is 5 and whose common ratio is $-\frac{1}{2}$.

39. Find S_6 for the geometric sequence $10, 10a, 10a^2, 10a^3, \ldots$.

40. Find S_8 for the geometric sequence $k, \frac{k}{b}, \frac{k}{b^2}, \frac{k}{b^3}, \ldots$.

In problems 41–48, find the indicated unknowns for each geometric sequence with the given characteristics.

41. $a_1 = 2$; $n = 3$; $S_n = 26$; r

42. $r = 2$; $n = 5$; $a_n = -48$; a_1 and S_n

43. $a_1 = 3$; $a_n = 192$; $n = 7$; r

44. $a_6 = 3$; $a_9 = -81$; r and a_1

45. $a_5 = \frac{1}{8}$; $r = -\frac{1}{2}$; a_9 and S_8

46. $a_1 = 1$; $r = (1.03)^{-1}$; $a_9 = (1.03)^{-8}$; S_8

47. $a_1 = \frac{1}{16}$; $r = 2$; $a_n = 32$; n and S_n

48. $a_1 = 250$; $r = \frac{1}{3}$; $a_n = \frac{98}{3}$; n and S_n

In problems 49–54, find the sum of the given geometric series.

49. $\frac{1}{3} + \frac{1}{9} + \frac{1}{27} + \cdots + (\frac{1}{3})^n + \cdots$

50. $\frac{2}{3} + \frac{4}{9} + \frac{8}{27} + \cdots + (\frac{2}{3})^n + \cdots$

51. $\frac{1}{5} + \frac{1}{25} + \frac{1}{125} + \cdots + (\frac{1}{5})^n + \cdots$

52. $\frac{9}{8} + \frac{9}{64} + \frac{9}{512} + \cdots + 9(\frac{1}{8})^n + \cdots$

53. $\sum_{k=1}^{\infty} \frac{4}{2^k}$

54. $\sum_{k=1}^{\infty} (0.7)^k$

In problems 55–60, use a geometric series to find the rational number represented by the given decimal.

55. $0.\overline{32}$

56. $0.04\overline{9}$

57. $0.\overline{46}$

58. $0.\overline{072}$

59. $3.\overline{561}$

60. $32.\overline{4218}$

9.3 Mathematical Induction and the Binomial Theorem

Frequently, we encounter properties and formulas that are valid for all positive integers, such as the basic properties of summation on page 507. It is impossible to prove such properties and formulas by checking *every* positive integer. For instance, suppose we were given the task to prove the following assertion:

$$1 + 2 + 3 + \cdots + n = \frac{n(n+1)}{2} \qquad \text{for any positive integer } n$$

Observe that if $n = 1$, then $1 = 1(1 + 1)/2$, which is true; if $n = 2$, then we have $1 + 2 = 2(2 + 1)/2$, which is also true; if $n = 3$, then $1 + 2 + 3 = 3(3 + 1)/2$, which is again true.

These tested values of n can only give us an *impression* that the statement is true. Simply testing these values is not adequate to establish a formal proof of the statement for all possible values of n.

Let us consider another example. Suppose that we want to prove the following statement:

$$1 + 4 + 7 + \cdots + (3n - 2) = \frac{n(3n - 1)}{2} \qquad \text{for any positive integer } n$$

We can begin to test the equation for specific values of n as we did in the first example.

If $n = 1$, then the statement becomes $1 = 1(3 \cdot 1 - 1)/2$, which is true; if $n = 2$, then we have $1 + 4 = 2(3 \cdot 2 - 1)/2$, which is also true; if $n = 3$, then we have $1 + 4 + 7 = 3(3 \cdot 3 - 1)/2$, which is true. Again, however, our testing process is not enough to give us a *generalized* proof. A formal proof of each of the above statements can be given by using the *principle of mathematical induction.*

Principle of Mathematical Induction

Suppose that $S_1, S_2, S_3, \ldots$ is a sequence of assertions; that is, suppose that for each positive integer n we have a corresponding assertion S_n. Assume that the following two conditions hold:

(i) S_1 is true.

(ii) For each fixed positive integer k, the truth of S_k implies the truth of S_{k+1}.

Then it follows that every assertion $S_1, S_2, S_3, \ldots$ is true; that is, S_n is true for all positive integers.

Let us see how this principle can be applied to prove the statement where S_n is the assertion:

$$1 + 2 + 3 + \cdots + n = \tfrac{1}{2}n(n + 1)$$

So far as we know, S_n may be true for certain values of n and false for other values of n. In order to show that S_n is, in fact, true for all values of n, we need to verify the following conditions:

(i) S_1 is true.

(ii) If S_k is true, then S_{k+1} is also true, where k is a fixed positive integer.

Condition (i) can be verified by direct computation, for S_1 is the assertion that $1 = (\tfrac{1}{2})(1)(1 + 1)$, which is clearly true.

To prove condition (ii), we must show that S_k implies S_{k+1}; that is, we must show that if S_k is assumed to be true, then S_{k+1} must be true. To this end, assume that S_k is true; that is, assume that the assertion

$$1 + 2 + 3 + \cdots + k = \tfrac{1}{2}k(k + 1)$$

is true. Since S_k is a true assertion, we can add $(k + 1)$ to both sides of this equation, to get

$$\begin{aligned} 1 + 2 + 3 + \cdots + k + (k + 1) &= \tfrac{1}{2}k(k + 1) + (k + 1) \\ &= (k + 1)(\tfrac{1}{2}k + 1) \\ &= (k + 1)\left(\frac{k + 2}{2}\right) \\ &= \tfrac{1}{2}(k + 1)(k + 2) \end{aligned}$$

But the latter assertion is precisely S_{k+1}. Hence, we have proved condition (ii); and, by the principle of mathematical induction, we conclude that S_n is true for any positive integer n; therefore,

$$1 + 2 + 3 + \cdots + n = \tfrac{1}{2}n(n + 1) \qquad \text{for any positive integer } n$$

Using sigma notation, this result can be written as

$$\sum_{k=1}^{n} k = \frac{n(n + 1)}{2}$$

In Examples 1 and 2, use mathematical induction to prove each assertion for any positive integer n. Express the result in sigma notation.

EXAMPLE 1 $1 + 4 + 7 + \cdots + (3n - 2) = \dfrac{n(3n - 1)}{2}$

SOLUTION Let S_n represent the above assertion. Using the principle of mathematical induction, we have

(i) S_1 becomes $1 = 1(3 \cdot 1 - 1)/2$, which is true.
(ii) Assume that S_k is true; that is, assume that, for any fixed positive integer k,

$$1 + 4 + 7 + \cdots + (3k - 2) = \frac{k(3k - 1)}{2}$$

We are to prove that S_{k+1} is true; that is, we must prove that, for each positive integer k,

$$1 + 4 + 7 + \cdots + (3k - 2) + [3(k + 1) - 2] = \frac{(k + 1)[3(k + 1) - 1]}{2}$$

$$1 + 4 + 7 + \cdots + (3k - 2) + (3k + 1) = \frac{(k + 1)(3k + 2)}{2}$$

Since S_k is true, we can add $(3k + 1)$ to each side of the equation expressing S_k to get

$$\begin{aligned} 1 + 4 + 7 + \cdots + (3k - 2) + (3k + 1) &= \frac{k(3k - 1)}{2} + (3k + 1) \\ &= \frac{3k^2 - k + 2(3k + 1)}{2} \\ &= \frac{3k^2 + 5k + 2}{2} \\ &= \frac{(k + 1)(3k + 2)}{2} \end{aligned}$$

Hence S_{k+1} is true and we have proved condition (ii). We may now conclude that S_n is true; that is, for any positive integer,

$$1 + 4 + 7 + \cdots + (3n - 2) = \frac{n(3n - 1)}{2}$$

This result can be written in sigma notation as

$$\sum_{k=1}^{n} (3k - 2) = \frac{n(3n - 1)}{2}$$

EXAMPLE 2 $1^2 + 2^2 + 3^2 + \cdots + n^2 = \frac{1}{6}n(n + 1)(2n + 1)$

SOLUTION Let S_n represent the above assertion. Using the principle of mathematical induction,

(i) S_1 becomes $1^2 = (\frac{1}{6})(1)(1 + 1)(2 \cdot 1 + 1) = (\frac{1}{6})(1)(2)(3) = 1$, which is true.
(ii) Assume that S_k is true; that is, assume that, for any fixed positive integer k,

$$1^2 + 2^2 + 3^2 + \cdots + k^2 = \frac{1}{6}k(k + 1)(2k + 1)$$

We must now prove that S_{k+1} is true, where S_{k+1} is the assertion

$$1^2 + 2^2 + 3^2 + \cdots + k^2 + (k + 1)^2 = \frac{1}{6}(k + 1)(k + 2)(2k + 3)$$

After adding $(k + 1)^2$ to both sides of the equation for S_k we have

$$\begin{aligned} 1^2 + 2^2 + 3^2 + \cdots + k^2 + (k + 1)^2 &= \frac{1}{6}k(k + 1)(2k + 1) + (k + 1)^2 \\ &= (k + 1)\left[\frac{1}{6}k(2k + 1) + (k + 1)\right] \\ &= (k + 1)\left(\frac{2k^2 + k + 6k + 6}{6}\right) \\ &= (k + 1)\left(\frac{2k^2 + 7k + 6}{6}\right) \\ &= \frac{1}{6}(k + 1)(k + 2)(2k + 3) \end{aligned}$$

Hence, S_{k+1} is true, and we have proved condition (ii). Thus, by the principle of mathematical induction, we conclude that S_n is true for any positive integer n. That is, for any positive integer n,

$$1^2 + 2^2 + 3^2 + \cdots + n^2 = \tfrac{1}{6}n(n+1)(2n+1)$$

We can use sigma notation to write the result as

$$\sum_{k=1}^{n} k^2 = \frac{n(n+1)(2n+1)}{6}$$

(See Property 5, page 508.)

Binomial Expansions

If we were asked to multiply out the binomial expression $(2x - y)^8$, the repetitive multiplication involved would be difficult. The *binomial theorem* provides a technique for expanding positive integer powers of binomial expressions. This theorem will be proved, using the principle of mathematical induction.

The pattern of the numerical coefficients of a binomial power expansion is easier to represent if we use *factorial notation* and *combinatorial notation,* which are defined as follows.

DEFINITION 1 **Factorial Notation**

The symbol $n!$ (read n factorial) is defined for all *nonnegative* integers as

$$0! = 1 \qquad \text{and} \qquad n! = n(n-1)(n-2)\ldots 2 \cdot 1$$

if n is a positive integer.

For instance,

$$5! = 5 \cdot 4 \cdot 3 \cdot 2 \cdot 1 = 120$$

and

$$7! = 7 \cdot 6 \cdot 5! = (42)(120) = 5040$$

DEFINITION 2 **Combinatorial Notation**

Assume k and n are integers such that $0 \leq k \leq n$. Then

$$\binom{n}{k} = \frac{n!}{k!(n-k)!}$$

For example,

$$\binom{5}{3} = \frac{5!}{3!(5-3)!} = \frac{5!}{3!2!} = 10$$

$$\binom{17}{14} = \frac{17!}{14!(17-14)!} = \frac{17!}{14!3!} = 680$$

$$\binom{5}{2} = \frac{5!}{2!(5-2)!} = \frac{5!}{2!3!} = 10$$

Note that since $0! = 1$, we have

$$\binom{n}{0} = \frac{n!}{0!n!} = 1 \qquad \text{and} \qquad \binom{n}{n} = \frac{n!}{n!0!} = 1$$

EXAMPLE 3 Simplify each expression.

(a) $\binom{n}{n-1}$ (b) $\frac{a_{n+1}}{a_n}$ if $a_k = \frac{3^k}{k!}$ (c) $\frac{a_{n+1}}{a_n}$ if $a_k = \frac{x^{2k}}{(2k)!}$

SOLUTION (a)

$$\binom{n}{n-1} = \frac{n!}{(n-1)![n-(n-1)]!} = \frac{n!}{(n-1)!1!}$$

$$= \frac{n\cancel{(n-1)!}}{\cancel{(n-1)!}} = n$$

(b) Since

$$a_n = \frac{3^n}{n!} \qquad \text{and} \qquad a_{n+1} = \frac{3^{n+1}}{(n+1)!}$$

it follows that

$$\frac{a_{n+1}}{a_n} = \frac{\frac{3^{n+1}}{(n+1)!}}{\frac{3^n}{n!}} = \frac{3^{n+1}}{(n+1)!} \cdot \frac{n!}{3^n}$$

$$= \frac{3^{n+1}}{3^n} \cdot \frac{n!}{(n+1)n!} = \frac{3}{n+1}$$

(c) Here

$$a_n = \frac{x^{2n}}{(2n)!} \qquad \text{and} \qquad a_{n+1} = \frac{x^{2n+2}}{(2n+2)!}$$

so that

$$\frac{a_{n+1}}{a_n} = \frac{x^{2n+2}}{(2n+2)!} \cdot \frac{(2n)!}{x^{2n}}$$

$$= \frac{x^{2n+2}}{x^{2n}} \cdot \frac{\cancel{(2n)!}}{(2n+2)(2n+1)\cancel{(2n)!}} = \frac{x^2}{4n^2 + 6n + 2}$$

Let us consider some examples of expanding the binomial $a + b$ to positive integer powers.

$$(a+b)^1 = a + b$$

$$(a+b)^2 = a^2 + 2ab + b^2$$

$$(a+b)^3 = a^3 + 3a^2b + 3ab^2 + b^3$$

$$(a+b)^4 = a^4 + 4a^3b + 6a^2b^2 + 4ab^3 + b^4$$

$$(a+b)^5 = a^5 + 5a^4b + 10a^3b^2 + 10a^2b^3 + 5ab^4 + b^5$$

Notice that the following pattern holds for the terms in these expansions of $(a+b)^n$.

1. There are $n + 1$ terms; the first term is a^n; the last term is b^n.

2. The power of a decreases by 1 for each term, and the power of b increases by 1 for each term. In any case, the sum of the exponents of a and b is n for each term. These patterns are generalized in the theorem, which we state using combinatorial notation.

THEOREM 1 **Binomial Theorem**

Let a and b be real numbers and let n be a positive integer; then

$$(a+b)^n = \binom{n}{0}a^n + \binom{n}{1}a^{n-1}b + \cdots + \binom{n}{k}a^{n-k}b^k + \cdots + \binom{n}{n}b^n$$

PROOF Use the principle of mathematical induction.

(i) S_1 is true since $(a+b)^1 = a^1 + b^1 = 1 \cdot a^1 + 1 \cdot b^1 = \binom{1}{0}a^1 + \binom{1}{1}b^1$.

(ii) We must show that if S_n is true, then S_{n+1} is also true. [Notice that we are using n instead of k.]

To this end, assume that S_n is true; that is, assume that

$$(a+b)^n = \binom{n}{0}a^n + \binom{n}{1}a^{n-1}b + \cdots + \binom{n}{k}a^{n-k}b^k + \cdots + \binom{n}{n}b^n$$

for n a positive integer. After multiplying both sides by $(a+b)$, we obtain

$$(a+b)^n(a+b) = (a+b)\left[\binom{n}{0}a^n + \binom{n}{1}a^{n-1}b + \cdots + \binom{n}{k}a^{n-k}b^k + \cdots + \binom{n}{n}b^n\right]$$

$$= \binom{n}{0}(a^{n+1} + a^nb) + \binom{n}{1}(a^nb + a^{n-1}b^2)$$

$$+ \cdots + \binom{n}{k}(a^{n-k+1}b^k + a^{n-k}b^{k+1}) + \cdots + \binom{n}{n}(ab^n + b^{n+1})$$

$$= \binom{n}{0}a^{n+1} + \left[\binom{n}{0} + \binom{n}{1}\right]a^n b + \left[\binom{n}{1} + \binom{n}{2}\right]a^{n-1}b^2$$

$$+ \cdots + \left[\binom{n}{k-1} + \binom{n}{k}\right]a^{n+1-k}b^k + \cdots + \binom{n}{n}b^{n+1}$$

However, $\binom{n}{0} = \binom{n+1}{0}$ (Problem 18a), $\binom{n}{n} = \binom{n+1}{n+1}$ (Problem 18b), and

$$\binom{n}{k-1} + \binom{n}{k} = \frac{n!}{(k-1)!(n-k+1)!} + \frac{n!}{k!(n-k)!}$$

$$= \frac{n!k + n!(n-k+1)}{k!(n-k+1)!} = \frac{n!(n+1)}{k!(n+1-k)!}$$

$$= \frac{(n+1)!}{k!(n+1-k)!} = \binom{n+1}{k}$$

so that

$$(a+b)^{n+1} = \binom{n+1}{0}a^{n+1} + \binom{n+1}{1}a^n b$$

$$+ \cdots + \binom{n+1}{k}a^{n+1-k}b^k + \cdots + \binom{n+1}{n+1}b^{n+1}$$

But the latter assertion is precisely S_{n+1}, and the proof is complete.

Using sigma notation, the binomial theorem asserts that if n is a positive integer, then

$$(a+b)^n = \sum_{k=0}^{n} \binom{n}{k} a^{n-k} b^k$$

Notice the "symmetry" of the values of the coefficients of the binomial expansions.

$$(a+b)^2 = 1a^2 + 2ab + 1b^2$$

$$(a+b)^3 = 1a^3 + 3a^2b + 3ab^2 + 1b^3$$

$$(a+b)^4 = 1a^4 + 4a^3b + 6a^2b^2 + 4ab^3 + 1b^4$$

In general,

$$(a+b)^n = \binom{n}{0}a^n + \binom{n}{1}a^{n-1}b + \binom{n}{2}a^{n-2}b^2 + \cdots + \binom{n}{n-2}a^2b^{n-2} + \binom{n}{n-1}ab^{n-1} + \binom{n}{n}b^n$$

where $\binom{n}{k} = \binom{n}{n-k}$ (Problem 12a)

In Examples 4–6, use the binomial theorem to expand the binomial.

EXAMPLE 4 $(x+y)^5$

SOLUTION

$$\begin{aligned}(x+y)^5 &= \sum_{k=0}^{5}\binom{5}{k}x^{5-k}y^k \\ &= \binom{5}{0}x^5 + \binom{5}{1}x^4y + \binom{5}{2}x^3y^2 + \binom{5}{3}x^2y^3 + \binom{5}{4}xy^4 + \binom{5}{5}y^5 \\ &= x^5 + \frac{5!}{1!4!}x^4y + \frac{5!}{2!3!}x^3y^2 + \frac{5!}{3!2!}x^2y^3 + \frac{5!}{4!1!}xy^4 + y^5 \\ &= x^5 + 5x^4y + 10x^3y^2 + 10x^2y^3 + 5xy^4 + y^5\end{aligned}$$

EXAMPLE 5 $(x-3)^4$

SOLUTION

$$\begin{aligned}(x-3)^4 &= [x+(-3)]^4 \\ &= \sum_{k=0}^{4} x^{4-k}(-3)^k \\ &= \binom{4}{0}x^4 + \binom{4}{1}x^3(-3) + \binom{4}{2}x^2(-3)^2 + \binom{4}{3}x(-3)^3 + \binom{4}{4}(-3)^4 \\ &= x^4 + 4x^3(-3) + 6x^2(-3)^2 + 4x(-3)^3 + (-3)^4 \\ &= x^4 - 12x^3 + 54x^2 - 108x + 81\end{aligned}$$

EXAMPLE 6 $\left(3x^2 - \frac{1}{2}\sqrt{y}\right)^4$

SOLUTION By the binomial theorem,

$$\begin{aligned}\left(3x^2 - \frac{1}{2}\sqrt{y}\right)^4 &= \sum_{k=0}^{4}(3x^2)^{4-k}\left(-\frac{1}{2}\sqrt{y}\right)^k \\ &= \binom{4}{0}(3x^2)^4 + \binom{4}{1}(3x^2)^3\left(-\frac{1}{2}\sqrt{y}\right) + \binom{4}{2}(3x^2)^2\left(-\frac{1}{2}\sqrt{y}\right)^2 \\ &\quad + \binom{4}{3}(3x^2)\left(-\frac{1}{2}\sqrt{y}\right)^3 + \binom{4}{4}\left(-\frac{1}{2}\sqrt{y}\right)^4\end{aligned}$$

$$= (3x^2)^4 - 4(3x^2)^3\left(\frac{1}{2}\sqrt{y}\right) + 6(3x^2)^2\left(\frac{1}{2}\sqrt{y}\right)^2$$
$$-4(3x^2)\left(\frac{1}{2}\sqrt{y}\right)^3 + \left(\frac{1}{2}\sqrt{y}\right)^4$$
$$= 81x^8 - 54x^6\sqrt{y} + \frac{27}{2}x^4y - \frac{3}{2}x^2y^{3/2} + \frac{1}{16}y^2$$

EXAMPLE 7 Find the sixth term of the expansion of $(2x - y^2)^8$.

SOLUTION The binomial theorem indicates that if $n = 8$, then the *sixth* term is of the form

$$\binom{8}{5}a^{8-5}b^5 = \binom{8}{5}a^3b^5$$

But in this situation, we write

$$(2x - y^2)^8 = [(2x) + (-y^2)]^8$$

so that $a = 2x$ and $b = -y^2$. Hence, the sixth term is

$$\binom{8}{5}(2x)^3(-y^2)^5 = -\binom{8}{5}8x^3y^{10}$$
$$= -\frac{8 \cdot 7 \cdot 6}{3 \cdot 2 \cdot 1}8x^3y^{10}$$
$$= -448x^3y^{10}$$

PROBLEM SET 9.3

In problems 1–10, use mathematical induction to prove the given assertion for all positive integers n. Express the result in sigma notation.

1. $1 + 3 + 5 + \cdots + (2n - 1) = n^2$
2. $1^3 + 2^3 + 3^3 + \cdots + n^3 = \frac{1}{4}n^2(n + 1)^2$
3. $2 + 4 + 6 + \cdots + 2n = n^2 + n$
4. $1^2 + 3^2 + 5^2 + \cdots + (2n - 1)^2 = \frac{1}{3}n(2n - 1)(2n + 1)$
5. $1 \cdot 2 + 2 \cdot 3 + 3 \cdot 4 + \cdots + n(n + 1) = \frac{1}{3}n(n + 1)(n + 2)$
6. $\frac{1}{1 \cdot 2} + \frac{1}{2 \cdot 3} + \frac{1}{3 \cdot 4} + \cdots + \frac{1}{n(n + 1)} = \frac{n}{n + 1}$
7. $4 + 4^2 + 4^3 + \cdots + 4^n = \frac{4}{3}(4^n - 1)$
8. $x^0 + x^1 + x^2 + \cdots + x^n = \frac{1 - x^{n+1}}{1 - x}$, for $x \neq 1$
9. $1 + 5 + 5^2 + \cdots + 5^{n-1} = \frac{1}{4}(5^n - 1)$
10. *DeMoivre's theorem* (see page 381): $[r(\cos\theta + i\sin\theta)]^n = r^n(\cos n\theta + i\sin n\theta)$.

11. Evaluate each of the given expressions.

(a) $\binom{15}{10}$ (b) $\binom{n}{n}$ (c) $\binom{6}{2}$ (d) $\binom{15}{5}$ (e) $\binom{n}{0}$ (f) $\binom{6}{5}$ (g) $\binom{15}{3}$ (h) $\binom{n}{1}$ (i) $\binom{6}{4}$

12. (a) Verify that

$$\binom{n}{k} = \binom{n}{n-k}$$

(b) Verify that

$$\binom{n}{r} + \binom{n}{r+1} = \binom{n+1}{r+1}$$

(c) Let $a = b = 1$ in the expansion of $(a + b)^n$, and find the sum

$$\binom{n}{0} + \binom{n}{1} + \binom{n}{2} + \cdots + \binom{n}{n}$$

In problems 13–17, simplify the expression.

13. $\binom{n+5}{n+4}$

14. $\dfrac{a_{n+1}}{a_n}$ if $a_k = \dfrac{x^k}{k!}$

15. $\dfrac{a_{n+1}}{a_n}$ if $a_k = \dfrac{(x-1)^k}{(2k)!}$

16. $\left|\dfrac{a_{n+1}}{a_n}\right|$ if $a_k = \dfrac{(-1)^k x^{k+1}}{(k+1)!}$

17. $\left|\dfrac{a_{n+1}}{a_n}\right|$ if $a_k = \dfrac{(-1)^k(k-1)!}{e^k}$

18. (a) Prove that $\binom{n}{0} = \binom{n+1}{0}$. (b) Prove that $\binom{n}{n} = \binom{n+1}{n+1}$.

In problems 19–26, use the binomial theorem to expand each expression as specified.

19. $(x + 3)^5$, all terms

20. $(2z + x)^4$, all terms

21. $(x - 2)^4$, all terms

22. $\left(\dfrac{1}{a} + \dfrac{x}{2}\right)^3$, all terms

23. $(x + y)^{12}$, first four terms

24. $(x - 3y)^7$, first four terms

25. $(a^{3/2} - 2x^2)^8$, first four terms

26. $(x + \frac{1}{2})^{10}$, first four terms

In problems 27–32, use the binomial theorem to find the indicated term.

27. $\left(\dfrac{x^2}{2} + a\right)^{15}$, fourth term

28. $(y^2 - 2z)^{10}$, sixth term

29. $\left(2x^2 - \dfrac{a^2}{3}\right)^9$, seventh term

30. $(x + \sqrt{a})^{12}$, middle term

31. $\left(a + \dfrac{x^2}{3}\right)^9$, term containing x^{12}

32. $\left(2\sqrt{y} - \dfrac{x}{2}\right)^{10}$, term containing y^4

33. The **combinations** of n objects, taken k at a time, include *all* possible collections of k of the objects without regard to the order of arrangement. For example, the combinations of the four letters a, b, c, and d, taken two at a time, are:

$$ab,\quad ac,\quad ad,\quad bc,\quad bd,\quad \text{and}\quad cd$$

In general, it can be shown that the *number* of combinations of n objects taken k at a time is given by $\binom{n}{k}$. Use this result to determine:

(a) The number of different card hands containing 5 cards that can be obtained from a deck containing 52 cards.

(b) The number of committees containing 3 members that can be formed from an organization that has 100 members.

REVIEW PROBLEM SET, CHAPTER 9

In problems 1–4, find the first five terms of each sequence.

1. $\{a_n\} = \left\{\dfrac{1}{n^2 + 3n}\right\}$ **2.** $\{a_n\} = \{4^{n/2}\}$ **3.** $\{a_n\} = \left\{\left(1 - \dfrac{3}{n}\right)^n\right\}$ **4.** $\{a_n\} = \{\cos n\pi\}$

In problems 5–8, find the indicated term of each arithmetic sequence.

5. The tenth term of 3, 7, 11, 15, . . .
6. The eleventh term of $2k - 3$, $2k + 2$, $2k + 7$, $2k + 12$, . . .
7. The ninth term of $\frac{9}{8}, \frac{13}{8}, \frac{17}{8}, \frac{21}{8}, \ldots$
8. The twelfth term of $-8, -5, -2, 1, \ldots$

In problems 9–12, find the indicated term of each geometric sequence.

9. The eighth term of $-\frac{2}{3}, -1, -\frac{3}{2}, -\frac{9}{4}, \ldots$
10. The seventh term of $\sqrt{3}, -3, 3\sqrt{3}, -9, \ldots$
11. The tenth term of $\sqrt{5}, -\sqrt{20}, \sqrt{80}, -\sqrt{320}, \ldots$
12. The ninth term of $1, (1.02)^{-1}, (1.02)^{-2}, (1.02)^{-3}, \ldots$

In problems 13–18, evaluate each sum.

13. $\sum_{k=1}^{5} k(2k - 1)$ **14.** $\sum_{k=1}^{4} 2k^2(k - 3)$ **15.** $\sum_{k=0}^{5} \dfrac{2}{3^k}$

16. $\sum_{k=2}^{6} (k + 1)(k + 2)$ **17.** $\sum_{k=5}^{10} (2k - 1)^2$ **18.** $\sum_{k=0}^{6} (2^{k+1} - 2^k)$

In problems 19–24, find the sum of the first n terms of each arithmetic sequence for the given value of n.

19. $8, 5, 2, -1, \ldots, n = 10$
20. $12, 16, 20, 24, \ldots, n = 15$
21. $\frac{3}{4}, \frac{1}{4}, -\frac{1}{4}, -\frac{3}{4}, \ldots, n = 12$
22. $\frac{10}{9}, \frac{7}{9}, \frac{4}{9}, \frac{1}{9}, \ldots, n = 8$
23. $2a + 3b, 3a + 2b, 4a + b, \ldots, n = 6$
24. $x - 6, x + 3, 2x, \ldots, n = 7$

In problems 25–28, find the indicated sum of the first n terms of each geometric sequence for the given value of n.

25. $\frac{1}{4}, \frac{1}{2}, 1, 2, \ldots, n = 8$
26. $3\sqrt{2}, 6, 6\sqrt{2}, \ldots, n = 10$
27. $\frac{1}{9}, \frac{1}{3}, 1, 3, \ldots, n = 10$
28. $\frac{3}{2}, \frac{1}{2}, \frac{1}{6}, \frac{1}{18}, \ldots, n = 6$

In problems 29–34, find the sum of each geometric series.

29. $\frac{1}{10} + \frac{1}{100} + \frac{1}{1000} + \cdots + (\frac{1}{10})^n + \cdots$

30. $\frac{3}{5} + \frac{9}{25} + \frac{27}{125} + \cdots + (\frac{3}{5})^n + \cdots$

31. $\frac{3}{4} + \frac{9}{16} + \frac{27}{64} + \cdots + (\frac{3}{4})^n + \cdots$

32. $(\frac{2}{3})^2 + (\frac{2}{3})^4 + (\frac{2}{3})^6 + \cdots + (\frac{2}{3})^{2n} + \cdots$

33. $\sum_{k=1}^{\infty} \frac{1}{8 \cdot 3^k}$

34. $\sum_{k=0}^{\infty} \frac{1}{5^{(k/2)+1}}$

In problems 35–38, use geometric series to find the rational number represented by the given decimal.

35. $0.\overline{14}$

36. $0.\overline{137}$

37. $4.6\overline{24}$

38. $7.3\overline{582}$

In problems 39–44, use mathematical induction to prove each formula. Express the result in sigma notation.

39. $2 + 2^2 + 2^3 + \cdots + 2^n = 2(2^n - 1)$

40. $1 \cdot 3 + 2 \cdot 4 + 3 \cdot 5 + \cdots + n(n + 2) = \frac{1}{6}n(n + 1)(2n + 7)$

41. $2 + 5 + 10 + \cdots + (n^2 + 1) = \dfrac{n(2n^2 + 3n + 7)}{6}$

42. $a + (a + d) + (a + 2d) + \cdots + [a + (n - 1)d] = \dfrac{n}{2}[2a + (n - 1)d]$

43. $\dfrac{1}{1 \cdot 4} + \dfrac{1}{4 \cdot 7} + \dfrac{1}{7 \cdot 10} + \cdots + \dfrac{1}{(3n - 2)(3n + 1)} = \dfrac{n}{3n + 1}$

44. $a + a^2 + a^3 + \cdots + a^n = \dfrac{a}{a - 1}(a^n - 1), \quad a \neq 1$

In problems 45–54, simplify each expression.

45. $\binom{12}{5}$

46. $\binom{50}{2}$

47. $\binom{81}{79}$

48. $\binom{34}{30}$

49. $\binom{n + 7}{n + 5}$

50. $\dfrac{a_{n+1}}{a_n}$ if $a_n = \dfrac{5^n}{n!}$

51. $\dfrac{a_{n+1}}{a_n}$ if $a_n = \dfrac{(n + 1)!}{7^n}$

52. $\dfrac{a_{n+1}}{a_n}$ if $a_n = \dfrac{3^n}{(2n)!}$

53. $\left|\dfrac{a_{n+1}}{a_n}\right|$ if $a_n = (-1)^n \dfrac{n^2 \cdot n!}{(2n)!}$

54. $\left|\dfrac{a_{n+1}}{a_n}\right|$ if $a_n = (-1)^n \dfrac{(n!)^2}{(2n)!}$

In problems 55–62, use the binomial theorem to expand each expression.

55. $(3x + y)^4$

56. $(3x + \sqrt{x})^5$

57. $\left(2x + \dfrac{1}{y}\right)^3$

58. $\left(x - \dfrac{1}{x}\right)^4$

59. $(a + \frac{3}{2})^5$

60. $(b^3 + 2)^6$

61. $(\sqrt{x} - 3)^4$

62. $\left(5\sqrt{3} - \dfrac{2}{x}\right)^5$

In problems 63 and 64, use the binomial theorem to find the indicated term.

63. $(3x - 2y)^6$, third term

64. $(3h - k^2)^4$, middle term

Appendix

APPENDIX

Tables

Table I Common Logarithms

n	0.00	0.01	0.02	0.03	0.04	0.05	0.06	0.07	0.08	0.09
1.0	.0000	.0043	.0086	.0128	.0170	.0212	.0253	.0294	.0334	.0374
1.1	.0414	.0453	.0492	.0531	.0569	.0607	.0645	.0682	.0719	.0755
1.2	.0792	.0828	.0864	.0899	.0934	.0969	.1004	.1038	.1072	.1106
1.3	.1139	.1173	.1206	.1239	.1271	.1303	.1335	.1367	.1399	.1430
1.4	.1461	.1492	.1523	.1553	.1584	.1614	.1644	.1673	.1703	.1732
1.5	.1761	.1790	.1818	.1847	.1875	.1903	.1931	.1959	.1987	.2014
1.6	.2041	.2068	.2095	.2122	.2148	.2175	.2201	.2227	.2253	.2279
1.7	.2304	.2330	.2355	.2380	.2405	.2430	.2455	.2480	.2504	.2529
1.8	.2553	.2577	.2601	.2625	.2648	.2672	.2695	.2718	.2742	.2765
1.9	.2788	.2810	.2833	.2856	.2878	.2900	.2923	.2945	.2967	.2989
2.0	.3010	.3032	.3054	.3075	.3096	.3118	.3139	.3160	.3181	.3201
2.1	.3222	.3243	.3263	.3284	.3304	.3324	.3345	.3365	.3385	.3404
2.2	.3424	.3444	.3464	.3483	.3502	.3522	.3541	.3560	.3579	.3598
2.3	.3617	.3636	.3655	.3674	.3692	.3711	.3729	.3747	.3766	.3784
2.4	.3802	.3820	.3838	.3856	.3874	.3892	.3909	.3927	.3945	.3962
2.5	.3979	.3997	.4014	.4031	.4048	.4065	.4082	.4099	.4116	.4133
2.6	.4150	.4166	.4183	.4200	.4216	.4232	.4249	.4265	.4281	.4298
2.7	.4314	.4330	.4346	.4362	.4378	.4393	.4409	.4425	.4440	.4456
2.8	.4472	.4487	.4502	.4518	.4533	.4548	.4564	.4579	.4594	.4609
2.9	.4624	.4639	.4654	.4669	.4683	.4698	.4713	.4728	.4742	.4757
3.0	.4771	.4786	.4800	.4814	.4829	.4843	.4857	.4871	.4886	.4900
3.1	.4914	.4928	.4942	.4955	.4969	.4983	.4997	.5011	.5024	.5038
3.2	.5051	.5065	.5079	.5092	.5105	.5119	.5132	.5145	.5159	.5172
3.3	.5185	.5198	.5211	.5224	.5237	.5250	.5263	.5276	.5289	.5302
3.4	.5315	.5328	.5340	.5353	.5366	.5378	.5391	.5403	.5416	.5428
3.5	.5441	.5453	.5465	.5478	.5490	.5502	.5514	.5527	.5539	.5551
3.6	.5563	.5575	.5587	.5599	.5611	.5623	.5635	.5647	.5658	.5670
3.7	.5682	.5694	.5705	.5717	.5729	.5740	.5752	.5763	.5775	.5786
3.8	.5798	.5809	.5821	.5832	.5843	.5855	.5866	.5877	.5888	.5899
3.9	.5911	.5922	.5933	.5944	.5955	.5966	.5977	.5988	.5999	.6010
4.0	.6021	.6031	.6042	.6053	.6064	.6075	.6085	.6096	.6107	.6117
4.1	.6128	.6138	.6149	.6160	.6170	.6180	.6191	.6201	.6212	.6222
4.2	.6232	.6243	.6253	.6263	.6274	.6284	.6294	.6304	.6314	.6325
4.3	.6335	.6345	.6355	.6365	.6375	.6385	.6395	.6405	.6415	.6425
4.4	.6435	.6444	.6454	.6464	.6474	.6484	.6493	.6503	.6513	.6522
4.5	.6532	.6542	.6551	.6561	.6571	.6580	.6590	.6599	.6609	.6618
4.6	.6628	.6637	.6646	.6656	.6665	.6675	.6684	.6693	.6702	.6712
4.7	.6721	.6730	.6739	.6749	.6758	.6767	.6776	.6785	.6794	.6803
4.8	.6812	.6821	.6830	.6839	.6848	.6857	.6866	.6875	.6884	.6893
4.9	.6902	.6911	.6920	.6928	.6937	.6946	.6955	.6964	.6972	.6981

n	0.00	0.01	0.02	0.03	0.04	0.05	0.06	0.07	0.08	0.09
5.0	.6990	.6998	.7007	.7016	.7024	.7033	.7042	.7050	.7059	.7067
5.1	.7076	.7084	.7093	.7101	.7110	.7118	.7126	.7135	.7143	.7152
5.2	.7160	.7168	.7177	.7185	.7193	.7202	.7210	.7218	.7226	.7235
5.3	.7243	.7251	.7259	.7267	.7275	.7284	.7292	.7300	.7308	.7316
5.4	.7324	.7332	.7340	.7348	.7356	.7364	.7372	.7380	.7388	.7396
5.5	.7404	.7412	.7419	.7427	.7435	.7443	.7451	.7459	.7466	.7474
5.6	.7482	.7490	.7497	.7505	.7513	.7520	.7528	.7536	.7543	.7551
5.7	.7559	.7566	.7574	.7582	.7589	.7597	.7604	.7612	.7619	.7627
5.8	.7634	.7642	.7649	.7657	.7664	.7672	.7679	.7686	.7694	.7701
5.9	.7709	.7716	.7723	.7731	.7738	.7745	.7752	.7760	.7767	.7774
6.0	.7782	.7789	.7796	.7803	.7810	.7818	.7825	.7832	.7839	.7846
6.1	.7853	.7860	.7868	.7875	.7882	.7889	.7896	.7903	.7910	.7917
6.2	.7924	.7931	.7938	.7945	.7952	.7959	.7966	.7973	.7980	.7987
6.3	.7993	.8000	.8007	.8014	.8021	.8028	.8035	.8041	.8048	.8055
6.4	.8062	.8069	.8075	.8082	.8089	.8096	.8102	.8109	.8116	.8122
6.5	.8129	.8136	.8142	.8149	.8156	.8162	.8169	.8176	.8182	.8189
6.6	.8195	.8202	.8209	.8215	.8222	.8228	.8235	.8241	.8248	.8254
6.7	.8261	.8267	.8274	.8280	.8287	.8293	.8299	.8306	.8312	.8319
6.8	.8325	.8331	.8338	.8344	.8351	.8357	.8363	.8370	.8376	.8382
6.9	.8388	.8395	.8401	.8407	.8414	.8420	.8426	.8432	.8439	.8445
7.0	.8451	.8457	.8463	.8470	.8476	.8482	.8488	.8494	.8500	.8506
7.1	.8513	.8519	.8525	.8531	.8537	.8543	.8549	.8555	.8561	.8567
7.2	.8573	.8579	.8585	.8591	.8597	.8603	.8609	.8615	.8621	.8627
7.3	.8633	.8639	.8645	.8651	.8657	.8663	.8669	.8675	.8681	.8686
7.4	.8692	.8698	.8704	.8710	.8716	.8722	.8727	.8733	.8739	.8745
7.5	.8751	.8756	.8762	.8768	.8774	.8779	.8785	.8791	.8797	.8802
7.6	.8808	.8814	.8820	.8825	.8831	.8837	.8842	.8848	.8854	.8859
7.7	.8865	.8871	.8876	.8882	.8887	.8893	.8899	.8904	.8910	.8915
7.8	.8921	.8927	.8932	.8938	.8943	.8949	.8954	.8960	.8965	.8971
7.9	.8976	.8982	.8987	.8993	.8998	.9004	.9009	.9015	.9020	.9025
8.0	.9031	.9036	.9042	.9047	.9053	.9058	.9063	.9069	.9074	.9079
8.1	.9085	.9090	.9096	.9101	.9106	.9112	.9117	.9122	.9128	.9133
8.2	.9138	.9143	.9149	.9154	.9159	.9165	.9170	.9175	.9180	.9186
8.3	.9191	.9196	.9201	.9206	.9212	.9217	.9222	.9227	.9232	.9238
8.4	.9243	.9248	.9253	.9258	.9263	.9269	.9274	.9279	.9284	.9289
8.5	.9294	.9299	.9304	.9309	.9315	.9320	.9325	.9330	.9335	.9340
8.6	.9345	.9350	.9355	.9360	.9365	.9370	.9375	.9380	.9385	.9390
8.7	.9395	.9400	.9405	.9410	.9415	.9420	.9425	.9430	.9435	.9440
8.8	.9445	.9450	.9455	.9460	.9465	.9469	.9474	.9479	.9484	.9489
8.9	.9494	.9499	.9504	.9509	.9513	.9518	.9523	.9528	.9533	.9538
9.0	.9542	.9547	.9552	.9557	.9562	.9566	.9571	.9576	.9581	.9586
9.1	.9590	.9595	.9600	.9605	.9609	.9614	.9619	.9624	.9628	.9633
9.2	.9638	.9643	.9647	.9652	.9657	.9661	.9666	.9671	.9675	.9680
9.3	.9685	.9689	.9694	.9699	.9703	.9708	.9713	.9717	.9722	.9727
9.4	.9731	.9736	.9741	.9745	.9750	.9754	.9759	.9763	.9768	.9773
9.5	.9777	.9782	.9786	.9791	.9795	.9800	.9805	.9809	.9814	.9818
9.6	.9823	.9827	.9832	.9836	.9841	.9845	.9850	.9854	.9859	.9863
9.7	.9868	.9872	.9877	.9881	.9886	.9890	.9894	.9899	.9903	.9908
9.8	.9912	.9917	.9921	.9926	.9930	.9934	.9939	.9943	.9948	.9952
9.9	.9956	.9961	.9965	.9969	.9974	.9978	.9983	.9987	.9991	.9996

Table II Natural Logarithms

t	0.00	0.01	0.02	0.03	0.04	0.05	0.06	0.07	0.08	0.09
1.0	0.0000	.0100	0.0198	0.0296	0.0392	0.0488	0.0583	0.0677	0.0770	0.0862
1.1	0.0953	0.1044	0.1133	0.1222	0.1310	0.1398	0.1484	0.1570	0.1655	0.1740
1.2	0.1823	0.1906	0.1989	0.2070	0.2151	0.2231	0.2311	0.2390	0.2469	0.2546
1.3	0.2624	0.2700	0.2776	0.2852	0.2927	0.3001	0.3075	0.3148	0.3221	0.3293
1.4	0.3365	0.3436	0.3507	0.3577	0.3646	0.3716	0.3784	0.3853	0.3920	0.3988
1.5	0.4055	0.4121	0.4187	0.4253	0.4318	0.4383	0.4447	0.4511	0.4574	0.4637
1.6	0.4700	0.4762	0.4824	0.4886	0.4947	0.5008	0.5068	0.5128	0.5188	0.5247
1.7	0.5306	0.5365	0.5423	0.5481	0.5539	0.5596	0.5653	0.5710	0.5766	0.5822
1.8	0.5878	0.5933	0.5988	0.6043	0.6098	0.6152	0.6206	0.6259	0.6313	0.6366
1.9	0.6419	0.6471	0.6523	0.6575	0.6627	0.6678	0.6729	0.6780	0.6831	0.6881
2.0	0.6931	0.6981	0.7031	0.7080	0.7130	0.7178	0.7227	0.7275	0.7324	0.7372
2.1	0.7419	0.7467	0.7514	0.7561	0.7608	0.7655	0.7701	0.7747	0.7793	0.7839
2.2	0.7885	0.7930	0.7975	0.8020	0.8065	0.8109	0.8154	0.8198	0.8242	0.8286
2.3	0.8329	0.8372	0.8416	0.8459	0.8502	0.8544	0.8587	0.8629	0.8671	0.8713
2.4	0.8755	0.8796	0.8838	0.8879	0.8920	0.8961	0.9002	0.9042	0.9083	0.9123
2.5	0.9163	0.9203	0.9243	0.9282	0.9322	0.9361	0.9400	0.9439	0.9478	0.9517
2.6	0.9555	0.9594	0.9632	0.9670	0.9708	0.9746	0.9783	0.9821	0.9858	0.9895
2.7	0.9933	0.9969	1.0006	1.0043	1.0080	1.0116	1.0152	0.0188	1.0225	1.0260
2.8	1.0296	1.0332	1.0367	1.0403	1.0438	1.0473	1.0508	1.0543	1.0578	1.0613
2.9	1.0647	1.0682	1.0716	1.0750	1.0784	1.0818	1.0852	1.0886	1.0919	1.0953
3.0	1.0986	1.1019	1.1053	1.1086	1.1119	1.1151	1.1184	1.1217	1.1249	1.1282
3.1	1.1314	1.1346	1.1378	1.1410	1.1442	1.1474	1.1506	1.1537	1.1569	1.1600
3.2	1.1632	1.1663	1.1694	1.1725	1.1756	1.1787	1.1817	1.1848	1.1878	1.1909
3.3	1.1939	1.1970	1.2000	1.2030	1.2060	1.2090	1.2119	1.2149	1.2179	1.2208
3.4	1.2238	1.2267	1.2296	1.2326	1.2355	1.2384	1.2413	1.2442	1.2470	1.2499
3.5	1.2528	1.2556	1.2585	1.2613	1.2641	1.2669	1.2698	1.2726	1.2754	1.2782
3.6	1.2809	1.2837	1.2865	1.2892	1.2920	1.2947	1.2975	1.3002	1.3029	1.3056
3.7	1.3083	1.3110	1.3137	1.3164	1.3191	1.3218	1.3244	1.3271	1.3297	1.3324
3.8	1.3350	1.3376	1.3403	1.3429	1.3455	1.3481	1.3507	1.3533	1.3558	1.3584
3.9	1.3610	1.3635	1.3661	1.3686	1.3712	1.3737	1.3762	1.3788	1.3813	1.3838
4.0	1.3863	1.3888	1.3913	1.3938	1.3962	1.3987	1.4012	1.4036	1.4061	1.4085
4.1	1.4110	1.4134	1.4159	1.4183	1.4207	1.4231	1.4255	1.4279	1.4303	1.4327
4.2	1.4351	1.4375	1.4398	1.4422	1.4446	1.4469	1.4493	1.4516	1.4540	1.4563
4.3	1.4586	1.4609	1.4633	1.4656	1.4679	1.4702	1.4725	1.4748	1.4770	1.4793
4.4	1.4816	1.4839	1.4861	1.4884	1.4907	1.4929	1.4952	1.4974	1.4996	1.5019
4.5	1.5041	1.5063	1.5085	1.5107	1.5129	1.5151	1.5173	1.5195	1.5217	1.5239
4.6	1.5261	1.5282	1.5304	1.5326	1.5347	1.5369	1.5390	1.5412	1.5433	1.5454
4.7	1.5476	1.5497	1.5518	1.5539	1.5560	1.5581	1.5602	1.5623	1.5644	1.5665
4.8	1.5686	1.5707	1.5728	1.5748	1.5769	1.5790	1.5810	1.5831	1.5851	1.5872
4.9	1.5892	1.5913	1.5933	1.5953	1.5974	1.5994	1.6014	1.6034	1.6054	1.6074
5.0	1.6094	1.6114	1.6134	1.6154	1.6174	1.6194	1.6214	1.6233	1.6253	1.6273
5.1	1.6292	1.6312	1.6332	1.6351	1.6371	1.6390	1.6409	1.6429	1.6448	1.6467
5.2	1.6487	1.6506	1.6525	1.6544	1.6563	1.6582	1.6601	1.6620	1.6639	1.6658
5.3	1.6677	1.6696	1.6715	1.6734	1.6752	1.6771	1.6790	1.6808	1.6827	1.6845
5.4	1.6864	1.6882	1.6901	1.6919	1.6938	1.6956	1.6974	1.6993	1.7011	1.7029

t	0.00	0.01	0.02	0.03	0.04	0.05	0.06	0.07	0.08	0.09
5.5	1.7047	1.7066	1.7084	1.7102	1.7120	1.7138	1.7156	1.7174	1.7192	1.7210
5.6	1.7228	1.7246	1.7263	1.7281	1.7299	1.7317	1.7334	1.7352	1.7370	1.7387
5.7	1.7405	1.7422	1.7440	1.7457	1.7475	1.7492	1.7509	1.7527	1.7544	1.7561
5.8	1.7579	1.7596	1.7613	1.7630	1.7647	1.7664	1.7682	1.7699	1.7716	1.7733
5.9	1.7750	1.7766	1.7783	1.7800	1.7817	1.7834	1.7851	1.7867	1.7884	1.7901
6.0	1.7918	1.7934	1.7951	1.7967	1.7984	1.8001	1.8017	1.8034	1.8050	1.8066
6.1	1.8083	1.8099	1.8116	1.8132	1.8148	1.8165	1.8181	1.8197	1.8213	1.8229
6.2	1.8245	1.8262	1.8278	1.8294	1.8310	1.8326	1.8342	1.8358	1.8374	1.8390
6.3	1.8406	1.8421	1.8437	1.8453	1.8469	1.8485	1.8500	1.8516	1.8532	1.8547
6.4	1.8563	1.8579	1.8594	1.8610	1.8625	1.8641	1.8656	1.8672	1.8687	1.8703
6.5	1.8718	1.8733	1.8749	1.8764	1.8779	1.8795	1.8810	1.8825	1.8840	1.8856
6.6	1.8871	1.8886	1.8901	1.8916	1.8931	1.8946	1.8961	1.8976	1.8991	1.9006
6.7	1.9021	1.9036	1.9051	1.9066	1.9081	1.9095	1.9110	1.9125	1.9140	1.9155
6.8	1.9196	1.9184	1.9199	1.9213	1.9228	1.9242	1.9257	1.9272	1.9286	1.9301
6.9	1.9315	1.9330	1.9344	1.9359	1.9373	1.9387	1.9402	1.9416	1.9430	1.9445
7.0	1.9459	1.9473	1.9488	1.9502	1.9516	1.9530	1.9544	1.9559	1.9573	1.9587
7.1	1.9601	1.9615	1.9629	1.9643	1.9657	1.9671	1.9685	1.9699	1.9713	1.9727
7.2	1.9741	1.9755	1.9769	1.9782	1.9796	1.9810	1.9824	1.9838	1.9851	1.9865
7.3	1.9879	1.9892	1.9906	1.9920	1.9933	1.9947	1.9961	1.9974	1.9988	2.0001
7.4	2.0015	2.0028	2.0042	2.0055	2.0069	2.0082	2.0096	2.0109	2.0122	2.0136
7.5	2.0149	2.0162	2.0176	2.0189	2.0202	2.0215	2.0229	2.0242	2.0255	2.0268
7.6	2.0282	2.0295	2.0308	2.0321	2.0334	2.0347	2.0360	2.0373	2.0386	2.0399
7.7	2.0412	2.0425	2.0438	2.0451	2.0464	2.0477	2.0490	2.0503	2.0516	2.0528
7.8	2.0541	2.0554	2.0567	2.0580	2.0592	2.0605	2.0618	2.0631	2.0643	2.0665
7.9	2.0669	2.0681	2.0694	2.0707	2.0719	2.0732	2.0744	2.0757	2.0769	2.0782
8.0	2.0794	2.0807	2.0819	2.0832	2.0844	2.0857	2.0869	2.0882	2.0894	2.0906
8.1	2.0919	2.0931	2.0943	2.0956	2.0968	2.0980	2.0992	2.1005	2.1017	2.1029
8.2	2.1041	2.1054	2.1066	2.1078	2.1090	2.1102	2.1114	2.1126	2.1138	2.1150
8.3	2.1163	2.1175	2.1187	2.1199	2.1211	2.1223	2.1235	2.1247	2.1258	2.1270
8.4	2.1282	2.1294	2.1306	2.1318	2.1330	2.1342	2.1353	2.1365	2.1377	2.1389
8.5	2.1401	2.1412	2.1424	2.1436	2.1448	2.1459	2.1471	2.1483	2.1494	2.1506
8.6	2.1518	2.1529	2.1541	2.1552	2.1564	2.1576	2.1587	2.1599	2.1610	2.1622
8.7	2.1633	2.1645	2.1656	2.1668	2.1679	2.1691	2.1702	2.1713	2.1725	2.1736
8.8	2.1748	2.1759	2.1770	2.1782	2.1793	2.1804	2.1815	2.1827	2.1838	2.1849
8.9	2.1861	2.1872	2.1883	2.1894	2.1905	2.1917	2.1928	2.1939	2.1950	2.1961
9.0	2.1972	2.1983	2.1994	2.2006	2.2017	2.2028	2.2039	2.2050	2.2061	2.2072
9.1	2.2083	2.2094	2.2105	2.2116	2.2127	2.2138	2.2148	2.2159	2.2170	2.2181
9.2	2.2192	2.2203	2.2214	2.2225	2.2235	2.2246	2.2257	2.2268	2.2279	2.2289
9.3	2.2300	2.2311	2.2322	2.2332	2.2343	2.2354	2.2364	2.2375	2.2386	2.2396
9.4	2.2407	2.2418	2.2428	2.2439	2.2450	2.2460	2.2471	2.2481	2.2492	2.2502
9.5	2.2513	2.2523	2.2534	2.2544	2.2555	2.2565	2.2576	2.2586	2.2597	2.2607
9.6	2.2618	2.2628	2.2638	2.2649	2.2659	2.2670	2.2680	2.2690	2.2701	2.2711
9.7	2.2721	2.2732	2.2742	2.2752	2.2762	2.2773	2.2783	2.2793	2.2803	2.2814
9.8	2.2824	2.2834	2.2844	2.2854	2.2865	2.2875	2.2885	2.2895	2.2905	2.2915
9.9	2.2925	2.2935	2.2946	2.2956	2.2966	2.2976	2.2986	2.2996	2.3006	2.3016

Table III Trigonometric Functions—Degree Measure

Degrees	Sin	Csc	Tan	Cot	Sec	Cos	
0° 0′	.0000	—	.0000	—	1.000	1.0000	90° 0′
10′	029	343.8	029	343.8	000	000	50′
20′	058	171.9	058	171.9	000	000	40′
30′	.0087	114.6	.0087	114.6	1.000	1.0000	30′
40′	116	85.95	116	85.94	000	0.9999	20′
50′	145	68.76	145	68.75	000	999	10′
1° 0′	.0175	57.30	.0175	57.29	1.000	.9998	89° 0′
10′	204	49.11	204	49.10	000	998	50′
20′	233	42.98	233	42.96	000	997	40′
30′	.0262	38.20	.0262	38.19	1.000	.9997	30′
40′	291	34.38	291	34.37	000	996	20′
50′	320	31.26	320	31.24	001	995	10′
2° 0′	.0349	28.65	.0349	28.64	1.001	.9994	88° 0′
10′	378	26.45	378	26.43	001	993	50′
20′	407	24.56	407	24.54	001	992	40′
30′	.0436	22.93	.0437	22.90	1.001	.9990	30′
40′	465	21.49	466	21.47	001	989	20′
50′	494	20.23	495	20.21	001	988	10′
3° 0′	.0523	19.11	.0524	19.08	1.001	.9986	87° 0′
10′	552	18.10	553	18.07	002	985	50′
20′	581	17.20	582	17.17	002	983	40′
30′	.0610	16.38	.0612	16.35	1.002	.9981	30′
40′	640	15.64	641	15.60	002	980	20′
50′	669	14.96	670	14.92	002	978	10′
4° 0′	.0698	14.34	.0699	14.30	1.002	.9976	86° 0′
10′	727	13.76	729	13.73	003	974	50′
20′	756	13.23	758	13.20	003	971	40′
30′	.0785	12.75	.0787	12.71	1.003	.9969	30′
40′	814	12.29	816	12.25	003	967	20′
50′	843	11.87	846	11.83	004	964	10′
5° 0′	.0872	11.47	.0875	11.43	1.004	.9962	85° 0′
10′	901	11.10	904	11.06	004	959	50′
20′	929	10.76	934	10.71	004	957	40′
30′	.0958	10.43	.0963	10.39	1.005	.9954	30′
40′	.0987	10.13	.0992	10.08	005	951	20′
50′	.1016	9.839	.1022	9.788	005	948	10′
6° 0′	.1045	9.567	.1051	9.514	1.006	.9945	84° 0′
10′	074	9.309	080	9.255	006	942	50′
20′	103	9.065	110	9.010	006	939	40′
30′	.1132	8.834	.1139	8.777	1.006	.9936	30′
40′	161	8.614	169	8.556	007	932	20′
50′	190	8.405	198	8.345	007	929	10′
7° 0′	.1219	8.206	.1228	8.144	1.008	.9925	83° 0′
10′	248	8.016	257	7.953	008	922	50′
20′	276	7.834	287	7.770	008	918	40′
30′	.1305	7.661	.1317	7.596	1.009	.9914	30′
40′	334	7.496	346	7.429	009	911	20′
50′	363	7.337	376	7.269	009	907	10′
8° 0′	.1392	7.185	.1405	7.115	1.010	.9903	82° 0′
	Cos	Sec	Cot	Tan	Csc	Sin	Degrees

Degrees	Sin	Csc	Tan	Cot	Sec	Cos	
8° 0′	.1392	7.185	.1405	7.115	1.010	.9903	82° 0′
10′	421	7.040	435	6.968	010	899	50′
20′	449	6.900	465	6.827	011	894	40′
30′	.1478	6.765	.1495	6.691	1.011	.8980	30′
40′	507	6.636	524	6.561	012	886	20′
50′	536	6.512	554	6.435	012	881	10′
9° 0′	.1564	6.392	.1584	6.314	1.012	.9877	81° 0′
10′	593	277	614	197	013	872	50′
20′	622	166	644	6.084	013	868	40′
30′	.1650	6.059	.1673	5.976	1.014	.9863	30′
40′	679	5.955	703	871	014	858	20′
50′	708	855	733	769	015	853	10′
10° 0′	.1736	5.759	.1763	5.671	1.015	.9848	80° 0′
10′	765	665	793	576	016	843	50′
20′	794	575	823	485	016	838	40′
30′	.1822	5.487	.1853	5.396	1.017	.9833	30′
40′	851	403	883	309	018	827	20′
50′	880	320	914	226	018	822	10′
11° 0′	.1908	5.241	.1944	5.145	1.019	.9816	79° 0′
10′	937	164	.1974	5.066	019	811	50′
20′	965	089	.2004	4.989	020	805	40′
30′	.1994	5.016	.2035	4.915	1.020	.9799	30′
40′	.2022	4.945	065	843	021	793	20′
50′	051	876	095	773	022	787	10′
12° 0′	.2079	4.810	.2126	4.705	1.022	.9781	78° 0′
10′	108	745	156	638	023	775	50′
20′	136	682	186	574	024	769	40′
30′	.2164	4.620	.2217	4.511	1.024	.9763	30′
40′	193	560	247	449	025	757	20′
50′	221	502	278	390	026	750	10′
13° 0′	.2250	4.445	.2309	4.331	1.026	.9744	77° 0′
10′	278	390	339	275	027	737	50′
20′	306	336	370	219	028	730	40′
30′	.2334	4.284	.2401	4.165	1.028	.9724	30′
40′	363	232	432	113	029	717	20′
50′	391	182	462	061	030	710	10′
14° 0′	.2419	4.134	.2493	4.011	1.031	.9703	76° 0′
10′	447	086	524	3.962	031	696	50′
20′	476	4.039	555	914	032	689	40′
30′	.2504	3.994	.2586	3.867	1.033	.9681	30′
40′	532	950	617	821	034	674	20′
50′	560	906	648	776	034	667	10′
15° 0′	.2588	3.864	.2679	3.732	1.035	.9659	75° 0′
10′	616	822	711	689	036	652	50′
20′	644	782	742	647	037	644	40′
30′	.2672	3.742	.2773	3.606	1.038	.9636	30′
40′	700	703	805	566	039	628	20′
50′	728	665	836	526	039	621	10′
16° 0′	.2756	3.628	.2867	3.487	1.040	.9613	74° 0′
	Cos	Sec	Cot	Tan	Csc	Sin	Degrees

Table III Trigonometric Functions—Degree Measure

Degrees	Sin	Csc	Tan	Cot	Sec	Cos	
16° 0′	.2756	3.628	.2867	3.487	1.040	.9613	74° 0′
10′	784	592	899	450	041	605	50′
20′	812	556	931	412	042	596	40′
30′	.2840	3.521	.2962	3.376	1.043	.9588	30′
40′	868	487	.2994	340	044	580	20′
50′	896	453	3026	305	045	572	10′
17° 0′	.2924	3.420	.3057	3.271	1.046	.9563	73° 0′
10′	952	388	089	237	047	555	50′
20′	.2979	357	121	204	048	546	40′
30′	.3007	3.326	.3153	3.172	1.048	.9537	30′
40′	035	295	185	140	049	528	20′
50′	062	265	217	108	050	520	10′
18° 0′	.3090	3.236	.3249	3.078	1.051	.9511	72° 0′
10′	118	207	281	047	052	502	50′
20′	145	179	314	3.018	053	492	40′
30′	.3173	3.152	.3346	2.989	1.054	.9483	30′
40′	201	124	378	960	056	474	20′
50′	228	098	411	932	057	465	10′
19° 0′	.3256	3.072	.3443	2.904	1.058	.9455	71° 0′
10′	283	046	476	877	059	446	50′
20′	311	3.021	508	850	060	436	40′
30′	.3338	2.996	.3541	2.824	1.061	.9426	30′
40′	365	971	574	798	062	417	20′
50′	393	947	607	773	063	407	10′
20° 0′	.3420	2.924	.3640	2.747	1.064	.9397	70° 0′
10′	448	901	673	723	065	387	50′
20′	475	878	706	699	066	377	40′
30′	.3502	2.855	.3739	2.675	1.068	.9367	30′
40′	529	833	772	651	069	356	20′
50′	557	812	805	628	070	346	10′
21° 0′	.3584	2.790	.3839	2.605	1.071	.9336	69° 0′
10′	611	769	872	583	072	325	50′
20′	638	749	906	560	074	315	40′
30′	.3665	2.729	.3939	2.539	1.075	.9304	30′
40′	692	709	.3973	517	076	293	20′
50′	719	689	.4006	496	077	283	10′
22° 0′	.3746	2.669	.4040	2.475	1.079	.9272	68° 0′
10′	773	650	074	455	080	261	50′
20′	800	632	108	434	081	250	40′
30′	.3827	2.613	.4142	2.414	1.082	.9239	30′
40′	854	595	176	394	084	228	20′
50′	881	577	210	375	085	216	10′
23° 0′	.3907	2.559	.4245	2.356	1.086	.9205	67° 0′
10′	934	542	279	337	088	194	50′
20′	961	525	314	318	089	182	40′
30′	.3987	2.508	.4348	2.300	1.090	.9171	30′
40′	.4014	491	383	282	092	159	20′
50′	041	475	417	264	093	147	10′
24° 0′	.4067	2.459	.4452	2.246	1.095	.9135	66° 0′
	Cos	Sec	Cot	Tan	Csc	Sin	Degrees

Degrees	Sin	Csc	Tan	Cot	Sec	Cos	
24° 0′	.4067	2.459	.4452	2.246	1.095	.9135	66° 0′
10′	094	443	487	229	096	124	50′
20′	120	427	522	211	097	112	40′
30′	.4147	2.411	.4557	2.194	1.099	.9100	30′
40′	173	396	592	177	100	088	20′
50′	200	381	628	161	102	075	10′
25° 0′	.4226	2.366	.4663	2.145	1.103	.9063	65° 0′
10′	253	352	699	128	105	051	50′
20′	279	337	734	112	106	038	40′
30′	.4305	2.323	.4770	2.097	1.108	.9026	30′
40′	331	309	806	081	109	013	20′
50′	358	295	841	066	111	.9001	10′
26° 0′	.4384	2.281	.4877	2.050	1.113	.8988	64° 0′
10′	410	268	913	035	114	975	50′
20′	436	254	950	020	116	962	40′
30′	.4462	2.241	.4986	2.006	1.117	.8949	30′
40′	488	228	.5022	1.991	119	936	20′
50′	514	215	059	977	121	923	10′
27° 0′	.4540	2.203	.5095	1.963	1.122	.8910	63° 0′
10′	566	190	132	949	124	897	50′
20′	592	178	169	935	126	884	40′
30′	.4617	2.166	.5206	1.921	1.127	.8870	30′
40′	643	154	243	907	129	857	20′
50′	669	142	280	894	131	843	10′
28° 0′	.4695	2.130	.5317	1.881	1.133	.8829	62° 0′
10′	720	118	354	868	134	.816	50′
20′	746	107	392	855	136	802	40′
30′	.4772	2.096	.5430	1.842	1.138	.8788	30′
40′	797	085	467	829	140	774	20′
50′	823	074	505	816	142	760	10′
29° 0′	.4848	2.063	.5543	1.804	1.143	.8746	61° 0′
10′	874	052	581	792	145	732	50′
20′	899	041	619	780	147	718	40′
30′	.4924	2.031	.5658	1.767	1.149	.8704	30′
40′	950	020	696	756	151	689	20′
50′	.4975	010	735	744	153	675	10′
30° 0′	.5000	2.000	.5774	1.732	1.155	.8660	60° 0′
10′	025	1.990	812	720	157	646	50′
20′	050	980	851	709	159	631	40′
30′	.5075	1.970	.5890	1.698	1.161	.8616	30′
40′	100	961	930	686	163	601	20′
50′	125	951	.5969	675	165	587	10′
31° 0′	.5150	1.942	.6009	1.664	1.167	.8572	59° 0′
10′	175	932	048	653	169	557	50′
20′	200	923	088	643	171	542	40′
30′	.5225	1.914	.6128	1.632	1.173	.8526	30′
40′	250	905	168	621	175	511	20′
50′	275	896	208	611	177	496	10′
32° 0′	.5299	1.887	.6249	1.600	1.179	.8480	58° 0′
	Cos	Sec	Cot	Tan	Csc	Sin	Degrees

Table III Trigonometric Functions—Degree Measure

Degrees	Sin	Csc	Tan	Cot	Sec	Cos	
32° 0′	.5299	1.887	.6249	1.600	1.179	.8480	58° 0′
10′	324	878	289	590	181	465	50′
20′	348	870	330	580	184	450	40′
30′	.5373	1.861	.6371	1.570	1.186	.8434	30′
40′	398	853	412	560	188	418	20′
50′	422	844	453	550	190	403	10′
33° 0′	.5446	1.836	.6494	1.540	1.192	.8387	57° 0′
10′	471	828	536	530	195	371	50′
20′	495	820	577	520	197	355	40′
30′	.5519	1.812	.6619	1.511	1.199	.8339	30′
40′	544	804	661	501	202	323	20′
50′	568	796	703	1.492	204	307	10′
34° 0′	.5592	1.788	.6745	1.483	1.206	.8290	56° 0′
10′	616	781	787	473	209	274	50′
20′	640	773	830	464	211	258	40′
30′	.5664	1.766	.6873	1.455	1.213	.8241	30′
40′	688	758	916	446	216	225	20′
50′	712	751	.6959	437	218	208	10′
35° 0′	.5736	1.743	.7002	1.428	1.221	.8192	55° 0′
10′	760	736	046	419	223	175	50′
20′	783	729	089	411	226	158	40′
30′	.5807	1.722	.7133	1.402	1.228	.8141	30′
40′	831	715	177	393	231	124	20′
50′	854	708	221	385	233	107	10′
36° 0′	.5878	1.701	.7265	1.376	1.236	.8090	54° 0′
10′	901	695	310	368	239	073	50′
20′	925	688	355	360	241	056	40′
30′	.5948	1.681	.7400	1.351	1.244	.8039	30′
40′	972	675	445	343	247	021	20′
50′	.5995	668	490	335	249	.8004	10′
37° 0′	.6018	1.662	.7536	1.327	1.252	.7986	53° 0′
10′	041	655	581	319	255	969	50′
20′	065	649	627	311	258	951	40′
30′	.6088	1.643	.7673	1.303	1.260	.7934	30′
40′	111	636	720	295	263	916	20′
50′	134	630	766	288	266	898	10′
38° 0′	.6157	1.624	.7813	1.280	1.269	.7880	52° 0′
10′	180	618	860	272	272	862	50′
20′	202	612	907	265	275	844	40′
30′	.6225	1.606	.7954	1.257	1.278	.7826	30′
40′	248	601	.8002	250	281	808	20′
50′	271	595	050	242	284	790	10′
39° 0′	.6293	1.589	.8098	1.235	1.287	.7771	51° 0′
10′	316	583	146	228	290	753	50
20′	338	578	195	220	293	735	40′
30′	.6361	1.572	.8243	1.213	1.296	.7716	30′
40′	383	567	292	206	299	698	20′
50′	406	561	342	199	302	679	10′
40° 0′	.6428	1.556	.8391	1.192	1.305	.7660	50° 0′
	Cos	Sec	Cot	Tan	Csc	Sin	Degrees

Degrees	Sin	Csc	Tan	Cot	Sec	Cos	
40° 0′	.6428	1.556	.8391	1.192	1.305	.7660	50° 0′
10′	450	550	441	185	309	642	50′
20′	472	545	491	178	312	623	40′
30′	.6494	1.540	.8541	1.171	1.315	.7604	30′
40′	517	535	591	164	318	585	20′
50′	539	529	642	157	322	566	10′
41° 0′	.6561	1.524	.8693	1.150	1.325	.7547	49° 0′
10′	583	519	744	144	328	528	50′
20′	604	514	796	137	332	509	40′
30′	.6626	1.509	.8847	1.130	1.335	.7490	30′
40′	648	504	899	124	339	470	20′
50′	670	499	.8952	117	342	451	10′
42° 0′	.6691	1.494	.9004	1.111	1.346	.7431	48° 0′
10′	713	490	057	104	349	412	50′
20′	734	485	110	098	353	392	40′
30′	.6756	1.480	.9163	1.091	1.356	.7373	30′
40′	777	476	217	085	360	353	20′
50′	799	471	271	079	364	333	10′
43° 0′	.6820	1.466	.9325	1.072	1.367	.7314	47° 0′
10′	841	462	380	066	371	294	50′
20′	862	457	435	060	375	274	40′
30′	.6884	1.453	.9490	1.054	1.379	.7254	30′
40′	905	448	545	048	382	234	20′
50′	926	444	601	042	386	214	10′
44° 0′	.6947	1.440	.9657	1.036	1.390	.7193	46° 0′
10′	967	435	713	030	394	173	50′
20′	.6988	431	770	024	398	153	40′
30′	.7009	1.427	.9827	1.018	1.402	.7133	30′
40′	030	423	884	012	406	112	20′
50′	050	418	.9942	006	410	092	10′
45° 0′	.7071	1.414	1.000	1.000	1.414	.7071	45° 0′
	Cos	Sec	Cot	Tan	Csc	Sin	Degrees

Table IV Trigonometric Functions—Radian Measure

t	$\sin t$	$\cos t$	$\tan t$	$\cot t$	$\sec t$	$\csc t$
.00	.0000	1.0000	.0000	—	1.000	—
.01	.0100	1.0000	.0100	99.997	1.000	100.00
.02	.0200	.9998	.0200	49.993	1.000	50.00
.03	.0300	.9996	.0300	33.323	1.000	33.34
.04	.0400	.9992	.0400	24.987	1.001	25.01
.05	.0500	.9988	.0500	19.983	1.001	20.01
.06	.0600	.9982	.0601	16.647	1.002	16.68
.07	.0699	.9976	.0701	14.262	1.002	14.30
.08	.0799	.9968	.0802	12.473	1.003	12.51
.09	.0899	.9960	.0902	11.081	1.004	11.13
.10	.0998	.9950	.1003	9.967	1.005	10.02
.11	.1098	.9940	.1104	9.054	1.006	9.109
.12	.1197	.9928	.1206	8.293	1.007	8.353
.13	.1296	.9916	.1307	7.649	1.009	7.714
.14	.1395	.9902	.1409	7.096	1.010	7.166
.15	.1494	.9888	.1511	6.617	1.011	6.692
.16	.1593	.9872	.1614	6.197	1.013	6.277
.17	.1692	.9856	.1717	5.826	1.015	5.911
.18	.1790	.9838	.1820	5.495	1.016	5.586
.19	.1889	.9820	.1923	5.200	1.018	5.295
.20	.1987	.9801	.2027	4.933	1.020	5.033
.21	.2085	.9780	.2131	4.692	1.022	4.797
.22	.2182	.9759	.2236	4.472	1.025	4.582
.23	.2280	.9737	.2341	4.271	1.027	4.386
.24	.2377	.9713	.2447	4.086	1.030	4.207
.25	.2474	.9689	.2553	3.916	1.032	4.042
.26	.2571	.9664	.2660	3.759	1.035	3.890
.27	.2667	.9638	.2768	3.613	1.038	3.749
.28	.2764	.9611	.2876	3.478	1.041	3.619
.29	.2860	.9582	.2984	3.351	1.044	3.497
.30	.2955	.9553	.3093	3.233	1.047	3.384
.31	.3051	.9523	.3203	3.122	1.050	3.278
.32	.3146	.9492	.3314	3.018	1.053	3.179
.33	.3240	.9460	.3425	2.920	1.057	3.086
.34	.3335	.9428	.3537	2.827	1.061	2.999
.35	.3429	.9394	.3650	2.740	1.065	2.916
.36	.3523	.9359	.3764	2.657	1.068	2.839
.37	.3616	.9323	.3879	2.578	1.073	2.765
.38	.3709	.9287	.3994	2.504	1.077	2.696
.39	.3802	.9249	.4111	2.433	1.081	2.630
.40	.3894	.9211	.4228	2.365	1.086	2.568
.41	.3986	.9171	.4346	2.301	1.090	2.509
.42	.4078	.9131	.4466	2.239	1.095	2.452
.43	.4169	.9090	.4586	2.180	1.100	2.399
.44	.4259	.9048	.4708	2.124	1.105	2.348

t	$\sin t$	$\cos t$	$\tan t$	$\cot t$	$\sec t$	$\csc t$
.45	.4350	.9004	.4831	2.070	1.111	2.299
.46	.4439	.8961	.4954	2.018	1.116	2.253
.47	.4529	.8916	.5080	1.969	1.122	2.208
.48	.4618	.8870	.5206	1.921	1.127	2.166
.49	.4706	.8823	.5334	1.875	1.133	2.125
.50	.4794	.8776	.5463	1.830	1.139	2.086
.51	.4882	.8727	.5594	1.788	1.146	2.048
.52	.4969	.8678	.5726	1.747	1.152	2.013
$\frac{\pi}{6}$	.5000	.8660	.5774	1.732	1.155	2.000
.53	.5055	.8628	.5859	1.707	1.159	1.978
.54	.5141	.8577	.5994	1.668	1.166	1.945
.55	.5227	.8525	.6131	1.631	1.173	1.913
.56	.5312	.8473	.6269	1.595	1.180	1.883
.57	.5396	.8419	.6410	1.560	1.188	1.853
.58	.5480	.8365	.6552	1.526	1.196	1.825
.59	.5564	.8309	.6696	1.494	1.203	1.797
.60	.5646	.8253	.6841	1.462	1.212	1.771
.61	.5729	.8196	.6989	1.431	1.220	1.746
.62	.5810	.8139	.7139	1.401	1.229	1.721
.63	.5891	.8080	.7291	1.372	1.238	1.697
.64	.5972	.8021	.7445	1.343	1.247	1.674
.65	.6052	.7961	.7602	1.315	1.256	1.652
.66	.6131	.7900	.7761	1.288	1.266	1.631
.67	.6210	.7838	.7923	1.262	1.276	1.610
.68	.6288	.7776	.8087	1.237	1.286	1.590
.69	.6365	.7712	.8253	1.212	1.297	1.571
.70	.6442	.7648	.8423	1.187	1.307	1.552
.71	.6518	.7584	.8595	1.163	1.319	1.534
.72	.6594	.7518	.8771	1.140	1.330	1.517
.73	.6669	.7452	.8949	1.117	1.342	1.500
.74	.6743	.7385	.9131	1.095	1.354	1.483
.75	.6816	.7317	.9316	1.073	1.367	1.467
.76	.6889	.7248	.9505	1.052	1.380	1.452
.77	.6961	.7179	.9697	1.031	1.393	1.437
.78	.7033	.7109	.9893	1.011	1.407	1.422
$\frac{\pi}{4}$	.7071	.7071	1.000	1.000	1.414	1.414
.79	.7104	.7038	1.009	.9908	1.421	1.408
.80	.7174	.6967	1.030	.9712	1.435	1.394
.81	.7243	.6895	1.050	.9520	1.450	1.381
.82	.7311	.6822	1.072	.9331	1.466	1.368
.83	.7379	.6749	1.093	.9146	1.482	1.355
.84	.7446	.6675	1.116	.8964	1.498	1.343

Table IV Trigonometric Functions—Radian Measure

t	$\sin t$	$\cos t$	$\tan t$	$\cot t$	$\sec t$	$\csc t$
.85	.7513	.6600	1.138	.8785	1.515	1.331
.86	.7578	.6524	1.162	.8609	1.533	1.320
.87	.7643	.6448	1.185	.8437	1.551	1.308
.88	.7707	.6372	1.210	.8267	1.569	1.297
.89	.7771	.6294	1.235	.8100	1.589	1.287
.90	.7833	.6216	1.260	.7936	1.609	1.277
.91	.7895	.6137	1.286	.7774	1.629	1.267
.92	.7956	.6058	1.313	.7615	1.651	1.257
.93	.8016	.5978	1.341	.7458	1.673	1.247
.94	.8076	.5898	1.369	.7303	1.696	1.238
.95	.8134	.5817	1.398	.7151	1.719	1.229
.96	.8192	.5735	1.428	.7001	1.744	1.221
.97	.8249	.5653	1.459	.6853	1.769	1.212
.98	.8305	.5570	1.491	.6707	1.795	1.204
.99	.8360	.5487	1.524	.6563	1.823	1.196
1.00	.8415	.5403	1.557	.6421	1.851	1.188
1.01	.8468	.5319	1.592	.6281	1.880	1.181
1.02	.8521	.5234	1.628	.6142	1.911	1.174
1.03	.8573	.5148	1.665	.6005	1.942	1.166
1.04	.8624	.5062	1.704	.5870	1.975	1.160
$\frac{\pi}{3}$	.8660	.5000	1.732	.5774	2.000	1.155
1.05	.8674	.4976	1.743	.5736	2.010	1.153
1.06	.8724	.4889	1.784	.5604	2.046	1.146
1.07	.8772	.4801	1.827	.5473	2.083	1.140
1.08	.8820	.4713	1.871	.5344	2.122	1.134
1.09	.8866	.4625	1.917	.5216	2.162	1.128
1.10	.8912	.4536	1.965	.5090	2.205	1.122
1.11	.8957	.4447	2.014	.4964	2.249	1.116
1.12	.9001	.4357	2.066	.4840	2.295	1.111
1.13	.9044	.4267	2.120	.4718	2.344	1.106
1.14	9086	.4176	2.176	.4596	2.395	1.101
1.15	.9128	.4085	2.234	.4475	2.448	1.096
1.16	.9168	.3993	2.296	.4356	2.504	1.091
1.17	.9208	.3902	2.360	.4237	2.563	1.086
1.18	.9246	.3809	2.427	.4120	2.625	1.082
1.19	.9284	.3717	2.498	.4003	2.691	1.077
1.20	.9320	.3624	2.572	.3888	2.760	1.073
1.21	.9356	.3530	2.650	.3773	2.833	1.069
1.22	.9391	.3436	2.733	.3659	2.910	1.065
1.23	.9425	.3342	2.820	.3546	2.992	1.061
1.24	.9458	.3248	2.912	.3434	3.079	1.057
1.25	.9490	.3153	3.010	.3323	3.171	1.054
1.26	.9521	.3058	3.113	.3212	3.270	1.050
1.27	.9551	.2963	3.224	.3102	3.375	1.047
1.28	.9580	.2867	3.341	.2993	3.488	1.044
1.29	.9608	.2771	3.467	.2884	3.609	1.041

t	$\sin t$	$\cos t$	$\tan t$	$\cot t$	$\sec t$	$\csc t$
1.30	.9636	.2675	3.602	.2776	3.738	1.038
1.31	.9662	.2579	3.747	.2669	3.878	1.035
1.32	.9687	.2482	3.903	.2562	4.029	1.032
1.33	.9711	.2385	4.072	.2456	4.193	1.030
1.34	.9735	.2288	4.256	.2350	4.372	1.027
1.35	.9757	.2190	4.455	.2245	4.566	1.025
1.36	.9779	.2092	4.673	.2140	4.779	1.023
1.37	.9799	.1994	4.913	.2035	5.014	1.021
1.38	.9819	.1896	5.177	.1931	5.273	1.018
1.39	.9837	.1798	5.471	.1828	5.561	1.017
1.40	.9854	.1700	5.798	.1725	5.883	1.015
1.41	.9871	.1601	6.165	.1622	6.246	1.013
1.42	.9887	.1502	6.581	.1519	6.657	1.011
1.43	.9901	.1403	7.055	.1417	7.126	1.010
1.44	.9915	.1304	7.602	.1315	7.667	1.009
1.45	.9927	.1205	8.238	.1214	8.299	1.007
1.46	.9939	.1106	8.989	.1113	9.044	1.006
1.47	.9949	.1006	9.887	.1011	9.938	1.005
1.48	.9959	.0907	10.983	.0910	11.029	1.004
1.49	.9967	.0807	12.350	.0810	12.390	1.003
1.50	.9975	.0707	14.101	.0709	14.137	1.003
1.51	.9982	.0608	16.428	.0609	16.458	1.002
1.52	.9987	.0508	19.670	.0508	19.695	1.001
1.53	.9992	.0408	24.498	.0408	24.519	1.001
1.54	.9995	.0308	32.461	.0308	32.476	1.000
1.55	.9998	.0208	48.078	.0208	48.089	1.000
1.56	.9999	.0108	92.620	.0108	92.626	1.000
1.57	1.0000	.0008	1255.8	.0008	1255.8	1.000
$\frac{\pi}{2}$	1.0000	.0000	—	.0000	—	1.000

Table V Exponential Function Values

x	e^x	e^{-x}	x	e^x	e^{-x}
0.00	1.0000	1.0000	3.0	20.086	0.0498
0.05	1.0513	0.9512	3.1	22.198	0.0450
0.10	1.1052	0.9048	3.2	24.533	0.0408
0.15	1.1618	0.8607	3.3	27.113	0.0369
0.20	1.2214	0.8187	3.4	29.964	0.0334
0.25	1.2840	0.7788	3.5	33.115	0.0302
0.30	1.3499	0.7408	3.6	36.598	0.0273
0.35	1.4191	0.7047	3.7	40.447	0.0247
0.40	1.4918	0.6703	3.8	44.701	0.0224
0.45	1.5683	0.6376	3.9	49.402	0.0202
0.50	1.6487	0.6065	4.0	54.598	0.0183
0.55	1.7333	0.5769	4.1	60.340	0.0166
0.60	1.8221	0.5488	4.2	66.686	0.0150
0.65	1.9155	0.5220	4.3	73.700	0.0136
0.70	2.0138	0.4966	4.4	81.451	0.0123
0.75	2.1170	0.4724	4.5	90.017	0.0111
0.80	2.2255	0.4493	4.6	99.484	0.0101
0.85	2.3396	0.4274	4.7	109.95	0.0091
0.90	2.4596	0.4066	4.8	121.51	0.0082
0.95	2.5857	0.3867	4.9	134.29	0.0074
1.0	2.7183	0.3679	5.0	148.41	0.0067
1.1	3.0042	0.3329	5.1	164.02	0.0061
1.2	3.3201	0.3012	5.2	181.27	0.0055
1.3	3.6693	0.2725	5.3	200.34	0.0050
1.4	4.0552	0.2466	5.4	221.41	0.0045
1.5	4.4817	0.2231	5.5	244.69	0.0041
1.6	4.9530	0.2019	5.6	270.43	0.0037
1.7	5.4739	0.1827	5.7	298.87	0.0033
1.8	6.0496	0.1653	5.8	330.30	0.0030
1.9	6.6859	0.1496	5.9	365.04	0.0027
2.0	7.3891	0.1353	6.0	403.43	0.0025
2.1	8.1662	0.1225	6.5	665.14	0.0015
2.2	9.0250	0.1108	7.0	1,096.6	0.0009
2.3	9.9742	0.1003	7.5	1,808.0	0.0006
2.4	11.023	0.0907	8.0	2,981.0	0.0003
2.5	12.182	0.0821	8.5	4,914.8	0.0002
2.6	13.464	0.0743	9.0	8,103.1	0.0001
2.7	14.880	0.0672	9.5	13,360	0.00007
2.8	16.445	0.0608	10.0	22,026	0.00004
2.9	18.174	0.0550			

Answers to Selected Problems

CHAPTER 1

PROBLEM SET 1.1 page 8

1. (a) $-1, 8, 0, 9$; (b) $-3, -1, 1, 3, 5$; (c) $-6, -5, -4, -3, -2, -1$ $0, 1, 2$; (d) $4, 6$
3. (a) F; (b) T; (c) T; (d) T; (e) T; (f) F; (g) F; (h) T; (i) T; (j) T; (k) F; (l) F **5.** 0.75
7. $-0.8\overline{3}$ **9.** $2.\overline{3}$ **11.** 17.25 **13.** $0.0\overline{21}$ **15.** (a) $\frac{27}{100}$; (b) $\frac{171}{100}$; (c) $-\frac{1}{8}$; (d) $-\frac{1}{125}$
17. (a) $-\frac{32}{9}$; (b) $\frac{170}{99}$; (c) $\frac{858}{37}$; (d) 1 **19.** 5 **21.** 23 **23.** 21 **25.** 0 **27.** $\pi - \sqrt{2}$
29. -1 **31.** 4 **33.** $|b + 4|$ **35.** $|2a - 9|$ or $|9 - 2a|$ **37.** 1.273 **39.** -1.441
41. -0.8660 **43.** 57.30 **45.** 403.2 **47.** 2.957 **49.** Closure under addition
51. Multiplication is commutative **53.** Addition is associative **55.** Distributive **57.** Distributive
59. Negation and closure of multiplication **61.** Cancellation of multiplication **63.** Zero factor

PROBLEM SET 1.2 page 16

1. 6^6 **3.** $2x^3y^2 - 3x^3$ **5.** Polynomial; degree 2; coefficients: 4, -1, -7; trinomial **7.** No
9. Polynomial; degree 4; coefficient: -11; monomial **11.** Polynomial; degree 5; coefficients: $\frac{1}{2}$, π; binomial
13. Polynomial; degree 20; coefficient: 16; monomial **15.** -352 **17.** 0.5946 **19.** $3xy^2 - 2x^2y^2$
21. $-4x^3 + 5x^2 - 3x - 5$ **23.** $2x^2 + 4x - 14$ **25.** $x^2 - 4x + 2$ **27.** $21x^3 - 34x^2 + 19x - 4$
29. $-22x^{11}$ **31.** $30x^5y^{10}$ **33.** $27x^6y^3z^9$ **35.** $8575x^{11}y^{10}$ **37.** $-288x^{11}y^{15}$ **39.** $6x^2 + 7x - 3$
41. $96x^3 + 64x^2y - 66xy^2 - 45y^3$ **43.** $-30x^3 - 32x^2 + 14x$ **45.** $2x^3 + 5x^2 + 11x - 7$
47. $x^4 - x^3 - 5x^2 - 21x + 54$ **49.** $4x^2 + 4xy + y^2$ **51.** $64y^2 - 128yz + 64z^2$
53. $8p^3 - 12p^2q^2 + 6pq^4 - q^6$ **55.** $25r^2 - 49s^2$ **57.** $27 + y^3$ **59.** $3x(3x + 1)$
61. $3x^2y(27x - 4y^4)$ **63.** $(x + y)(m + x + y)$ **65.** $(3y - 1)(3y + 1)$ **67.** $(4x + 5y)(4x - 5y)$
69. $(x^2 + 9y^2)(x + 3y)(x - 3y)$ **71.** $(x - 2)(x^2 + 2x + 4)$ **73.** $(5y + 4)(25y^2 - 20y + 16)$
75. $(w + 4)(w^2 - w + 7)$ **77.** $(x - 9)(x - 7)$ **79.** $(3x - 1)(x + 2)$ **81.** $(4x - 1)(3x + 5)$
83. $-x(2x - 3)(x + 4)$ **85.** $-2x^2(x + 6)(x - 2)$ **87.** $(2 - 3x)(y - 2)$
89. $(5x - 2y)(5x + 2y + 1)$

PROBLEM SET 1.3 page 24

1. $x \neq 7$ **3.** $x \neq -8$ **5.** $x \neq -2, -3$ **7.** $x \neq 0, 1, -2$ **9.** $x \neq y$ **11.** $\dfrac{y^5}{5xa^2}$

13. $-\dfrac{2x + 3}{3x}$ **15.** $\dfrac{1}{7x + 5y}$ **17.** $\dfrac{x + 4}{x - 2}$ **19.** $\dfrac{z + y}{y + w}$ **21.** $-\dfrac{1}{x^2}$ **23.** $\dfrac{-2(x + 2)}{(x - 1)^3}$

25. $\dfrac{h}{2x + h + 5}$ **27.** $\dfrac{3}{5(x + 3)}$ **29.** $7a + 2$ **31.** $\dfrac{4x(x - 1)}{(3x - 5)(x + 2)}$ **33.** $\dfrac{2(x + 2)}{3(x - 2)(x + 1)}$

35. $-\dfrac{x^2+4}{2x(x-3)}$ **37.** $x-3$ **39.** $\dfrac{2x-5}{(x-5)(x+5)}$ **41.** $\dfrac{-4x+3}{(x-3)(x+3)(x-2)}$

43. $\dfrac{7y+11}{(2y-1)(y+1)(y+2)}$ **45.** $\dfrac{-x(4+x)}{(x-2)(x+2)}$ **47.** $\dfrac{12x^3-x^2-2x-1}{4x^3(x+1)}$ **49.** $\dfrac{x+5}{x(x-1)(x+1)}$

51. x **53.** $\dfrac{x^2-y^2}{x^2+y^2}$ **55.** $-\dfrac{10}{3}$ **57.** $\dfrac{-2x-h}{x^2(x+h)^2}$ **59.** $\dfrac{-3x^2-3xh-h^2}{6x^3(x+h)^3}$

PROBLEM SET 1.4 page 34

1. $\dfrac{3}{32}$ **3.** $\dfrac{1}{320}$ **5.** 4096 **7.** $\dfrac{64}{729}$ **9.** $\dfrac{y^8}{x^2}$ **11.** x **13.** $\dfrac{-1}{ab}$ **15.** $\dfrac{b+a}{b-a}$ **17.** $x^{1/3}$

19. $a^{101/24}$ **21.** $\dfrac{x}{y^{1/2}}$ **23.** $\dfrac{x^3y^4}{3}$ **25.** $\dfrac{8(4x+1)^2(x+4)}{(2x+3)^3}$ **27.** $\dfrac{-16x+1}{(4x+5)^{5/2}(2x-1)^{3/2}}$

29. (a) $2\sqrt{x}$; (b) $4\sqrt{x}$; (c) $x+y+2\sqrt{xy}$; (d) $\sqrt{x^2+y^2}$; (e) $\sqrt[3]{(x-y)^2}$; (f) $\dfrac{1}{\sqrt{3x^2-5x+7}}$;

(g) $\dfrac{1}{(x^3+2y^3)\sqrt[3]{x^3+2y^3}}$; (h) $(x^2+9)\sqrt{x^2+9}$; (i) $\dfrac{x+4}{5-x}\sqrt[3]{\left(\dfrac{x+4}{5-x}\right)^2}$ **31.** $-2\sqrt[5]{2}$

33. $5x\sqrt{x}$ **35.** $2x^2y^3\sqrt[4]{x^3y}$ **37.** $5xy^2\sqrt[3]{2xy}$ **39.** $2v^2/u$ **41.** $\dfrac{t}{2}\sqrt[3]{9}$ **43.** $\dfrac{3y^2-2y^4}{(1-y^2)^{3/2}}$

45. $-11\sqrt{2}$ **47.** $21x\sqrt{3}-5x$ **49.** $-9x\sqrt[3]{x}$ **51.** $4x^2+15x+29$ **53.** $6\sqrt{3}-4\sqrt{7}+2$

55. $-20+5\sqrt{6}+4\sqrt{3}-3\sqrt{2}$ **57.** $53-10\sqrt{6}$ **59.** 17 **61.** $9x-y$ **63.** $\sqrt{2}$

65. $2\sqrt{5x}/3$ **67.** $-\frac{1}{4}(5\sqrt{3}+5\sqrt{7}+\sqrt{6}+\sqrt{14})$ **69.** $6+3\sqrt{2}-\sqrt{6}-6\sqrt{3}$ **71.** $\sqrt{t}-\sqrt{t-1}$

73. $6+8i$ **75.** $-2-2i$ **77.** $27+23i$ **79.** 53 **81.** $-1-3i$ **83.** $8-6i$

85. $18\sqrt{x}\,i$ **87.** $-\frac{3}{5}-\frac{7}{5}i$ **89.** $-\frac{3}{25}+\frac{29}{25}i$

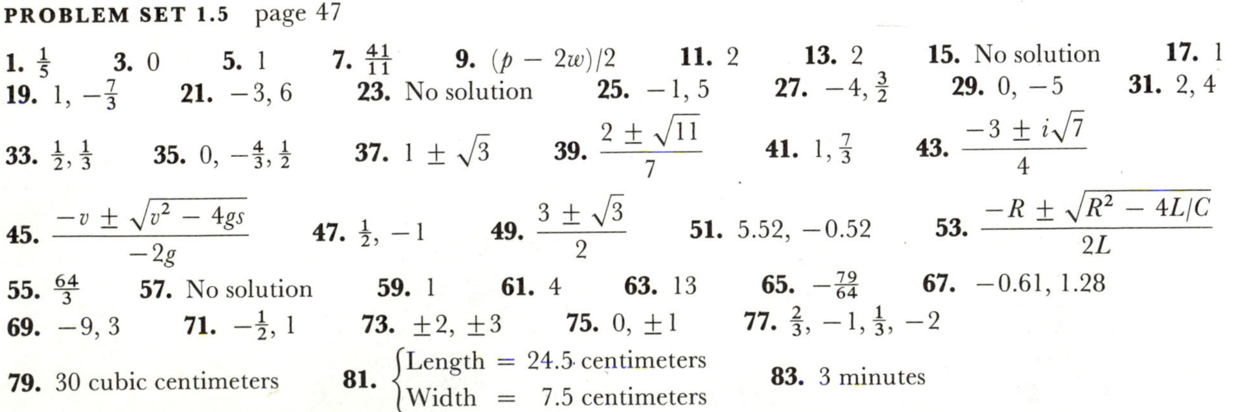

PROBLEM SET 1.5 page 47

1. $\frac{1}{5}$ **3.** 0 **5.** 1 **7.** $\frac{41}{11}$ **9.** $(p-2w)/2$ **11.** 2 **13.** 2 **15.** No solution **17.** 1

19. $1, -\frac{7}{3}$ **21.** $-3, 6$ **23.** No solution **25.** $-1, 5$ **27.** $-4, \frac{3}{2}$ **29.** $0, -5$ **31.** 2, 4

33. $\frac{1}{2}, \frac{1}{3}$ **35.** $0, -\frac{4}{3}, \frac{1}{2}$ **37.** $1 \pm \sqrt{3}$ **39.** $\dfrac{2 \pm \sqrt{11}}{7}$ **41.** $1, \frac{7}{3}$ **43.** $\dfrac{-3 \pm i\sqrt{7}}{4}$

45. $\dfrac{-v \pm \sqrt{v^2-4gs}}{-2g}$ **47.** $\frac{1}{2}, -1$ **49.** $\dfrac{3 \pm \sqrt{3}}{2}$ **51.** $5.52, -0.52$ **53.** $\dfrac{-R \pm \sqrt{R^2-4L/C}}{2L}$

55. $\frac{64}{3}$ **57.** No solution **59.** 1 **61.** 4 **63.** 13 **65.** $-\frac{79}{64}$ **67.** $-0.61, 1.28$

69. $-9, 3$ **71.** $-\frac{1}{2}, 1$ **73.** $\pm 2, \pm 3$ **75.** $0, \pm 1$ **77.** $\frac{2}{3}, -1, \frac{1}{3}, -2$

79. 30 cubic centimeters **81.** Length = 24.5 centimeters; Width = 7.5 centimeters **83.** 3 minutes

85. (a) 3 seconds; (b) 6 seconds

PROBLEM SET 1.6 page 60

1. $[-1, 3)$

3. $(\frac{3}{4}, \infty)$ or $(-\infty, 0)$

5. $\left(\frac{1}{\pi}, \pi\right)$ or $[4, \infty)$

7. $0 \leq x \leq \frac{1}{2}$

9. $x < 1$

11. $-2 < x < 4$ or $5 \leq x < 8$

13. Addition property **15.** Addition property **17.** Transitive property **19.** Multiplication property

21. $x < 3, (-\infty, 3)$

23. $x \geq \frac{9}{4}, [\frac{9}{4}, \infty)$

25. $t \geq 4, [4, \infty)$

27. $x > -\frac{18}{13}, (-\frac{18}{13}, \infty)$

29. $x \geq 1, [1, \infty)$

31. $x \leq -\frac{1}{2}, (-\infty, -\frac{1}{2}]$

33. $-14 < t < 14, (-14, 14)$

35. $x \geq 0, [0, \infty)$

37. $x \leq -\frac{24}{25}, (-\infty, -\frac{24}{25}]$

39. $[-4, 4]$

41. $[-2, 4]$

43. $(-2, -1)$

45. $(-\infty, -3]$ or $[3, \infty)$

47. $(-\infty, -7]$ or $[3, \infty)$

49. $(-\infty, -\frac{3}{2})$ or $(3, \infty)$

51. $(-\infty, \frac{1}{2}]$ or $[\frac{7}{2}, \infty)$

53. $[-\frac{7}{2}, -\frac{3}{2}]$

55. $(-3, 1)$

57. $(-\infty, -6]$ or $[1, \infty)$

59. $[-3, \frac{3}{2}]$

61. $(-\frac{1}{2}, \frac{4}{3})$

63. $(-\infty, -0.90]$ or $[2.58, \infty)$

65. $(-\infty, -3)$ or $(1, 2)$

67. $[-2, \frac{1}{3}]$ or $[7, \infty)$

69. $(-2, -\frac{1}{2})$

71. $(1, 4)$

73. $(-\infty, 0)$ or $[\frac{1}{2}, \infty)$

75. $(-\infty, -3)$ or $(4, \infty)$

77. $(-1, \frac{11}{9}]$ or $(3, \infty)$

79. $(-\infty, 2]$ or $[\frac{14}{3}, \infty)$

81. 2572 message units **83.** At most 33.23 feet **85.** After 8 years

REVIEW PROBLEM SET, CHAPTER 1 page 62

1. (a) 0.171; (b) -0.182; (c) 3.125; (d) 1.167; (e) -5.955; (f) 15.395; (g) 2.088; (h) 0.170; (i) 0.318; (j) 0.0005 **3.** (a) $\frac{7}{50}$; (b) $\frac{-7}{2000}$; (c) $\frac{31}{99}$; (d) $-\frac{1256}{495}$; (e) $\frac{4}{75}$; (f) $\frac{784}{2395}$ **5.** Closure for addition **7.** Commutativity of multiplication **9.** Multiplicative identity **11.** Cancellation property for multiplication **13.** Zero factor **15.** $3x^2 - 2x + 3$ **17.** $4x^2 + 6x - 4$ **19.** $2x^4 - 5x^3 + 7x^2 - 5x + 2$ **21.** $28 + xy - 15x^2y^2$ **23.** $x^4 - 16$ **25.** $x^6 - y^3$ **27.** $13x^2y^2(2x + 3x^3y^2 - 4y)$ **29.** $(y - 11)(y + 11)$ **31.** $(5y - 9z)(5y + 9z)$ **33.** $(x + 4)(x^2 - 4x + 16)$ **35.** $(x - 8)(x + 7)$ **37.** $\frac{x-4}{x}$ **39.** $\frac{x+1}{x-3}$ **41.** $\frac{x+3}{(x-4)(x-3)(x-1)}$ **43.** $\frac{1}{x-1}$ **45.** $\frac{y-x}{x^2y^2(x+y)}$ **47.** x^6y^9 **49.** $\frac{1}{x^2y^{4/3}}$ **51.** $-16x\sqrt{2}$ **53.** $4x^2y^5z^7\sqrt{2xz}$ **55.** $2x^2y^4\sqrt[6]{2xy}$ **57.** $13 - 2\sqrt{42}$ **59.** $10 - 5i$ **61.** $-1 + 12i$ **63.** $30 - 16i$ **65.** $\frac{5}{7} - \frac{6\sqrt{3}}{7}i$ **67.** $-2y\sqrt{xy}\,i$ **69.** $145 + 58i$ **71.** $\frac{11}{5}$ **73.** 25 **75.** 3 **77.** $-1 \pm \sqrt{7}$ **79.** $-\frac{1}{2}, 1$ **81.** $\frac{-3 \pm \sqrt{13}}{4}$ **83.** $\frac{-1 \pm i}{2}$ **85.** 11 **87.** 0 **89.** 37 calls **91.** (a) 0 second or $\frac{11}{2}$ seconds; (b) $\frac{11}{4}$ seconds **93.** $\frac{1}{512}, -1$ **95.** (a) F; (b) F; (c) T; (d) T **97.** $(4, \infty)$ **99.** $[\frac{4}{3}, \infty)$ **101.** $(\frac{10}{3}, \frac{13}{3}]$ 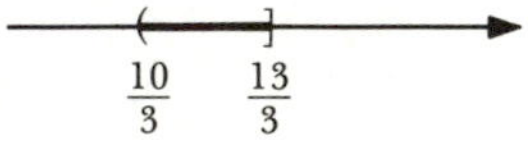 **103.** At most 7 tickets **105.** $\frac{8}{3}, -\frac{16}{3}$ **107.** $[-7, 2]$  **109.** $(-\infty, -3)$ or $(\frac{7}{3}, \infty)$  **111.** $\frac{1}{3}$ **113.** $\mathbb{R}$ **115.** $[-1, 5]$ **117.** $(-\infty, -1)$ or $(\frac{5}{3}, \infty)$ 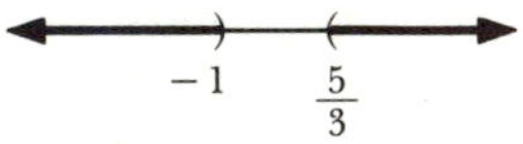 **119.** $[-1, 0]$ or $[3, \infty)$ **121.** $(-\infty, -1)$ or $[\frac{1}{5}, \infty)$ **123.** $(-25, -5)$ or $(-5, 5)$

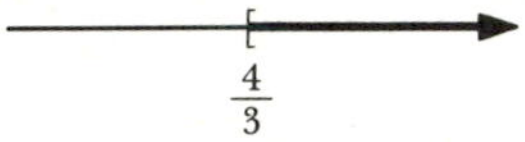

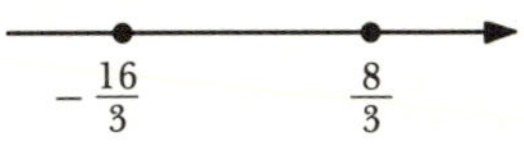

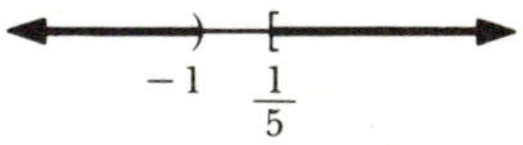

CHAPTER 2

PROBLEM SET 2.1 page 73

1. (a) Q_{I}; (b) Q_{II}; (c) Q_{IV}; (d) Q_{IV}; (e) y axis; (f) y axis; (g) x axis; (h) Q_{III} **3.** $Q = (1, -4)$; $R = (-1, 4)$; $S = (-1, -4)$ **5.** $Q = (-3, -2)$; $R = (3, 2)$; $S = (3, -2)$ **7.** $Q = (2, 3)$; $R = (-2, -3)$; $S = (-2, 3)$ **9.** 10 **11.** $\sqrt{17}$ **13.** 1 **15.** $\sqrt{41}/2$ **17.** $2\sqrt{1 + t^2}$ **19.** $\sqrt{13 + 2\sqrt{6} - 4\sqrt{3}}$ **21.** 51.80

27. (a) 7, −3; (b)

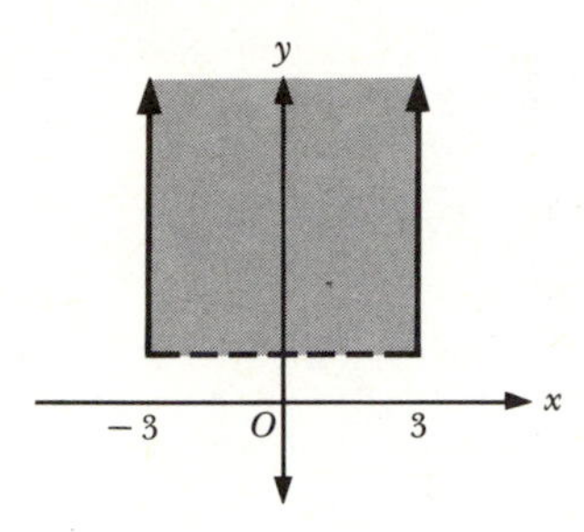

29. $18 + 2\sqrt{10}$ **31.** Collinear **33.** Noncollinear
35. (a) $(-\frac{11}{2}, \frac{1}{2})$; (b) $(\frac{5}{2}, 3)$; (c) $(3, \frac{5}{2})$; (d) $(-\frac{3}{2}, -3)$
37. $(x - 3)^2 + (y + 2)^2 = 25$
39. $(x + 3)^2 + (y - 1)^2 = 16$ **41.** $(1, -2)$, 5
43. $(\frac{3}{2}, -2)$, $\frac{3}{2}$ **45.** $(-3, 1)$, 2 **47.** $(1, -\frac{2}{3})$, $\frac{11}{3}$

PROBLEM SET 2.2 page 85

1. $\frac{5}{3}$

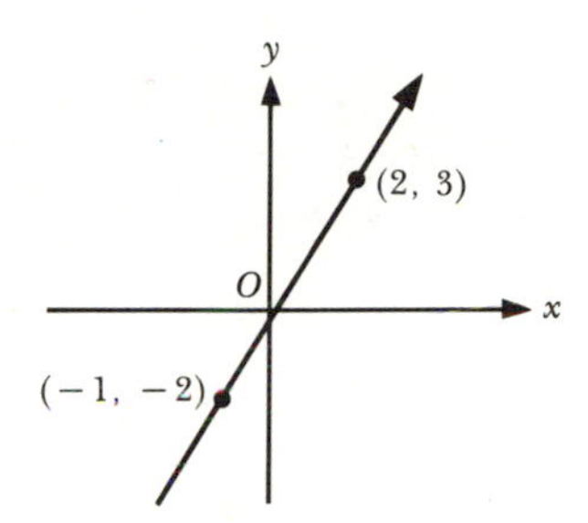

3. −2

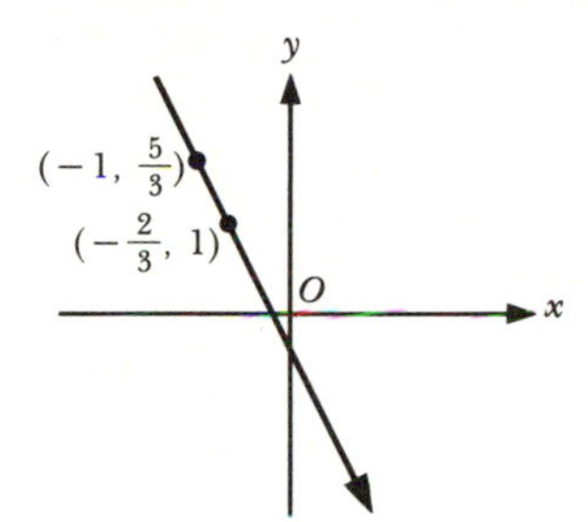

5. $\frac{7}{3}$

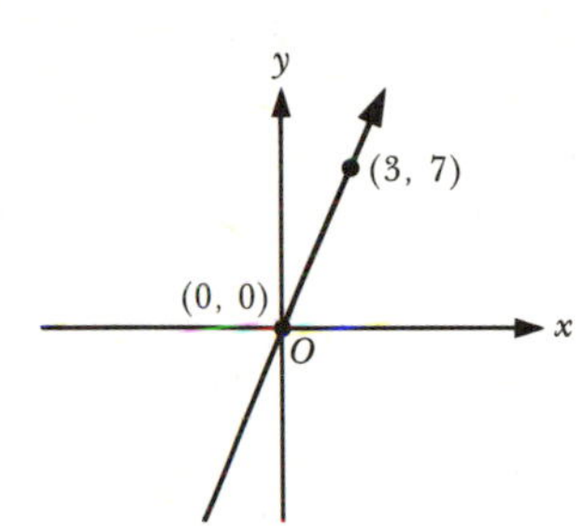

7. 0

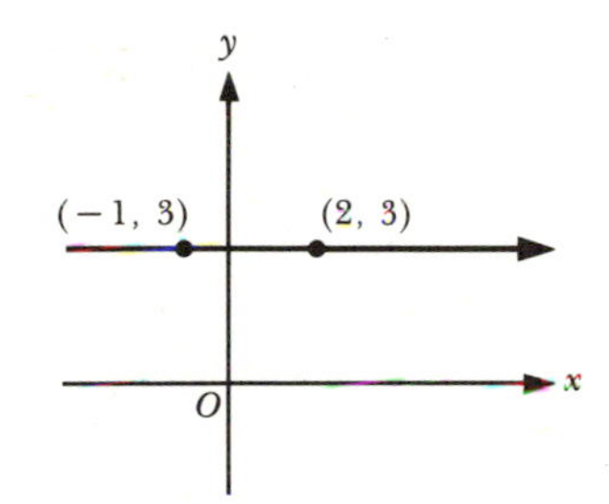

9. Undefined

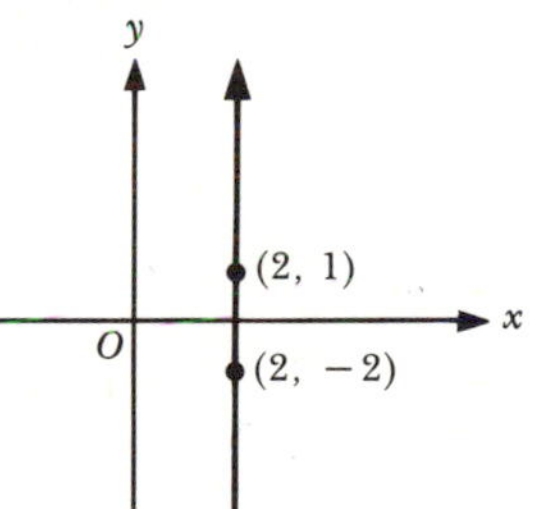

11.

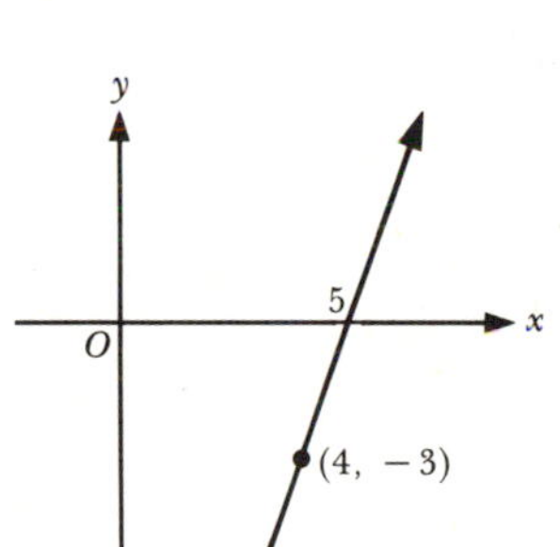

13.

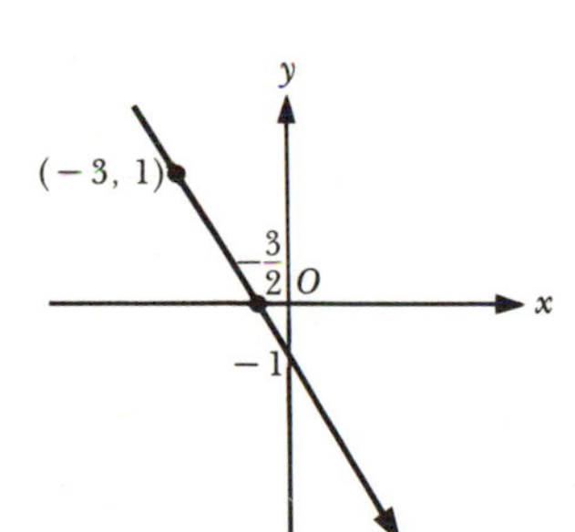

15.

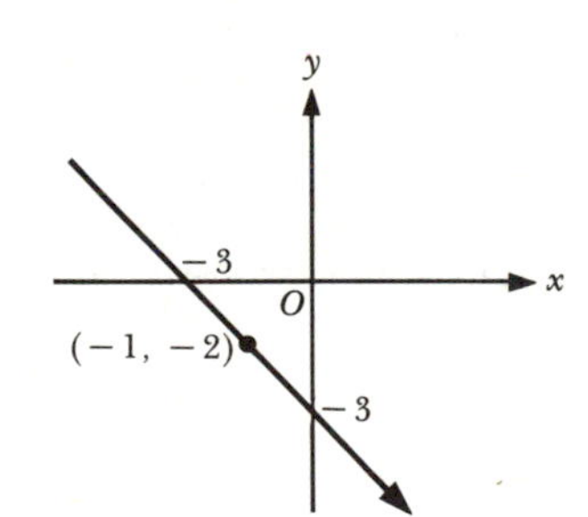

17. (a) 1; (b) $\frac{5}{2}$ **19.** $y - 1 = -3(x + 4)$ **21.** $y - 4 = 9(x - 3)$ or $y + 5 = 9(x - 2)$

23. $y = \frac{3}{5}x$; $m = \frac{3}{5}$; y intercept $= 0$; x intercept $= 0$

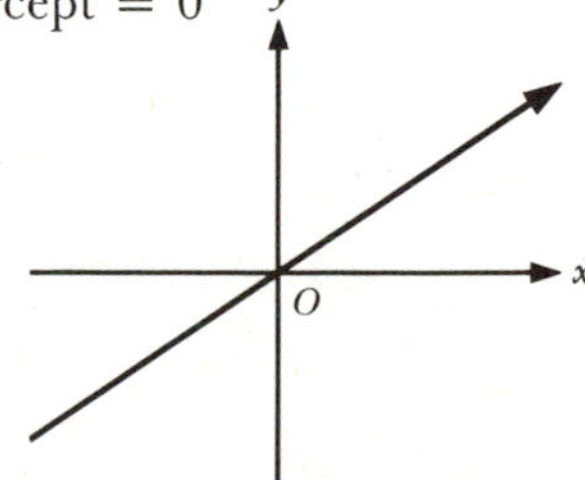

25. $y = -2x - 3$; $m = -2$; y intercept $= -3$; x intercept $= -\frac{3}{2}$

27. $y = -\frac{4}{5}x - \frac{7}{5}$; $m = -\frac{4}{5}$; y intercept $= -\frac{7}{5}$; x intercept $= -\frac{7}{4}$

29. $y = -\frac{5}{12}x + 3$; $m = -\frac{5}{12}$; y intercept $= 3$; x intercept $= \frac{36}{5}$

31. $y = 9x - 23$; $m = 9$; y intercept $= -23$; x intercept $= \frac{23}{9}$

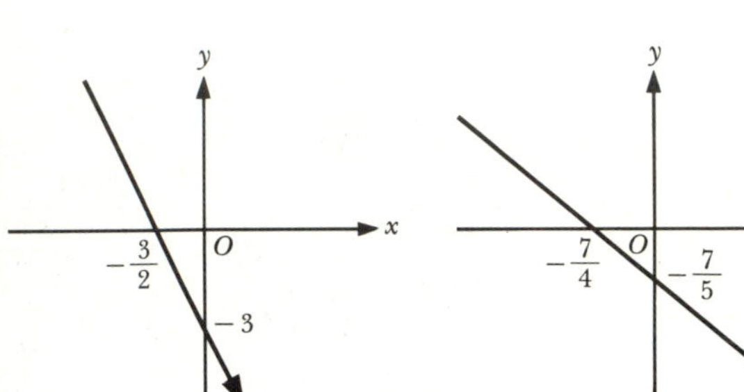

Figure Problem 25

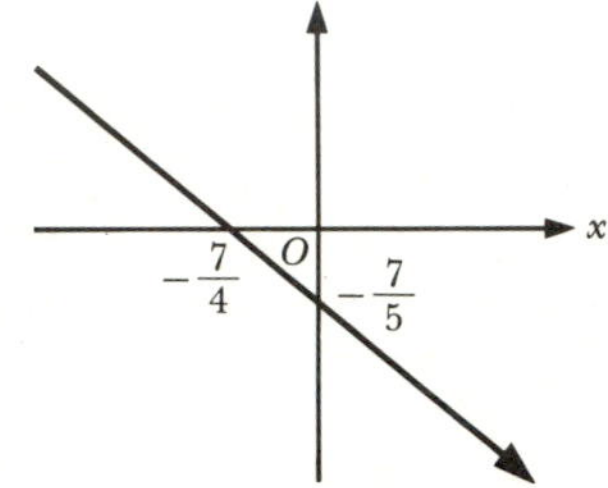

Figure Problem 27

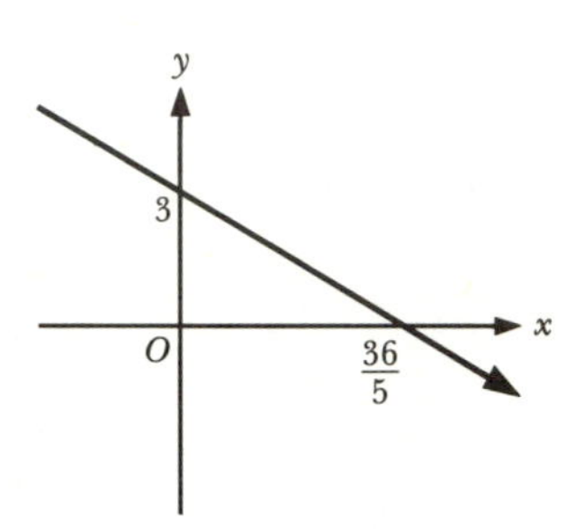

Figure Problem 29

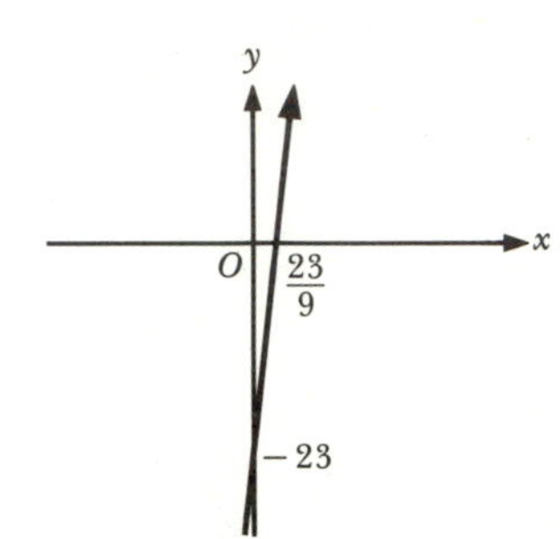

Figure Problem 31

33. (a) $y - 1 = 4(x - 1)$; (b) $y = 4x - 3$; (c) $4x - y - 3 = 0$
35. (a) $y - 2 = -3x$; (b) $y = -3x + 2$; (c) $3x + y - 2 = 0$
37. (a) $y - 3 = \frac{5}{2}(x - 3)$; (b) $y = \frac{5}{2}x - \frac{9}{2}$; (c) $5x - 2y - 9 = 0$
39. (a) $y - 4 = -\frac{3}{4}(x + 1)$; (b) $y = -\frac{3}{4}x + \frac{13}{4}$; (c) $3x + 4y - 13 = 0$
41. (a) $y + 5 = \frac{2}{3}(x + 3)$; (b) $y = \frac{2}{3}x - 3$; (c) $2x - 3y - 9 = 0$ **43.** $x = -3$
45. $F = \frac{9}{5}C + 32$, 113°F **47.** $T = 0.07I - 210$, \$15,000 **49.** 2°C

PROBLEM SET 2.3 page 94

1. 5 **3.** $-\frac{3}{2}$ **5.** 2 **7.** 11 **9.** 6 **11.** 5 **13.** 0 **15.** 4 **17.** −3 **19.** $\dfrac{3a - 1}{a - 2}$

21. $2a + 9$ **23.** $a + 3$ **25.** $4|x|$ **27.** $2b$ **29.** $2ab + b^2 - 2b$ **31.** $\dfrac{7x + 3}{7x - 2}$ **33.** $14x + 35$

35. $7x^2 - 14x + 56$ **37.** $\dfrac{1 + 3x + 3a}{1 - 2x - 2a}$ **39.** $-2x + 5$ **41.** $x^2 + 2x + 8$ **43.** $|3 - 4x^2|$

45. $4x^2 + 20x + 25$ **47.** $x^4 - 4x^3 + 20x^2 - 32x + 64$ **49.** 29.203 **51.** 0.086 **53.** 3

55. −5 **57.** $\dfrac{-1}{x(x + h)}$ **59.** $\dfrac{-2x - h}{x^2(x + h)^2}$ **61.** (a) −2; (b) 6; (c) −7 **63.** (a) $\frac{1}{2}$; (b) 5; (c) −1

65. (a) $f(-2) = 14$; $f(-1) = 4$; $f(0) = 0$; $f(1) = 2$; $f(2) = 10$
(b) Function notation—see (a) $f\colon -2 \mapsto 14, f\colon -1 \mapsto 4, f\colon 0 \mapsto 0, f\colon 1 \mapsto 2, f\colon 2 \mapsto 10$;
$(-2, 14), (-1, 4), (0, 0), (1, 2), (2, 10)$
67. $\mathbb{R}$ **69.** $(-\infty, -1)$ together with $(-1, \infty)$ **71.** $[\frac{2}{3}, \infty)$
73. $(-\infty, -1)$ together with $(-1, 1)$ and $(1, \infty)$

75.

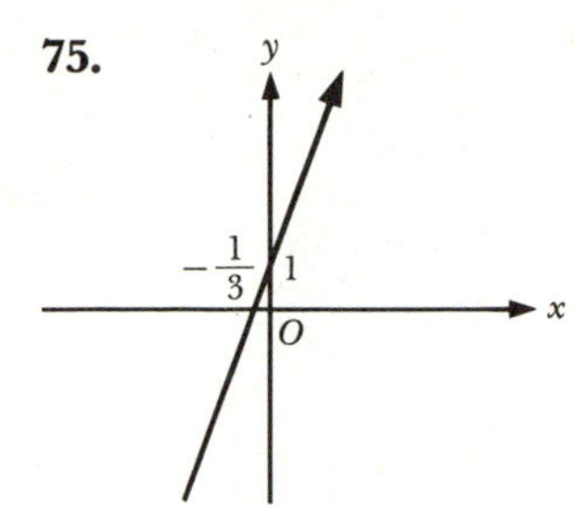

77.

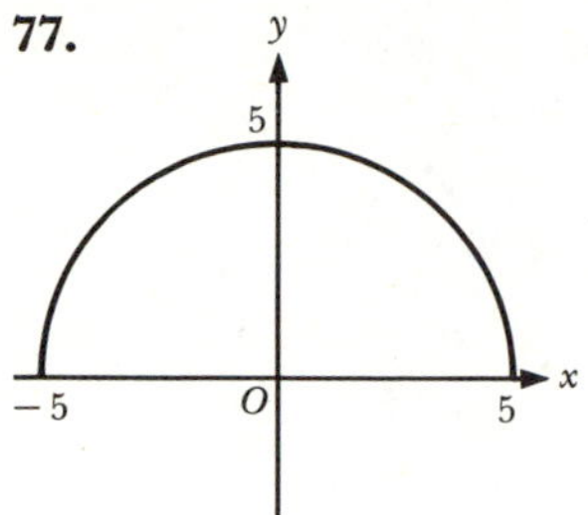

79. Not a function

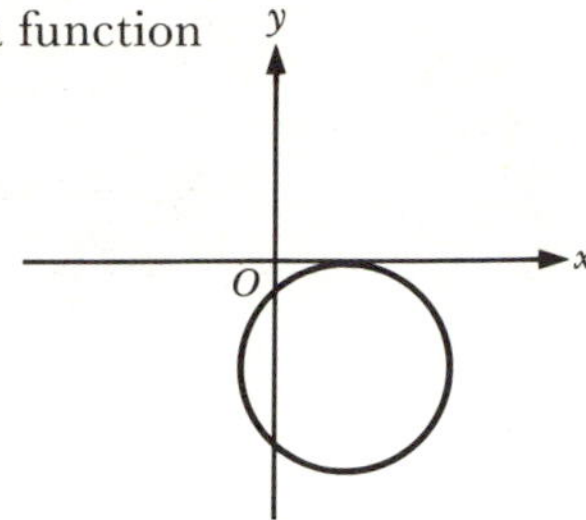

81. (c), (d) **83.** (a) 27 meters; (b) 21.28 meters; (c) 0.19 meter

85. (a) 0°C; (b) 30°C; (c) $(\frac{-160}{9})$°C; (d) −25°C

87. $\dfrac{(8+\pi)P^2}{4(\pi+4)^2}$ **89.** $V = x(10-2x)(14-2x)$ or $V = 4x^3 - 48x^2 + 140x$

91. $V = 64\pi h$, 640π cubic feet

PROBLEM SET 2.4 page 110

1.

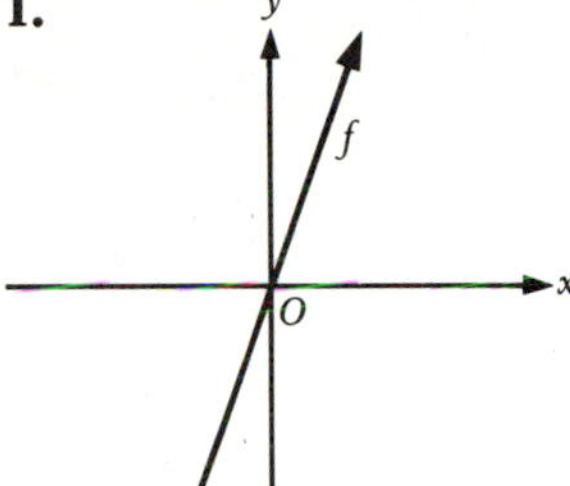

3.

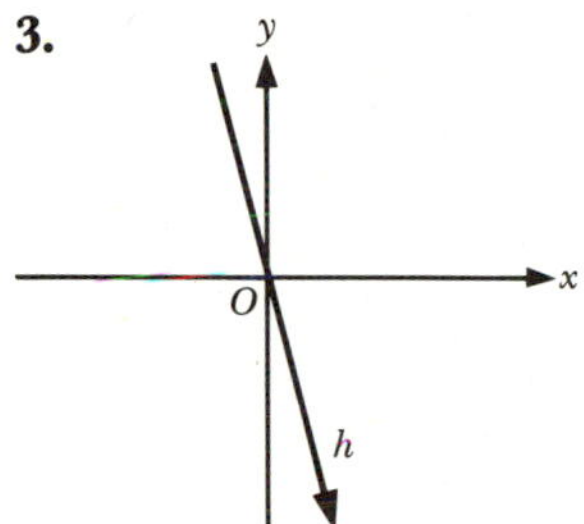

5.

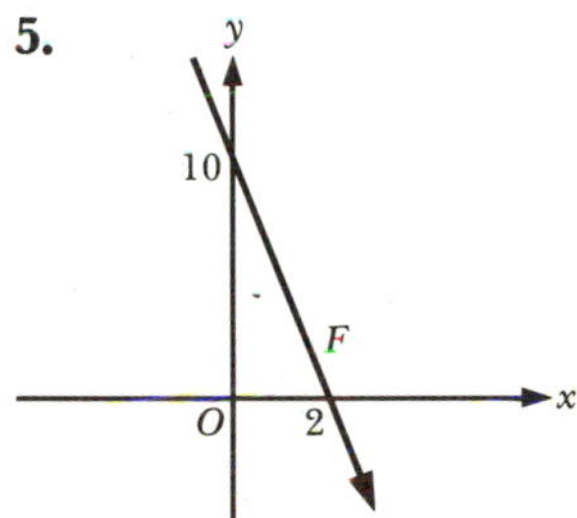

7.

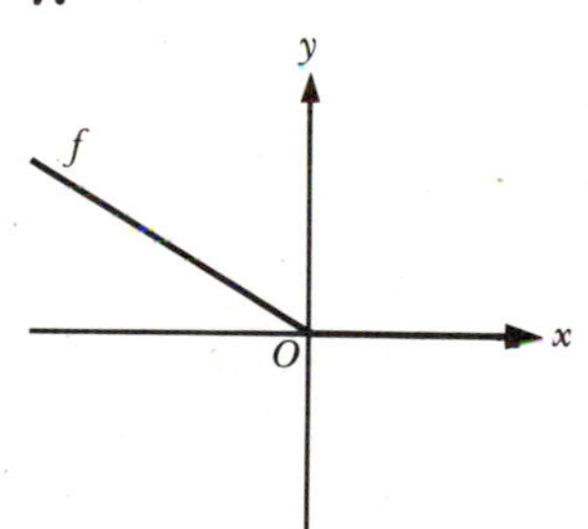

9.

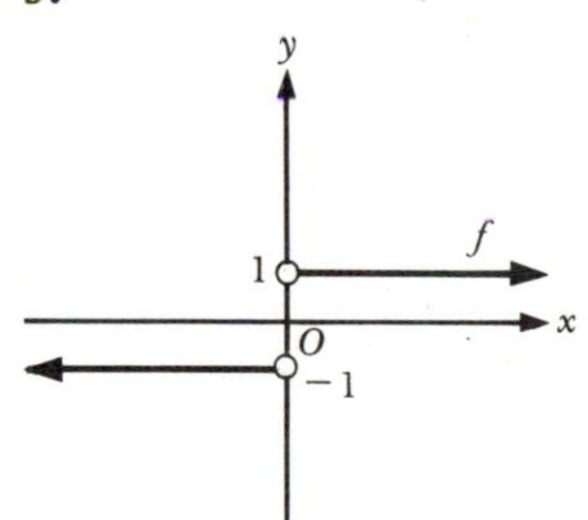

11.

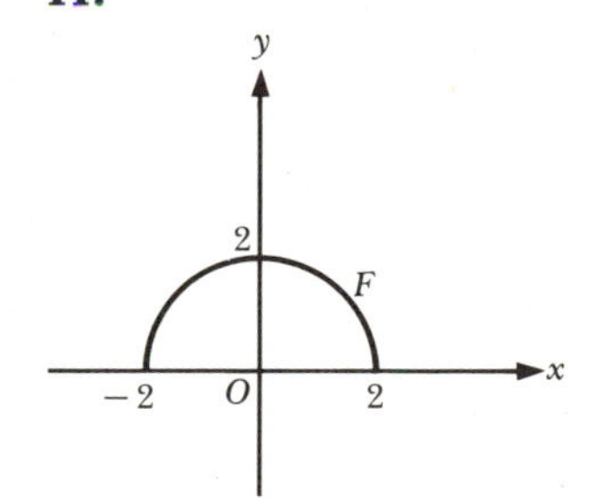

13.

Even-Odd	Domain	Range	Increasing	Decreasing	Constant
(a) Even	$\mathbb{R}$	$[-2, \infty)$	$[-2, 0]$ and $[2, \infty)$	$(-\infty, -2]$ and $[0, 2]$	None
(b) Neither	$\mathbb{R}$	$(-\infty, 3]$	$(-\infty, -3]$ and $[-1, 2]$	$[2, 3]$	$[-3, -1]$ and $[3, \infty)$
(c) Even	$[-6, 6]$	$[0, 2]$	$[-6, -4]$, $[-2, -1]$, $[0, 1]$, and $[2, 4]$	$[-4, -2]$, $[-1, 0]$, $[1, 2]$, and $[4, 6]$	None
(d) Neither	$\mathbb{R}$	$(-\infty, 2]$	$[-4, -2]$ and $[1, \frac{5}{2}]$	$[-2, 1]$ and $[\frac{5}{2}, \infty)$	$(-\infty, -4]$

15. (a) f is symmetric with respect to y axis; g is symmetric with respect to origin;

(b)

Point	Reflection
$(2, 16)$	$(-2, 16)$
$(-3, 81)$	$(3, 81)$
$(8, 2)$	$(-8, -2)$
$(-1, -1)$	$(1, 1)$

(c) f is decreasing on $(-\infty, 0]$; increasing on $[0, \infty)$; g is always increasing on $\mathbb{R}$

17. f is even; symmetric with respect to y axis
19. h is even; symmetric with respect to y axis
21. f is odd; symmetric with respect to origin
23. f is even; symmetric with respect to y axis

25.

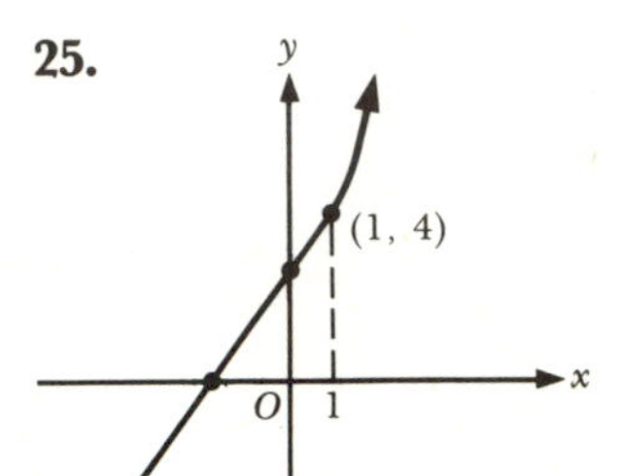

27.

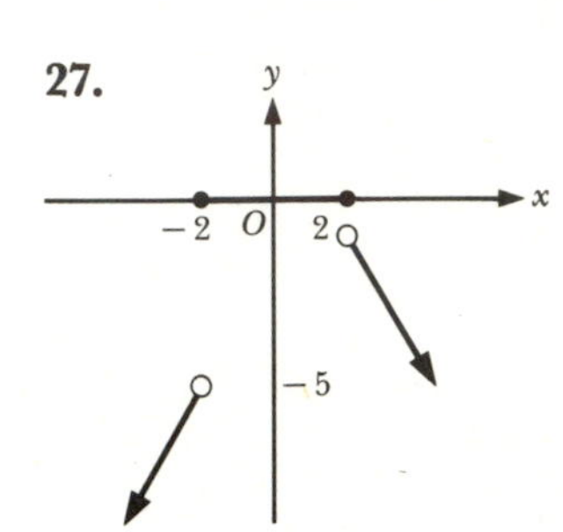

29.

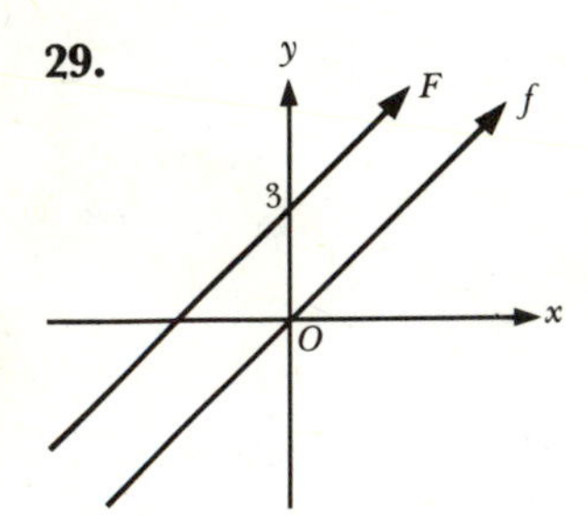

31.

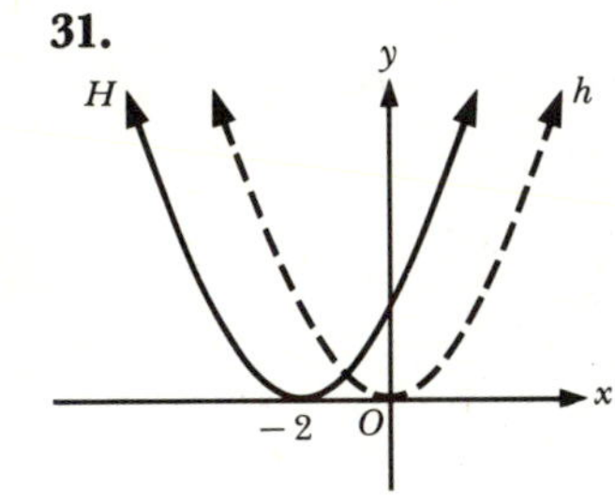

33.

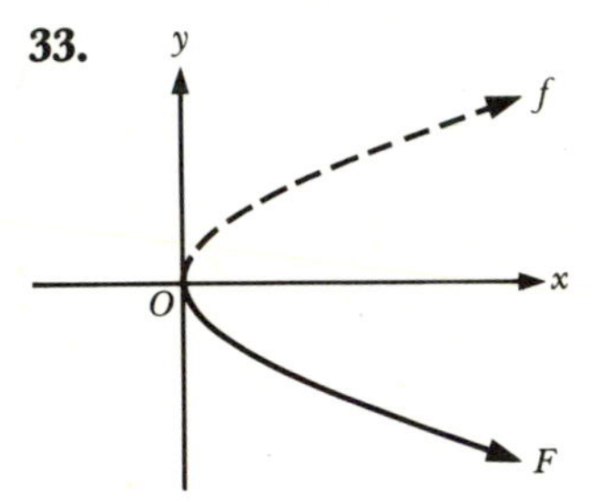

35.

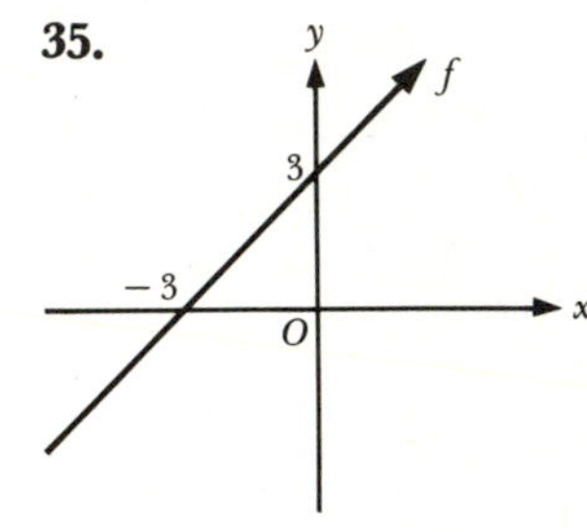

37.

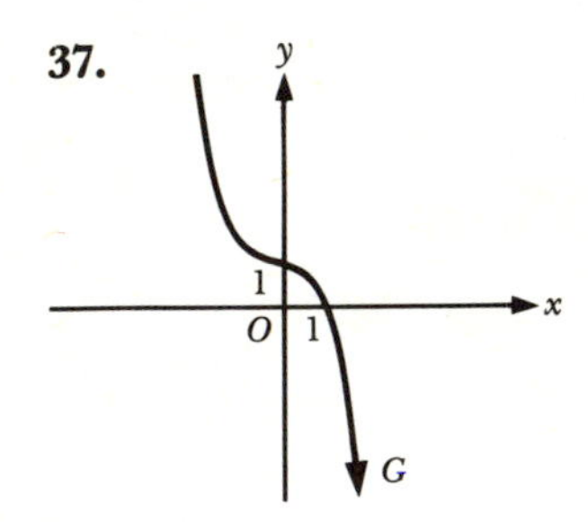

39.

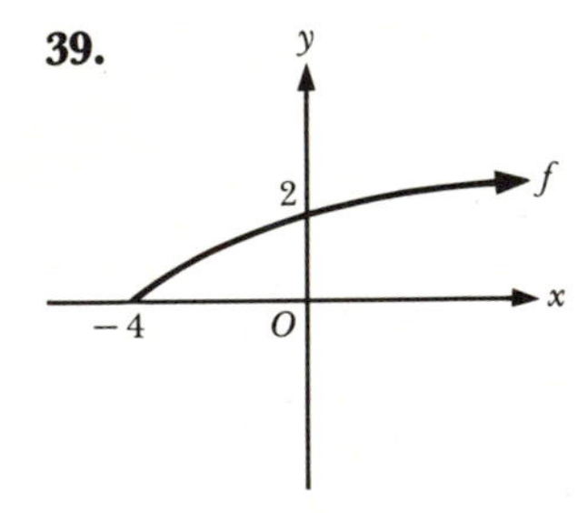

41.

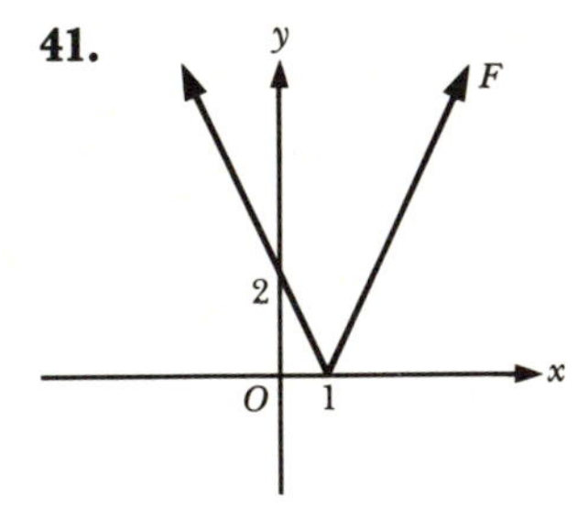

43.

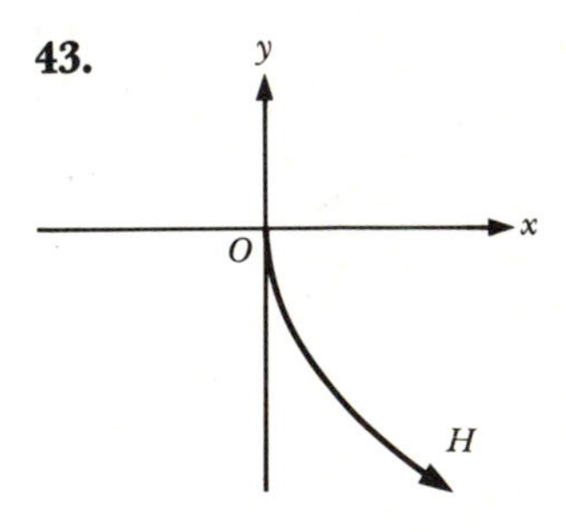

45.

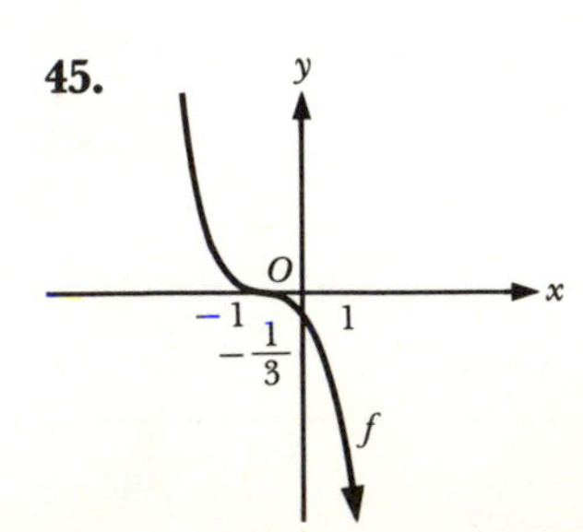

47.

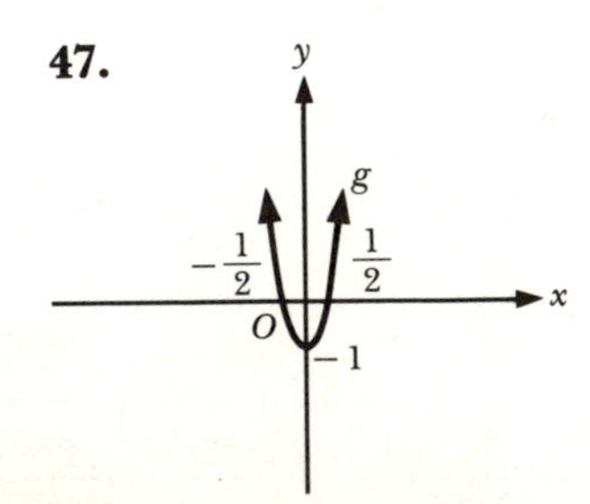

49.

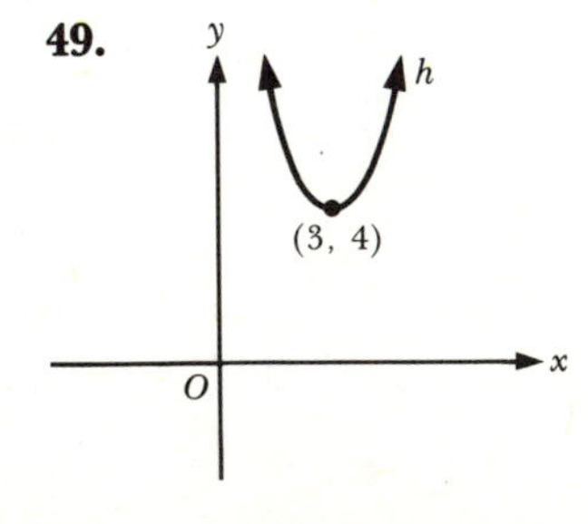

PROBLEM SET 2.5 page 118

1. (a) $6x - 6$; (b) 8; (c) $9x^2 - 18x - 7$; (d) $\frac{3x + 1}{3x - 7}$, all reals but $\frac{7}{3}$

3. (a) $5x + 1$; (b) $3x - 11$; (c) $4x^2 + 19x - 30$; (d) $\frac{4x - 5}{x + 6}$, all reals but -6

5. (a) $\frac{-x^2 + 48x - 7}{3x^2 - 14x - 5}$; (b) $\frac{13x^2 - 26x + 13}{3x^2 - 14x - 5}$; (c) $\frac{-14x^2 - 17x + 6}{3x^2 - 14x - 5}$;
(d) $\frac{6x^2 + 11x + 3}{-7x^2 + 37x - 10}$, all reals but $\frac{2}{7}$, 5, $-\frac{1}{3}$

7. (a) $4x + 9$; (b) $10x - 7$; (c) $-21x^2 + 53x + 8$; (d) $\frac{7x + 1}{-3x + 8}$, all reals but $\frac{8}{3}$

9. (a) $5 + x$; (b) $2x^2 - x + 5$; (c) $-x^4 + x^3 - 5x^2 + 5x$; (d) $\frac{x^2 + 5}{-x^2 + x}$, all reals but 0, 1

11. 518 **13.** 398 **15.** 268 **17.** 2744 **19.** 134

21. (a) $10x - 6$, domain $= \mathbb{R}$; (b) $10x - 3$, domain $= \mathbb{R}$

23. (a) $-9x$, domain $= \mathbb{R}$; (b) $-9x$, domain $= \mathbb{R}$ **25.** (a) x, domain $= \mathbb{R}$; (b) x, domain $= \mathbb{R}$

27. (a) $\frac{x - 1}{13x + 12}$, domain = all reals except $-\frac{3}{2}$ and $-\frac{12}{13}$; (b) $\frac{2x - 5}{11x - 15}$, domain = all reals except $\frac{5}{3}$ and $\frac{15}{11}$

29. $f(x) = x^3$, $g(x) = 5x - 3$ **31.** $f(t) = t^{-2}$, $g(t) = t^2 - 2$ **33.** $f(x) = \sqrt[3]{x}$, $g(x) = x + x^{-1}$

35. $-\frac{7}{2}$ **37.** $(f \circ g)(x) = (x + 1)^4 + 1$, $(g \circ f)(x) = (x^2 + 2)^2$ **39.** (a) $-15x + 7$; (b) $\frac{1}{3}$

41. (a) $x + 9$; (b) 4 **43.** -1 **45.** $P(x) = 140x - \frac{x^2}{30} - 72{,}000$; \$63,000 **47.** (a) $\frac{2\pi}{2 + t^2}$; (b) $\frac{\pi}{9}$

PROBLEM SET 2.6 page 125

7. (a) Yes; (b) No; (c) No; (d) Yes

9. (a)

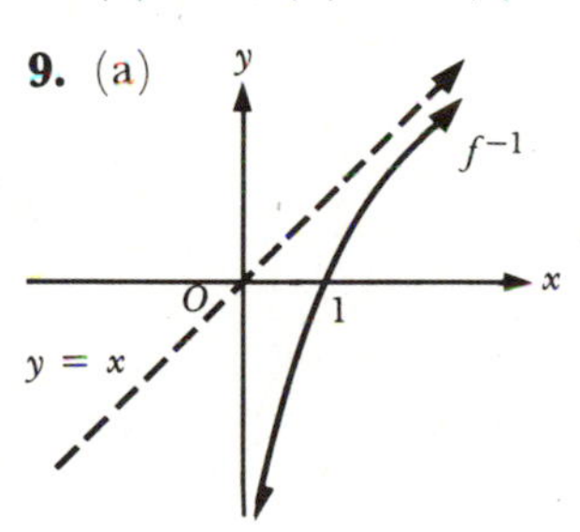

(b)

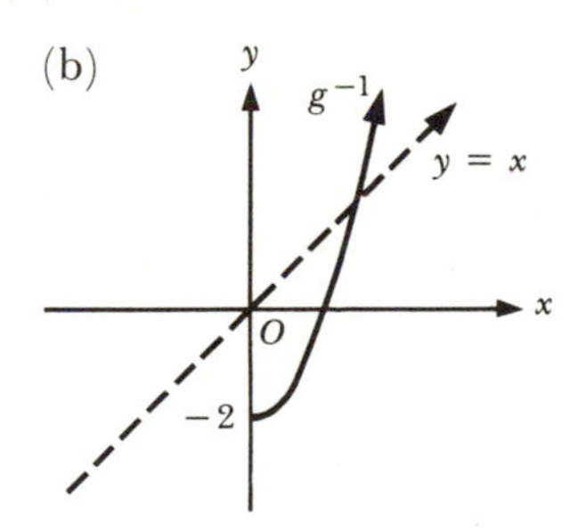

11. $f^{-1}(x) = \frac{x - 5}{7}$

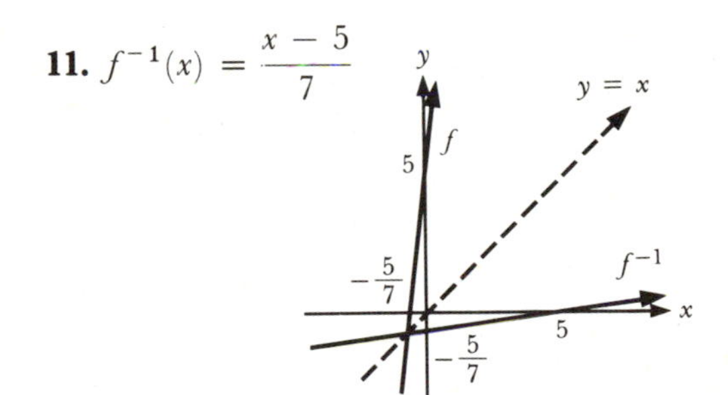

13. $f^{-1}(x) = \frac{1 - x}{3}$

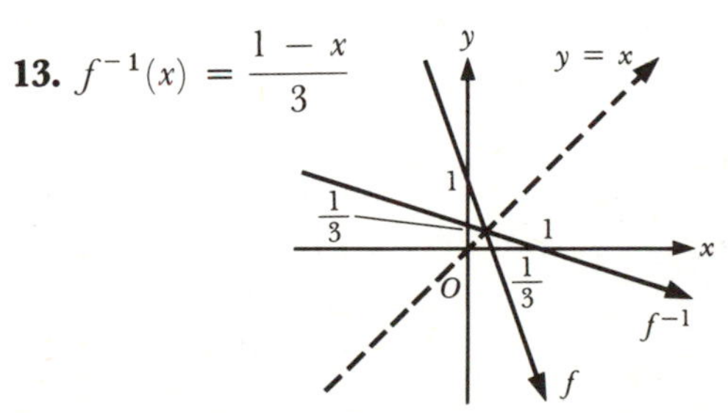

15. $f^{-1}(x) = \sqrt[3]{x + 8}$

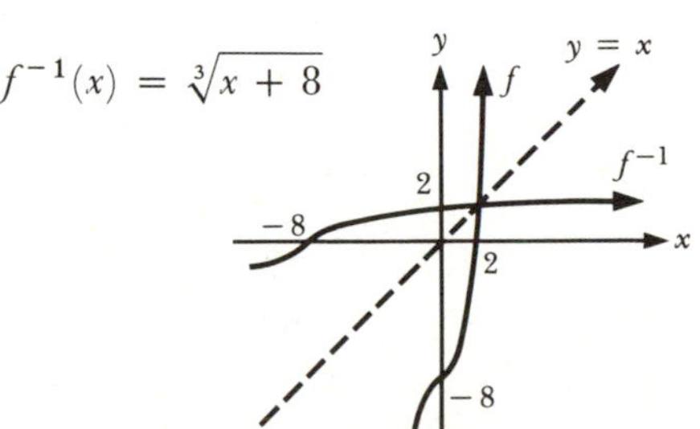

17. $f^{-1}(x) = \sqrt{x}$

19. (a) $f^{-1}(x) = \dfrac{x+7}{3}$ **21.** (a) $f^{-1}(x) = \frac{4}{3}(x-5)$

23. (a) $f^{-1}(x) = \dfrac{1}{1+x}$ **25.** (a) $f^{-1}(x) = -\dfrac{5}{x}$

27. (a) $f^{-1}(x) = x^2 - 1, x \geq 0$

31. No **35.** Show $f[g(x)] = g[f(x)] = x$

REVIEW PROBLEM SET, CHAPTER 2 page 128

1. (a) Q_{I}; (b) Q_{II}; (c) Q_{IV}; (d) Q_{III}; (e) Q_{IV}; (f) x axis; (g) x axis; (h) y axis **3.** (a) $2\sqrt{10}$; (b) $(3, -2)$

5. (a) $4\sqrt{5}$; (b) $(-2, -1)$ **7.** $|\overline{P_2P_3}| = |\overline{P_1P_3}| = \sqrt{17}$

9. $(x-3)^2 + (y+2)^2 = 25$; center $= (3, -2)$, $r = 5$

11. $m = -\dfrac{1}{3}$, $y - 3 = -\frac{1}{3}(x-2)$ or $y - 4 = -\frac{1}{3}(x+1)$

13. $m = \frac{11}{5}$, $y - 6 = \frac{11}{5}(x-4)$ or $y + 5 = \frac{11}{5}(x+1)$

15.

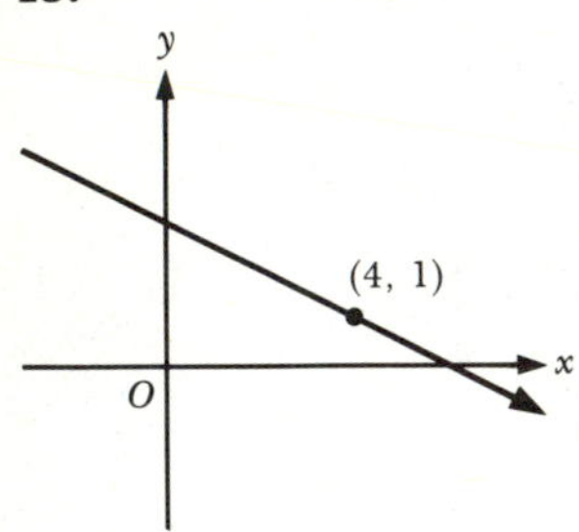

17. $y = 3x - 7$

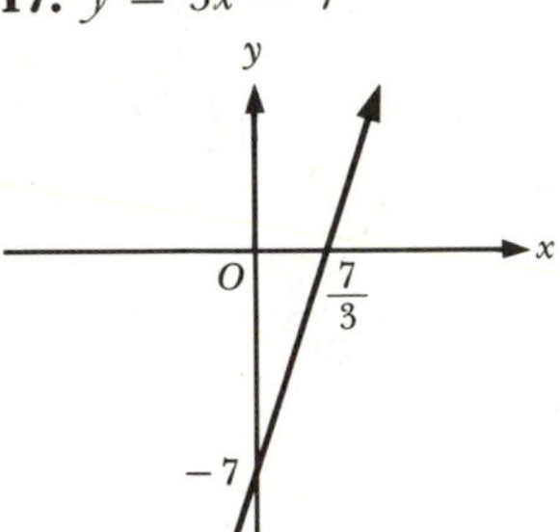

19. $y = 4x - 14$

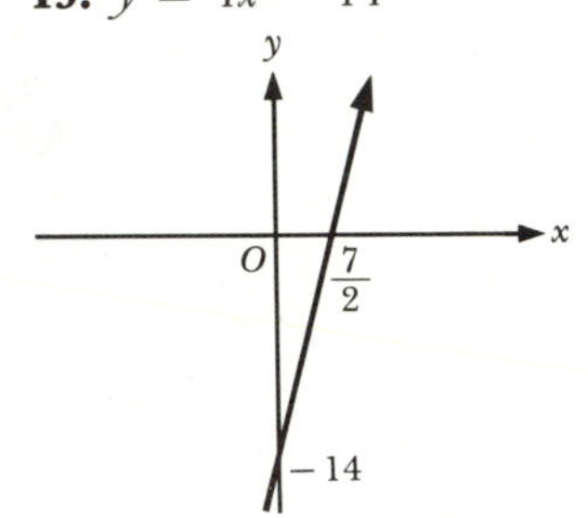

21. $y = -3x - 3$

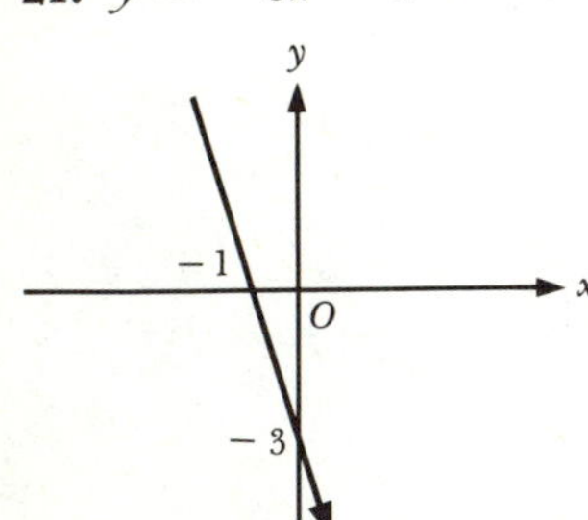

23. $y = \frac{7}{8}x - \frac{23}{4}$

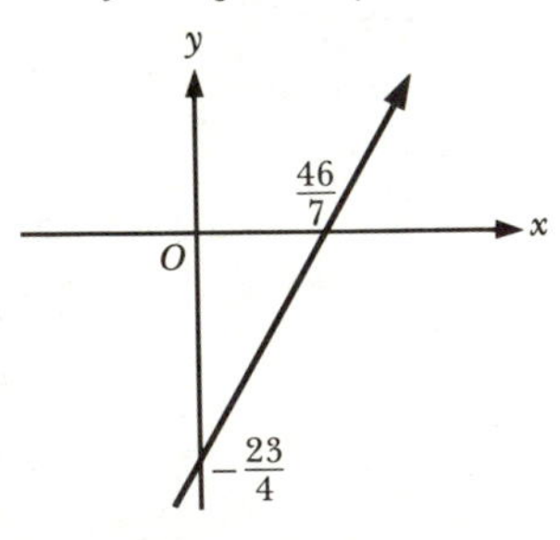

25. $(5, 9)$ **27.** -1 **29.** 50 **31.** 2 **33.** $\frac{1}{27}$ **35.** 2 **37.** x^3 **39.** All reals, $\mathbb{R}$ except 2

41. (a), (b), and (c) **43.** (a) $f(-2) = f(2) = 3, f(0) = \sqrt{5}$;
(b) Function notation—see (a); $f\colon -2 \mapsto 3, f\colon 2 \mapsto 3, f\colon 0 \mapsto \sqrt{5}$; $(-2, 3), (2, 3), (0, \sqrt{5})$

45. $f(x) = 350x - 55$ **47.** $\dfrac{\sqrt{3}}{4}x^2$ **49.** $y = \sqrt{9 - x^2}$; $A = 4x\sqrt{9 - x^2}$ **51.** $[1, \infty)$, neither

53. $\mathbb{R}$, even

55. Domain = $\mathbb{R}$; range = $\mathbb{R}$; neither odd nor even; decreases on $\mathbb{R}$
57. Domain = $\mathbb{R}$; range = $[0, \infty)$; neither odd nor even; graph decreases on $(-\infty, 0]$
59. Domain = $\mathbb{R}$; range = $\{-1\}$; even; symmetric with respect to y axis
61. Domain = $\mathbb{R}$; range = $(-\infty, 1]$; neither odd nor even; graph increases on $(-\infty, 1]$, decreases on $[1, \infty)$
63. Domain = $\mathbb{R}$; range = $\mathbb{R}$; odd; symmetric with respect to origin; decreases on $\mathbb{R}$
65. Domain = $\mathbb{R}$; range = $[0, \infty)$; neither odd nor even; graph decreases on $(-\infty, 0]$, increases on $[0, \infty)$

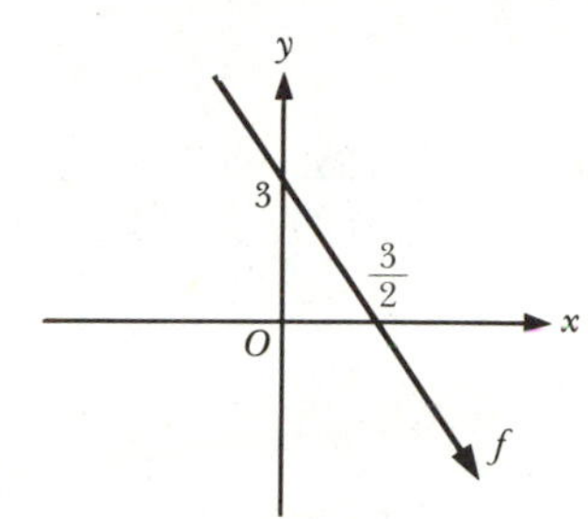

Figure Problem 55

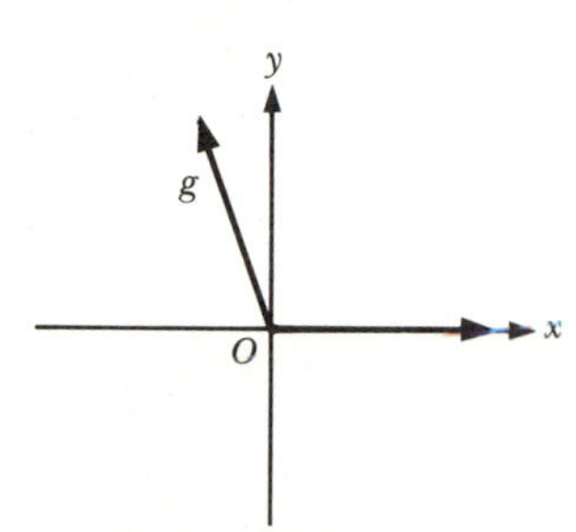

Figure Problem 57

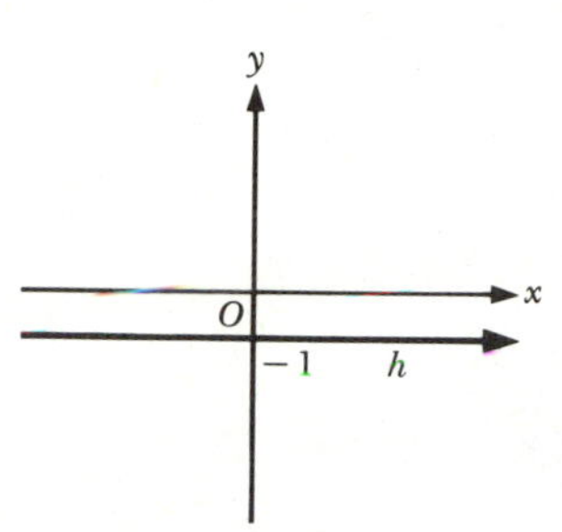

Figure Problem 59

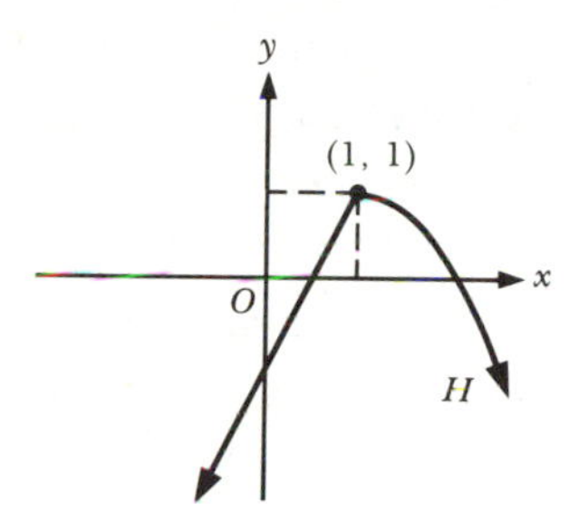

Figure Problem 61

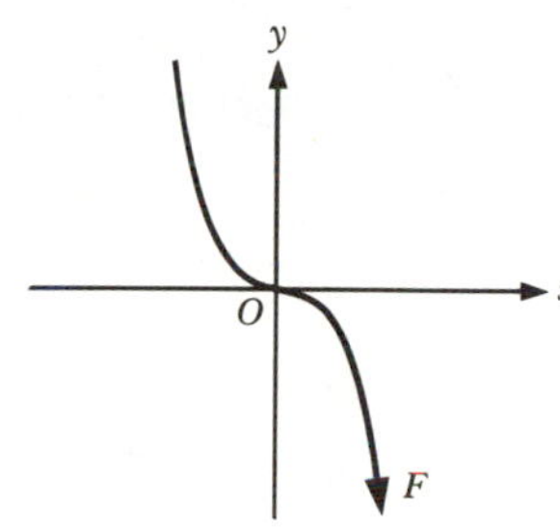

Figure Problem 63

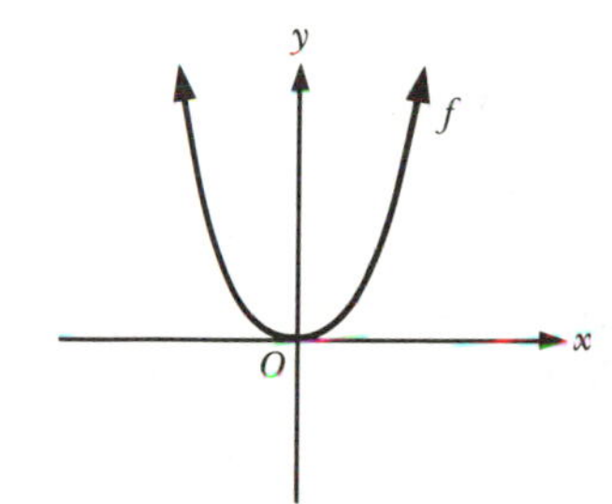

Figure Problem 65

67.

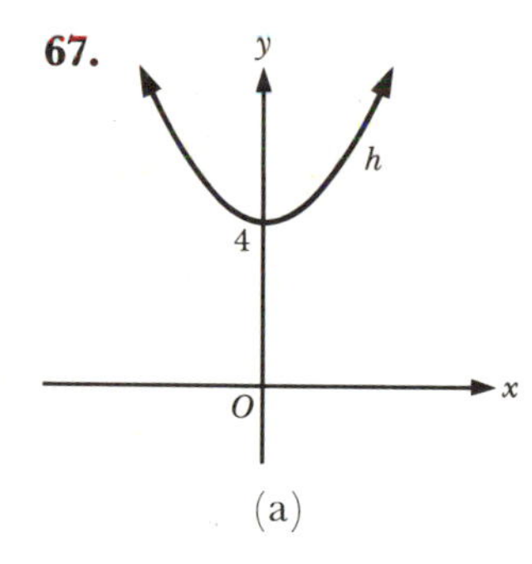

(a)

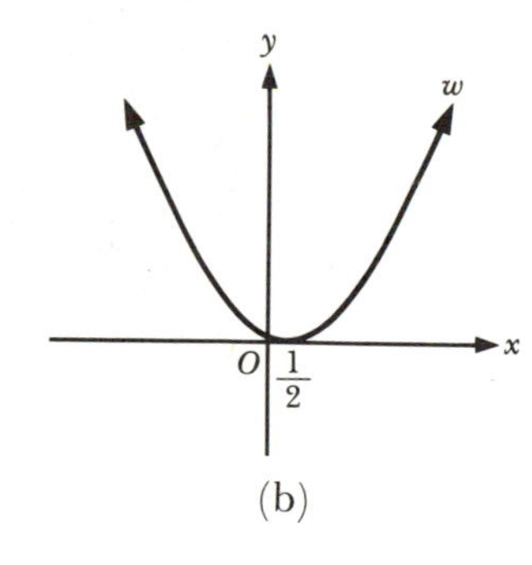

(b)

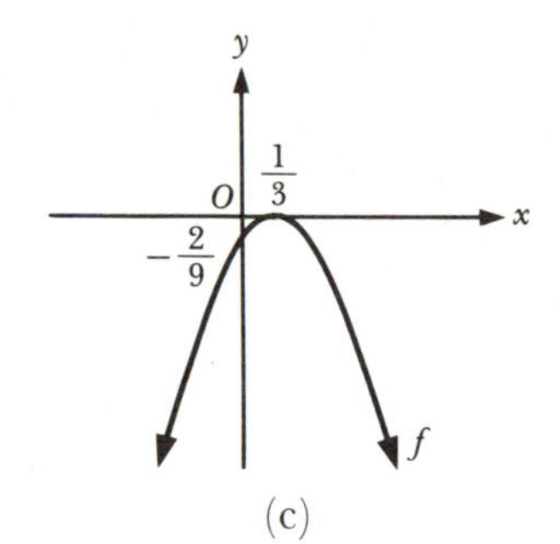

(c)

69. (a) $2x + 1$; (b) 3; (c) $x^2 + x - 2$; (d) $x + 1$; (e) $x + 1$; (f) $\dfrac{x+2}{x-1}$, domain is $\mathbb{R}$, except 1

71. (a) $x^2 + 2x + 1$; (b) $x^2 - 2x - 1$; (c) $2x^3 + x^2$; (d) $(2x+1)^2$; (e) $2x^2 + 1$
(f) $\dfrac{x^2}{2x+1}$, domain is $\mathbb{R}$, except $-\frac{1}{2}$

73. (a) $\dfrac{9x+3}{(1-x)(1+5x)}$; (b) $\dfrac{11x+1}{(1-x)(1+5x)}$;
(c) $\dfrac{2}{(1-x)(1+5x)}$; (d) $\dfrac{2(1+5x)}{5x}$; (e) $\dfrac{1-x}{11-x}$; (f) $\dfrac{2+10x}{1-x}$, domain is $\mathbb{R}$, except 1

75. (a) $8 + 5x - x^2$; (b) $6 - 5x - x^2$; (c) $7 - x^2 + 35x - 5x^3$; (d) $6 - 10x - 25x^2$; (e) $36 - 5x^2$; (f) $\dfrac{7 - x^2}{1 + 5x}$, domain is $\mathbb{R}$ except $-\dfrac{1}{5}$ **77.** (a) $|x| + |x - 3|$; (b) $|x| - |x - 3|$; (c) $|x^2 - 3x|$; (d) $|x - 3|$; (e) $||x| - 3|$; (f) $\left|\dfrac{x}{x - 3}\right|$, domain is $\mathbb{R}$, except 3

79. 4.3095 **81.** (a) $f(x) = x^5$, $g(x) = 7x + 2$; (b) $f(t) = \sqrt{t}$, $g(t) = t^2 + 17$ **85.** $g(x) = \dfrac{5 - x}{4}$

87. $f^{-1}(x) = \dfrac{7 - x}{13}$

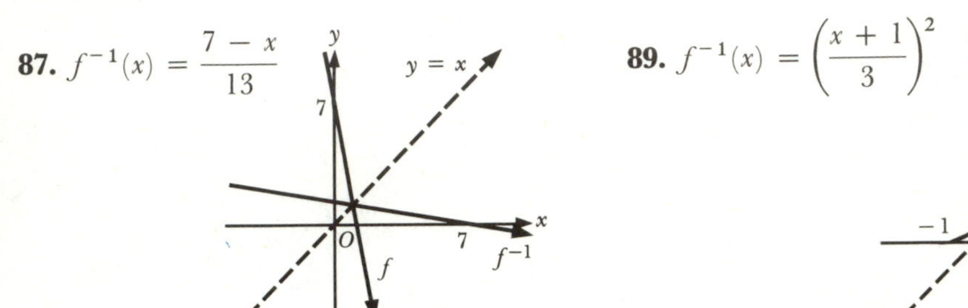

89. $f^{-1}(x) = \left(\dfrac{x + 1}{3}\right)^2$

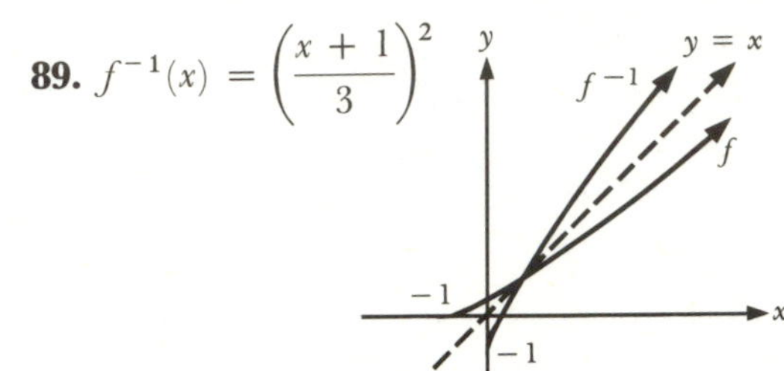

91. No inverse

CHAPTER 3

PROBLEM SET 3.1 page 142

1. (a) $a = \frac{1}{3}, 1, 4, 7$; (b) $a = -\frac{1}{3}, -1, -4, -7$; (c) $\frac{1}{3}$; (d) 4, 7

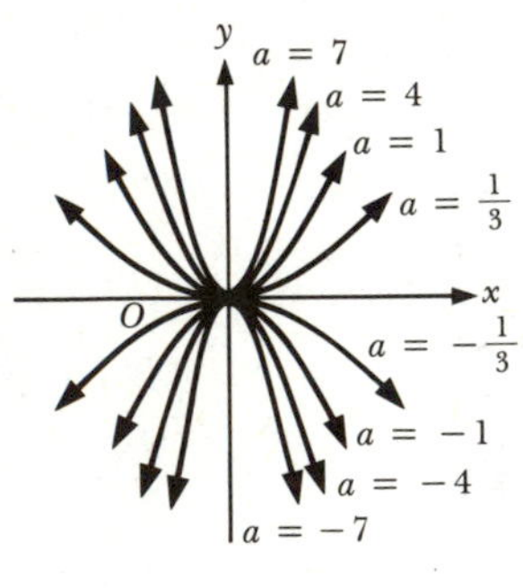

3.

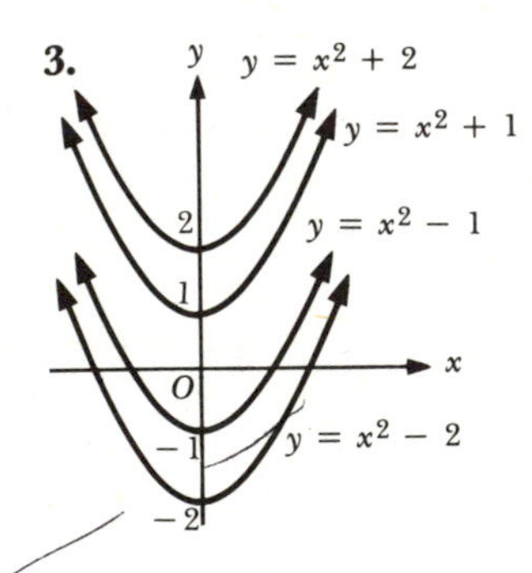

5.

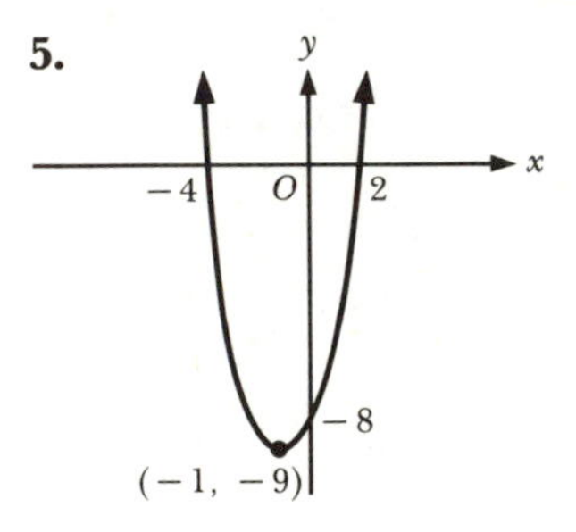

7.

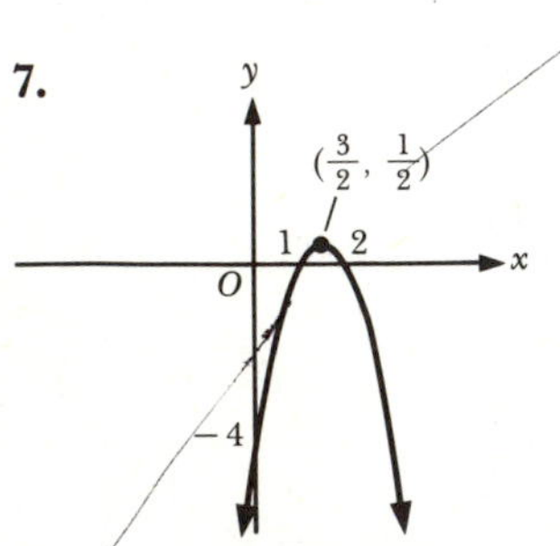

9.

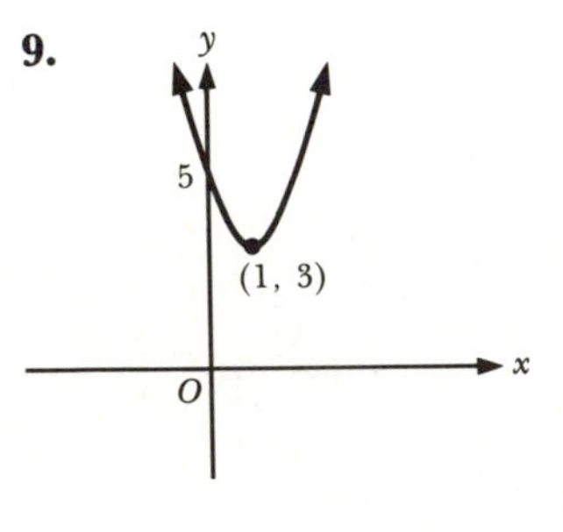

11. Minimum point: $(2, -9)$ **13.** Maximum point: $(3, 9)$ **15.** Minimum point: $\left(\frac{1}{10}, -\frac{61}{20}\right)$

17. Range $[-\frac{81}{4}, \infty)$
Line of symmetry: $x = \frac{9}{2}$

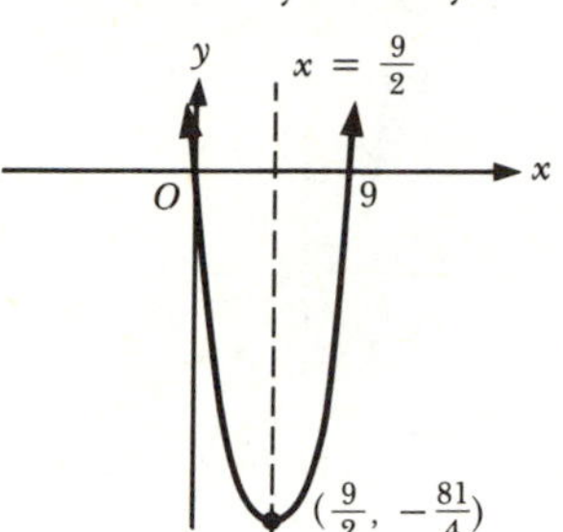

19. Range $[-\frac{81}{4}, \infty)$
Line of symmetry: $x = -\frac{5}{2}$

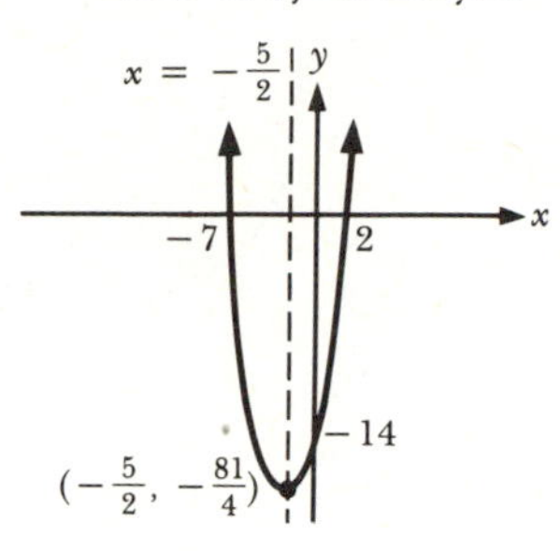

21. Range $(-\infty, 0]$
Line of symmetry: $x = -2$

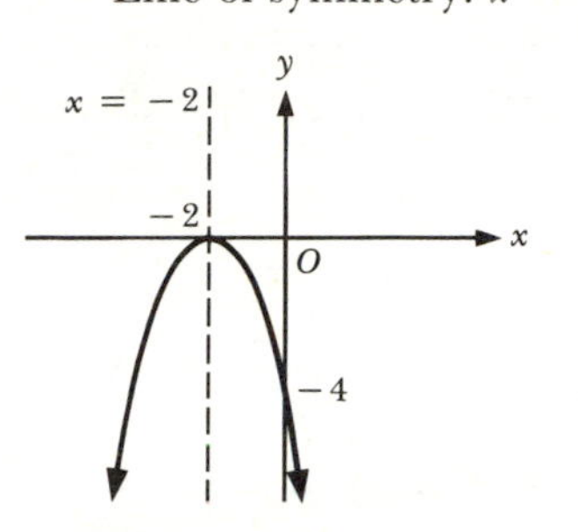

23. Range $(-\infty, 4]$
Line of symmetry: $x = \frac{1}{2}$

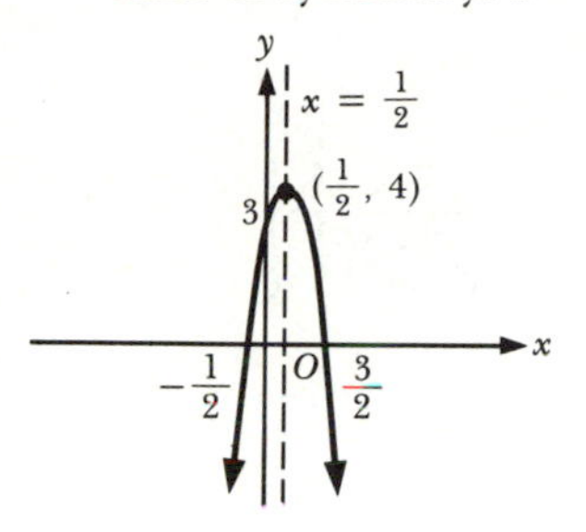

25. Range $[-5, \infty)$
Line of symmetry: $x = -1$

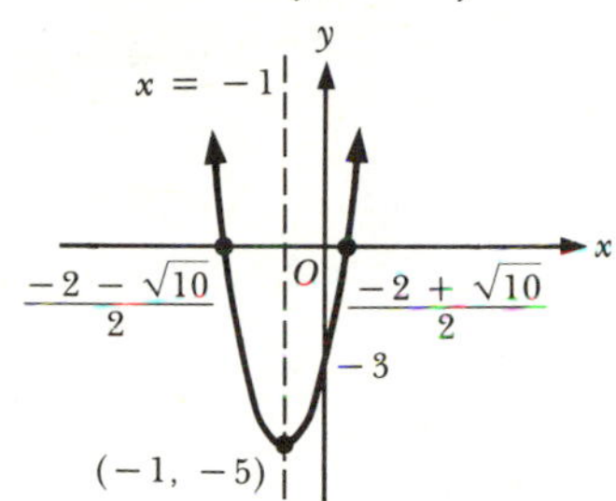

27. Range $(-\infty, \frac{1}{8}]$
Line of symmetry: $x = \frac{3}{4}$

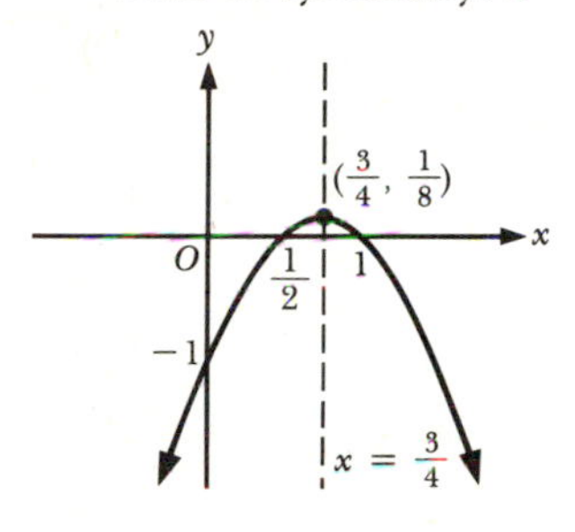

29. Range $[-10.65, \infty)$
Line of symmetry: $x = 2.94$

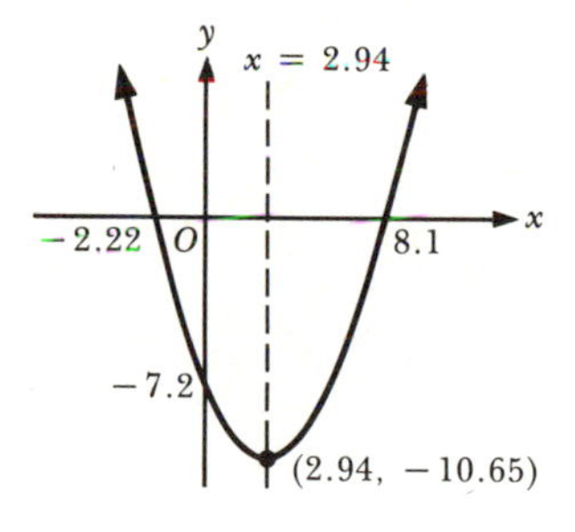

31. $(-\infty, 2)$ or $(7, \infty)$

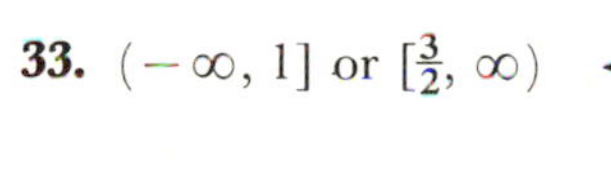

33. $(-\infty, 1]$ or $[\frac{3}{2}, \infty)$

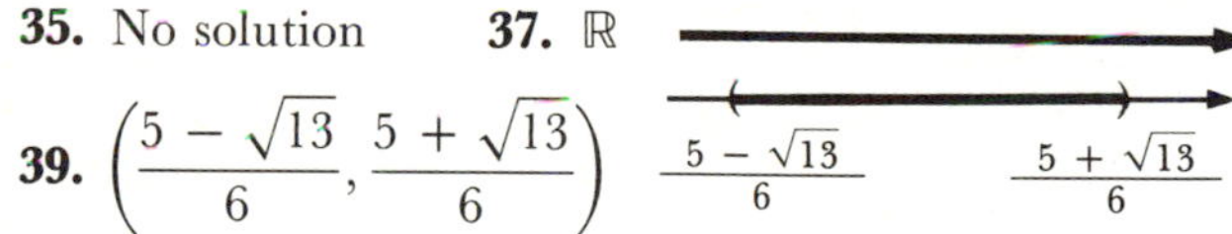

35. No solution **37.** $\mathbb{R}$

39. $\left(\dfrac{5 - \sqrt{13}}{6}, \dfrac{5 + \sqrt{13}}{6}\right)$

41. (a) 256 feet; (b) 400 feet; (c) 8 seconds

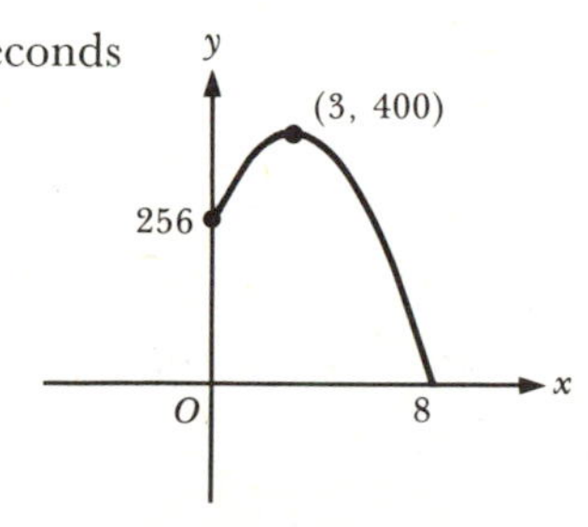

43. 312.5 square yards

PROBLEM SET 3.2 page 149

	Polynomial	Degree	Coefficients
1.	Yes	3	$4, -2, 1, -5$
3.	Yes	0	$\frac{1}{3}$
5.	Yes	4	$5, \frac{1}{2}, -2$
7.	No	—	—
9.	Yes	5	$-3, 1, -1, 8$

11.

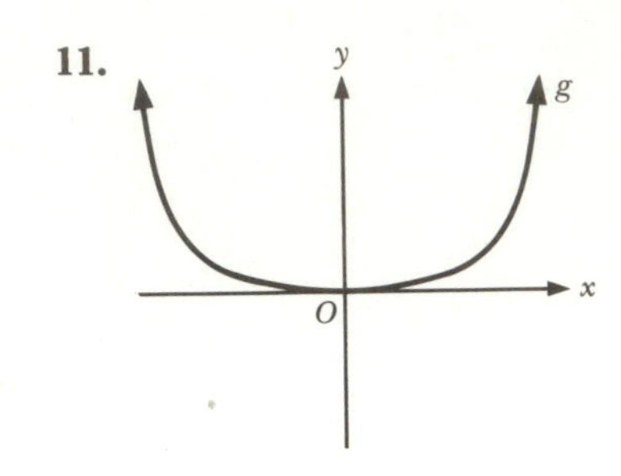

13.

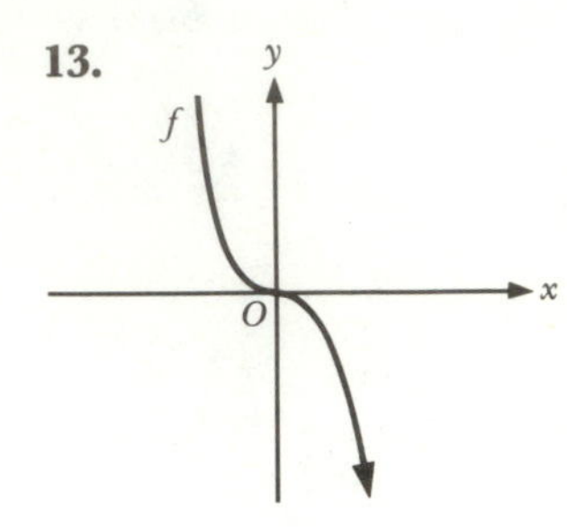

15.

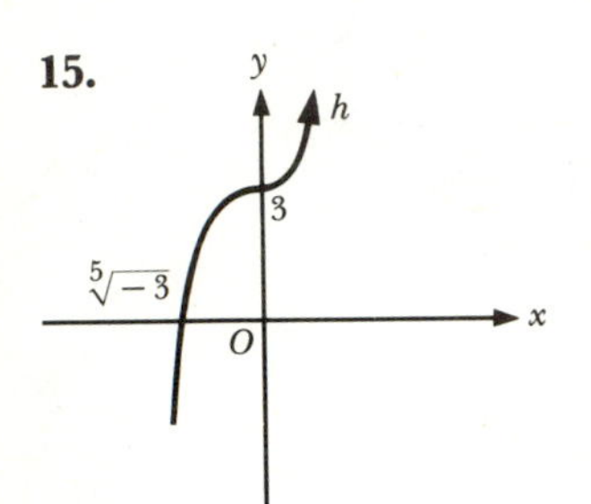

17.

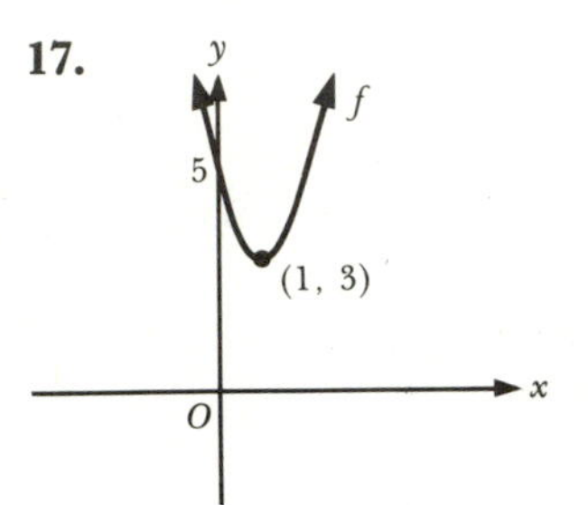

19.

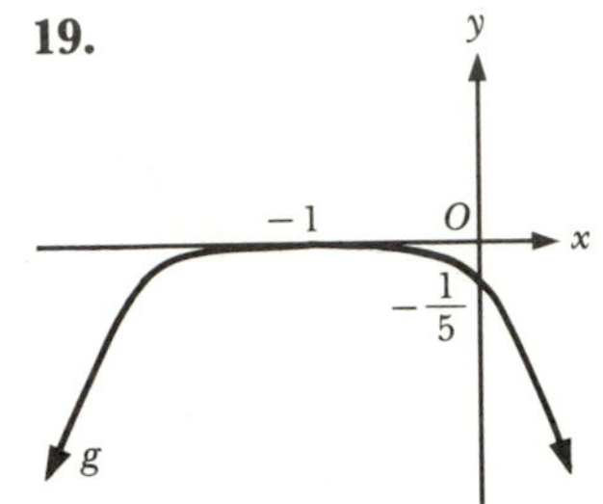

21.

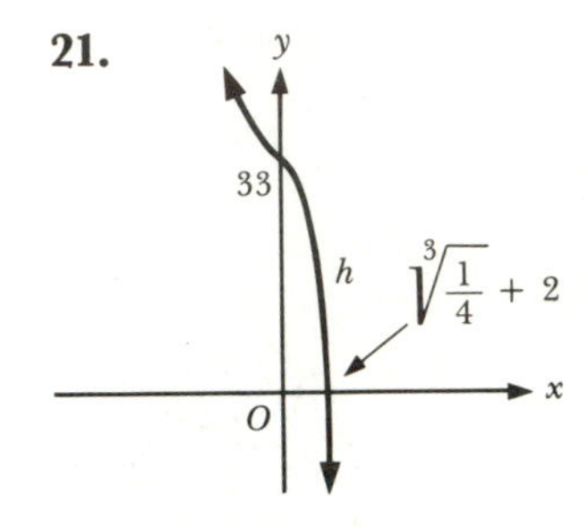

23.

	Zero	Multiplicity
25.	$\frac{2}{3}$	1
27.	1	3
	2	2
29.	2	1
	-2	1

	Zero	Multiplicity
31.	0	1
	3	1
	-3	1
33.	-1	1
	5	3
	-5	1

35. $(-2, 0)$ or $(1, \infty)$ **37.** $(-\infty, -3]$ or $[-2, -1]$ **39.** $(-\infty, 2)$ or $(2, 3)$
41. $(-\infty, 4]$ or $[5, \infty)$ **43.** $\mathbb{R}$

PROBLEM SET 3.3 page 158

1. $Q(x) = 4x + 6$; $R(x) = 7x - 19$ **3.** $Q(x) = 5x - 8$; $R(x) = 42x - 23$
5. $Q(x) = x^2 + 2x + \frac{1}{2}$; $R(x) = -\frac{9}{2}x^2 - \frac{7}{2}x - 5$ **7.** $x^3 - 4x^2 + 5x + 7 = (x - 2)(x^2 - 2x + 1) + 9$
9. $2x^4 - x^3 + 5x - 3 = (x + 1)(2x^3 - 3x^2 + 3x + 2) - 5$
11. $3.8x^2 - 7.3x - 2.1 = (x - 2.3)(3.8x + 1.44) + 1.212$ **13.** $Q(x) = x - 5$; $R = 0$
15. $Q(x) = 4x^3 + x^2 + 3x + 3$; $R = 8$ **17.** $Q(x) = 5x^2 + 13x + 42$; $R = 122$
19. $Q(x) = 5x^4 - 10x^3 + 20x^2 - 39x + 78$; $R = -153$
21. $Q(x) = -4x^5 - 8x^4 - 16x^3 - 37x^2 - 71x - 141$; $R = -275$ **23.** $x + 7 + \dfrac{15}{x - 2}$

25. $4x^2 - 10x + 21 + \dfrac{-47}{x+2}$ **27.** $5(x^4 + x^3 + x^2 + x + 1) + \dfrac{0}{x-1}$ **29.** 29 **31.** 2

33. 35 **35.** $\dfrac{25}{2}$ **37.** -20

45. $f(-5) = 0, f(-4) = 30, f(-3) = 40, f(-2) = 36, f(-1) = 24,$
$f(1) = 0, f(2) = 0, f(3) = 16, f(4) = 54, f(5) = 120;$
$f(x) = (x+5)(x-2)(x-1)$

PROBLEM SET 3.4 page 166

	Number of Positive Zeros	Number of Negative Zeros	Number of Complex Zeros
1.	1	1	0
3.	2	1	0
	0	1	2
5.	1	2	2
	1	0	4
7.	1	1	2
9.	4	0	0
	2	0	2
	0	0	4

11. $1, 2, -2$ **13.** 1 **15.** $\frac{1}{3}$, **17.** $\frac{2}{5}, \frac{1}{2}, -1$ **19.** $1, -1, -\frac{1}{2}, \frac{3}{2}$ **21.** No rational zeros
23. $-\frac{1}{2}$ **25.** $-\frac{1}{2}, -\frac{1}{4}$

27. $f(x) = (x+1)(x-4)(x+2)$; x intercepts $-1, 4, -2$; y intercept -8

29. $h(x) = (x-3)(x-2)(x+1)$; x intercepts $3, 2, -1$; y intercept 6

31. $g(x) = (x-1)(3x-2)(3x+4)(x+2)$; x intercepts $1, \frac{2}{3}, -\frac{4}{3}, -2$; y intercept 16

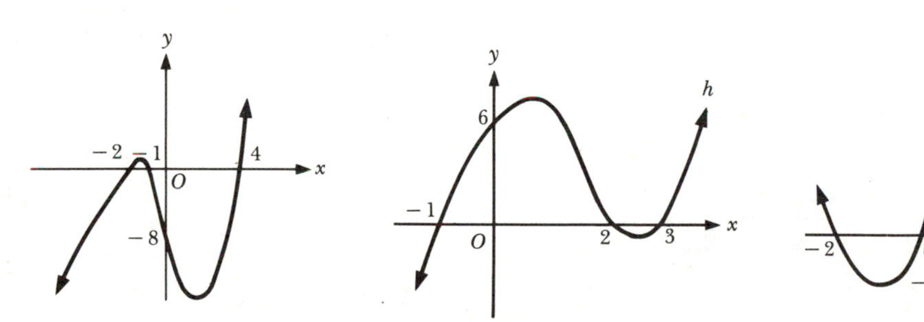

Figure Problem 27 *Figure Problem 29* *Figure Problem 31*

33. $h(x) = (x + 2)(2x - 3)(x - 4)$; x intercepts $-2, \frac{3}{2}, 4$; y intercept 24

35. $f(x) = x^2(3x - 4)(x - 5)$; x intercepts $0, \frac{4}{3}, 5$; y intercept 0

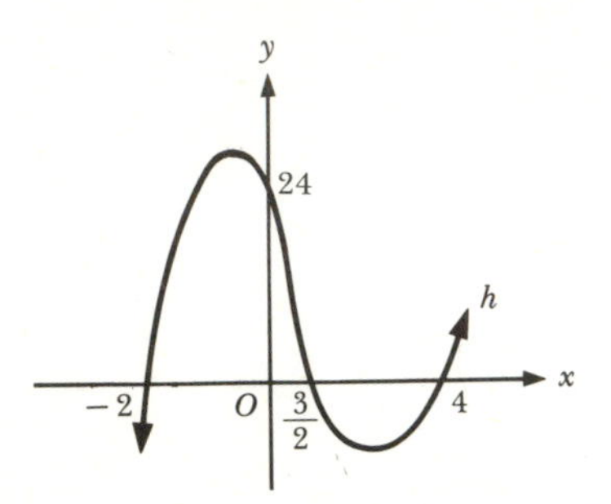

Figure Problem 33

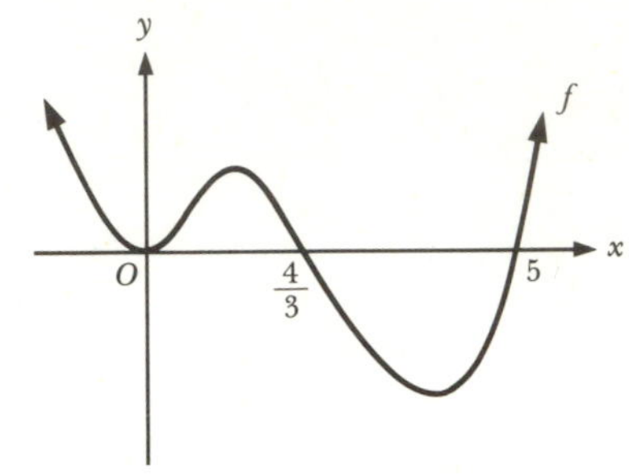

Figure Problem 35

PROBLEM SET 3.5 Page 173

1. $f(x) = x^2 - 2x + 2$ **3.** $f(x) = x^3 - 3x^2 + x + 5$ **5.** $f(x) = x^6 - x^5 - 4x^4 - 2x^3 - 11x^2 - x - 6$
7. $f(x) = x^4 - 4x^3 + 24x^2 - 40x + 100$ **9.** $f(x) = x^4 - 3x^3 + x^2 + 4$
11. $f(x) = (x - 2i)(x + 2i)(x + 3i)(x - 3i)$ **13.** $f(x) = (x + 1 - i)(x + 1 + i)(x + \sqrt{2})(x - \sqrt{2})$
15. $-1, 4$ **17.** $-3, 3$ **19.** $-4, 5$ **21.** -1.2012 **23.** 1.4141

25. 1.2129 **27.** $-1.4142, 1.4142$

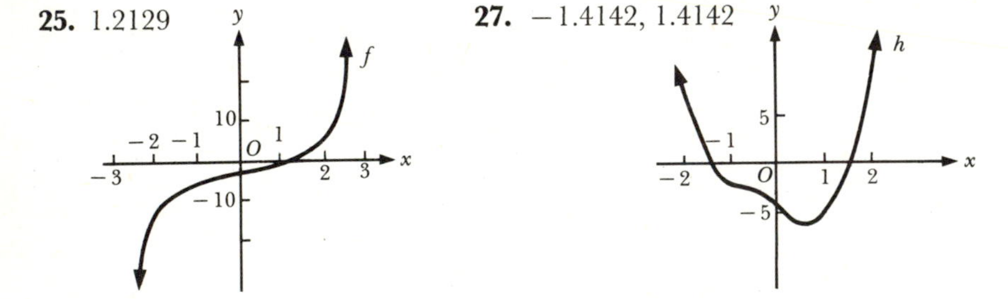

PROBLEM SET 3.6 page 181

Domain	Vertical Asymptotes
1. All real numbers except -7	$x = -7$
3. All real numbers except -7 and 4	$x = 4$ and $x = -7$
5. All real numbers except -5 and 1	$x = -5$ and $x = 1$
7. All real numbers except 3	None
9. $\mathbb{R}$	None

11. Vertical asymptote $x = -4$
Horizontal asymptote $y = 0$

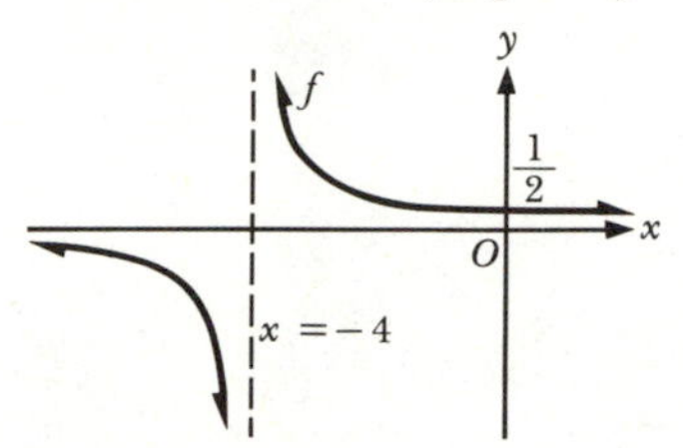

13. Vertical asymptote $x = 3$
Horizontal asymptote $y = 0$

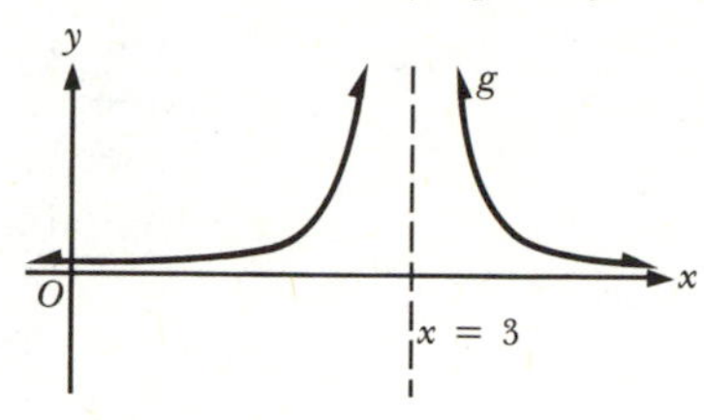

15. Vertical asymptote $x = -1$
Horizontal asymptote $y = 0$

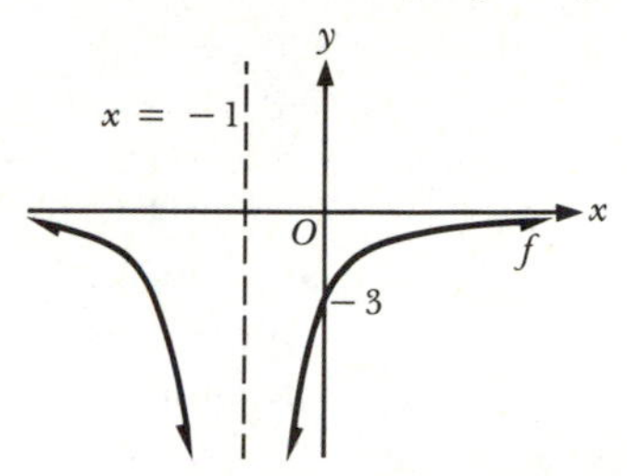

17. Vertical asymptote $x = 0$
Horizontal asymptote $y = 3$

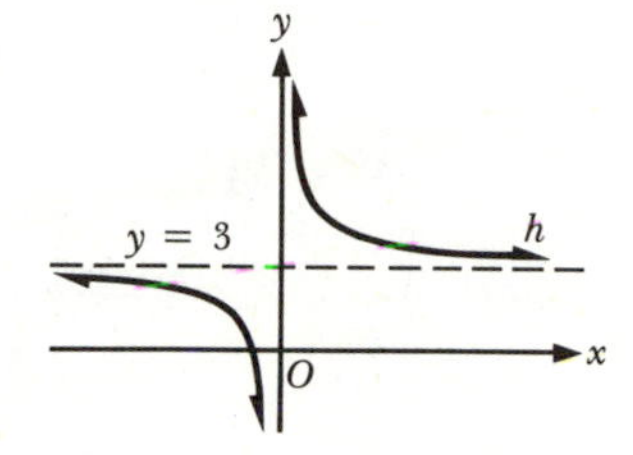

19. Vertical asymptote $x = 0$
Horizontal asymptote $y = 1$

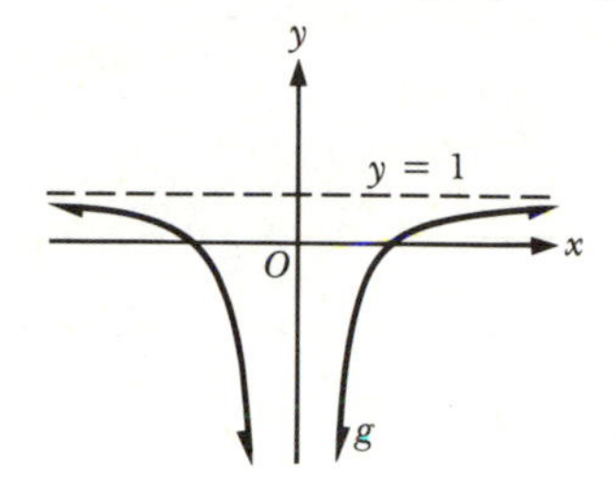

21. Horizontal asymptote $y = -3$
Vertical asymptote $x = -3$

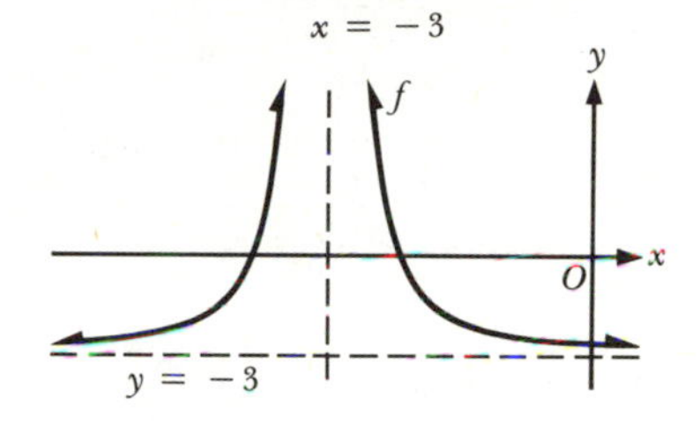

23. Vertical asymptote $x = 2$
Horizontal asymptote $y = 2$

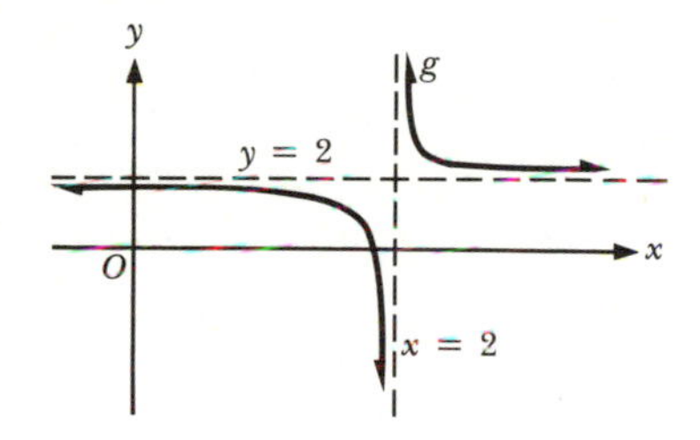

25. Vertical asymptote $x = -1$
Horizontal asymptote $y = 5$

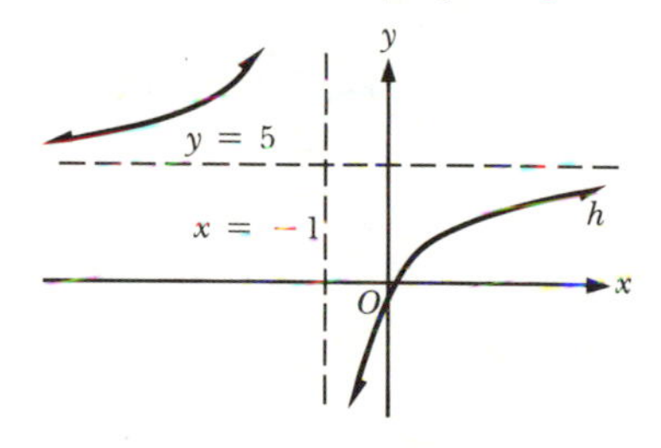

27. Domain: all reals but 3; no vertical or horizontal asymptotes

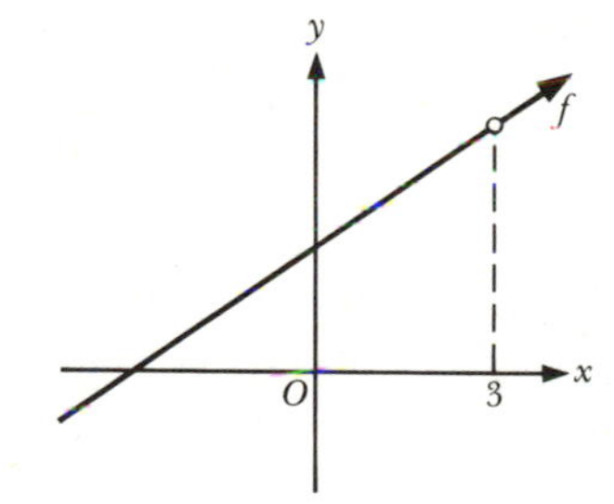

29. Domain: all reals but -1;
horizontal asymptote $y = 2$
vertical asymptote $x = -1$;

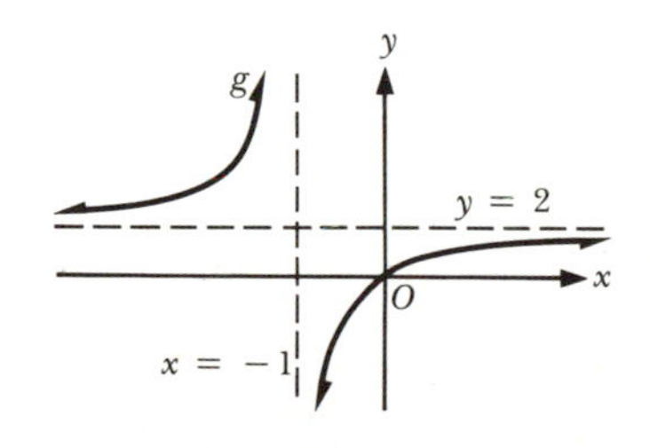

31. Domain $= \mathbb{R}$; no vertical asymptote;
horizontal asymptote $y = 1$

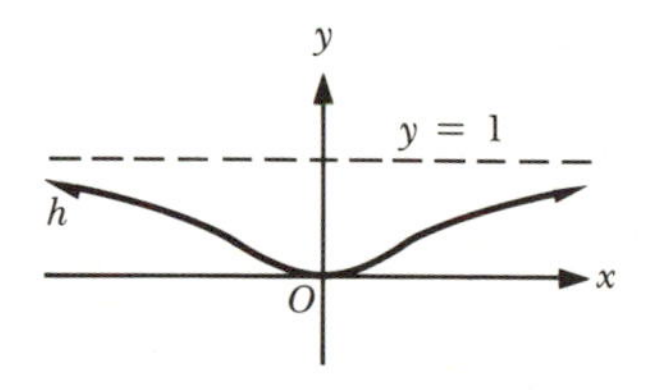

33. Domain: all reals but $0, -1$;
vertical asymptotes $x = 0$, $x = -1$;
horizontal asymptote $y = 1$

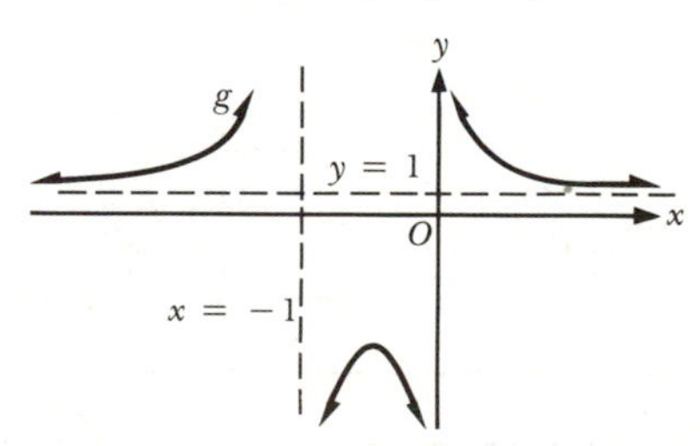

35. Domain: all reals but 1 or -1; vertical asymptotes $x = 1$, $x = -1$; horizontal asymptote $y = 1$

37. Domain: all reals but $\frac{3}{2}$, -5; vertical asymptotes $x = \frac{3}{2}$, $x = -5$; horizontal asymptote $y = 0$

39. Domain: all reals but 1, -2; vertical asymptotes $x = 1$, $x = -2$; horizontal asymptote $y = 0$

41. Domain: all reals but 2, -1; vertical asymptote $x = 2$; horizontal asymptote $y = 1$

43. (a) Domain of f: all reals but 0; domain of g: all reals but 0;
(b) The graph of f is symmetric with respect to origin; graph of g is symmetric with respect to y axis;
(c) $x = 0$ is a vertical asymptote for f and g; $y = 0$ is a horizontal asymptote for f and g

45. Horizontal asymptote $y = 3$
Vertical asymptote $x = 0$

47. Horizontal asymptote $y = 0$
Vertical asymptote $x = 0$

49. Vertical asymptote $x = 2$
Horizontal asymptote $y = 1$

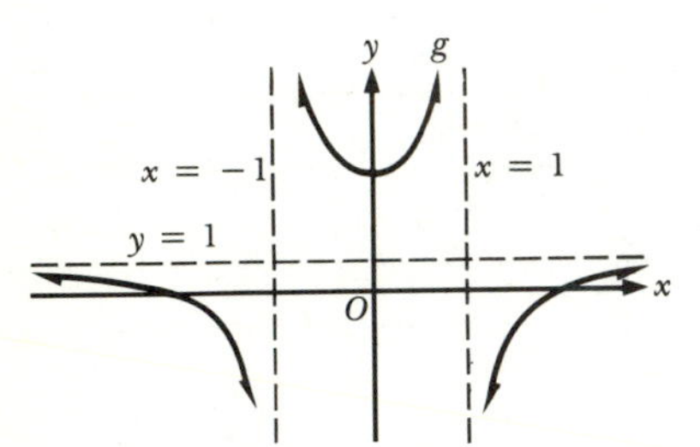

Figure Problem 35

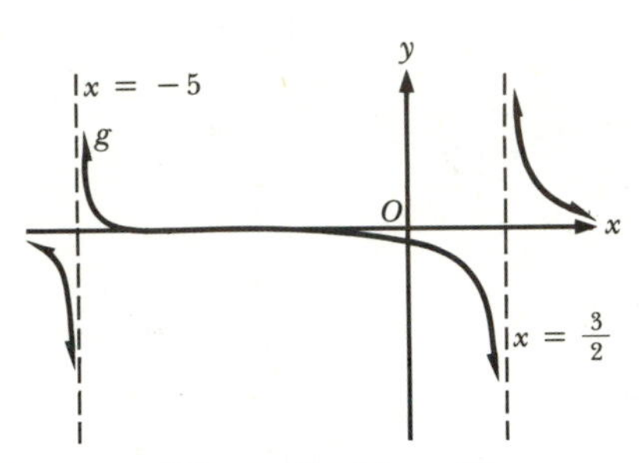

Figure Problem 37

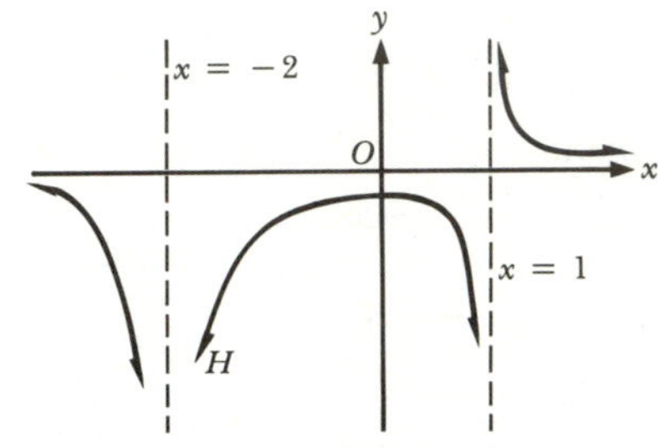

Figure Problem 39

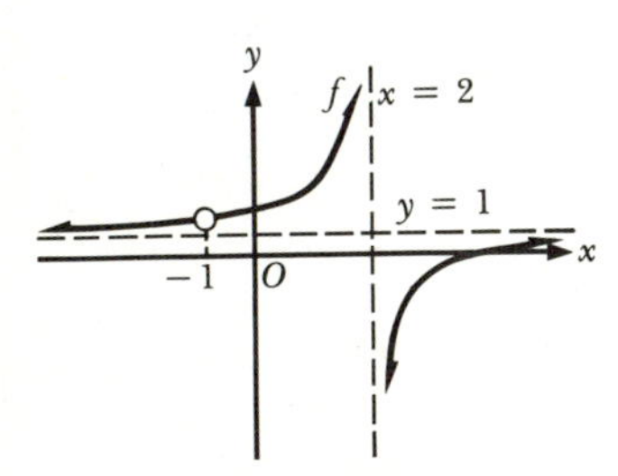

Figure Problem 41

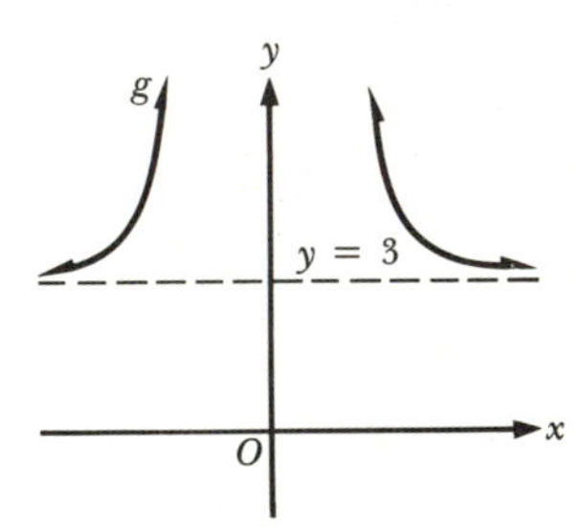

Figure Problem 45

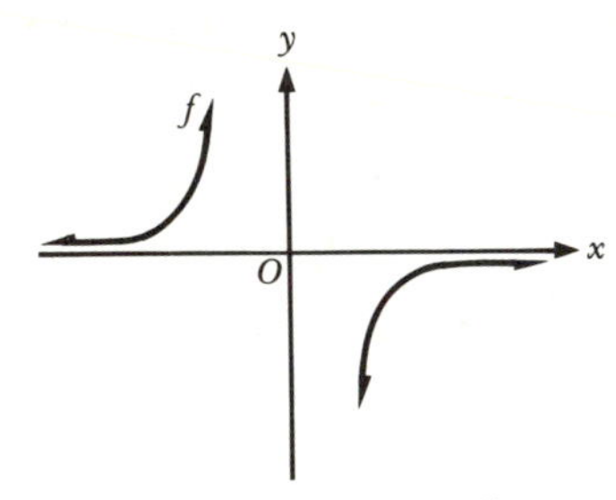

Figure Problem 47

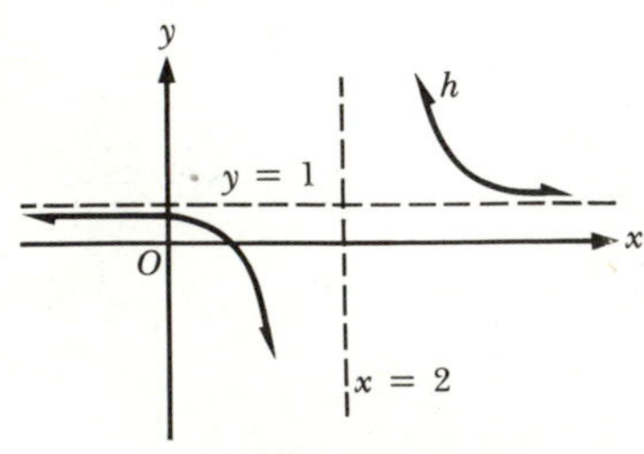

Figure Problem 49

51.

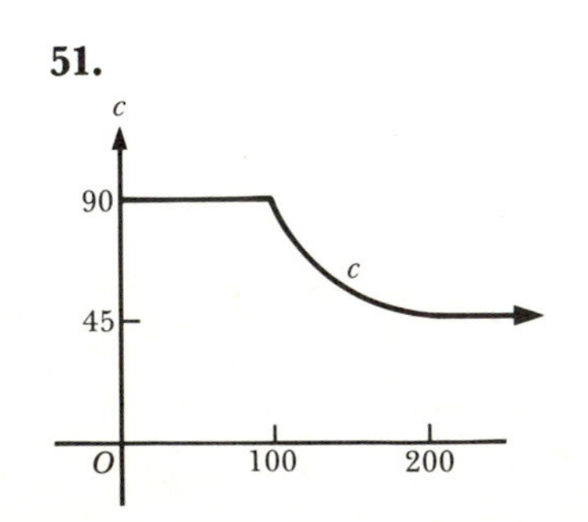

53.

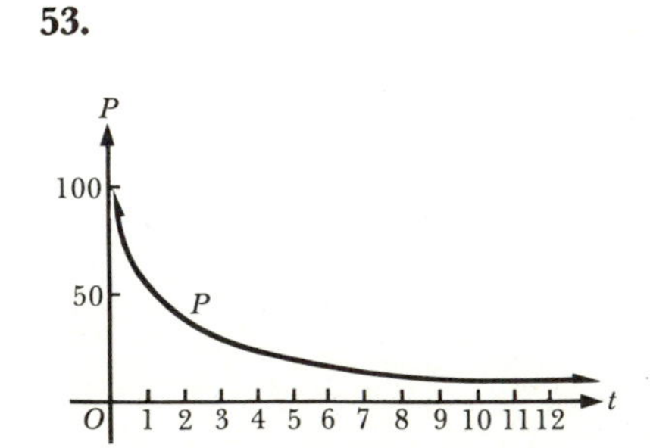

REVIEW PROBLEM SET, CHAPTER 3 page 183

1. (a)

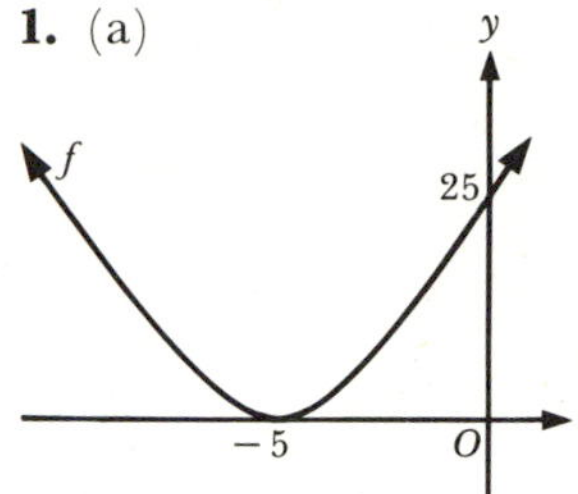

(b)

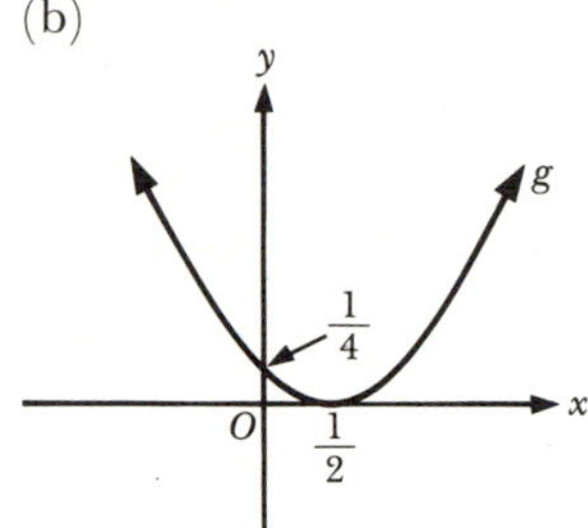

(c)

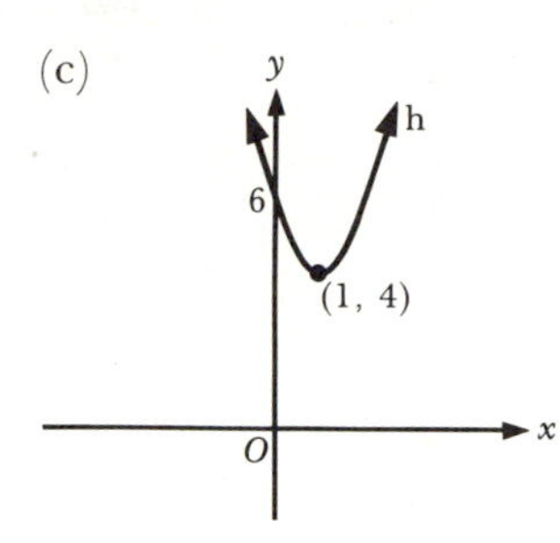

(d)

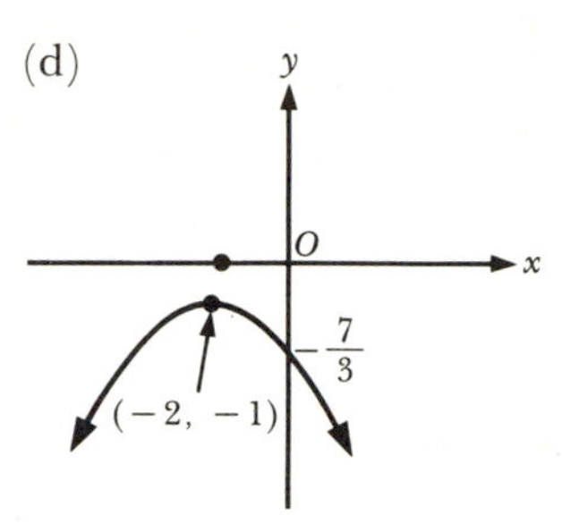

3. $x = -7$; $[0, \infty)$

5. $x = \frac{1}{4}$; $(-\infty, \frac{25}{8}]$

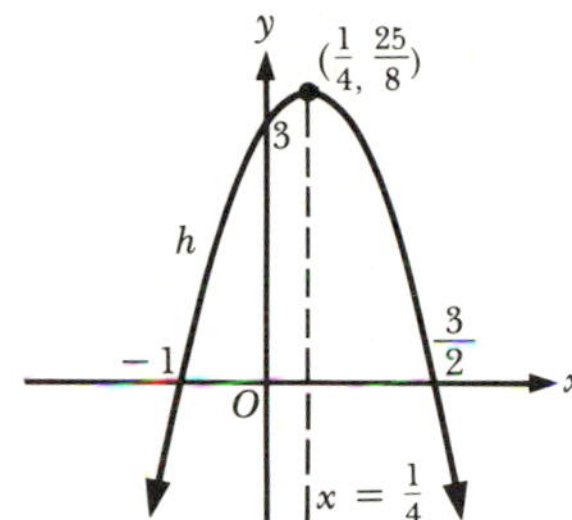

7. $x = -\frac{3}{2}$; $(-\infty, \frac{5}{4}]$

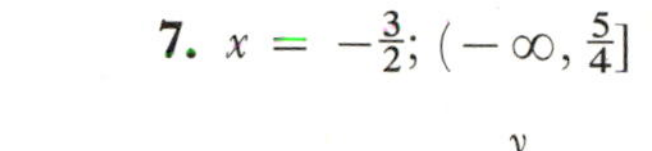

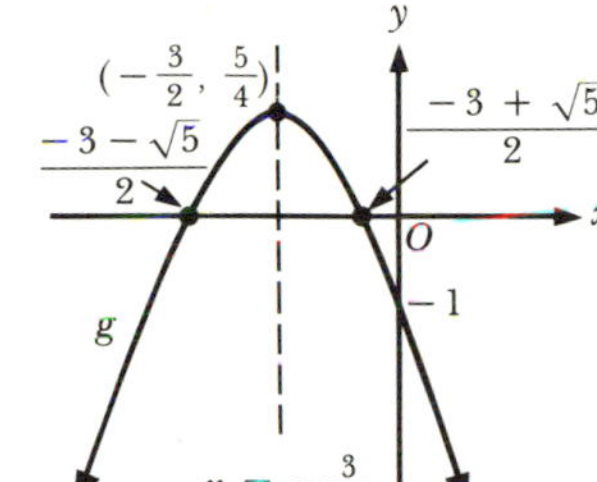

9. $x = -\frac{7}{10}$; $[-\frac{9}{20}, \infty)$

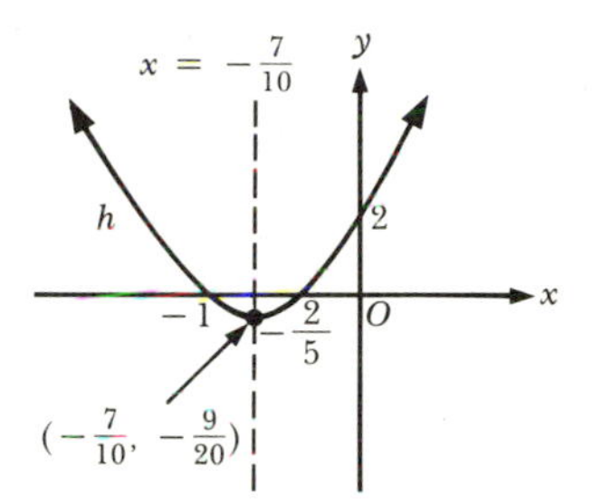

11. $[-\sqrt{2}, \sqrt{2}]$ **13.** $[-\frac{1}{2}, \frac{4}{3}]$ **15.** All real numbers except 1 **17.** 600

19.

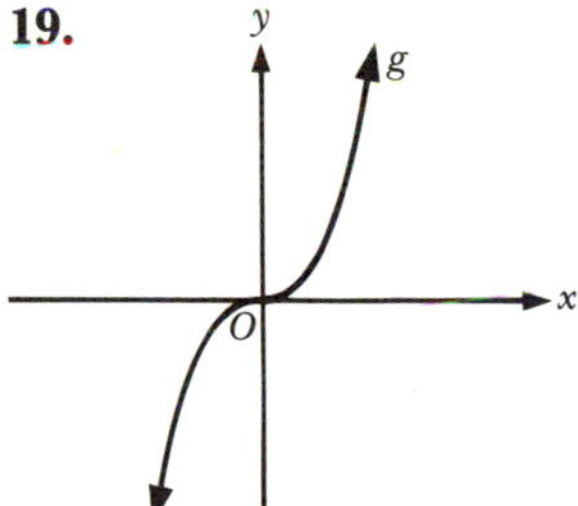

21.

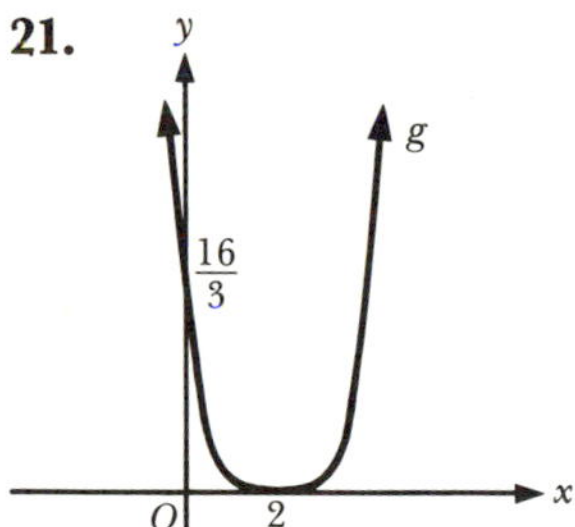

23.

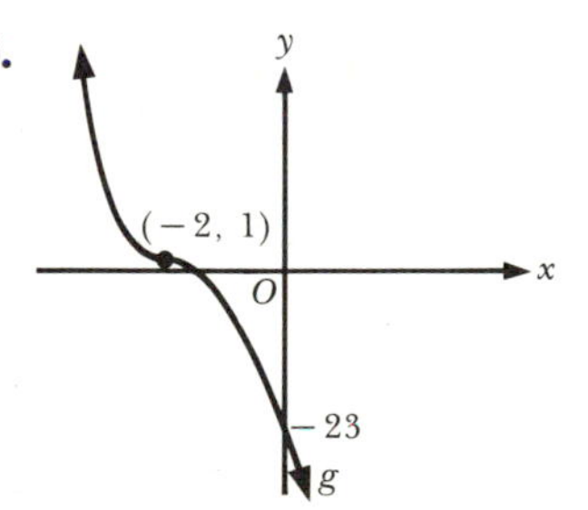

25. $-\frac{3}{2}$ has multiplicity 1 **27.** 1 has multiplicity 2; 4 has multiplicity 1 **29.** 2 has multiplicity 2; -2 has multiplicity 1, 1 has multiplicity 1

31. $(-\infty, -3)$ or $(0, 5)$ **33.** $[-2, 0]$ or $x = 1$ **35.** $Q(x) = 3x^2 + 11x + 29$; $R = 55$

37. $Q(x) = 2x^2 + 2x + 1$; $R = -19$ **39.** (a) 13; (b) 208; (c) 4; (d) -7; (e) $\frac{191}{32}$

	Number of Positive Zeros	Number of Negative Zeros	Number of Complex Zeros
41.	1	2	0
	1	0	2
43.	3	1	0
	1	1	2

45. $-1, 2, 3$ **47.** $3, -3, -1, \frac{1}{2}$ **49.** $f(x) = (x + 3)(x - 3)(x + 1)(x - 1)$
51. $h(x) = (x - 3)(x + 1)^2$ **53.** $f(x) = x^4 + 4x^3 - 5x^2 - 36x - 36$
55. $f(x) = x^3 - 6x^2 + 13x - 10$ **57.** $-2, 5$ **59.** $-2, 4$ **61.** 0.5352
63. Domain: all reals except -1; vertical asymptote $x = -1$
65. Domain: all reals except 2 and -2; vertical asymptotes $x = 2$, $x = -2$
67. Domain: all reals except 6 and -1; vertical asymptotes $x = 6$, $x = -1$
69. See figures below
71. Domain: all reals except 5; vertical asymptote $x = 5$; horizontal asymptote $y = 0$
73. Domain: all reals except $3, -3$; vertical asymptotes $x = 3$, $x = -3$; horizontal asymptote $y = 0$
75. Domain $= \mathbb{R}$; no vertical asymptote; horizontal asymptote $y = 1$

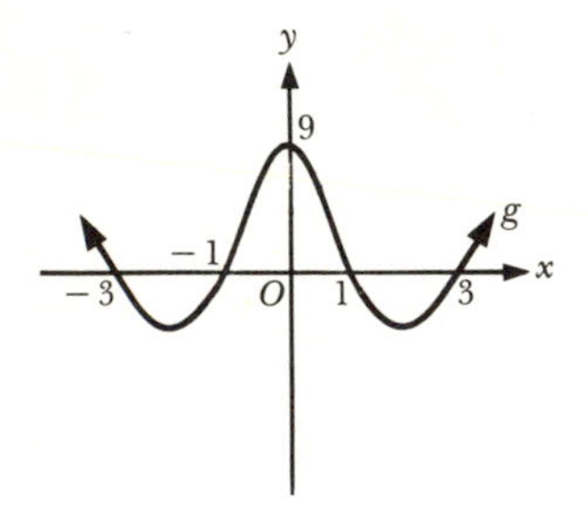

Figure Problem 49

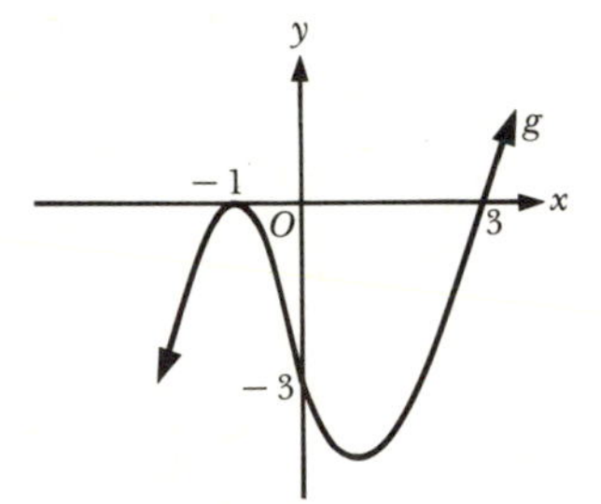

Figure Problem 51

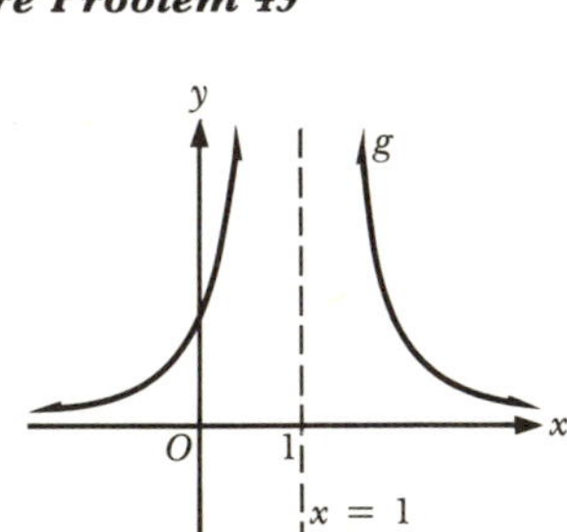

Figure Problem 69a

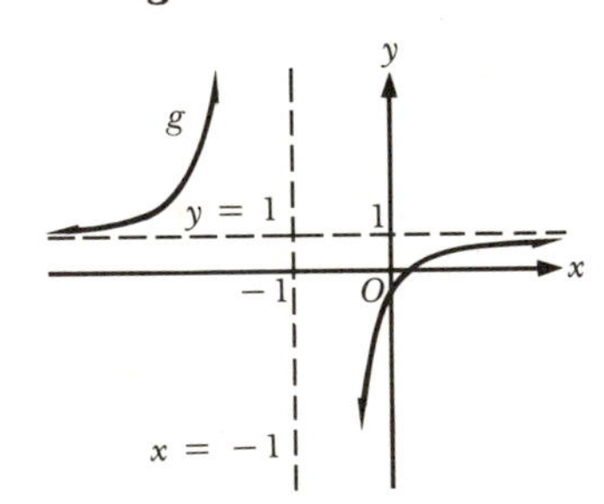

Figure Problem 69b

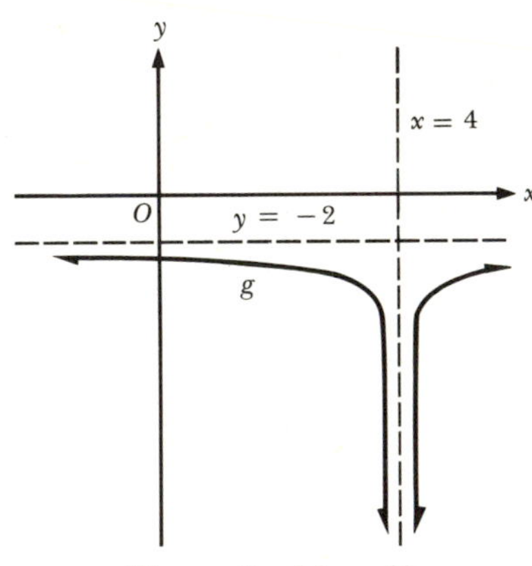

Figure Problem 69c

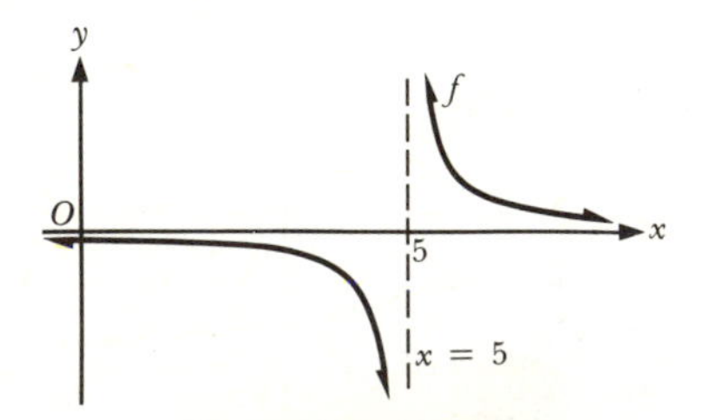

Figure Problem 71

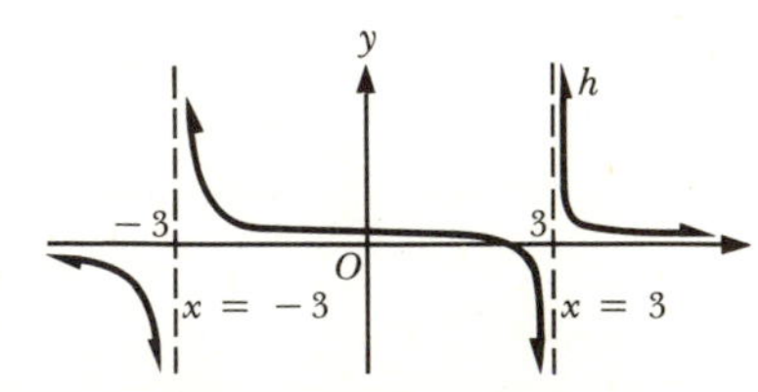

Figure Problem 73

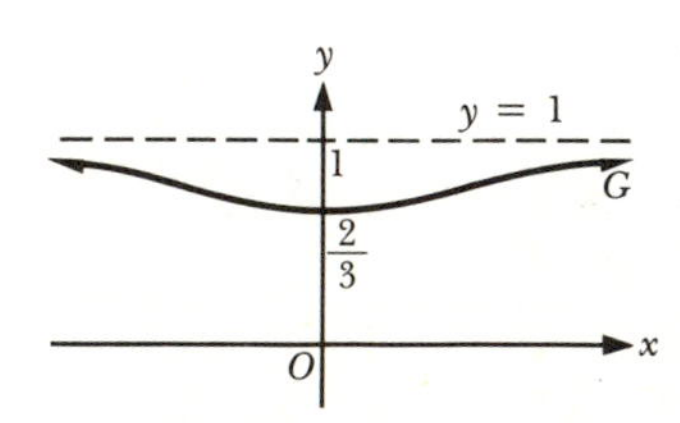

Figure Problem 75

CHAPTER 4

PROBLEM SET 4.1 page 194

1. (a) 2; (b) 4; (c) $\frac{1}{4}$; (d) 0.8123; (e) 2.2974; (f) 0.3010

3. Domain $= \mathbb{R}$
Range $= (0, \infty)$
Increasing
Horizontal asymptote: $y = 0$

5. Domain $= \mathbb{R}$
Range $= (0, \infty)$
Decreasing
Horizontal asymptote: $y = 0$

7. Domain $= \mathbb{R}$
Range $= (0, \infty)$
Decreasing
Horizontal asymptote: $y = 0$

9. Domain $= \mathbb{R}$
Range $= (0, \infty)$
Increasing
Horizontal asymptote: $y = 0$

11. Domain $= \mathbb{R}$
Range $= (-\infty, 0)$
Decreasing
Horizontal asymptote: $y = 0$

13. Domain $= \mathbb{R}$
Range $= (0, \infty)$
Increasing
Horizontal asymptote: $y = 0$

15. Domain $= \mathbb{R}$
Range $= (3, \infty)$
Increasing
Horizontal asymptote: $y = 3$

17. Domain $= \mathbb{R}$
Range $= (0, \infty)$
Increasing
Horizontal asymptote: $y = 0$

19. Domain $= \mathbb{R}$
Range $= (-3, \infty)$
Decreasing
Horizontal asymptote: $y = -3$

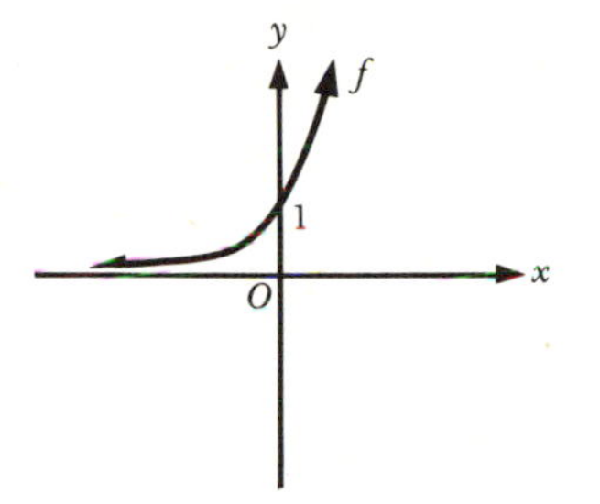

Figure Problem 3

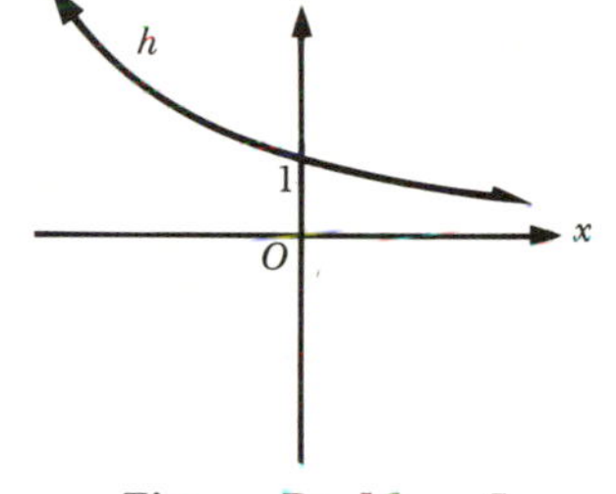

Figure Problem 5

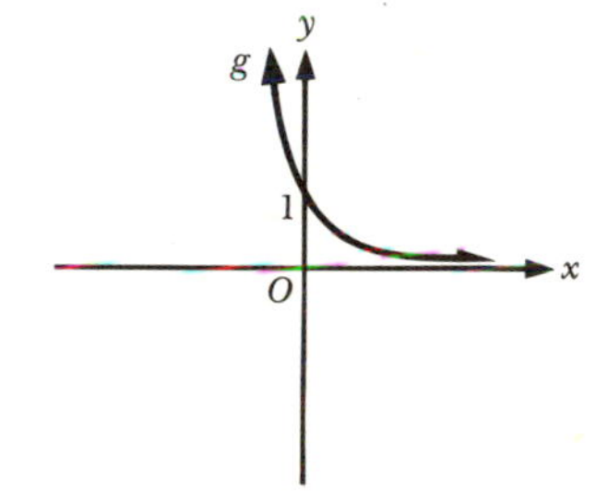

Figure Problem 7

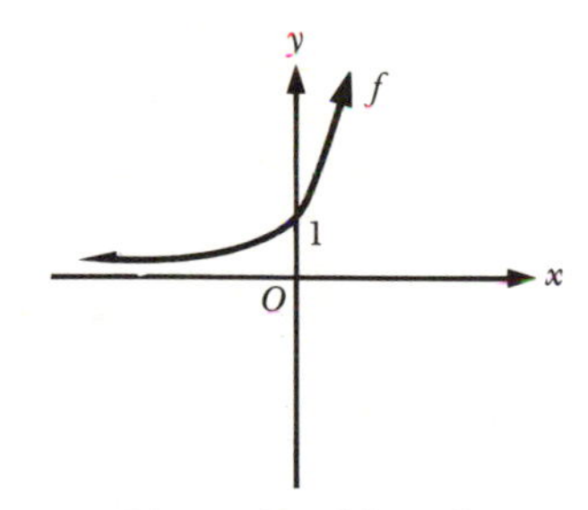

Figure Problem 9

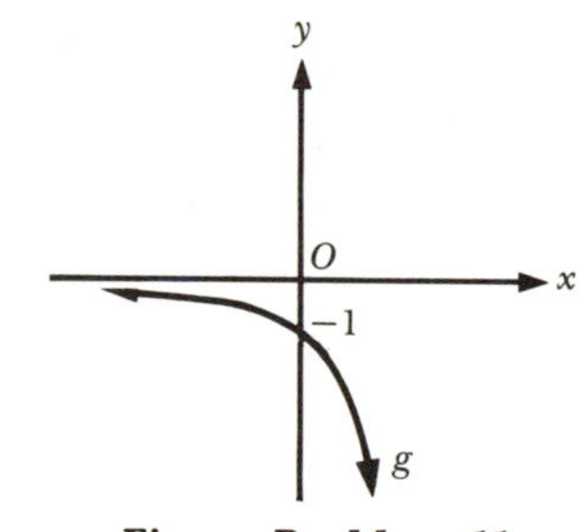

Figure Problem 11

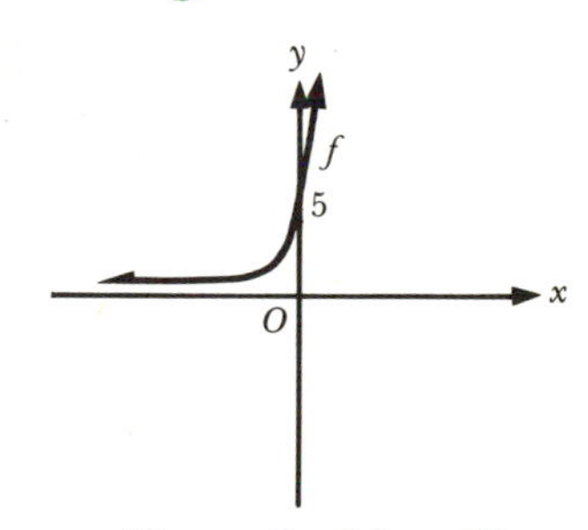

Figure Problem 13

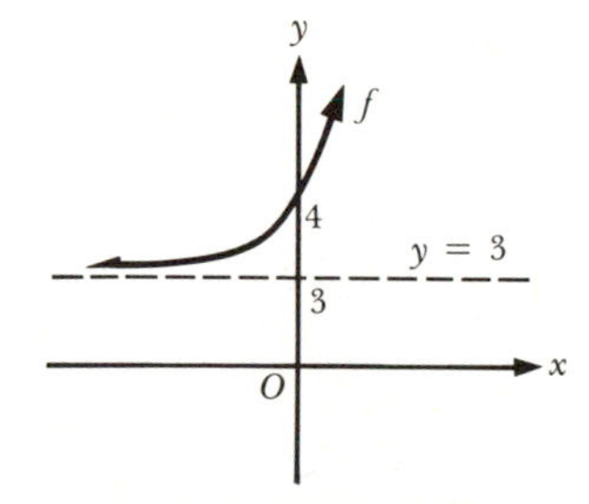

Figure Problem 15

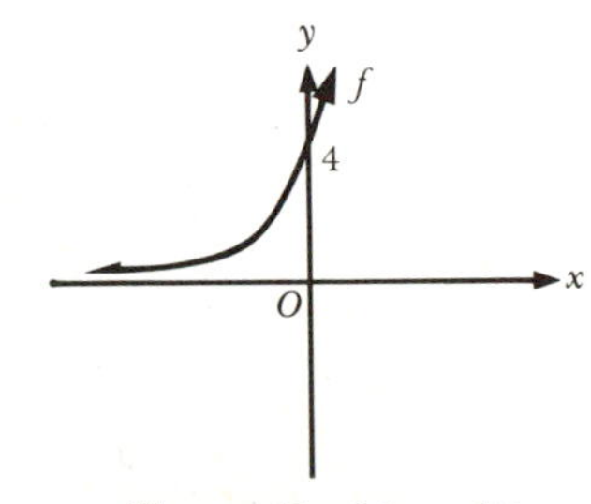

Figure Problem 17

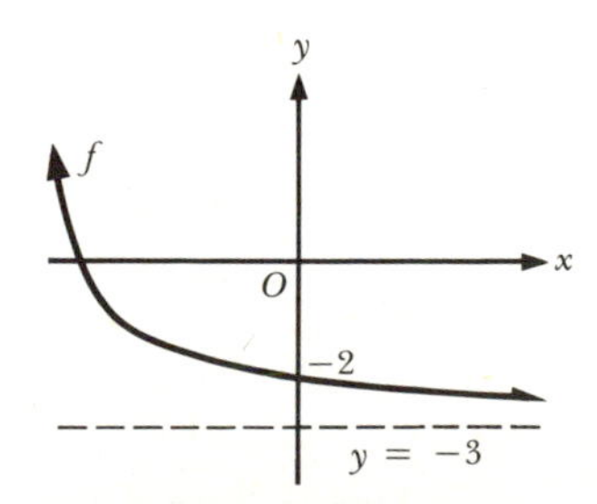

Figure Problem 19

21. $c = 1, b = 3$ **23.** $c = -1, b = e$ **29.** $-\frac{4}{3}, -1$

31. Domain = $\mathbb{R}$
Range = $(1, \infty)$
Increasing

33. Domain = $\mathbb{R}$
Range = $(-\infty, 2)$
Decreasing

35. $4/(e^x + e^{-x})^2$
37. (a) 29.96410; (b) 23.14069; (c) 4.113250; (d) 15.15426; (e) 0.0820850; (f) 0.6514391; (g) 0.1769212; (h) 0.4930687 **39.** 1 **41.** -4 **43.** $-\frac{1}{2}$ **45.** $-4, 1$ **47.** $0, -1$ **49.** \$4105.71 **51.** \$25,809.23 **53.** \$19,401.48 **55.** \$6478.74 **57.** \$6734.57 **59.** 9.20% **61.** 10.38% **63.** 91 **65.** 389 **67.** 16.26 horsepower

---

PROBLEM SET 4.2 page 204

1. $\log_5 125 = 3$ **3.** $\log_9 3 = \frac{1}{2}$ **5.** $\log_{32} 2 = \frac{1}{5}$ **7.** $9^2 = 81$ **9.** $(\frac{1}{3})^{-2} = 9$ **11.** $36^{3/2} = 216$
13. -3 **15.** -1 **17.** 1 **19.** 0 **21.** -4 **23.** 2 **25.** -3 **27.** -2 **29.** 4
31. $\frac{1}{81}$ **33.** $(-1, \infty)$ **35.** $(-\infty, 1)$ **37.** $\mathbb{R}$ **39.** $(-\infty, 1)$ and $(2, \infty)$

41. Domain = $(-1, \infty)$
Range = $\mathbb{R}$
Increasing
Vertical asymptote: $x = -1$

43. Domain = $(0, \infty)$
Range = $\mathbb{R}$
Increasing
Vertical asymptote: $x = 0$

45. Domain = $(0, \infty)$
Range = $\mathbb{R}$
Decreasing
Vertical asymptote: $x = 0$

47. Domain = all reals except 0
Range = $\mathbb{R}$
Decreasing on $(-\infty, 0)$ and
Increasing on $(0, \infty)$
Vertical asymptote: $x = 0$

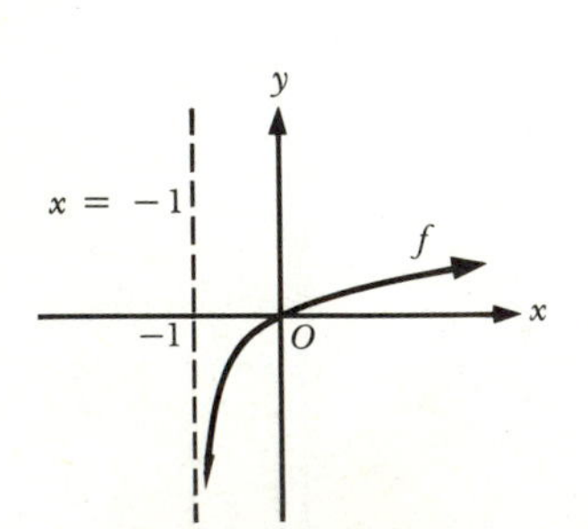

Figure Problem 41

Figure Problem 43

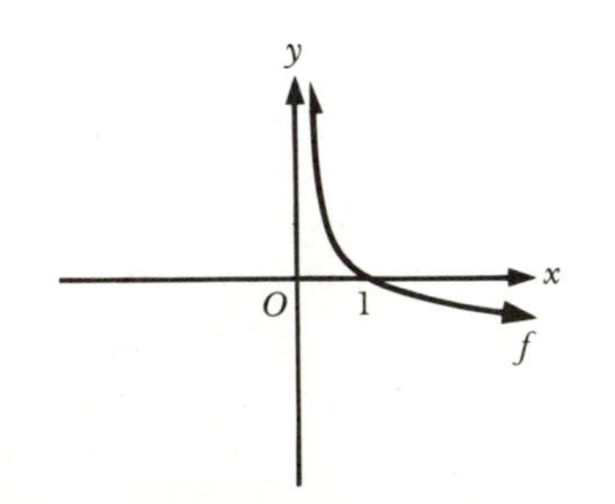

Figure Problem 45

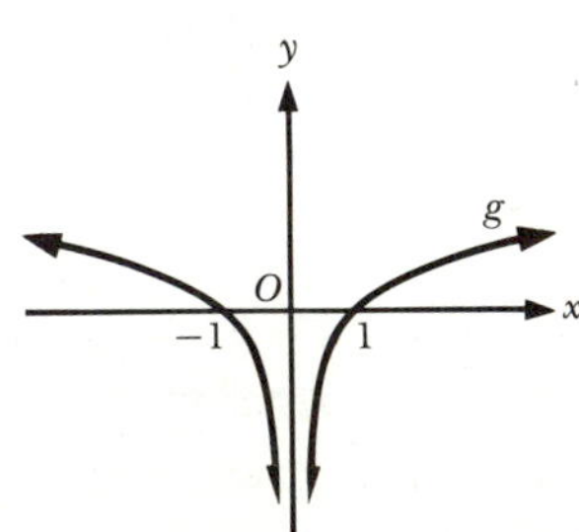

Figure Problem 47

49. Domain = $(0, \infty)$
Range = $\mathbb{R}$
Decreasing
Vertical asymptote: $x = 0$

51. 1.593286 **53.** 3.238748
55. -2.235674 **57.** 6.867974
59. -0.2850190 **61.** -5.475055
63. 2.622 **65.** 13.66 **67.** 0.0035

71.

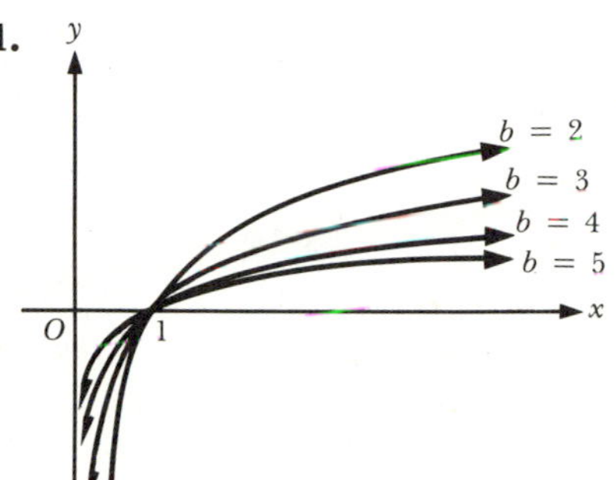

73.

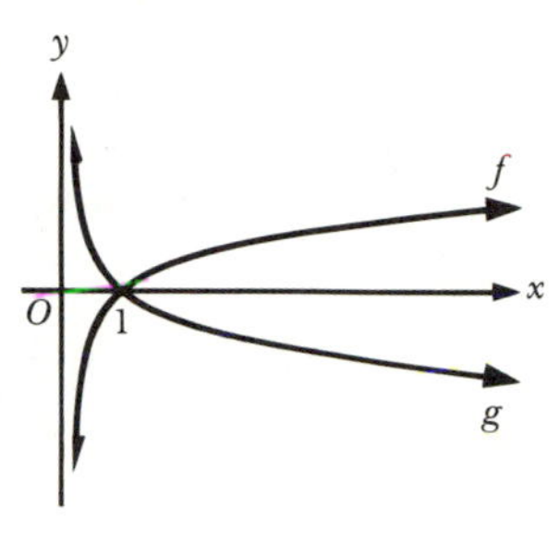

75. (a) $0 < x < 1$
(b) $x > 1$
(c) $x = 1$
(d) $\log_{10} x_1 < \log_{10} x_2$

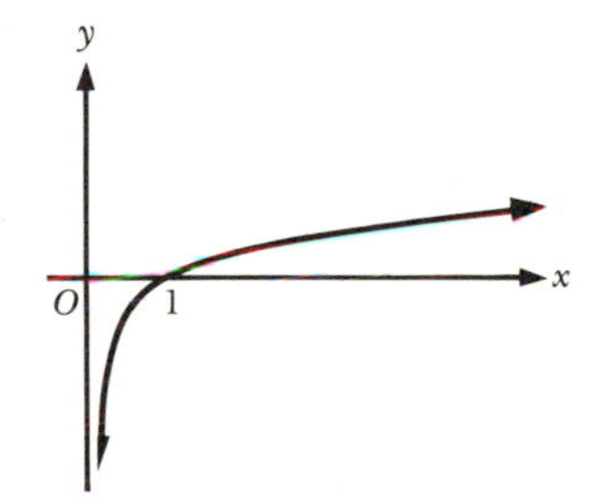

77. (a) \$260,332.77
(b) \$269,247.93
(c) \$279,080.85

PROBLEM SET 4.3 page 212

1. $\log_3 x + \log_3(x + 1)$ **3.** $2 \log_b x + 3 \log_b y$ **5.** $4 \log_b(x + 3)$ **7.** $\frac{5}{6} \log_b x - \frac{1}{3} \log_b y$
9. $\frac{1}{3} \log x + \frac{1}{9} \log y$ **11.** $\ln y + \frac{2}{3} \ln(3x + 1)$ **13.** $\frac{1}{3} \ln(x + y) + \frac{1}{3} \ln(x - y)$ **15.** $\frac{1}{3} \ln(4x + 3) - \frac{4}{3} \ln x$
17. $\log_b 5 + \log_b x + 2 \log_b(x^2 + 1) - \log_b(x + 1) - \frac{1}{2} \log_b(7x + 3)$ **19.** $-\frac{1}{7} \ln(3x + 1) - \frac{1}{7} \ln(2x - 3)$
21. $\log_5 \frac{8}{7}$ **23.** $\log_3 2$ **25.** $\log_c \frac{a}{7}$ **27.** $\ln\left(\frac{m + 1}{m - 1}\right)$ **29.** $\ln\left(\frac{2u + 3}{4u^2 + 6u + 9}\right)$ **31.** $\log_7\left(\frac{x + 2}{2x}\right)$
33. $\log_e \frac{b^7}{4}$ **35.** 4 **37.** $\frac{3}{4}$ **39.** $\frac{8}{5}$ **41.** $\frac{23}{10}$
43. (a) 1.00; (b) 1.70; (c) 1.99; (d) 0.58; (e) 0.23; (f) 0.70; (g) 0.37 **45.** 9 **47.** $\frac{1}{2}, -\frac{1}{2}$
49. 8 **51.** 28 **53.** $\frac{26}{125}$ **55.** $-1, -2$ **57.** 3 **59.** 4 **61.** 7 **63.** $\frac{1}{3}$ **65.** 22
67. 2.322 **69.** 1.730 **71.** 1.39 **73.** 0.61 **75.** 1.87 **77.** -0.39 **79.** -3.30
81. (a) About 14; (b) About 13 years, 10 months; (c) About 13 years, 9 months; (d) About 13 years, 9 months

PROBLEM SET 4.4 page 220

1. (a) 2.50; (b) 7.80; (c) 6.40 **3.** 114.77 decibels **5.** 3 **7.** 5628.36 meters **9.** 24.65 volts
11. (a) 600; (b) 437,400 **13.** (a) 252,128,085; (b) 281,276,067 **15.** About 63 and 4 months
17. 149.36 grams **19.** 77.97 milligrams **21.** (a) About 42; (b) Second day
23. (a) 65°C; (b) After 48 minutes and 11 seconds **25.** 2.4 amperes

PROBLEM SET 4.5 page 230

1. (a) 2.9395; (b) 0.9395; (c) -0.0605; (d) -1.0605 **3.** (a) 1.6749; (b) 0.6749; (c) -0.3251; (d) -3.3251
5. (a) 5.7404; (b) 1.7404; (c) -2.2596; (d) -0.2596 **7.** (a) 3.0238; (b) -3.0238; (c) 0.75595
9. 4.8947 **11.** -1.1288 **13.** 0.7880 **15.** 1.0917 **17.** -0.2618 **19.** 5.6747 **21.** 8.20
23. 55.2 **25.** 157 **27.** 0.0556 **29.** 0.0004 **31.** 3.653 **33.** 15.77 **35.** 0.1668
37. 0.007527 **39.** 0.02456 **41.** 1.6094 **43.** 2.4849 **45.** 6.1225 **47.** 1.4476 **49.** 8.9257
51. 9.007 **53.** 3.511

REVIEW PROBLEM SET, CHAPTER 4 page 231

1. (a) 4; (b) 4; (c) $\frac{1}{8}$ (d) $-\frac{1}{27}$

3. Domain = $\mathbb{R}$
Range = $(0, \infty)$
Decreasing
Horizontal asymptote: $y = 0$

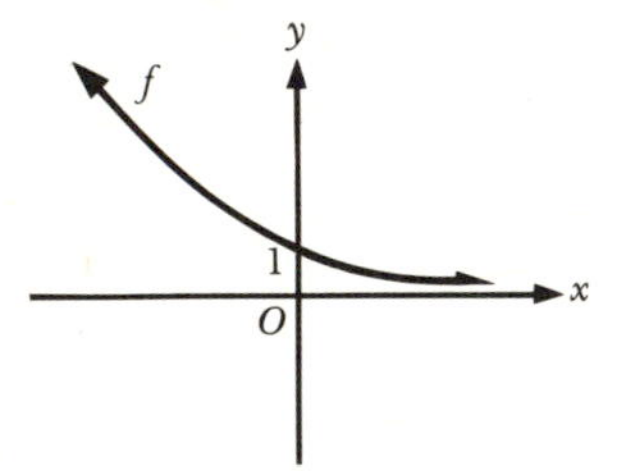

Figure Problem 3

5. Domain = $\mathbb{R}$
Range = $(0, \infty)$
Increasing
Horizontal asymptote: $y = 0$

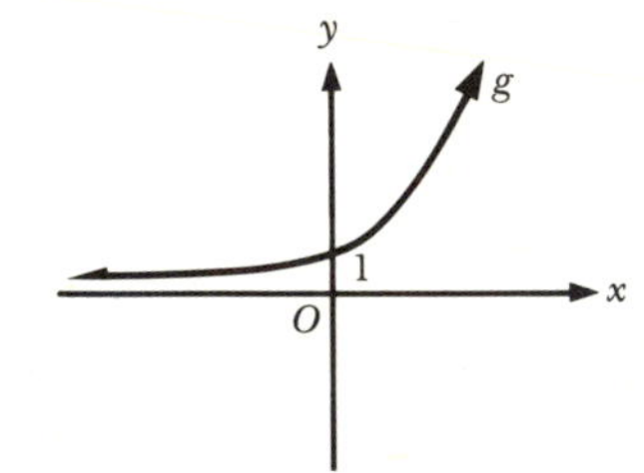

Figure Problem 5

7. Domain = $\mathbb{R}$
Range = $(0, \infty)$
Increasing
Horizontal asymptote: $y = 0$

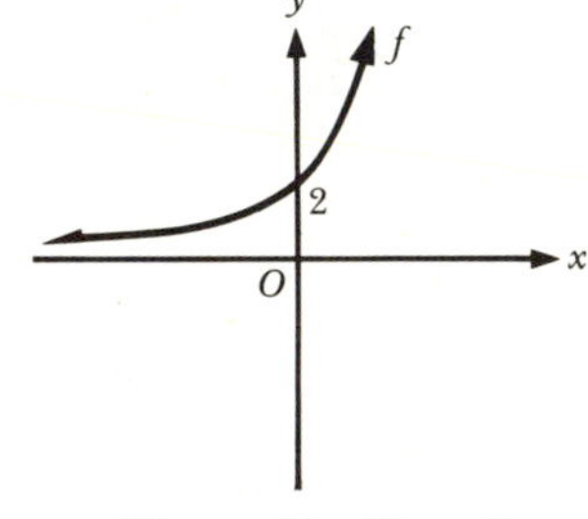

Figure Problem 7

9. (a) \$1655.00; (b) \$1668.17; (c) \$1675.01; (d) \$1679.67; (e) \$1682.03
11. 8.34% compounded continuously **13.** Domain = $\mathbb{R}$; Range = $(-\infty, 2)$
15. (a) 61.75072; (b) 0.0446010; (c) 0.0000002 **17.** (a) \$8000; (b) \$15,616.71 **19.** $\frac{3}{2}$ **21.** $-2, 1$

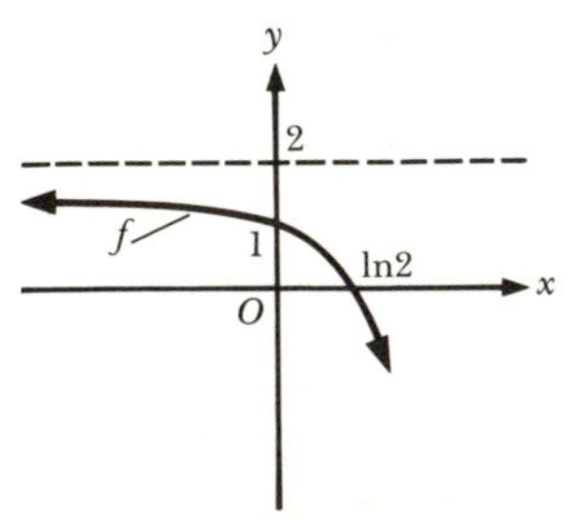

Figure Problem 13

23. $-\frac{8}{3}$ **25.** $-5, 1$ **27.** $c = 7, b = 2$ **29.** $(\frac{1}{2}, \infty)$ **31.** All reals except -2
33. Domain $= (0, \infty)$; Range $= \mathbb{R}$ **35.** Domain $= (0, \infty)$; Range $= \mathbb{R}$
37. Domain = all reals except 0;
Range $= \mathbb{R}$

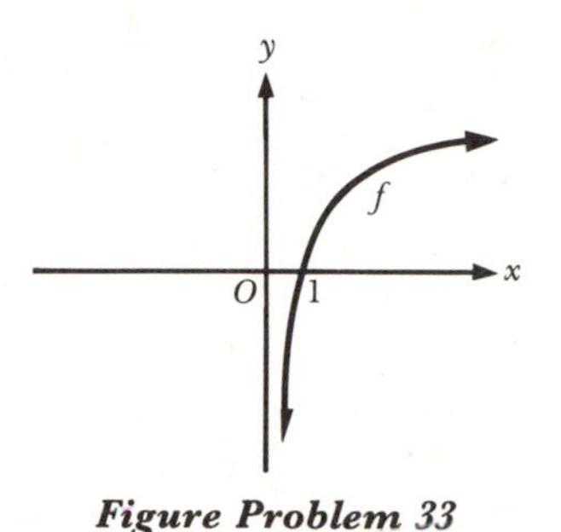

Figure Problem 33

Figure Problem 35

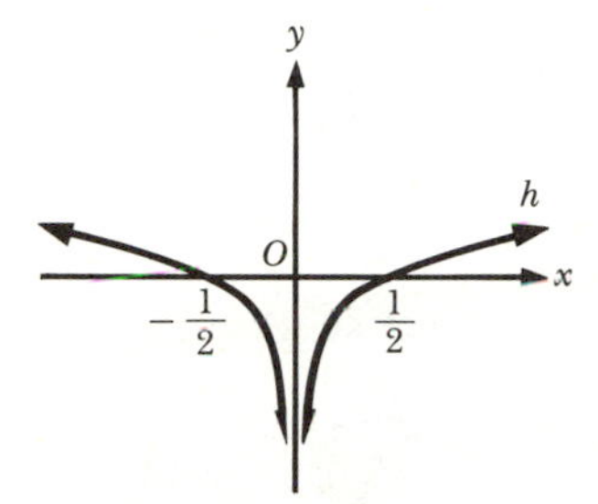

Figure Problem 37

39. (a) 2; (b) $\frac{3}{2}$; (c) 0; (d) -2; (e) $-\frac{3}{2}$; (f) $-\frac{1}{2}$
41. (a) 4.436163; (b) 1.871981; (c) 1.084223; (d) 2.252628; (e) 1.741130; (f) 2.182111; (g) -0.1469665; (h) 0.6826352 **43.** $\log_3(y \pm \sqrt{y^2 - 1})$ **45.** $\frac{1}{2}\ln\left(\frac{1+y}{1-y}\right)$ **47.** 0.7925 **49.** 1.869
51. 1.404 **53.** $6 \log_2 3 + 7 \log_2 5$ **55.** $\log_2 x + 5 \log_2 y$ **57.** $2 \log_a x - 4 \log_a y$ **59.** $\log_2 \frac{2}{9}$
61. $\log_5 \frac{1}{6}$ **63.** $\log_a(x^5/y^3)$ **65.** 13.2 **67.** 0.92 **69.** -1.64 **71.** 3.1 **73.** $-\frac{11}{4}$
75. $(25, \infty)$ **77.** 5 **79.** $\sqrt{6}\ 2$ **81.** $\frac{3}{4}$ **83.** (a) -3; (b) -3 **85.** 7.89 **87.** 1.46
89. -1.10 **91.** (a) About 7 years and 9 months; (b) About 7 years and 8 months
93. (a) 2.80; (b) 10.50 **95.** 1.63 hours **97.** About 99,037 years **99.** 4.9736 **101.** -0.4660
103. 29.7764 **105.** 99.3 **107.** 0.002525 **109.** 10.9430 **111.** 2.576

CHAPTER 5

PROBLEM SET 5.1 page 246

1. $-2\pi/3$ **3.** $5\pi/2$ **5.** Q_{III} **7.** Q_{II} **9.** x axis **11.** Q_{II} **13.** Q_{II} **15.** Q_{II}
17. Q_{IV} **19.** Q_{III} **21.** Q_{IV} **23.** Q_{IV} **25.** Q_{II} **27.** Q_{II}

	$\sin t$	$\cos t$	$\tan t$	$\cot t$	$\sec t$	$\csc t$
29.	$\frac{1}{2}$	$-\sqrt{3}/2$	$-\sqrt{3}/3$	$-\sqrt{3}$	$-2\sqrt{3}/3$	2
31.	$-\frac{12}{13}$	$\frac{5}{13}$	$-\frac{12}{5}$	$-\frac{5}{12}$	$\frac{13}{5}$	$-\frac{13}{12}$
33.	$2\sqrt{13}/13$	$-3\sqrt{13}/13$	$-\frac{2}{3}$	$-\frac{3}{2}$	$-\sqrt{13}/3$	$\sqrt{13}/2$
35.	$\sqrt{3}/2$	$\frac{1}{2}$	$\sqrt{3}$	$\sqrt{3}/3$	2	$2\sqrt{3}/3$
37.	0.8899	-0.4561	-1.9511	-0.5125	-2.1925	1.1237
39.	0	-1	0	undefined	-1	undefined
41.	-1	0	undefined	0	undefined	-1

43. (a) $(\frac{1}{2}, -\sqrt{3}/2)$; (b) $(-\frac{1}{2}, \sqrt{3}/2)$; (c) $(-\frac{1}{2}, -\sqrt{3}/2)$
45. (a) $(-\sqrt{3}/2, \frac{1}{2})$; (b) $(\sqrt{3}/2, -\frac{1}{2})$; (c) $(\sqrt{3}/2, \frac{1}{2})$
47. (a) $(\sqrt{2}/2, \sqrt{2}/2)$; (b) $(-\sqrt{2}/2, -\sqrt{2}/2)$; (c) $(-\sqrt{2}/2, \sqrt{2}/2)$
49. $(-\sqrt{2}/2, -\sqrt{2}/2)$; $\sin(-3\pi/4) = -\sqrt{2}/2$; $\cos(-3\pi/4) = -\sqrt{2}/2$
51. $(\sqrt{2}/2, -\sqrt{2}/2)$; $\sin(7\pi/4) = -\sqrt{2}/2$; $\cos(7\pi/4) = \sqrt{2}/2$
53. $(\frac{1}{2}, \sqrt{3}/2)$; $\sin(-5\pi/3) = \sqrt{3}/2$; $\cos(-5\pi/3) = \frac{1}{2}$ **55.** $(\sqrt{3}/2, -\frac{1}{2})$; $\sin(11\pi/6) = -\frac{1}{2}$; $\cos(11\pi/6) = \sqrt{3}/2$
57. -1 **59.** Undefined **61.** $\sqrt{3}/3$ **63.** $-\frac{1}{2}$ **65.** $\sqrt{3}$ **67.** $\sqrt{2}$ **69.** $-\sqrt{3}/2$ **71.** $-\frac{1}{2}$

PROBLEM SET 5.2 page 255

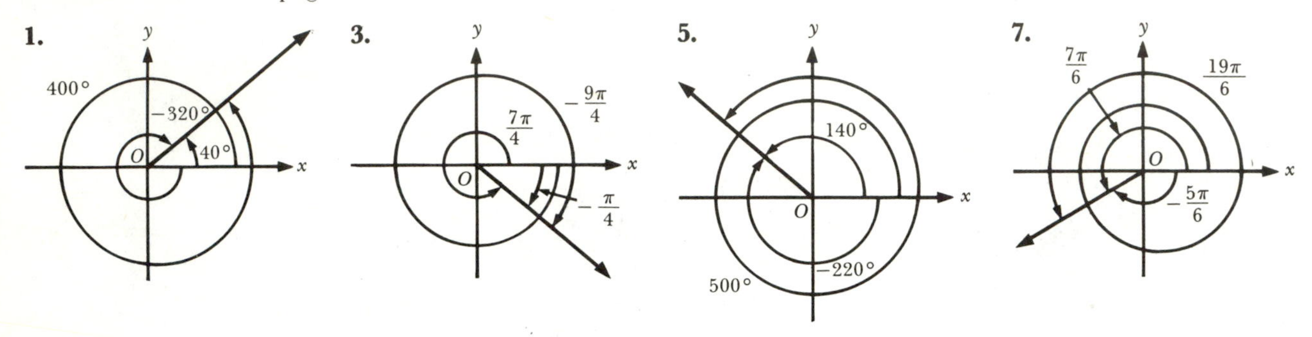

9. 87°21′ **11.** −25°33′ **13.** −65°22′12″ **15.** 18.7011° **17.** −920.4167° **19.** 70.5878°
21. $3\pi/2$ centimeters **23.** 0.6 radian **25.** 4 inches
27. (a) $2\pi/9$; (b) $5\pi/12$; (c) $4\pi/3$; (d) $11\pi/6$; (e) $-19\pi/36$; (f) $-11\pi/9$; (g) 6π; (h) $-305\pi/18$; (i) $-7\pi/3$; (j) $37\pi/15$; (k) $3\pi/8$; (l) $-\pi/24$
29. (a) 120°; (b) 330°; (c) 70°; (d) 60.5°; (e) 105°; (f) 1290°; (g) −80°; (h) −900°; (i) −390°; (j) 1845°; (k) −67.5°; (l) −(90/7)°
31. $11\pi/12$ **33.** $20\pi/3$ feet **35.** $\frac{55}{6}$ units **37.** $21\pi/4$ square centimeters
39. $750/\pi$ square feet **41.** $1225\pi/72$ square centimeters **43.** 30π meters
45. (a) 2.4 radians per second; (b) 11.67 feet per second **47.** 2250π meters per minute
49. (a) $\frac{10}{3}$ radians per second; (b) 6 feet per second

PROBLEM SET 5.3 page 264

	$\sin\theta$	$\cos\theta$	$\tan\theta$	$\cot\theta$	$\sec\theta$	$\csc\theta$
1.	$\frac{3}{5}$	$\frac{4}{5}$	$\frac{3}{4}$	$\frac{4}{3}$	$\frac{5}{4}$	$\frac{5}{3}$
3.	$-\frac{4}{5}$	$-\frac{3}{5}$	$\frac{4}{3}$	$\frac{3}{4}$	$-\frac{5}{3}$	$-\frac{5}{4}$
5.	-1	0	undefined	0	undefined	-1
7.	-0.90	0.44	-2.04	-0.49	2.27	-1.11
9.	$-\frac{15}{17}$	$\frac{8}{17}$	$-\frac{15}{8}$	$-\frac{8}{15}$	$\frac{17}{8}$	$-\frac{17}{15}$
11.	$\sqrt{2}/2$	$\sqrt{2}/2$	1	1	$\sqrt{2}$	$\sqrt{2}$
13.	1	0	undefined	0	undefined	1

	$\sin\theta$	$\cos\theta$	$\tan\theta$	$\cot\theta$	$\sec\theta$	$\csc\theta$
15.	-1	0	undefined	0	undefined	-1
17.	$\frac{3}{5}$	$\frac{4}{5}$	$\frac{3}{4}$	$\frac{4}{3}$	$\frac{5}{4}$	$\frac{5}{3}$
19.	$2\sqrt{29}/29$	$5\sqrt{29}/29$	$\frac{2}{5}$	$\frac{5}{2}$	$\sqrt{29}/5$	$\sqrt{29}/2$
21.	$\frac{15}{17}$	$\frac{8}{17}$	$\frac{15}{8}$	$\frac{8}{15}$	$\frac{17}{8}$	$\frac{17}{15}$

23. $\frac{1}{2}$ **25.** $\sqrt{3}$ **27.** $\sqrt{2}$ **29.** -1 **31.** $-\sqrt{3}/2$ **33.** $2\sqrt{3}/3$ **35.** $-\sqrt{3}/2$ **37.** $\sqrt{2}/2$
39. $\sqrt{2}/2$ **41.** -2 **43.** 2 **45.** $\sqrt{3}/3$ **47.** $-\sqrt{3}/3$
49. (a) Angles are coterminal, (b) Angles are coterminal

PROBLEM SET 5.4 page 276

1. $-\sqrt{2}/2$ **3.** $-\sqrt{3}/2$ **5.** $-\sqrt{3}/3$ **7.** 1 **9.** $2\sqrt{3}/3$ **11.** $\sqrt{2}$ **13.** $\sqrt{2}/2$
15. -1 **17.** -1 **19.** $-\sqrt{3}$ **21.** $-2\sqrt{3}/3$ **23.** $-2\sqrt{3}/3$ **25.** $\frac{1}{2}$ **27.** $\sqrt{2}/2$
29. -1 **31.** $\sqrt{3}/3$ **33.** $-\sqrt{2}$ **35.** $\sqrt{2}/2$ **37.** -1 **39.** $-2\sqrt{3}/3$ **41.** Q_{I}
43. Q_{IV} **45.** Q_{II} **47.** Q_{II}

	$\sin\theta$	$\cos\theta$	$\tan\theta$	$\cot\theta$	$\sec\theta$	$\csc\theta$
49.	—	—	$-\frac{8}{15}$	$-\frac{15}{8}$	$-\frac{17}{15}$	$\frac{17}{8}$
51.	—	$\frac{8}{17}$	$\frac{15}{8}$	$\frac{8}{15}$	$\frac{17}{8}$	$\frac{17}{15}$
53.	$\frac{12}{13}$	$-\frac{5}{13}$	$-\frac{12}{5}$	$-\frac{5}{12}$	—	$\frac{13}{12}$
55.	$-\frac{24}{25}$	$-\frac{7}{25}$	$\frac{24}{7}$	—	$-\frac{25}{7}$	$-\frac{25}{24}$
57.	$-\frac{7}{25}$	$\frac{24}{25}$	—	$-\frac{24}{7}$	$\frac{25}{24}$	$-\frac{25}{7}$
59.	$\frac{1}{2}$	$-\sqrt{3}/2$	$-\sqrt{3}/3$	$-\sqrt{3}$	$-2\sqrt{3}/3$	—
61.	$-\frac{24}{25}$	—	$\frac{24}{7}$	$\frac{7}{24}$	$-\frac{25}{7}$	$-\frac{25}{24}$
63.	$\frac{12}{13}$	—	$\frac{12}{5}$	$\frac{5}{12}$	$\frac{13}{5}$	$\frac{13}{12}$
65.	$2\sqrt{5}/5$	$-\sqrt{5}/5$	-2	$-\frac{1}{2}$	$-\sqrt{5}$	—
67.	—	$\frac{3}{4}$	$-\sqrt{7}/3$	$-3\sqrt{7}/7$	$\frac{4}{3}$	$-4\sqrt{7}/7$
69.	0.84	−0.55	—	−0.65	−1.83	1.19
71.	—	−0.85	−0.62	−1.61	−1.18	1.89
73.	—	$\sqrt{1-u^2}$	$\frac{u\sqrt{1-u^2}}{1-u^2}$	$\frac{\sqrt{1-u^2}}{u}$	$\frac{\sqrt{1-u^2}}{1-u^2}$	$\frac{1}{u}$
75.	$\frac{u\sqrt{1+u^2}}{1+u^2}$	$\frac{\sqrt{1+u^2}}{1+u^2}$	—	$\frac{1}{u}$	$\sqrt{1+u^2}$	$\frac{\sqrt{1+u^2}}{u}$

77. $-\tan t$ **79.** $-\cot\theta$ **81.** $-\cot\frac{1}{t}$ **83.** $\frac{\sqrt{2}}{2}$ **85.** $\frac{\sqrt{3}}{3}$ **87.** $\frac{-2\sqrt{3}}{3}$ **89.** $\frac{-2\sqrt{3}}{3}$

PROBLEM SET 5.5 page 285

	θ	$\sin\theta$	$\cos\theta$	$\tan\theta$	$\cot\theta$	$\sec\theta$	$\csc\theta$
1.	43°	0.68200	0.73135	0.93252	1.07236	1.36733	1.46628
3.	119.23°	0.87267	−0.48832	−1.78709	−0.55957	−2.04785	1.14591
5.	326.21°	−0.55615	0.83108	−0.66919	−1.49435	1.20325	−1.79807
7.	−103°8′12″	−0.97383	−0.22727	4.28482	0.23338	−4.39996	−1.02687
9.	$\pi/5$	0.58779	0.80902	0.72655	1.37638	1.23607	1.70130
11.	$-3\pi/7$	−0.97493	0.22252	−4.38129	−0.22824	4.49396	−1.02572
13.	1.588	0.99985	−0.01720	−58.12139	−0.01721	−58.12999	1.00015
15.	12.407	−0.15870	0.98733	−0.16073	−6.22147	1.01284	−6.30132
17.	−8.764	−0.61373	−0.78952	0.77735	1.28642	−1.26660	−1.62938
19.	16.574	−0.76177	−0.64785	1.17584	0.85046	−1.54357	−1.31274

21. 57° **23.** 31° **25.** 84°30′ **27.** 85.2° **29.** 80.2° **31.** 1.12 **33.** 1.22 **35.** 0.60
37. 0.70 **39.** 1.31 **41.** 0.8572 **43.** 0.0729 **45.** −2.971 **47.** 0.1790 **49.** −3.747
51. 16.458 **53.** 0.8192 **55.** 6.314 **57.** 2.366 **59.** 0.7660 **61.** 0.2079 **63.** −8.144
65. −1.305 **67.** −0.9902 **69.** −2.176 **71.** −0.9949 **73.** 2.572 **75.** −0.0168
77. 0.8873 **79.** −0.6682 **81.** 1.165 **83.** 1.202 **85.** 0.9870

PROBLEM SET 5.6 page 298

1. Period = 2π; Amplitude 10
Phase shift = 0

3. Period = 2π; Amplitude = 2
Phase shift = 0

5. Period = 2π; Amplitude = 1
Phase shift = 0

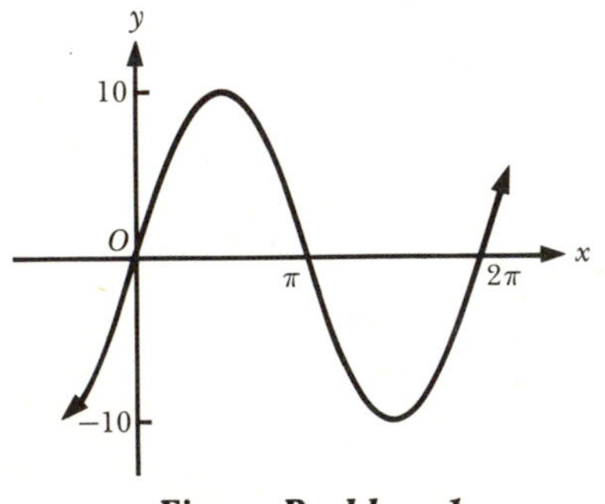

Figure Problem 1

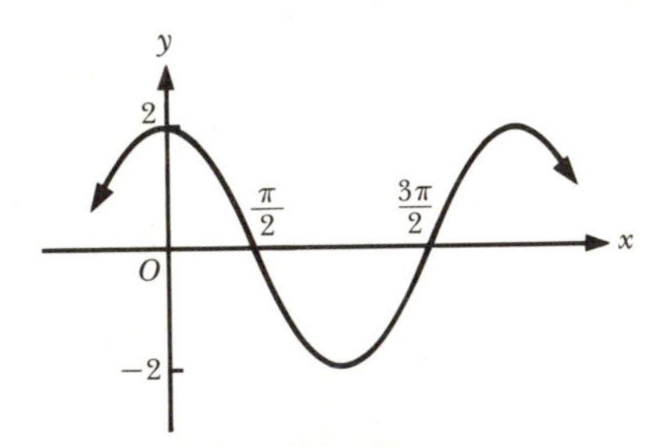

Figure Problem 3

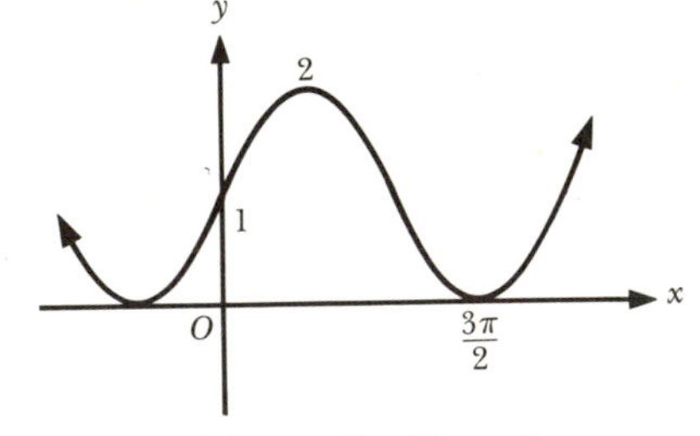

Figure Problem 5

7. Period = 2π; Amplitude = 1
Phase shift = 0

9. Period = 6; Amplitude = 1
Phase shift = 0

11. Period = $\pi/5$; Amplitude = 5
Phase shift = 0

13. Period = $\frac{2}{3}$; Amplitude = 3
Phase shift = 0

15. Period $\pi/3$; Amplitude = $\frac{5}{3}$
Phase shift = 0

17. Period = $2\pi/5$; Amplitude = π
Phase shift = 0

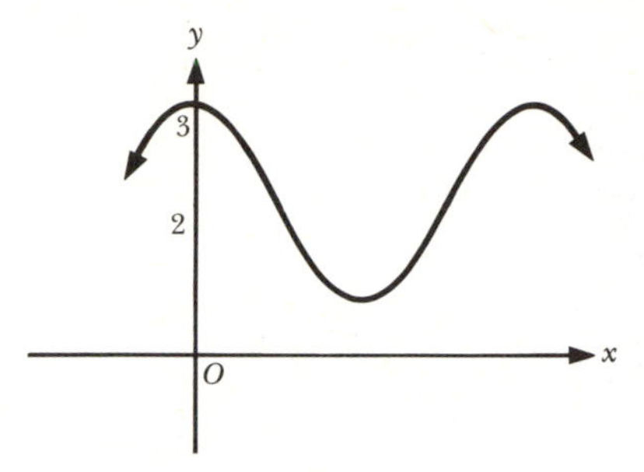

Figure Problem 7

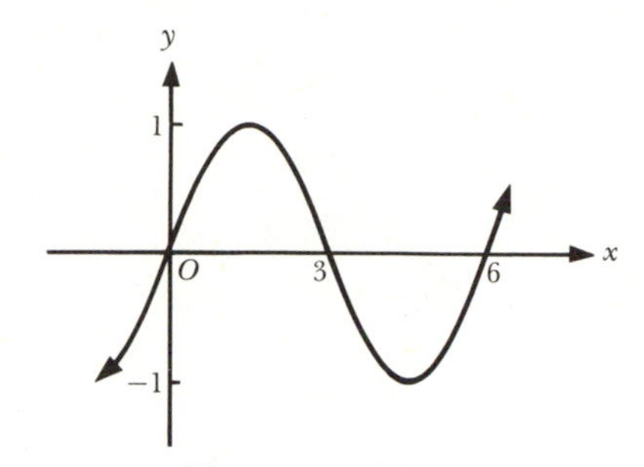

Figure Problem 9

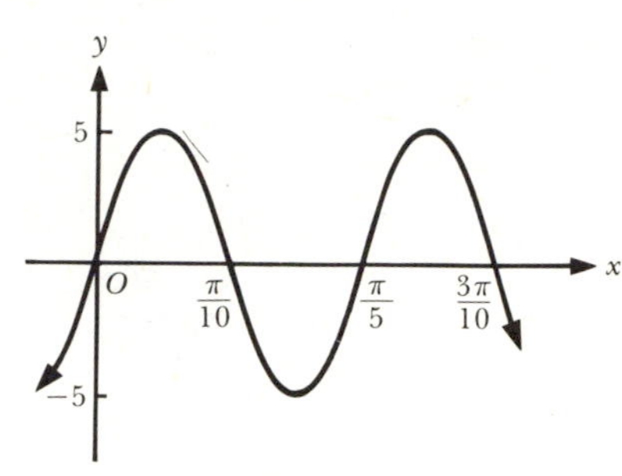

Figure Problem 11

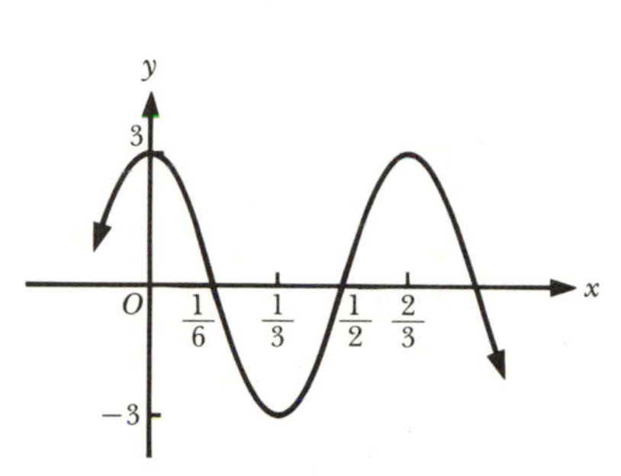

Figure Problem 13

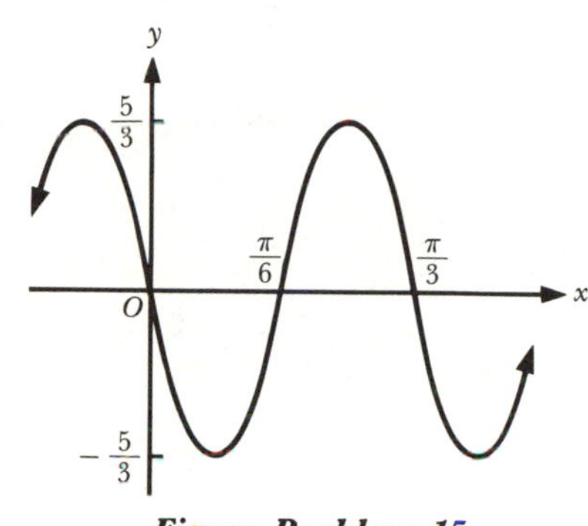

Figure Problem 15

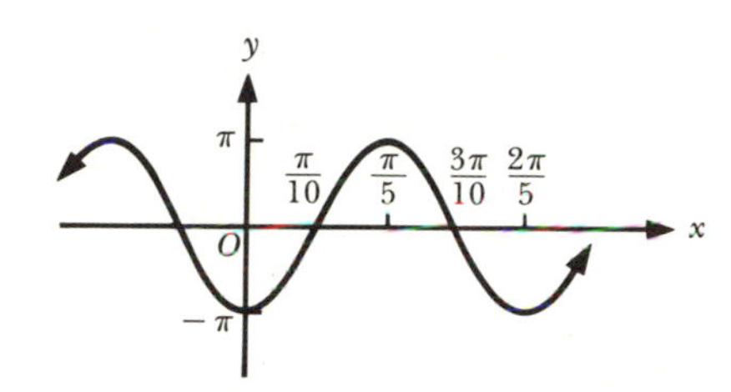

Figure Problem 17

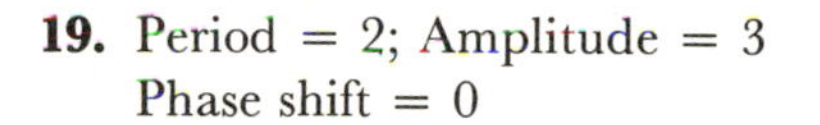

19. Period = 2; Amplitude = 3
Phase shift = 0

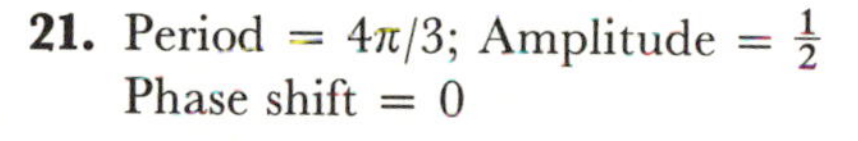

21. Period = $4\pi/3$; Amplitude = $\frac{1}{2}$
Phase shift = 0

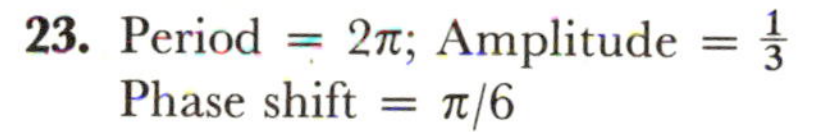

23. Period = 2π; Amplitude = $\frac{1}{3}$
Phase shift = $\pi/6$

25. Period = 4π; Amplitude = 3
Phase shift = $-\pi/2$

27. Period = $2\pi/3$; Amplitude = 2
Phase shift = $\frac{5}{3}$

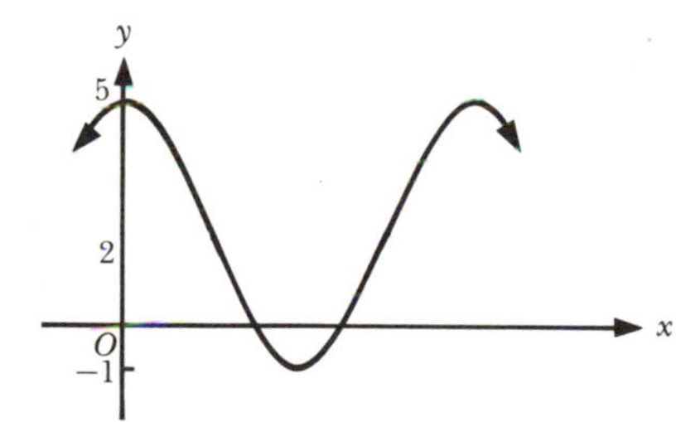

Figure Problem 19

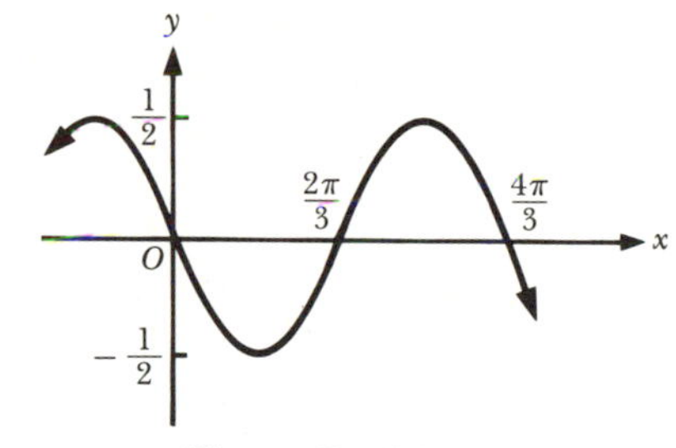

Figure Problem 21

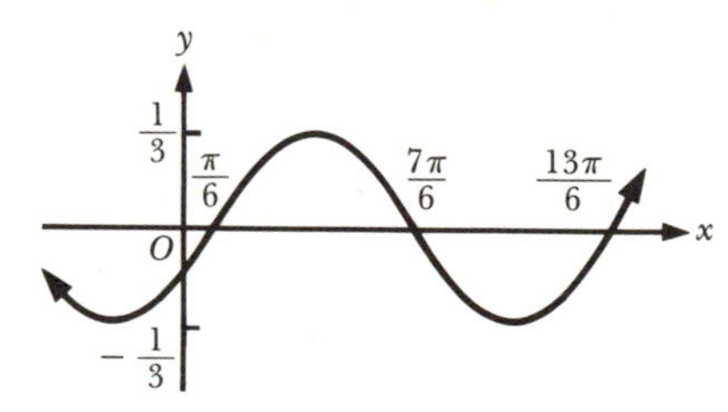

Figure Problem 23

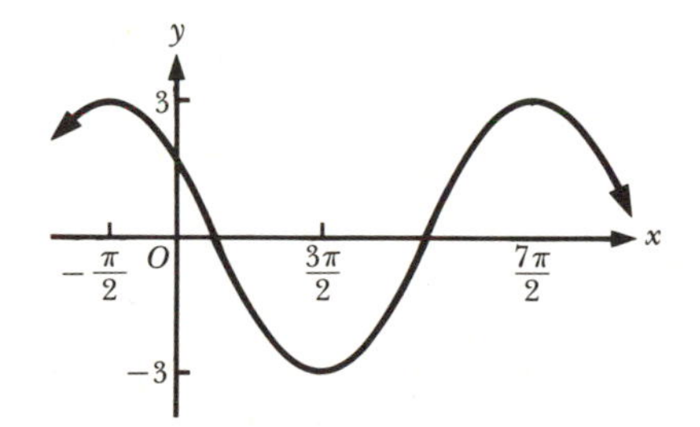

Figure Problem 25

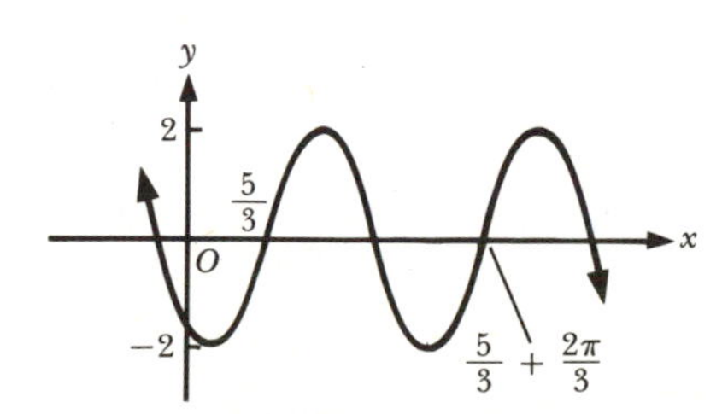

Figure Problem 27

29.

$y = 2\sin(2x - \frac{\pi}{2})$ $y = \sin 2x$ $y = \sin x$ $y = \sin(2x - \frac{\pi}{2})$

31.

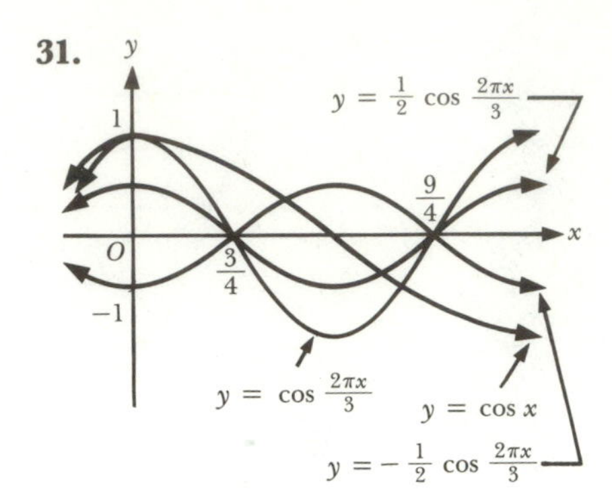

37. Amplitude = 1; period = $\frac{12}{5}$; phase angle = 0; phase shift = 0; angular frequency = $5\pi/6$.

39. Amplitude = 4; period = 8; phase angle = $\frac{\pi}{2}$; phase shift = 2; angular frequency = $\pi/4$.

41. (a) 3; (b) 5; (c) 1; (d) $2\pi/5$; (e) $\frac{1}{5}$ **43.** (a) 1.57; (b) 0.64; (c) 4

45. (a) 1.00; (b) 1.00; (c) 6.26 **47.** (a) 0.01885 second; (b) 53.05; (c) 0.0047; (d) 0.0126

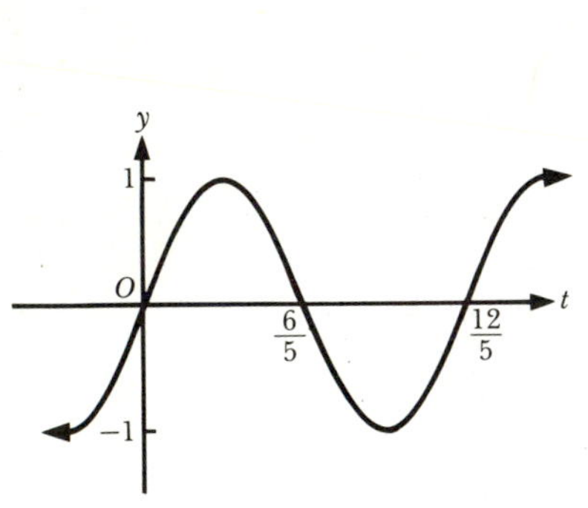

Figure Problem 37

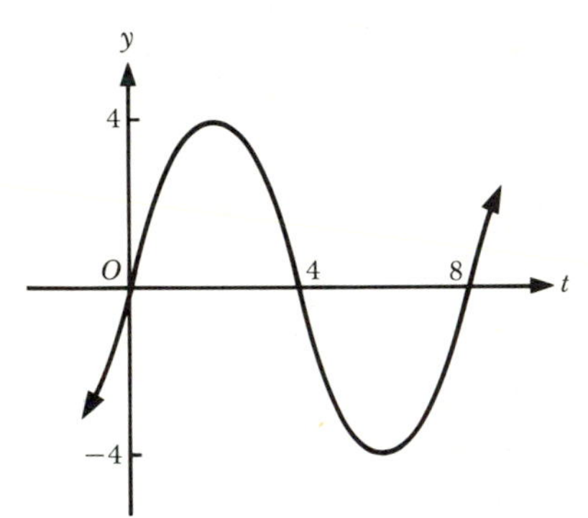

Figure Problem 39

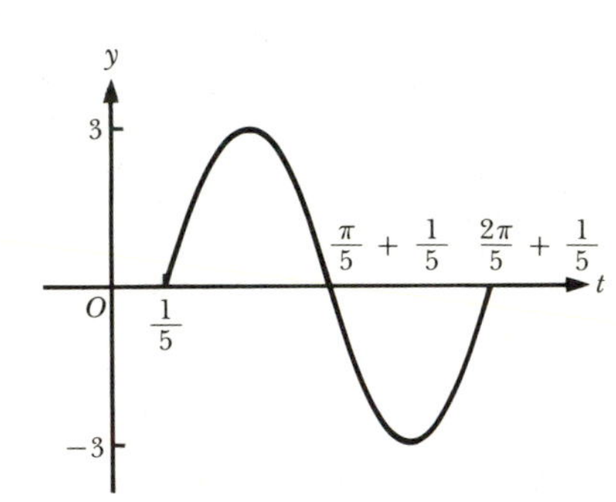

Figure Problem 41 (f)

PROBLEM SET 5.7 page 308

1.

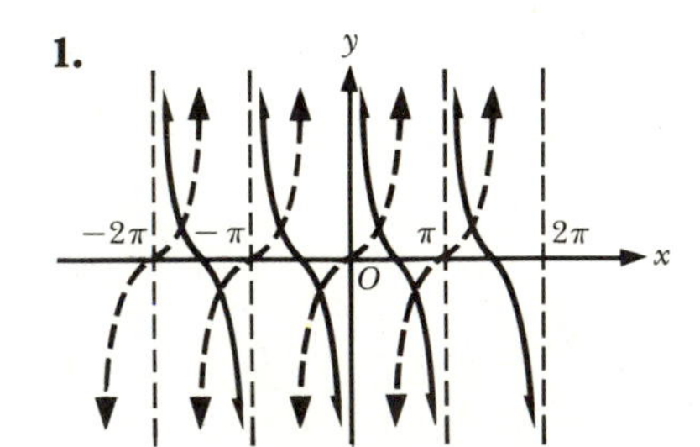

	(i)	(ii)	(iii)	(iv)
cot x	decreasing	decreasing	decreasing	decreasing

3.

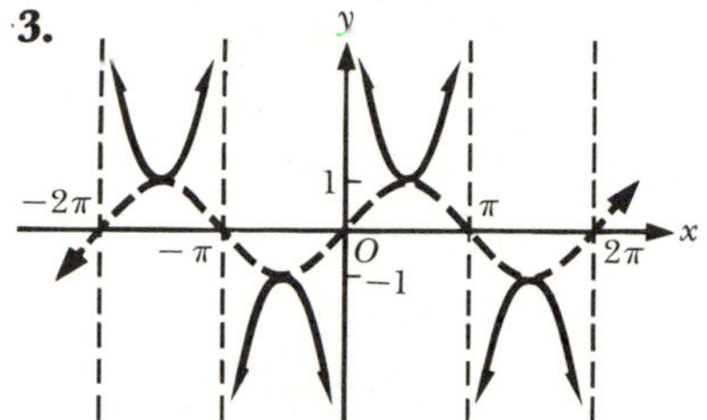

	(i)	(ii)	(iii)	(iv)
csc x	decreasing	increasing	increasing	decreasing

5. π

7.

	0 to $\frac{\pi}{2}$	$\frac{\pi}{2}$ to π	π to $\frac{3\pi}{2}$	$\frac{3\pi}{2}$ to 2π
tan x	—	increasing	—	increasing
cot x	decreasing	—	decreasing	decreasing
sec x	increasing	increasing	decreasing	—
csc x	decreasing	increasing	increasing	decreasing

9.

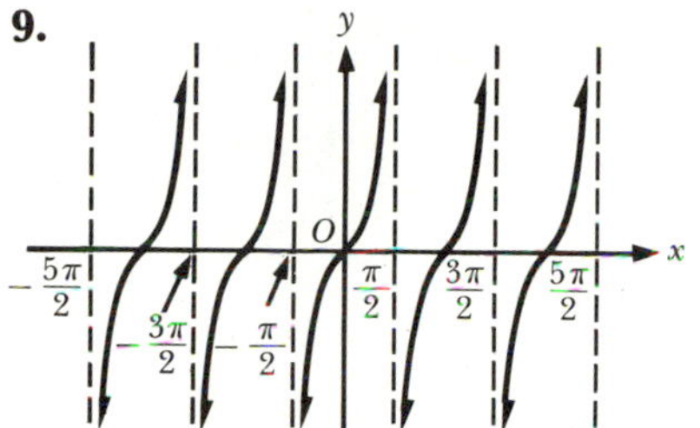

11.

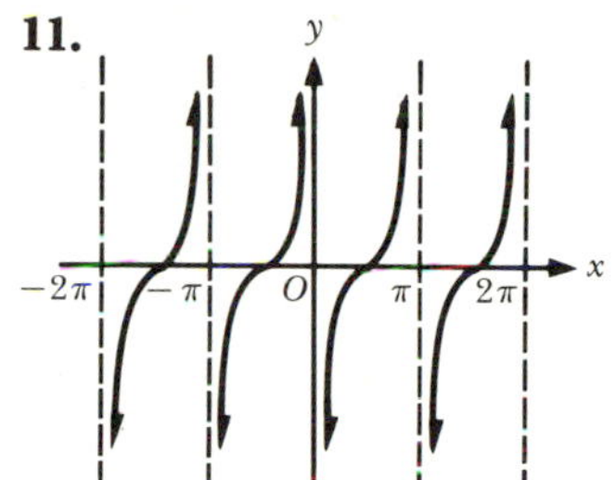

13.

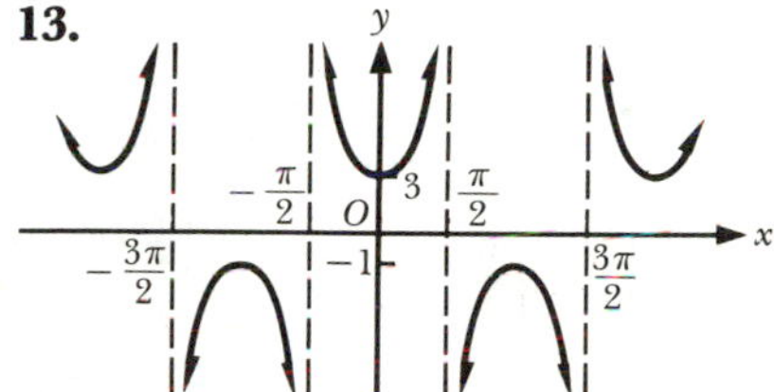

15.

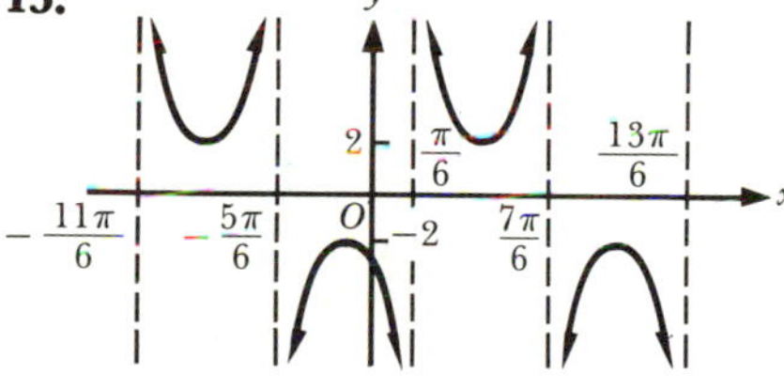

17.

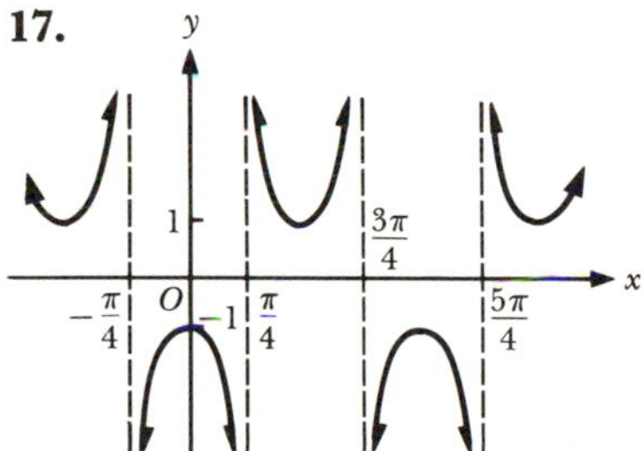

19.

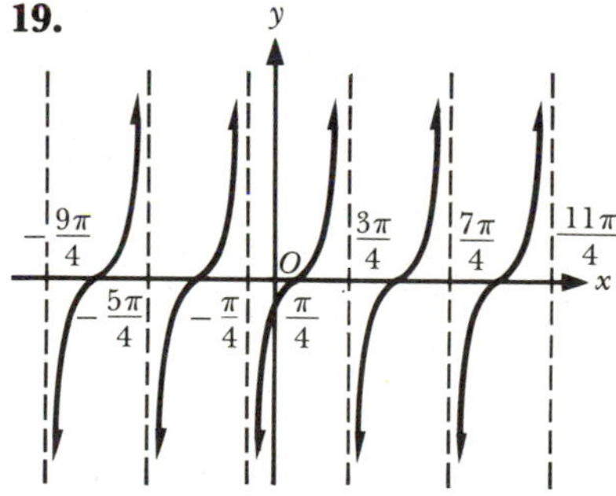

21.

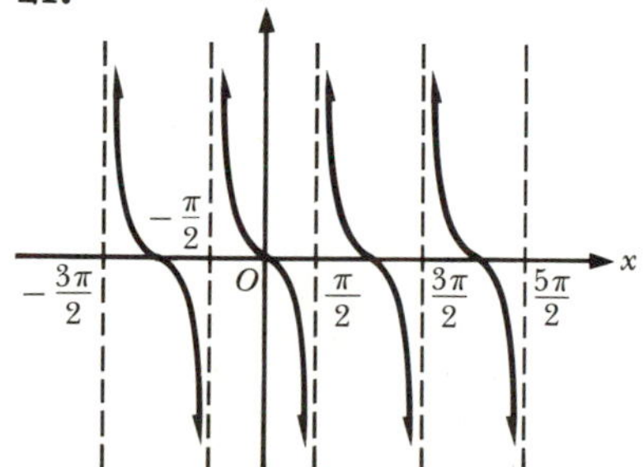

27.

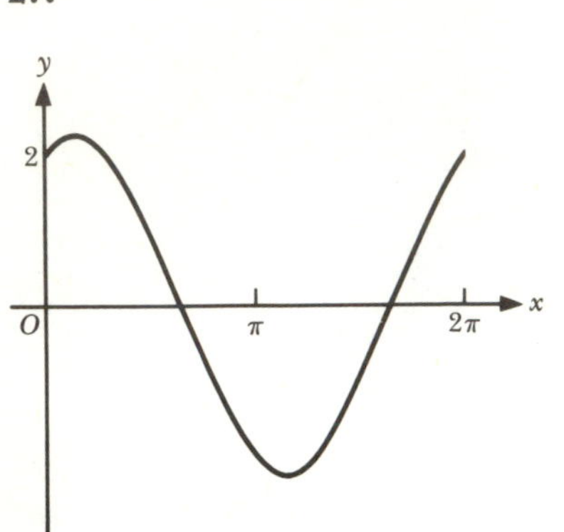

29.

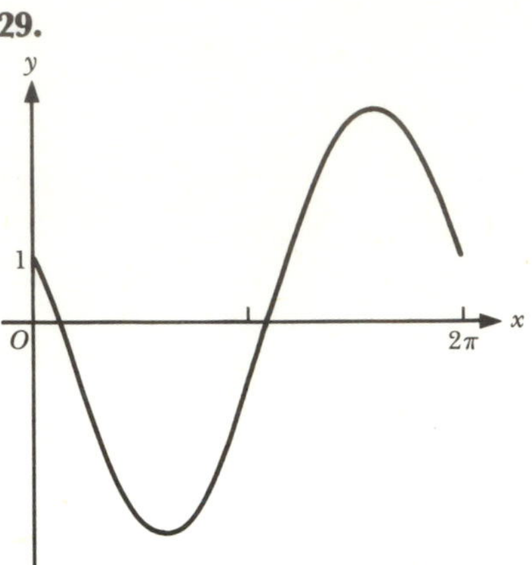

31.

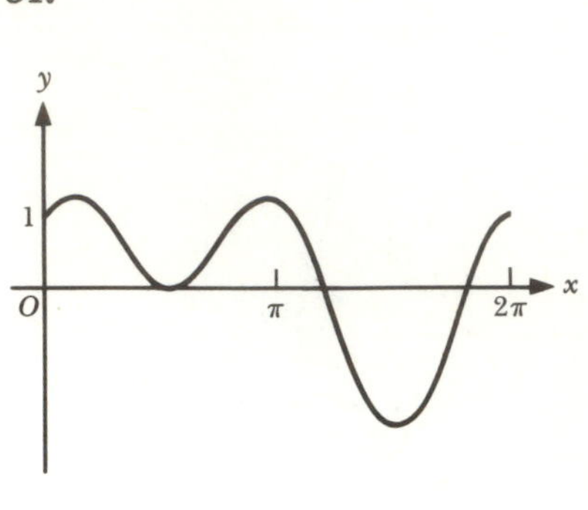

33.

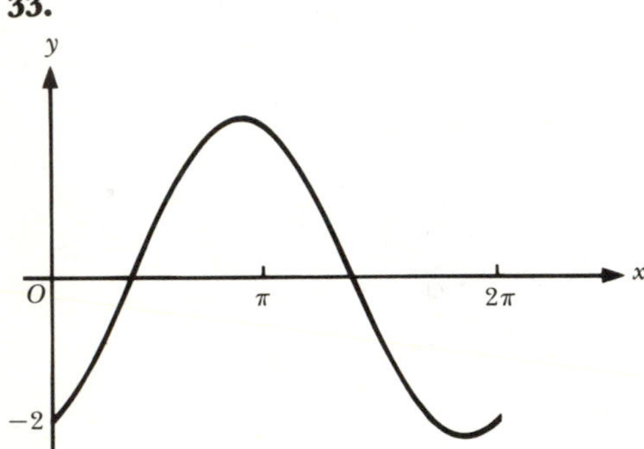

REVIEW PROBLEM SET, CHAPTER 5 page 309

1. y axis **3.** Q_{II} **5.** Q_{III}

	$\sin t$	$\cos t$	$\tan t$	$\cot t$	$\sec t$	$\csc t$
7.	$-\frac{1}{2}$	$-\sqrt{3}/2$	$\sqrt{3}/3$	$\sqrt{3}$	$-2\sqrt{3}/3$	-2
9.	$-\sqrt{10}/10$	$3\sqrt{10}/10$	$-\frac{1}{3}$	-3	$\sqrt{10}/3$	$-\sqrt{10}$

11. $29\pi/18$ **13.** $41\pi/9$ **15.** $-7\pi/18$ **17.** $19\pi/18$ **19.** $300°$ **21.** $-585°$ **23.** $-462.86°$
25. $333.63°$ **27.** $35\pi/12$, $245\pi/24$ **29.** 2π, 10π **31.** 2.5, $143.31°$, 20
33. 12π centimeters per second, 4π radians per second

	$\sin\theta$	$\cos\theta$	$\tan\theta$	$\cot\theta$	$\sec\theta$	$\csc\theta$
35.	$-\frac{4}{5}$	$-\frac{3}{5}$	$\frac{4}{3}$	$\frac{3}{4}$	$-\frac{5}{3}$	$-\frac{5}{4}$
37.	$-6\sqrt{61}/61$	$5\sqrt{61}/61$	$-\frac{6}{5}$	$-\frac{5}{6}$	$\sqrt{61}/5$	$-\sqrt{61}/6$
39.	0.83	−0.56	−1.47	−0.68	−1.79	1.20
41.	$\frac{7}{25}$	$\frac{24}{25}$	$\frac{7}{24}$	$\frac{24}{7}$	$\frac{25}{24}$	$\frac{25}{7}$

43. 2 **45.** $\sqrt{3}/2$ **47.** $-\sqrt{3}$ **49.** $-\sqrt{3}/2$ **51.** 1 **53.** $-\sqrt{3}$ **55.** -2 **57.** 2
59. $\sqrt{3}$ **61.** $-\sqrt{2}/2$ **63.** Q_{III}

	$\sin\theta$	$\cos\theta$	$\tan\theta$	$\cot\theta$	$\sec\theta$	$\csc\theta$
65.	—	$-\sqrt{5}/3$	$-2\sqrt{5}/5$	$-\sqrt{5}/2$	$-3\sqrt{5}/5$	$\frac{3}{2}$
67.	$\frac{1}{5}$	$2\sqrt{6}/5$	$\sqrt{6}/12$	$2\sqrt{6}$	$5\sqrt{6}/12$	—
69.	-0.92	-0.38	2.42	0.41	—	-1.09
71.	$-6\sqrt{61}/61$	$-5\sqrt{61}/61$	—	$\frac{5}{6}$	$-\sqrt{61}/5$	$-\sqrt{61}/6$
73.	$7\sqrt{674}/674$	$-25\sqrt{674}/674$	$-\frac{7}{25}$	—	$-\sqrt{674}/25$	$\sqrt{674}/7$

75. 0.14112 **77.** -0.30902 **79.** -0.56962 **81.** 0.48055 **83.** 1.00982
85. -1.05832 **87.** -0.92004 **89.** 0.91402 **91.** -1.83922 **93.** 66° **95.** $2\pi/11$

97. Amplitude $= \frac{1}{3}$; period $= 2\pi$; phase shift $= 0$

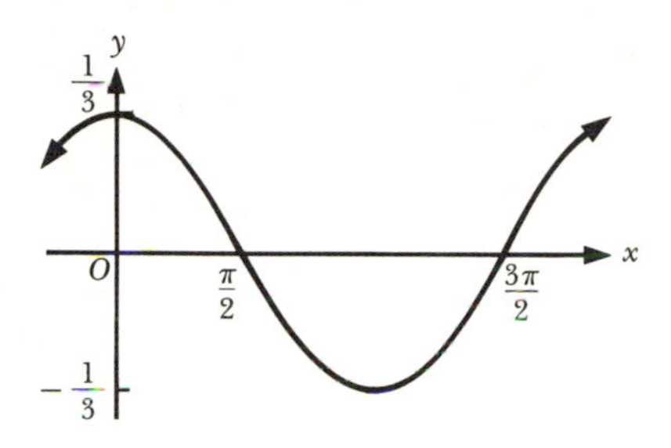

99. Amplitude $= 2$; period $= \pi$; phase shift $= 0$

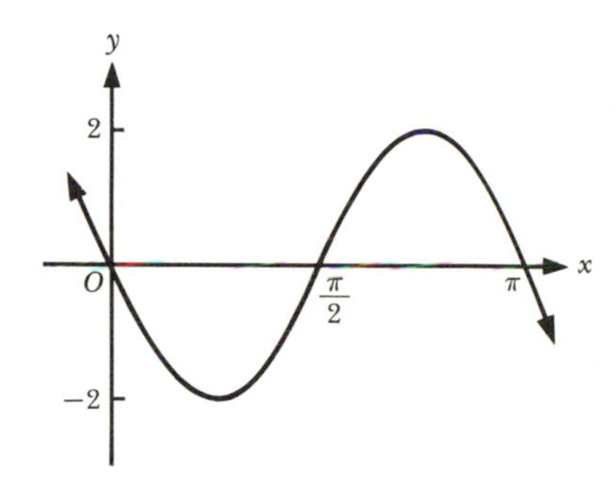

101. Amplitude $= \frac{1}{5}$; Period $= \pi$; phase shift $= -\pi/16$

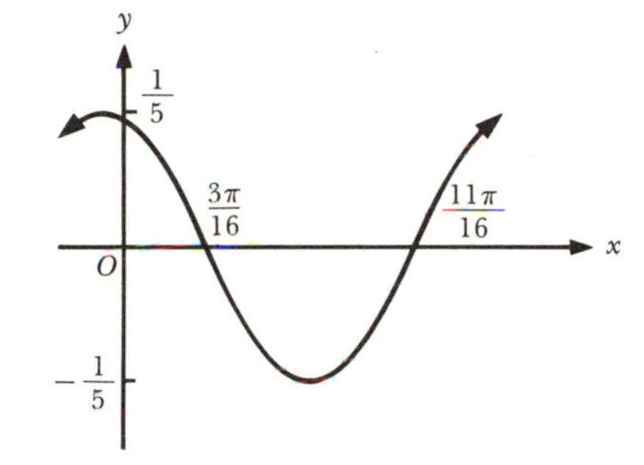

103.

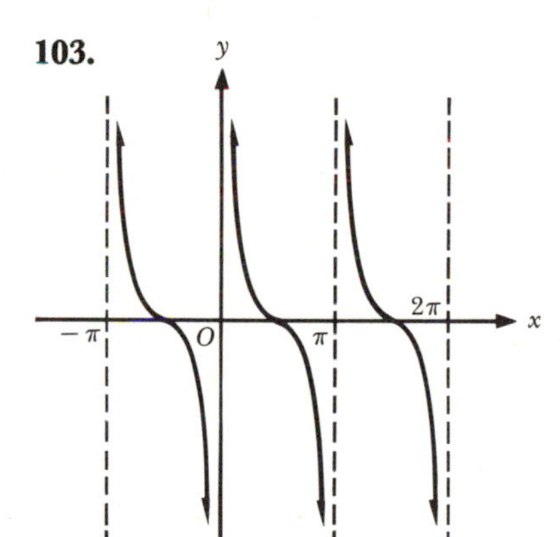

105.

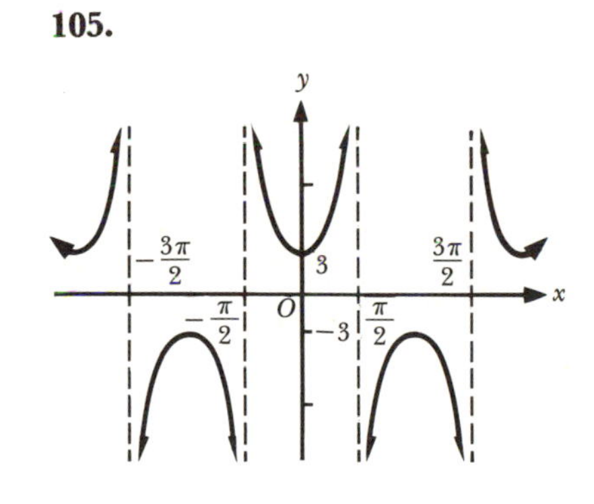

107.

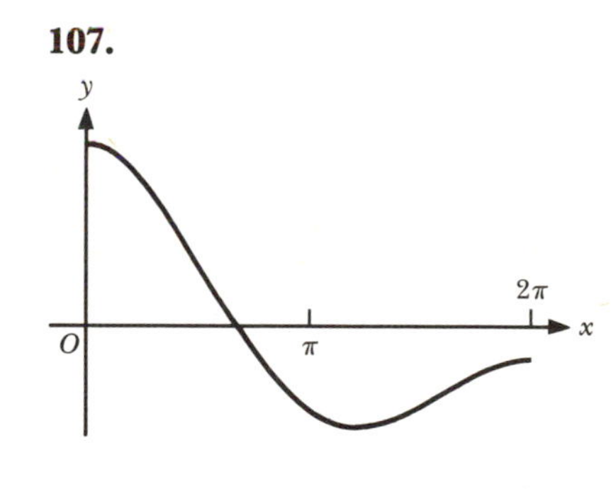

109. $\dfrac{300}{\pi}$ **111.** (a) 2; (b) 0.05; (c) 3; (d) 40π; (e) 60

113. $y = 12 \sin 1.1\pi t$

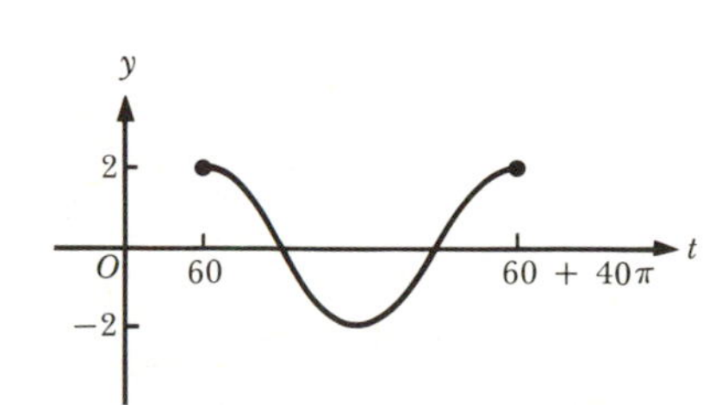

Figure Problem 111 (f)

CHAPTER 6

PROBLEM SET 6.1 page 317

1. 1 **3.** $\cos^2\theta$ **5.** 1 **7.** 1 **9.** 0 **11.** 1 **13.** $\sin^2\theta$ **15.** 1 **17.** $\sin t$
19. $\sec\theta$ **21.** 1 **23.** $\cos\theta$ **25.** $\dfrac{1}{(1-\cos\theta)\cos\theta}$ **27.** $2\csc t$ **29.** $\dfrac{1}{\cos\theta}$ **31.** $\dfrac{\sin\theta}{\cos\theta}$
33. $\cos t$ **35.** $2\cos\theta$ **37.** $3\sec\theta$ **39.** $5\tan\theta$ **41.** $27\sec^3\theta$ **43.** $2\sec\theta$

PROBLEM SET 6.2 page 328

1. $\dfrac{\sqrt{6}+\sqrt{2}}{4}$ **3.** $-\dfrac{\sqrt{6}+\sqrt{2}}{4}$ **5.** $-\sqrt{3}-2$ **7.** $\dfrac{\sqrt{6}+\sqrt{2}}{4}$
9. $-\dfrac{\sqrt{2}+\sqrt{6}}{4}$ **11.** $\cos\theta$ **13.** $\tan\theta$ **15.** $\sin\theta$ **17.** $\sec\theta$ **19.** $-\cot\theta$
21. $\sqrt{3}/2$ **23.** $\sqrt{3}/2$ **25.** 0 **27.** $\sin t$ **29.** $\sqrt{3}$
31. (a) $-\frac{63}{65}$; (b) $-\frac{16}{65}$; (c) $-\frac{33}{65}$; (d) $\frac{56}{65}$; (e) $\frac{63}{16}$; (f) $\frac{65}{56}$
33. (a) $\frac{304}{425}$; (b) $-\frac{297}{425}$; (c) $\frac{416}{425}$; (d) $\frac{-87}{425}$; (e) $\frac{-304}{297}$; (f) $-\frac{425}{87}$ **35.** $\sin(s+t)$ **37.** $-\csc t$
55. $y=\sqrt{2}\cos[s-(\pi/4)]$ **57.** $\frac{1}{2}(\cos 3w-\cos 5w)$ **59.** $\frac{1}{2}(\cos 4v+\cos 12v)$ **61.** $\frac{1}{2}(\cos 2t+\cos 2s)$
63. $2\sin(5\theta/2)\cos(\theta/2)$ **65.** $2\cos 3t\cos t$

PROBLEM SET 6.3 page 336

1. $2\sin 40^\circ\cos 40^\circ$ **3.** $\cos^2 10^\circ-\sin^2 10^\circ$ or $2\cos^2 10^\circ-1$ or $1-2\sin^2 10^\circ$ **5.** $\dfrac{2\tan(\pi/9)}{1-\tan^2(\pi/9)}$
7. $\sin 6t$ **9.** $\cos 14t$ **11.** $\cos 8t$ **13.** $2\cos\theta$ **15.** $2\cos\theta$ **17.** $\tan 8\theta$
19. (a) $\frac{24}{25}$; (b) $-\frac{7}{25}$; (c) $-\frac{24}{7}$ **21.** (a) $\frac{336}{625}$; (b) $-\frac{527}{625}$; (c) $-\frac{336}{527}$ **23.** (a) $\frac{-120}{169}$; (b) $\frac{119}{169}$; (c) $-\frac{120}{119}$;
25. $\dfrac{\sqrt{2+\sqrt{2}}}{2}$ **27.** $-\dfrac{\sqrt{2+\sqrt{3}}}{2}$ **29.** $\dfrac{\sqrt{2+\sqrt{3}}}{2}$ **31.** $\sqrt{7-4\sqrt{3}}$ **33.** (a) $\frac{3}{5}$; (b) $\frac{4}{5}$; (c) $\frac{3}{4}$
35. (a) $\dfrac{3\sqrt{10}}{10}$; (b) $\dfrac{\sqrt{10}}{10}$; (c) 3 **37.** $\cos 40^\circ$ **39.** $-\sin 70^\circ$ **41.** $\tan(\pi/5)$

PROBLEM SET 6.4 page 345

1. $\pi/6$ or 30° **3.** $2\pi/3$ or 120° **5.** $-\pi/2$ or -90° **7.** $-\pi/4$ or -45° **9.** $\pi/4$ or 45°
11. $-\pi/6$ or -30° **13.** 0.22 **15.** 1.36 **17.** -0.89 **19.** 2.58 **21.** 1.25 **23.** $\pi/3$ or 60°
25. $\sqrt{3}/2$ **27.** $\frac{1}{2}$ **29.** $\frac{3}{5}$ **31.** $\frac{4}{5}$ **33.** $\dfrac{\sqrt{6}+\sqrt{2}}{4}$ **35.** $x=\dfrac{\cos y}{2}$ **37.** $x=\tan y-1$
39. $\sqrt{3}/2$ **41.** -1 **43.** $-\frac{24}{7}$ **45.** $\cos t=\frac{3}{5}$; $\cos 2t=-\frac{7}{25}$ **47.** $\sec v=\sqrt{5}/2$; $\tan 2v=-\frac{4}{3}$

55.

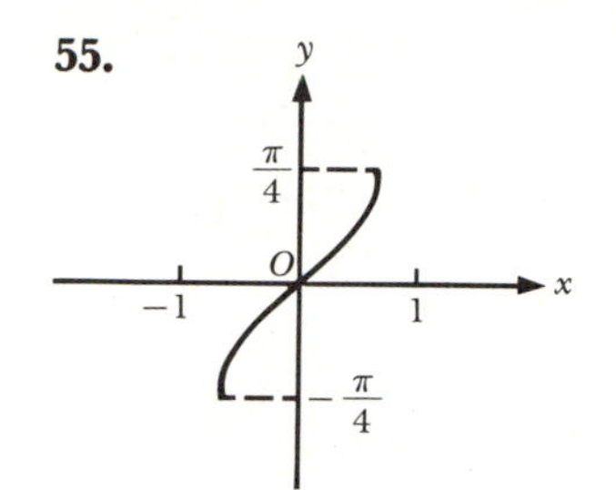

57.

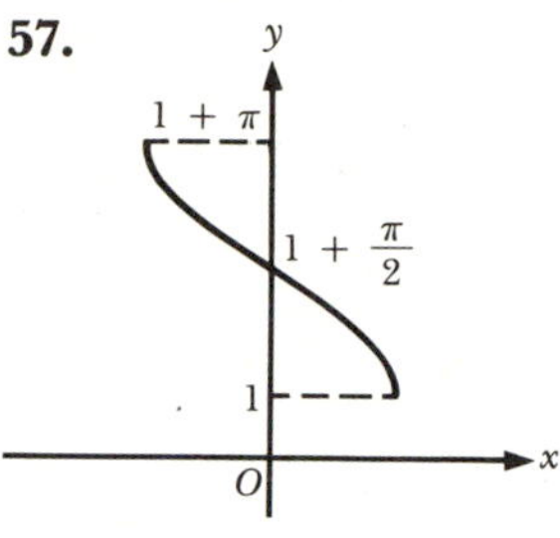

59. True **61.** False
63. Domain of $y = \sin^{-1} x$ is $[-1, 1]$, and π is not in this interval

65. (a)

Domain is $[1, \infty)$ and $(-\infty, -1]$; range is $0 \leq y \leq \pi$, $y \neq \pi/2$

PROBLEM SET 6.5 page 352

1. (a) $\pi/3$; (b) $\pi/3$; $5\pi/3$; (c) $(\pi/3) + 2n\pi$, $(5\pi/3) + 2n\pi$, n is any integer **3.** $\pi/3$, $2\pi/3$ **5.** 135°, 315° **7.** $\pi/4$, $7\pi/4$ **9.** 90°, 270° **11.** 60°, 240° **13.** $\pi/4$, $3\pi/4$ **15.** $2\pi/3$, $4\pi/3$ **17.** 240°, 300° **19.** $\pi/3$, $2\pi/3$ **21.** $2\pi/3$, $4\pi/3$ **23.** 0.95, 2.19 **25.** 114.68°, 245.32° **27.** 0.69, 3.83 **29.** 8.59°, 188.59° **31.** 45°, 75°, 165°, 195°, 285°, 315° **33.** $\pi/8$, $7\pi/8$, $9\pi/8$, $15\pi/8$ **35.** $\pi/2$ **37.** 60°, 120°, 240°, 300° **39.** 60°, 120°, 240°, 300° **41.** 0, π **43.** 90°, 240°, 300°, 270° **45.** 30°, 90°, 270°,150° **47.** $\pi/4$, $3\pi/4$, $5\pi/4$, $7\pi/4$ **49.** 0°, 120°, 240° **51.** 60°, 90°, 270°, 300° **53.** 45°, 225° **55.** 0, $\pi/6$, $5\pi/6$, π **57.** 90°, 210°, 330° **59.** $\pi/4$, $5\pi/4$ **61.** 30°, 150°, 270° **63.** 60°, 120°, 240°, 300° **65.** 0°, 270° **67.** 0.20, 0.34, 2.80, 2.94 **69.** 19.47°, 41.81°, 138.19°, 160.53° **71.** 0.64, 2.50, 4.71 **73.** 11.92° **75.** 62.04°

PROBLEM SET 6.6 page 358

1. $c = 5.831$; $\alpha = 59.04°$; $\beta = 30.96°$ **3.** $a = 6.428$; $b = 7.660$; $\alpha = 40°$ **5.** $b = 7.216$; $c = 18.47$; $\alpha = 67°$ **7.** $b = 30.28$; $c = 37.43$; $\beta = 54°$ **9.** $b = 98.90$; $\alpha = 36.79°$; $\beta = 53.21°$ **11.** $a = 14.03$; $b = 43.95$; $\beta = 72.3°$ **13.** $a = 25.21$; $b = 96.77$; $\beta = 75.4°$ **15.** 10.68 inches **17.** 2.113 meters **19.** 46.06 feet **21.** 75.96° **23.** 170.2 kilometers **25.** 1.339 kilometers **27.** 346.4 feet **29.** 7.52° **31.** 42.26 meters **33.** 13,740 meters **35.** 8507 meters **37.** 188.4 meters **39.** 2.285 kilometers **41.** 3990.6 feet

PROBLEM SET 6.7 page 367

1. 12.25 **3.** 3.264 **5.** no triangle, since $\beta > 1$
7. $\beta_1 = 36.82°$; $\gamma_1 = 121.18°$; $c_1 = 11.42$; $\beta_2 = 143.18°$; $\gamma_2 = 14.82°$; $c_2 = 3.414$

9. $\beta = 22.48°$; $\gamma = 122.52°$; $c = 13.23$ **11.** 17.78 **13.** 26.38° **15.** 15.62 **17.** 38.08 **19.** 104.48° **21.** 61.77 **23.** 0.8144 kilometer; 1.081 kilometers **25.** 414.6 meters **27.** 258.6 meters **29.** 23.49 miles **31.** 193.6 miles **33.** 756.2 meters **35.** 42.19 nautical miles **37.** 16.25 square centimeters **39.** 0.6681 square kilometers

PROBLEM SET 6.8 page 376

1. $(3\sqrt{3}, 3)$ **3.** $\left(-\frac{7}{2}, \frac{7\sqrt{3}}{2}\right)$ **5.** $(2\sqrt{2}, -2\sqrt{2})$ **7.** $(2.67, 1.36)$ **9.** $(-2\sqrt{3}, -2)$ **11.** $(0.76, -1.85)$ **13.** $(2, 120°)$ **15.** $(5, 36.87°)$ **17.** $(5\sqrt{2}, 45°)$ **19.** $(6, 120°)$ **21.** $(5\sqrt{2}, 225°)$ **23.** $(7.07, 188.13°)$ **25.** $y = x$ **27.** $x^2 + (y - \frac{3}{2})^2 = \frac{9}{4}$ **29.** $x = 3$ **31.** $r = 5$ **33.** $\theta = \tan^{-1} 3$ **35.** $r = 6\cos\theta$

37.

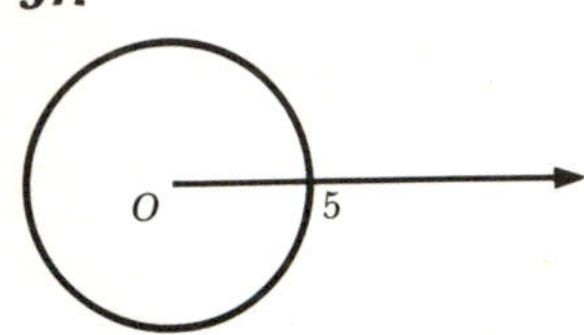

39.

41.

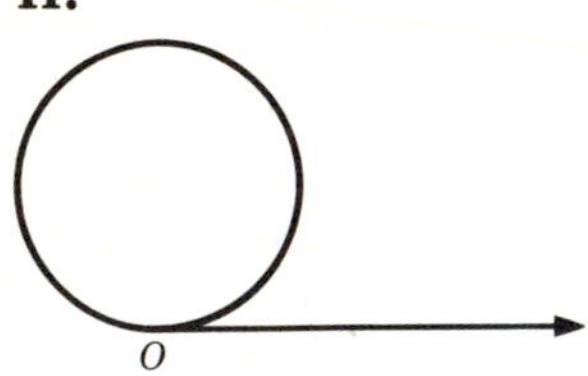

43.

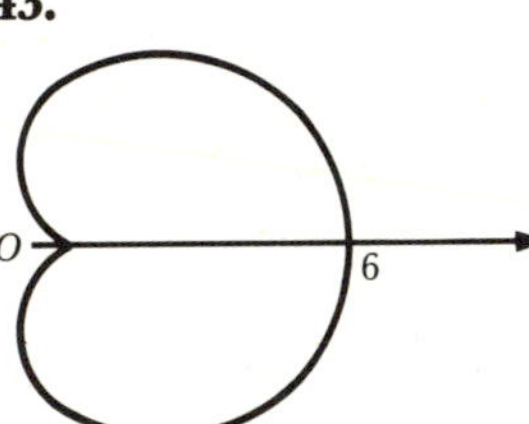

45.

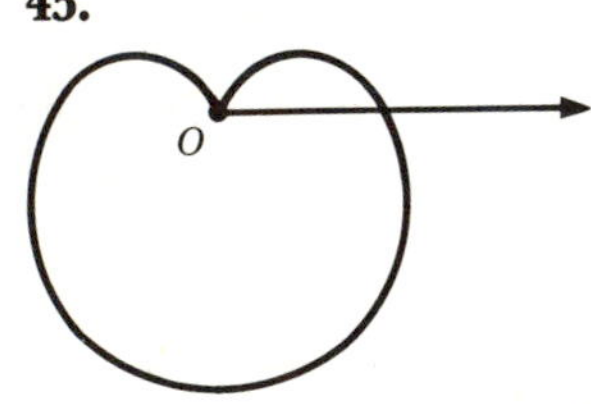

47.

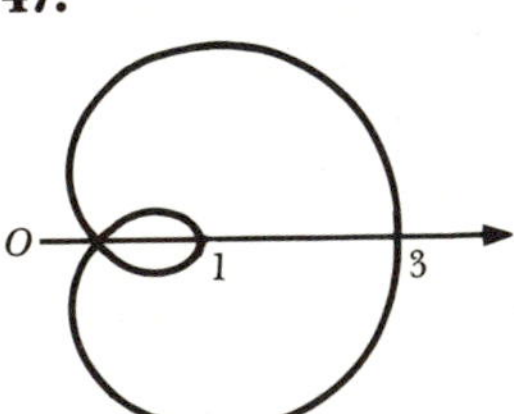

PROBLEM SET 6.9 page 385

1. $\sqrt{53}$ **3.** $\sqrt{1537}$ **5.** $\frac{\sqrt{1537}}{29}$ **7.** 53 **11.** $\sqrt{2}\left(\cos\frac{5\pi}{4} + i\sin\frac{5\pi}{4}\right)$ **13.** $2\left(\cos\frac{3\pi}{2} + i\sin\frac{3\pi}{2}\right)$ **15.** $2\left(\cos\frac{7\pi}{6} + i\sin\frac{7\pi}{6}\right)$ **17.** $1\left(\cos\frac{2\pi}{3} + i\sin\frac{2\pi}{3}\right)$ **19.** $1.9696 + 0.3473i$ **21.** $0.7764 - 2.8977i$ **23.** $4 + 0i$

	(a) $z_1 z_2$	(b) z_1/z_2
25.	$5(\cos 225° + i\sin 225°) = -\frac{5\sqrt{2}}{2} - \frac{5\sqrt{2}}{2}i$	$5(\cos 115° + i\sin 115°) = -2.1130 + 4.5315i$
27.	$6(\cos 90° + i\sin 90°) = 0 + 6i$	$\frac{2}{3}(\cos 10° + i\sin 10°) = 0.6565 + 0.1157i$
29.	$8\left(\cos\frac{7\pi}{6} + i\sin\frac{7\pi}{6}\right) = -4\sqrt{3} - 4i$	$2\left(\cos\frac{\pi}{2} + i\sin\frac{\pi}{2}\right) = 0 + 2i$
31.	$2(\cos 180° + i\sin 180°) = -2 + 0i$	$1[\cos(-90°) + i\sin(-90°)] = 0 - i$
33.	$12\sqrt{2}[\cos(-195°) + i\sin(-195°)] = -16.3923 + 4.3923i$	$\frac{2\sqrt{2}}{3}[\cos(-285°) + i\sin(-285°)] = 0.2440 - 0.9107i$

35. $-\frac{\sqrt{3}}{2} - \frac{1}{2}i$ **37.** $512 + 512\sqrt{3}i$ **39.** $256 + 0i$ **41.** $-125{,}000i$ **43.** $-8 - 8\sqrt{3}i$
45. -2^{30} **47.** -1 **49.** $\frac{\sqrt{2}}{2} + \frac{\sqrt{2}}{2}i;\ -\frac{\sqrt{2}}{2} - \frac{\sqrt{2}}{2}i$ **51.** $2 + 0i;\ -1 + \sqrt{3}i;\ -1 - \sqrt{3}i$
53. $0.8536 + 0.7686i;\ -0.4672 + 1.0494i;\ -1.1424 - 0.1201i;\ -0.2388 - 1.1236i;\ 0.9948 - 0.5743i$
55. $1 + i\sqrt{3};\ -\sqrt{3} + i;\ -1 - \sqrt{3}i;\ \sqrt{3} - i$ **57.** $1 + i\sqrt{3};\ -2 + 0i;\ \sqrt{3} - i$
59. $\frac{3\sqrt{2}}{2} + \frac{3\sqrt{2}}{2}i;\ \frac{-3\sqrt{2}}{2} + \frac{3\sqrt{2}}{2}i;\ \frac{-3\sqrt{2}}{2} - \frac{3\sqrt{2}}{2}i;\ \frac{3\sqrt{2}}{2} - \frac{3\sqrt{2}}{2}i$
61. $\sqrt{3} + i;\ 0 + 2i;\ -\sqrt{3} + i;\ -\sqrt{3} - i;\ \sqrt{3} - i;\ 0 - 2i$
63. $4096\left(\cos\frac{5\pi}{6} + i\sin\frac{5\pi}{6}\right) = -2048\sqrt{3} + 2048i$ **65.** (a) $-\frac{\sqrt{3}}{2} - \frac{1}{2}i$; (b) $\frac{1}{32} - \frac{1}{32}i$

PROBLEM SET 6.10 page 394

5. $\langle 3, 0\rangle$ **7.** $\langle 5\sqrt{2}/2, 5\sqrt{2}/2\rangle$ **9.** $\langle -\sqrt{3}/2, -\frac{1}{2}\rangle$ **11.** $\langle -0.0883, 0.4922\rangle$ **13.** $2\sqrt{2}$; 135°
15. 5; 180° **17.** 10; 216.87° **19.** $\sqrt{61}$; 320.19° **21.** $\langle -5, -2\rangle$; $\sqrt{29}$ **23.** $\langle 5, -11\rangle$; $\sqrt{146}$
25. $\langle -6, -6\rangle$; $6\sqrt{2}$ **27.** $\langle 0, 20\rangle$ **29.** $5 + 2\sqrt{5}$ **31.** $\langle 8, 16\rangle$ **33.** $\langle 12, -4\rangle$ **35.** 15
37. (a) $(\frac{10}{3}, 2)$; (b) $(1, -1)$ **39.** (a) 3; (b) $-5\sqrt{3}/2$ **41.** (a) 0; (b) -5; (c) -2; (d) 0
43. 55.71 pounds; 20.49° **45.** (a) 12.63; (b) 17.29°
47. (a) 540.0 miles per hour; (b) 3.66°; (c) N81.34°E
49. (a) 51.22 kilometers per hour; (b) 38.66°; (c) N1.34°E

REVIEW PROBLEM SET, CHAPTER 6 page 396

1. $\cot\theta$ **3.** $-\tan\theta$ **5.** 1 **7.** 1 **9.** $216\cos^3\theta$ **19.** $\sin 5t$ **21.** $\tan^2 3t$ **23.** $\tan 6t$
25. $-\cos 8\theta$ **27.** $\frac{\sin 14t}{2}$ **29.** $-\sin t$ **31.** $\frac{\sqrt{2 - \sqrt{2}}}{2}$ **33.** $-\frac{\sqrt{2 + \sqrt{3}}}{2}$ **35.** $\sqrt{3} + 2$
37. $-\frac{240}{289}$ **39.** $-\frac{77}{85}$ **41.** $-\frac{84}{85}$ **43.** $\sqrt{10}/10$ **45.** $-\frac{36}{77}$ **61.** $\frac{1}{2}[\sin 3\theta + \sin 2\theta]$

63. $2\cos\frac{5t}{2}\sin\frac{3t}{2}$ **65.** $\pi/4$ or $45°$ **67.** $-\pi/6$ or $-30°$ **69.** $\pi/4$ or $45°$ **71.** 1.3416

73. 0.7884 **75.** -1.4802 **77.** (a) $\sqrt{2}/2$; (b) $45°$ **79.** $\frac{3}{4}$ **81.** $-\sqrt{3}/3$ **83.** $4\sqrt{5}/9$

87. $x = 2\sin y$ **89.** $x = \frac{1}{4}\sec(y-3)$ **91.** $270°$ **93.** $60°, 240°$ **95.** $210°, 330°$ **97.** $\pi/3, 2\pi/3$

99. $\pi/4, 3\pi/4$ **101.** $0, 2\pi/3, 4\pi/3$ **103.** $0, \pi, \pi/6, 7\pi/6$ **105.** $c = 13$; $\alpha = 22.62°$; $\beta = 67.38°$

107. $\alpha = 25°$; $a = 11.66$; $c = 27.58$ **109.** $\alpha = 39°$; $b = 20.99$; $c = 27.01$

111. $b = 16$; $c = 20$; $\alpha = 36.86°$; $\beta = 53.13°$ **113.** $c = 3.658$; $\alpha = 43.16°$; $\beta = 106.84°$

115. $\gamma = 60°$; $b = 11.15$; $a = 8.165$

117. $\alpha_1 = 101.41°$; $a_1 = 7.842$; $\gamma_1 = 48.59°$; $\alpha_2 = 18.59°$; $a_2 = 2.550$; $\gamma_2 = 131.41°$

119. $\gamma = 54.28°$; $\beta = 10.72°$; $b = 0.9851$ **121.** $\alpha = 87°$; $b = 6.700$; $c = 7.782$ **123.** 1319.43 meters

125. 480.71 feet **127.** 513.48 feet **129.** 31.72 feet **131.** $(5\sqrt{2}/2, 5\sqrt{2}/2)$ **133.** $(-1, -1)$

135. $(0, -3)$ **137.** $(3, 180°)$ **139.** $(10, 210°)$ **141.** $(14, 270°)$ **143.** $x^2 + y^2 + 4y = 0$

145. $x^2(x^2 + y^2) = 16y^2$ **147.** $r = 4$ **149.** $r = 4\cot\theta\csc\theta$

151.

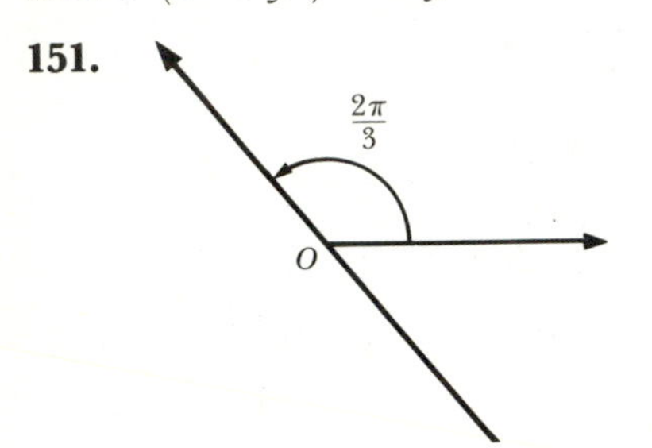

153.

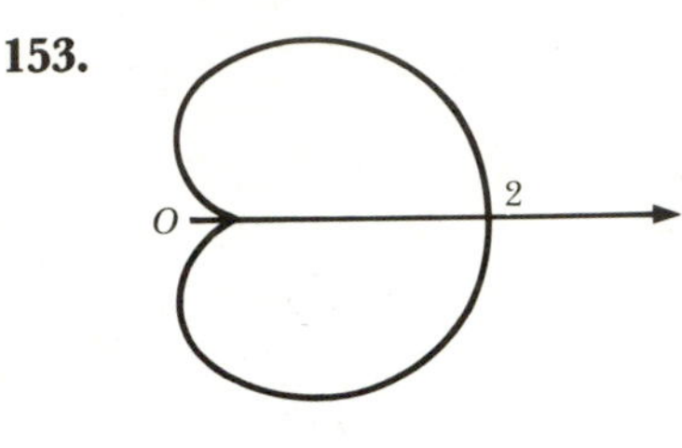

155. $5\sqrt{2}\left(\cos\frac{\pi}{4} + i\sin\frac{\pi}{4}\right)$ **157.** $12\left(\cos\frac{\pi}{6} + i\sin\frac{\pi}{6}\right)$ **159.** $2\sqrt{2}\left(\cos\frac{\pi}{4} + i\sin\frac{\pi}{4}\right)$

161. $5(\cos 53.13° + i\sin 53.13°)$

	(a) z_1z_2	(b) z_1/z_2
163.	$6\left(\cos\frac{3\pi}{2} + i\sin\frac{3\pi}{2}\right) = 0 - 6i$	$\frac{2}{3}\left(\cos\frac{\pi}{2} + i\sin\frac{\pi}{2}\right) = 0 + \frac{2}{3}i$
165.	$12(\cos 322° + i\sin 322°) = 9.4561 - 7.3879i$	$3[\cos(-102°) + i\sin(-102°)] = -0.6237 - 2.9343i$

167. $\cos 300° + i\sin 300° = \frac{1}{2} - \frac{\sqrt{3}}{2}i$ **169.** $1(\cos 25\pi + i\sin 25\pi) = -1 + 0i$

171. $2^{30}(\cos 5\pi + i\sin 5\pi) = -2^{30}$ **173.** $0.8680 + 4.9240i$ **175.** $2 + 2\sqrt{3}i$; -4; $2 - 2\sqrt{3}i$

177. $\sqrt[8]{2}\left(\cos\frac{\pi}{16} + i\sin\frac{\pi}{16}\right)$; $\sqrt[8]{2}\left(\cos\frac{9\pi}{16} + i\sin\frac{9\pi}{16}\right)$; $\sqrt[8]{2}\left(\cos\frac{17\pi}{16} + i\sin\frac{17\pi}{16}\right)$; $\sqrt[8]{2}\left(\cos\frac{25\pi}{16} + i\sin\frac{25\pi}{16}\right)$

179. $\langle 9, 12\rangle$ **181.** $\langle -2, -5\rangle$ **183.** $\left\langle \frac{5\sqrt{3}}{2}, \frac{5}{2}\right\rangle$ **185.** $\langle -4\sqrt{3}, 4\rangle$

187. (a) $\langle 10, -5\rangle$; $5\sqrt{5}$; (b) $\langle 10, -7\rangle$; $\sqrt{149}$ **189.** $93.73°$

CHAPTER 7

PROBLEM SET 7.1 page 409

1. Consistent **3.** Inconsistent **5.** Consistent **7.** $(4,3)$ **9.** $(1,-1)$ **11.** $(\frac{7}{3}, -\frac{4}{3})$
13. $(2,3,-1)$ **15.** $(4,3,2)$ **17.** $(1,1)$ **19.** $(3,2)$ **21.** $(-\frac{3}{2},\frac{3}{2})$ **23.** $(2,-1)$ **25.** $(1,1,1)$
27. $(-5,\frac{5}{3},\frac{28}{3})$ **29.** $(5,2,1)$ **31.** 7 and 5 **33.** 1137 at $4.25 and 1863 at $5.50
35. $28,000 at 10%; $22,000 at 8% **37.** 50 miles per hour and 60 miles per hour
39. Car at 40 miles per hour: 5 hours; car at 55 miles per hour: 3 hours
41. 2 pounds of meat; 1 pound of beans **43.** The first is 90% pure and the second is 80% pure
45. $4000 at 6%, $10,000 at 7%, and $6000 at 8%
47. 8 tons of 5% phosphate, 18% nitrogen; 4 tons of 6% phosphate, 10% nitrogen; 8 tons of 9% phosphate, 7% nitrogen

PROBLEM SET 7.2 page 423

1. $(3,5)$ **3.** $(\frac{23}{4},\frac{13}{2})$ **5.** $(x, 4-3x)$, x in $\mathbb{R}$ **7.** $(1,2,3)$ **9.** $(1,1,1)$ **11.** $(4,5,6)$
13. $(1,1)$ **15.** $(3,0,5)$ **17.** $(x,4,3-x)$, x in $\mathbb{R}$
19. $(x_2+2x_4+1, x_2, -3x_4+1, x_4, 1)$, x_2 and x_4 in $\mathbb{R}$ **21.** $(2,3)$ **23.** $(5,9)$ **25.** $(1,1,0)$
27. Inconsistent **29.** Dependent

31. (a) $\begin{bmatrix} 8 & 11 \\ -2 & 7 \end{bmatrix}$, (b) $\begin{bmatrix} -2 & -1 \\ 0 & 1 \end{bmatrix}$; (c) $\begin{bmatrix} 12 & 20 \\ -4 & 16 \end{bmatrix}$; (d) $\begin{bmatrix} -25 & -30 \\ 5 & -15 \end{bmatrix}$; (e) $\begin{bmatrix} -13 & -10 \\ 1 & 1 \end{bmatrix}$

33. (a) $\begin{bmatrix} 8 & 12 & 8 \\ 2 & 3 & 0 \end{bmatrix}$; (b) $\begin{bmatrix} -2 & -10 & -4 \\ -4 & 3 & 2 \end{bmatrix}$; (c) $\begin{bmatrix} 12 & 4 & 8 \\ -4 & 12 & 4 \end{bmatrix}$;
(d) $\begin{bmatrix} -25 & -55 & -30 \\ -15 & 0 & 5 \end{bmatrix}$; (e) $\begin{bmatrix} -13 & -51 & -22 \\ -19 & 12 & 9 \end{bmatrix}$

35. (a) $\begin{bmatrix} -9 & 8 & 3 \\ -3 & 3 & 7 \\ 6 & -2 & 4 \end{bmatrix}$; (b) $\begin{bmatrix} -1 & -4 & -1 \\ -1 & -1 & -3 \\ 0 & 0 & 0 \end{bmatrix}$; (c) $\begin{bmatrix} -20 & 8 & 4 \\ -8 & 4 & 8 \\ 12 & -4 & 8 \end{bmatrix}$;
(d) $\begin{bmatrix} 20 & -30 & -10 \\ 5 & -10 & -25 \\ -15 & 5 & -10 \end{bmatrix}$; (e) $\begin{bmatrix} 0 & -22 & -6 \\ -3 & -6 & -17 \\ -3 & 1 & -2 \end{bmatrix}$

37. (c) $\begin{bmatrix} 0 & 0 \\ 0 & 0 \end{bmatrix}$ and $\begin{bmatrix} 0 & 0 \\ 0 & 0 \end{bmatrix}$; (d) Yes; $D = \begin{bmatrix} -1 & -5 \\ 2 & -8 \end{bmatrix}$ **39.** (a) $\begin{bmatrix} 19 & -1 \\ 2 & 1 \end{bmatrix}$; (b) $\begin{bmatrix} 6 & 9 \\ 7 & 14 \end{bmatrix}$

41. (a) $\begin{bmatrix} -9 & -29 \\ 4 & -14 \end{bmatrix}$; (b) $\begin{bmatrix} -7 & 10 & -12 \\ 0 & -1 & 11 \\ 5 & -5 & -15 \end{bmatrix}$ **43.** (a) A; (b) A **51.** $(2,1)$ **53.** $(4,-2)$
55. $(2,-1,3)$

PROBLEM SET 7.3 page 433

1. 17 **3.** 63 **5.** 0 **7.** $\frac{1}{4}$ **9.** $-\frac{5}{7}$ **11.** Theorem 2 **13.** Theorem 1
15. Theorems 1 and 3 **17.** 0 **19.** -20 **21.** 3 **23.** Yes **25.** No **27.** No
29. $(\frac{1}{3}, \frac{2}{3})$ **31.** $(0, 0)$ **33.** $(\frac{19}{4}, -\frac{3}{4}, -\frac{29}{4})$ **35.** $(1, 1, 1)$ **37.** $(3, -1, 2)$ **39.** 16.5 square units

PROBLEM SET 7.4 page 440

1. $\frac{1}{x-1} + \frac{2}{x-3}$ **3.** $\frac{\frac{19}{5}}{x-3} + \frac{\frac{21}{5}}{x+2}$ **5.** $\frac{3}{3x+2} + \frac{-5}{2x+1}$ **7.** $\frac{\frac{1}{3}}{x-1} + \frac{-4}{x+1} + \frac{\frac{14}{3}}{x+2}$

9. $\frac{2}{x} + \frac{7}{(x-2)^2}$ **11.** $\frac{3}{x+2} + \frac{2}{x-3} + \frac{-1}{(x-3)^2}$ **13.** $2y + \frac{1}{y-2} + \frac{3}{(y-2)^2}$ **15.** $\frac{2}{x} + \frac{-2x}{x^2+3}$

17. $\frac{4}{2x-1} + \frac{-8x-4}{4x^2+1}$ **19.** $1 + \frac{\frac{24}{5}}{2p-1} + \frac{\frac{13}{5}p + \frac{9}{5}}{p^2+1}$ **21.** $\frac{-5}{2x-1} + \frac{3x+1}{x^2+4}$

23. $\frac{1}{x+1} + \frac{-x+1}{x^2+1} + \frac{2x-2}{(x^2+1)^2}$ **25.** $\frac{4}{x} + \frac{-2x-2}{x^2+1} + \frac{-3x-4}{(x^2+1)^2}$ **27.** $\frac{2}{x^2+4} + \frac{5x+8}{(x^2+4)^2}$

29. $\frac{-1}{x-1} + \frac{-4x-1}{x^2+4} + \frac{-4x}{(x^2+4)^2}$

PROBLEM SET 7.5 page 443

1. $(-1, -2)$; $(2, 1)$ **3.** $(-3, \frac{9}{2})$; $(2, 2)$ **5.** $(3, 4)$; $(2, 6)$ **7.** $(2, 4)$; $(-\frac{4}{5}, -\frac{22}{5})$ **9.** No solution
11. $(\frac{7}{2}, \frac{5}{2})$; $(\frac{11}{4}, \frac{5}{4})$ **13.** No solution **15.** $(0, 0)$; $(4, 4)$ **17.** $(4, 2)$; $(4, -2)$
19. $(3, 2)$; $(-3, 2)$; $(3, -2)$; $(-3, -2)$ **21.** $(-2\sqrt{6}, 1)$; $(-2\sqrt{6}, -1)$; $(2\sqrt{6}, 1)$; $(2\sqrt{6}, -1)$
23. $(-2, 2\sqrt{3})$; $(-2, -2\sqrt{3})$; $(2, 2\sqrt{3})$; $(2, -2\sqrt{3})$ **25.** $(4, 2)$; $(-4, 2)$; $(4\sqrt{15}/3, -\frac{10}{3})$; $(-4\sqrt{15}/3, -\frac{10}{3})$
27. $(3, 2)$; $(-3, 2)$; $(-3, -2)$; $(3, -2)$ **29.** $(-2, 1)$; $(2, 1)$ **31.** No solution
33. $(\frac{41}{10}, 3\sqrt{91}/10)$; $(\frac{41}{10}, -3\sqrt{91}/10)$ **35.** $(a, b) = (16, 2)$ **37.** $(r, s) = (0, -3), (\sqrt{5}, 2), (-\sqrt{5}, 2)$
39. $(u, v) = (1.568, 1.625), (3.125, 0.815)$ **41.** Numbers are 9 and 6, or -9 and -6, or 9 and -6, or -9 and 6
43. 15 centimeters, 20 centimeters, and 25 centimeters
45. One chemical is 140 kilograms, and the other is 170 kilograms

PROBLEM SET 7.6 page 450

1.

3.

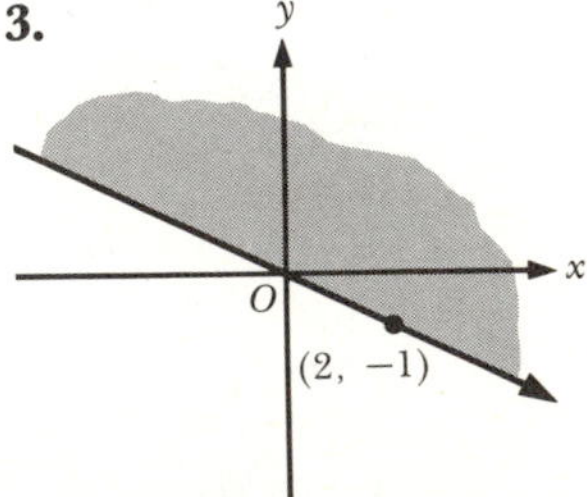

5.

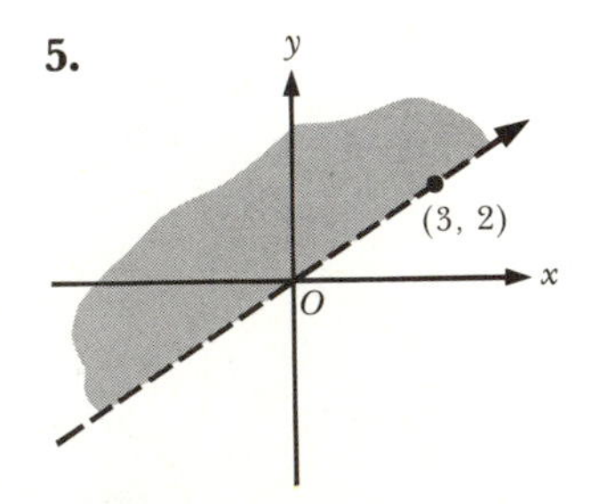

7.

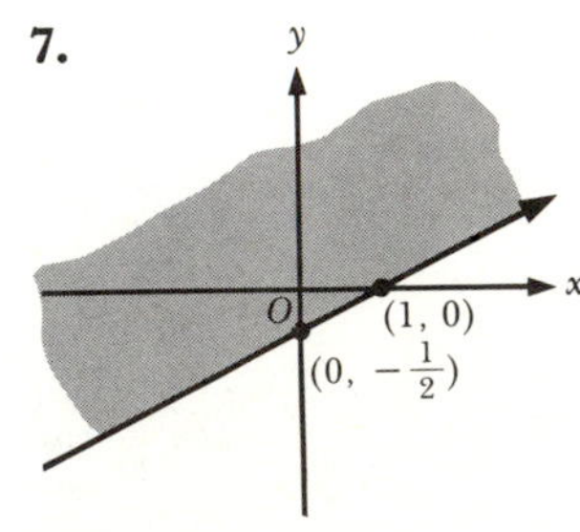

9.

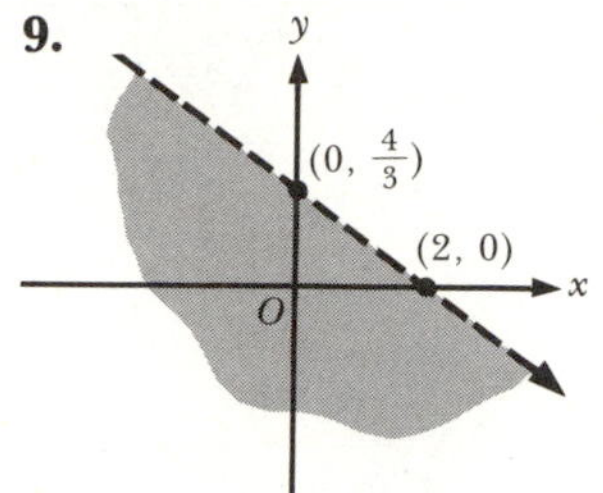

11.

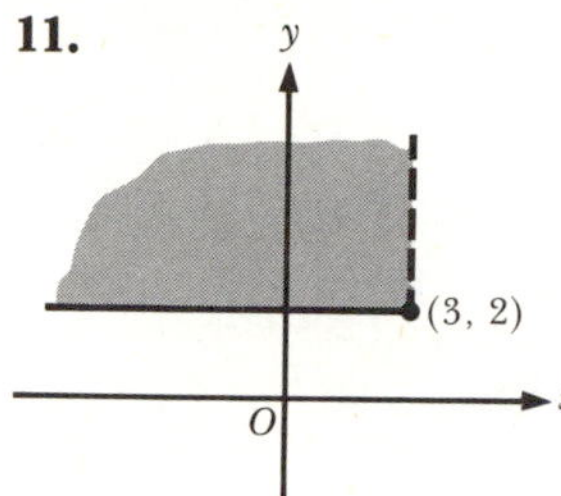

13.

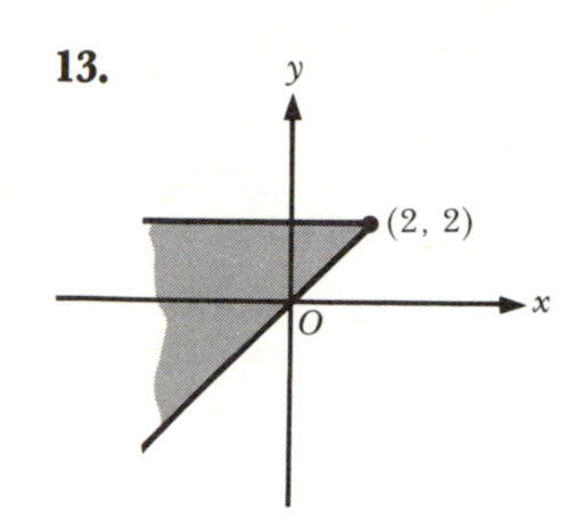

15.

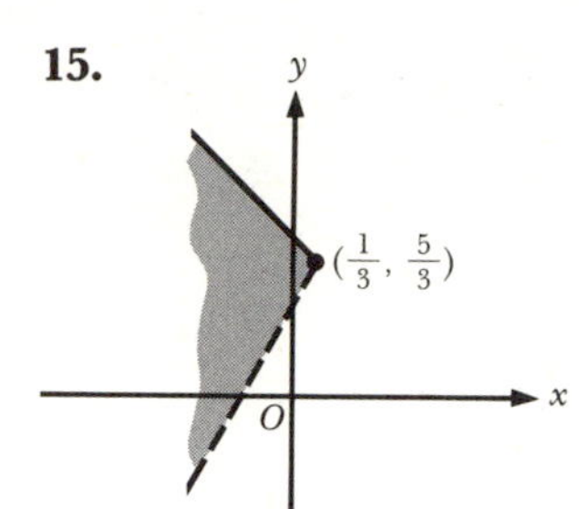

17.

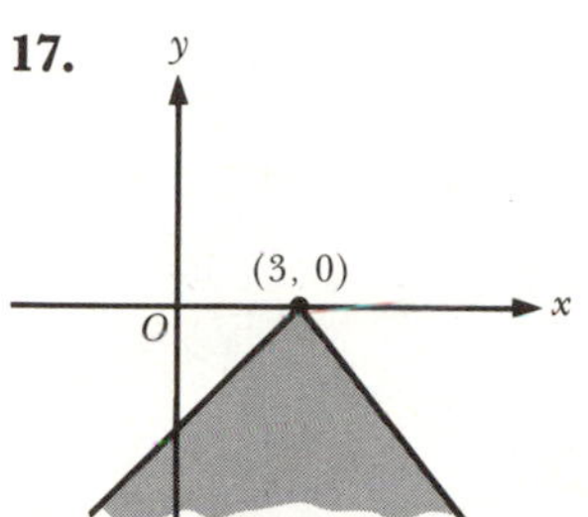

19.

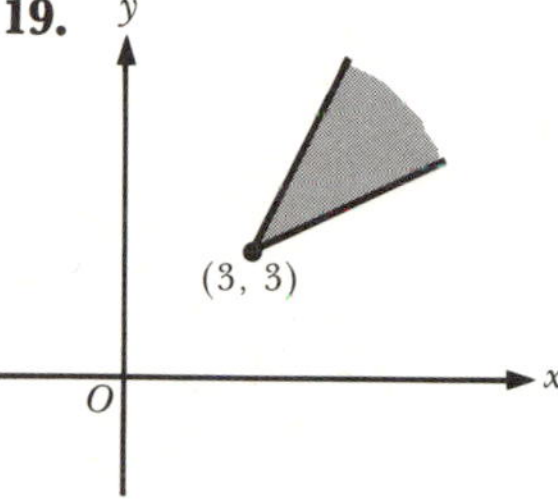

21. maximum = 30
minimum = 0

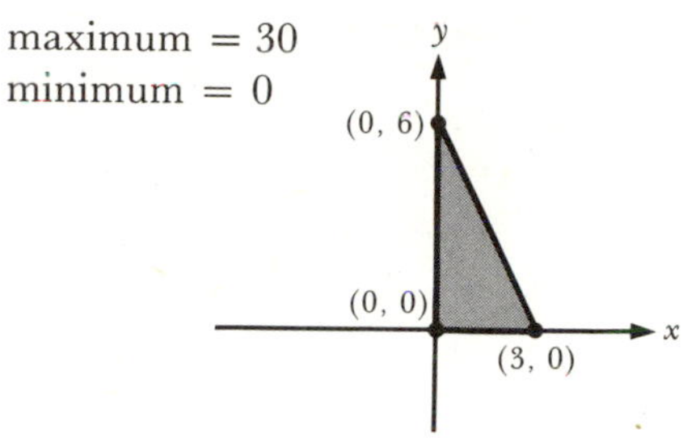

23. maximum = 24
minimum = 0

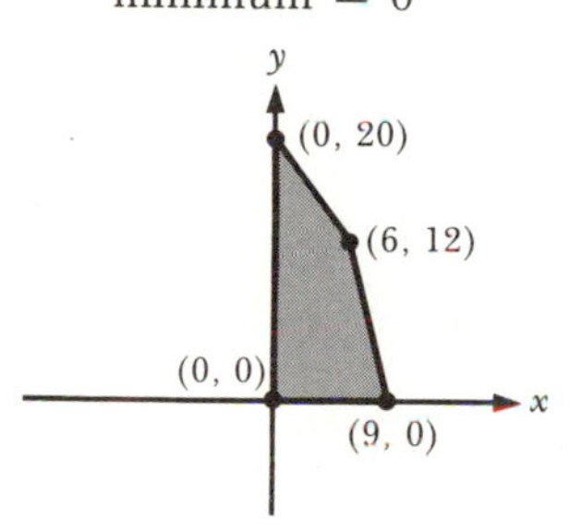

25. maximum = 23
minimum = −12

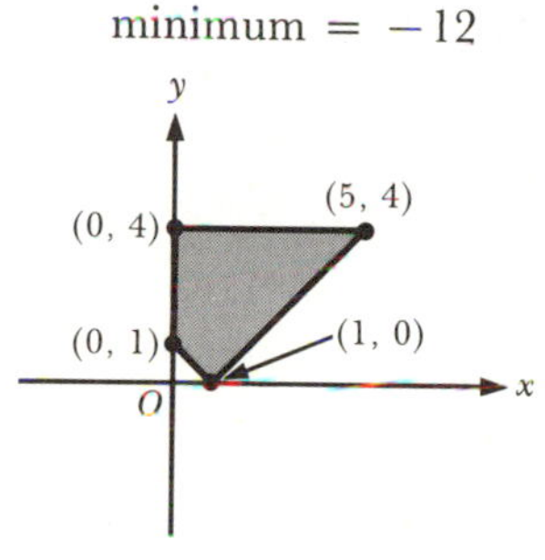

27. maximum = 25
minimum = 6

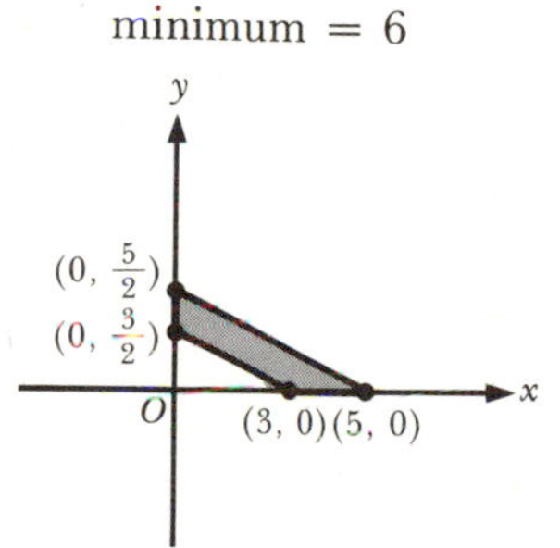

29. maximum = 8
minimum = 14/3

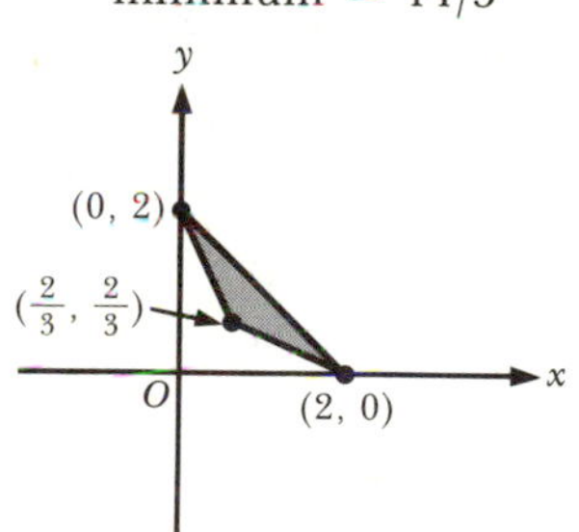

31. 450 hamburgers and 200 tacos **33.** 33 regular and 66 deluxe, with leftover material
35. 50 tons of 80/20 type and 100 tons of 60/40 type

REVIEW PROBLEM SET, CHAPTER 7 page 452

1. Consistent; $(2, -2)$ **3.** Consistent; $(\frac{1}{2}, -\frac{1}{4})$ **5.** Consistent; $(\frac{26}{5}, -\frac{36}{5})$ **7.** $(-1, -2)$
9. $(3, -2, 5)$ **11.** $(2, -1)$ **13.** $(-1, 3, 2)$ **15.** $(-\frac{2}{9}, \frac{28}{9})$ **17.** $(0, 1, 1)$ **19.** $(-9, 5)$

21. $(-7, 4, 3)$ **23.** (a) $\begin{bmatrix} 12 & -3 \\ 7 & 7 \end{bmatrix}$; (b) $\begin{bmatrix} 14 & -11 \\ 9 & 19 \end{bmatrix}$ **25.** (a) $\begin{bmatrix} 7 & 15 \\ 2 & -4 \\ 7 & 2 \end{bmatrix}$; (b) $\begin{bmatrix} -11 & -26 \\ -2 & 7 \\ -10 & -1 \end{bmatrix}$

27. (a) $\begin{bmatrix} -4 & 5 \\ 17 & 9 \end{bmatrix}$; (b) $\begin{bmatrix} 9 & 5 \\ 17 & -4 \end{bmatrix}$ **29.** (a) $\begin{bmatrix} 0 & 5 \\ 11 & -18 \\ 1 & -7 \end{bmatrix}$; (b) Not possible **35.** $(-1, 2)$

37. $(1, -1, 5)$ **39.** 5 **41.** 58.931 **43.** 2 **45.** $4 \pm \sqrt{21}$ **47.** 0 **49.** 10 **51.** 0

53. $(3, 2)$ **55.** $(1.298, 2.392)$ **57.** $(\frac{1}{2}, -\frac{1}{2}, \frac{3}{2})$ **59.** $\dfrac{\frac{9}{8}}{x-7} + \dfrac{-\frac{1}{8}}{x+1}$ **61.** $\dfrac{1}{x} + \dfrac{-1}{x-1} + \dfrac{1}{(x-1)^2}$

63. $\dfrac{-7x+34}{x^2+3} + \dfrac{-15}{x^2+2}$ **65.** $(3, -4)$ **67.** $(6, 0)$; $(-5, \sqrt{11})$; $(-5, -\sqrt{11})$

69. $(p, q) = (2, 3), (2, -3), (-2, 3), (-2, -3)$ **71.** $(x, y) = (3, 2), (2, 3)$

73. maximum = 21
minimum = 0

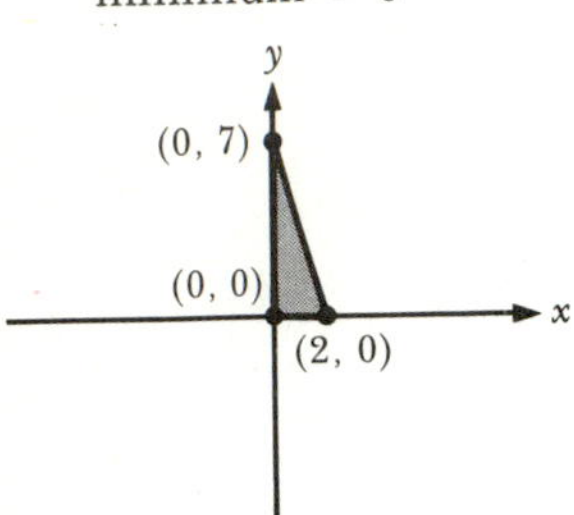

75. maximum = 20
minimum = 0

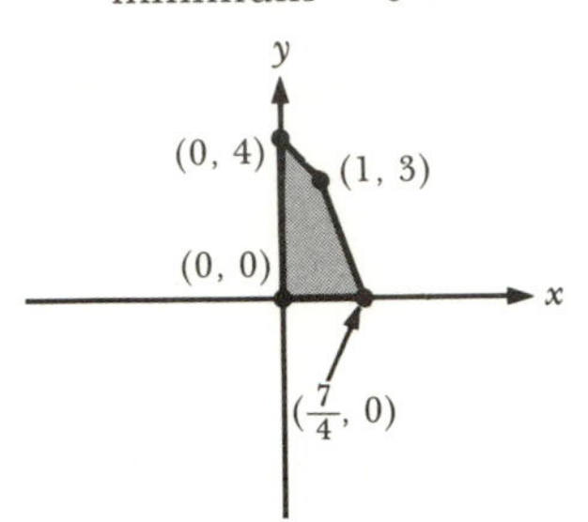

77. maximum = 990
minimum = 0

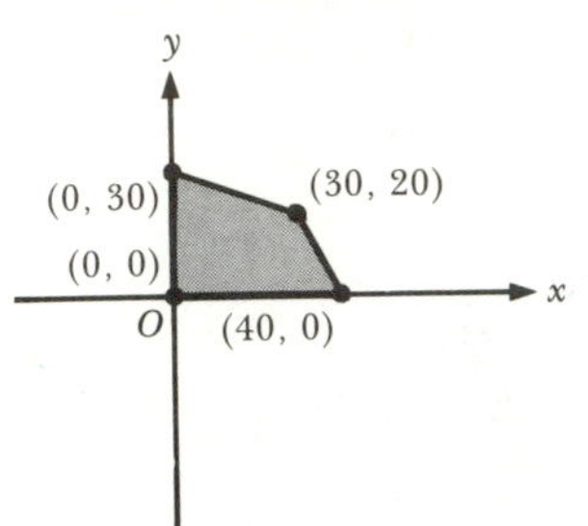

79. 10 quarters, 5 dimes, and 20 nickels

81. State income tax = \$2453
City income tax = \$1387
Federal tax = \$10,410

83. 7 and 8; 7 and -8; -7 and 8; -7 and -8

CHAPTER 8

PROBLEM SET 8.1 page 465

1. $(x-2)^2 + (y+5)^2 = 25$ **3.** $x^2 + (y-3)^2 = 25$

5. $(x-6)^2 + (y+2)^2 = 10$ and $x^2 + (y+4)^2 = 10$

7. $(x+1)^2 + (y-2)^2 = 1$; $(x-1)^2 + (y+2)^2 = 1$; $(x+1)^2 + (y+2)^2 = 1$; $(x-1)^2 + (y-2)^2 = 1$

9. $(x - \frac{1}{4})^2 + (y - \frac{1}{4})^2 = \frac{1}{8}$ **11.** $x^2 + y^2 = 9$ **13.** $(x-2)^2 + (y-10)^2 = 106$

	Vertices	Foci
15.	$(4, 0), (-4, 0), (0, -3), (0, 3)$	$(-\sqrt{7}, 0), (\sqrt{7}, 0)$
17.	$(0, -4), (0, 4), (2, 0), (-2, 0)$	$(0, -2\sqrt{3}), (0, 2\sqrt{3})$
19.	$(-4, 0), (4, 0), (0, -2), (0, 2)$	$(-2\sqrt{3}, 0), (2\sqrt{3}, 0)$

	Center	Vertices	Foci
21.	$(1, -2)$	$(9, -2), (-7, -2), (1, 4\sqrt{3} - 2), (1, -4\sqrt{3} - 2)$	$(5, -2), (-3, -2)$
23.	$(-1, 1)$	$(-3, 1), (1, 1), (-1, 2), (-1, 0)$	$(-1 - \sqrt{3}, 1)(-1 + \sqrt{3}, 1)$
25.	$(-1, 2)$	$(-1, 5), (-1, -1), (1, 2), (-3, 2)$	$(-1, 2 + \sqrt{5}), (-1, 2 - \sqrt{5})$
27.	$(-2, 1)$	$(-7, 1), (3, 1), (-2, 5), (-2, -3)$	$(1, 1), (-5, 1)$

29. $\frac{(x-3)^2}{4} + \frac{(y+2)^2}{25} = 1$ **31.** $\frac{(x-3)^2}{4} + \frac{(y-1)^2}{25} = 1$ **33.** $\frac{(x-2)^2}{4} + \frac{(y-4)^2}{3} = 1$
35. $\frac{(x+3)^2}{1} + \frac{(y-1)^2}{25} = 1$ **37.** Circle with center $(0, 0)$ and radius $|a|$ **39.** $\frac{x^2}{2500} + \frac{y^2}{625} = 1$
41. 0.62 kilometer

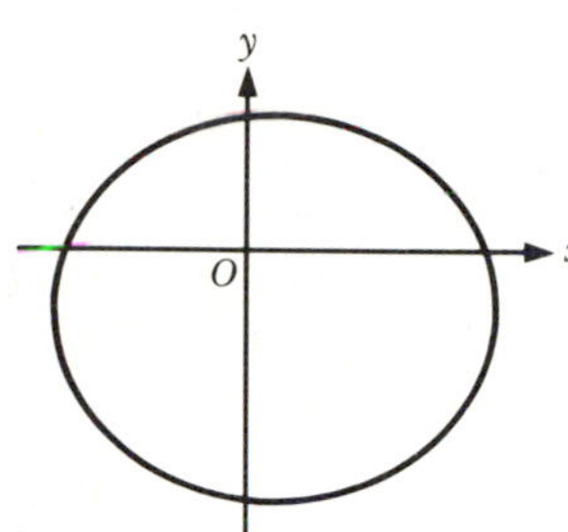

Figure Problem 21

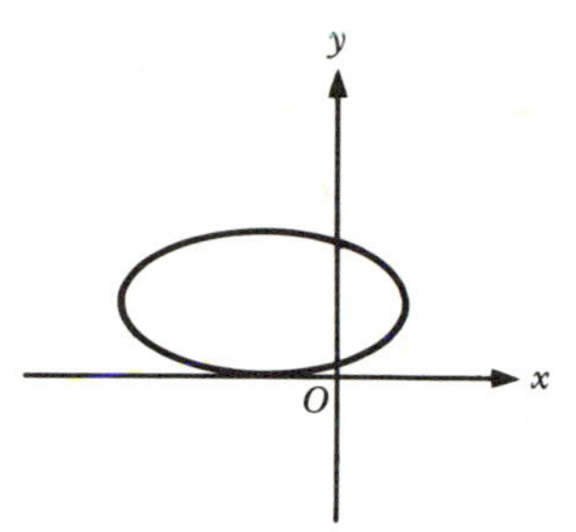

Figure Problem 23

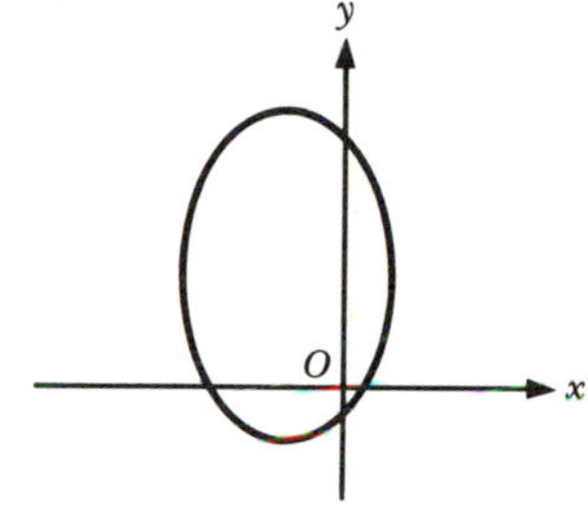

Figure Problem 25

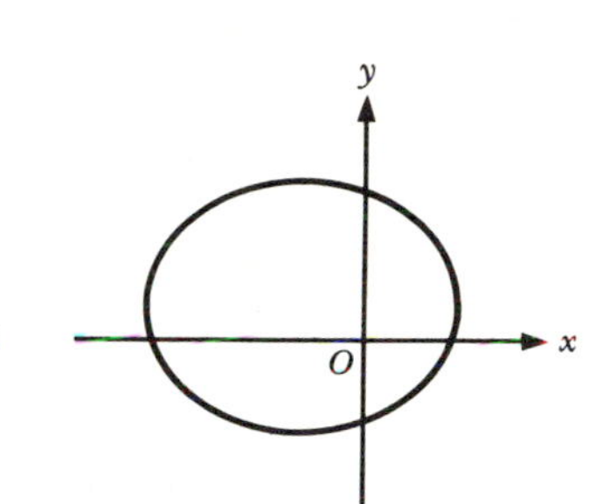

Figure Problem 27

PROBLEM SET 8.2 page 472

	Vertex	Focus	Axis of Symmetry	Directrix
1.	$(0, 0)$	$(2, 0)$	x axis	$x = -2$
3.	$(0, 0)$	$(0, -1)$	y axis	$y = 1$
5.	$(0, 0)$	$(\frac{5}{4}, 0)$	x axis	$x = -\frac{5}{4}$
7.	$(0, 0)$	$(0, \frac{1}{14})$	y axis	$y = -\frac{1}{14}$
9.	$(0, 0)$	$(-\frac{3}{4}, 0)$	x axis	$x = \frac{3}{4}$

11.

(a) (b) (c) (d)

	Vertex	Focus	Axis of Symmetry	Directrix
13.	$(2, -1)$	$(2, -\frac{5}{2})$	$x = 2$	$y = \frac{1}{2}$
15.	$(1, 2)$	$(\frac{1}{2}, 2)$	$y = 2$	$x = \frac{3}{2}$
17.	$(-3, -1)$	$(-\frac{7}{2}, -1)$	$y = -1$	$x = -\frac{5}{2}$
19.	$(-2, -\frac{1}{6})$	$(-2, -\frac{2}{3})$	$x = -2$	$y = \frac{1}{3}$
21.	$(\frac{3}{2}, 2)$	$(3, 2)$	$y = 2$	$x = 0$

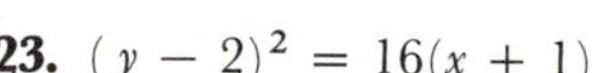

23. $(y - 2)^2 = 16(x + 1)$ **25.** $x^2 = 4(y - 3)$ **27.** $(x - 1)^2 = 4(y - 3)$ **29.** $x^2 = 2y$
31. $y^2 = -4x$ **33.** $(3, 2), (3, -2)$ **35.** 100, 58, 28, 10, 4, 10, 28, 58, 100 meters **37.** $\frac{371}{30}$ centimeters
39. 0.22 meter

PROBLEM SET 8.3 page 480

	Center	Vertices	Foci	Asymptotes
1.		$(0, 3), (0, -3)$	$(0, 3\sqrt{2}), (0, -3\sqrt{2})$	$y = \pm x$
3.		$(3, 0), (-3, 0)$	$(\sqrt{10}, 0), (-\sqrt{10}, 0)$	$y = \pm \frac{1}{3}x$
5.		$(0, \sqrt{3}), (0, -\sqrt{3})$	$(0, \sqrt{15}), (0, -\sqrt{15})$	$y = \pm \frac{1}{2}x$
7.		$(2, 0), (-2, 0)$	$(3, 0), (-3, 0)$	$y = \pm \frac{\sqrt{5}}{2}x$
9.		$\left(\frac{\sqrt{2}}{2}, 0\right), \left(\frac{-\sqrt{2}}{2}, 0\right)$	$\left(\frac{\sqrt{42}}{6}, 0\right), \left(\frac{-\sqrt{42}}{6}, 0\right)$	$y = \pm \frac{2\sqrt{3}}{3}x$
	Center	Vertices	Foci	Asymptotes
11.	$(5, 4)$	$(10, 4), (0, 4)$	$(5 - \sqrt{29}, 4), (5 + \sqrt{29}, 4)$	$y - 4 = \pm \frac{2}{5}(x - 5)$
13.	$(-2, 3)$	$(-2, 7), (-2, -1)$	$(-2, 3 - 4\sqrt{2}), (-2, 3 + 4\sqrt{2})$	$y - 3 = \pm (x + 2)$
15.	$(4, 1)$	$(4, 3), (4, -1)$	$(4, 1 + \sqrt{7}), (4, 1 - \sqrt{7})$	$y - 1 = \pm \frac{2\sqrt{3}}{3}(x - 4)$
17.	$(-2, -4)$	$(1, -4), (-5, -4)$	$(-2 + \sqrt{34}, -4)$, $(-2 - \sqrt{34}, -4)$	$y + 4 = \pm \frac{5}{3}(x - 5)$

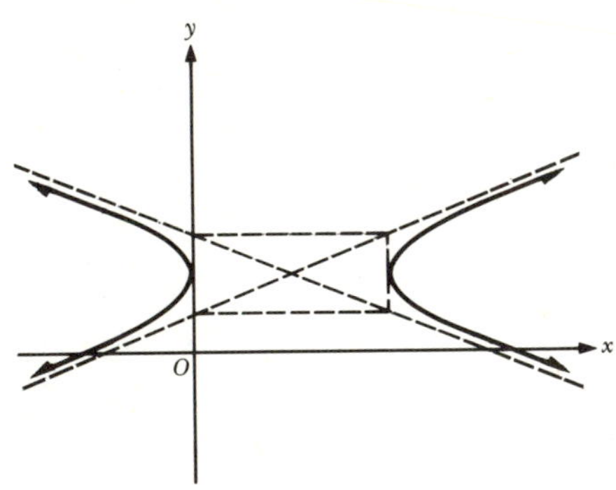

Figure Problem 11

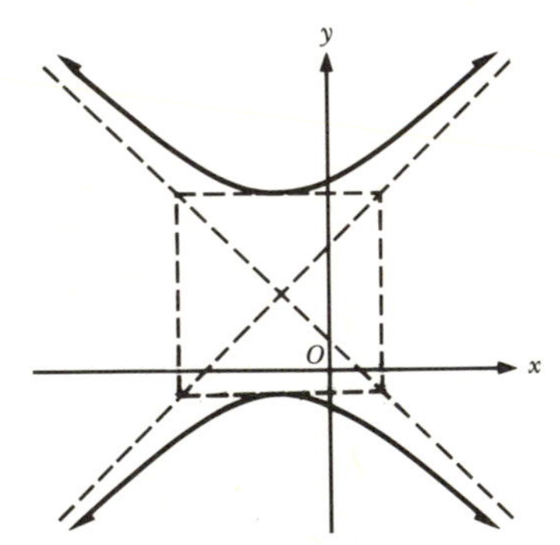

Figure Problem 13

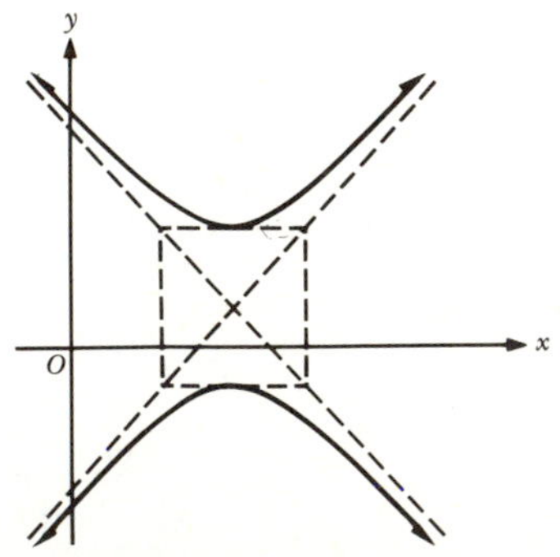

Figure Problem 15

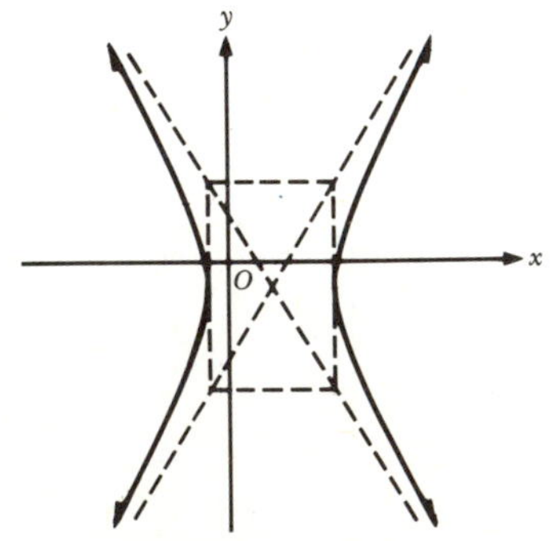

Figure Problem 17

	Center	Vertices	Foci	Asymptotes
19.	$(-1, -1)$	$(-1, -1 + \sqrt{3})$, $(-1, -1 - \sqrt{3})$	$\left(-1, -1 + \frac{\sqrt{15}}{2}\right)$, $\left(-1, -1 - \frac{\sqrt{15}}{2}\right)$	$y + 1 = \pm 2(x + 1)$
21.	$(-\frac{1}{2}, 2)$	$\left(-\frac{1}{2} + \frac{\sqrt{33}}{6}, 2\right)$, $\left(-\frac{1}{2} - \frac{\sqrt{33}}{6}, 2\right)$	$\left(-\frac{1}{2} + \frac{\sqrt{231}}{12}, 2\right)$, $\left(-\frac{1}{2} - \frac{\sqrt{231}}{12}, 2\right)$	$y - 2 = \pm\frac{\sqrt{3}}{2}\left(x + \frac{1}{2}\right)$

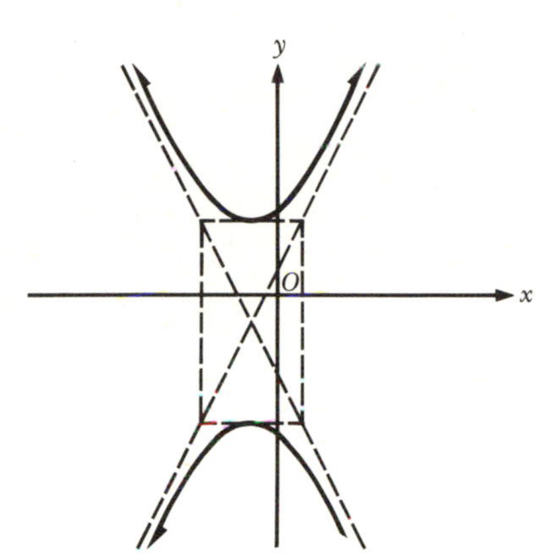

Figure Problem 19

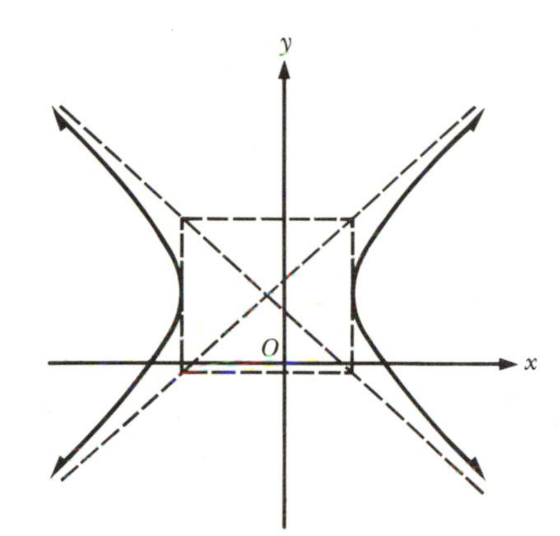

Figure Problem 21

23. $\frac{x^2}{256} - \frac{y^2}{400} = 1$ **25.** $\frac{(y-5)^2}{1} - \frac{(x+1)^2}{3} = 1$ **27.** $\frac{x^2}{9} - \frac{y^2}{7} = 1$ **29.** $\frac{3x^2}{7} - \frac{y^2}{35} = 1$

31. $(1, 2), (1, -2), (-1, 2), (-1, -2)$

PROBLEM SET 8.4 page 488

1. (i) (a) $(2, -1)$; (b) $(1, 5)$; (c) $(4, 10)$; (d) $(2, -2)$; (e) $(-1, 3)$; (f) $(0, 0)$
(ii) (a) $(9, -1)$; (b) $(8, 5)$; (c) $(11, 10)$; (d) $(9, -2)$; (e) $(6, 3)$; (f) $(7, 0)$

3. (i) (a) $(3, -10)$; (b) $(-4, 0)$; (c) $(-2, 1)$; (d) $(\frac{11}{4}, -\frac{25}{4})$; (e) $(-6, -8)$; (f) $(4, -10)$
(ii) (a) $(-1, -11)$; (b) $(-8, -1)$; (c) $(-6, 0)$; (d) $(-\frac{5}{4}, -\frac{29}{4})$; (e) $(-10, -9)$; (f) $(0, -11)$

5. $\bar{x} = x + 5, \bar{y} = y - 3$ **7.** $\bar{x} = x + 3, \bar{y} = y - 2$ **9.** $\bar{y} = y - 2, \bar{x} = x + \frac{1}{2}$

11. $\bar{x} = x + 4, \bar{y} = y - \frac{7}{4}$ **13.** $(-\sqrt{3} - \frac{5}{2}, 1 - \frac{5}{2}\sqrt{3})$ **15.** $\left(\frac{11\sqrt{2}}{2}, -\frac{9\sqrt{2}}{2}\right)$ **17.** $(-5.16, 0.21)$

19. $\frac{\bar{x}^2}{2} - \frac{\bar{y}^2}{2} = 1$ **21.** $\frac{\bar{x}^2}{8} + \frac{\bar{y}^2}{(8/5)} = 1$ **23.** $31\bar{x}^2 + 10\sqrt{3}\bar{x}\bar{y} + 21\bar{y}^2 = 144$ **25.** $\bar{x} = \sqrt{2}/2$

27. $\frac{\bar{x}^2}{3} + \frac{\bar{y}^2}{(9/7)} = 1$; 45° **29.** $\frac{\bar{x}^2}{36} + \frac{\bar{y}^2}{16} = 1$; 63.4° **31.** $\frac{\bar{x}^2}{1} + \frac{\bar{y}^2}{2} = 1$; 26.6°

PROBLEM SET 8.5 page 495

1. $(y - \frac{15}{4})^2/(\frac{125}{16}) - (x - 3)^2/(\frac{25}{4}) = 1$ **3.** $(x^2/25) + (y^2/16) = 1$
5. $y^2 = -4(x - 3)$ **7.** $e = \frac{3}{5}$
9. $e = \sqrt{6}/3$ **11.** $e = \frac{5}{3}$ **13.** $e = \sqrt{6}/3$
15. $C = (-1, -\frac{3}{2})$; $V{:}(-1, -\frac{13}{2}), (-1, \frac{7}{2})$; $F{:}(-1, -\frac{3}{2} + \sqrt{29}), (-1, -\frac{3}{2} - \sqrt{29})$
Asymptotes: $y + \frac{3}{2} = \pm\frac{5}{2}(x + 1)$; $e = \sqrt{29}/5$
17. (a) $\frac{1}{2}$; (b) $\sqrt{3}a$; $2a$ **19.** $e = 1$; parabola **21.** $e = \frac{1}{2}$; ellipse **23.** $e = 3$; hyperbola

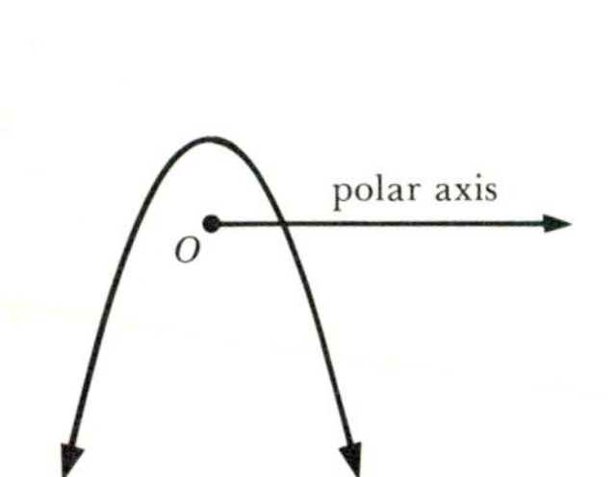

Figure Problem 19

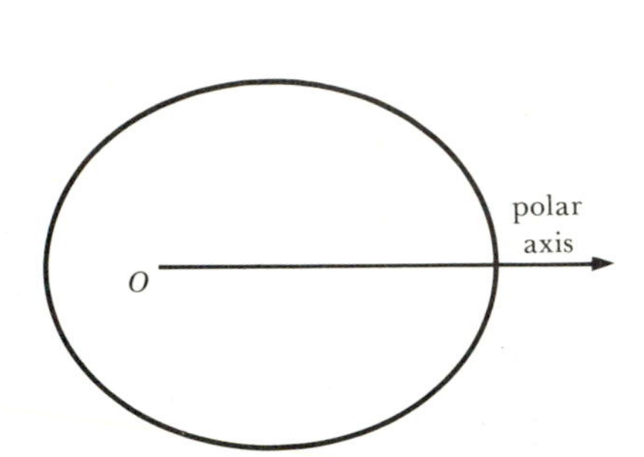

Figure Problem 21

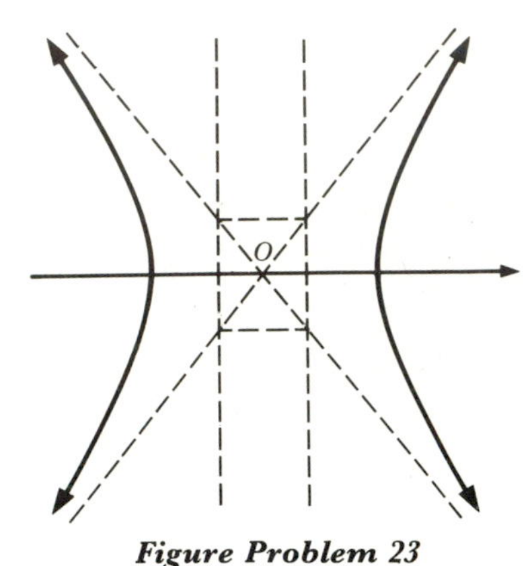

Figure Problem 23

25. $x^2 + y^2 = 16$ **27.** $y = x^2 - 2x, \ -1 \le x \le 3$ **29.** $y = \sqrt{1 - x^2}$
31. $\dfrac{(x - 3)^2}{4} + \dfrac{(y - 5)^2}{4} = 1$ **33.** $x^2 - y^2 = 1, \ x \ge 1$ or $x \le -1$

REVIEW PROBLEM SET, CHAPTER 8 page 497

1. $(x - 4)^2 + (y + 3)^2 = 16$ **3.** $(x - 1)^2 + (y - 2)^2 = 5$ **5.** $(x - 1)^2 + (y + \frac{5}{6})^2 = \frac{325}{36}$
7. $(x^2/25) + (y^2/16) = 1$ **9.** $\dfrac{(x - 1)^2}{1} + \dfrac{(y - 2)^2}{9} = 1$ **11.** $(x^2/18) + (y^2/2) = 1$

	Center	Vertices	Foci
13.	$(0, 0)$	$(0, 2\sqrt{3}), (0, -2\sqrt{3}), (2\sqrt{2}, 0), (-2\sqrt{2}, 0)$	$(0, 2), (0, -2)$
15.	$(-1, 1)$	$(-6, 1), (4, 1), (-1, 4), (-1, -2)$	$(3, 1), (-5, 1)$
17.	$(-4, 6)$	$(-4, 12), (-4, 0), (0, 6), (-8, 6)$	$(-4, 6 + 2\sqrt{5}), (-4, 6 - 2\sqrt{5})$

19. $(x + 4)^2 = -6(y - \frac{3}{2})$ **21.** $(x - 1)^2 = 3(y + \frac{13}{4})$ **23.** $(y + 3)^2 = -12(x + 5)$

	Vertex	Focus	Axis of Symmetry
25.	$(0,0)$	$(0,-2)$	y axis
27.	$(-2,0)$	$(-\frac{13}{4},0)$	x axis
29.	$(-1,3)$	$(-1,7)$	$x=-1$

31. $(x^2/4)-(y^2/5)=1$ **33.** $(y^2/16)-(x^2/9)=1$

	Center	Vertices	Foci	Asymptotes
35.	$(0,0)$	$(6\sqrt{2},0),(-6\sqrt{2},0)$	$(4\sqrt{5},0),(-4\sqrt{5},0)$	$y=\pm\frac{1}{3}x$
37.	$(-2,3)$	$(2,3),(-6,3)$	$(-2-2\sqrt{5},3),(-2+2\sqrt{5},3)$	$y-3=\pm\frac{1}{2}(x+2)$

39. $y=\pm\frac{3}{2}x$ **41.** $\dfrac{x^2}{302{,}500}-\dfrac{y^2}{697{,}500}=1$ **43.** (a) $(-3,-2)$; (b) $(7,-7)$; (c) $(-5,-5)$
45. $\bar{x}=x-1, \bar{y}=y+3$
47. (a) $\bar{x}=\sqrt{2}, \bar{y}=2\sqrt{2}$; (b) $\bar{x}=9\sqrt{2}/2, \bar{y}=\sqrt{2}/2$; (c) $\bar{x}=-5\sqrt{2}/2, \bar{y}=3\sqrt{2}/2$; (d) $\bar{x}=-\sqrt{2}, \bar{y}=-3\sqrt{2}$; (e) $\bar{x}=\sqrt{2}, \bar{y}=-\sqrt{2}$
49. $3\bar{y}^2-\bar{x}^2=6$ **51.** $\bar{x}^2+3\bar{y}^2=9$; $18.4°$ **53.** $(x^2/9)+(y^2/5)=1$ **55.** $(x^2/64)-(y^2/36)=1$
57. $e=\sqrt{6}$; hyperbola **59.** $e=\sqrt{2}/2$; ellipse
61. $e=\frac{1}{3}$; ellipse **63.** $e=2$; hyperbola **65.** $(x-5)^2+(y+2)^2=1$ **67.** $xy=1, x>0$

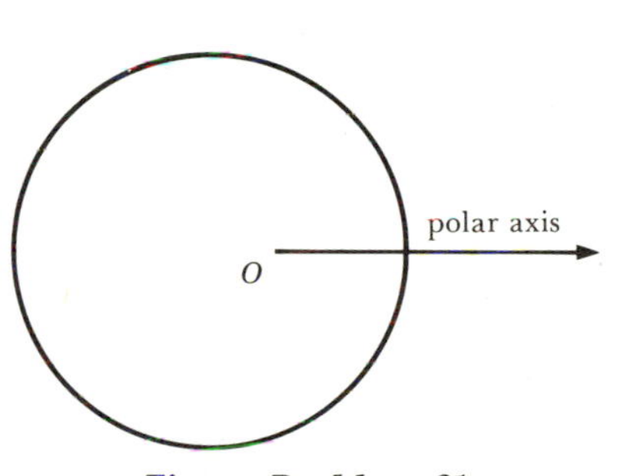

Figure Problem 61

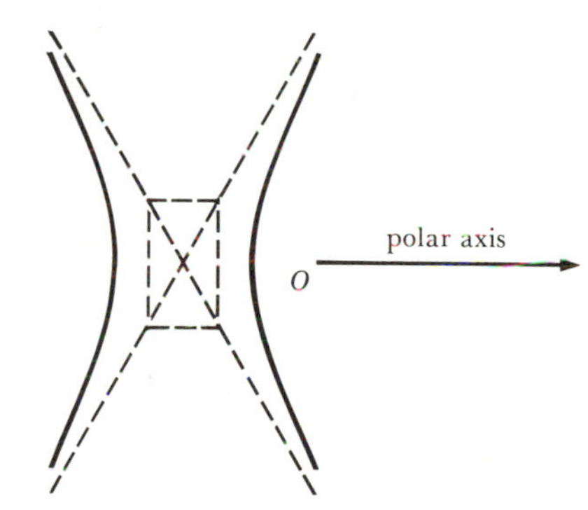

Figure Problem 63

CHAPTER 9

PROBLEM SET 9.1 page 505

1. $\frac{3}{2}, 4, \frac{15}{2}, 12, \frac{35}{2}$ **3.** $-1, -1, 0, 2, 5$ **5.** 2, 4, 2, 4, 2 **7.** Arithmetic; $d=3$
9. Arithmetic; $d=5$ **11.** Arithmetic; $d=-13$ **13.** Arithmetic; $d=1.2$ **15.** 50, 85
17. $16a+4b$; $25a-8b$ **19.** Geometric; $r=3$ **21.** Geometric; $r=-2$ **23.** Geometric: $r=\frac{2}{3}$
25. Geometric; $r=-\frac{2}{3}$ **27.** $\frac{1}{128}$ **29.** 2 **31.** $(1+a)^{n-1}$ **33.** 192; 3072 **35.** 48
37. Sixth

PROBLEM SET 9.2 page 513

1. 120 **3.** 660 **5.** 15 **7.** 83 **9.** 1562 **11.** 1,691,450 **13.** $\sum_{k=1}^{5} (3k - 2)$

15. $\sum_{k=1}^{4} (\frac{3}{5})^k$ **17.** True **19.** False **21.** True **23.** 145 **25.** -28 **27.** $63b - 63$

29. $a_{10} = 33$; $S_{10} = 195$ **31.** $a_{18} = \frac{719}{3}$; $d = \frac{668}{51}$ **33.** $n = 32$; $d = \frac{21}{31}$ **35.** $\frac{189}{2}$ **37.** 5460

39. $\dfrac{10(a^6 - 1)}{a - 1}$ **41.** 3 or -4 **43.** 2 or -2 **45.** $a_9 = \frac{1}{128}$; $S_8 = \frac{85}{64}$ **47.** $n = 10$; $S_{10} = \frac{1023}{16}$

49. $\frac{1}{2}$ **51.** $\frac{1}{4}$ **53.** 4 **55.** $\frac{32}{99}$ **57.** $\frac{46}{99}$ **59.** $\frac{1186}{333}$

PROBLEM SET 9.3 page 523

1. $\sum_{k=1}^{n} (2k - 1) = n^2$ **3.** $\sum_{k=1}^{n} (2k) = n^2 + n$ **5.** $\sum_{k=1}^{n} k(k + 1) = \frac{1}{3}n(n + 1)(n + 2)$

7. $\sum_{k=1}^{n} 4^k = \frac{4}{3}(4^n - 1)$ **9.** $\sum_{k=1}^{n} 5^{k-1} = \frac{1}{4}(5^n - 1)$

11. (a) 3003; (b) 1; (c) 15; (d) 3003; (e) 1; (f) 6; (g) 455; (h) n; (i) 15

13. $n + 5$ **15.** $\dfrac{x - 1}{(2n + 1)(2n + 2)}$ **17.** $\dfrac{n}{e}$ **19.** $x^5 + 15x^4 + 90x^3 + 270x^2 + 405x + 243$

21. $x^4 - 8x^3 + 24x^2 - 32x + 16$ **23.** $x^{12} + 12x^{11}y + 66x^{10}y^2 + 220x^9y^3$

25. $a^{12} - 16a^{21/2}x^2 + 112a^9x^4 - 448a^{15/2}x^6$ **27.** $\frac{455}{4096}x^{24}a^3$ **29.** $\frac{224}{243}x^6a^{12}$ **31.** $\frac{28}{243}a^3x^{12}$

33. (a) 2,598,960; (b) 161,700

REVIEW PROBLEM SET, CHAPTER 9 page 525

1. $\frac{1}{4}, \frac{1}{10}, \frac{1}{18}, \frac{1}{28}, \frac{1}{40}$ **3.** $-2, \frac{1}{4}, 0, \frac{1}{256}, \frac{32}{3125}$ **5.** 39 **7.** $\frac{41}{8}$ **9.** $\frac{-729}{64}$ **11.** $-512\sqrt{5}$

13. 95 **15.** $\dfrac{728}{243}$ **17.** 1246 **19.** -55 **21.** -24 **23.** $27a + 3b$ **25.** $\frac{225}{4}$

27. $\frac{29,524}{9}$ **29.** $\frac{1}{9}$ **31.** 3 **33.** $\frac{1}{16}$ **35.** $\frac{14}{99}$ **37.** $\frac{763}{165}$ **39.** $\sum_{k=1}^{n} 2^k = 2(2^n - 1)$

41. $\sum_{k=1}^{n} (k^2 + 1) = \dfrac{n(2n^2 + 3n + 7)}{6}$ **43.** $\sum_{k=1}^{n} \dfrac{1}{(3k - 2)(3k + 1)} = \dfrac{n}{3n + 1}$ **45.** 792

47. 3240 **49.** $\dfrac{(n + 7)(n + 6)}{2}$ **51.** $\dfrac{n + 2}{7}$ **53.** $\dfrac{(n + 1)}{2(2n + 1)n^2}$

55. $81x^4 + 108x^3y + 54x^2y^2 + 12xy^3 + y^4$ **57.** $8x^3 + \dfrac{12x^2}{y} + \dfrac{6x}{y^2} + \dfrac{1}{y^3}$

59. $a^5 + \frac{15}{2}a^4 + \frac{45}{2}a^3 + \frac{135}{4}a^2 + \frac{405}{16}a + \frac{243}{32}$ **61.** $x^2 - 12x\sqrt{x} + 54x - 108\sqrt{x} + 81$ **63.** $4860x^4y^2$

Index

VERMONT COLLEGE LIBRARY
0 00 10 0045543 4

Trigonometric Functions

Acute Angles

$$\sin\theta = \frac{\text{opp}}{\text{hyp}} \qquad \csc\theta = \frac{\text{hyp}}{\text{opp}}$$

$$\cos\theta = \frac{\text{adj}}{\text{hyp}} \qquad \sec\theta = \frac{\text{hyp}}{\text{adj}}$$

$$\tan\theta = \frac{\text{opp}}{\text{adj}} \qquad \cot\theta = \frac{\text{adj}}{\text{opp}}$$

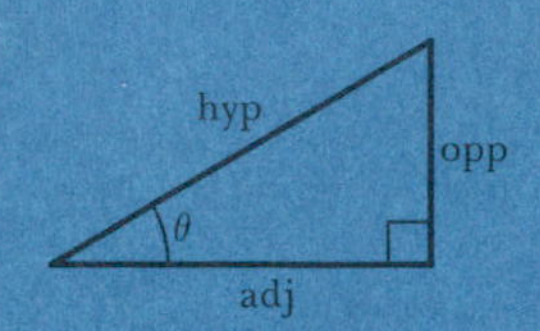

General Angles

$$\sin\theta = \frac{y}{r} \qquad \csc\theta = \frac{r}{y}$$

$$\cos\theta = \frac{x}{r} \qquad \sec\theta = \frac{r}{x}$$

$$\tan\theta = \frac{y}{x} \qquad \cot\theta = \frac{x}{y}$$

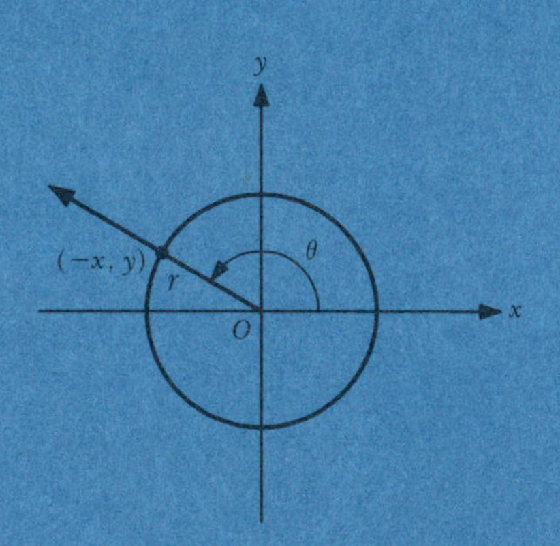

Trigonometric Identities

Fundamental Identities

1. $\csc\theta = \dfrac{1}{\sin\theta}$
2. $\sec\theta = \dfrac{1}{\cos\theta}$
3. $\cot\theta = \dfrac{1}{\tan\theta}$
4. $\tan\theta = \dfrac{\sin\theta}{\cos\theta}$
5. $\cot\theta = \dfrac{\cos\theta}{\sin\theta}$
6. $\cos^2\theta + \sin^2\theta = 1$
7. $1 + \tan^2\theta = \sec^2\theta$
8. $1 + \cot^2\theta = \csc^2\theta$

Even-Odd Identities

1. $\sin(-\theta) = -\sin\theta$
2. $\cos(-\theta) = \cos\theta$
3. $\tan(-\theta) = -\tan\theta$
4. $\cot(-\theta) = -\cot\theta$
5. $\sec(-\theta) = \sec\theta$
6. $\csc(-\theta) = -\csc\theta$

Addition Formulas

1. $\sin(\alpha + \beta) = \sin\alpha\cos\beta + \sin\beta\cos\alpha$
2. $\cos(\alpha + \beta) = \cos\alpha\cos\beta - \sin\alpha\sin\beta$
3. $\tan(\alpha + \beta) = \dfrac{\tan\alpha + \tan\beta}{1 - \tan\alpha\tan\beta}$

Subtraction Formulas

1. $\sin(\alpha - \beta) = \sin\alpha\cos\beta - \sin\beta\cos\alpha$
2. $\cos(\alpha - \beta) = \cos\alpha\cos\beta + \sin\alpha\sin\beta$
3. $\tan(\alpha - \beta) = \dfrac{\tan\alpha - \tan\beta}{1 + \tan\alpha\tan\beta}$